# Neutrons in Soft Matter

# Neutrons in Soft Matter

Edited by

**Toyoko Imae, Toshiji Kanaya, Michihiro Furusaka, and Naoya Torikai**

WILEY

A JOHN WILEY & SONS, INC., PUBLICATION

Published by John Wiley & Sons, Inc., Hoboken, New Jersey
Published simultaneously in Canada



***Library of Congress Cataloging-in-Publication Data:***

Neutron in Soft Matter / [edited by] Toyoko Imae ...
[et al.].
p. cm.
Includes bibliographical references and index.
ISBN 978-0-470-40252-8 (hardback)
1. Neutrons–Scattering. 2. Soft condensed matter. I. Imae, Toyoko.
QC793.5.N4628E86 2011
539.7'58–dc22

2010030994

# Contents

# Preface

Toward peaceful and safe human life, technology science, including nanotechnology, biotechnology, and information technology, is one of the key sciences in the twenty-first century, besides the environmental and energy sciences. Such technology science is complementary to materials science, analytical methodology, and related sciences. One of the innovations in analytical methodology is the development of neutron and synchrotron research in a category of "big science." Neutron and synchrotron facilities, which are out-of-laboratory level, have been improved in scale and quality with the support of national projects in several countries. Considering the situation that new neutron sources will lead to, a steep increase in the number of users of neutron facilities cannot be ruled out. Accordingly, there is a need of an adequate guidebook or textbook on neutron science.

Neutron beam used in a neutron facility is of short wavelength. Besides, the analysis of neutron research gives us information of small range like nanoscale. Thus, new research for chemical and biological objects will be undertaken because of the demand for an adequate tool for micro- and nanostructure research and for fast dynamics research of atomic location in materials. Considering such scientific requirement, we seek to publish a specialized book on neutron research. Different from already published professional books on neutron, this book focuses on instrumentation as well as theory and/or applications; each of the sections of theory, instrumentation, and applications is well described by contributors with deep knowledge and expertise in the field.

In Chapter I, the basic concepts of neutron scattering are briefly discussed. Chapter II meticulously describes instrumentation such as small-angle neutron scattering, neutron reflectometry, quasi and inelastic neutron scattering, and neutron imaging. Chapter III elucidates data treatment and sample environment for convenience of the users. Some practical applications are exemplified for soft matters like small molecules, linear polymers, dendritic polymers, gels, and proteins in Chapter IV. Finally, Chapter V deals with the current facilities based on pulsed neutron source and reactor.

This book on neutron research is useful for chemists, particularly those in the soft matter field; however, it is also valuable for physicists and biologists as they always look for a blow-by-blow account of neutron research. This book also includes the basic technological terms related to the field. It is expected that such a comprehensive book will prove useful to many scientists and engineers, who are already utilizing or will utilize neutron facilities, as well as readers who are interested in neutron research. In addition, it is a highly informative textbook for postgraduate students and researchers of neutron science.

The editors greatly wish to acknowledge all contributors for their enormous contributions. We also appreciate Ms. Hanako Ishida at the Institute for Chemical Research, Kyoto University, for designing the cover picture of the book. It is a pleasure to thank all the staff in our laboratory and the colleagues in the institution who helped us in bringing out this book. We acknowledge Japan Atomic Energy Agency for kind transfer permission of the aerial photograph of J-PARC. We are particularly indebted to our family for their emotional support and patience showed during the compiling of this book. We thank our publishers for their great support for this project. Finally, Toyoko Imae would like to thank other editors and Dr. Koji Mitamura for showing their unlimited perseverance and untiring energy during the editing process.

Toyoko Imae
Toshiji Kanaya
Michihiro Furusaka
Naoya Torikai

# Contributors

**Michael Agamalian**, Neutron Scattering Science Division, Oak Ridge National Laboratory, Oak Ridge, TN, USA (Chapter II.1.3.1)

**Masatoshi Arai**, J-PARC Centre, Japan Atomic Energy Agency, Tokai-mura, Japan (Chapter V.1)

**David G. Bucknall**, Materials Science and Engineering, Georgia Institute of Technology, Atlanta, GA, USA (Chapter IV.6)

**Richard A. Campbell**, Institut Laue-Langevin, Grenoble, France (Chapter III.5)

**Colin J. Carlile**, Lund University, ESS Scandinavia Secretariat, Lund, Sweden (Chapter V.2)

**Bernhard Frick**, Institut Laue-Langevin, Grenoble, France (Chapter II.3.2)

**Barbara J. Gabrys**, Department of Materials, University of Oxford, Oxford, UK (Chapter IV.3)

**Mitsuhiro Hirai**, Department of Physics, Gunma University, Maebashi, Japan (Chapter III.4)

**Toyoko Imae**, National Taiwan University of Science and Technology, Honors College, Graduate Institute of Engineering, Taipei, Taiwan (Chapter IV.2)

**Toshiji Kanaya**, Institute for Chemical Research, Kyoto University, Uji, Japan (Chapter IV.3)

**Mikio Kataoka**, Graduate School of Materials Science, Nara Institute of Science and Technology, Ikoma, Japan (Chapter IV.5)

**Satoshi Koizumi**, Strongly Correlated Supermolecule Group, Quantum Beam Science Directorate, Japan Atomic Energy Agency, Tokai-mura, Japan (Chapter II.1.3.2)

**Peter V. Konarev**, EMBL c/o DESY, Hamburg, Germany (Chapter III.3)

**Ruep E. Lechner**, Guest at Helmholtz-Zentrum Berlin, Berlin, Germany (Chapter II.3.3)

**Tsang-Lang Lin**, Department of Engineering and System Science, National Tsing Hua University, Hsinchu, Taiwan (Chapter IV.1)

**Peter Lindner**, Institut Laue-Langevin, Grenoble, France (Chapter III.5)

**Ferenc Mezei**, LANSCE, Los Alamos National Laboratories, Los Alamos, NM, USA (Chapter I.1)

**Koji Mitamura**, Japan Science and Technology Agency, Exploratory Research for Advanced Technology (JST/ERATO), Takahara Soft Interfaces Project, Fukuoka, Japan (Chapter IV.2)

**Michael Monkenbusch**, Institut für Festkörperforschung, Forschungszentrum Jülich, Jülich, Germany (Chapter II.3.1)

**Kell Mortensen**, Department of Natural Sciences, Faculty of Life Sciences, University of Copenhagen, Frederiksberg, Denmark (Chapter II.1.1)

**Geoffrey R. Mitchell**, Centre for Advanced Microscopy, University of Reading, Reading, UK (Chapter IV.7)

**Hiroshi Nakagawa**, Neutron Biophysics Group, Japan Atomic Energy Agency, Tokai-mura, Japan (Chapter IV.5)

**Dan Neumann**, NIST Center for Neutron Research, National Institute of Standards and Technology, Gaithersburg, MD, USA (Chapter II.3.2)

**Toshiya Otomo**, Institute of Materials Structure Science, High Energy Accelerator Research Organization (KEK), Tsukuba, Japan (Chapter II.1.2)

**Dieter Richter**, Institut für Festkörperforschung, Forschungszentrum Jülich, Jülich, Germany (Chapter II.3.1)

**Ralf Schweins**, Institut Laue-Langevin, Grenoble, France (Chapter III.5)

**Hideki Seto**, Institute of Materials Structure Science, High Energy Accelerator Research Organization (KEK), Tsukuba, Japan (Chapter III.2)

**Mitsuhiro Shibayama**, Neutron Science Laboratory, Institute for Solid State Physics, The University of Tokyo, Kashiwa, Japan (Chapter IV.4)

**Dmitri I. Svergun**, EMBL c/o DESY, Hamburg, Germany (Chapter III.3)

**Nobuyuki Takenaka**, Department of Mechanical Engineering, Graduate School of Engineering, Kobe University, Kobe, Japan (Chapter II.4)

**Naoya Torikai**, Department of Chemistry for Materials, Graduate School of Engineering, Mie University, Tsu, Japan (Chapter II.2)

**George D. Wignall**, Neutron Scattering Sciences Division, Oak Ridge National Laboratory, Oak Ridge, TN, USA (Chapter III.1)

# I Neutron Scattering

# I.1 Basic Concepts

*Ferenc Mezei*

## I.1.1 INTRODUCTION

The observation of scattering of radiation is one of the most common ways to see objects in everyday life: The light from the sun or a lamp falls on the objects around us and is partially absorbed and partially scattered. It is the observation of the scattered radiation by the eyes that allows us to "see" things around us, primarily the surface of the objects. Some materials such as glasses, gases, and many liquids neither scatter nor absorb much of light, so they are largely transparent or nearly invisible. The partial reflection of light on glass surfaces is also a form of scattering, and analogous processes with neutrons are widely used to study optically flat interfaces.

The way eyes and similarly operating cameras detect and process scattered light is called optical imaging, which is based on the capability to have focusing devices (lenses) with a reasonably large angular acceptance, which is related to the power to deviate the radiation by a substantial angle in the range of 10° or more. This is not the only way to extract information from scattered radiation. The information carried by the modified sound reaching us from objects behind obstacles or from moving objects gives a vague and limited idea of what can be inferred from the observation of more subtle properties of scattered radiation without the capability of image formation. The example of bats using sonar-like technology by capturing ultrasound scattered by flying insects is a particularly telling example.

The basic physical properties of the radiation determine the features of the objects we can observe. Light has a wavelength $\lambda$ in the range of a few tenths of a micron, and this leads to an unavoidable lower limit of geometrical details, which can be observed by using light. Although the optical capabilities of our eyes cannot fully exploit these confines, with the help of microscopes we can look for finer details down to the limit set by the wavelength of the light radiation used. Other limit is set by the capability of the radiation to make its way inside the materials observed. Light is absorbed in most materials within a very short distance, so the scattered radiation

*Neutrons in Soft Matter* Edited by Toyoko Imae, Toshiji Kanaya, Michihiro Furusaka, and Naoya Torikai

originates from the immediate vicinity of the surface of the objects; hence, it is not adequate to gain information from the inside of such materials.

X-rays, which essentially are very short wavelength light, have wavelength in the Å range (1 Å = 0.1 nm), just as the distance between atoms in chemical bonds or in condensed matter. Therefore, they can offer the capability to observe the atomic structure of matter, for which purpose X-ray diffraction is the oldest and most common method of exploration. Here we already meet a major difference compared to vision based on light: there is no way to directly observe images. This is due, on the one hand, to the fact that we do not know materials that could be used to produce X-ray optical devices, which could deviate this radiation by a substantial angle in the range of several degrees. (Even if this would not be the case, it would still be of quite some challenge to build optical devices with stability and precision of an Å or so. This, however, could be ingeniously achieved with scanning tunneling microscopes, atomic force microscopes, and similar devices for the study of surfaces with atomic resolution!) For this reason, the actual three-dimensional atomic image building has to be achieved by another way: by complex mathematical reconstruction from the scattering patterns arising due to interference of the radiation scattered on various atoms, more or less close to each other in the sample. With the advent of very intense future X-ray sources, such as free electron lasers, it is hoped that such reconstruction of the 3D image of the atoms will also become possible for samples consisting of a single large molecule.

Neutron radiation can also provide us with a probe with wavelength in the Å range, and thus make possible to observe atomic structures similarly to X-rays. Indeed, this has become a standard tool in the study of condensed matter since the groundbreaking discoveries by C. Shull and coworkers around 1950. In this type of research, the scarcer and more expensive neutron radiation is less commonly used than one or the other forms of X-ray sources, including tabletop laboratory equipment. However, neutrons offer a number of special capabilities for collecting information fully inaccessible to X-rays: (a) neutrons can observe light atoms (including hydrogen) in the presence of heavier ones; (b) neutrons can penetrate inside complex equipment for *in situ* studies or inside bulky samples to reveal structures far from the surface; and (c) neutrons allow to mark selected atoms within a given species in the sample by isotopic replacement. The large difference in the neutron scattering characteristics between hydrogen and deuterium is of particular importance in the study of biological matter.

Radiation in general is characterized not only by its wavelength, but also by its frequency. Here there is another very important difference between neutrons and X-rays. The relation between wavelength $\lambda$ and frequency $f$ for X-rays is given as

$$f = c/\lambda,$$

where $c$ is the velocity of light, 299,792,458 m/s, and for radiation with wavelength of 1 Å we have $f \cong 3 \times 10^{18}$ Hz. This is a very high frequency that corresponds to 12.5 keV energy or to the thermal energy $k_B T$ at a temperature of about 150 million K.

In contrast to X-rays, the neutron velocity $v$ depends on the wavelength, according to the de Broglie relation

$$mv = 2\pi\hbar/\lambda, \tag{I.1.1}$$

where $m$ is the neutron mass and $\hbar$ is the Planck constant. With this, the frequency $f$ of the neutron radiation becomes

$$f = v/\lambda = \frac{2\pi\hbar}{m\lambda^2}, \tag{I.1.2}$$

which for $\lambda = 1$ Å gives $f \cong 1.97 \times 10^{13}$ Hz, corresponding to 81.8 meV or to the thermal energy at 950K.

The huge difference between the frequencies of X-ray and neutron radiations of the wavelength comparable to atomic radii has the major consequence that atomic motions in materials under everyday conditions can be readily traced by neutron radiation, and this can be only partially accomplished by X-rays, or by any other radiation known to us, for the same matter. The scattering of radiation on atoms in motion is a quantum mechanical process, in which quanta of energy $\hbar\omega$ can be exchanged between the radiation and the sample, where $f \equiv 2\pi/\omega$ is a characteristic frequency of the motion of the atoms in the sample. In such a so-called "inelastic scattering" process, the energy of the scattered radiation is different from the incoming one by the amount $\hbar\omega$: either larger (called "energy gain scattering") or lower (called "energy loss scattering").

The motion of atoms in matter under ordinary conditions is thermally excited, and the possible frequencies of various kinds ("modes") of atomic motions in condensed matter range from essentially 0 (e.g., sound waves of a few tens or hundreds of Hz) to the equivalent of thermal energies at which the materials disintegrate, for example, a solid melts or a liquid evaporates. For ordinary materials, this corresponds to a few hundred to 1000K, pretty close to the equivalent of the energy of neutrons with $\lambda = 1$ Å wavelength. Therefore, we conclude that in ordinary matter inelastic scattering processes of this kind change the neutron energy by a substantial, easily detectable amount, while for X-rays with similar wavelength the change is in the range of a part in a million. This difference becomes decisive for the study of slower atomic or molecular motions. At the end, inelastic neutron scattering allows us to explore microscopic process in the time domain of $10^{-15}$ to $10^{-6}$ s (i.e., energies from neV to eV), while inelastic X-ray scattering cannot access times longer than $10^{-11}$ s (energies less than about a few tenths of a meV). This capability of exploring slow processes, together with its high sensitivity to look at hydrogen atoms, makes neutrons one of the particularly valuable tools for the microscopic investigation of soft matter.

It is worth emphasizing that the important feature of these radiations in the study of condensed matter is the direct exploration of the microscopic and nanoscopic time and length domains in both of its dimensions. This is made possible by being able to detect appropriate changes in frequency in the scattering process for radiation with wavelength comparable to the size of atoms or molecules. This offers invaluable additional insight into information on dynamics one can obtain by macroscopic spectroscopies, such as light scattering, dielectric response, and so on, exploring the average behavior of bulk matter, or by local probes, such as NMR, μSR, and so on, which directly test single atoms. The understanding of cooperative, collective aspects of the motion of atoms and molecules requires the additional dimension offered by neutron and to a lesser extent X-ray radiation: to directly observe the evolution of microscopic motion in the domain between the local atomic and the bulk macroscopic length scales.

Neutrons are particularly formidable microscopic probes of magnetism, due to their magnetic moment that can scrutinize the internal magnetic fields **B** on the atomic level inside materials. Magnetism is less common in soft matter; however, the sizable magnetic moment of neutrons—in contrast to their lack of electric charge—is a valuable additional experimental handle one can also make unique use of in experimental exploration of nonmagnetic materials, and as such also of prime importance in soft matter studies.

## I.1.2 RADIATION OF PARTICLES AND WAVES

The quantum mechanical duality between the particle and wave aspects of radiations used in scattering experiments is of fundamental importance in the practice of neutron scattering. As a rule of thumb, any radiation (even familiar waves as light) behaves as an ensemble of point-like classical particles when propagating between obstacles (structures) very large compared to its wavelength. In contrast, when the structure has details comparable in size to the wavelength of the radiation, wave propagation needs to be considered, with all the complexities of interference effects.

The physical reason for this behavior is that if optical path differences in wave propagation span a range much larger than the wavelength, interference effects will occur with a wide spread of phase differences in excess of $2\pi$. Therefore, these effects, the single experimental signature of wave propagation, will average to zero, and the wave looks like an ensemble of classical point-like particles following uniquely determined, precise trajectories, for example, as in geometrical optics.

A more practical and quantitative criterion for particle versus wave behavior can be formulated by considering the way radiation is collimated for the purposes of scattering experiments. For neutron beams, we have typically about $10^6$–$10^8$ particles/cm$^2$/s. If we would need to study a sample, for example, a crystalline grain of say 0.1 mm diameter, this would mean mere 100–10000 neutron/s impinging on the sample, far from enough for a reasonable scattering experiment. This illustrates the point that neutron scattering is an "intensity-limited" technique and we need most often to use beam cross sections and sample areas in the range of cm$^2$ or more. With such sample size, in order to achieve a well-defined beam direction a few meter flight path is quite common. How an obstacle of a $d = 1$ cm diameter hole in $L = 10$ m distance relates to the wavelength of the radiation can be evaluated by considering the possible path differences between wave/particle paths from one point to another along the beam propagation. Figure I.1.1 illustrates possible paths of beam

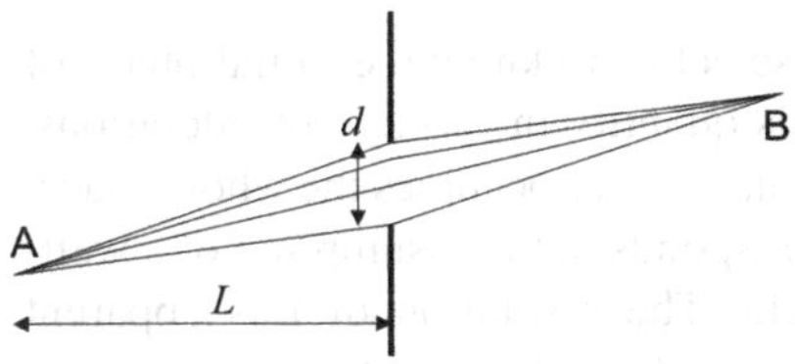


**Figure I.1.1** Potential optical paths for wave propagation from point A to point B through a diaphragm as obstacle.

propagation defined by a diaphragm as obstacle. The maximum possible difference in optical path lengths is given by

$$\delta = 2\left(\sqrt{L^2 + (d/2)^2} - L\right) \cong \frac{d^2}{4L}. \tag{I.1.3}$$

With $L \sim 10$ m and $d \sim 1$ cm, we get $\delta \sim 0.0025$ mm = 2500 nm, that is, about four orders of magnitude larger than the neutron wavelength. Thus, here the neutron can be considered as a classical point-like particle. For visible light, however, with $\lambda \sim 500$ nm one would be better advised to use wave mechanics to precisely evaluate an experiment, as it was well known nearly 200 years ago. At synchrotron X-ray sources, the typical beam diameter $d \sim 0.1$ mm, that is, we have $\delta \sim 2.5$ Å, which is indeed comparable to the wavelength of the radiation we are interested in.

Eq. (I.1.3) also implies that if we consider the propagation of neutrons inside materials, where the "obstacle" structure means atoms or molecules at distances $d$ from each other in the range of a few Å or nm, we will always encounter wave behavior, independently of the value of $L$ (which practically cannot be less than atomic distances). In practical terms, this leads to the basic rule of neutron scattering experimentation (and implies a great practical simplification of data reduction): the beam propagation in the spectrometers outside materials can be exactly described by classical mechanics of point-like particles. This is in contrast to synchrotron X-ray scattering, which fact is commonly and rather imprecisely referred to by calling synchrotron X-rays "coherent" and neutrons "incoherent" radiation. In actual fact, there is no difference in the coherence of these radiations as far as their interaction with the samples is concerned; the difference is instead in their interaction with the spectrometers operating with very different beam cross sections.

This classical particle ("incoherent") behavior of neutron beams as shaped and observed in scattering instruments paradoxically leads to a simplification of the description of the scattering process of neutrons on the atomic scale details in materials. Since the state of a classical particle impinging on a scattering objects can be described by a well-defined velocity vector $\vec{v}$, it will be characterized as a wave by a single, well-defined wavelength $\lambda$, as given by eq. (I.1.1), and the corresponding exact plane wave will thus be given as

$$\phi = e^{i(\vec{k}\vec{r} - \omega t + \varphi_0)}, \tag{I.1.4}$$

where the wave vector $\vec{k}$ is defined as $\vec{k} = m\vec{v}/\hbar$ (i.e., the absolute value of $\vec{k}$ equals $2\pi/\lambda$) and $\omega = 2\pi f$. While we can select radiation with a more or less well-defined wave vector $\vec{k}$ by some kind of a monochromator, which is part of any neutron experiment, there is fundamentally no way to select or to know the initial phase of the wave, $\varphi_0$. For this reason, we will drop this quantity in further considerations.

The paradox here appears to be that such a plane wave occupies the whole space, instead of being point-like, while it exactly corresponds to the assumption of a well-defined, unique velocity of a classical particle. The resolution of this apparent contradiction is quite fundamental for the understanding of the scattering processes on actual samples. Except for perfect crystals (the only practical examples of which are semiconductor crystals such as Si grown for use in electronic chips), real samples have a limited coherence range, that is, the range of exactly ordered arrangement of atoms. Thus, when the neutron wave interacts with the sample, its wave properties are tested only over the extent of the coherence volume in the sample. For ordinary practical materials, including usual (nonperfect) single crystals, this volume gets as large as crystalline grains can be, that is, usually a small fraction of a millimeter. Thus, ordinary samples in actual fact are a collection of a large number of small, independent samples. Here the scattered particle will act as "point-like" with the precision of the point as defined by the coherence volume in the sample. If this volume is much smaller than the beam dimensions, we have the paradoxical classical particle propagation together with scattering on the sample as extended plane wave according to eq. (I.1.4). If the "points" defined by the coherence volumes of the sample are comparable to the beam dimensions—which never happens in soft matter samples—the classical point-like particle propagation becomes an unsatisfactory approximation.

Thus, we can conclude that (with the exception of the exotic case of large perfect single crystal samples) neutron radiation plays out the famous quantum mechanical duality in a very convenient and practical manner for us: Neutrons can be considered as perfect, exactly point-like classical mechanical particles when they travel through usual neutron scattering instruments and as infinitely extended perfect plane waves when they meet atoms inside materials. This is a very convenient picture, which simplifies the understanding of neutron scattering to a great deal.

## I.1.3 NEUTRON SPIN AND OTHER PROPERTIES

Beyond its mass of $1.675 \times 10^{-27}$ g, the only significant feature of a neutron particle for our purposes is its magnetic moment of $\mu = -1.913$ Bohr magneton related to its $s = 1/2$ quantum mechanical spin. The minus sign indicates that the neutron magnetic moment and its spin are oppositely oriented. The existence of a magnetic moment implies that the neutron feels magnetic fields $\vec{B}$ via the Zeeman energy term $-\vec{\mu}\vec{B}$. This interaction can be quite substantial for magnetic fields in the range of several tesla, which neutrons most frequently encounter inside magnetic materials. It is important to keep in mind that the neutrons do not interact directly with the magnetic moments of atoms; they only see the magnetic field created by the atoms or by any other source, such as current carrying conductors.

The electric charge and electric dipole moment of the neutron are zero for all practical purposes (i.e., ongoing efforts to find them remained without success by now), which confer the neutrons the unique property for a radiation with low energy that they can traverse large chunks of most materials, in some cases up to several meters. This also implies that the neutron radiation is very largely destruction free and biological and other soft matter samples suffer no significant radiation damage even under long irradiation in neutron scattering experiments at the most intense neutron sources.

The only mechanism for the neutron itself to damage materials is knocking out atoms from their bound position in materials. This requires that the neutrons possess sufficient kinetic energy, typically in the range of several eV, which is much more than the energy of the neutrons ideally suited for soft matter research. Nevertheless, this process, the so-called Compton scattering of neutrons, is of some interest for the investigation for fluids of light elements, such as $H_2$ or He. This will not be further discussed here. For the neutron energies with which we are concerned, the neutron collision with atoms inside soft matter does not break the bonds of the atoms with their neighbors, so from the point of view of mechanics the collision is actually between the neutron and the whole sample. Thus, the process is recoil free for the bound atom, with momentum change in the collision being taken up by the quasi-infinitely large mass of the sample compared to that of the neutron.

Inside matter, neutrons interact only with the nuclei of the atoms, beyond, of course, the gravity and the magnetic field. Important here is that they do not interact (for any practical purpose) with the electrons and the large electric fields inside atoms. For this reason, the neutron itself does not ionize. However, one form of the interaction with the nuclei is the absorption capture of the neutron by a nucleus, by which a new isotope is formed. If this isotope is stable, no further damage to the sample occurs. In most cases, the isotope is a short-lived "intermediate nucleus" and leads to immediate decay with emission of ionizing particles (such as $\alpha$, $\beta$, or $\gamma$ radiation) or the sample becomes radioactive with the decay process taking place over an extended period, which can range from hours to years. The two practical consequences of neutron absorption by certain nuclei for neutron scattering work are that (a) the high neutron absorption of a few elements (actually usually some isotopes of the element) can be a difficulty for the study of samples containing such elements and (b) samples can become radioactive in neutron scattering experiments and need to be treated with caution. On the other hand, the radiation damage to the sample by the secondary radiation after neutron absorption capture is still negligible in virtually all cases and presents no concern for the study of soft matter, in contrast to some other kinds of radiations.

## I.1.4 NEUTRON INTERACTION WITH MATTER

The other form of interaction between neutrons and atomic nuclei is scattering, which in the exact terminology of nuclear physics is called "elastic scattering," as opposed to the nuclear reactions (including absorption) between neutrons and nuclei, which are called "inelastic" in nuclear physics. We stress here, in order to avoid confusion, that the notions of "elastic" and "inelastic" scattering are very different in our field of interest, the study of condensed matter by slow neutron scattering, which entirely falls into the domain of elastic processes (i.e., with no change to the particles themselves) in nuclear physics. This scattering process can be classically envisaged as the rebound of a billiard ball on another, while we must stress from the outset that this process inside condensed matter must be considered in terms of the wave properties of the particles. This is clear from our above introductory analysis, since atoms in a matter will form an ensemble of "obstacles" on the way of the neutron propagation showing a structure on a scale comparable to the neutron wavelength, that is, leading to neutron path differences between possible neutron trajectories comparable to its wavelength.

The radius of nuclei is many orders of magnitude smaller than the nm scale of atomic distances and the neutron wavelengths we are concerned with, so the nuclei can be considered as structureless points, described in space by delta functions $\delta(\vec{r}-\vec{r}_i(t))$, where $\vec{r}_i(t)$ is the position of the nucleus of atom $i$ at time $t$. If we consider an atom at rest ($r_i = 0$), its interaction with the neutron wave will be described by the Fermi pseudopotential

$$V(\vec{r}) = \frac{2\pi\hbar^2}{m} b_i \delta(\vec{r}), \tag{I.1.5}$$

where $b_i$ is the so-called neutron scattering length, which characterizes the neutron interaction with the given atomic nucleus $i$, and its value can be found in the tables for all common isotopes of the elements (http://www.ncnr.nist.gov/resources/n-lengths/). Here we have to stress that the potential equation, eq. (I.1.5), is assumed to be rigid in the sense that the object that brings it about is bound to position $r = 0$ and will not recoil when hit by the neutron. As discussed above, this is exactly the situation we are concerned with here for atoms embedded in condensed matter and for this reason the relevant scattering lengths one can find in the tables are specified as "bound scattering lengths."

It is important to observe that the scattering length $b$ not only varies from one isotope of one element to the other (in contrast to X-rays), but can also depend on the nuclear spin:

$$b_i = b_i[1 + c_i(\vec{\sigma}\vec{I}_i)], \tag{I.1.6}$$

where $\vec{\sigma}$ and $\vec{I}_i$ are the neutron and nuclear spins, respectively, and $c_i$ is a constant. For practical reasons that will become clear later, the scattering lengths of elements and isotopes are specified in the tables by their mean value and root mean square deviation over all isotopes in natural elements and/or over the different relative orientations of the neutron and nuclear spins. The first one

$$\bar{b}_i = \langle b_i \rangle \tag{I.1.7}$$

is referred to as "coherent scattering length" and the second one

$$b_i^{\mathrm{inc}} = \sqrt{\langle b_i^2 \rangle - \bar{b}_i^2} \tag{I.1.8}$$

is referred to as "incoherent scattering length."

It is particularly relevant for soft matter research that both the coherent and the incoherent scattering lengths of the two isotopes of hydrogen, $^1$H and $^2$H (or D), are substantial compared to the other elements and quite different from each other (−3.74 fm versus 6.67 fm for the coherent one and 25.27 fm versus 4.04 fm for the incoherent one). Thus, on the one hand and contrary to X-rays, with neutron scattering hydrogen atoms are well observable even if the sample contains heavier elements, and on the other hand, the substitution of a proton in a chemically well-defined position by deuteron allows us to mark and single out particular hydrogen atoms in an ensemble of many.

As pointed out above, in view of the nanoscale structure presented by atomic arrangements, the propagation of neutron radiation inside matter needs to be analyzed in terms of wave mechanics. For the point-like potential equation, eq. (I.1.5), represented by a single atom, the exact wave mechanical answer is that the incoming plane wave $\phi$ in eq. (I.1.4) will be transformed into a sum $\phi'$ of the original plane wave and a spherical wave emitted by the scattering center at $r = 0$ (cf. Figure I.1.2):

$$\phi' = e^{i(\vec{k}\vec{r}-\omega t)} + b\frac{e^{i(kr-\omega t)}}{r}. \tag{I.1.9}$$

Compared to the infinitely extended plane wave, the spherical wave represents negligible particle density, so the incoming plane wave continues without

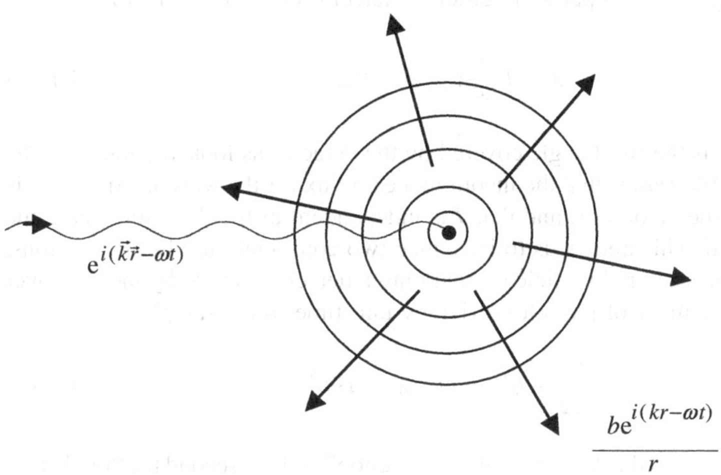


**Figure I.1.2** Scattering of a neutron plane wave on a point-like fixed object generates a spherical wave with an amplitude characterized by the scattering length $b$.

attenuation. This is a fundamental assumption in the commonly used analysis of neutron scattering processes (the so-called first Born approximation).

This approximation is indeed perfectly sufficient for the analysis of soft matter samples, as we will see later in more detail. However, considering the propagation of the beam through a sample containing some $10^{20}$ atoms or more, one needs to consider the attenuation of the beam by the cumulative effects of such a large number of atoms. This can be, however (and most fortunately in view of the close to untreatable complexity it would imply otherwise), perfectly well treated in the classical point-like particle description of the neutron beam—of course again with the exception of large perfect single crystal samples, which we will simply exclude for all what follows.

## I.1.5 SCATTERING CROSS SECTIONS

In order to proceed, we need to introduce the concept of scattering cross section, which tells us what a neutron detector will see in a scattering experiment, for example, in the one on a single atom as given in eq. (I.1.9). The neutron current density described by the wave function $\phi'$ (most commonly called "flux" and measured in units of particles/cm$^2$/s) is given as product of the particle density and the velocity of propagation $v$:

$$j = |\phi'|^2 v. \tag{I.1.10}$$

The detector can only be placed at a large distance from the scattering center compared to atomic dimensions, and at a position where the incoming original wave is masked out by diaphragms defining the beam. Therefore, we only need to consider at the detector the spherical wave component in eq. (I.1.9) and we will arrive at a neutron counting rate for a perfectly efficient detector of effective area $F$:

$$J = F\frac{b^2}{r^2}v = b^2 v\,\mathrm{d}\Omega, \tag{I.1.11}$$

where $\mathrm{d}\Omega = F/r^2$ is the solid angle covered by the detector as looked upon from the sample at $r=0$. We realize that the incoming beam flux for the wave in eq. (I.1.4) is just identical to the velocity $v$, and that $J$ is independent of the direction where the detector is placed. This allows us to introduce two commonly used cross sections, which express the scattered particle current onto a detector surface by the area over which the same number of particles arrive by unit time to the sample:

$$\frac{\mathrm{d}\sigma}{\mathrm{d}\Omega} = b^2 \quad \text{and} \quad \sigma = 4\pi b^2. \tag{I.1.12}$$

The first of these is the "differential cross section" and the second the "total cross section." The first one characterizes the measurable scattered beam intensity around a given detector position and the second one the scattered beam intensity integrated over all directions. Of course, these expressions give the cross sections only for a single bound atom, described by the potential equation, eq. (I.1.4). The cross sections for an ensemble of atoms will be considered below, and we will see that it delivers a lot of information on the atomic structure and atomic motion in the ensemble.

The total cross sections $\sigma$, as defined by eq. (I.1.12), are also given in the usual tables (http://www.ncnr.nist.gov/resources/n-lengths/) in units of barn $= 10^{-24}\,\mathrm{cm}^2$, for both coherent and incoherent scattering lengths. In view of eq. (I.1.8), the sum of coherent and incoherent scattering cross sections gives the grand total per atom in average over all nuclear spin states and all isotopes in the natural element. It is worth recalling that the typical diameter of an atom is in the range of Å, so the geometrical cross section presented by the atom to the neutron beam is in the $10^{-16}\,\mathrm{cm}^2$ range, while neutron scattering cross sections are less than $100\,\mathrm{barn} = 10^{-22}\,\mathrm{cm}^2$. This illustrates the good validity of eq. (I.1.9): only a most tiny fraction of the incoming beam hitting an atom is scattered. The cross section, on the other hand, is quite comparable to the geometrical cross section of the nuclei themselves.

The cross section is also a practical way to describe the probability of absorption of neutron by atoms, with the help of the so-called absorption cross section $\sigma^{\mathrm{abs}}$, defined analogously to the one introduced in the text before eq. (I.1.12). It is a good illustration of the intensity of the strong interaction (which acts between neutrons and nuclei, in form of both scattering and absorption) that $\sigma^{\mathrm{abs}}$ can be much larger than the geometrical cross section of the nucleus. The absorption of neutrons happens independently, without correlation between atoms; the process can be evaluated by adding up the cross sections for the atoms in a sample. For example, if a sample contains $n_i$ atoms/cm$^3$ with an average cross section $\bar{\sigma}_i^{\mathrm{abs}}$, the attenuation of the neutron beam will be given as

$$J(x) = J_0 \exp(-x n_i \bar{\sigma}_i^{\mathrm{abs}}), \tag{I.1.13}$$

where $x$ is the distance in the direction of the neutron beam propagation and $1/n_i\bar{\sigma}_i^{\mathrm{abs}}$ is the penetration range of neutrons for the material.

Note that in the general quantum mechanical formalism the scattering length $b$ is treated as a complex number and the absorption cross section is then related to its imaginary part. This imaginary part will also result in eq. (I.1.9) in a small shift of the phase of the spherical wave. However, the imaginary part of $b$ is in practice only large enough to become measurable for a couple of very highly absorbing isotopes, and for most materials the cross-section tables only cite the real part of $b$ and the absorption cross section itself. Thus, in what follows we implicitly assume that the neutron scattering length $b$ is a real number for all the atoms we have to do with in soft matter. Nevertheless, in order to keep some formulas in the usual more general form, whenever it does not make matter more complicated, we will formally keep the distinction between $b$ and its complex conjugate $b^*$ (and also for related quantities).

The absorption cross section for most isotopes is inversely proportional to the neutron velocity for the slow neutrons we are interested in (as it depends on the period of time the neutron spends near the absorbing nucleus), with the notable exception of a few very highly absorbing isotopes, such as $^{157}$Gd .

There is a final notion related to cross sections we need to introduce: that of elastic and inelastic scattering processes. Eq. (I.1.9) describes an elastic scattering process, since the wave number (momentum) of the outgoing scattered radiation, which becomes in good approximation a plane wave with wave number $k$ at macroscopic distances from the sample, is the same as the wave number of the incoming beam. Thus, incoming and outgoing neutrons have the same velocity, that is, energy. This would not be the case, for example, if we would consider scattering on an object moving with respect to the neutron beam direction. In this case, Doppler effect will occur; for example, if the scattering object moves opposite to the neutron beam with velocity $v_0$, the rebounding neutron will have higher velocity than the incoming one, and this velocity change will reach a maximum $2v_0$ in the direction of backward scattering. This can be easily demonstrated by considering a reference frame in which the scattering object is at rest. Atoms do move in soft matter objects, either around a quasi-equilibrium position or in a diffusive manner or in more complex ways. So the scattered neutron energy $E' = \hbar^2 k'^2/2m$ can be different from the incoming neutron energy $E$. Analogously to the differential cross section, this process can be characterized by the so-called "double differential cross section"

$$\frac{d^2\sigma}{dE'\,d\Omega}, \qquad \text{(I.1.14)}$$

which accounts for the neutrons that reach the detector of solid angle $d\Omega$ with an outgoing final neutron energy within a given energy interval with width $dE'$.

Note that the scattering cross section in the tables (http://www.ncnr.nist.gov/resources/n-lengths/) usually refers to a single atom. In practice, cross sections are more often evaluated for one chemical formula unit of the sample, instead of one atom.

## I.1.6 BEAM PROPAGATION THROUGH SAMPLES

Let us now consider in more detail what happens if instead of just one atom, neutrons scatter on ensemble of bound atoms, each of them described by a Fermi pseudo-potential function similar to eq. (I.1.5). For our purpose, we will be mostly concerned with the superposition of the scattered spherical waves, which will produce a complex pattern of interference between the waves scattered from the individual atomic nuclei, following the many centuries old Huygens principle. Before turning our focus to this final point, we will complete the discussion of the first term in eq. (I.1.9), that is, the fate of the nonscattered part of the incoming beam.

On the one hand, the neutron beam will get attenuated by the absorption when going through substantial amount of matter, as discussed above, and also by the scattering, even if this attenuation could be perfectly neglected when considering only one atom, as in eq. (I.1.9). Although, in contrast to absorption, the total scattering cross section of materials is not simply the sum of the scattering cross sections of the atoms it contains (as we will see in detail below), this is a good rough approximation, in particular for noncrystalline matter and for neutrons with wavelength not greater than about 5 Å. This implies to replace in eq. (I.1.13) the average absorption cross section by the sum of the average absorption and scattering cross sections $\bar{\sigma}_i^{\text{abs}} + \bar{\sigma}_i^{\text{scatt}}$ (cf. http://www.ncnr.nist.gov/resources/n-lengths/).

The attenuation range $\Lambda = 1/n_i(\bar{\sigma}_i^{\text{abs}} + \bar{\sigma}_i^{\text{scatt}})$ is a characteristic parameter for different materials, and it can range from a few μm for highly absorbing materials to a substantial fraction of a meter. Due to the high incoherent cross section of the proton (80.3 barn), organic materials containing high density of hydrogen show strong attenuation ($\Lambda \sim 0.1$–$0.2$ mm), while heavy materials, such as steel, can be quite transparent for neutrons with $\Lambda \sim 1$–$10$ cm, depending on the specific composition of the alloy; for example, Fe displays low absorption, while Co absorbs more than 10 times stronger than Fe. For X-rays, the contrast is quite the opposite: organic matter is generally quite transparent compared to highly absorbing metals.

The attenuation of the neutron beams in matter is being used in neutron radiography, where essentially the neutron shadow of objects is recorded with high-resolution position-sensitive neutron imaging detectors (usually optically recorded scintillator plates), which can achieve above 20 μm resolution. This method is widely used to detect holes, cracks, or highly absorbing organic materials inside metallic parts. The most widespread application is the routine check of airplane jet engine turbine blades for the presence of oil coolant in the cavities inside the blades.

Neutron tomography came of age more recently. It consists of taking the radiographic image of an object in different directions and creating—using well-established computer tomography algorithms—a three-dimensional picture of the object, which can reveal internal structures not accessible to other observation tools without destroying the sample. This method is particularly precious in engineering and in the study of valuable archeological artifacts, as also shown in Figure I.1.3 (Kardjilov et al., 2006).

## I.1.7 REFRACTION AND REFLECTION

On the other hand, the volume average of the potential equation, eq. (I.1.4), of all atomic nuclei will represent a small, but not always negligible contribution to the potential energy felt by the nonscattered part of the wave in eq. (I.1.9) or for the same matter, the classical particle it corresponds to. This potential will depend on the so-called "scattering length density" of the sample $\rho_{\text{scatt}}$, which is the product of the atomic density $\rho_{\text{a}}$ and the average scattering length over all atoms in the sample $\bar{b}$. In inhomogeneous, layered, and other nanostructured materials, $\rho_{\text{scatt}}$ can be a function of the position $\vec{r}$. Thus, we have a total locally averaged potential relevant for the propagation of the nonscattered beam:

$$U(\vec{r}) = mgh - \mu B(\vec{r}) + \frac{2\pi\hbar^2}{m}\rho_{\text{scatt}}(\vec{r}). \qquad \text{(I.1.15)}$$

Here the first term is the gravitational energy, which turns the neutron trajectories into free fall parabolas, in some cases significantly different from straight lines. The other two terms have been discussed in detail above. Note that in ferromagnetic materials

(a)

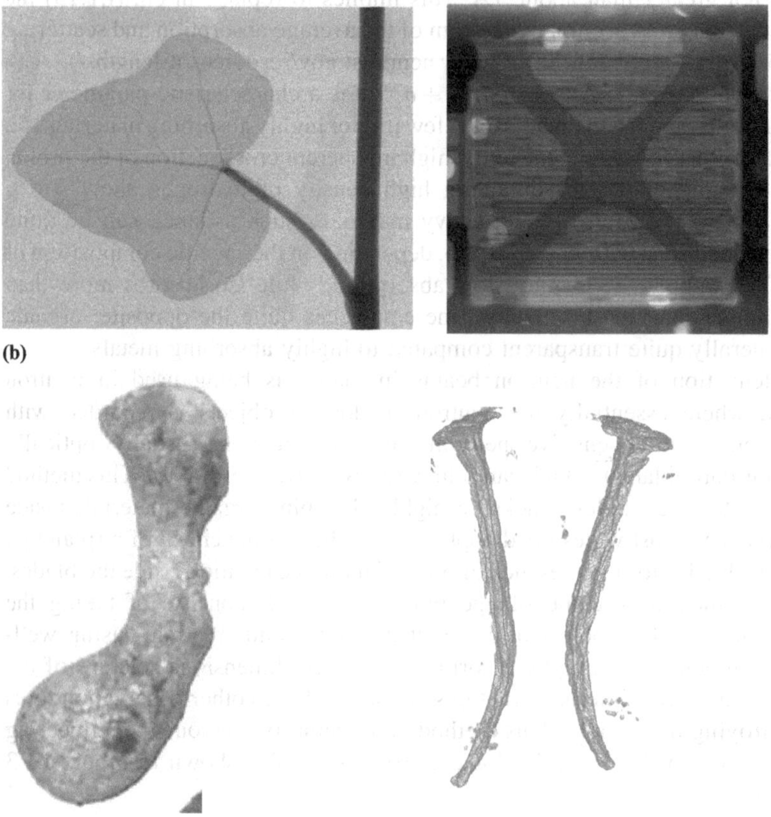

(b)

**Figure I.1.3** Examples of neutron imaging. (a) Neutron radiography (shadow image) of a plant leaf (left) and a fuel cell (right). In the latter image, the dark horizontal lines show the distribution of water inside the closed cell. (b) Neutron computer tomography reconstruction in three dimensions of an object (nail in different views, right) hidden inside an archaeological calcareous concretion shell (left).

the saturation magnetization $\vec{M}$ contributes to the local value of the magnetic field $\vec{B}$ (averaged over interatomic atomic distances) by the substantial amount of $4\pi\vec{M}$.

For most materials, the potential in eq. (I.1.15) corresponds to a potential barrier in the range of up to some 100 neV; that is, the neutrons need to exceed a critical velocity $v_c$ of up to about 5 m/s in order to be able to enter the sample. This is quite a small velocity in general, but it comes to prominent significance in the case the neutrons approach an optically flat surface at a small grazing angle. If the component of the neutron velocity perpendicular to the surface is inferior to $v_c$, the neutron will be totally reflected. This leads to a critical angle of $\theta_c = \arcsin(v_c/v)$, which can amount to about 0.5° for 5 Å wavelength neutrons.

At grazing angles ranging between $\theta_c$ and few times this value, we observe a range of significant partial reflectivity, whose investigation leads to the fast growing field of neutron reflectometry in the past three decades. This chapter of modern neutron research is based on the fact that the reflectivity curve $R(\theta)$ for $\theta > \theta_c$ is determined by the variation of the scattering length density function in the direction $z$ perpendicular to the optically flat sample surface, $\rho_{scatt}(z)$. For example, in a periodic multilayer system one can observe sizable peaks in the reflectivity curve $R(\theta)$, analogous to the Bragg peaks in crystalline matter. For magnetic multilayer structures, the reflectivity might depend very strongly on whether the neutron magnetic moment is oriented parallel or antiparallel to the magnetic field $B$ applied in order to align the magnetization in the layers, cf. eq. (I.1.15). This way magnetic multilayers can serve as powerful neutron beam polarizers by reflecting neutrons with one spin direction with much higher probability than those with the other (Mezei and Dagleish, 1977). A detailed description of the art of neutron reflectometry can be found in Chapter II.2.

## I.1.8 SCATTERING, INTERFERENCE, AND COHERENCE

We now turn our attention to the evaluation of the interference processes that can occur between neutron waves scattered by different atoms. This is the central subject for the study of matter on the atomic scale by actually any radiation (even including electron microscopy, where the atomic scale resolution often apparently achieved by direct imaging is also the result of shrewd processing and reconstruction). When particle waves scattered on different atoms are superposed, the interference between these waves is determined by the optical path length differences between waves scattered at different atoms. The optical path is defined as the real geometrical distance in the direction of wave propagation multiplied by the wave number of the radiation, cf. $\vec{k}\vec{r}$ in eq. (I.1.4). A change of the optical path comparable to unity implies substantial modification of the wave by a shift of its phase by a radian. Hence, the high sensitivity of neutron wave interference to distances comparable to $1/k = \lambda/2\pi$ offers us a spatial resolution capability not accessible by direct imaging methods.

Let us consider the optical paths for scattering on a point-like object (nucleus) at a position $\vec{r}_i$, by comparing it to the optical path for a scattering object at $r = 0$. The incoming radiation will be considered as a plane wave with wave vector $\vec{k}$ and the outgoing radiation as spherical waves from each scattering center with a wave number $k'$, cf. eq. (I.1.9). Here we assume that the detection point is at a very large distance compared to the distance between the scattering points; therefore, the outgoing radiation arrives to the detector practically as a plane wave with wave vector $\vec{k}'$, with a direction determined by the position of the neutron detection. As shown in Figure I.1.4, the optical path difference between the two spherical waves scattered at the origin and $\vec{r}_i$ will be given by

$$\delta s = |\vec{k}\vec{r}_i| + |\vec{k}'\vec{r}_i| = -\vec{q}\vec{r}_i, \tag{I.1.16}$$

where $\vec{q} = \vec{k}' - \vec{k}$ is the so-called momentum transfer vector. If we have to do with elastic scattering, that is, the absolute value of $\vec{k}'$ is equal to that of $\vec{k}$ (i.e., $k$), then we

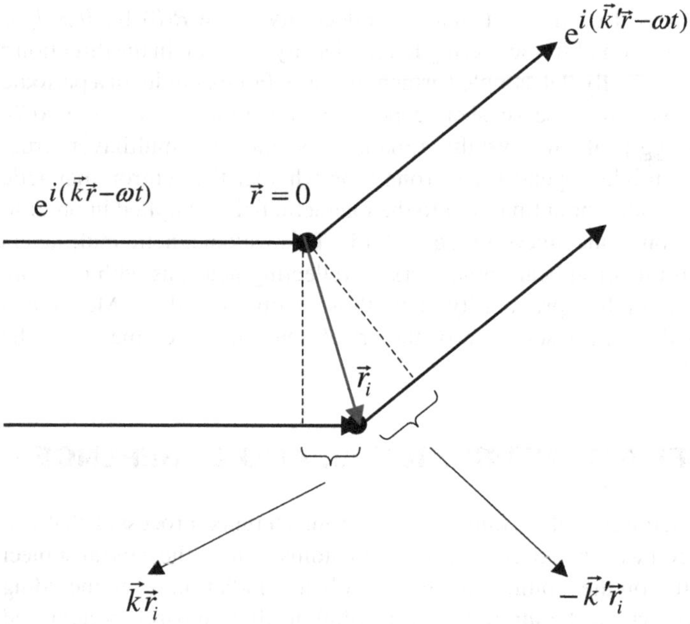


**Figure I.1.4** Optical path difference for scattering on a point-like object at position $\vec{r}_i$ compared to scattering at $r=0$.

will find that $q = 2k\sin(\theta/2)$, where $\theta$ is the angle between the incoming and outgoing radiations, the so-called scattering angle.

Thus, analogously to eq. (I.1.9), the scattered beam wave function becomes (omitting in what follows the nonscattered part of the incoming beam)

$$\phi' = b\frac{e^{i(k'r-\omega' t)}}{r} + b_i\frac{e^{i(k'|\vec{r}-\vec{r}_i|-\omega' t)}}{|\vec{r}-\vec{r}_i|} \cong \frac{e^{i(k'r-\omega' t)}}{r}(b+b_i e^{-i\vec{q}\vec{r}_i}), \tag{I.1.17}$$

where we took into account that the detector is at a distance $r$ very much larger than the distance between the scattering nuclei $r_i$. Note that eq. (I.1.9) is valid for a fixed, rigid scattering center and implies elastic scattering with equal absolute values $k=k'$, so this will also apply for the ensembles of scattering objects considered in the next paragraphs. The square of absolute value of the wave function (i.e., the finding probability) thus becomes

$$|\phi'|^2 \cong \frac{1}{r^2}|(b+b_i e^{-i\vec{q}\vec{r}_i})|^2 = \frac{1}{r^2}|b+b_i\cos(\vec{q}\vec{r}_i) - ib_i\sin(\vec{q}\vec{r}_i)|^2 \tag{I.1.18}$$

and following the definition of cross section in eqs. (I.1.10)–(I.1.12) we arrive at the scattering cross section

$$\frac{d\sigma}{d\Omega} = \left[b^2 + 2bb_i\cos(\vec{q}\vec{r}_i) + b_i^2\right]. \tag{I.1.19}$$

Eq. (I.1.19) illustrates how we can observe atomic scale structures by observing the neutron scattering cross section as a function of the momentum transfer vector $\vec{q}$. From its definition in connection with eq. (I.1.16), we can see that the simplest way to vary $\vec{q}$ is to observe the scattering at different scattering angles and we can also change the direction of $\vec{r}_i$ by rotating the sample. Thus, in this simplest example given in eq. (I.1.19), one will be able to determine both the distance between the two atoms with a good precision on the Å length scale and their scattering lengths $b$ and $b_i$, which could help to identify them. As a matter of fact, diatomic molecules in a low-density gas are a practical example for this trivial model, if the motion of these molecules is slow enough that it can be neglected.

Two fundamental assumptions have been implicitly used in deriving eqs. (I.1.16)–(I.1.19): (a) the incoming radiation for both scattering objects is the same plane wave with wave vector $\vec{k}$ and (b) the outgoing radiation for both scattering objects has the same wave vector $\vec{k}'$ (which also implies the same frequency). Point (b) is a simple practical requirement: the optical path length difference between radiations with different wave vectors would depend on the exact position of the neutron detection. Since neutron detection happens randomly within a macroscopic volume in the detectors, such a dependence of neutron paths leads to an averaging over a broad distribution of relative phase differences of the superposed waves; thus, no interference effect will be observable.

Point (a) is a more fundamental restriction: there is no way to prepare an incoming neutron beam in which wave components with different wave vectors $\vec{k}$ would display a correlation of the initial phases $\varphi_0$ ("random phase theorem"). Although one can mathematically easily write down wave functions with such phase correlation (coherence) between different components with different wave numbers (as it is done in textbooks discussing quantum mechanical "wave packets" of different "coherence lengths"), with real-life elementary particles such as neutrons there is no method known to reproduce such a wave packet with identical relative phases between components with different wave numbers. This is one of the most fundamental points in understanding the radiation created by some kind of thermodynamic processes (as opposed to macroscopically generated radiation, like a resonating cavity driven by a radio frequency power supply): every wave is only coherent with "itself"; that is, there is no coherence between radiations with different wave vectors $\vec{k}$. Therefore, all considerations in neutron scattering make physical sense only if they refer to a perfectly well-defined neutron wave vector (and hence neutron velocity), and all radiation must be considered as an incoherent, classical ensemble of particles with infinitely well-defined velocities/wave vectors. In view of our introductory discussion, coherence between different neutron wave components can practically only be generated by the scattering process in the sample, for example, in eq. (I.1.9) between the plane and spherical wave components.

## I.1.9 CROSS SECTIONS AND PAIR CORRELATION FUNCTIONS

Eq. (I.1.18) can be readily generalized to derive the general expression of the cross section for elastic scattering on an arbitrary ensemble of atoms. Let us describe the ensemble of atoms by a microscopic scattering length density function $\rho(\vec{r})$, in which each nucleus will appear as a delta function in view of eq. (I.1.5). Thus, in contrast to the average scattering length density function $\rho_{\text{scatt}}$ introduced above, the microscopic scattering length density function reflects the atomic structure in all details. Then the sum over the two scattering objects in eq. (I.1.18) needs to be extended to an integral over the whole volume of the sample, and similarly to eq. (I.1.17) we get for the scattering cross section

$$\begin{aligned}\frac{d\sigma}{d\Omega} &= \left|\int_V \rho(\vec{R})e^{-i\vec{q}\vec{R}}\, d\vec{R}\right|^2 = \int_V \rho^*(\vec{R})e^{i\vec{q}\vec{R}}\, d\vec{R} \int_V \rho(\vec{R})e^{-i\vec{q}\vec{R}}\, d\vec{R} \\ &= \iint_V \rho^*(\vec{R})\rho(\vec{R}')e^{i\vec{q}\vec{R}-i\vec{q}\vec{R}'}\, d\vec{R}\, d\vec{R}',\end{aligned} \tag{I.1.20}$$

where the integrations are made over the volume of the sample, $V$. We introduce as new variable the distance between points $\vec{R}$ and $\vec{R}'$, that is, $\vec{r} = \vec{R}' - \vec{R}$:

$$\frac{d\sigma}{d\Omega} = \iint_V \rho^*(\vec{R})\rho(\vec{R}+\vec{r})e^{-i\vec{q}\vec{r}}\, d\vec{R}\, d\vec{r} = \int_V d\vec{R} \int_V \rho^*(\vec{R})\rho(\vec{R}+\vec{r})e^{-i\vec{q}\vec{r}}\, d\vec{r}. \tag{I.1.21}$$

Here the first integration is a summation over the volume of the sample, which effectively means an averaging of the second integration over the whole volume of the sample:

$$\frac{d\sigma}{d\Omega} = \int_V \langle\rho^*(\vec{R})\rho(\vec{R}+\vec{r})\rangle e^{-i\vec{q}\vec{r}}\, d\vec{r} = \int_V g(\vec{r})e^{-i\vec{q}\vec{r}}\, d\vec{r} = S(\vec{q}). \tag{I.1.22}$$

As mentioned above, the cross section is usually expressed per some essentially freely defined unit amount of the sample matter, usually chemical formula unit, characteristic atoms, and elementary structural unit, among others. This normalization is included here in the definition of the averaging under the $\langle\cdots\rangle$ sign. To calculate the total scattering power of the amount of material in the actual sample, one needs to determine the number of reference units contained in the sample volume. Fundamentally, the total cross section of the sample should be proportional to its total volume, which is indeed the case if the beam does not suffer essential attenuation within the sample. To correct for such "self-shielding" effect, we have to consider the beam attenuation factors discussed above.

The average indicated by $\langle\rho^*(\vec{R})\rho(\vec{R}+\vec{r})\rangle = g(\vec{r})$ is called the pair correlation function and gives the probability of finding a particle in the sample at position $\vec{R}+\vec{r}$ if there is one at position $\vec{R}$. So this function describes the correlations between atoms in the sample. In noncrystalline materials, all directions are equivalent; thus, $g(r)$ becomes a function of the absolute value of the distance $r$. If, for example, there are no atoms (i.e., atomic nuclei) in the sample at distance $r$ from each other, $g(r)$ will be zero. This is the case for distances shorter than the minimum separation of nearest neighbors. In contrast, in amorphous matter $g(r)$ has a maximum at around the average distance of nearest neighbors.

The integrals in eq. (I.1.22) are, paradoxically, independent of the sample volume. In particular in noncrystalline matter, but also in polycrystalline samples or nonperfect single crystals, there is a characteristic correlation length $\xi$ in the sample structure: there will be no correlation between the positions of atoms further apart than this distance. It is easiest to envisage this in a polycrystalline material. There is a periodic order between atoms within one crystalline grain, but beyond that the relative positions are random. This is expressed by behavior of $g(r)$ for $r > \xi$. It will tend to a constant, the square of the mean scattering density $\bar{\rho}^2$, and the integration of the periodic function $e^{-iqr}$ will average to zero for $r > \xi$. Thus, the so-called "structure factor" $S(\vec{q})$, defined in eq. (I.1.22) as the Fourier transform of the correlation function $g(\vec{r})$, will have a value independent of the integration volume considered, if it extends over the range of correlation $\xi$.

This observation contains one of the other fundamental facts about neutron scattering and similar experiments. A macroscopic sample in fact consists of a multitude of independent samples with the dimension of the sample correlation length $\xi$ and what we measure is the average behavior of these myriads of subsamples within the typically cm$^3$ large sample volume. The real physical sample volume, the coherence volume of the order of $\xi^3$ in soft matter or polycrystalline samples, typically ranges from $10^{-22}$ to some $10^{-6}$ cm$^3$. So indeed, the observed cross sections are good averages over the many possible atomic configurations and orientations to realize the more or less short-range local atomic order characterizing the material we are concerned with.

We will come back to the exact interpretation of the cross section formulas, eqs. (I.1.20)–(I.1.22), we were able to deduce in rather conspicuous matter: simple mindedly, adding the spherical waves emitted by all atoms in the sample, á la Huygens. This simple approach shows clearly the mechanism involved, how the interference between these individual spherical waves carry information on the atomic arrangement, cf. the example of eqs. (I.1.16)–(I.1.19). Nevertheless, there is an important underlying approximation: the atoms are not rigid in a real sample; there is always some degree of motion. The exact theory, which indeed contains no approximation with the one exception that the sample has to be small enough to avoid substantial beam attenuation, is due to van Hove (1954). It is mathematically more complex and less evident to visualize. Its result is, however, just a very plausible generalization of eq. (I.1.22).

## I.1.10 DYNAMIC STRUCTURE FACTOR

Let us define a more general pair correlation function similar to $g(\vec{r})$ in eq. (I.1.22):

$$g(\vec{r},t) = \langle \rho^*(\vec{R},0)\rho(\vec{R}+\vec{r},t)\rangle. \tag{I.1.23}$$

This van Hove correlation function expresses the probability that, if there was an atom at position $\vec{R}$ at time 0, we will find an atom at a location displaced by $\vec{r}$ at a later time $t$. Then the double differential cross section will be determined by the space–time Fourier transform of correlation function, eq. (I.1.23):

$$\frac{\mathrm{d}^2\sigma}{\mathrm{d}\Omega\,\mathrm{d}E'} = \frac{k'}{k}\int_V \mathrm{d}\vec{r}\,\frac{1}{2\pi\hbar}\int_{-\infty}^{\infty} g(\vec{r},t)e^{-i(\vec{q}\vec{r}-\omega t)}\,\mathrm{d}t = \frac{k'}{k}S(\vec{q},\omega). \qquad \text{(I.1.24)}$$

$S(\vec{q},\omega)$ is the so-called "scattering function" or "dynamic structure factor" and the origin of the prefactor $k'/k$ is the fact that the neutron current density is proportional to the beam velocity, cf. eq. (I.1.10). One important feature of this exact result is that the scattering function as defined in the expression of the double differential cross section in eq. (I.1.24) is a function of two parameters describing the scattering process only: the momentum transfer $\vec{q}$, as introduced in eq. (I.1.16), and the neutron energy transfer $\hbar\omega = E'-E$, where $E$ and $E'$ are the incoming and outgoing neutron energies $mv^2/2$, respectively. The often considered intermediate scattering function is

$$I(\vec{q},t) = \int_{-\infty}^{\infty} S(\vec{q},\omega)e^{-i\omega t}\,\mathrm{d}\hbar\omega \equiv \int_V g(\vec{r},t)e^{-i\vec{q}\vec{r}}\,\mathrm{d}\vec{r}. \qquad \text{(I.1.25)}$$

It is of particular practical significance in model calculations and in neutron spin echo (NSE) spectroscopy (cf. later in this book).

With the exact results, eqs. (I.1.24) and (I.1.25), at hand, we can now return to the interpretation of the time-independent correlation function $g(\vec{r})$ involved in "elastic" scattering experiments, cf. eq. (I.1.22). If it is defined to describe the correlations at a given moment of time, say $t=0$ ("equal time" or "static" correlation function), then it will correspond to the intermediate correlation function at $t=0$, that is,

$$I(\vec{q},t=0) = \int_{-\infty}^{\infty} S(\vec{q},\omega)\,\mathrm{d}\hbar\omega = \int_{-\infty}^{\infty} \frac{k}{k'}\frac{\mathrm{d}^2\sigma}{\mathrm{d}\Omega\,\mathrm{d}E'}\mathrm{d}E' = S(\vec{q}) = \int_V g_{\mathrm{stat}}(\vec{r})e^{-i\vec{q}\vec{r}}\,\mathrm{d}\vec{r}. \qquad \text{(I.1.26)}$$

The integration of the double differential cross section here over all energy changes is feasible only within a more or less fair approximation. In the so-called "elastic" neutron scattering experiments, in which the final neutron energy is not analyzed, only the angular dependence of the scattered neutron intensity is determined (e.g., diffraction). The expression "elastic" is fully misleading here, actually what one tries to accomplish is just to integrate over all energy transfers in order to obtain the differential cross section $\mathrm{d}\sigma/\mathrm{d}\Omega$. The rough approximation here is that for neutron energy changes corresponding to the frequencies of atomic motion in soft matter the wave number of the scattered neutrons $k'$ will span a substantial range and a given scattering angle will not correspond to scattering with a given momentum transfer $q$, cf. the comments after eq. (I.1.16). The kinematic prefactor $k/k'$ is also not taken care of in such an "elastic" experiment. The smaller the deviations from an apparent $q$ (Placek corrections), the higher the incoming neutron energy. As a matter of fact, with the very high energies of X-rays in the 10 keV range, these deviations become negligible and the static X-ray structure factor $S_X(\vec{q})$ can indeed be directly observed in view of the inherently correct integration over the energy in "elastic" X-ray experiments, cf. eq. (I.1.26).

On the other hand, eq. (I.1.22) will exactly hold, corresponding to its derivation for a rigid array of atoms, if we define the correlation function $g$ as the one for infinite times $g_{\text{avr}}(\vec{r})$, that is, looking at the correlations in the long time average of the scattering density $\rho_\infty(\vec{r})$. This will correspond to the real elastic scattering with energy change $E=0$, which of course can only be exactly determined experimentally by using an energy discrimination method to single out the elastic scattering contribution. This can be equated to the $t=\infty$ limit of the intermediate scattering function. However, in solid samples and in particular at low temperatures, much of the scattering is elastic (e.g., Bragg peaks in crystalline matter), so the "elastic" scattering experimentation (i.e., without analyzing the scattered beam energy) indeed corresponds to mostly elastic scattering, and there is little difference between the static ($t=0$ "equal time") correlations and the "time-averaged" ($t=\infty$) correlations.

## I.1.11 DEBYE–WALLER FACTOR: COHERENT AND INCOHERENT SCATTERING

It is also worth evaluating eq. (I.1.20) for the case when the neutron scattering density $\rho(\vec{r})$ corresponds just to the sum of $\delta$ functions corresponding to different nuclei $i$ with scattering lengths $b_i$ at locations $\vec{r}_i$, that is, $\sum b_i\delta(\vec{r}-\vec{r}_i(t))$. We readily get

$$\frac{d\sigma}{d\Omega} = \left|\sum_i b_i e^{-i\vec{q}\vec{r}_i}\right|^2 = \sum_{i,j} b_i^* b_j e^{i\vec{q}(r_j - r_i)}. \tag{I.1.27}$$

This case applies only to the static $t=0$ equal time correlations, since only at a well-defined time instant can the atomic nuclei be assigned to $\delta$ functions at exact positions.

In contrast, the long time average even for nuclei stationarily sitting at equilibrium positions will be, instead of a $\delta$ function, a density distribution function with a width corresponding to the amplitude of the vibrational atomic motions around the equilibrium positions. In the simplest approximation, these displacements in time will lead to a Gaussian distribution of the time-averaged probability of finding an atom around its equilibrium position, for example,

$$\rho_{i,\infty}(\vec{u}) \propto e^{-x^2/2\sigma_{i,x}^2 + y^2/2\sigma_{i,y}^2 + z^2/2\sigma_{i,z}^2}, \tag{I.1.28}$$

where $x$, $y$, and $z$ are the coordinates of the displacement $\vec{u} = \vec{R} - \vec{r}_i$ and $\sigma_{i,x}$ is the root mean square displacement of atom $i$ from the equilibrium position $\vec{r}_i$ in the $x$-direction. This then leads to the time-averaged density function for the whole sample:

$$\rho_\infty(\vec{R}) = \sum_i b_i \rho_{i,\infty}(\vec{R} - \vec{r}_i) \tag{I.1.29}$$

and the integration by $\vec{R}$ in eq. (I.1.20) will result in Fourier transforming $\rho_{i,\infty}(\vec{r})$ into the so-called structure factor $f_i(\vec{q})$ (or thermal factor), and eq. (I.1.27) becomes for the exactly elastic scattering representing the time-averaged structure

$$\left(\frac{\mathrm{d}\sigma}{\mathrm{d}\Omega}\right)_{\text{elastic}} = \left|\sum_i b_i f_i(\vec{q}) e^{-i\vec{q}\vec{r}_i}\right|^2 = \sum_{i,j} b_i^* b_j f_i^*(\vec{q}) f_j(\vec{q}) e^{i\vec{q}(r_j - r_i)}. \tag{I.1.30}$$

In a system that is symmetric to the change of the direction of the coordinates (e.g., $x$ to $-x$), the structure factors $f_i$ have real values. Also note that $f_i(0) = 1$ by definition, which is the consequence of the conditions that $\int \rho_{i,\infty}(\vec{r})\mathrm{d}\vec{r} = b_i$, since $\rho_{i,\infty}(\vec{r})$ describes how the time average of the finding probability of the one atom $i$ is distributed in space.

Let us assume that all atoms in the sample perform isotropic thermal vibrations around their equilibrium positions by about the same amplitude, which can be characterized by the mean square displacement $\langle u^2 \rangle$ around the equilibrium positions. Then the Gaussian distribution in eq. (I.1.28) takes the form $\exp(-r^2/2\langle u^2/3\rangle)$ and its Fourier transform becomes $f_i(\vec{q}) = \exp(-q^2\langle u^2\rangle/6)$, that is, $f_i^*(\vec{q})f_j(\vec{q}) = \exp(-q^2\langle u^2\rangle/3)$. Finally, eq. (I.1.30) can be written as

$$\left(\frac{\mathrm{d}\sigma}{\mathrm{d}\Omega}\right)_{\text{elastic}} = e^{-2W}\left|\sum_i b_i e^{-i\vec{q}\vec{r}_i}\right|^2 = e^{-2W}\sum_{i,j} b_i^* b_j e^{i\vec{q}(r_j - r_i)}. \tag{I.1.31}$$

Here the so-called Debye–Waller factor $e^{-2W} = \exp(-q^2\langle u^2\rangle/3)$ can be approximated at small mean square displacements (moderate temperatures) as $e^{-2W} = 1 - q^2\langle u^2\rangle/3$.

Note that for magnetic atoms the structure factor plays a more fundamental role and it is only marginally related to the thermal vibrations of the atoms. As we have pointed out above, the neutrons by their magnetic moment also see the magnetic fields, cf. eq. (I.1.15). The magnetism of atoms comes from their electrons, not from the point-like nuclei, so the strong microscopic magnetic field characterizing each magnetic atom extends practically over the whole atomic diameter. For this reason, the atomic magnetic interaction for the neutron corresponds to a density distribution over the whole electron cloud of the atom (i.e., it reaches out for far larger distances than the atomic vibrations in a solid) and consequently the corresponding magnetic scattering length will need to be characterized including a structure factor $f_\mathrm{M}(\vec{q})$ defined with a normalization such that $f_\mathrm{M}(0) = 1$. Thus, in eq. (I.1.27) for the magnetic atoms we have to replace $b_i$ by $b_i + p_i^{\mathrm{eff}} f_{\mathrm{M}}(\vec{q})$, where the magnetic scattering length of the atom $p_i^{\mathrm{eff}}$ depends in an intricate manner on the neutron spin and atomic magnetization, including their relative orientation. These more special aspects of magnetic neutron scattering will not be discussed here; the interested reader is referred to the literature (Squires, 1978).

We will conclude this introduction to the basic features of neutron scattering by a discussion of coherent and incoherent scattering. It was pointed out above that the scattering length $b_i$ for a given atomic species can depend on further details of the nuclei involved, for example, many elements have different isotopes and many isotopes have nuclear spin and hence different nuclear spin states. We will examine the consequence of this in mathematical detail for the simple expression of the cross section, eq. (I.1.27), and will generalize the result to the van Hove cross-section formula by analogy. Let $\bar{b}_i$ denote the average scattering length for a given atomic species (e.g., Ni, which has many isotopes, or $^1$H, which can display different spin states with respect the neutron spin), as discussed in connection with eqs. (I.1.7) and (I.1.8). We will then explicitly split the terms in the sum in eq. (I.1.27) into the average values and the deviations from the average:

$$\frac{d\sigma}{d\Omega} = \sum_{i,j} (\bar{b}_i^* \bar{b}_j + b_i^* b_j - \bar{b}_i^* \bar{b}_j) e^{i\vec{q}(r_j - r_i)} = \sum_{i,j} \bar{b}_i^* \bar{b}_j e^{i\vec{q}(r_j - r_i)} + \sum_{i,j} (b_i^* b_j - \bar{b}_i^* \bar{b}_j) e^{i\vec{q}(r_j - r_i)}. \tag{I.1.32}$$

The key of the matter now is to remember that in any sample (except perfect crystals) we have a correlation length $\xi$ beyond which the atoms are randomly placed and therefore do not contribute to the sums involved here. So in actual fact each sample has to be considered as an ensemble of a large number of independent, distinct "subsamples" of the material studied, and the cross section will be an average over all these subsamples. The first sum on the right-hand side of eq. (I.1.32) is the same for all subsamples, and it is called "coherent" scattering. Considering the average of any element $(i, j)$ in the second sum over all subsamples, we need to keep in mind the definition of the average value $\bar{b}_j = \langle b_j \rangle$. Since under ordinary conditions isotopes and nuclear spins states are fully randomly distributed in the sample, there is no correlation between the occupations of site $i$ and site $j$ by these possible choices. Therefore, if $i \neq j$

$$\langle b_i^* b_j - \bar{b}_i^* \bar{b}_j \rangle = \langle b_i^* \rangle \langle b_j \rangle - \bar{b}_i^* \bar{b}_j \equiv 0 \tag{I.1.33}$$

while for $i = j$

$$\langle b_i^* b_i - \bar{b}_i^* \bar{b}_i \rangle = \langle b_i^* b_i \rangle - \bar{b}_i^* \bar{b}_i = (b_i^{\mathrm{inc}})^2 \tag{I.1.34}$$

in view of the definition of the incoherent scattering length in eq. (I.1.8), which is introduced under the practical assumption that $b_i$ are real numbers (http://www.ncnr.nist.gov/resources/n-lengths/). Thus, eq. (I.1.32) becomes

$$\frac{d\sigma}{d\Omega} = \sum_{i,j} \bar{b}_i^* \bar{b}_j e^{i\vec{q}(r_j - r_i)} + \frac{1}{4\pi} \sum_i \sigma_i^{\text{inc}}. \quad \text{(I.1.35)}$$

Similarly we can rewrite eq. (I.1.33):

$$\left(\frac{d\sigma}{d\Omega}\right)_{\text{elastic}} = \sum_{i,j} \bar{b}_i^* \bar{b}_j f_i^*(\vec{q}) f_j(\vec{q}) e^{i\vec{q}(r_j - r_i)} + \frac{1}{4\pi} \sum_i |f_i(\vec{q})^2| \sigma_i^{\text{inc}}. \quad \text{(I.1.36)}$$

In practical terms, in particular in soft matter or matter in other than solid phase, the elastic scattering is defined by the resolution of the spectrometer used to distinguish between elastic and inelastic scattering contributions. So the structure factors and, for the same matter, the mean square displacements of atoms depend on the timescale selected to build the averages $\bar{\rho}_i(\vec{r})$ and $\langle u_i^2 \rangle$, as determined by the experimental resolution $\delta E$ of the spectrometer by the relation $t_{\text{avr}} = \hbar/\delta E$. With this definition of the averaging timescale kept in mind, the experimentally determined factor $|f_i(\vec{q})^2|$ in eq. (I.1.36) is called the incoherent elastic structure factor (EISF). It describes the Fourier transform of the probability distribution $\bar{\rho}_i(\vec{r})$ of finding atom $i$ at different locations around its average position over the averaging time $t_{\text{avr}}$. For example, for a hydrogen bond with a proton tunneling back and forth between two positions, this distribution will consist of two spots (broadened by vibrations) where the atom can be.

The plausible generalization of eqs. (I.1.35) and (I.1.36) to the case of the double differential cross sections as expressed by the van Hove correlation functions can be achieved by considering two correlation functions. The first one gives rise to the coherent scattering by considering the correlation between all particles (summation over all pairs of atoms $i$ and $j$) weighted by their coherent scattering length for each atomic species in the sample (i.e., averaged over all isotopes and nuclear spin states for each atomic species):

$$g_{\text{coh}}(\vec{r},t) = \langle \bar{\rho}^*(\vec{R},0) \bar{\rho}(\vec{R}+\vec{r},t) \rangle \quad \text{with} \quad \bar{\rho}(\vec{r},t) = \sum_i \bar{b}_i \delta(r - r_i(t)), \quad \text{(I.1.37)}$$

$$\left(\frac{d^2\sigma}{dE'\, d\Omega}\right)_{\text{coh}} = \frac{k'}{k} \int_V d\vec{r} \frac{1}{2\pi\hbar} \int_{-\infty}^{\infty} g_{\text{coh}}(\vec{r},t) e^{-i(\vec{q}\vec{r} - \omega t)}\, dt = \frac{k'}{k} S_{\text{coh}}(\vec{q},\omega). \quad \text{(I.1.38)}$$

The second one corresponds to considering only the self-correlations of the atoms, that is, the correlation between the positions of the same atom at time $t = 0$ and at an arbitrary time $t$, weighted by the incoherent scattering length for each atomic species (i.e., the root square deviation from the average for that atomic species):

$$g_{\text{self}}(\vec{r},t) = \left\langle \sum_i \rho_i^{\text{inc}}(\vec{R},0) \rho_i^{\text{inc}}(\vec{R}+\vec{r},t) \right\rangle \quad \text{with} \quad \rho_i^{\text{inc}}(\vec{r},t) = b_i^{\text{inc}} \delta(r - r_i(t)),$$

$$\text{(I.1.39)}$$

$$\left(\frac{\mathrm{d}^2\sigma}{\mathrm{d}E'\,\mathrm{d}\Omega}\right)_{\mathrm{inc}} = \frac{k'}{k}\int\limits_V \mathrm{d}\vec{r}\,\frac{1}{2\pi\hbar}\int\limits_{-\infty}^{\infty} g_{\mathrm{inc}}(\vec{r},t)e^{-i(\vec{q}\vec{r}-\omega t)}\,\mathrm{d}t = \frac{k'}{k}S_{\mathrm{inc}}(\vec{q},\omega). \tag{I.1.40}$$

For mathematical completeness, note that the incoherent scattering length as defined in eq. (I.1.8) and given in the tables (http://www.ncnr.nist.gov/resources/n-lengths/) can only be a real, positive number.

The simplest example of a self-correlation function often encountered in soft matter is the one describing conventional diffusion of atoms, such as hydrogen, which can be readily observed by inelastic neutron scattering due to the high incoherent cross section of the proton. If a particle was found at $r=0$ at $t=0$, at a later time $t$ its finding probability will be described by a Gaussian distribution around its original position with the mean square distance for displacement in the diffusion process linearly increasing with the time: $\sigma^2 \propto t$. Thus,

$$g_{\mathrm{self}}(\vec{r},t) \propto (b^{\mathrm{inc}})^2 e^{-3r^2/2\sigma^2}. \tag{I.1.41}$$

By Fourier transformation in the space variable $r$, we get the intermediate scattering function, which will also be a Gaussian function of the form $\exp(-q^2\sigma^2/6)$. Keeping in mind that $\sigma^2 \propto t$, we can rewrite the exponent here by introducing an appropriate proportionality constant $D$ and we get

$$I_{\mathrm{inc}}(\vec{q},t) = (b^{\mathrm{inc}})^2 e^{-q^2 D|t|}. \tag{I.1.42}$$

This intermediate scattering function satisfies the requirement that at $t = 0$ it has to be equal to the differential cross section $\mathrm{d}\sigma/\mathrm{d}\Omega$. The observation of this scattering function by NSE spectroscopy is an efficient way to study diffusion processes. In soft and glassy matter, one often finds unconventional "sublinear" diffusion, which implies that the mean square displacement evolves with time slower than linearly, that is, $\sigma^2 \propto t^{\beta}$, with $\beta < 1$. This leads in eq. (I.1.42) to the so-called stretched exponential form for the time dependence of the intermediate scattering function: $e^{-(t/\tau)^{\beta}}$ instead of the conventional $e^{-t/\tau}$, with $\tau$ being the characteristic time constant. By inspecting the right-hand side of eq. (I.1.42), we find that $\tau$ varies as $q^{-2}$ in the case of conventional diffusion. In contrast, we find $\tau \propto q^{-2/\beta}$ for the unconventional sublinear diffusion.

The double differential cross section will be obtained by Fourier transformation of the intermediate scattering function in eq. (I.1.42) with respect to the time variable. This results in the well-known scattering function for the double differential inelastic cross section consisting of a Lorentzian line around zero energy transfer $\hbar\omega$ (the so-called quasielastic scattering):

$$S_{\mathrm{inc}}(q,\omega) = \frac{(b^{\mathrm{inc}})^2}{\pi}\,\frac{\Gamma^2}{\Gamma^2 + (\hbar\omega)^2}, \tag{I.1.43}$$

where the full width at half maximum (FWHM) of the Lorentzian line is $\Gamma = Dq^2$. This behavior is the well-known signature of conventional diffusion processes, which is often explored by incoherent quasielastic neutron scattering.

We have obtained this result by using a classical mechanical picture of particle diffusion, which leads to the Gaussian correlation function, eq. (I.1.41). This is an excellent approximation at high sample temperatures, that is, for times $t$ that are much longer than the characteristic time of thermal fluctuations given by the thermal energy $k_B T$ as $t_{th} = \hbar / k_B T$. In other terms, this condition will translate into the relation $\Gamma = Dq^2 \ll k_B T$.

### I.1.12 DETAILED BALANCE, BOSE FACTOR

Inelastic neutron scattering involves energy changes of the neutron: either the incoming neutron transfers energy to the sample and loses energy in the scattering process or the neutron can take up energy from the sample in the scattering process and go out with an energy gain. The probabilities of both processes are exactly implied in the mathematical properties of the van Hove correlation functions. We will recall here two significant aspects, which are important quantum mechanical requirements. We will use plausibility arguments to introduce them.

The fundamental difference between classical and quantum notions of processes involving exchange of energy between objects is that quantum mechanics revealed that this exchange can happen only in quanta; that is, in a process with angular frequency $\omega$, the energy change can only be a multiple of $E = \hbar\omega$. In the exchange of energy between the sample and the neutron, this becomes a prominent effect, when the thermal energy of the sample $k_B T$ is smaller than or comparable to $\hbar\omega$. Here $k_B$ is the Boltzmann constant and $T$ is the temperature. Since the thermal fluctuations provide the energy for the sample that can be communicated to the neutron, it takes a long time—that is, will become a process of lower probability—to wait for a thermal fluctuation that can deliver energy more than $k_B T$. Actually, this probability is known to be proportional to $\exp(-E/k_B T)$. This leads to an expression in the detailed balance property of the dynamic structure factor $S(\vec{q}, \omega)$:

$$S(\vec{q}, \omega) = e^{-\hbar\omega/k_B T} S(\vec{q}, -\omega), \qquad \text{(I.1.44)}$$

where $\omega > 0$ means energy gain for the neutron in the scattering and $\omega < 0$ means neutron energy loss. Mathematically, we also need to assume here that the reversal of the direction of $\vec{q}$ has no effect on the sample (i.e., it is inherently microscopically symmetric), which generally holds for soft matter. This so-called detailed balance condition implies that the probability of the scattering process with neutron energy gain $E = \hbar\omega > 0$ is lower than that of the process with equal neutron energy loss (i.e., energy transferred to the sample) just by the thermal probability for the sample to be in a higher energy initial state by this amount.

Detailed balance is a very general requirement and it is exactly obeyed in nature, if the sample is in thermal equilibrium, that is, if it can be characterized by a unique temperature. Beyond spurious inhomogeneities of the sample temperature, there can also be a physically founded breakdown of thermal equilibrium, if the temperature of one part of the system develops extremely long thermal equilibration times. A practical case of interest is the temperature of nuclear spin systems, which occasionally are very weakly coupled to the rest of the sample. A prominent case is molecular hydrogen, where the thermal equilibrium between the *ortho* and *para* states at low temperatures can take several days to be reached.

The other property of $S(\vec{q}, \omega)$ we will consider is related to the temperature dependence of the inelastic scattering probabilities. Let us assume, as an example, that the sample has a spectrum of excitations, which can be characterized by the density of states function $Z(\omega)$. It tells us how many excited states are between energy $\hbar\omega$ and $\hbar(\omega + \delta\omega)$, where by definition $\omega$ is positive. We can envisage the state of the system at a given temperature, by assuming that these excited states are occupied by a factor $n$, which is given by the Bose statistical occupation number

$$n = \frac{1}{e^{\hbar\omega/k_B T} - 1}, \tag{I.1.45}$$

where by definition $\omega > 0$ must hold. For high temperatures, this expression becomes in good approximation $n = \hbar\omega/k_B T$. The probability for the neutron to gain $\hbar\omega$ energy in the scattering will be proportional to the occupation number $n$ of the states with this energy. In contrast, the probability for the neutrons to lose energy (i.e., excite an excitation with energy $\hbar\omega$) is required by quantum mechanics to be $n + 1$. This is summarized by the Bose factor $n_B(\omega)$ for the temperature dependence of the cross section for neutron to scatter on this spectrum of excitation $Z(\omega)$:

$$n_B(\omega) = \begin{cases} n + 1, & \omega < 0, \\ n, & \omega > 0. \end{cases} \tag{I.1.46}$$

It is to be stressed that the sign of $\omega$ here refers to the sign of the neutron energy change, and $|\omega|$ is to be used for the calculation of the occupation number $n$ according to eq. (I.1.45). Note that with this definition $n_B(\omega)$ satisfies the detailed balance condition, eq. (I.1.44).

Measured scattering functions $S(\vec{q}, \omega)$ often show a temperature dependence proportional to $n_B(\omega)$, which suggest that indeed the sample displays a spectrum of excitations $Z(\omega)$, which itself is independent of the temperature, that is, only the occupation numbers change. Such a temperature dependence governed by the Bose factor is called harmonic. It indicates that the sample is in a robust state that in itself is little influenced by the temperature.

When the temperature dependence of $S(\vec{q}, \omega)$ cannot be accounted for by the Bose factor in eq. (I.1.46), it becomes "anharmonic." This is an indication of changes of the state and functioning of the sample. Such anharmonic evolutions are frequent in soft matter and deserve particular attention. For example, the onset of diffusion (i.e., particles in the sample become mobile to move over substantial distances) is a process that typically does not obey Bose factor temperature dependence. It is rather controlled by the evolution of the diffusion constant $D$, cf. eq. (I.1.43), which can follow Arrhenius-type activation instead. Indeed, the onset of diffusion indicates a substantial change in the function of the atoms: they do not stay anymore around an equilibrium position. Note that eq. (I.1.43) derived from a classical model obviously violates detailed balance, as all classical model calculations principally do, since the scattering function, eq. (I.1.43), is symmetric, the same for $\omega$ and $-\omega$. This is a good approximation if $k_B T \gg \hbar\omega$, which usually holds for the slow diffusion processes, and there are phenomenological approaches to reconcile classical model calculations with detailed balance (Squires, 1978).

## REFERENCES

KARDJILOV, N., FIORI, F., GIUNTA, G., HILGER, A., RUSTICHELLI, F., STROBL, M., BANHART, J., and TRIOLO, R. *J. Neutron Res.* **2006**, *14*, 29.

MEZEI, F. and DAGLEISH, P.A. *Commun. Phys.* **1977**, *2*, 41.

SQUIRES, G.L. *Introduction to the Theory of Thermal Neutron Scattering*, Cambridge University Press, Cambridge, 1978.

van HOVE, L. *Phys. Rev.* **1954**, *95*, 249.

# II Instrumentation

# II.1 Small-Angle Neutron Scattering

## II.1.1 Small-Angle Neutron Scattering at Reactor Sources

*Kell Mortensen*

### II.1.1.1 INTRODUCTION

Small-angle scattering is principally a very simple technique, as schematically illustrated in Figure II.1.1.1: a sample is placed in a collimated, monochromatic beam, and the beam scattering by the sample is monitored. This scattering pattern reflects structural properties of the sample. Small-angle neutron scattering (SANS) is an ideal technique for studying bulk structures on the 1–500 nm length scale, that is, nanoscale structures. The SANS technique is thereby complementary to other scattering techniques such as small-angle X-ray scattering and static and dynamic light scattering. Other complementary methods include imaging techniques such as electron microscopy (TEM and SEM) and atomic tunneling techniques (AFM) and spectroscopic techniques such as NMR.

The scattered beam depends on two terms: (i) a contrast factor determined by the ability of individual atoms to interact with the neutron and (ii) the structure factor resulting from interference effects between radiation originating from the different sites in the sample, thus giving information on structural properties of the sample. The fact that the interaction with neutrons, the scattering length, is sensitive to the nuclear isotopes is the reason for particular interests within soft matter research. Specifically, the scattering cross section of hydrogen is large and comparable to that of other isotopes, in contrast to X-rays, and the scattering power of normal hydrogen, $^{1}H$, and of deuterium, $D = {}^{2}H$, is significantly different. By specific labeling, it is thereby possible to highlight specific structural properties. The fact that neutrons are magnetic spin-1/2 particles gives the possibility of studying magnetic structures and fluctuations, which is, however, usually not

*Neutrons in Soft Matter* Edited by Toyoko Imae, Toshiji Kanaya, Michihiro Furusaka, and Naoya Torikai

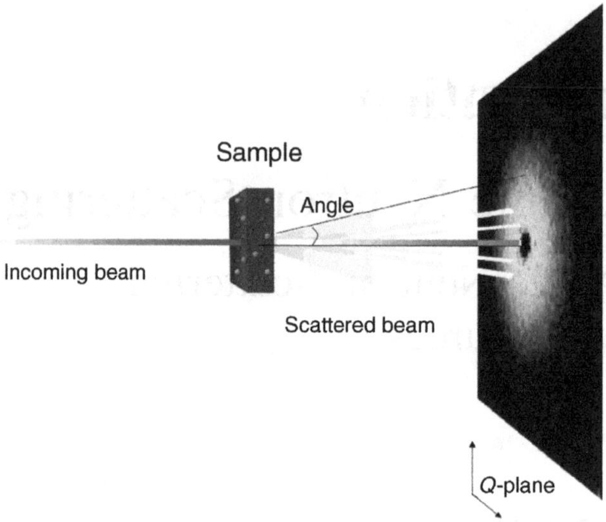


**Figure II.1.1.1** Illustration of small-angle scattering.

relevant to soft matter research. The spin-1/2 property also leads to incoherent scattering from systems where the individual nuclei possess magnetic moment, leading to significant background. Hydrogen is unfortunately an example that in this way causes large background. Another feature of SANS is a result of the relative weak interaction between neutrons and matters, which makes it easy to penetrate most materials. It is thereby feasible to make relative complex sample environments for *in situ* studies of large-scale structures.

## II.1.1.2 SMALL-ANGLE SCATTERING

The SANS method relies on the wave character of neutrons, which in site $\mathbf{r}$ may be expressed as

$$A(\mathbf{r}) = A_0 \exp[-i(\mathbf{k} \cdot \mathbf{r})], \qquad \text{(II.1.1.1)}$$

where $A_0$ represents the neutron beam amplitude and $\mathbf{k}$ is the wave vector parallel to the direction of the beam and with size $k = 2\pi/\lambda$, $\lambda$ being the neutron wavelength given by the de Broglie relation (Figure II.1.1.2).

The beam scattered from a site $\mathbf{r}_i$ into a direction $\mathbf{k}_\theta$ given by the scattering angle $2\theta$ is expressed as

$$A(\mathbf{k}_\theta)_{\mathbf{r}_i} = A \cdot b(\mathbf{r}_i)\exp[-i(\mathbf{k}_\theta \cdot \mathbf{r})], \qquad \text{(II.1.1.2)}$$

where $b(\mathbf{r}_i)$ denotes the ability of the atom at site $\mathbf{r}_i$ to scatter neutrons, the so-called scattering length. The scattering length of neutrons is dominated by the interaction with the nucleus. Values important for soft matter studies are given in Table II.1.1.1.

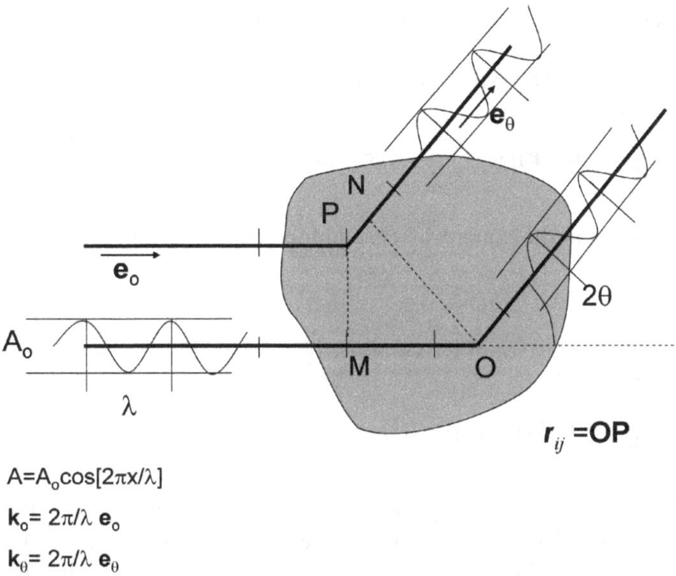


**Figure II.1.1.2** Illustration of scattering of a neutron beam.

More extended tables may be found, for example, at http://www.ncnr.nist.gov/resources/n-lengths/.

It is convenient to rewrite the scattered wave vector $\mathbf{k}_\theta$ as $\mathbf{k}_0 + \mathbf{Q}$, where $\mathbf{Q}$ is the scattering vector, or momentum transfer, given as a function of scattering angle:

$$Q = |\mathbf{Q}| = \frac{4\pi}{\lambda}\sin\theta. \tag{II.1.1.3}$$

Equation (II.1.1.2) is thereby rewritten as

$$A(\mathbf{Q})_{\mathbf{r}_i} = A \cdot b(\mathbf{r}_i)\exp[-i(\mathbf{Q}\cdot\mathbf{r})]\exp[-i(\mathbf{k}_0\cdot\mathbf{r})]. \tag{II.1.1.4}$$

**Table II.1.1.1** Scattering Length and Incoherent Scattering Cross Section of Typical Nuclei of Soft Matter Materials

| Nuclei | Coherent scattering length, $b$ (cm) | Incoherent cross section, $\sigma_{ic}$ (cm$^2$) |
|---|---|---|
| H | $-0.3739 \times 10^{-12}$ | $80.26 \times 10^{-24}$ |
| $^1$H | $-0.3741 \times 10^{-12}$ | $80.27 \times 10^{-24}$ |
| $^2$H = D | $0.6671 \times 10^{-12}$ | $2.05 \times 10^{-24}$ |
| C | $0.6646 \times 10^{-12}$ | $0.001 \times 10^{-24}$ |
| N | $0.936 \times 10^{-12}$ | $0.5 \times 10^{-24}$ |
| O | $0.5803 \times 10^{-12}$ | $0.0 \times 10^{-24}$ |
| Si | $0.4149 \times 10^{-12}$ | $0.004 \times 10^{-24}$ |

Nuclei with no index represent natural, mixed isotopes (http://www.ncnr.nist.gov/resources/n-lengths).

The total amplitude scattered from a sample is the simple sum of these terms:

$$
\begin{aligned}
A(\mathbf{Q}) &= \sum_{\mathbf{r}_i \in \text{sample}} A_0 \cdot b(\mathbf{r}_i)\exp[-i(\mathbf{Q}\cdot\mathbf{r})]\exp[-i(\mathbf{k}_0\cdot\mathbf{r})] \\
&= A_0 \exp[-i(\mathbf{k}_0\cdot\mathbf{r})] \cdot \sum_{\mathbf{r}_i \in \text{sample}} b(\mathbf{r}_i)\exp[-i(\mathbf{Q}\cdot\mathbf{r})].
\end{aligned}
\qquad \text{(II.1.1.5)}
$$

The measured intensity is the square of ensemble-averaged amplitude of the scattered beam, that is,

$$
\begin{aligned}
I_{\mathrm{m}}(\mathbf{Q}) &= A_0^2 \cdot \left\langle \sum_{\mathbf{r}_i \in \text{sample}} b(\mathbf{r}_i)\exp[-i(\mathbf{Q}\cdot\mathbf{r})] \sum_{\mathbf{r}_i \in \text{sample}} b(\mathbf{r}_i)\exp[+i(\mathbf{Q}\cdot\mathbf{r})] \right\rangle \\
&= I_0 \sum_i \sum_j \langle b(\mathbf{r}_i) b(\mathbf{r}_j)\rangle \exp[-i(\mathbf{Q}\cdot\mathbf{r}_{ij})],
\end{aligned}
\qquad \text{(II.1.1.6)}
$$

where $\mathbf{r}_{ij} = \mathbf{r}_j - \mathbf{r}_i$. Equation (II.1.1.6) expresses how the wave character of the neutron beam gives rise to interference phenomena, when the beam is scattered from different sites within a sample. In the following, we will consider the intensity normalized with the incoming beam: $I(\mathbf{Q}) = I_{\mathrm{m}}(\mathbf{Q})/I_0$.

Small-angle neutron scattering experiments do not give atomic length-scale resolution. In the scattering function expression for SANS, the scattering length $b$ characterizing individual atoms (nuclei) is therefore conveniently replaced by a continuous *scattering length density* function

$$
\rho(\mathbf{r}) = \sum_V b_i/V = \sum_V b_i N_{\mathrm{A}} \delta / M_V \qquad \text{(II.1.1.7)}
$$

averaged over an appropriate volume $V$. $M_V$ is the molar mass within the volume $V$, $\delta$ is the mass density, and $N_{\mathrm{A}}$ is Avogadro's number. In polymers, an appropriate volume may be that of one monomer, and in solvents that of the solvent molecule. Figure II.1.1.3 gives the scattering length density of different biological and synthetic macromolecules, shown in a diagram versus $D_2O/H_2O$ fraction. It is clear from the figure that contrast match for proteins appears within the range of 40–50% $D_2O$; that is, the proteins are invisible to the neutrons within the given scattering length density value. DNA, on the other hand, is invisible at ~60% $D_2O$. In DNA–protein complexes, it is thereby possible to measure the individual protein and DNA components making the appropriate $D_2O/H_2O$ mixtures.

Substituting the scattering length with scattering length densities, we can reformulate the scattering function, eq. (II.1.1.6), into integral form expressed as

$$
I(\mathbf{Q}) = \int_i \int_j \langle \rho(\mathbf{r}_i)\rho(\mathbf{r}_j)\exp[-i(\mathbf{Q}\cdot\mathbf{r}_{ij})]\rangle \mathrm{d}\mathbf{r}_i \, \mathrm{d}\mathbf{r}_j. \qquad \text{(II.1.1.8)}
$$

In an "isotropic medium," the averaged correlation function $\langle \rho(\mathbf{r}_i)\rho(\mathbf{r}_j)\rangle$ cannot depend on specific sites, but only on the distance $\mathbf{r}_{ij}$. One of the integrals in

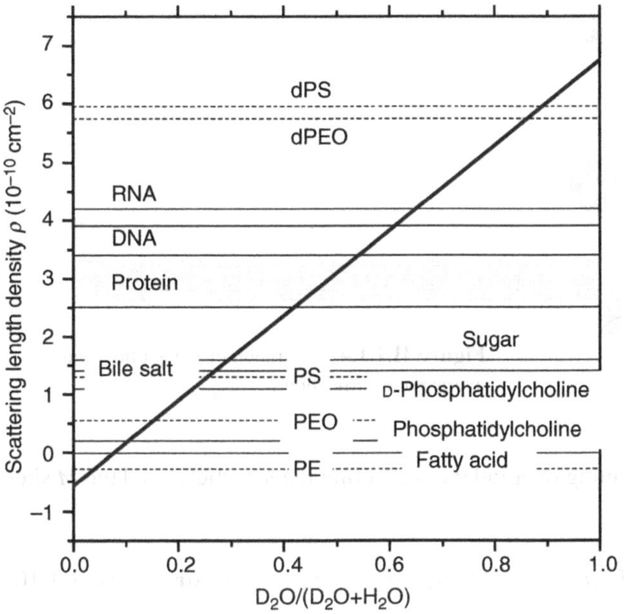


**Figure II.1.1.3** Scattering length densities of water mixtures and different biological and synthetic molecules.

eq. (II.1.1.8) can thereby be eliminated, giving

$$
\begin{aligned}
I(\mathbf{Q}) &= V \int_{\text{sample}} \langle \rho(\mathbf{r}')\rho(\mathbf{r}'+\mathbf{r})\exp[-i(\mathbf{Q}\cdot\mathbf{r})]\rangle d\mathbf{r} \\
&= V \int_{\text{sample}} \langle \gamma(\mathbf{r})\exp[-i(\mathbf{Q}\cdot\mathbf{r})]\rangle d\mathbf{r}.
\end{aligned}
\qquad \text{(II.1.1.9)}
$$

We thereby see that the scattering function is the Fourier transform of the ensemble-averaged correlation function $\gamma(\mathbf{r}) = \rho(\mathbf{r}')\rho(\mathbf{r}'+\mathbf{r})$, correlating densities separated by a distance $\mathbf{r}$.

In a typical sample, the correlations expressed by $\gamma(\mathbf{r})$ relate to both intraparticle correlations and interparticle correlations; in polymers, for example, these are represented by correlations within individual chains (reflecting the chain conformation) and between different chains (reflecting the distribution of chains), respectively.

In idealized systems where all scattering objects are identical and that on average can be assumed to have spherical symmetry, it is possible to split the scattering function into a product of two terms: one term relating to the intraparticle correlations (the form factor) and the other term relating to the interparticle correlations (the structure factor). To see this, let us assume a sample consisting of $n$ identical particles, each composed of $M$ scattering sites with excess scattering length density $\Delta\rho$, as illustrated in Figure II.1.1.4. To calculate the scattering function, we must thereby

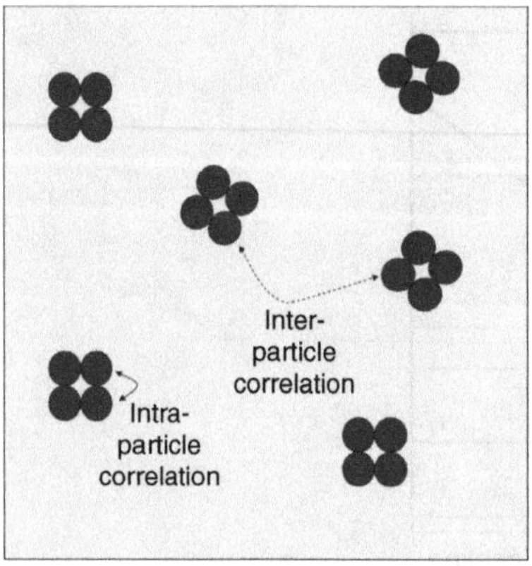


**Figure II.1.1.4** Illustration of intraparticle and interparticle correlations.

make a double integral running over indexes reflecting the $n$ particles and the $M$ sites (subparticles):

$$I(\mathbf{Q}) = \iint_{(i,j)\in n} \iint_{(p,q)\in M} \langle (\Delta\rho)^2 \exp[-i\mathbf{Q}\cdot(\mathbf{r}_{pi}-\mathbf{r}_{qj})]\rangle d\mathbf{r}_{pi}\, d\mathbf{r}_{qi}\, d\mathbf{r}_{pj}\, d\mathbf{r}_{qj}, \qquad \text{(II.1.1.10)}$$

where $p$ and $q$ represent sites within each of the $n$ particles, and $i$ and $j$ represent different particles. Mathematically, eq. (II.1.1.10) may be rewritten into the form

$$I(\mathbf{Q}) = (\Delta\rho)^2 \iint_{(i,j)\in n} \iint_{(p,q)\in M} \langle \exp[-i\mathbf{Q}\cdot(\mathbf{r}_{pi}-\mathbf{r}_{qi})]\exp[-i\mathbf{Q}\cdot(\mathbf{r}_{qi}-\mathbf{r}_{qj})]\rangle d\mathbf{r}_{pi}\, d\mathbf{r}_{qi}\, d\mathbf{r}_{pj}\, d\mathbf{r}_{qj}, \qquad \text{(II.1.1.11)}$$

where we by adding and subtracting $\mathbf{Q}\cdot\mathbf{r}_{qi}$ in the exponent have split the term into two: one term representing intraparticle correlations ($i=j$) and the other term representing interparticle correlations ($p=q$). Using that all particles are exactly similar, and assuming that the interparticle correlations are only relative weak, it is possible to split the sum in eq. (II.1.1.11) into

$$\begin{aligned} I(\mathbf{Q}) &= n(\Delta\rho)^2 M^2 \cdot \frac{1}{M^2} \iint_{(p,q)\in M} \langle \exp[-i\mathbf{Q}\cdot(\mathbf{r}_i-\mathbf{r}_j)]\rangle d\mathbf{r}_i\, d\mathbf{r}_j \\ &\quad \cdot \left[ 1 + \frac{1}{n} \iint_{(i\neq j)\in n} \langle \exp[-i\mathbf{Q}\cdot(\mathbf{r}_p-\mathbf{r}_q)]\rangle d\mathbf{r}_p\, d\mathbf{r}_q \right], \end{aligned} \qquad \text{(II.1.1.12)}$$

which we will rewrite into the usual form

$$I(\mathbf{Q}) = nM^2(\Delta\rho)^2 \cdot P(\mathbf{Q})S(\mathbf{Q}). \qquad \text{(II.1.1.13)}$$

We see that the scattering function is split into three terms: The first term is the prefactor that is proportional to the number concentration of particles, $n$, and the squared number of intraparticle scattering sites, that is, molar mass, $M$. The prefactor is further proportional to the contrast factor $(\Delta\rho)^2 = (\rho_2 - \rho_1)^2$, where $\rho_1$ and $\rho_2$ in the example expressed in Figure II.1.1.4 represent the scattering length densities of the particle and the solvent, respectively.

The second term in eq. (II.1.1.13) is the *form factor*, $P(Q)$, representing the form and size of the scattering object. $P(Q)$ is normalized to 1 at forward scattering: $P(Q=0)=1$. The third term is the structure factor reflecting mutual distribution of the particles. For very dilute systems with no particular interparticle interaction, $S(Q)$ approaches unity, as evident from eq. (II.1.1.12).

Equation (II.1.1.12) is an approximation that principally holds only for dilute, isotropic systems of ideal identical particles. In practice, eq. (II.1.1.12) is often applied even though the systems are known to be somewhat polydisperse in size and/or form. The split into a product of form factor and structure factor makes it generally easy to analyze the scattering data and the degree of polydispersity is relatively easily incorporated into the form factor by an appropriate convolution. A number of approaches have been formulated to correct for the nonideal system. Kotlarchyk and Chen (1983) proposed the "decoupling approximation" where it is assumed that the position of the particles is independent of their size and that the structure factor can be calculated based on an average size. Pedersen (1994) suggested a different approach assuming that the system can be divided into subsystems of monodisperse ensembles.

### II.1.1.2.1 Form Factor

The form factor expresses structural details of the molecules or aggregates to be studied. To analyze experimental data, one usually needs to compose appropriate model functions that represent the molecular object. Alternatively, one may make a Fourier transform of the data to get the model-independent pair correlation function.

In the case of a dilute sample of identical, randomly oriented particles, the scattering function, eq. (II.1.1.12), may be reduced to a function of the absolute value of $|\mathbf{r}| = r$:

$$I(\mathbf{Q}) = n_v(\Delta\rho)^2 4\pi \int r^2\gamma(r) \cdot \frac{\sin(Qr)}{Qr} \mathrm{d}r, \qquad \text{(II.1.1.14)}$$

where $n_v$ is the number density of particles and where it is assumed that the structure factor term of eq. (II.1.1.12) can be set to unity, $S(Q) = 1$. At $Q$-values small compared to the inverse of the characteristic length of the scattering molecules, the $\sin(Qr)/(Qr)$ term may be expanded into a series in $(Qr)$, thereby giving the Guinier approximation

$$I(Q) \propto \exp(-R_g^2 Q^2/3), \qquad \text{(II.1.1.15)}$$

where $R_g$ is the radius of gyration of the scattering object. Thus, by plotting $\ln(I)$ versus $Q^2$, the slope directly gives the overall size of the molecular system studied.

Generally, only the most simple structural form can be represented by analytical expressions for the form factor. This includes linear polymer chains when obeying Gaussian chain statistics, given by the Debye formula

$$I(Q) = \frac{2}{Q^4 R_g^4}\left[\exp(-Q^2R_g^2) - 1 + Q^2R_g^2\right]. \qquad \text{(II.1.1.16)}$$

Figure II.1.1.5 shows an experimental example of SANS measurements of a polystyrene polymer melt composed of mixed hydrogenated and deuterated chains, giving contrast to resolve the structure of an individual coil.

Another example where $P(Q)$ is expressed analytically is a system that may be represented by simple dense sphere of radius $R$,

$$P(Q) = \left[\frac{3}{Q^3R^3}[\sin(QR) - qR\cos(QR)]\right]^2. \qquad \text{(II.1.1.17)}$$

The form factor of ellipsoidal and cylinder-shaped particles is also expressed analytically. From these simple geometrically formed units, it is possible to construct analytical expressions for even more complex molecules or molecular aggregates. A simple example is a system that can be approached to composition of concentric spherical shells. The form factor can then be expressed as

$$I(Q) = \frac{\rho_1 V(R_1) P_{\text{sphere}}(Q, R_1) + \sum(\rho_i - \rho_{i-1}) V(R_i) P_{\text{sphere}}(Q, R_i)}{\rho_1 V(R_1) + \sum(\rho_i - \rho_{i-1}) V(R_i)}, \qquad \text{(II.1.1.18)}$$

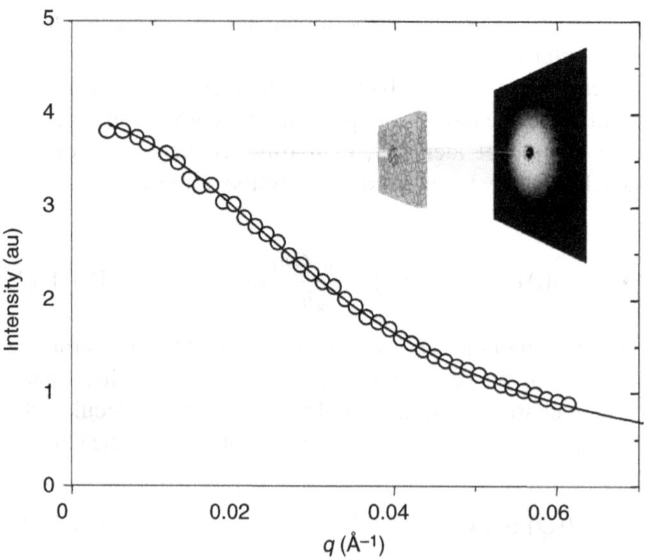


**Figure II.1.1.5** SANS experiment of a polymer melt composed of mixed hydrogenated and deuterated polystyrene chains. The solid line represents best fit to the Debye function, eq. (II.1.1.16).

where the index $i=1$ refers to the innermost dense sphere. Such composition of concentric shells may be used to simulate the structure of spherical micelles. Another approach for describing the form factor of spherical polymer micelles assumes composition of a dense spherical core (form factor $P_s(Q)$) surrounded by Gaussian polymer chains (form factor $P_c(Q)$) (Pedersen and Gerstenberg, 1996):

$$\begin{aligned} I(Q) &= N_{agg}^2 \rho_s^2 P_s(Q,R_s) + N_{agg}\rho_c^2 P_c(Q,R_g) \\ &\quad + N_{agg}(N_{agg}-1)\rho_c^2 S_{cc}(Q) + 2N_{agg}^2 \rho_s \rho_c S_{sc}(Q). \end{aligned} \tag{II.1.1.19}$$

$N_{agg}$ represents the micellar aggregation number, the index "s" represents properties of the spherical core, and the index "c" represents properties of the chain. $S_{cc}(Q)$ accounts for the interference term between two chains:

$$S_{cc}(Q) = \left[\frac{\sin(QR_s)}{(QR_s)^2}\right]^2 \left[\frac{1-\exp(-Q^2R_g^2)}{(QR_g)^2}\right]^2 \tag{II.1.1.20}$$

and $S_{sc}(q)$ accounts for the interference term between the spherical core and the chain:

$$S_{sc}(Q) = 3\frac{\sin(QR_s)-QR_s\cos(QR_s)}{Q^3R_s^3}\frac{\sin^2(QR_s)}{Q^2R_s^2}\frac{1-\exp(-Q^2R_g^2)}{Q^2R_g^2}. \tag{II.1.1.21}$$

Figure II.1.1.6 shows examples of experimental SANS data of a suspension of polymer micelles and the fits using eq. (II.1.1.19).

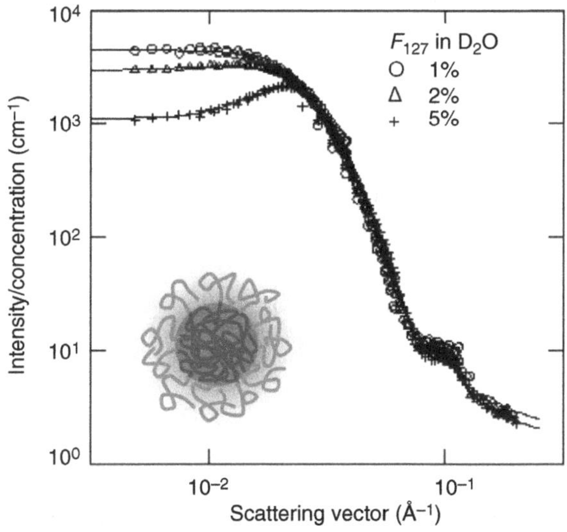


**Figure II.1.1.6** Experimental SANS data on PEO–PPO–PEO type of block copolymer micelles and fits using the micellar form factor, eq. (II.1.1.19), and a hard-sphere structure factor, eq. (II.1.1.22) (Mortensen and Talmon, 1995).

### II.1.1.2.2 Structure Factor

The scattering function of particles in solution becomes increasingly dominated by interparticle correlations as the concentration is increased. This is given by the structure factor. In very dense systems, this term becomes the dominating part and evolves into resolution-limited Bragg peaks if the system forms ordered structure. There exist a few analytical expressions for the structure factor of dilute suspensions. Here we will only discuss the one based on the Ornstein–Zernike (OZ) approximation and applying the Percus–Yevick closure with hard-sphere interaction potential. Other approaches also use the OZ approximation, but use other closures. Examples are the hypernetted chain approximation (HNCA), which has been used successfully for charged particles. HNCA is, however, a nonlinear theory that must be solved by numerical methods. A linearized version of HNCA is the mean spherical approximation (MSA), which is less accurate, but can be solved analytically for specific interaction potentials, such as the Derjaguin–Landau–Verwey–Overbeek (DLVO) double-layer repulsive potential (Hayter and Penfold, 1981). $S(Q)$ is in the hard-sphere Percus–Yevick approximation given analytically by two parameters only, the volume fraction $\phi$ and the hard-sphere interaction distance $R_{hs}$ (Kinning and Thomas, 1984) (Figure II.1.1.7):

$$S(Q) = \frac{1}{1 + 24\phi G(QR_{hs}, \phi)/(2QR_{hs})}, \qquad \text{(II.1.1.22)}$$

where $G$ is a trigonometric function of $QR_{hs}$ and $\phi$.

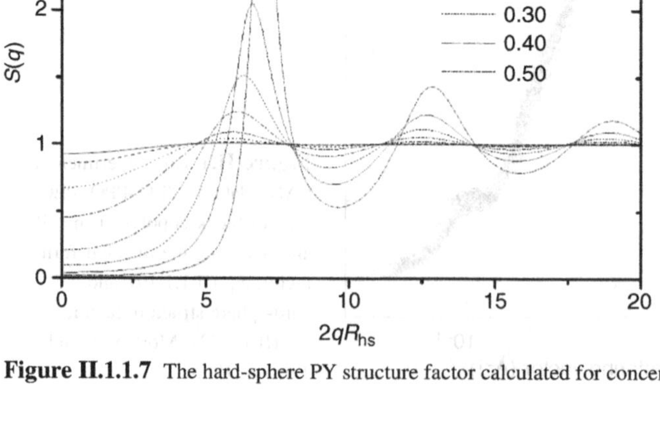


**Figure II.1.1.7** The hard-sphere PY structure factor calculated for concentration in the range 1–50%.

We see that for relatively small volume fractions, the main effect of the structure factor is to decrease the intensity at the lowest angles, while at higher volume fractions, oscillations in the structure factor will become a dominating factor in the scattering data, eventually evolving into real Bragg reflections if the suspension forms an ordered system.

## II.1.1.3 INSTRUMENTATION

From the presentation above, it appears that to make small-angle scattering one needs well-monochromatized neutron beam with a given neutron wavelength $\lambda$ (neutron energy), and a highly collimated beam toward the sample, to get unique value for the scattering angle $2\theta$ (Schmatz et al., 1974; Koehler and Hendricks, 1978, 1979). In this chapter, we will describe in some detail SANS instruments as installed at reactor sources. For more details on specific instruments, refer to web pages or scientific reports (Ibel, 1976; Schelten and Hendricks, 1978; Hofmeyr et al., 1979; Child and Spooner, 1980; Mildner et al., 1981; Wignall, 1986, 1997; Glinka et al., 1986, 1998; Schwahn et al., 1991; Rekveldt, 1996; Kohlbrecher and Wagner, 2000; Aswal and Goyal, 2000; Wiedenmann, 2001; Lynn et al., 2003; Strunz et al., 2004; Okabe et al., 2005; Gilbert et al., 2006; Strobl et al., 2007; Han et al., 2007; Giri et al., 2007; Lieutenant et al., 2007; Desert et al., 2007; Dewhurst, 2008).

Small-angle neutron scattering instruments are in principle relatively simple facilities, consisting of only five principal elements: the neutron source, the neutron monochromator selecting neutrons with a given energy/wavelength, the collimation section specifying divergence of the neutron beam directing toward the sample, the sample environment, and finally the flight tank for the scattered beam with the neutron detector. In addition to these units, the instrument may include a variety of additional units such as choppers for time-of-flight measurements and neutron polarizers. Furthermore, the instrument includes detector electronics for data acquisition and electronics and computer for determining the instrumental setting (Figure II.1.1.8).

Even though small-angle neutron scattering has been known as a technique since the early days of neutron diffraction, the method did not become the present-day major techniques within soft matter sciences until few decades ago. A major breakthrough was the development of area-sensitive neutron detectors.

### II.1.1.3.1 The Neutron Source

Neutrons are elementary particles that constitute about half of the nucleus mass, or more. Each neutron has a mass of $m_n = 1.675 \times 10^{-14}$ g. In fission of specific nuclei, neutrons might escape and can be utilized for materials research. The application of neutron scattering technique is based on the wave character of such neutrons, as determined through the de Broglie relation

$$\lambda = h/m_n v_n, \tag{II.1.1.23}$$

where $h$ is the Planck constant, $m_n$ is the neutron mass, and $v_n$ is the neutron velocity.

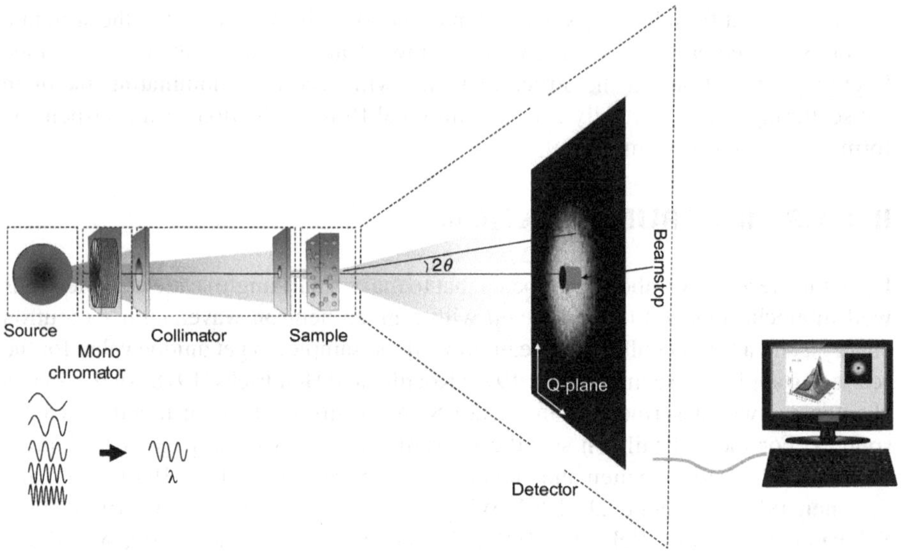


**Figure II.1.1.8** Schematic illustration of a SANS instrument and its various components, including neutron source, monochromator system, collimator, sample environment, and flight tube with area-sensitive detector and beamstop.

Two types of applied neutron sources exist: the reactor type and the spallation type. Both types produce, at best, a beam of about $10^7$ neutrons/s/cm$^2$ at the sample position of a SANS instrument. This chapter focuses on SANS instruments at reactors, but most of the SANS characteristics are common independent of the source (Mildner, 1984). As a consequence of the special needs and expenses involved in operation and building neutron sources, sources for neutron scattering are *mega science facilities* that rely on international user groups.

In the reactor core, fast neutrons with energy $E_n$ of the order of 1–2 MeV are produced in a fission chain reaction. Usually, enriched uranium $^{235}U$ is used as a fuel. Almost all reactor facilities act as continuous source, providing a constant neutron flux. Examples of major reactor facilities are ILL in France, FRM2 in Germany, NIST in the United States, and JAEA in Japan. Most national neutron sources are of the reactor type; examples include facilities at institutes in the United States, Canada, Germany, the Netherlands, Norway, Korea, Malaysia, Indonesia, and China. In the reactor, the core is surrounded by a moderator that is usually based on $D_2O$, $H_2O$, or graphite. During the interaction with the moderator material, the neutrons are thermalized to the energy given by the temperature of the moderator, for example, $T = 50°C$. The resulting Maxwell–Boltzmann distribution of such thermal neutrons has a maximum flux for neutron energies $E_n = k_B T = 30$ eV, $k_B$ being the Boltzmann constant, corresponding to a neutron velocity

$$v_n = \sqrt{2k_B T/m_n} \tag{II.1.1.24}$$

of the order of 3000 m/s, and thereby a neutron wavelength of the order of $\lambda = 1.4$ Å, according to the de Broglie relation, eq. (II.1.1.23).

In small-angle scattering facilities, which are designed for structural studies on the mesoscopic length scale of 1–100 nm, it is desirable to use low-energy neutrons with wavelength typically in the 3–20 Å range. In an attempt to move the peak flux to the energy corresponding to such wavelengths, cooled moderators are inserted next to the reactor core in the neutron beam hole. Such units are usually named *cold sources.* The cold source typically works with deuterium close to the supercritical point at 28K and 15 bar, giving a peak flux of neutrons in the range of 4.6 Å.

Further advances for cold beam small-angle scattering facilities are gained if the instrument is moved away from the reactor source using neutron guides. This ensures reduction of stray neutrons, high-energy neutrons, and gamma-radiation, thus resulting in a significantly improved signal-to-background ratio.

In spallation sources, discussed in detail in Chapter II.I.2, the neutron beam is obtained by splintering heavy nuclei by high-energy particles such as 800 MeV protons. The method of accelerating protons is often optimized to bursts of high-energy protons, and hence pulsed neutron beam. But continuous, or quasicontinuous, spallation sources exist as well, for example, the SINQ facility in Switzerland. The instrumentation and application of such facilities is principally like that applicable to reactor sources, and is thus included in this chapter. Pulsed reactors also exist and are from the SANS instrumental point of view more like that of spallation sources. An example is the IBR2 pulsed reactor in JINR Dubna.

The advantages and disadvantages of various types of sources depend on the type of problem to be studied by neutron scattering. Small-angle scattering is generally best performed at reactor sources. This is mainly a consequence of higher degree of flux stability and better reliability in determining the background in the form of incoherent scattering from the sample itself. The pulsed sources have, on the other hand, possibilities for better high-resolution data. An optimal SANS instrument may be that using a reliable pulsed spallation source, where the beam is monochromatized. Such instruments have been proposed for the European ESS project.

### II.1.1.3.2 Monochromator

Neutrons with a particular wavelength can be extracted from the "white" beam of the neutron source by monochromators. In pulsed facilities, the obvious way is to use time of flight, even though additional monochromatization may be useful as discussed above. For continuous sources, monochromatization is essential.

Two basic different methods are used, given by the two dual properties of the neutron, wave character and particle character. Using the wave nature of the neutrons, a particular wavelength can be selected via Bragg scattering from a crystal with suitable crystal lattice spacing, *d*. A high reflectivity is needed, and to match the general resolution of the instrument, a relatively high mosaic spread of the monochromator crystal is favored, that is, a crystal with some spread in the crystal orientation, fulfilling the Bragg condition for neutron wavelength within a given band. Typically, pyrolytic graphite may be used as a crystal monochromator, where wavelength spread up to about 6% centered on the nominal value is possible.

Alternatively, one can use the particle nature of the neutrons to choose the appropriate neutron velocity, and thereby given wavelength. The velocity of 6 Å neutrons, for example, is 600 m/s. This relatively low speed allows the use of mechanical selectors made up of rotating plates of neutron absorbing materials (Clark et al., 1966). The neutrons with too low speed are absorbed by the selector from the back, while those with too high speed hit the selector plate in the front. Mechanical selectors are usually made as slightly screwed turbine-like devices. The screw angle, $\varphi$, defines the wavelength spread

$$\frac{\Delta\lambda}{\lambda} = A/\varphi, \tag{II.1.1.25}$$

while the speed of revolution,$\omega$, gives the nominal wavelength:

$$\lambda = B\frac{\varphi}{\omega}. \tag{II.1.1.26}$$

$A$ and $B$ are device constants that depend on the distance between the plates. By changing the axis of rotation relative to the beam direction, there is an additional possibility for variation of wavelength spread. Typical wavelength spread is about 10%, but 5–30% spread is feasible and regularly used.

#### II.1.1.3.3 Chopper

Choppers are usually not standard modules in reactor SANS instruments, even though such devices may allow significantly improved $Q$-resolution, and in particular may be used for high-resolution time-resolved experiments. The application of choppers, however, usually causes major decrease in neutron flux.

High-resolution time-resolved SANS experiments (TITANE) may be performed using stroboscopic techniques adding a fast chopper near the beginning of the instrument, typically just after the mechanical velocity selector (Kipping et al., 2008). Spectroscopic time-resolved SANS experiments require in addition a device that modulates the scattering properties of the sample. The TITANE technique has been used for studying dynamics of magnetic colloids, applying an oscillating magnetic field on the sample (Wiedemann et al., 2006).

#### II.1.1.3.4 Collimation

The neutron beam direction is defined by the collimator, which in principle can be given by just two pinholes, one near the source point and the other near the sample. In practice, the collimator system is incorporated with additional pinholes between the two defining ones, in an attempt to avoid stray neutrons reaching the detector. Furthermore, reactor SANS facilities are usually equipped with a variable length of the collimator as well as variable pinhole sizes to optimize flux with appropriate resolution. The effective length is varied by inserting sections of neutron guides between the monochromator and the sample. Such neutron guides are glass channels coated with Ni or supermirrors reflecting neutrons with wavelength above a certain critical value, which typically is slightly more than 3 Å, depending on the Ni isotope that is used or the design of the super mirror. As an example, a 10 m collimator with 2 cm diameter pinhole at the entrance and 1 cm diameter pinhole near the sample corresponds to a beam divergence of 0.1°.

## II.1.1.3.5 Focusing Devices

Other more sophisticated collimators with focusing elements may be applied, to allow the application of much smaller sample volume than otherwise typically needed in SANS experiments. Alternatively, focusing on the detector allows measurements to lower scattering momentum, and thereby possibility to study structures of larger sizes. In some SANS instruments, the collimator is composed of ensembles of smaller pinholes. A challenge with such multiple-pinhole collimator design is to avoid crosstalks between the different beam passes. Examples are the LLB design of a multibeam pinhole collimator converging onto a high-resolution detector (Brulet et al., 2008) and the trumpet guide design proposed for the upgraded D11 instrument at ILL (Lieutenant et al., 2007).

A number of studies have been made to apply optical devices for neutron focusing. Since refractive indices for cold neutrons ($\lambda \sim 10$ Å) differ from unity by at most a few parts in $10^5$, grazing incidence reflection optics have long been considered as the most promising means for focusing neutrons in small-angle neutron scattering instruments (Alefeld et al., 1997; Kentzinger et al., 2004). The focusing of a cold neutron beam by multiple biconcave lenses has, however, more recently been demonstrated using $MgF_2$ material (Eskildsen et al., 1998). Such compound refractive lenses (CRL) have been proposed as a practical means of improving the minimum $Q$-value of conventional SANS instruments that use pinhole collimation with circular apertures separated by distances of several meters (Eskildsen et al., 1998; Choi et al., 2000; Mildner and Hammouda, 2005; Mildner, 2005). It is likely that most future SANS instruments will be facilitated with such focusing CRL optics.

For most materials, the neutron refractive index, $n$, is less than unity. Therefore, in contrast to light where $n$ is greater than unity, a concave lens is convergent while a convex lens is divergent. To get an appropriate focal length, a relatively large number of lenses must be applied. Since the focal length of a lens depends on $\lambda^{-2}$, a lens system is strongly chromatic. Therefore, neutrons with well-defined wavelength are required for good focusing. With typical SANS wavelength spread $\Delta\lambda/\lambda \sim 10\%$, a certain amount of chromatic aberration is expected.

Recently, proposals have been forwarded for neutron focusing using magnetic neutron lens based on a sextupole magnet (Oku et al., 2007, 2008; Koizumi et al., 2007). Such a device requires polarized neutrons, since the magnetic neutron lens functions as both focusing and defocusing lens depending on the neutron polarity: negative polarity neutrons are defocused by the lens and spread over the detector of the instrument. Therefore, a neutron polarizing device with a very high polarizing efficiency needs to be employed together with a magnetic neutron lens.

### II.1.1.3.6 Neutron Polarization

In soft matter research, there is only seldom use for the application of polarized neutrons. This is in contrast to other condensed matter research, where polarization is an important part of the instrument, even though still not standard option in most SANS facilities. Polarization may be obtained using multilayer mirrors of magnetic materials or gases that absorb only one polarization (Gentile et al., 2000; Wiedemann, 2005).

### II.1.1.3.7 Neutron Detector

Most SANS facilities are equipped with an area-sensitive neutron detector that is positioned in a large evacuated flight path tank. Most detectors in SANS instruments are gas-filled proportional counters, typically large-area multiwire detectors filled with $^3$He as the detection medium and $CF_4$ or $CH_4$–Ar gas mixture used as a stopping gas (Kopp et al., 1981). The neutron absorption by a target nucleus such as $^3$He induces a fission reaction and emission of two charged particles, one $^3$H tritium and one $^1$H proton, with a total energy of 760 keV,

$$n + {}^3\text{He} \rightarrow {}^3\text{H} + {}^3\text{H} + 0.76\,\text{MeV}, \qquad \text{(II.1.1.27)}$$

inducing the primary ionization of the gas. Alternative active gases may be used based on either $^{10}$B or $^6$Li detection system.

A plane of wires mounted in the center of the detector, the anode plane, is based at a high tension (3–5 kV). The electrons are accelerated to get more ionization and to amplify the signal. The primary protons and tritium nuclei produce a trace of ionized atoms and electrons. The electrons are accelerated toward this anode plane, thereby producing an amplified charge signal. The charges are absorbed in the anode and distributed toward an electronic system, which determines the one coordinate of the position of the incoming neutron. The stopping gas reduces the path length of the electron for a good position resolution. An additional plane, the cathode plane next to the anode, gets partly by inductive coupling and partly by the ions produced by the electrons an electronic pulse of opposite sign, which with appropriate design of the planes is used to determine the second coordinate. The output from the neutron detector is analyzed electronically and saved in a histogram memory for further data treatment.

The gas detector has a limited capability in terms of neutron flux and spatial sensitivity. For SANS detectors with diameter of the order of 50–100 cm, the typical spatial resolution is within the 0.5–1.0 cm range and maximum count rates are up to the order of $10^6$ neutron counts per second. Dead times in the detectors are at best of the order of $\tau = 10-100$ ns, representing a loss of 10% at $10^6-10^7$ n/s count rate. If the count rate is distributed over the whole detector, one may easily recalculate the real count rate $C_{\text{real}}$ based on the measured count rate $C_{\text{meas}}$ and known dead time:

$$C_{\text{real}} \approx \frac{C_{\text{meas}}}{1-\tau C_{\text{meas}}}. \qquad \text{(II.1.1.28)}$$

Often, however, some pixels are exposed significantly more than others. The simple equation should then be used only with caution.

New modern detectors will likely be based on scintillation units or other solid-state devices, ensuring not only larger dynamic ranges, but also better spatial resolution (Heiderich et al., 1991; Cheng et al., 2000).

### II.1.1.3.8 Sample Environment: Auxiliary Units Applied Together with SANS

The sample environment depends naturally on the specific system to be studied. Typical scattering experiments on polymer melts and solutions concern investigation of structure and structural changes versus temperature, pressure, shear, or other external fields. The investigation can involve detailed examination of an ordered mesophase with the aim of determining the crystal structure or obtaining details on the phase behavior near a critical point.

In general, a high-quality temperature control needs to be adapted to the scattering facility. Typical temperature ranges are within the −20 to 300°C range, but depending on the examined system, of course. Another thermodynamic parameter, which in recent years has proven important for the understanding of polymer phase behavior, is a pressure device possibly combined with temperature regulation. Pressure devices, which are presently available, are typically limited to approximately 3000 bar, but pressures beyond this limit would certainly be interesting for future studies.

Shear and stretching devices have proven to be extremely important in studies of surfactant systems, gels, and polymer melts, reflecting properties during processing and applications. Shear can cause dramatic changes in phase behavior, texture, and so on. Figure II.1.1.9 shows schematically a Couette-type shear device used for SANS experiments. The two commonly used configurations are included: the beam going radially through the cups giving the $(e, v)$-scattering plane, and the beam tangential to the shearing cups, giving the $(e, \nabla)$-scattering plane, where the index $v$ denotes parallel to the flow velocity, $\nabla$ parallel to the shear gradient, and $e$ parallel to the neutral vorticity axis.

Figure II.1.1.10 shows a commercial parallel plate shear instrument (RSA-II from Rheometrics) modified for *in situ* SANS experiments.

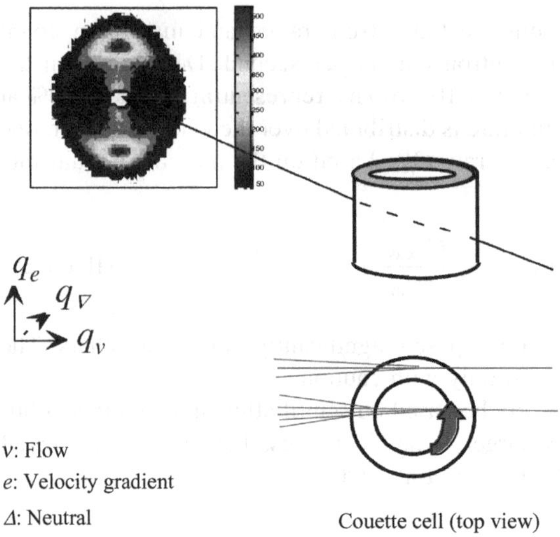


**Figure II.1.1.9** Schematic illustration of a Couette-type shear cell used for SANS experiments in radial and tangential modes.

**Figure II.1.1.10** Modified commercial RSA-II parallel plate shear device installed at a SANS facility for *in situ* rheology and scattering experiments. The bottom part shows schematically the parallel plate device, the neutron beam, and the scattering pattern.

## II.1.1.4 SANS FACILITIES

Table II.1.1.2 lists available SANS facilities at reactor sources around the world. The information listed in the table is taken mainly from the home pages of the different facilities. The list is not complete, neither concerning the existing SANS instruments available, nor the auxiliary apparatus listed. The given parameters should also be taken with some reservation. We refer to the home pages for updated instrument information. The home pages also give information on access programs for the different instruments.

**Table II.1.1.2** List of Major SANS Facilities at Reactor Sources and Some of Their Main Characteristics

| Reactor sources | Instrument | Wavelength range (nm) | $Q$-range ($nm^{-1}$) | High temperature | Cryogenic temperature | High magnetic field | Shear – strain | Polarization | Chopper | Web home pages |
|---|---|---|---|---|---|---|---|---|---|---|
| Europe | | | | | | | | | | |
| ILL Grenoble | D22 | 0.45–4 | 0.01–10 | x | x | x | | | x | http://www.ill.eu/d22 |
| ILL Grenoble | D11 | | 0.006–3.0 | x | x | x | | | x | http://www.ill.eu/d11 |
| FRM II JCNS Garching/ München | KWS2 | 0.45–2 | 0.001–3 | | | | | | x | http://www.frm2.tum.de/en/ science/diffractometer/kws-2 |
| FRM II JCNS Garching/ München | KWS1 | 0.45–2 | 0.01–2.0 | | | x | | x | x | http://www.frm2.tum.de/en/ science/diffractometer/kws-1 |
| FRM II JCNS Garching/ München | KWS3 | 0.8–2 | 0.005–0.4 | | | | | | | http://www.frm2.tum.de/en/ science/diffractometer/kws-3 |
| FRM II GKSS Garching/ München | SANS-1 | 0.45–2 | 0.001–20 | | | | | x | x | http://www.frm2.tum.de/en/ science/diffractometer/sans-1 |
| PSI Villigen/ Zürich | SANS-I | 0.45–2 | 0.006–5.4 | | | | x | x | | http://kur.web.psi.ch/sans1 |
| PSI Villigen/ Zürich | SANS-II | 0.45–2 | 0.02–3.5 | x | x | x | | x | | http://sans2.web.psi.ch |
| HMI | VSANS[a] | 0.35–3 | 0.01–8.5 | x | x | x | | x | | http://www.helmholtz-berlin.de/ userservice/neutrons/ instrumentation/neutron-instruments/v16_en.html |
| BNC Budapest | SANS | 0.4 | 0.1–5.0 | x | x | x | | | | http://www.bnc.hu/modules.php? name=News&file=article&sid=3 |

(*continued*)

**Table II.1.1.2** (*Continued*)

| Reactor sources | Instrument | Wavelength range (nm) | $Q$-range ($nm^{-1}$) | High temperature | Cryogenic temperature | High magnetic field | Shear – strain | Polarization | Chopper | Web home pages |
|---|---|---|---|---|---|---|---|---|---|---|
| GKSS Hamburg | SANS-1 | | | | | | | x | | http://www.gkss.de/central_departments/genf/instruments/003096/index_0003096.html.en |
| GKSS Hamburg | SANS-2 | | –25 | | | | | x | | http://www.gkss.de/central_departments/genf/instruments/003124/index_0003124.html.en |
| IFE Kjeller/Oslo | SANS | 0.45–1 | 0.08–3.2 | | | | | | | http://www.ife.no/departments/physics/sections/neutronlab/view?set_language=en |
| LLB Paris | PACE | 0.4–2 | 0.02–5 | | | | | | | http://www-llb.cea.fr/spectros/pdf/pace-llb.pdf |
| LLB Paris | PAXY | 0.4–2 | 0.03–10 | | | | | | | http://www-llb.cea.fr/spectros/pdf/paxy-llb.pdf |
| LLB Paris | PAXE | 0.4–2 | 0.05–5 | | | | | | | http://www-llb.cea.fr/spectros/pdf/paxe-llb.pdf |
| LLB Paris | TPA | 0.7–1.6 | 0.002–0.2 | | | | | | | |
| LLB Paris | Papyrus | 8 | 0.008–0.24 | | | | | | | http://www-llb.cea.fr/papirus/PAPYRUSSTART.html |
| IRI Delft | SESANS | | | | | | | | | http://www.tnw.tudelft.nl/live/pagina.jsp?id=68715280-da24-4949-a251-7e53c4e11695&lang=en |
| Asia and Pacifics | | | | | | | | | | |
| ANSTO Lucas Heights | QUOKKA | 0.45–2.0 | 0.015–1 | x | x | x | | x | | http://www.ansto.gov.au/opal.html |
| JAEA Tokai | SANS-J | | 0.01–6.0 | | | | | | | |

| | | | | | | | | | |
|---|---|---|---|---|---|---|---|---|---|
| JAEA Tokai | SANS-U | | 0.005–5.0 | x | x | | x | | |
| KAERI Taejon | SANS | 0.4–1.0 | 0.06–6.0 | | | | | | |
| NSL-BATAN Serpong | SANS | 0.27–0.56 | 0.02–6.0 | | | | | | |
| Dhruva Barc Trombay | SANS | | | x | | | | | |
| North America | | | | | | | | | |
| ORNL | BIO-SANS[a] | 0.6–3 | 0.002–1 | x | x | x | | | http://neutrons.ornl.gov/hfir_instrument_systems/CG-3.shtml |
| ORNL | CG-2 | 4–30 | 0.004–0.5 | | | | | | http://neutrons.ornl.gov/hfir_instrument_systems/CG-2.shtml |
| NIST Gaithersburg | NG-7 | 0.5–2 | 0.007–7.0 | x | x | x | x | | http://www.ncnr.nist.gov/instruments/ng7sans/ |
| NIST Gaithersburg | NG-3 | 0.5–2 | 0.008–6.0 | x | x | x | x | x | http://www.ncnr.nist.gov/instruments/ng3sans/ |

The parameters and the listed auxiliary sample environments are taken mostly from the instrument home pages and are not complete.

[a] Under construction.

## II.1.1.5 EXPERIMENTS

The major advances using reactor sources for SANS experiments are the high degree of both stability and reliability in the neutron flux. This allows easy treatment of background, which is essential for high-quality data of weak scattering samples. The background does not need to be measured almost simultaneously with the experiments, but can be separated in time, making better planning for optimal use of beam time and use of maximum environmental parameters. This has particular interests in connection with time-resolved experiments, where artifacts coming from time-dependent changes in neutron flux should not be a matter to take into account.

### II.1.1.5.1 The Sample

Optimal sample thickness depends on absorption and coherent scattering cross section. The interaction between neutron and materials is generally very weak compared to that of X-rays. With the relatively low flux in neutron spectrometers compared to X-ray instruments, one needs rather large sample volume. The optimal thickness is a compromise between absorption and scattering. The absorption cross section, and thereby the transmission, depends on the neutron wavelength. Optimal conditions are typically given by a transmission factor, $t$, of roughly $1/e$:

$$t = \exp(-\Sigma_{\text{total}} d) \approx 1/e \sim 37\%, \qquad (\text{II.1.1.29})$$

where $\Sigma_{\text{total}}$ is the total scattering cross section, including absorption and scattering terms, and $d$ is the sample thickness.

For an aqueous sample measured with 5 Å neutrons, the transmission of a 1 mm thick sample is about 52%, while that of a 3 mm sample is only 14%. For the 1 mm thick sample, the transmission is reduced to 34% using 15 Å neutrons. Often, a compromise is needed in the choice of sample design: one needs to consider optimal scattering power and the need to use the same sample for different instrumental configurations.

### II.1.1.5.2 SANS Measurements

A typical SANS measurement includes a large number of measurements beyond that of the sample itself. The scattering from the sample container or other relevant sample background, for example, a filled quartz cuvette, must be measured, as well as the electronic and stray radiation noise. Further, one may need to measure a standard sample used to normalize the data and correct for possible detector nonlinearity, by measuring the corresponding background spectra. Finally, one needs to measure the transmission factors for sample, sample container, and so on. Based on these measurements, one can determine the absolute scattering function of a given sample, according to

$$S(Q) = \frac{d\Sigma}{dQ} = \frac{1}{d_S}\frac{1}{t_S}\frac{I_S - t_S/t_B \cdot I_B - (1-t_S/t_B)\cdot I_E}{I_N - t_N/t_{NB}\cdot I_{NB} - (1-t_N/t_{NB})\cdot I_{NE}}C, \qquad \text{(II.1.1.30)}$$

where $S(Q) = d\Sigma/dQ$ is the scattering cross section of interest, $d_S$ is the sample thickness, and $I_S$, $I_B$, and $I_E$ are measured $Q$-dependent intensities. $I_S$ is the intensity of the sample including container, $I_B$ is the intensity of an appropriate background, and $I_E$ is the measured intensity resulting from integrated noise arising from sources not part of the sample environment (electronic noise, stray neutrons, etc.). $t_S$ and $t_B$ are the transmission factors measured for the sample (including container) and background, respectively. $I_N$, $I_{NB}$, and $I_{NE}$ are the corresponding measured intensities with a normalization standard mounted as a sample, and $t_N$ and $t_{NB}$ are the corresponding transmission factors. $C$ is a constant that brings the data to absolute value. Often, water is used for absolute normalization, since hydrogen ensures well-defined isotropic scattering. $C$ is then given by

$$C_{H_2O} = \frac{(1-t_{H_2O})}{4\pi}g, \qquad \text{(II.1.1.31)}$$

where the first term reflects that the fraction $(1-t_{H_2O})$ is scattered into $4\pi$. The last term, $g$, is a factor close to unity, which corrects for nonperfect scattering into $4\pi$. Table II.1.1.3 gives values of $g$ versus neutron wavelength. The correction factor depends slightly on exact sample geometry and properties of the SANS instrument.

After the SANS data have been corrected for background and possibly reduced from two-dimensional to one-dimensional data, the results must be interpreted using model fit or other tools. Typically, one will try to fit an appropriate model function, $I_{\text{model}}(Q,\ldots)$, to the experimental scattering data, $I_{\exp}(Q)$, measured at $N$ $Q$-values, using least square methods:

$$\chi^2 = \sum_{i=1}^{N}\frac{(I_{\exp}(Q_i) - I_{\text{model}}(Q_i))^2}{\sigma_i^2} \qquad \text{(II.1.1.32)}$$

minimizing $\chi^2$. $\sigma_i$ is the statistical error at the given measurement. The model function may be one of the expressions discussed above.

SANS data are always incorporated with some nonnegligible smearing from instrumental properties as apparent from the discussions above, giving details on wavelength spread from the monochromator ($\Delta\lambda/\lambda \sim 10\%$), collimation giving beam divergence of the order of tenths of a degree, and the detector resolution of the order of 0.5–1.0 cm (Antonimi et al., 1970; Wignall et al., 1988; Wignall, 1991; Pedersen et al., 1990; Barker and Pedersen, 1995). Gravity has some effect in large instrument when using long-wavelength neutrons (Boothroyd, 1989). Although applying time of flight will typically reduce $\Delta\lambda/\lambda$, it should be incorporated (Grabcev, 2007). In the data evaluation, these terms should be taken into account.

**Table II.1.1.3** Correction Parameter $g$ Correcting for Nonideal Isotropic Scattering from 1 mm Thick Water Sample

| $\lambda$ (Å) | $g$ |
|---|---|
| 2 | 1.84 |
| 3 | 1.63 |
| 4 | 1.48 |
| 5 | 1.33 |
| 6 | 1.24 |
| 7 | 1.15 |
| 8 | 1.11 |
| 9 | 1.05 |
| 10 | 1.0 |

This is most effectively done by smearing the model function by an appropriate smearing function $R(Q)$:

$$\tilde{I}_{\mathrm{model}}(\langle Q\rangle) = \int R(\langle Q\rangle, Q) I_{\mathrm{model}}(Q)\mathrm{d}Q, \qquad \text{(II.1.1.33)}$$

where $\langle Q\rangle$ is the nominal scattering vector. In the least square fit, $I_{\mathrm{model}}$ should be replaced by $\tilde{I}_{\mathrm{model}}$. Another effect that may influence the scattering function is multiple scattering events (Mazumder et al., 1993; Allen and Berk, 1994) and the effect of inelastic scattering (Heenan, 1993). A variety of program packages for data analysis are available at the SANS facilities (Ghosh and Rennie, 1990; Strunz and Wiedenmann, 1997; Kline, 2006; Kohlbrecher, 2003).

Alternatively, data treatment may be based on transforming the scattering data into real space by Fourier transform of the data. This is highly relevant for samples of dilute suspensions of molecules or molecular aggregates. Since the measured data are limited to a given $Q$-range and are affected by statistical noise, a simple mathematical Fourier transform is not practically applicable. Glatter proposed an alternative method to determine the pair distribution function, namely, the so-called indirect method based on a simple fitting procedure, where the pair correlation functions are composed as a sum of appropriate model functions (e.g., cubic spline functions) that is easily transformed (Glatter, 1997). Fitting routine results in the parameters that determine the Fourier transform. For realistic transformation, one needs to incorporate appropriate damping parameters in the calculations.

## II.1.1.6 SOFT MATTER SCIENCE AT SANS FACILITIES

The type of scientific topics that are treated using SANS facilities at reactor sources is very wide, including molecular biology, pharmaceutical and food sciences, colloids, and polymer sciences.

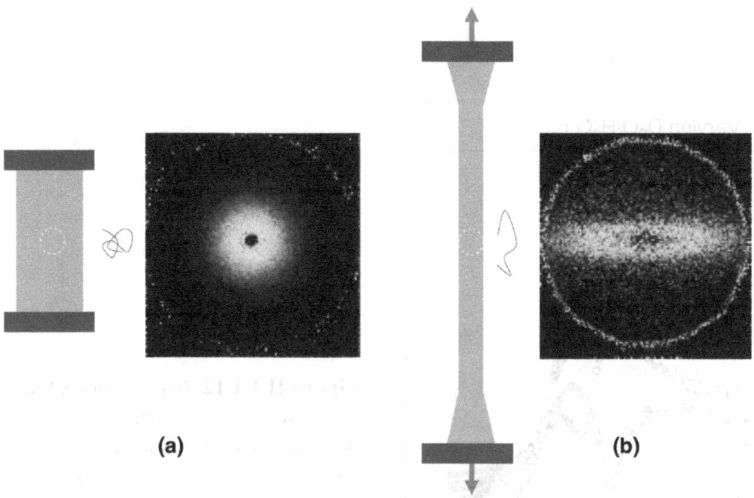


**Figure II.1.1.11** Experimental SANS pattern of a polystyrene sample (PS/PSd) when relaxed (a) and when exposed to elongational flow (b) (Hassager et al., 2009).

In soft matter sciences, labeling using hydrogen–deuterium exchange is most relevant. Systematic studies labeling single molecules or parts of a molecule have led to detailed structural insight. In polymer sciences, the polymer coil conformation within the polymer melt state is among the most groundbreaking results of polymer physics, showing Gaussian coil conformation as proposed by Flory. An experimental example is shown in Figure II.1.1.5. Corresponding labeling is applied when studying the molecular response to macroscopic deformation, elongation, or shear. Figure II.1.1.11 shows an example of the scattering pattern of a polystyrene sample of mixed hydrogenated and deuterated chains, when relaxed (a) and when exposed to elongational flow (b) (Hassager et al., 2009). Analysis of the scattering pattern provides details on the polymer coil deformation as resulting from the flow field. Such data are important for understanding rheological properties and features during polymer processing.

Systematic studies based on labeling different units of complex synthetic molecules or biomolecules have led to detailed structural insight. Examples include the studies of the ribosome structure, biomembranes, and other complex molecular aggregates, which require involvement of a large number of scientists with interdisciplinary expertises including biology, biochemistry, chemistry, and scattering physics. A more simple example of such complex molecular aggregate structure is that of spherical micelles. Figure II.1.1.12 shows the scattering of polymeric micelles, using different contrast conditions such as different mixing ratios of $H_2O$ and $D_2O$. The difference in the apparent form factor is quite dramatic, indicating that such contrast studies provide significantly improved structural insight relative to the information that is gained from an experiment using a single contrast condition only.

Shear devices have proven to be extremely important in studies of surfactant systems, gels, and polymer melts, reflecting properties during processing and

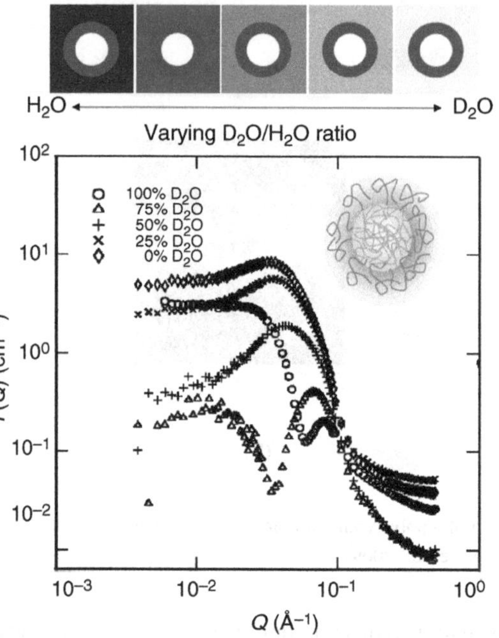


**Figure II.1.1.12** Experimental SANS pattern of polymer micelles (PPO–dPEO–PPO) in various $H_2O/D_2O$ mixtures ranging from pure $H_2O$ to pure $D_2O$. The top row illustrates schematically the contrast condition for the micelles in the measured mixtures, where the micelles are shown as simple core–shell particles.

applications. Shear can cause dramatic changes in phase behavior, texture, and so on. This is, for example, the situation in extruders used in the polymer industry. In the scattering experiment, shear can be used to study both hydrodynamic changes of polymer coil conformation related to shear and stress and possible changes in miscibility parameters and thereby phase transition temperatures. Shear has in addition proven extremely useful for making monodomain ordered structures suitable for crystallographic studies, for example, after mounted on a goniometer. Figure II.1.1.13 shows experimental scattering data of a polymer micellar gel that

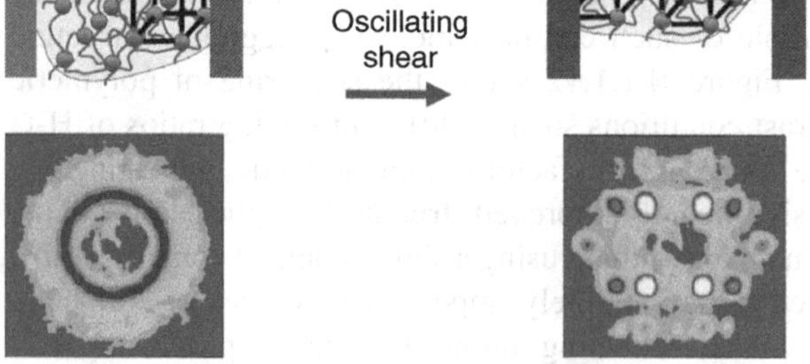


**Figure II.1.1.13** Experimental SANS pattern of an ordered polymer micellar gel, as formed and after exposure to large-amplitude oscillatory shear. The top part shows schematically the alignment of ordered domains (Mortensen, 2004).

forms a cubic ordered system (Mortensen, 2004). As formed, the sample shows broad, isotropic Debye–Scherrer rings that are difficult to separate. Upon large-amplitude oscillatory shear, the sample forms a near monodomain (or twin-domain) structure with highly resolved Bragg reflections.

## REFERENCES


Alefeld, B., Hayes, C., Mezei, F., Richter, D., and Springer, T. *Physica B* **1997**, *234–236*, 1052.

Allen, A.J. and Berk, N.F. *J. Appl. Crystallogr.* **1994**, *27*, 878.

Antonimi, M., Daneri, A., and Toselli, G. *J. Appl. Crystallogr.* **1970**, *3*, 145.

Aswal, V.K. and Goyal, P.S. *Curr. Sci.* **2000**, *79*, 947–953.

Barker, J.G. and Pedersen, J.S. *J. Appl. Crystallogr.* **1995**, *28*, 105.

Boothroyd, A.T. *J. Appl. Crystallogr.* **1989**, *22*, 252.

Brulet, A., Thevenot, V., Lairez, D., Lecommandoux, S., Agut, W., Armes, S.P., Du, J., and Desert, S. *J. Appl. Crystallogr.* **2008**, *41*, 161.

Cheng, Y.T., Mildner, D.F.R., Chen-Mayer, H.H., Sharov, V.A., and Glinka, C.J. *J. Appl. Crystallogr.* **2000**, *33*, 1253.

Child, H.R. and Spooner, S. *J. Appl. Crystallogr.* **1980**, *13*, 259.

Choi, S.M., Barker, J.G., Glinka, C.J., Cheng, Y.T., and Gammel, P.L. *J. Appl. Crystallogr.* **2000**, *33*, 793.

Clark, C.D., Mitchell, E.W.J., Palmer, D.W., and Wilson, I.H. *J. Sci. Instrum.* **1966**, *43*, 1–5.

Desert, S., Thevenot, V., Oberdisse, J., and Brulet, A. *J. Appl. Crystallogr.* **2007**, *40*, s471.

Dewhurst, C.D. *Meas. Sci. Technol.* **2008**, *19*, 034007.

Eskildsen, M.R., Gammel, P.L., Isaacs, E.D., Detlefs, C., Mortensen, K., and Bishop, D.J. *Nature* **1998**, *391*, 563.

Gentile, T.R., Jones, G.L., Thompson, A.K., Barker, J., Glinka, C.J., Hammouda, B., and Lynn, J.W. *J. Appl. Crystallogr.* **2000**, *33*, 771.

Ghosh, R.E. and Rennie, A.R. *Inst. Phys. Conf. Ser.* **1990**, *107*, 233.

Gilbert, E.P., Schulz, J.C., and Noakes, T.J. *Physica B* **2006**, *385–386*, 1180.

Giri, E., Putra, R., Ikram, A., Santoso, E., and Bharoto, B. *J. Appl. Crystallogr.* **2007**, *40*, 447.

Glatter, O. *J. Appl. Crystallogr.* **1977**, *10*, 415–421.

Glinka, C.J., Rowe, J.M., and LaRock, J.G. *J. Appl. Crystallogr.* **1986**, *19*, 427.

Glinka, C.J., Barker, J.G., Hammouda, B., Krueger, S., Moyer, J.J., and Orts, W.J. *J. Appl. Crystallogr.* **1998**, *31*, 430.

Grabcev, B. *J. Appl. Crystallogr.* **2007**, *40*, 40.

Han, Y.-S., Choi, S.-M., Kim, T.-H., Lee, S.-H., and Kim, H.-R. *J. Appl. Crystallogr.* **2007**, *40*, s442.

Hassager, O., Mortensen, K., Bach, A., Almdal, K., Rasmussen, H.K., and Pyckhout-Hintzen, W. Unpublished.

Hayter, J.P. and Penfold, J. *J. Mol. Phys.* **1981**, *42*, 109.

Heenan, R.K. *Proceedings of the International Collaboration on Advanced Neutron Sources*, ICANS XII, RAL Report 94-025, Vol. I, May 1993, p. 241.

Heiderich, M., Reinartz, R., Kurz, R., and Schelten, J. *Nucl. Instrum. Methods Phys. Res.* **1991**, *305*, 423.

Hofmeyr, C., Mayer, R.M., and Tillwick, D.L. *J. Appl. Crystallogr.* **1979**, *12*, 192.

Ibel, K. *J. Appl. Crystallogr.* **1976**, *9*, 296.

Kentzinger, E., Dohmen, L., Alefeld, B., Rucker, U., Stellbrink, J., Ioffe, A., Richter, D., and Bruckel, T. *Physica B* **2004**, *350*, e779.

Kinning, D.J. and Thomas, E.L. *Macromolecules* **1984**, *17*, 1712.

Kipping, D., Gahler, R., and Habicht, K. *Phys. Lett. A* **2008**, *372*, 1541.

Kline, S.R. *J. Appl. Crystallogr.* **2006**, *39*, 895.

Koehler, W.C. and Hendricks, R.W. *Acta Crystallogr. A* **1978**, *34*, s346.

Koehler, W.C. and Hendricks, R.W. *J. Appl. Phys.* **1979**, *50*, 1951.

Kohlbrecher, J. http://num.web.psi.ch/reports/2003/ASQ/ASQ17 Joachim SANSpol.pdf, 2003.

Kohlbrecher, J. and Wagner, W. *J. Appl. Crystallogr.* **2000**, *33*, 804.

Koizumi, S., Iwase, H., Suzuki, J., Oku, T., Motokawa, R., Sasao, H., Tanaka, H., Yamaguchi, D., Shimizu, H.M., and Hashimoto, T. *J. Appl. Crystallogr.* **2007**, *40*, s474.

Kopp, M.K., Valentine, K.H., Christophorou, L.G., and Carter, J.G. *Nucl. Instrum. Methods Phys. Sci.* **1981**, *201*, 395.

Kotlarchyk, M. and Chen, S.H. *J. Chem. Phys.* **1983**, *79*, 2461.

Lieutenant, K., Lindner, P., and Gahler, R. *J. Appl. Crystallogr.* **2007**, *40*, 1056.

Lynn, G.W., Buchanan, M.V., Butler, P.D., Magid, L.J., and Wignall, G.D. *J. Appl. Crystallogr.* **2003**, *36*, 829.

Mazumder, S., Sequeira, A., Roy, S.K., and Biswas, A.B. *J. Appl. Crystallogr.* **1993**, *26*, 357.

Mildner, D.F.R. *J. Appl. Crystallogr.* **1984**, *17*, 293.

Mildner, D.F.R. *J. Appl. Crystallogr.* **2005**, *38*, 488.

Mildner, D.F.R., Berliner, R., Pringle, O.A., and King, J.S. *J. Appl. Crystallogr.* **1981**. *14*, 370.

Mildner, D.F.R., Hammouda, B., and Kline, S.R. *J. Appl. Crystallogr.* **2005**, *38*, 979.

Mortensen, K. *J. Polym. Sci. Polym. Phys.* **2004**, *42*, 3095.

Mortensen, K. and Talmon, Y. *Macromolecules* **1995**, *28*, 8829.

Okabe, S., Nagao, M., Karino, T., Watanabe, S., Adachi, T., Shimizu, H., and Shibayama, M. *Appl. Crystallogr.* **2005**, *38*, 1035.

Oku, T., Iwase, H., Shinohara, T., Yamada, S., Hirota, K., Koizumi, S., Suzuki, J., Hashimoto, T., and Shimizu, H.M. *J. Appl. Crystallogr.* **2007**, *40*, s408.

Oku, T., Shinohara, T., Kikuchi, T., Oba, Y., Iwase, H., Koizumi, S., Suzuki, J., and Shimizu, H.M. *Meas. Sci. Technol.* **2008**, *19*, 034011.

Pedersen, J.S. *J. Appl. Crystallogr.* **1994**, *27*, 595.

Pedersen, J.S. Gerstenberg, M. *Macromolecules* **1996**, *29*, 1363.

Pedersen, J.S., Posselt, D., and Mortensen, K. *J. Appl. Crystallogr.* **1990**, *23*, 321.

Rekveldt, M.T. *Nucl. Instrum. Methods Phys. Res. B* **1996**, *114*, 366.

Schelten, J. and Hendricks, R.W. *J. Appl. Crystallogr.* **1978**, *11*, 297.

Schmatz, W., Springer, T., Schelten, J., and Ibel, K. *J. Appl. Crystallogr.* **1974**, *7*, 96.

Schwahn, D., Meier, G., and Springer, T. *J. Appl. Crystallogr.* **1991**, *24*, 568.

Strobl, M., Treimer, W., Ritzoulis, C., Wagh, A.G., Abbas, S., and Manke, I. *J. Appl. Crystallogr.* **2007**, *40*, 463.

Strunz, P. and Wiedenmann, A. *J. Appl. Crystallogr.* **1997**, *30*, 1132.

Strunz, P., Mortensen, K., and Janssen, S. *Physica B* **2004**, *350*, e783.

Wiedenmann, A. *Physica B* **2001**, *297*, 226.

Wiedemann, A. *Physica B* **2005**, *356*, 246.

Wiedemann, A., Keiderling, U., Habicht, K., Russina, M., and Gahler, R. *Phys. Rev. Lett.* **2006**, *97*, 057202.

Wignall, G.D. *Polym. Eng. Sci.* **1986**, *26*, 695.

Wignall, G.D. *J. Appl. Crystallogr.* **1991**, *24*, 479.

Wignall, G.D. *J. Appl. Crystallogr.* **1997**, *30*, 884.

Wignall, G.D., Christen, D.K., and Ramakrishnan, V. *J. Appl. Crystallogr.* **1988**, *21*, 438.

# II Instrumentation

# II.1 Small-Angle Neutron Scattering

## II.1.2 SANS Instruments at Pulsed Neutron Sources

*Toshiya Otomo*

### II.1.2.1 INTRODUCTION

An accelerator-driven neutron source produces neutrons with a wide range of wavelength. With the time-of-flight method, momentum transfer is obtained by the arrival time of the neutron:

$$Q = \frac{4\pi \sin\theta}{\lambda} = \frac{4\pi \sin\theta}{ht/mL}, \qquad \text{(II.1.2.1)}$$

where $m$ is the neutron mass, $h$ is the Planck's constant, $t$ is the arrival time (time-of-flight (TOF)) of a neutron at a detector since its generation and $L$ is the flight path length of neutrons. The wide-$Q$ coverage with the wide range of wavelength is the benefit to install small-angle neutron scattering (SANS) at pulsed sources (hereafter, it is referred to as TOF-SANS). The wide-$Q$ coverage enables us to observe both small-angle scattering and powder diffractions simultaneously. On the other hand, corrections of wavelength-dependent factors are critical to obtain accurate scattering functions.

Figure II.1.2.1 shows a TOF diagram of a pulsed neutron source that produces neutron with a repetition of accelerator (Hz), $R$. The time interval between the pulses is $1/R$ s and one interval is called as "one frame." At each frame, neutrons of a wide range of wavelength are generated and shorter wavelength neutron may pass longer wavelength neutrons generated in the previous frame. The longest wavelength of neutron, $\lambda_{max}$, which arrives in $1/R$ s, is calculated by the following equation:

*Neutrons in Soft Matter* Edited by Toyoko Imae, Toshiji Kanaya, Michihiro Furusaka, and Naoya Torikai

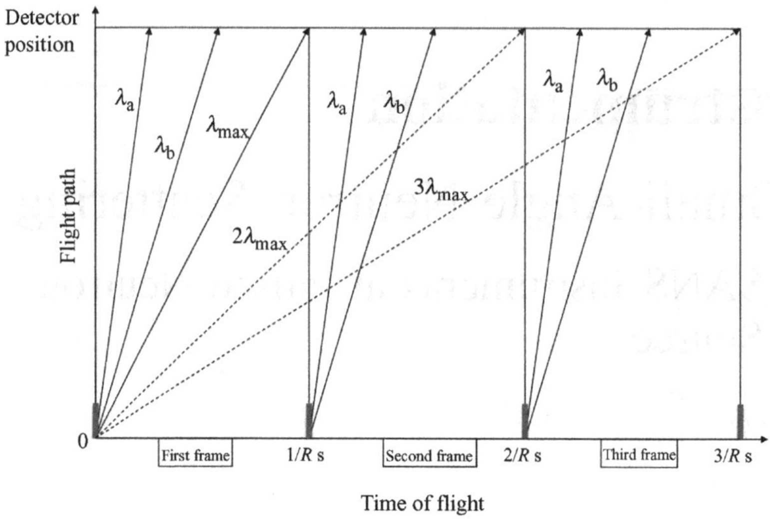


**Figure II.1.2.1** Time-of-flight diagram without choppers. Neutrons within the bandwidth such as $\lambda_a$ and $\lambda_b$ are accessible in the first frame. As the repetition rate, $R$, is larger or total neutron flight is shorter, $\lambda_{max}$ becomes larger. Since neutrons are produced every $1/R$ s, there are longer wavelength neutrons ($>\lambda_{max}$) that arrive at the detector in the following time frames.

$$\lambda_{max} = \frac{h}{mv_{min}} = \frac{h}{m}\frac{1}{L_{total}R} = \frac{3956}{L_{total}/R}, \tag{II.1.2.2}$$

where $L_{total}$ is total flight path (m). Let us put a frame as first frame, then wavelength range next frame (second frame) is from $\lambda_{max}$ to $2\lambda_{max}$ and one of the $n$th frame is from $(n-1)\lambda_{max}$ to $n\lambda_{max}$. Since $\lambda$ is linear to time (TOF), $\lambda_{max}$ also indicates an accessible bandwidth ($\lambda_{width} = \lambda_{max} - \lambda_{min}$) in a frame. $\lambda_{max}$ is very important parameter to design SANS at pulsed source because it decides the value of $Q_{min}$.

## II.1.2.2 RESOLUTION AND WAVELENGTH BANDWIDTH OF PINHOLE-TYPE SANS FOR PULSED NEUTRON SOURCE

In this section, basic parameters to design SANS at pulsed sources will be described by taking the pinhole-type collimation as an example. The most important parameter of SANS is momentum transfer ($Q$) range ($Q_{min} < Q < Q_{max}$) to be measured by the instrument. This parameter should be decided by the scientific motivation at the first step of the instrumentation.

$Q_{\min}$ is determined with a detector at minimum scattering angle ($2\theta_{\min}$) and $\lambda_{\max}$ using eqs. (II.1.2.1) and (II.1.2.2):

$$Q_{\min} = \frac{4\pi}{\lambda_{\max}} \sin\theta_{\min}$$

$$\propto L_{\text{total}} \sin\theta_{\min}. \tag{II.1.2.3}$$

It is obvious that smaller $L_{\text{total}}$ realizes smaller $Q_{\min}$. The other parameter of $Q_{\min}$ is the minimum scattering angle, $2\theta_{\min}$, and $\theta_{\min}$ is determined by a beam divergence, $\Delta\theta$, taking into account the required $Q$-resolution and the instrument geometry in the case of pinhole collimation.

The $Q$-resolutions of a TOF-SANS with a point-collimator system are calculated approximately by

$$\frac{\Delta Q}{Q} = \left[\left(\frac{\Delta\lambda}{\lambda}\right)^2 + (\cot\theta \cdot \Delta\theta)^2\right]^{1/2}. \tag{II.1.2.4}$$

At a small-angle region, the $Q$-resolution can be approximated as the angular resolution,

$$\frac{\Delta Q}{Q} \approx \frac{\Delta\theta}{\theta}. \tag{II.1.2.5}$$

This is because the wavelength resolution $\Delta\lambda/\lambda$ is few percent and it is small enough when compared to the angular resolution. $\theta_{\min}$ at a small-angle region with a certain $Q$-resolution can be calculated approximately with

$$\theta_{\min} \approx \frac{\Delta\theta}{\Delta Q/Q}, \tag{II.1.2.6}$$

and then

$$Q_{\min} \cong \frac{4\pi}{\lambda_{\max}} \times \frac{\Delta\theta}{\Delta Q/Q}. \tag{II.1.2.7}$$

From eq. (II.1.2.7), $Q_{\min}$ of TOF-SANS of pinhole collimation type determined by three parameters of $\lambda_{\max}$, $\Delta\theta$, and $\Delta Q/Q$. These parameters, however, depend on each other.

The beam divergence $\Delta\theta$ is calculated by the following equation in the case of pinhole-type collimation (Mildner and Carpenter, 1984),

$$\Delta\theta = \frac{1}{2}\sqrt{\frac{(A_0/2)^2}{L_1^2} + \left(\frac{A_S}{2}\right)^2\left(\frac{1}{L_1} + \frac{1}{L_2}\right)^2\frac{A_{\text{dr}}^2}{L_2^2}}, \tag{II.1.2.8}$$

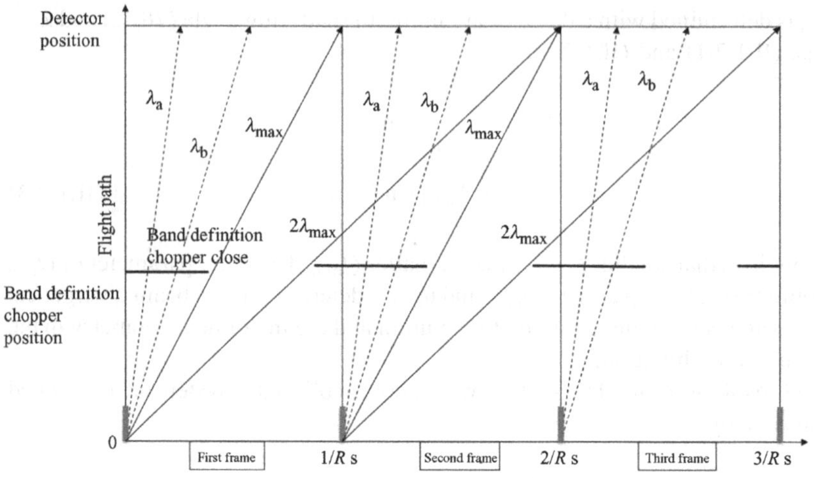


**Figure II.1.2.2** Wavelength definition by a band definition chopper. Shorter wavelength neutrons ($<\lambda_{max}$) are shielded by the chopper and "second frame" in the range from $\lambda_{max}$ to $2\lambda_{max}$ is accessible.

where $L_1$ is the moderator-to-sample distance, $L_2$ is the sample-to-detector distance, $A_0$ is the moderator width, $A_S$ is a sample width, $A_d$ is a detector width, and $A_{dr}$ is a detector resolution, respectively. In case of pinhole-type collimation, the beam divergence, $\Delta\theta$, can be smaller if longer $L_1$ and $L_2$, that is, longer $L_{total}$ ($=L_1 + L_2$) is chosen. On the other hand, longer $L_{total}$ leads to smaller $\lambda_{max}$ and larger $Q_{min}$ from eq. (II.1.2.2). This effect is from the wavelength bandwidth of neutron and is a typical constrain of TOF-SANS. One of the ways to use longer wavelength at certain $L_{total}$ is to use a so-called band definition choppers to select second or third frame. A band definition chopper is rotating disk made with a neutron-absorbing area and a transparent area. Neutron can only transmit the disk through the deficit. Once the distance between the neutron source and the chopper is decided, a usable wavelength range of neutron is calculated with the opening time and phase of the chopper (see Figure II.1.2.2). Curved neutron guide (or so-called "T0 chopper") can eliminate fast neutrons and wider bandwidth can be realized. In this case, at least one frame will be lost then the total number of neutron at sample is less than half. Therefore, simple pinhole-type collimation is not suitable to pursue $Q_{min}$ for SANS at pulsed sources.

Table II.1.2.1 shows the basic parameters of TOF-SANS in the world. Basic parameters are similar with these instruments. In the case of lower $Q$ region, that is, smaller $Q_{min}$, is required, neutron optical devices are installed to utilize longer wavelength neutrons.

In the case of SWAN (Otomo et al., 1999), which had been operated at KENS until 2006, $Q$-range was decided as $Q_{min} = 0.01\,\text{Å}^{-1}$ and $Q_{max} = 20\,\text{Å}^{-1}$. The $Q$-resolutions ($\Delta Q/Q$) of SWAN (small- and wide-angle neutron diffractomter) at KENS are shown in Figure II.1.2.3 with $A_0 = 0.06$ m; $A_s = 0.02$ m; $L_1 = 13.2$ m; and $L_2 = 4.3$ m for small angle, $L_2 = 3.3$ m for second small-angle bank, $L_2 = 1.15$ m for

**Table II.1.2.1** Instrument Parameters of TOF-SANS

| | LOQ | LQD | SAND | SWAN |
|---|---|---|---|---|
| Facility | ISIS, Rutherford Appleton Laboratory | Lujan Center, Los Alamos National Laboratory | IPNS, Argonne National Laboratory | KENS, High Energy Accelerator Research Organization (KEK) |
| Repetition rate | 50 | 20 | 30 | 20 |
| Moderator | Liquid hydrogen (25K) | Partially coupled liquid hydrogen (20K) | Coupled solid methane (24K) | Solid methane (20K) |
| Primary flight path | Solar supermirror bender 24 mrad | | Focusing solar collimators 3.4 mrad @FWHM | |
| L1 | 11.3 | 12.72 | 7.0 | 13.2 |
| L2 | 3.85 | | 2.0, 1.524 | 4.3, 3.3, 1.15, 0.5 |
| Sample size | 8 mm $\phi$ (2 mm $\phi$–20 mm $\phi$) | 10 × 13 mm | | 20 × 20 mm |
| Wavelength | >2 (5.2) | 1.5 ~ 15 | 1 ~ 14 | 0.5 ~ 12 |
| Detector pixel | 6 × 6 mm | | 4 ~ 6 mm, 1 ~ 2 cm | 1 ~ 2 cm |
| $Q$-range | 0.006 ~ 1.4 | 0.003 ~ 0.5 | 0.0035 ~ 2 | 0.01 ~ 20 |
| User program | Available | Available | Closed (2008) | Closed (2006) |

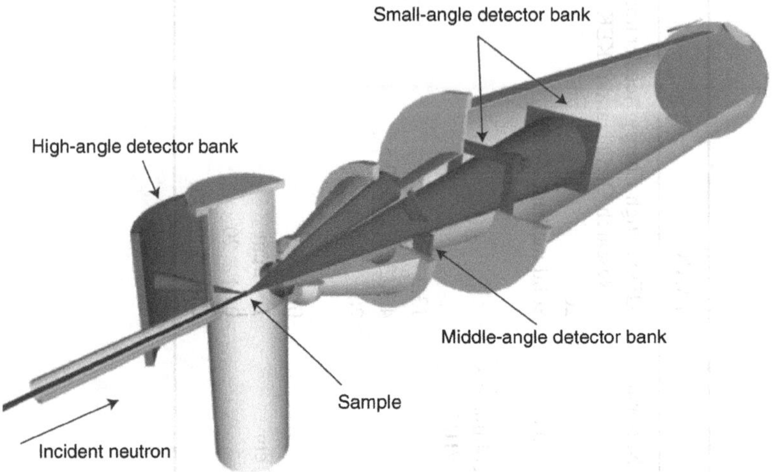


**Figure II.1.2.3** Schematic view of SWAN operated at KENS, KEK. It consisted of several detector banks: two small-angle bank, middle-angle bank, and high-angle bank. As the scattering angle is larger, sample–detector distance can be shorter to realize reasonable resolution. It helps to realize larger solid angle of detectors and hence higher counting rate.

middle-angle bank, and $L_2 = 0.5$ m for high-angle bank. $A_D$ is 0.012 m for the small- and high-angle bank and is 0.025 for the medium-angle bank, respectively. These $A_D$ values depend on the selection of the neutron detector. In case of SWAN, $^3$He gas detector was chosen: linear position-sensitive detector (0.5 in. diameter, 60 cm of effective length, and 10 atm) for small- and high-angle bank and normal detector (1 in. diameter and 10 atm) for middle angle bank. The normal detectors were aligned so that the axis of the detectors was perpendicular to the direction of neutron beam. $Q$-resolutions of higher angle bank can be calculated with eq. (II.1.2.4). In this chapter, selection of neutron detector, optimization of spatial resolution of detector will not be discussed, but it is closely related to the instrument parameters. The selection for SANS at pulsed source is rather complicated than the reactor since a neutron detection efficiency in a wide wavelength range is required. The calculated $Q_{min}$ of SWAN using eq. (II.1.2.4) is 0.005 Å$^{-1}$ with $\Delta Q/Q = 0.3$. Because of the cot $\theta$ effect, the resolution is better at a higher $Q$-range. The resolution of the medium-angle bank is lower than that of the small-angle bank. This is caused by the size of the detector element, $A_D$. The resolution of a high-angle bank is less than 1% even at short $L_2$, for example, $L_2 = 0.5$ m.

The scattered intensity of neutrons with wavelength between $\lambda$ and $\lambda + \Delta\lambda$, $I(\lambda)$ $\Delta\lambda$, from a sample with cross section d$\sigma$/d$\Omega$ is given by (Ishikawa et al., 1986)

$$I(\lambda)\Delta\lambda = I_0(\lambda)\left[\frac{d\sigma}{d\Omega}\right]A(\lambda)VN\Delta\Omega\eta(\lambda)\Delta\lambda, \qquad \text{(II.1.2.9)}$$

where $I_0(\lambda)$ $\Delta\lambda$ is the intensity of incident neutrons with wavelength between $\lambda$ and $\lambda + \Delta\lambda$ at the sample position. With a range of neutron wavelength, one detector element at certain scattering angle can observe a range of $Q$. On the other hand, scattering intensity at certain $Q$, $I(Q)$, of TOF-SANS is the sum of detector elements of different scattering angles. The neutron intensity observed in $\Delta Q$ at $Q$ is expressed as follows (Furusaka et al., 1990):

$$I_{\text{obs}}(Q)\Delta Q = \int_{k_z^{\min}}^{k_z^{\max}} \mathrm{d}k_z I_0(k_z)\eta(k_z)\mathrm{VNA}(k_z)\frac{\mathrm{d}\sigma}{\mathrm{d}\Omega}(Q)\frac{\Delta\Omega(k_Z, Q)}{\Delta Q}\Delta Q, \qquad \text{(II.1.2.10)}$$

where, $k_z$ is a wave number vector along the neutron beam direction, $k_z^{\max}$ is the largest $k_z$ available, $\frac{\Delta\Omega(k_z, Q)}{\Delta Q}$ is the solid-angle covered by the detector system in $\Delta Q$ for a fixed $k_z$. $F(Q)$ in eq. (II.1.2.11), which is called as figure of merit of TOF-SANS, is the instrument-dependent factors in eq. (II.1.2.10). The ideal line in Figure II.1.2.4 corresponds to the ideal detector arrangements that can use full range of wavelength. As depicted in Figure II.1.2.5, $Q$-resolution is variable according to $Q$.

$$F(Q) = \int_{k_z^{\min}}^{k_z^{\max}} \mathrm{d}k_z I_0(k_z)\eta(k_z)\frac{\Delta\Omega(k_z, Q)}{\Delta Q}. \qquad \text{(II.1.2.11)}$$

The intensity of $Q_{\min}$ is determined by a monochromatic neutron flux (neutron/s) of $\lambda_{\max}$. In order to overcome such prospect of pinhole geometry, optical devices are

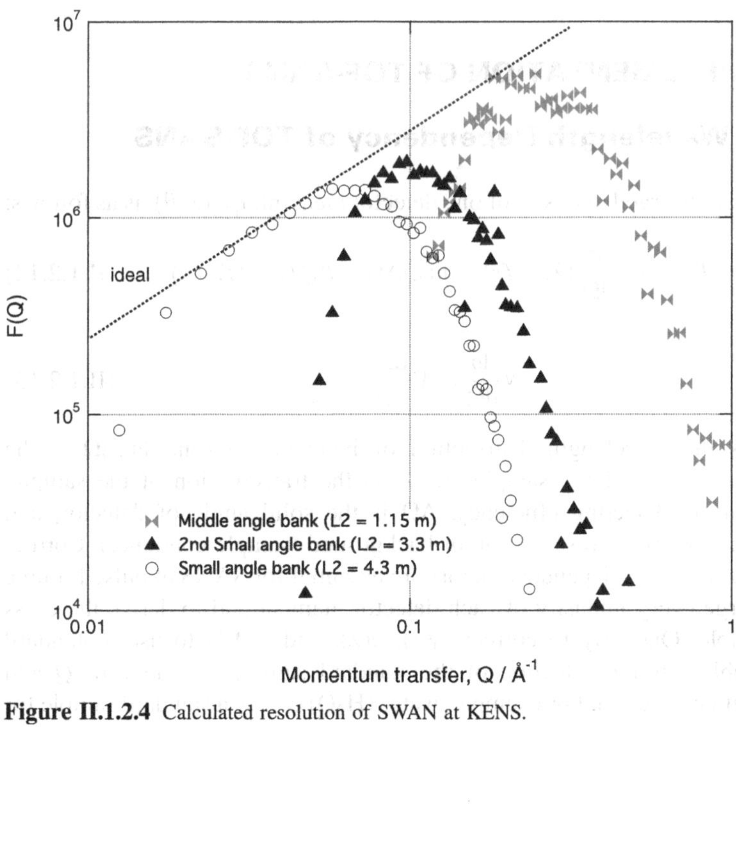


**Figure II.1.2.4** Calculated resolution of SWAN at KENS.

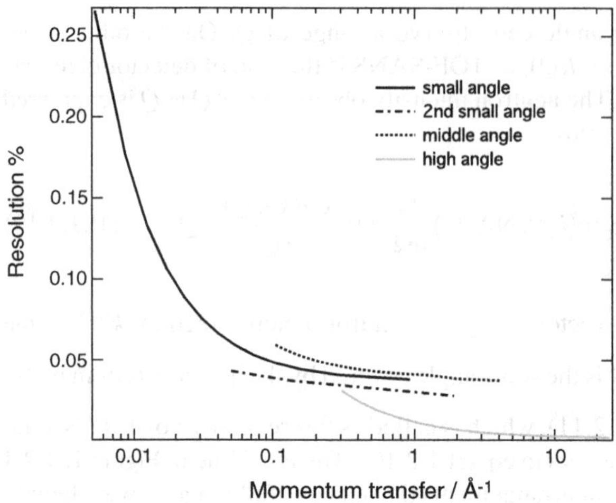


**Figure II.1.2.5** Calculated $F(Q)$ of small-angle bank, second small-angle bank, and middle-angle bank of SWAN at KENS. Since the small-angle bank covered wide scattering angle, $F(Q)$ in the $Q$-range from 0.02 to 0.06 Å$^{-1}$ is almost ideal. This means that all wavelength ranges were used to measure the $Q$-range.

aggressively developing and became essential for SANS at pulsed source (Carpenter et al., 2003; Iwashita et al., 2009; Littrell, 2004; Oku et al., 2009; Shinohara et al., 2009). Intensity gains with such optical devices are significant and these devices will be implemented in the next-generation TOF-SANS. The gains are estimated with Monte Carlo calculations. Next-generation TOF-SANS such as SANS2D (ISIS), EQ-SANS (SNS), and Taikan (J-PARC) will be available for user program in the near future.

## II.1.2.3 IMPLEMENTATION OF TOF-SANS

### II.1.2.3.1 Wavelength Dependency of TOF-SANS

Simplest form of observed intensity of one detector element, $I^{s}_{obs}(\lambda, \theta)$, is as follows:

$$I^{s}_{obs}(\lambda, \theta) = I^{s}_{obs}(\lambda) N \frac{d\sigma^{s}}{d\Omega}(\lambda, \theta) Tr^{s}(\lambda) \eta(\lambda) \Delta\Omega + I^{bck}_{obs}(\lambda, \theta) Tr^{s}(\lambda) \qquad \text{(II.1.2.12)}$$

$$N \frac{d\sigma}{d\Omega} = V \frac{d\Sigma}{d\Omega} \qquad \text{(II.1.2.13)}$$

where $I^{s}_{0}(\lambda)$ is the wavelength distribution of incident neutron, $d\sigma^{s}/d\Omega$ is the scattering cross section of the sample, $Tr^{s}(\lambda)$ is the transmission of the sample, $\eta(\lambda)$ is the neutron detecting efficiency, $\Delta\Omega$ is the solid angle of detector, and $I^{bck}_{obs}(\lambda, \theta)$ is the scattering cross section of background (sample container). Corrections of three wavelength-dependent factors are essential for SANS at pulsed source in order to merge every intensity of each detector element and to derive the cross section of sample. One way to correct $I_0(\lambda)$, $\eta(\lambda)$, and $\Delta\Omega$ is to use a standard sample. A problem is the selection of the standard sample for the wide-$Q$ and wide-wavelength correction. For example, water ($H_2O$) can be a standard sample for small-angle but not for high-angle because of its recoil effect. The other deficit of the standard sample correction is intrinsic small solid angle in the small-angle region—it takes longer time to measure the intensity from the standard sample until enough statistics is secured. From this point of view, the instrument should prepare to measure each wavelength dependency. Multiple scattering and inelastic effect, most difficult terms for corrections, are not described in eq. (II.1.2.12).

## II.1.2.3.2 Implementations for Wavelength Dependency Correction

Again, the corrections of $I_0(\lambda)$, $Tr(\lambda)$, and $\eta(\lambda)$ are the key to use TOF-SANS. The instrument implementation for the correction is varies depend on instruments. Here, it is tried to demonstrate a concrete example by showing the implementation of SWAN at KENS.

### *II.1.2.3.2.1 Monitor for Incident Neutron*

Typical wavelength dependence of incident neutron at pulsed source is shown in Figure II.1.2.6. This was observed on SWAN, i.e. neutron flux of solid methane moderator at KENS. This consists of epithermal region (<2 Å) and thermal region (>2 Å). It is essential to know wavelength-dependent distribution neutrons radiated

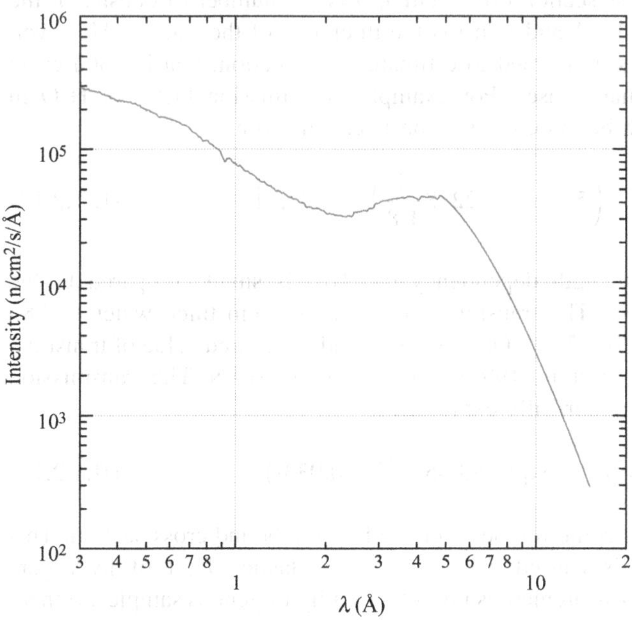


**Figure II.1.2.6** Wavelength dependency of incident neutron of SWAN at KENS measured by the incident monitor and the efficiency corrected.

to the sample in absolute value. There are different ways to observe incident neutron with TOF. In case of SWAN, $^{3}He$ monitor with $10^{-3}$ efficiency was installed in the incident neutron beam at 1 m upstream from the sample position. This low efficiency was realized by low $^{3}He$ gas pressure. The neutron-counting efficiency of the monitor is calculated by

$$\eta_{mon}(\lambda) = 0.00008309 + 0.0005263 \times \lambda. \qquad \text{(II.1.2.14)}$$

This equation was simply calculated by assuming that the efficiency is $10^{-3}$ at the wavelength 1.8 Å. Note that the efficiency is linear to $\lambda$ and good for the correction of wavelength dependency.

#### II.1.2.3.2.2 Monitor for Neutron Transmission Through Sample

The transmission factor, $Tr(\lambda)$, can be calculated if the total cross section is known:

$$Tr(\lambda) = \exp(\Sigma_{total}(\lambda) \times t), \qquad \text{(II.1.2.15)}$$

where,

$$\Sigma_{total}(\lambda) = \left(\sigma_{sea} + \sigma_{abs}\frac{\lambda}{1.8}\right)\rho \qquad \text{(II.1.2.16)}$$

is the total cross section of the sample, $\sigma_{sca}$ is the scattering cross section ($10^{-24}$ cm$^2$), $\sigma_{abs}$ is the absorption cross section ($10^{-24}$ cm$^2$), $\rho$ is the number of density of the sample (1/cm$^3$) with $\lambda = 1.8$ Å and $t$(cm) is the thickness of the sample. However, scattering measurement is performed to estimate cross section, that is, total cross section is unknown in many cases. For example, transmission factor of $H_2O$ in ambient temperature with bound cross section is calculated as

$$Tr(\lambda) = \exp\left[-\left(56.09 + 0.2218\frac{\lambda}{1.8}\right) \times 0.033t\right]. \qquad \text{(II.1.2.17)}$$

It is clear that wavelength dependency (0.2218) is small compared to the independent factor (56.09). The transmission factor of 1 mm thick water is 0.83 with $\lambda = 4.75$ Å from eq. (II.1.2.17). On the other hand, measured value of transmission is 0.539 (Thiyagarajan et al., 1997) and is 0.56 by SWAN. The transmission measured at SWAN was empirically expressed as

$$Tr(\lambda) = \exp(-88.45\lambda^{0.45} \times 0.033t). \qquad \text{(II.1.2.18)}$$

It is obvious that this could not be calculated by the bound cross section. This wavelength dependency is caused by incoherent inelastic effect of hydrogen. Therefore, transmission measurement is important for hydrogenous samples. If there is an incident monitor and detector element at transmitted beam position (direct beam monitor), the transmission can be obtained by taking the ratio between sample and empty (nothing at sample position) measurements.

$$\frac{[I_d^s(\lambda)/\eta_d(\lambda)]/[I_{mon}^s(\lambda)/\eta_{mon}(\lambda)]}{[I_d^{empty}(\lambda)/\eta_d(\lambda)]/[I_{mon}^{empty}(\lambda)/\eta_{mon}(\lambda)]} = \frac{[I_d^s(\lambda)/I_{mon}^s(\lambda)]}{[I_d^{empty}(\lambda)/I_{mon}^{empty}(\lambda)]}. \qquad (II.1.2.19)$$

Although implementation of the direct beam monitor has not been established, the most popular way is to put a direct beam monitor when a transmission measurement is performed. The other way is to put neutron absorber at the direct beam position of the small-angle detector in order to prevent counting loss. Benefits of these are safer in signal-to-noise ratio in the former case and real time measurement of transmission in the latter case.

#### *II.1.2.3.2.3 Detector-Counting Efficiency*

It is difficult to know absolute and accurate efficiency of every detector. For the case that one large area detector is used for small-angle scattering measurement, a correction method to use the center of the detector was proposed (Heenan et al., 1997). The center of the detector counts quickly because the direct neutron beam is radiated. If the neutron flux at sample is high enough, single crystal of vanadium is good incoherent scatterer to estimate detector efficiency experimentally. Rough estimation is possible by an empirical method in the case of a tube detector: Area integration of absorption cross section of detector by following equation:

$$\eta(\lambda) = \int_0^{r_2} \exp\left[-\Sigma_{tube} \times \left(\sqrt{r_1^2 - x^2} - \sqrt{r_1^2 - x^2}\right)\right] \exp\left[1 - \Sigma_{He} \times 2\left(\sqrt{r_2^2 - x^2}\right)\right] dx, \qquad (II.1.2.20)$$

where $\Sigma_{tube}$ is the total cross section of detector tube (container), $\Sigma_{He}$ is the total cross section of He gas, $r_1$ and $r_2$ are the outer diameter of the tube, respectively. Figure II.1.2.7 shows the calculated detector efficiency of SWAN. At longer $\lambda$, efficiencies decrease because neutrons are scattered by the stainless tube.

#### *II.1.2.3.2.4 Absolute Correction by Vanadium*

In principle, if we could estimate every parameters in eq. (II.1.2.12), absolute value of the sample cross section can be obtained. Unfortunately, it is not normal case, because experiment time is finite. By utilizing wide-$Q$ coverage of TOF-SANS, polycrystal vanadium is useful to normalize the intensity as eq. (II.1.2.21) and obtain an absolute value:

$$I_{sample}^{abs}(Q) = \frac{I_{sample}^{abs}(Q)}{V_{sample}} \times \left(\frac{d\Sigma}{d\Omega}\right)_{van} \times \frac{V_{van}}{I_{van}^{obs}(Q)} = \frac{V_{van}}{V_{sample}} \times \frac{I_{sample}^{abs}(Q)}{I_{van}^{abs}(Q)} \times \left(\frac{d\Sigma}{d\Omega}\right)_{van}. \qquad (II.1.2.21)$$

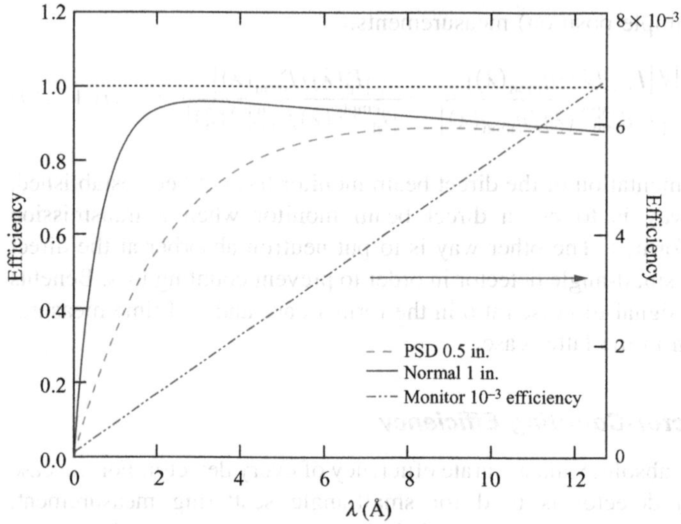


**Figure II.1.2.7** Calculated wavelength-dependent counting efficiencies of neutron detectors of SWAN at KENS. "PSD 0.5 in" is the detector used at small-, second small bank, and high-angle bank and "normal 1 in" is used at middle-angle bank. 0.5 and 1 in. are the diameters of detectors. As the diameter is larger, the probability to capture neutron by $^3$He gas, that is, counting efficiency, increases.

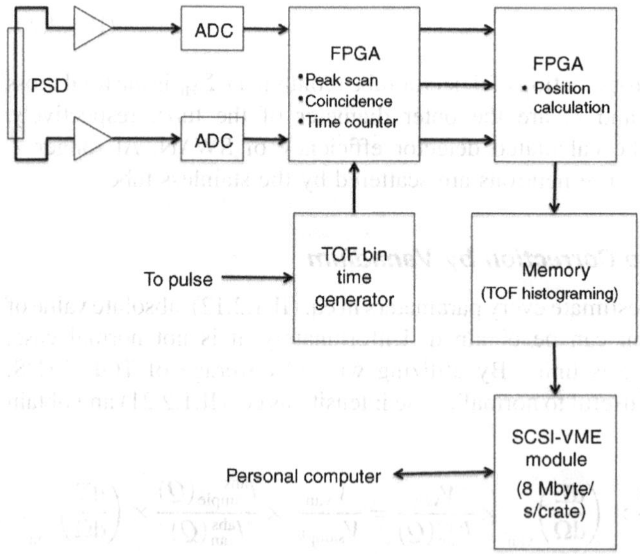


**Figure II.1.2.8** Block diagram of TOF measurement electronics for a PSD.

### *II.1.2.3.2.5 Data Acquisition System*

Figure II.1.2.8 shows a brief block diagram of TOF measurement electronics for a position-sensitive $^3$He -gas detector (PSD). Details were described in Ref. (Satoh et al., 2004, 2009). Linear PSD produces signals from both ends of the PSD ($Ph_1$, $Ph_2$). These signals will be amplified and converted to digital signal by "ADC." Then a field programmable gate array (FPGA) searches peaks corresponding to neutron detection, and checks the timing coincidence of the two signals. If it is coincidence, the signals are send to another FPGA with the TOF value and the detected position is calculated by $L_d Ph_1/(Ph_1 + Ph_2)$ where $L_d$ is the length of the detection area of the

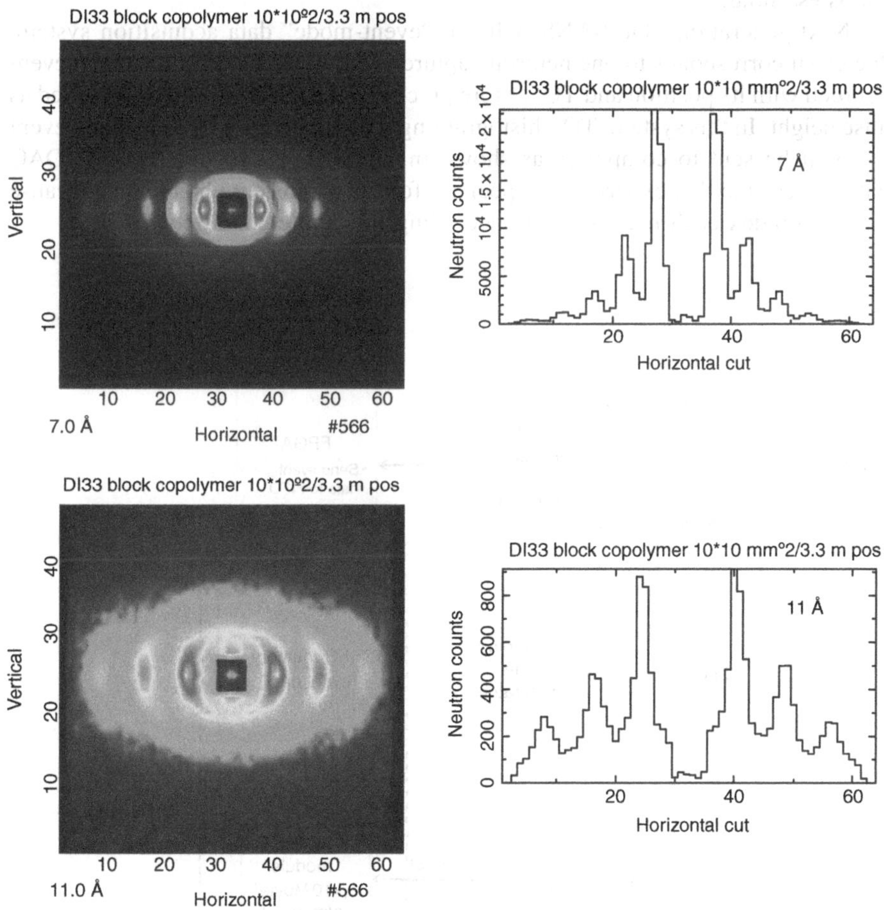


**Figure II.1.2.9** Bragg peaks from styrene-$d_8$-isoprene diblock copolymer clearly observed by 7 and 11 Å neutrons. These two wavelengths were chosen for the demonstration. By TOF method, wavelength dependency is continuously measured. This intensity map is a typical example to show wide-wavelength measurement by a TOF-SANS.

PSD. The calculated data are stored on the on-board memory as TOF histogram. The basic components are similar to reactor's one but there is "TOF bin time generator." Since the observed data contain many wavelength dependencies, "on-line" analysis should be implemented to perform efficient experiment.

Observed small-angle intensities of block copolymer (styrene-$d_8$-isoprene diblock copolymer) are shown in Figure II.1.2.9. Bragg peaks from two-dimensional lamellar structure (periodicity is ~430 Å) were observed. By using position-sensitive detectors, intensity map can be observed as well as reactor–source SANS. Besides the position information, wavelength can be selected from the observed data. Intensities of 7 and 11 Å are depicted. Since $Q_{max}$ is lower at 11 Å, the intensity map is zooming into lower $Q$-range than the case of 7 Å. This is an example showing the volumetric data of TOF-SANS. Another important aspect is that software to handle the measured data is essential.

Next-generation TOF-SANS will use "event-mode" data acquisition system. One event corresponds to one neutron capture event at the detector and each event recorded with its position and TOF. More precisely, the position may be recorded as pulse height. In this system, TOF histograming will not be done by FPGA and event data will be sent to computers as shown in Figure II.1.2.10. Event-mode DAQ realizes very flexible experiment especially for time-transient phenomena because users can index each neutron event according to sample conditions.

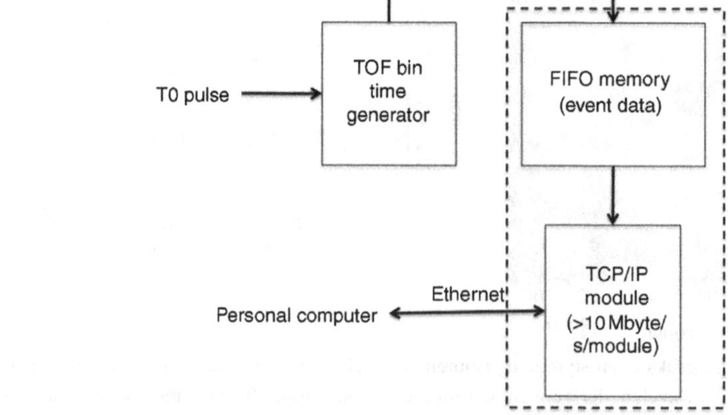


**Figure II.1.2.10** Block diagram of TOF measurement electronics for an event mode data acquisition.

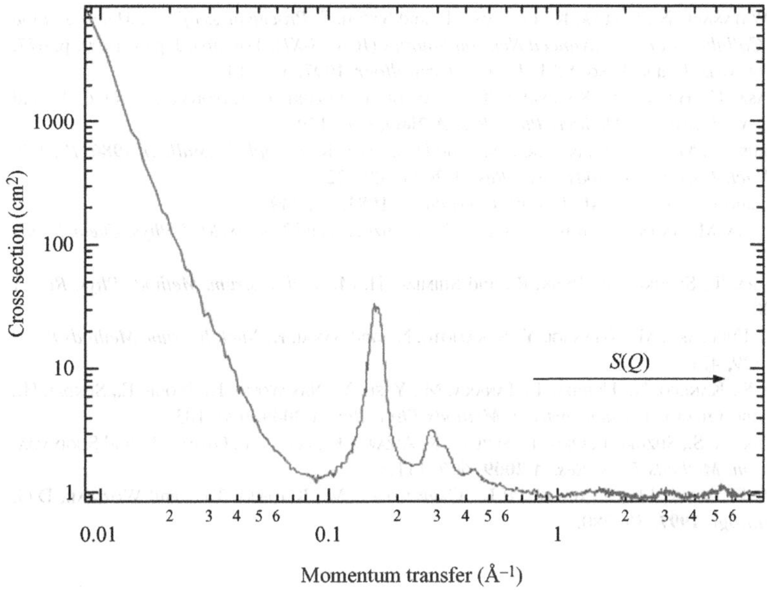


**Figure II.1.2.11** Wide-$Q$ structure factor of mesoporous silica measured by SWAN at KENS.

### *II.1.2.3.2.6 Analyzed Example*

Figure II.1.2.11 shows one of the examples: wide-$Q$ data of mesoporous silica, which has 3 nm pores aligned hexagonally and walls of the pore are made with silica glass, measured by SWAN. This sample shows Bragg peaks of hexagonal periodicity at $0.1 < Q < 0.5$ and structure factor $S(Q)$ of glass structure above $Q = 0.5$ Å. In this case, the profile was obtained by the following procedures: (1) detector pixel were merged by radial averaging, (2) corrections of wavelength-dependent factors, (3) merging of radial-averaged intensities to obtain a scattering function with a wide-$Q$ range, (4) final absolute correction by a standard sample (vanadium). Absolute value of $S(Q)$ was obtained at first and then lower $Q$-range intensities were connected to $S(Q)$. This is useful because intensity of $S(Q)$ is oscillating around constant value, 1.

## II.1.2.4 SUMMARY

Instrumentations of TOF-SANS were introduced. Full utilization of pulsed neutron source provides wide-$Q$ measurement and useful information for variety of material science. Corrections of wavelength dependencies are not trivial, but there is a way to reach reliable absolute value by step-by-step corrections.

## REFERENCES

Carpenter, J.M., Agamalian, M., Littrell, K.C., Thiyagarajan, P., and Rehm, C. *J. Appl. Crystallogr.* **2003**, *36*, 76.

FURUSAKA, M., WATANABE, N., SUZUYA, K., FUJIKAWA, I., and SATOH, S. *Proceedings of 11th Meeting of the International Collaboration on Advanced Neutron Sources (ICANS-XI)*, Tsukuba, Japan, 1990, p. 677.
HEENAN, R.K., PENFOLD, J., and KING, S.M. *J. Appl. Crystallogr.* **1997**, *30*, 1140.
IWASHITA, H., IWASA, H., HIRAGA, F., KAMIYAMA, T., KIYANAGI, Y., SUZUKI, J., SHINOHARA, T., OKU, T., and SHIMIZU, H.M. *Nucl. Instrum. Methods Phys. Res. A* **2009**, *600*, 129.
ISHIKAWA, Y., FURUSAKA, M., NIIMURA, N., ARAI, M., and HASEGAWA, K. *J. Appl. Crystallogr.* **1986**, *19*, 229.
LITTRELL, K.C. *Nucl. Instrum. Methods Phys. Res. A* **2004**, *529*, 22.
MILDNER, D.F.R. and CARPENTER, J.M. *J. Appl. Crystallogr.* **1984**, *17*, 249.
OTOMO, T., FURUSAKA, M., SATOH, S., ITOH, S., ADACHI, T., SHIMIZU, S., and TAKEDA, M. *J. Phys. Chem. Solids* **1999**, *60*, 1579.
OKU, T., SHINOHARA, T., SUZUKI, J.-i., PYNN, R., and SHIMIZU, H. M. *Nucl. Instrum. Methods Phys. Res. A* **2009**, *600*, 100.
SATOH, S., INO, T., FURUSAKA, M., KIYANAGI, Y., SAKAMOTO, N., and SAKAI, K. *Nucl. Instrum. Methods Phys. Res. A* **2004**, *529*, 421.
SATOH, S., MUTO, S., KANEKO, N., UCHIDA, T., TANAKA, M., YASU, Y., NAKAYOSHI, K., INOUE, E., SENDAI, H., NAKATANI, T., and OTOMO, T. *Nucl. Instrum. Methods Phys. Res. A* **2009**, *600*, 103.
SHINOHARA, T., TAKATA, S., SUZUKI, J., OKU, T., SUZUYA, K, AIZAWA, K., ARAI, M., OTOMO, T., and SUGIYAMA, M. *Nucl. Instrum. Methods Phys. Res. A* **2009**, *600*, 111.
THIYAGARAJAN, P., EPPERSON, J.E., CRAWFORD, R.K., CARPENTER, J.M., KLIPPERT, T.E., and WOZNIAK, D.G. *J. Appl. Crystallogr.* **1997**, *30*, 280.

# II Instrumentation

# II.1 Small-Angle Neutron Scattering

## II.1.3 Ultra-Small-Angle Neutron Scattering

### II.1.3.1 Bonse–Hart USANS Instrument

*Michael Agamalian*

#### II.1.3.1.1 INTRODUCTION

Small-angle neutron scattering became significant since the mid-1970s when the focus of structural investigations in biology, industrial, materials, polymer, food and environmental sciences, geology, colloids, complex fluids, and organic chemistry shifted to the supra-atomic level of structural organization. Obtaining reliable structure information at different levels of morphology, including not only the molecular and the nanometer but also the micron range, is nowadays becoming more and more important; however, a complete SANS analysis cannot be accomplished using just one instrument. Most of the conventional high-resolution pinhole geometry SANS machines described in Chapter II.1.1 (see also Glinka et al., 1998; Dewhurst, 2008) operate effectively in the $Q$-range, $1 \times 10^{-3}\,\text{Å}^{-1} < Q < 1\,\text{Å}^{-1}$ and thus are capable of measuring the maximum diffraction distance $D_{max} \sim 2\pi/Q_{min} \sim 6000\,\text{Å}$. The other type of a small-angle diffractometer, an ultra-small-angle neutron scattering (USANS) instrument, with much higher $Q$-resolution is required to extend the SANS investigations to the micron range of the length scale. The combined USANS/SANS technique has already become a powerful tool of structural investigations in the field of materials sciences (Schaefer, 2004) and Earth sciences (Triolo and Agamalian, 2009), that is, in studies of objects exhibiting hierarchical structures with complex morphology (Agamalian, 2005). Particularly, important scientific results were obtained examining the μm-scale structures of rocks

*Neutrons in Soft Matter* Edited by Toyoko Imae, Toshiji Kanaya, Michihiro Furusaka, and Naoya Torikai

(Radliňski et al., 1999), polymer blends (Agamalian et al., 1999a), colloidal silica gels (Muzny et al., 1999), gels of attractive block copolymer micelles (Crichton and Bhatia, 2005), and porous Vycor glass (Kim and Glinka, 2006). In all mentioned studies, the Bonse–Hart USANS instruments have been used in conjunction with the conventional high-resolution pinhole SANS machines and USANS data were corrected for the slit geometry collimation smearing before combining them with SANS profiles. The aspects of slit geometry collimation correction are discussed in Chapter 1 and also in the textbook (Feigin and Svergun, 1987). The Bonse–Hart USANS technique has also been successfully tested as an autonomous tool in several investigations, for example, in the studies of colloidal crystals (Matsuoka et al., 1999) and hydrating cement paste (Sabine, 1999). This chapter describes the USANS instruments based on the neutron double-crystal diffractometers (DCDs) with multi-bounce channel-cut crystals. The alternative ultrahigh-resolution SANS instruments are considered in Chapter II.1.3.2.

It is known that the reflectivity function of a single crystal and as a consequence the rocking curve of the DCD has extremely narrow angular resolution. This fact offers an opportunity for measuring ultrasmall-angle scattering from a sample placed in between the monochromator and the analyzer crystals; the scattered radiation can be obtained as the difference between the two rocking curves measured with and without the sample under study. However, the single-bounce reflectivity function contains also tails (or wings) significantly diminishing the sensitivity of the DCD to small-angle scattering because the intensity of SAS occurring in the range of the wings is usually weaker than the intensity of wings. U. Bonse and M. Hart found a way to solve this problem by pioneering the remarkable "tailless single-crystal reflection" technique based on application of channel-cut multi-bounced crystals (Bonse and Hart, 1965). This technique, originally developed for X-rays, immediately gave rise to ultrasmall-angle X-ray scattering (USAXS) measurements (Bonse and Hart, 1966); however, the first attempts to adapt it for neutrons, made in the mid-1980s at Jülich, Germany, had only limited success (Schwahn et al., 1985). The Bonse–Hart technique was properly applied for USANS studies in 1996–1997 following several successful neutron dynamical diffraction experiments (Agamalian et al., 1997; Agamalian et al., 1998a, 1998c) carried out at the high-flux isotope reactor (HFIR), Oak Ridge National Laboratory (ORNL). As a result of those experiments, the sensitivity of the ORNL neutron DCD, equipped with triple-bounce Si(111) channel-cut crystals, increased by more than three orders of magnitude compared to that reported by Schwahn et al. (1985). Nowadays, the Bonse–Hart DCDs are in use for USANS investigations at the National Institute of Standards and Technology (NIST), USA (Barker et al., 2005), as well as in Japan (JRR-3M Research Reactor) (Aizawa and Tomimitsu, 1995; Takahashi et al., 1999), Germany (GKSS-Forschungszentrum Geesthacht GmbH) (Bellmann et al., 2000), (Berlin Neutron Scattering Center) (Treimer et al., 2000), and France (Institute of Laue-Langevin, Grenoble) (Hainbuchner et al., 2000). A new adjustable time-of-flight USANS (TOF-USANS) instrument (Carpenter et al., 2003; Agamalian et al., 2005) is currently under construction at the ORNL Spallation Neutron Source (SNS).

## II.1.3.1.2 DCD WITH SINGLE-BOUNCE CRYSTALS

Conventional reactor-based SANS instruments (Chapter II.1.1) use narrowly collimated monochromatic beams of radiation and area detectors that record a two-dimensional image of the scattering function $I(Q)$ (Glinka et al., 2007a; Dewhurst, 2008). Measurements cover the $Q$-range $\sim 1 \times 10^{-3}\,\text{Å}^{-1} < Q < 1\,\text{Å}^{-1}$, typically, spanning the interval in several overlapping runs. The instrument resolution $\Delta Q$ is defined by the angular collimation of the incident beam, the range of wavelengths, the wavelength resolution, the sample size, and the detector pixel size. The conventional reactor-based SANS instruments usually use long-wavelength neutrons, $3\,\text{Å} < \lambda < 15\,\text{Å}$. Typical reactor-based SANS instruments employ a pin-hole geometry collimation scheme with wavelength-variable monochromatic flux, $\Delta\lambda/\lambda \sim 5$–10%, and variable source-to-sample and sample-to-detector collimation distances. Time-of-flight SANS instruments at pulsed spallation neutron sources (Chapter II.I.2) have similar collimation schemes with fixed collimators and detectors, but gain $Q$-range by using a broad band of wavelengths approximately in the range 0.5–15 Å (Carpenter and Faber, 1978). All SANS instruments with geometrical collimation are limited in the $Q$-resolution, $\Delta Q = Q_{\min} \geq 5 \times 10^{-4}\,\text{Å}^{-1}$. Therefore, a different approach is required to extend the $Q$-range of SANS to significantly smaller $Q$ values.

A way to overcome the practical limitations of geometrical collimation stems from the fact that, according to dynamical diffraction theory, the Bragg reflection from a single crystal is mostly confined to a very narrow, several seconds of arc, angular range. The reflectivity function $R(y)$ derived by Darwin (see, for example, Zachariasen, 1967) for an infinitely thick and transparent crystal $R(y) \equiv R_D(y)$ is

$$
\begin{aligned}
R_D(y) &= 1, \quad |y| = 1; \\
R_D(y) &= [|y| - (y^2 - 1)^{0.5}]^2, \quad |y| > 1,
\end{aligned}
\qquad \text{(II.1.3.1.1)}
$$

where $y = (\theta - \theta_B)/\delta\theta_D$ is the dimensionless angular parameter of dynamical diffraction theory, $\theta$ is the angle of incidence, $\theta_B$ is the Bragg angle, and $\delta\theta_D$ is the half-width of the Darwin plateau (DP). The function (II.1.3.1.1) has a typical plateau with $R_D(y) = 1$, the Darwin plateau, in the range $|y| \leq 1$; the total width of the DP $2\delta\theta_D$ is

$$
2\delta\theta_D = 2\left(\text{DWF} \times \frac{\lambda_{hkl}^2}{\sin 2\theta_B}\right) \times \left(\frac{|F(h,k,l)|}{\pi V_0}\right), \qquad \text{(II.1.3.1.2)}
$$

where $\lambda_{hkl} = 2d_{hkl} \times \sin\theta_B$ is the wavelength of the Bragg reflection, $d_{hkl}$ is the $d$-spacing, $V_0$ is the volume of the crystallographic unit cell, DWF is the Debye–Waller factor and $|F(h, k, l)|$ is the magnitude of the structure factor for a given family of crystallographic planes. The term $[|F(h, k, l)|/\pi V_0]$ in eq. (II.1.3.1.2) depends only on the parameters of a chosen crystallographic system. The Debye–Waller factor is

$$
\text{DWF} = \exp\left[-B\left(\frac{\sin\theta_B}{\lambda_{hkl}}\right)^2\right], \qquad \text{(II.1.3.1.3)}
$$

where $B = 8\pi^2 u^2$ and $u$ is the root mean square displacement of an atom perpendicular to the Bragg crystallographic planes; for example, for Si, $B \approx 0.45$ Å$^2$ (Sears and Shelley, 1991). The Darwin reflectivity function (II.1.3.1.3) coincides in the range of DP with that calculated by Ewald for a finite-thickness, transparent crystal, but departs in the range of wings. The Ewald reflectivity function, $R_E(y)$, is

$$R_E(y) = 1, \quad |y| \le 1;$$
$$R_E(y) = 1-(1-y^{-2})^{0.5}, \quad |y| > 1, \qquad \text{(II.1.3.1.4)}$$

T. Takahashi and M. Hashimoto (Takahashi and Hashimoto, 1995) have derived the relationship between $R_D(y)$ and $R_E(y)$, which is

$$R_E(y) = R_D(y) + [1-R_D(y)]^2 \times R_D(y) \times \left[1 + R_D^2(y) + R_D^4(y) + \cdots\right] = \frac{2R_D(y)}{1+R_D(y)}. \qquad \text{(II.1.3.1.5)}$$

The terms $R_D(y)$ and $[1 - R_D(y)]^2 \times R_D(y) = R_{BF1}(y)$ correspond to the first front-face (FF) and back-face (BF) reflections, respectively; the remaining terms in eq. (II.1.3.1.5) are successive FF–BF reflections inside a thick transparent crystal. In the range of far wings where $y \gg 1$ the Ewald function $R_E(y) \approx 2R_D(y)$ (Takahashi and Hashimoto, 1995) and the first BF reflection $R_{BF1}(y) = [1 - R_D(y)]^2 \times R_D(y) \approx R_D(y)$ (see Figure II.1.3.1.1).

Therefore, the first FF and BF reflections give nearly equal contributions to $R_E(y)$ in the range of the wings. This result following from the Takahashi–Hashimoto theory is very important for the optimization of sensitivity of the Bonse–Hart USANS instruments with triple-bounce channel-cut crystals.

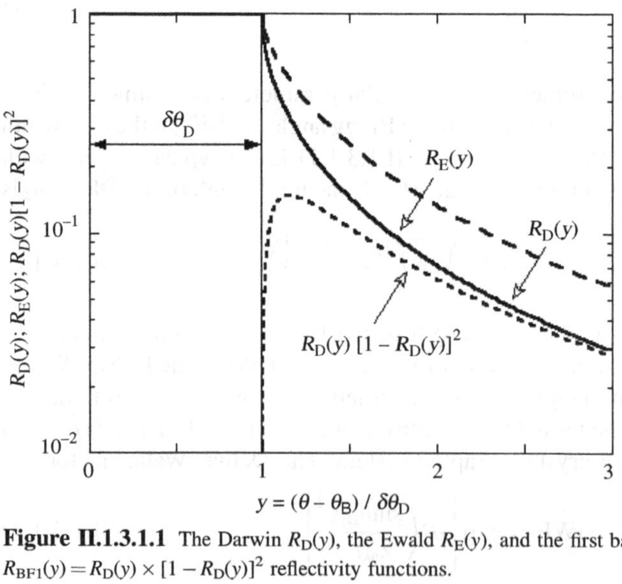


**Figure II.1.3.1.1** The Darwin $R_D(y)$, the Ewald $R_E(y)$, and the first back-face $R_{BF1}(y) = R_D(y) \times [1 - R_D(y)]^2$ reflectivity functions.

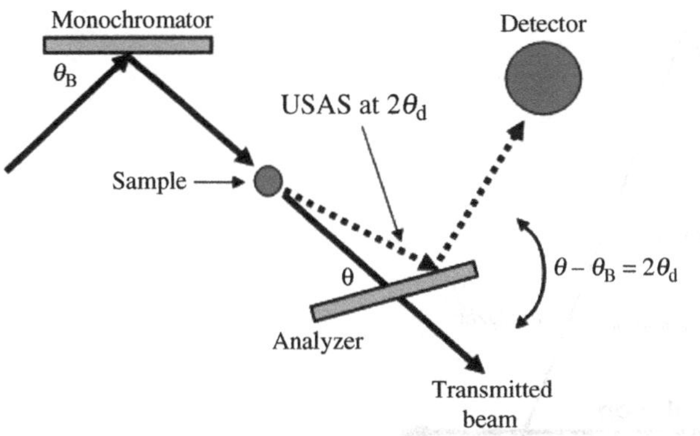


**Figure II.1.3.1.2** The optical scheme of a DCD with single-bounce crystals and the geometry of USAS experiments.

The Darwin and Ewald reflectivity functions cannot be observed in a direct diffraction experiment but can be measured using a parallel double-crystal arrangement or a DCD. Figure II.1.3.1.2 shows the optical scheme of the simplest DCD with identical single-bounce crystals.

The main unit of a typical DCD consists of two perfect crystals; the upstream immovable crystal reflecting radiation at the exact Bragg angle $\theta_B$ is called the monochromator and the downstream one is the analyzer. The angular scan of the analyzer crystal in horizontal plane, $(\theta - \theta_B)$, where $\theta$ is the angle of incidence for the analyzer with respect to the transmitted beam (solid arrow in Figure II.1.3.1.2), corresponds to the convolution $I(y)$ of reflectivity functions of the monochromator, $R_1(y)$, and the analyzer, $R_2(y)$,

$$I(\Delta) = \int R_1(y) \times R_2(y+\Delta)\mathrm{d}y. \qquad \text{(II.1.3.1.6)}$$

The function $I(y)$ in eq. (II.1.3.1.6) is also known as the theoretical rocking curve. The reflectivity functions $R_1(y)$ and $R_2(y)$ can be determined by fitting $I(y)$ to the experimental rocking curve, which can be measured by means of a DCD.

Numerous publications have demonstrated that a DCD based on perfect single-bounce crystals can be used to measure ultrasmall-angle scattering (USAS) with X-rays and neutrons; Figure II.1.3.1.2 shows the geometry of USAS (dashed arrows) measurements with a DCD. Two rocking curves of the analyzer crystal measured with and without a sample placed between them provide USAS data when the second rocking curve (without a sample) is subtracted from the first one. The FWHM of the experimental rocking curve of an empty DCD determines the angular resolution of this instrument. Because this curve has a nearly triangular shape in the vicinity of the exact Bragg reflection, it is convenient to simulate it by the convolution of $R_1(y)$ and $R_2(y)$ approximated as rectangles with the width equal to $2\delta\theta_D$ (see Figure II.1.3.1.3):

$$\begin{aligned} R_1(y) &= R_2(y) = 1, \quad |y| = 1; \\ R_1(y) &= R_2(y) = 0, \quad |y| > 1. \end{aligned} \qquad \text{(II.1.3.1.7)}$$

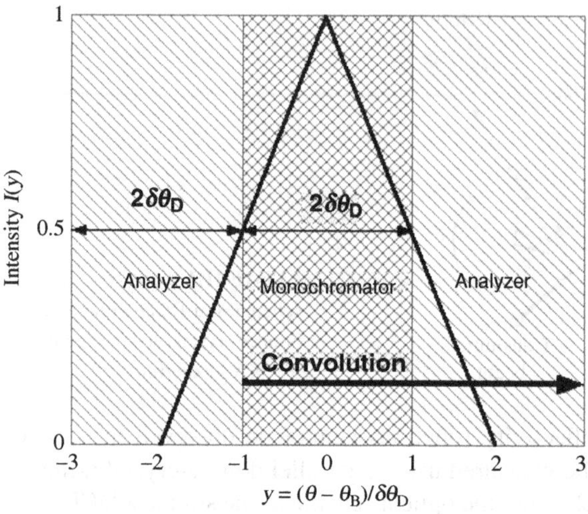


**Figure II.1.3.1.3** The triangular approximation of the resolution function of a DCD with perfect crystals.

The rectangular model, eq. (II.1.3.1.7), is practical because it simplifies the relationship between the smallest accessible scattering angle from an object under study $2\theta_{d,min}$, where $\theta_d$ is the diffraction angle, and the width of the DP. The diagram shown in Figure II.1.3.1.3 demonstrates that the triangular resolution function of a DCD starts at $y=(\theta-\theta_B)/\delta\theta_D=\pm 2$. Assuming that $|y|=2$ corresponds to the minimum scattering angle, $2\theta_{d,min}$, measurable in the DCD with perfect crystals, then according to Figure II.1.3.1.3, $2\theta_{d,min}=2\delta\theta_D$.

Consequently, $Q$-resolution and $Q_{min}$ of the DCD is defined as

$$\Delta Q = Q_{min} = \frac{4\pi \sin\theta_{d,min}}{\lambda_{hkl}} = \frac{4\pi \sin\delta\theta_D}{\lambda_{hkl}} \approx 4\pi \times \frac{\delta\theta_D}{\lambda_{hkl}}. \qquad \text{(II.1.3.1.8)}$$

The value of $Q_{min}$ calculated with eq. (II.1.3.1.8) for a neutron DCD can be very small, for example, for the ORNL DCD with Si(111) crystals set up for the Bragg angle $\theta_B=24.4^{\circ}$, the $Q$-resolution calculated by eq. (II.1.3.1.8) is $\Delta Q = Q_{min} \approx 2 \times 10^{-5}$ Å$^{-1}$ (Agamalian et al., 1997). This value of $\Delta Q$ is about two orders of magnitude smaller than that for a conventional SANS instrument.

Another advantage of the DCD with perfect crystals is the effect of optical (but not geometrical) collimation leading to a considerable neutron flux gain, achievable without losing $Q$-resolution. The dynamical diffraction reflectivity functions $R_D(y)$ and $R_E(y)$ are calculated assuming theoretically absolute monochromaticity and zero divergence of the incident radiation. In reality, the incident neutron flux is polychromatic and divergent; however, the beam reflected from a perfect crystal is "encoded" in terms of angular–wavelength correlation ($\theta$–$\lambda$ correlation) set up with the accuracy determined by the width of the DP. Therefore, the upstream crystal-monochromator of the DCD (see Figure II.1.3.1.2) is an "encoder" and the identical downstream crystal–analyzer is a "decoder," which decodes the $\theta$–$\lambda$ correlation set up by the monochromator and preserves the ultra-high angular resolution. That is why the rocking curve of the analyzer crystal $I(y)$ has angular resolution $2\delta\theta_D$, which is equal to that for the reflectivity function of a perfect crystal $R_D(y)$ (see Figure II.1.3.1.3). For example, the horizontal angular divergence of the neutron beam in the NIST USANS instrument is as great as approximately $\pm 0.8°$ (Barker et al., 2005); however, the FWHM of the experimental rocking curve is in good agreement with the theoretical value, $2\delta\theta_D \sim 1.5$ arcs, calculated with equation (II.1.3.1.2). Therefore, the geometrical divergence of the beam in horizontal plane exceeds the angular resolution of the NIST USANS instrument by more than three orders of magnitude. The vertical divergence and the horizontal and the vertical spatial spread of the beam also do not affect the angular resolution in horizontal plane. This effect offers an opportunity to focus the incoming primary beam not only in the vertical but also in the horizontal plane (Freund, 1983) achieving the significant flux-at-sample gain (Barker et al., 2005). In addition, it allows varying the sample cross-sectional area without changing the resolution, which is another way to enhance the neutron flux. However, application of the DCDs based on single-bounce perfect crystals for USAS investigations is handicapped by the wings of the reflectivity functions $R_D(y)$ and $R_E(y)$ falling off as $y^{-2}$, which dramatically decrease its sensitivity. The wings produce an intense (compared to a typical intensity of small-angle scattering) background beyond the Darwin plateau and thus only USAS from strongly scattering objects (preferably exhibiting periodical structure generating well-pronounced Bragg peaks) can be examined with these instruments.

## II.1.3.1.3 IMPLEMENTATION OF THE BONSE–HART TECHNIQUE FOR USANS

In 1965, U. Bonse and M. Hart (Bonse and Hart, 1965) pioneered a way to overcome the wings of $R_D(y)$ using multibounce channel-cut crystals. According to the Bonse–Hart concept of a "tailless" reflectivity function, the classical Darwin function $R_D(y)$ transforms into $R_D^m(y)$ after $m$ consecutive Bragg reflections inside the channel-cut crystal (Figure II.1.3.1.4), which is typically cut from a massive single crystal of Si or Ge with diffractive faces parallel to a chosen crystallographic plane. The X-ray beam enters the channel from one end and, after $m$ successive FF Bragg reflections, leaves the crystal from the other end. With symmetric reflections inside a channel of the width $w$, the maximum width of the beam $a$ is given by $a = 2w\cos\theta_B$ and the total length $s$ of the crystal is $s = (m+1)w\cos\theta_B$. The quintuple-bounce channel-cut crystal designed by U. Bonse and M. Hart (Figure II.1.3.1.4) should theoretically suppress the wings of the Darwin reflectivity function as $R_D^5(y)$, while the resolution and the peak reflectivity are preserved.

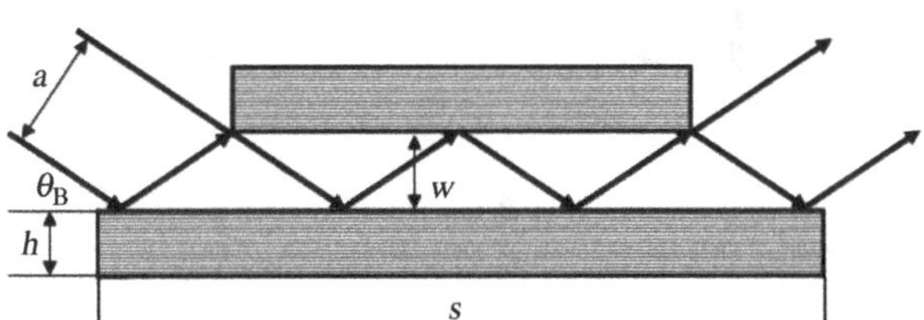


**Figure II.1.3.1.4** The Bonse–Hart quintuple-bounce channel-cut crystal designed for X-ray radiation.

Originally Si and Ge channel-cut crystals were tested with X-rays (Bonse and Hart, 1965) using the parallel double-crystal arrangement (Figure II.1.3.1.2) with quintuple-bounce monochromator and analyzer crystals (see Figure II.1.3.1.4). The experimental results qualitatively proved the Bonse–Hart concept; however, the quantitative agreement appeared satisfactory only for the resolution and the peak reflectivity. The suppression of wings was many orders of magnitude less than the theoretical prediction due to the appearance of an unexpected background falling off as $y^{-2}$ in the range $y > 4$; this parasitic effect was experimentally observed in many X-ray studies (see, for example, Lambard and Zemb, 1991; Matsuoka et al., 1991; Diat et al., 1994; Ilavsky et al., 2002). The deviation of experimental results from the $R_{\mathrm{D}}^{n}(y)$ model still is not understood completely and is known in dynamical diffraction physics as the long-standing "wings problem." In spite of this dramatic departure from the theoretical prediction, U. Bonse and M. Hart demonstrated that the X-ray DCD on multi-bounce channel-cut crystals can be used for USAXS studies much more effectively than that on single-bounce crystals (Bonse and Hart, 1966). The Bonse–Hart USAXS instruments found broad application in structural studies of condensed matter and thus are routinely available at many X-ray laboratories worldwide (Lambard and Zemb, 1991; Matsuoka et al., 1991; Diat et al., 1994; Ilavsky et al., 2002).

Fast progress in the development and the application of the Bonse–Hart USAXS technique has been made due to the strong absorption of X-rays in Si and Ge used as channel-cut crystals, which significantly simplifies the optimization of multi-bounce crystal optics for this type of radiation. However, the neutron version of the Bonse–Hart DCD on nearly transparent Si channel-cut crystals suffered from contamination of the rocking curve wings from back-face reflections. The dynamical diffraction studies performed at ORNL demonstrated how the first BF reflection, $R_{\mathrm{BF1}}(y) = [1 - R_{\mathrm{D}}(y)]^2 \times R_{\mathrm{D}}(y)$, comes into view in the neutron DCD with Si triple-bounce channel-cut crystals (Agamalian et al., 1997). Also some of the other sources of parasitic scattering in Si crystals, edge-face diffraction (Shull, 1973) and surface-induced scattering (Agamalian et al., 1998a) were examined with neutrons in the vicinity of the first Bragg reflection. Finally, M. Agamalian modified the classical Bonse–Hart design (see Figure II.1.3.1.4) by introducing a slot for a cadmium absorber in the middle of the long wall of a triple-bounce channel-cut crystal (Agamalian et al., 1997). In this modification (see Figure II.1.3.1.5), cadmium plates are incorporated to shield the triple bouncing beam (solid arrows), against contamination with the parasitic scattering originating from the gray triangular zones of the diffractive walls. Therefore, only the useful neutron radiation can pass

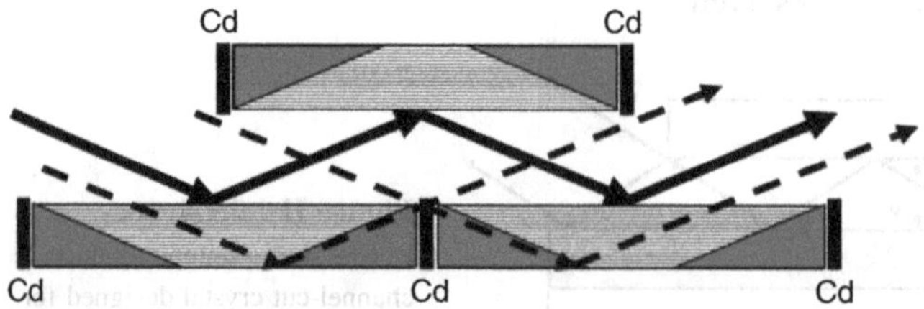


**Figure II.1.3.1.5** The triple-bounce channel-cut crystal modified for neutron radiation.

throughout the trapezoidal zones reflecting from the front and the back diffractive surfaces.

The triple-bounce reflectivity function $R_{TB}(y)$ of the channel-cut crystal shown in Figure II.1.3.1.5 can be calculated with the formula

$$R_{TB}(y) \approx \{R_D(y) + B(\theta_B, a, h) \times R_{BF1}(y)\}^3, \qquad \text{(II.1.3.1.9)}$$

where $R_{BF1}(y) = [1 - R_D(y)]^2 \times R_D(y)$ and $0 \le B(\theta_B, a, h) \le 1$ is the geometrical coefficient depending on the Bragg angle $\theta_B$, the width of the triple-bounce beam $a$, and the thickness of the diffractive wall $h$ (see Figure II.1.3.1.5). If $B(\theta_B, a, h) = 0$, $R_{TB}(y) = R_D^3(y)$, the Darwin solution, and $R_{TB}(y) \approx R_E^3(y)$, the Ewald solution, when $B(\theta_B, a, h) = 1$.

The strongest parasitic effect is the first single-bounce BF reflection $R_{BF1}(y)$ (see the dashed arrows in Figure II.1.3.1.5) originating inside the gray triangular zone in the middle of the long wall. The purpose of the Cd strip in the middle of the long wall is to block this parasitic diffraction. Cadmium plates covering the edges of the channel-cut crystal eliminate unwanted edge-face diffraction (Shull, 1973) and thus the neutron beam can pass throughout the channel-cut crystal only by undergoing triple FF and BF Bragg reflections (solid arrows in Figure II.1.3.1.5 show the triple FF reflection). Careful mechanical and chemical (polishing and etching) treatment of the front and the back diffractive surfaces suppresses surface-induced scattering (Agamalian et al., 1998a); the edge refraction from Cd strips and slits studied in (Treimer et al., 2002) can be significantly reduced by rounding the edges of the Cd plates.

Figure II.1.3.1.6 shows the rocking curves normalized by the peak intensity, $I(\theta - \theta_B)/I(0)$, measured in the ORNL neutron DCD ($\theta_B = 24.4°$) equipped with differently fabricated Si(111) crystals; the edges of the crystals under study were covered with Cd plates blocking the edge-face diffraction (Shull, 1973).

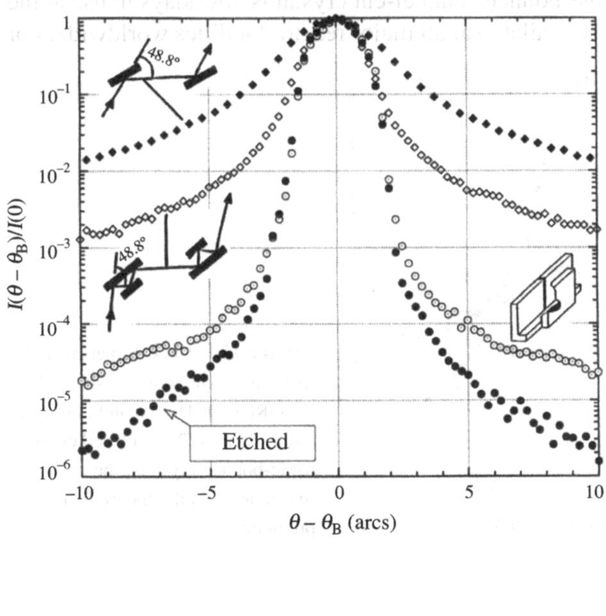


**Figure II.1.3.1.6** Rocking curves obtained in the ORNL DCD with Si(111) crystals: single-bounce (closed diamonds), triple-bounce (open diamonds), triple-bounce with the central Cd absorber (open circles), and deeply etched triple-bounce with the central Cd absorber (closed circles).

Comparison of the two upper rocking curves in Figure II.1.3.1.6 (closed and opened diamonds) demonstrates that replacement of the Si(111) single bounce crystals with the Bonse–Hart triple-bounce channel-cut crystals (Figure II.1.3.1.4) reduces the intensity of wings at $|\theta - \theta_B| = 10$ arcs by an order of magnitude. However, the Cd blocker inserted in the middle of the long wall (see Figure II.1.3.1.5) improves this result by additional two orders of magnitude (compare the rocking curves shown with opened diamonds and opened circles of Figure II.1.3.1.6). It is remarkable to note that the rocking curve in Figure II.1.3.1.6 shown with open circles practically coincides with that measured in the X-ray DCD with quintuple-bounce crystals (see Figure 7 in the study by Agamalian et al., 1975). Finally, the best result was achieved after additional deep etching of the modified triple-bounce channel-cut crystals (closed circles). Figure II.1.3.1.7 demonstrates the fit of the best experimental neutron rocking curves, obtained in the ORNL DCD with Si(111) (opened circles) and Si(220) (closed circles) triple-bounce channel-cut crystals, to the theoretical prediction calculated by formula (II.1.3.1.6) for $R_1(y) = R_2(y) = R_E^3(y)$.

Analyzing the experimental and the theoretical rocking curves presented in Figures II.1.3.1.6 and II.1.3.1.7, one can conclude that the Si triple-bounce channel-cut crystals optimized for neutron application effectively suppress the wings of the rocking curve and thus the Bonse–Hart technique is fully adoptable for neutrons. On the other hand, the experimental neutron rocking curves show departure from the theoretical prediction in the range $|y| > 4$, which reaches approximately two orders of magnitude at $|y| = 12$.

## II.1.3.1.4 THE BONSE–HART USANS INSTRUMENT WITH ADJUSTABLE RESOLUTION

The ORNL design of the triple-bounce channel-cut crystal is nowadays in use at the USANS instruments routinely available at all major reactor facilities worldwide. For

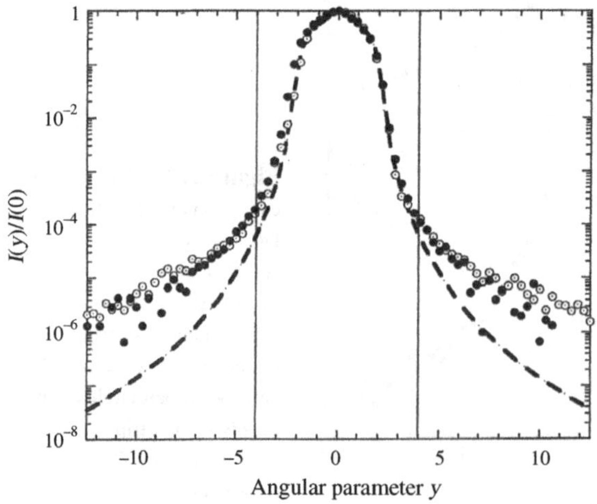


**Figure II.1.3.1.7** Experimental neutron rocking curves obtained in the ORNL DCD with Si(111) (open circles) and Si(220) (closed circles) triple-bounce crystals; the dashed line represents the theoretical prediction.

example, the NIST Bonse–Hart DCD fully optimized for USANS measurements is equipped with Si(220) triple-bounce channel-cut crystals set up at the Bragg angle $\theta_B = 38.4^o$ (Barker et al., 2005). The typical $Q$-range of this instrument, operating at the wavelength $\lambda_{220} = 2.38$ Å, is $3 \times 10^{-5}$ Å$^{-1} < Q < 5 \times 10^{-3}$ Å$^{-1}$ and its sensitivity, or signal-to-noise ratio (SNR), is SNR $\sim 2.5 \times 10^6$ for $Q \geq 5 \times 10^{-4}$ Å$^{-1}$. The total neutron flux-at-sample, $\sim 5.6 \times 10^4$ n/s, is achieved after incorporating a 2D focusing PG(002) premonochromator into the optical scheme of the instrument following the theoretical calculations (Freund, 1983). The estimated flux gain reached with this optical device is a factor of 6, while the resolution, the sensitivity, and the $Q$-range are preserved. The Bonse–Hart DCD at NIST allows accommodating sample environment (such as temperature, magnetic/electric field, pressure, and mechanical tension) similarly to the conventional SANS instrument. Therefore, its performance is nearly optimal; however, the reactor-based version has a serious limitation related to the fixed wavelength $\lambda_{hkl}$ originating from the first Bragg reflection. As a result, the $Q$-resolution $\Delta Q = Q_{\min} \approx 4\pi \times \delta\theta_D/\lambda_{hkl}$ is also fixed and thus the neutron flux-at-sample optimized for the high $Q$-resolution ($Q_{\min} \sim 2 \times 10^{-5}$ Å$^{-1}$) cannot be adjusted. The Bonse–Hart USANS instruments at steady-state neutron sources typically cover the range $2 \times 10^{-5}$ Å$^{-1} < Q < 5 \times 10^{-3}$ Å$^{-1}$ overlapping with that for the conventional SANS machine ($Q_{\min} \sim 2 \times 10^{-3}$ Å$^{-1}$) only for intensely scattering objects. In the case of weakly scattering samples, the combined USANS/SANS data usually have a gap in the range $\sim 8 \times 10^{-4}$ Å$^{-1} < Q < 4 \times 10^{-3}$ Å$^{-1}$. This disadvantage seriously compromises the quality of USANS/SANS studies of the objects having peculiarities of the diffraction patterns in this $Q$-range.

With the advent of powerful pulsed neutron sources, a new multiwavelength time-of-flight (TOF) concept is possible—TOF-USANS (Carpenter et al., 2003; Agamalian et al., 2005). The key postulate of this concept is based on the fact that the Darwin plateau width and consequently the $Q$-resolution of a DCD with perfect crystals defined by eq. (II.1.3.1.8) and as a result the neutron flux depend on the order of Bragg reflection $n$ belonging to the same family of crystallographic planes $(h, k, l)$:

$$\Delta Q_n = Q_{\min,n} \approx 4\pi \times \frac{\delta\theta_{D,n}}{\lambda_{hkl}}. \qquad \text{(II.1.3.1.10)}$$

At pulsed sources, $n$ different monochromatic wavelengths $\lambda_{hkl}$, Bragg reflected from a perfect crystal at the same angle $\theta_B$, appear at the detector at different times, which provide simultaneous USANS measurements at $n$ different $Q$-ranges by making a single $(\theta - \theta_B)$ scan of the analyzer crystal. Because the Bragg angle stays the same for a chosen family of crystallographic planes, $\lambda_{hkl} = 2d_{hkl} \times \sin\theta_B$, the main unit of the TOF-DCD (Figure II.1.3.1.8) is identical to the classical one and only the configuration of the upstream optics is different.

The presence of the high-order Bragg reflections (starting from the third order) in the spectrum of a reactor-based DCD is usually ignored assuming that the contribution to the scattering data from these wavelengths is negligibly small. Mostly, Si(111) and Si(220) channel-cut crystals are in use in the USANS instruments and a typical

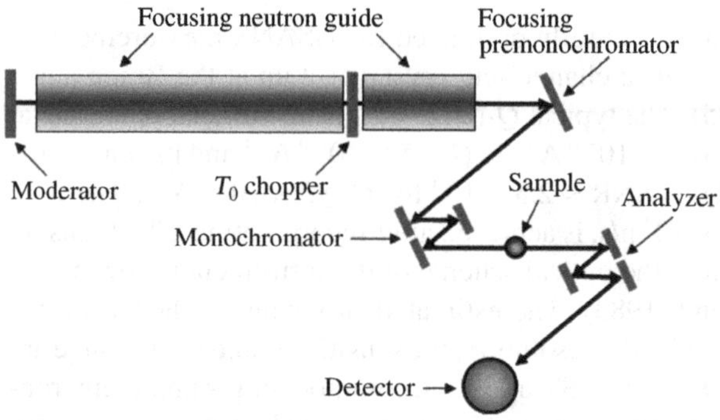


**Figure II.1.3.1.8** The optical scheme of the TOF-USANS instrument.

value of the Bragg angle is in the range $24^\circ < \theta_B < 45^\circ$. The wavelength spectrum of the DCDs with Si(111) crystals do not contain the second (forbidden) order of Bragg reflections and it is usually filtered in the DCDs with Si(220) crystals (Barker et al., 2005); the summed intensity of the remaining high-order Bragg peaks is much smaller than that of the first order. Therefore, the Bragg angles, $\theta_B$, of reactor-based DCDs are set up for the parameters $\delta\theta_{D,1}$ and $\lambda_{hkl}$ of the first Bragg reflection, calculated to reach the optimal neutron flux for the chosen $Q$-resolution. In contrast, the discrete multi-wavelength spectrum of the TOF-DCD offers an opportunity to optimize the flux-at-sample with the first order of Bragg reflection and simultaneously obtain extremely high $Q$-resolution with the high-order reflections. This approach is based on the effect of the increase of the Darwin plateau width $2\delta\theta_{D,n}$ with the increase of $\theta_B$ for a given Bragg reflection. Figure II.1.3.1.9 shows the angular dependence of $2\delta\theta_{D,1}$ in the range $20^\circ < \theta_B < 80^\circ$ calculated for the first Bragg reflection from the Si(220) family of crystallographic planes.

Because the flux-at-sample is proportional to $2\delta\theta_{D,n}$ of the monochromator, the ratio of the DP widths for the Bragg angles $\theta_B = 70^\circ$, $2\delta\theta_{D,1} = 5.35$ arcs

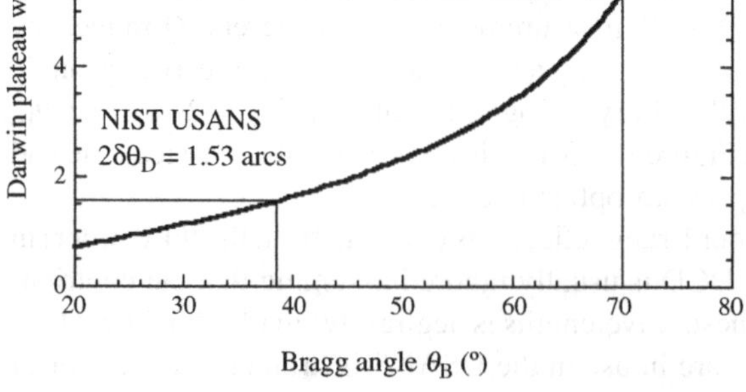


**Figure II.1.3.1.9** The DP width versus the Bragg angle $\theta_B$ calculated for Si(220).

(SNS TOF-USANS instrument configuration) and $\theta_B = 38.4°$, $2\delta\theta_{D,1} = 1.53$ arcs (NIST USANS instrument configuration) gives the flux-at-sample gain factor, 5.35/1.53 = 3.48, assuming that the primary beam entering the double-crystal arrangement is identical for these two instruments. The intensity of neutrons scattering from a sample is $I(Q) \sim (\delta\theta_{D,n})^2$ because the angular acceptance of the analyzer crystal is also proportional to the width of the DP. Therefore, for given source intensity the flux gain of the SNS TOF-USANS instrument over the NIST DCD related to the back-scattering geometry is a factor of 12. Finally, accounting for the differing source intensities, the total flux gain factor calculated for the SNS liquid hydrogen moderator and the source power 2 MW is ~30 (Agamalian et al., 2005). This significant gain factor is presumably helpful to eliminate the gap between USANS and SANS data in the case of weekly scattering samples.

Table II.1.3.1.1 contains the diffraction parameters of the first four Bragg reflections from the Si(220) family of crystallographic planes at $\theta_B = 70°$ chosen for the SNS TOF-USANS instrument; the last column of the table gives also the $Q$-resolution calculated by eq. (II.1.3.1.10) for the each wavelength.

It is obvious that the back-scattering geometry ($\theta_B = 70°$) flux gain is achieved for $\lambda_{220}$ by sacrificing the resolution ($Q_{\min,1} \sim 4.5 \times 10^{-5}$ Å$^{-1}$). However, the discrete multiwavelength spectrum gives a chance not only to reach the $Q$-resolution of the conventional (reactor-based) DCD for the second order of Bragg reflection ($Q_{\min,2} \sim 2 \times 10^{-5}$ Å$^{-1}$) but also considerably improve it for the fourth order ($Q_{\min,4} \sim 7 \times 10^{-6}$ Å$^{-1}$). Therefore, the discrete multiwavelength spectrum of Bragg diffraction in the back-scattering geometry dramatically improves the flux-at-sample in the range $Q_{\min} > 4.5 \times 10^{-5}$ Å$^{-1}$ and at the same time extends the USANS $Q$-range to the smaller $Q$ values. The TOF approach makes the USANS instrument as adjustable as the conventional pinhole geometry SANS machine but in contrast, it covers the total $Q$-range (four subranges) in one run (angular scan of the analyzer crystal).

Discrete multiwavelength spectra formed by Bragg reflections from single crystals can be examined in a time-of-flight powder diffractometer; this technique offers an opportunity to measure, separated in time-of-flight, orders of Bragg reflections (elastic scattering) and inelastic thermal diffuse scattering (TDS) from a chosen family of crystallographic planes. Figure II.1.3.1.10 shows the 2D (angles versus wavelengths) diffraction data in the angular, $18° < 2\theta < 156°$, and

**Table II.1.3.1.1** Bragg Diffraction Parameters for the Si(220) Family of Crystallographic Planes for $\theta_B = 70°$ and the TOF-USANS Instrument Resolution $Q_{\min,n}$ as a Function of $\lambda_{hkl}$ and $\delta\theta_{D,n}$.

| $n$ | Si($h$, $k$, $l$) | $d$ (Å) | $\lambda_{hkl}$ (Å) | $2\delta\theta_{D,n}$ (arcs) | $Q_{\min,nl}$ $10^{-6}$ (Å$^{-1}$) |
|---|---|---|---|---|---|
| 1 | 220 | 1.92 | 3.609 | 5.347 | 45.1 |
| 2 | 440 | 0.96 | 1.804 | 1.219 | 20.6 |
| 3 | 660 | 0.64 | 1.203 | 0.466 | 11.8 |
| 4 | 880 | 0.48 | 0.902 | 0.211 | 7.14 |

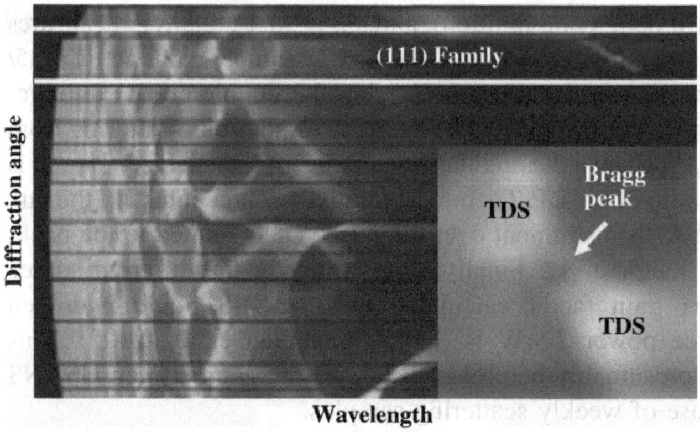


**Figure II.1.3.1.10** The 2D representation of Bragg diffraction and TDS from Si(111) crystal at $\theta_B = 45°$. (See the color version of this figure in Color Plates section.)

wavelength, $0.2 < \lambda < 4.0$ Å, ranges obtained from a slab-shaped single-bounce Si(111) crystal setup at the Bragg angle $\theta_B = 45°$. The general purpose powder diffractometer (GPPD) at the Intense Pulsed Neutron Source, Argonne National Laboratory was used for these experiments.

Several families of Bragg reflections are detected as tiny bright spots surrounded by TDS; the (111) family is highlighted and the insert shows the magnified image of the third-order (333) peak. The continuous intensity distribution of TDS is quasiperiodic, coincident with the reciprocal lattice of the Si single crystal. The Bragg peaks mark the centers of Brillouin zones. Appearance of TDS in the diffraction data is not a surprise because it is well known that thick Si crystals generate TDS in the vicinity of Bragg reflections (see, for example, Graf et al., 1981). The contribution of TDS to the integrated intensity of Bragg reflections increases at short wavelengths as $I(\lambda)_{TDS}/I(\lambda)_B \sim \lambda^{-3}$ (Popa and Willis, 1997); thus, TDS originating in the vicinity of high-order Bragg reflections is significantly stronger than that generated near the first Bragg reflection (see Figure II.1.3.1.10). TDS compromises the performance of the TOF-USANS instrument at high-order reflections and thus has been studied in more detail (Agamalian et al., 2009).

Two Si(111) crystals identical in material and thickness, a slab-shaped single-bounce, and a channel-cut triple-bounce shielded with Cd strips (see Figure II.1.3.1.5) were measured in GPPD. This experiment revealed the undetected-so-far mechanism of parasitic scattering by comparing the time-of-flight spectra obtained on the two crystals. The measurements span the wavelength range $0.2 < \lambda < 4.0$ Å of the first seven allowed Bragg reflections from Si(111) at $\theta_B = 24.4°$. The data summed at $2\theta_B = 48.8 \pm 0.75°$ were normalized by the monitor rate and the spectrum of the primary beam. The spectrum measured from the slab-shaped crystal (upper curve in Figure II.1.3.1.11) shows TDS appearing nearby all of the single-bounce reflections.

The shapes of the triple-bounce (111), (333), and (444) reflections (lower spectrum in Figure II.1.3.1.11) are very different from those in the upper spectrum;

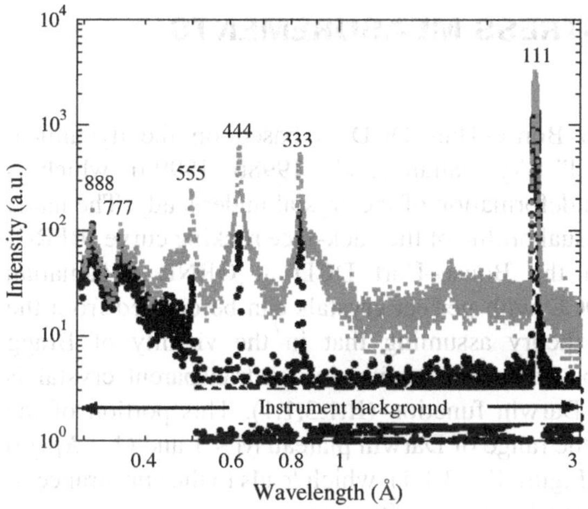


**Figure II.1.3.1.11** Bragg reflections and TDS from the slab-shaped (upper curve) and the triple-bounce (lower curve) Si(111) crystals.

also the peak intensity of these triple-bounce reflections, $I(\lambda_{111})_{\mathrm{TB,peak}}$, $I(\lambda_{333})_{\mathrm{TB,peak}}$, and $I(\lambda_{444})_{\mathrm{TB,peak}}$, is proportional to $2\delta\theta_{\mathrm{D},n} \sim \lambda_{hkl}^{2}$ indicating that the TDS in the range $0.6 < \lambda < 3.0$ Å is below the detection limit. However, the intensity of TDS grows rapidly in the range of (555), (777), and (888) triple-bounce reflections, reaching the level of TDS registered for (888) single-bounce reflection (compare upper and lower spectra in Figure II.1.3.1.11). The abnormal increase of TDS in the range of wavelength $0.3 < \lambda < 0.6$ Å is due to contamination of the triple-bounce (555), (777), and (888) reflections with single-bounce Bragg diffraction and TDS penetrating through the Cd shielding for $\lambda < 0.6$ Å (Agamalian et al., 2009). The increase of TDS in the triple-bounce spectrum at $0.3 < \lambda < 0.6$ Å correlates with the increase of Cd transmission $T(\lambda)$ in this range of wavelength. In the vicinity of the eighth order (888) $T(\lambda) \approx 0.9$ and that is why the single- and triple-bounce spectra are nearly identical in this range of wavelength. The Cd shielding of the triple-bounce crystal (see Figure II.1.3.1.5) is ineffective at short wavelengths, $\lambda < 0.6$ Å, and thus the parasitic effects (mostly the single-bounce BF reflection), blocked with Cd for $\lambda > 0.6$ Å, become observable.

It is practically impossible to separate this parasitic scattering from the triple-bounce reflection in DCDs at steady-state neutron beam lines except by the use of highly curved neutron guides (Wagh et al., 2001). Otherwise, the first triple-bounce Bragg reflection is usually contaminated with the single-bounce high-order back-face reflections and TDS originating from the wavelength range $\lambda < 0.6$ Å (see the dashed arrows in Figure II.1.3.1.5). This additional source of parasitic scattering is already taken into consideration in the design of the SNS TOF-USANS instrument, which will use the wavelength range $0.8 < \lambda < 4.0$ Å where the Cd shielding operates properly and the triple-bounce channel-cut crystal effectively suppresses TDS (see Figure II.1.3.1.11). The range $\lambda < 0.6$ Å where Cd shielding becomes transparent may be eliminated from the neutron spectrum entering the double-crystal arrangement with the T0 chopper (see Figure II.1.3.1.8).

## II.1.3.1.5 RESIDUAL STRESS MEASUREMENTS IN THIN FILMS

This application of the neutron Bonse–Hart DCD is based on the dynamical diffraction effect "neutron camel" (Agamalian et al., 1998b, 1999b), which is extremely sensitive to ultra-small deformation of the crystal under study. The name "neutron camel" relates to an unusual profile of the back-face rocking curve (BFRC) discovered experimentally with the Bonse–Hart DCD at ORNL (Agamalian et al., 1997). The BFRC for a DCD with perfect crystals can be derived from the classical dynamical diffraction theory assuming that in the vicinity of Bragg reflection the part of the neutron beam penetrating into a transparent crystal is $(1 - R_D(y))$, where $R_D(y)$ is the Darwin function (II.1.3.1.1). This portion of the neutron beam $(1 - R_D(y)) = 0$ in the range of Darwin plateau $|y| \leq 1$ and $(1 - R_D(y)) > 0$ for $y > +1$ and $y < -1$ (see Figure II.1.3.1.1), which leads to the appearance of two peaks in the BFRC, $I_{BF}(y)$, which can be written as

$$I_{BF}(\Delta) = \int R^m(y) \times R_{BF1}(y + \Delta)\mathrm{d}y, \qquad \text{(II.1.3.1.11)}$$

where $R_{BF1}(y) = 1 - R_D(y)]^2 \times R_D(y)$ is the reflectivity function of the first BF reflection and $R^m(y)$ is the reflectivity function of a multibounce monochromator. The BFRC is symmetrical with respect to $y = 0$ and contains two sharp peaks, the resolution of which improves significantly when the number of reflections in the monochromator is increased. The theoretical BFRCs, calculated from eq. (II.1.3.1.11), agree well with the experimental results obtained from thick, slab-shaped perfect Si crystals (Agamalian et al., 1998b). However, the function $R_{BF1}(y) = 1 - R_D(y)]^2 \times R_D(y)$, based on the classical dynamical diffraction theory, does not describe the asymmetry of the experimental BFRCs observed on lightly deformed crystals and therefore the dynamical two-wave diffraction theory (Bragg case) of deformed crystals (see, for example, Pinsker, 1978) requires to explain this effect.

The asymmetry of BFRCs has been modeled theoretically and verified by neutron diffraction experiments with controlled ultra-small static bending of the crystal under study (Agamalian et al., 2008b). These experiments revealed the extremely high sensitivity of the BFRC to ultra-small deformation strain (bending) in single crystals, which cannot be detected with conventional (front-face) rocking curve measurements. In the vicinity of the exact Bragg reflection the neutron beam propagating inside a transparent perfect crystal partly reflects from the BF (the first BF reflection) and partly leaves the crystal; the process continues and the beam propagates inside the crystal in the direction parallel to the diffractive surfaces (BF-FF-BF-... mode). When the crystal is lightly deformed, the second mode of propagation (see Figure II.1.3.1.12), created by so-called garland reflections only from the front-face (FF-FF-FF-... mode), appears in addition to the BF-FF-BF-... mode (Chukhovskii and Petrasheń, 1988; Agamalian et al., 2001).

A chosen model of cylindrical deformation contains one independent dimensionless parameter of deformation $b$, which is proportional to the gradient of the lattice constant.

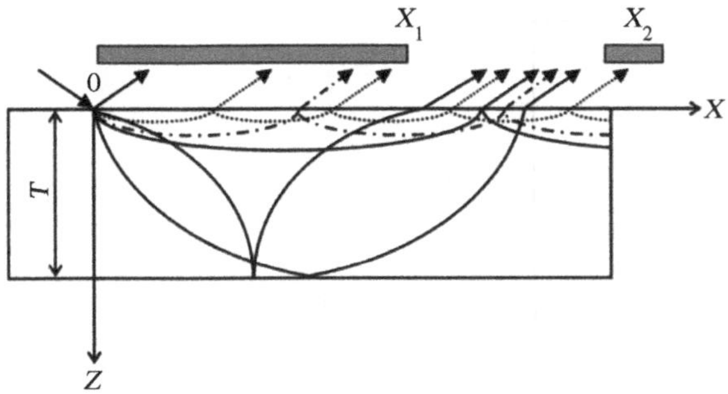


**Figure II.1.3.1.12** Trajectories of FF-BF-FF... and FF-FF-FF-... reflections inside a deformed crystal. X1, X2 are the coordinates of the Cd slit and $T$ is the crystal thickness.

$$b \sim \frac{\partial^2(\mathbf{QU})}{\partial s_0 \partial s_h}, \qquad \text{(II.1.3.1.12)}$$

where $\mathbf{Q}$ is the vector of scattering, $Q = 4\pi \sin\theta_B/\lambda$; $\mathbf{U}$ is the displacement of nuclei under the deformation force; and $s_0 = (X/\cos\theta_B + Z/\sin\theta_B)/2$ and $s_h = (X/\cos\theta_B - Z/\sin\theta_B)/2$ are the coordinates directed along the incident and the diffracted beams, respectively. The intensity of the first BF reflection of the lightly deformed crystal $R_{BF1}(y)$ is derived for $b > 0$, $y < -1$, and $b > 0$, $y > 1 + 2bT$, where $T_1 = (T/\tau) \times \pi \times ctg\theta_B$ is the dimensionless crystal thickness, $T$ is the crystal thickness in $\mu$m, and $\tau$ is the extinction length.

$$R_{BF1}(y) = \exp\{-2\,\mathrm{arccos}h[|y| - 2bT_1 \mathrm{sign}(y)]\}\{1 - \exp[-2\,\mathrm{arccos}\,h(|y|)]\}^2 \times$$
$$\times H(X_r - X_1)H(X_2 - X_r);$$
$$X_r = \frac{\mathrm{sign}(y)}{b}\left[\sqrt{y^2 - 1} - \sqrt{(|y| - 2bT_1 \mathrm{sign}(y))^2 - 1}\right], \qquad \text{(II.1.3.1.13)}$$

where $y = (\theta - \theta_B)/\delta\theta_D$ and $H(X)$ is the Heaviside unit step function. Figure II.1.3.1.13 shows the theoretical BF reflectivity functions integrated over the width of the Cd slit ($X_1$, $X_2$), $R_{BF1}(\theta - \theta_B)$, calculated for the parameters $\delta\theta_D = 0.8$ arcs, $T = 8.19$ mm, and $b \sim 4 \times 10^{-4}$. This diagram demonstrates that the reflectivity function of the first BF reflection from a lightly deformed crystal is asymmetric and contains the garland reflections (see the sharp diffraction peaks in the vicinity of $(\theta - \theta_B) = -1.0$). This function differs dramatically from that calculated for a perfect crystal (see Figure II.1.3.1.1).

The neutron experiments were carried out on a perfect 8.1 mm thick Si(111) slab-shaped crystal with the 114 × 40 mm diffractive surfaces polished mechanically, etched, and finally polished chemically (Agamalian et al, 2001, 2002). The crystal was set up in the Bonse–Hart ORNL DCD as an analyzer (see Figure II.1.3.1.14); the neutron beam reflected from the triple-bounce monochromator was restricted with a stationary 1.8 mm wide cadmium slit Cd1. The second 4 mm-wide scanning slit Cd2

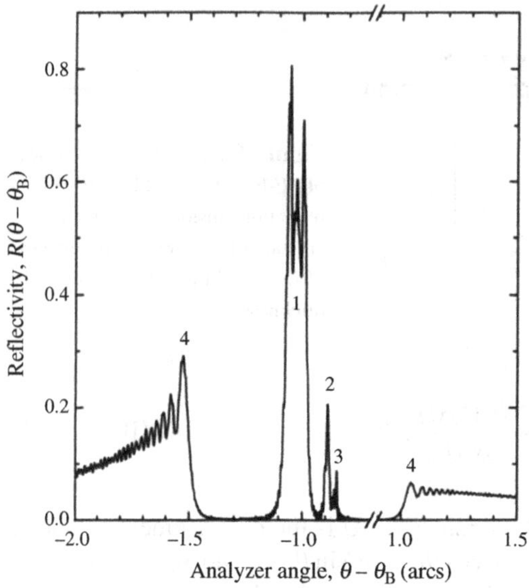


**Figure II.1.3.1.13** The $R_{BF1}(\theta - \theta_B)$ reflectivity function calculated for the deformed crystal with $T = 8.19$ mm and $b \sim 4 \times 10^{-4}$.

was mounted directly in front of the detector to map the neutron beam diffracted from the internal volume of the crystal; the transmitted beam was used as a monitor signal to determine the exact Bragg angle, $\theta_B$. The FF and BF rocking curves were measured on the crystal under study after the surface treatment; the same measurements were repeated after coating of one of the diffractive surfaces with a 2000 Å thick Ni film to introduce a strain field. In both cases, the slab-shaped Si(111) crystal was mounted on the rotation stage without external deformation strain.

Figure II.1.3.1.15 demonstrates the experimental BFRC obtained from the perfect (without Ni film) Si(111) crystal, symmetric with respect to the exact Bragg angle ($\theta - \theta_B = 0$) (open circles), together with the instrument background (closed circles). This result is consistent with the theoretical BFRC calculated by eq. (II.1.3.1.11) for $R_{BF1}(y) = 1 - R_D(y)]^2 \times R_D(y)$.

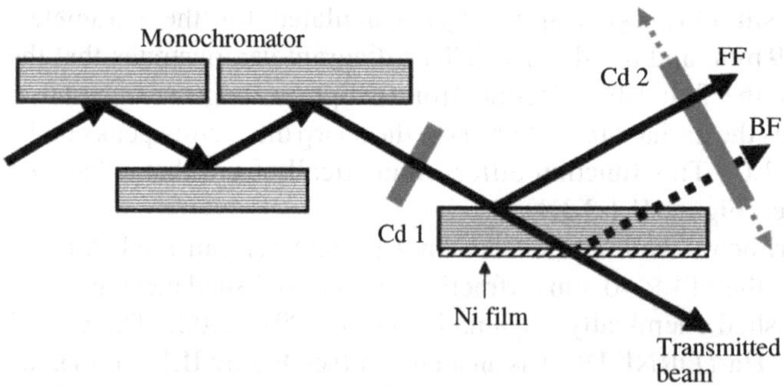


**Figure II.1.3.1.14** Optical scheme of the residual stress experiment. Cd 1 and Cd 2 are the stationary and the scanning Cd collimation slits correspondingly.

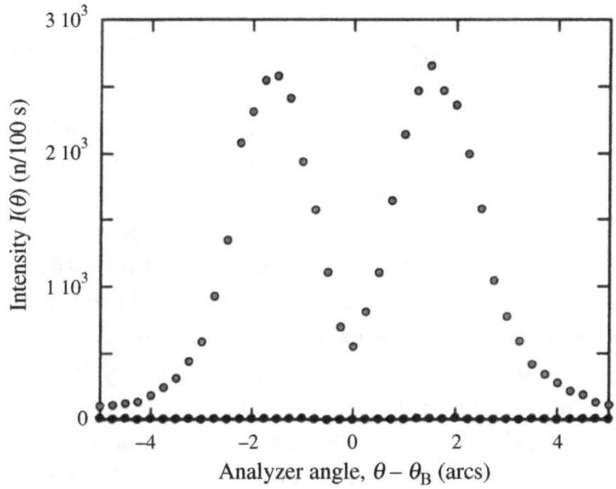


**Figure II.1.3.1.15** The experimental BFRC (open circles) measured from Si(111) crystal-substrate before coating with a Ni film. The instrument background is shown with closed circles.

The dramatic asymmetry of the BFRCs was observed in the experiments on the same crystal after coating one of the diffractive surfaces with a 2000 Å Ni film (Figure II.1.3.1.16). The BFRCs were measured for two orientations of the crystal: (1) the Ni coated surface is set up as the FF (Ni on FF) and (2) the Ni coated surface is set up as the BF (Ni on BF). The best-fit to the experimental BFRCs (see Figure II.1.3.1.16) was found for the parameters of deformation $b \approx 4 \times 10^{-4}$ and $b \approx 3.5 \times 10^{-4}$, which corresponds to the relative deformation of the Si crystallographic cells in the vicinity of diffractive surfaces, $|\partial u_z/\partial z| \sim 1.6 \times 10^{-6}$, and to the radius of bending $R_b \approx 19$ km. It is worthwhile to note that conventional FF Bragg diffraction is not sensitive to such light bending of the crystal (the FF rocking curves obtained on the Si(111) crystal before and after coating with a 2000 Å Ni film are identical).

The Stoney formula (see, for example, Noyan et al., 1995) converts the value of $R_b$ to the tension force $F$ applied to the film as a result of the substrate deformation:

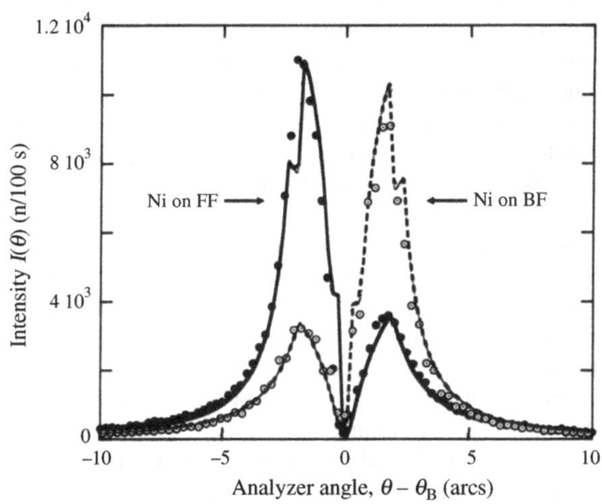


**Figure II.1.3.1.16** BFRCs measured after coating with the 2000 Å Ni film. Closed: the Ni film is on the BF. Open circles: the Ni film is on the FF. Solid and dashed lines are the simulation curves calculated for $b \approx 4 \times 10^{-4}$ and $b \approx 3.5 \times 10^{-4}$, respectively.

$$F = \frac{ET^2}{6(1-\nu^2)R_b}, \tag{II.1.3.1.14}$$

where $E \sim 10^{12}$ dyn/cm$^2$ is the modulus of elasticity, and $\nu$ is the Poisson constant for Si. The calculated value of the tensile force, $F \approx 90$ N/m, indicates that the Ni film is strongly strained (expanded along the X-axis shown in Figure II.1.3.1.12).

Optical devices consisting of thin reflecting layers deposited on silicon or silicon dioxide substrates have found wide application in light, X-ray, and neutron diffraction. A significant surface-induced residual stress usually remaining in the films as well as in the substrates after deposition creates a serious limitation of quality of these devices. The residual stress in crystalline films can be detected directly by the conventional X-ray diffraction technique (Noyan et al., 1995). The laser-based *in situ* technique, surface-stress-induced optical deflection (SSIOD), detects small deformation strains in substrates during the coating process (Bicker et al., 1998). The back-face diffraction (BFD) from a perfect Si crystal is capable of revealing residual stress in single crystals even when the relative deformation of the crystallographic cells is as small as $\sim 8 \times 10^{-7}$, which corresponds to the radius of bending of $\sim$40 km (Agamalian et al., 1998b). Because the BFD technique can be applied for residual stress measurements in thin films deposited on the diffractive surface of thick Si substrates, combined reflectometry (see Chapter II.2) and BFD measurements of the same samples are possible. The BFD technique works in principle similarly to the SSIOD *in situ* technique, detecting the deformation of the substrate, thus, it is capable to measure residual stress not only in crystalline, likewise the X-ray diffraction technique, but also in amorphous, polymer, colloidal, mono- and multilayer thin films deposited on the diffractive surface of Si single crystals.

## II.1.3.1.6 SUMMARY

The conventional (reactor-based) Bonse–Hart DCD with Si triple-bounce channel-cut crystals fully optimized for USANS investigations extends the SANS range by two orders of magnitude to the smaller $Q$ values. USANS and SANS profiles measured from identical samples can easily be joined; this feature is the most beneficial output of the breakthrough to ultra-small angles. Combined USANS/SANS measurements have been found extremely effective for examination of hierarchical structures common for many natural and man-made materials with multilevel morphology (atoms–molecules–aggregates–agglomerates). The examples of these studies can be found in the review articles (Schaefer and Agamalian, 2004; Triolo and Agamalian, 2009).

The Bonse–Hart TOF-USANS instrument for pulsed sources, which according to the theoretical predictions should further improve the $Q$-resolution ($\Delta Q = Q_{min} \sim 7 \times 10^{-6}$ Å$^{-1}$) and at the same time the intensity of scattering radiation in the USANS/SANS overlapping range $\sim 8 \times 10^{-4}$ Å$^{-1} < Q < 4 \times 10^{-3}$ Å$^{-1}$, is under construction at the SNS, ORNL.

The Bonse–Hart USANS instrument can be easily set up for residual stress measurements in thin films coated on the surface of thick Si substrates. This configuration of the neutron DCD with a multibounce channel-cut monochromator and a single-bounce analyzer offers an opportunity to conduct another category of combined neutron diffraction studies—reflectometry (see Chapter II.2) and residual stress measurements of the same samples.

## REFERENCES


AGAMALIAN, M. *Notiziario Neutroni e Lucedi Sincrotrone* **2005**, *10*, 22.

AGAMALIAN, M., WIGNALL, G.D., and TRIOLO, R. *J. Appl. Crystallogr.* **1997**, *30*, 345.

AGAMALIAN, M., CHRISTEN, D.K., DREWS, A.R., GLINKA, C.J., MATSUOKA, H., and WIGNALL, G.D. *J. Appl. Crystallogr.* **1998a**, *31*, 235.

AGAMALIAN, M., GLINKA, C.J., IOLIN, E., RUSEVICH, L., and WIGNALL, G.D. *Phys. Rev. Lett.* **1998b**, *81*, 602.

AGAMALIAN, M., WIGNALL, G.D., and TRIOLO, R. *Neutron News* **1998c**, *9*, 24.

AGAMALIAN, M., ALAMO, R.G., KIM, M.-H., LONDONO, J.D., MANDELKERN, L., and WIGNALL, G.D. *Macromolecules* **1999a**, *32*, 3093.

AGAMALIAN, M., IOLIN, E., WIGNALL, G.D. *Neutron News* **1999b**, *10*, 24.

AGAMALIAN, M, IOLIN, E., KAISER, H., REHM, Ch., and WERNER, S.A., *Phys. Rev. B* **2001**, *64*, 161402.

AGAMALIAN, M., IOLIN, E., KAISER, H., REHM, Ch., and WERNER, S.A. *Appl. Phys. A* **2002**, *74* (Suppl.), S1686.

AGAMALIAN, M., CARPENTER, J.M., LITTRELL, K.C., RICHARDSON, J.W., and STOICA, A. *Proceedings of the 17th Meeting of the International Collaboration on Advanced Neutron Sources* 2005, p. 762

AGAMALIAN, M., CARPENTER, J.M., and RICHARDSON, J. W. *Phys. Lett. A* **2009**, *373*, 292.

AIZAWA, K and TOMIMITSU, H. *Physica B* **1995**, *213–214*, 884.

BARKER, J.G., GLINKA, C.J., MOYER, J., KIM, M.-H., DREWS, A.R., and AGAMALIAN, M. *J. Appl. Crystallogr.* **2005**, *38*, 1004.

BELLMANN, D., STARON, P., and BECKER, P. *Physica B* **2000**, *276*, 124.

BICKER, M., von HŰLSEN, U., LAUDAHN, U., PUNDT, A., and GEYER, U., *Rev. Sci. Instrum.* **1998**, *69*, 460.

BONSE, U. and HART, M. *Appl. Phys. Lett.* **1965**, *7*, 238.

BONSE, U. and HART, M. *Zeitschrift für Physik* **1966**, *189*, 151.

CARPENTER, J.M. and FABER, J. *J. Appl. Crystallogr.* **1978**, *11*, 464.

CARPENTER, J.M., AGAMALIAN, M., LITTRELL, K.C., THIYAGARAJAN, P., and REHM, Ch. *J. Appl. Crystallogr.* **2003**, *36*, 763.

CHUKHOVSKII, F.N. and PETRASHEŃ, P.V. *Acta Crystallogr. A* **1988**, *44*, 8.

CRICHTON, M.A. and BHATIA, S.R. *Langmuir* **2005**, *23*, 10028.

DEWHURST, C.D. *Meas. Sci. Technol.* **2008**, *19*, 034007.

DIAT, O., BÖSENKE, P., FERRERO, C., FREUND, A.K., LAMBARD, J., and HEINTZMANN, R. *Nucl. Instrum. Methods. Phys. Res. A* **1994**, *365*, 566.

FEIGIN, L.A. and SVERGUN, D.I. Structural Analysis by Small-Angle X-ray and Neutron Scattering. Plenum Press, New York, 1987.

FREUND, A.K. *Nucl. Instrum. Methods Phys. Res.* **1983**, *216*, 269.

GLINKA, C.J., BARKER, J.G., HAMMOUDA, B., KRUEGER, S., MOYER, J., and ORTS, W.J. *J. Appl. Crystallogr.* **1998**, *31*, 430.

GRAF, H.A., SCHNEIDER, J.R., FREUND, A.K., and LEHMANN, M.S. *Acta Crysallogr. A* **1981**, *37*, 863.

HAINBUCHNER, M., VILLA, M., KROUPA, G., BRUCKNER, G., BARON, M., AMENITSCH, H., SEIDL, E., and RAUCH, H. *J. Appl. Crystallogr.* **2000**, *33*, 851.

ILAVSKY, J., ALLEN, A.J., LONG, G.G., and JEMIAN, P.R. *Rev. Sci. Instrum.* **2002**, *73*, 1660.

KIM, M.-H. and GLINKA, C.J. *Microporous Mesopoous. Mater.* **2006**, *91*, 305.

LAMBARD, J. and ZEMB, T. *J. Appl. Crystallogr.* **1991**, *24*, 555.

MATSUOKA, H., KAKIGAMI, K., ISE, K., KOBAYASHI, Y., MACHITANI, Y., KATO, T., and KIKUCHI, T. *Proc. Natl. Acad. Sci. USA* **1991**, *88*, 6618.

MATSUOKA, H., IKEDA, T., YAMAOKA, T., HASHIMOTO, M, TAKAHASHI, T, AGAMALIAN, M., and WIGNALL, G.D. *Langmuir* **1999**, *15*, 293.

MUZNY, C.D., BUTLER, B.D., HANLEY, H.J.M., and AGAMALIAN, M. *J. Phys: Condens. Matter* **1999**, *11*, L295.

NOYAN, I.C., HUANG, T.C., and YORK, B.R. *Crit. Rev. Solid State Mater. Sci.* **1995**, *20* (2), 125.

PINSKER, Z.G. *Dynamical Scattering of X-Rays in Crystals.* Springer-Verlag, Berlin, 1978.

POPA, N.C. and WILLIS T.M. *Acta Crystallogr A* **1997**, *53*, 537.

RADLIŃSKI, A.P., RADLIŃSKA, E.Z., AGAMALIAN, M., WIGNALL, G.D., LINDNER, P., and RANDL, O.G. *Phys. Rev. Lett.* **1999**, *82*, 3078.

SABINE, T.M. *Concrete in Australia* **1999**, *25*, 21.

SCHAEFER, D.W. and AGAMALIAN, M. *Curr. Opin. Solid State Mater. Sci.* **2004**, *8*, 39.
SCHWAHN, D., MIKŠOVSKY, A., RAUCH, H., SEIDL, E., and ZUGAREK, G. *Nucl. Instrum. Methods Phys. Res. A* **1985**, *239*, 229.
SEARS, V.F. and SHELLEY, S.A. *Acta Crystallogr. A* **1991**, *47*, 441.
SHULL C.G. *J. Appl. Crystallogr.* **1973**, *6*, 257.
TAKAHASHI, T. and HASHIMOTO, M. *Phys. Lett. A* **1995**, *200*, 73.
TAKAHASHI, T., HASHIMOTO, M., and NAKATAMI, S. *J. Phys. Chem. Solids* **1999**, *60*, 1591.
TREIMER, W., STROBL, M., and HILGER, A. *Phys. Lett. A* **2001**, *289*, 151.
TREIMER, W., STROBL, M., and HILGER, A. *Phys. Lett. A* **2002**, *305*, 87.
TRIOLO, R. and AGAMALIAN, M. The combined ultra-small- and small-angle neutron scattering (USANS/SANS) technique for Earth sciences. In: LIANG L., et al. (editors). Neutron Applications in Earth, Energy and Environmental Science, Springer, 2009, p. 567.
WAGH, A.G., RAKHECHA, V.C., and TREIMER, W. *Phys. Rev. Lett.* **2001**, *87*, 125504.
ZACHARIASEN, W.H. Theory of X-ray Diffraction in Crystals, Dover Publications, Inc., New York, 1967.

# II.1.3.2 Focusing USANS Instrument

*Satoshi Koizumi*

## II.1.3.2.1 UTILITY OF ULTRASMALL-ANGLE NEUTRON SCATTERING

Small-angle neutron scattering (SANS) is an extremely powerful method for performing *quantitative* and *in situ* observation of swollen, dispersed, and fluctuating tissue in solvent. The SANS method does not require any special treatment such as freezing, sectioning, or staining, in contrast to electron microscopy. First, we consider the question of how small-angle scattering is used to approach the whole cell, that is, a *hierarchically self-assembled system, composed of multiple components of macromolecules and small molecules.* When observing the inside of a cell, the organelles (e.g., nucleolus, mitochondria, chloroplast, and endoplasmic reticulum) are enclosed by a plasma membrane composed of lipids. A large number of proteins are present in the membrane and cytoplasm; these proteins play many roles including serving as channels, enzymes, or cytoskeleton. The size of the hierarchical structures ranges from 10 $\mu$m to several nanometers. Therefore, we employ a new technique for focusing ultrasmall-angle neutron scattering (USANS) to continuously cover a wide range of length scales from few angstroms to several tens of micrometers.

SANS detects the differential scattering cross section $d\Sigma/d\Omega(q)$ for the small magnitude of the scattering vector $q$, which is defined as $q = (4\pi/\lambda)\sin(\theta/2)$ where $\lambda$ and $\theta$ are the wavelength and the scattering angle, respectively. The measured intensity $I(q)$ is affected by the incident beam flux $I_0$, the detector window's solid angle $\Delta\Omega$ and the instrument constant $K_I$, which depends on the neutron source and the scattering spectrometer. Sample parameters of transmission $T_S$ and volume $V_S$, the latter of which is given by the sample cross section $A_S$ and thickness $D_S$, also affect $I(q)$. Thus, $I(q)$ is given by

$$I(q) = K_I I_0 (A_S D_S) T_S \frac{d\Sigma}{d\Omega}(q) \Delta\Omega. \qquad \text{(II.1.3.2.1)}$$

The key technical challenge to be overcome to apply the SANS method is effectively measuring $I(q)$ across a wide range of length scales from a few micrometers to several angstroms. The focusing ultra-SANS (USANS) technique is necessary to address this issue.

For a conventional pinhole SANS (Schmatz et al., 1974), the beam size (R in radius) at the detector position is decided by the first and the second apertures (diameters $S_1^p$ and $S_2^p$) as shown in Figure II.1.3.2.1a. $S_2^p$ sets $A_S$ in accordance with $S_2^{p2}$. For long $\lambda$ (several angstroms), long scattering path $L_S$ ($\sim$10 m) and small $R$, the minimum experimentally accessible $q$ ($q_{min}$) is given by

$$q_{min} = \frac{2\pi R}{\lambda L_S}. \qquad \text{(II.1.3.2.2)}$$

The usual choice of $L_s = 10$ m, $R = 20$ mm, and $\lambda = 6$ Å gives $q_{min} = 2 \times 10^{-3}$ Å$^{-1}$, which corresponds to a length scale of 600 nm ($\sim 2\pi/q_{min}$). The principle of eq. (II.1.3.2.2) is common for the focusing USANS method (Figure II.1.3.2.1b), as discussed in this section.

The double-crystal (Bonse–Hart) USANS method (Shwahn et al., 1985) was developed at the same time as the pinhole SANS spectrometer (see details in Chapter II.1.3.1). The smallest $q_{min}$ accessible by the double-crystal method is on

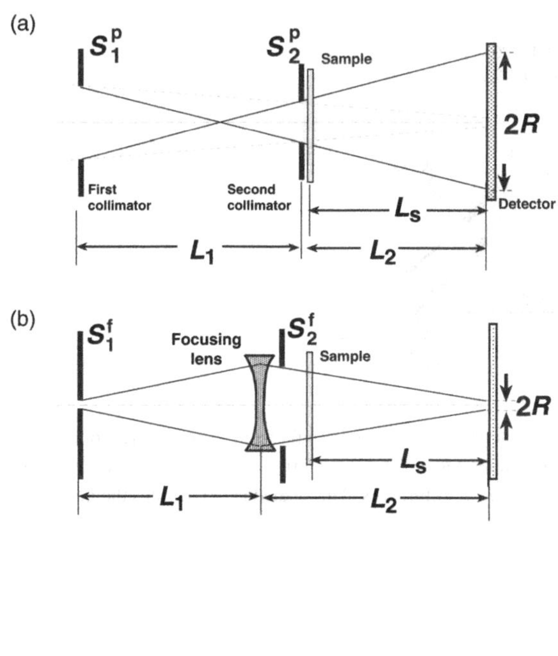


**Figure II.1.3.2.1** Schematic of optical geometries for (a) pinhole SANS and (b) focusing USANS. First and second apertures ($S_1$ and $S_2$) and detectors are allocated at distances $L_1$ and $L_2$. Sample–detector distance is $L_S$ ($\cong L_2$). Beam size at detector is indicated by $R$.

the order of $10^{-5}$ Å$^{-1}$, which is obtained by using thermal neutrons and grooved perfect crystals made from silicon. The first grooved crystal, which is placed before a sample in the path of the incident beam, selects only neutrons with small angular divergence (on the order of microradians) after triple diffraction. By rotating the second grooved crystal, USANS from the sample is analyzed again by triple diffraction. Because of the fine resolution, the neutron flux at the sample is decreased by too large an extent. A highly useful characteristic of the double-crystal method is that $q_{min}$ is decoupled from the beam size that irradiates the specimen. As a matter of fact, the beam size is on the order of 10 cm$^2$, which crucially contributes to the recovery of counting efficiency.

By combining the conventional pinhole SANS and double-crystal USANS methods, it is possible to cover a wide range of $q$ from 0.1 to $10^{-5}$ Å$^{-1}$. However, there is a gap in $q$ between these two methods. Figure II.1.3.2.2 shows the scattering curves obtained for poly(*N*-isopropyl acrylamide) gel swollen with $D_2O$ (Koizumi et al., 2004) that exhibits a large enhancement in scattering intensity as temperature increases across the volume phase transition temperature $T_V$. At low temperatures for the swollen state, the pinhole SANS detects diffuse scattering as given by an Ornstein–Zernike formalism, whereas at higher temperatures for a collapsed state,

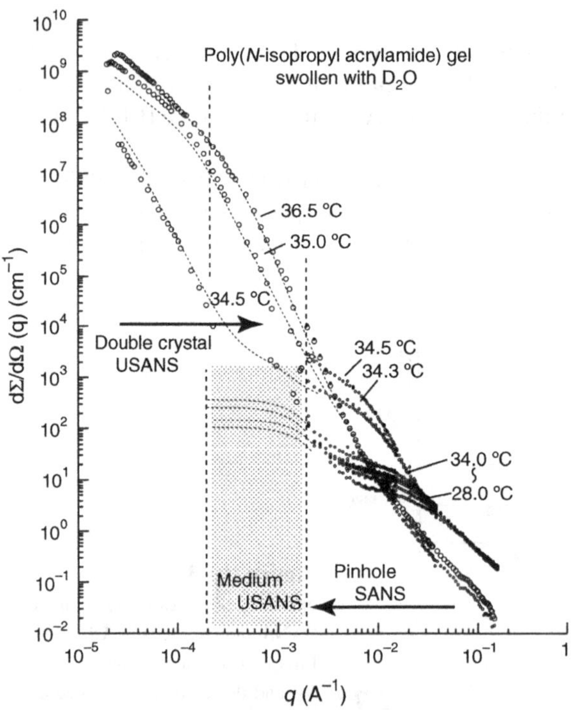


**Figure II.1.3.2.2** Ultrasmall-angle neutron scattering and small-angle neutron scattering from a polymer gel swollen with water at different temperatures. Whole scattering curves covering from 0.1 to $01^{-5}$ Å$^{-1}$ are detected by double-crystal and pinhole collimators. The shaded area indicates the invisible $q$-region, referred to as medium USANS.

the double-crystal USANS successfully observes domain scattering with a sharp interface boundary, which behaves in accordance with $\sim q^{-4}$ in $q = 10^{-5}$ Å$^{-1}$. An invisible $q$-region of $10^{-4}$ Å$^{-1}$ (the gap in $q$) appears between those covered by pinhole SANS and double-crystal USANS spectrometers. We refer to this invisible region as *medium USANS*.

## II.1.3.2.2 HISTORY OF NEUTRON FOCUSING LENS AND ATTEMPTS TO ACCESS MEDIUM USANS

A number of reflective and refractive devices have been developed and examined in order to focus a neutron beam and detect medium USANS. Since the refractive index ($n$) for cold neutrons is very small ($n$ is almost unity and slightly deviates on the order of $10^{-4}$), a reflective device, that is a mirror, was first thought to be more promising than refractive devices. After numerous trials (Maier-Leibnitz and Springer, 1963; Lartigue et al., 1995), a toroidal mirror was utilized at the USANS spectrometer KW3 at the FRM-II reactor, Germany (Alefeld et al., 1997). KW3 successfully covers medium USANS although the small-angle scattering from the mirror surface degrades the focusing behavior.

A refractive lens made of quartz was first proposed and fabricated by Gähler et al. (1980). Cold neutrons could be focused with a reasonable focal length ($f$) by using multiply stacked refractive lenses, which were made of a single $MgF_2$ crystal and biconcave in shape (Eskildsen et al., 1998). These biconcave refractive lenses were first installed on conventional pinhole SANS spectrometer at the National Institute of Standard and Technology (NIST) in the United States (Choi et al., 2000). By combining a higher resolution detector with a spatial resolution of 0.5 mm, the focusing USANS spectrometer was established at Tokai, Japan (SANS-J-II) (Koizumi et al., 2006, 2007), the details of which are reported in Section II.1.3.2.4. The biconcave lens with a parabolic surface shape was tested in order to improve a spherical aberration effect with a largely illuminated sample cross section.

The Stern–Gerlach apparatus (Estermann and Stern, 1933; Stern, 1934) uses an inhomogeneous magnetic field to split a neutron beam spatially into two beams of opposite polarization. The magnetic gradient system makes use of the sign of the force exerted on a neutron magnetic moment; this force depends on the sign of the spin component with respect to the magnetic field. The forces accelerate neutrons oppositely transverse to an inhomogeneous magnetic field. The sextuploe geometry of the magnets gives rise to a symmetrical force for focusing polarized neutrons (a magnetic lens) (Williams, 1988). This idea has been realized by using a permanent magnetic or superconducting magnetic lens (Shimizu et al., 1999). For the focusing USANS method, the magnetic lens requires highly polarized neutrons since those with an up spin component converge (focusing), whereas those with other spin components diverge (defocusing). This effect is discussed later in Section II. 1.3.2.5 with the experimental results.

Other techniques for tailoring neutron beams have been comprehensively reviewed by Crawford and Carpenter (1988). The multipinhole collimator selects neutron trajectories and focuses the neutrons on the detector. This technique was utilized to build the medium USANS spectrometer at Saclay France (Brûlet et al., 1998). By bending a single Si crystal and controlling its mosaic, the angular resolution of a Bonse–Hart spectrometer can be adjusted to the medium USANS range. A bent crystal combined with a position-sensitive detector, located along the bending direction, has been utilized for the multiangular scanning medium USANS spectrometer (Mikula et al., 1988; Strunz et al., 1997; Strobl et al., 2007). The principle of neutron spin echo, which is usually utilized to detect inelastic neutron scattering, was also applied to USANS (Bouwman et al., 2008). By the spin echo method, a pair correlation function $G(r)$ can be obtained in real space, which is advantageous for resolution calibration. A common idea underlying all these attempts, which successfully detect medium USANS, is "decoupling of the beam size at a sample and the $q$-resolution." The focusing USANS technique by using lenses is not an exception (see Figure II. 1.3.2.1 (b)).

## II.1.3.2.3 THEORETICAL BACKGROUND

### II.1.3.2.3.1 Basic Equations for Focal Properties

In this section, we summarize the basic concepts and equations that describe the focal properties of neutron focusing lenses (Sears, 1989). The index of refraction ($n$) is defined as

$$n^2 = 1 - \xi \qquad \text{(II.1.3.2.3)}$$

with

$$\xi = \frac{\lambda^2 \rho b_c}{\pi}, \qquad \text{(II.1.3.2.4)}$$

where $\rho$ is the atomic density, $b_c$ is the bound coherent length. For most materials, $b_c$ is positive; consequently, $n$ is smaller than unity. For a compound system such as $MgF_2$, $\rho b_c$ is the averaged bound coherent scattering length density,

$$\rho b_c = \sum_l \rho_l b_{C,l}, \qquad \text{(II.1.3.2.5)}$$

where $l$ denotes different atomic species in a compound.

Figure II.1.3.2.1b illustrates the focal properties of biconcave lens. The distances $L_1$ and $L_2$ from the center of the lens to the object and the image points, respectively, obey the following fundamental relation according to Gaussian optics (Born and Wolf, 1975),

$$\frac{1}{f} = \frac{1}{L_1} + \frac{1}{L_2}. \qquad \text{(II.1.3.2.6)}$$

From eq. (II.1.3.2.6), we obtain $L_2 = f$ when $L_1 = \infty$.

For a single biconcave lens, the focal length ($f_0$) is approximately derived as follows (Sears, 1989),

$$f_0 = \frac{r}{2(1-n)} \quad \text{or} \quad f_0 = \left(\frac{\pi r}{\rho b_c}\right)\frac{1}{\lambda^2}, \tag{II.1.3.2.7}$$

where $r$ is the radius of curvature of the lens surfaces. When lenses are stacked ($N$ is the number of lenses in the stack), $f$ is approximated by

$$f = \frac{f_0}{N}. \tag{II.1.3.2.8}$$

Since $n < 1$ for most materials, a concave lens is convergent (producing a real image), whereas a convex lens is divergent (producing a virtual image). This is opposite to the ordinary optics of light, where $n > 1$. From eq. (II.1.3.2.7), we find the important fact that $f \propto \lambda^{-2}$, indicating that (i) a long wavelength is necessary to obtain a small $f$ and (ii) the chromatic aberration is important for neutrons with a wavelength distribution ($\Delta\lambda$), which is true for SANS experiments. The chromatic aberration results in a broader focus, which is modeled and experimentally examined in Section I.1.3.2.5.

For a magnetic lens, $f$ is given by the lens power ($G$) and the lens length ($Z_m$) and the constant ($\alpha = 5.77$) (Shimizu et al., 1999), as given by

$$f = \frac{h^2}{G\alpha m_n^2}\frac{1}{\lambda^2 Z_m}, \tag{II.1.3.2.9}$$

where $h$ and $m_n$ denote Planck's constant and the mass of a neutron, respectively. We find that $f$ is inversely proportional to $G$, which determines the radial magnetic field in the lens; the magnetic field is designed to decrease according to the parabolic gradient from the edge of the lens to its center. $f$ is also inversely proportional to $\lambda^2$ and the length of lens $Z_m$. If we compare eqs. (II.1.3.2.8) and (II.1.3.2.9), we find that $N$ is equivalent to $Z_m$ and that $h^2/G\alpha\, m_n$ is comparable to $n r/\rho b_c$.

### II.1.3.2.3.2 Resolution and Minimum Accessible $q$ ($q_{min}$) with Refractive Focusing Optics

Mildner and Carpenter (1984) developed the resolution function of pinhole SANS. Mildner et al. (2005) and Hammouda and Mildner (2007) extended the arguments to the focusing USANS technique. In this section, we present their arguments. The smearing effect of isotropic scatterers and the fluctuations and, therefore, azimuthal isotropic small-angle scattering are given by the one-dimensional convolution integral,

$$\frac{d\Sigma(q_x)}{d\Omega} = \int_{-\infty}^{\infty} dq_x P(q'_x)\frac{d\Sigma(|q_x - q'_x|)}{d\Omega}. \tag{II.1.3.2.10}$$

$P(q)$ in eq. (II.1.3.2.10) is a resolution function approximately given by a Gaussian function as follows:

$$P(q'_x) = \left(\frac{1}{2\pi\sigma_q^2}\right)\exp\left(\frac{q_x^2}{2\sigma_q^2}\right). \qquad \text{(II.1.3.2.11)}$$

The $q$-resolution $\sigma_q$ is determined by two factors: (i) the wavelength distribution $\sigma_{q,w}$ and (ii) the instrument configuration geometry $\sigma_{q,g}$. We assume that there is no correlation between $\sigma_{q,w}$ and $\sigma_{q,g}$; thus, $\sigma_q$ is given by

$$\sigma_q^2 = \sigma_{q,w}^2 + \sigma_{q,g}^2, \qquad \text{(II.1.3.2.12)}$$

$\sigma_{q,w}$ is given as

$$\sigma_{q,w}^2 = q^2\frac{1}{6}\left[\frac{\Delta\lambda}{\lambda}\right]^2 \qquad \text{(II.1.3.2.13)}$$

and $\sigma_{q,g}$ is given by

$$\sigma_{q,g}^2 = \left(\frac{2\pi}{\lambda L_2}\right)^2 \sigma_g^2, \qquad \text{(II.1.3.2.14)}$$

where $\sigma_g$ describes the instrument configuration geometry.

According to Mildner and Carpenter (1984), for the conventional pinhole SANS, as shown in Figure II.1.3.2.1a, $\sigma_g$ is given by

$$\sigma_g^2 = \left(\frac{L_2}{L_1}\right)^2 \frac{S_1^{P^2}}{4} + \left(\frac{L_1 + L_2}{L_1}\right)^2 \frac{S_2^{P^2}}{4} + \frac{1}{3}\left(\frac{\Delta d}{2}\right)^2, \qquad \text{(II.1.3.2.15)}$$

where $L_1$ and $L_2$ ($\cong L_S$) are the source–sample and sample–detector distances, respectively, as shown in Figure II.1.3.2.1a. $\Delta d$ is the detector pixel size. This argument is based on the assumption that the source aperture $S_1^P$ is uniformly illuminated and the incident beam diverges uniformly. The contributions from the $x$- and the $y$-directions (horizontal and perpendicular) are given as

$$\sigma_q^2 = \sigma_{qx}^2 + \sigma_{qy}^2. \qquad \text{(II.1.3.2.16)}$$

$\sigma_{qy}$ for the perpendicular direction contains gravitational effects.

For the conventional pinhole SANS, the beam size at the detector position is given by

$$R = \frac{L_2}{L_1} S_1^P + \frac{L_1 + L_2}{L_1} S_2^P + \frac{\Delta d}{2}. \qquad \text{(II.1.3.2.17)}$$

$q_{min}$ is given by eqs. (II.1.3.2.2) and (II.1.3.2.17).

Next, we discuss the case of focusing USANS (Figure II.1.3.2.1b). We consider the case that focusing collimation is optimized for the principle wavelength $\lambda_0$ with a triangular wavelength distribution. According to Mildner et al. (2005) and Hammouda and Mildner (2007), $\sigma_g$ and $R$ are given by

$$\sigma_g^2 = \left(\frac{L_2}{L_1}\right)^2 \frac{S_1^{f\,2}}{4} + \left(\frac{L_1 + L_2}{L_1}\right)^2 \frac{2}{3}\left(\frac{\Delta\lambda}{\lambda_0}\right)^2 \frac{S_2^{f\,2}}{4} + \frac{1}{3}\left(\frac{\Delta d}{2}\right)^2 \qquad \text{(II.1.3.2.18)}$$

and

$$R = \frac{L_2}{L_1} S_1^{f} + \frac{L_1 + L_2}{L_1} 2\left(\frac{\Delta\lambda}{\lambda_0}\right) S_2^{f} + \frac{\Delta d}{2}. \qquad \text{(II.1.3.2.19)}$$

The second term of eq. (II.1.3.2.19) describes the chromatic aberration effect due to the wavelength distribution ($\Delta\lambda$), which is enhanced by $S_2^f$.

If we compare the pinhole and the focusing collimations, a gain factor ($g$) for neutron intensity at the sample position is defined by the slit area ratio as follows:

$$g = \left(\frac{S_1^{f\,2}}{S_1^{p2}}\right)\left(\frac{S_2^{f\,2}}{S_2^{p2}}\right) T_L. \qquad \text{(II.1.3.2.20)}$$

The focusing geometry employs the inverted collimation of a narrow first pinhole ($\sim S_1$) and a wider second pinhole $S_2$, as shown in Figure II.1.3.2.1. This wider slit size is critical for increasing $A_S$ in eq. (II.1.3.2.2), even as $R_f$ decreases (the principle of *decoupling beam size and q-resolution*). $T_L$ is the transmission of the lens.

## II.1.3.2.4 CONSTRUCTION OF FOCUSING USANS SPECTROMETER

In this section, we describe attempts to construct a *"focusing and polarized neutron ultrasmall-angle scattering spectrometer* (*SANS-J-II*) at JRR3, Tokai, Japan (Koizumi et al., 2006, 2007). We installed a focusing lens based on a conventional pinhole SANS spectrometer.

### II.1.3.2.4.1 Conventional Pinhole SANS Spectrometer (SANS-J)

First, we describe the conventional pinhole SANS spectrometer SANS-J (Figure II.1.3.2.3a), which has operated since 1991 at the guide beamhole of research reactor JRR3 (20 MW). The cold neutrons, moderated by liquid hydrogen, are transmitted by a Ni guide tube (20 mmW × 50 mmH) to the neutron guide beamhole. *SANS-J* is equipped at the end of the cold neutron guide (C-3-2) (neutron flux is $1.0 \times 10^8$ n/cm$^2$s). By using a disk-type velocity selector (provided by Central Research Institute, Hungary), the cold neutrons are monochromatically selected at $\lambda = 6.0$ Å, which is a maximum of the Maxwellian wavelength distribution.

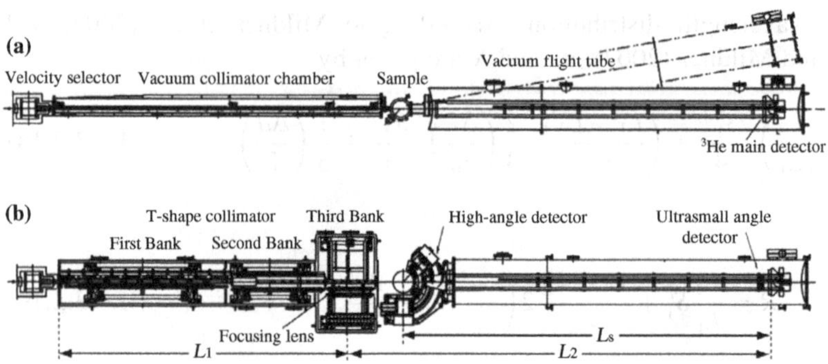


**Figure II.1.3.2.3** Schematic of (a) conventional SANS spectrometer (SANS-J) and (b) focusing USANS spectrometer (SANS-J-II).

The total spectrometer length (20 m) is symmetrically divided at the sample position (collimator and scattering flight tubes are about 10 m in length). In the vacuum flight tube, an $^3$He position-sensitive main-detector with a diameter of 60 cm and positional resolution of about 5 mm is equipped (provided by RISØ National Laboratory, Denmark). In the collimator, we combine Ni guides and pinhole sizes ($S_1$ and $S_2$ at the upper and the lower positions in the collimator), depending on the available sample–camera distance $L_s$, from 1.3 to 10 m. Using typical apparatus conditions to obtain $q_{\min}=3\times10^{-4}\,\text{Å}^{-1}$ ($S_1=20\,\text{mm}\phi$, $S_2=8\,\text{mm}\phi$, $L_S=10.2\,\text{m}$, and $\lambda=6.5\,\text{Å}$), we obtain a beam flux of about $5\times10^5\,\text{n/cm}^2\cdot\text{s}$ at the sample position. With the shortest camera length 1.3 m, we reach up to $q=0.2\,\text{Å}^{-1}$. Further extension toward higher $q$ ($\cong 0.4\,\text{Å}^{-1}$) is achieved by tilting a flight tube angle up to 10°.

## II.1.3.2.4.2 Focusing and Polarized Neutron Ultrasmall-Angle Scattering Spectrometer (SANS-J-II)

To construct the focusing USANS instrument, we retained the instrument base of *SANS-J*: (i) the total spectrometer length ($L_1+L_2=20\,\text{m}$), (ii) the velocity selector (providing $\lambda=6.5\,\text{Å}$ and $\Delta\lambda/\lambda=0.08\sim0.13$) and (iii) the $^3$He RISØ-type area detector. On this base, we newly constructed the following three items.

### *II.1.3.2.4.2.1 "T-Shape Collimator" with Focusing and Polarizing Devices*

We constructed the "*T-Shaped*" collimator chamber, partially made from nonmagnetic stainless steel (length: 11.372 m) (Figure II.1.3.2.3b). The polarizing devices (Fe/Si supermirrors and quadrupole permanent magnet for polarization (Oku et al., 2007a), and $\pi$-flippers) are installed upstream in the chamber. By sliding the optical benches, we aim to prepare a variety of collimations using polarized or unpolarized neutrons. The incident beam is collimated by fourfold beam narrowers composed of sintered $B_4C$ plates with a tapered edge; the position of the $B_4C$ plates is controlled by four ultrasonic motors.

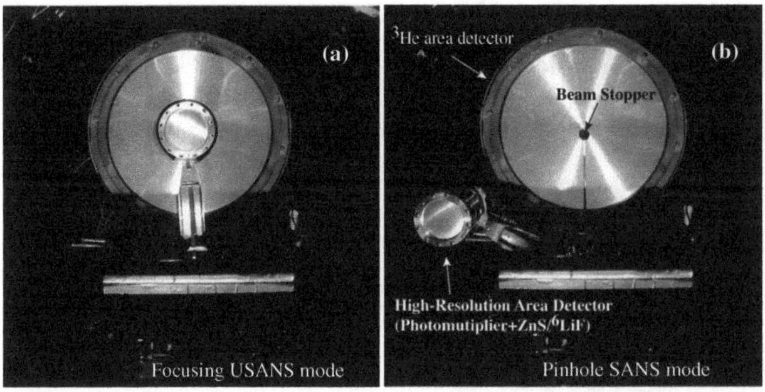


**Figure II.1.3.2.4** A hybrid detector system composed of an $^3$He 2-dimensional detector (60 cm in diameter) and high-resolution photomultiplier with ZnS scintillator (12.7 cm in diameter). For focusing USANS (a), the photomultiplier covers a direct beam position, whereas for conventional SANS (b), it moves out to the bottom.

#### *II.1.3.2.4.2.2 High-Resolution USAS Detector Combined with Main $^3$He Detector*

In front of the $^3$He area detector, we installed a second ultrasmall-angle area detector with higher positional resolution of 0.5 mm, specifically, a cross-wired position-sensitive photomultiplier (provided by Hamamatsu Photonics Co., Ltd.) with a thin plate of an ZnS/$^6$LiF scintillator (thickness: 0.2 mm). According to the charge division method, the R3239 photomultiplier has positional resolution of 0.5 mm (Hirota et al., 2005). The USAS detector slides in the direct beam position as shown in Figure II.1.3.2.4a ($L_s = 9.6$ m for the R3239 photomultiplier, which is slightly shorter than $L_S = 10.2$ m for the $^3$He main detector). During focusing USANS measurements, conventional SANS ($q > 3 \times 10^{-3}$ Å$^{-1}$) is simultaneously detected by the $^3$He main area detector, which is located behind the R3239 photomultiplier. For conventional pinhole SANS ($q > 3 \times 10^{-3}$ Å$^{-1}$), the R3239 photomultiplier can be moved to an empty space at the bottom left (Figure II.1.3.2.4b).

#### *II.1.3.2.4.2.3 Refractive and Magnetic Focusing Lenses*

The focusing lenses are positioned symmetrically ($L_1 \cong L_2$) in the T-shaped collimator. According to eq. (II.1.3.6), the focusing lenses are required to have $f = 5$ m with $\lambda = 6.5$ Å. For the refractive compound lens made of $MgF_2$, the stack number is estimated as $N = 70$, according to eqs. (II.1.3.2.7) and (II.1.3.2.8) with $\xi = 6.8 \times 10^{-5}$ for $\lambda = 6.5$ Å. Although a parabolic surface is optically ideal (Frielinghaus et al., 2009), a spherical surface, as designed by NIST (Choi et al., 2000), is employed for *SANS-J-II* (diameter: 30 mm; radius of curvature: 25 mm; center thickness: 1 mm) (Figure II.1.3.2.5).

For the sextupole permanent magnet lens, the lens power $G = 1.081 \times 10^4$ T/m$^2$ satisfies $f = 5$ m with $\lambda = 6.5$ Å. The magnetic lens has additional parameters of

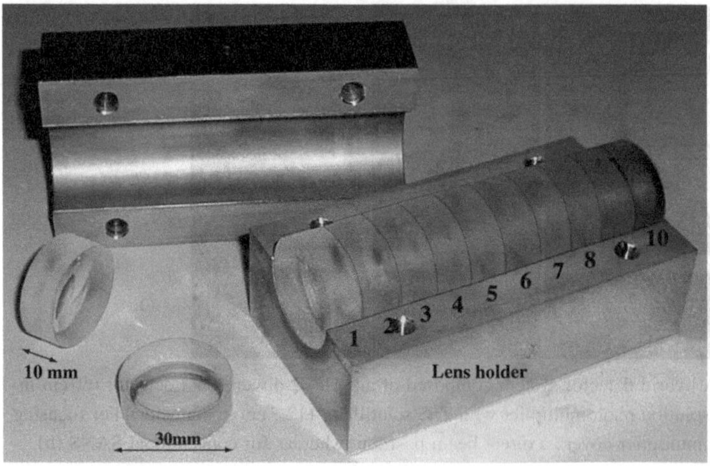


**Figure II.1.3.2.5** Biconcave refractive lenses, made from a single crystal of magnesium difluorides ($MgF_2$), which are aligned in an aluminum holder.

$Z_m = 1.26$ m, aperture size of 35 mm$\phi$, external diameter of 160 mm and weight of 357 kg (Oku et al., 2007b).

## II.1.3.2.5 FOCUSED NEUTRON BEAM

The focal properties for the refractive lens ($MgF_2$ and $N = 70$) and the sextupole magnetic lens were examined by varying $\lambda$. For this investigation, we chose $S_1^f$ and $S_2^f$ values of 2 and 20 mm$\phi$, respectively. The minimum $R$ was obtained for $\lambda_c = 6.63$ and 6.65 Å for the refractive and the magnetic lenses, respectively (Figure II.1.3.2.6). The parabolic change of $R$ is attributed to $\lambda^{-2}$ in eq. (II.1.3.2.7) or eq. (II.1.3.2.9).

The focused beam, obtained using the refractive lens at $\lambda_c = 6.63$ Å, was further examined by separately observing the different beam paths in a radial direction. In front of the refractive lens, we placed a Cd mask having the pinholes of 2 mm$\phi$ with pitch of 4 mm, as shown in Figure II.1.3.2.7a. Figure II.1.3.2.7b shows each spot at the focal point, corresponding to the different pinhole positions in the Cd mesh. The center spot (number 3 in Figure II.1.3.2.7a) is circularly symmetric, whereas the spots from the outer shell (number 1 or 5 in Figure II.1.3.2.7a) are elliptically elongated in the radial direction. This is the result of the second term in eq. (II.1.3.2.19), and the effect of chromatic aberration is more enhanced for large $R_2$. The elliptical elongation of the spot is attributed to broadening by chromatic aberration. It should be denoted that the transmission for the beam path through the outer shell is lower than that at the center because the lens thickness increases as $R_2$ increases, which determined experimentally (Iwase and Koizumi, 2010). The total transmission considering all beam paths (from 1 to 5) was determined to be $T_L = 0.5$ for the refractive lens ($MgF_2$ and $N = 70$). For the sextupole magnetic lens, the

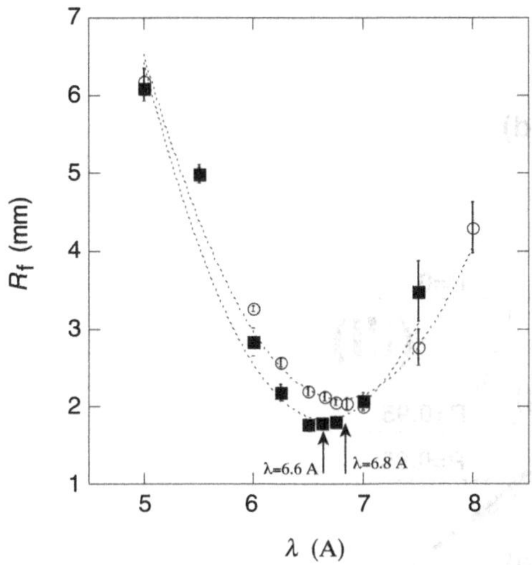


**Figure II.1.3.2.6** Focused beam size $R_f$, examined by changing wavelength $\lambda$ (filled squares and open circles correspond to $R_f$ of a refractive lens and magnetic lens, respectively).

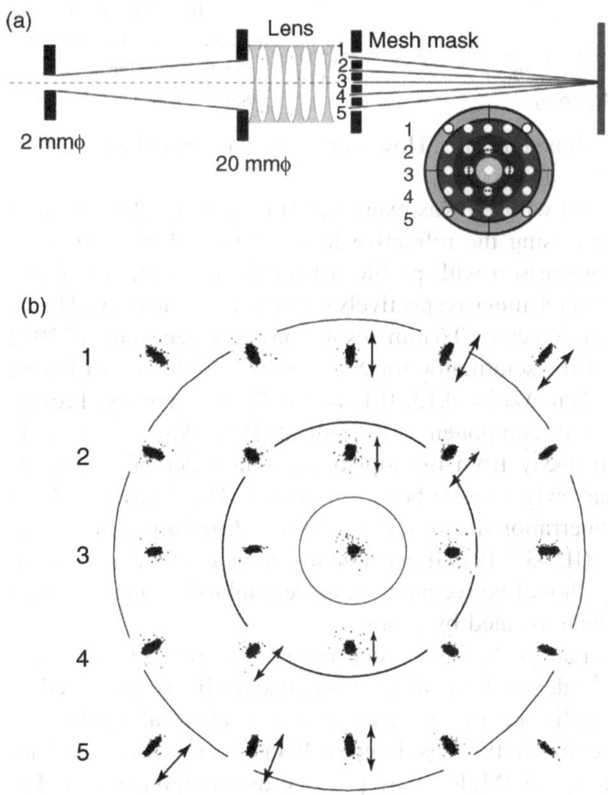


**Figure II.1.3.2.7** (a) Schematic diagram of pinhole beam mask after refractive lens. (b) Focused beams detected for different beam paths along radial direction of refractive lens.

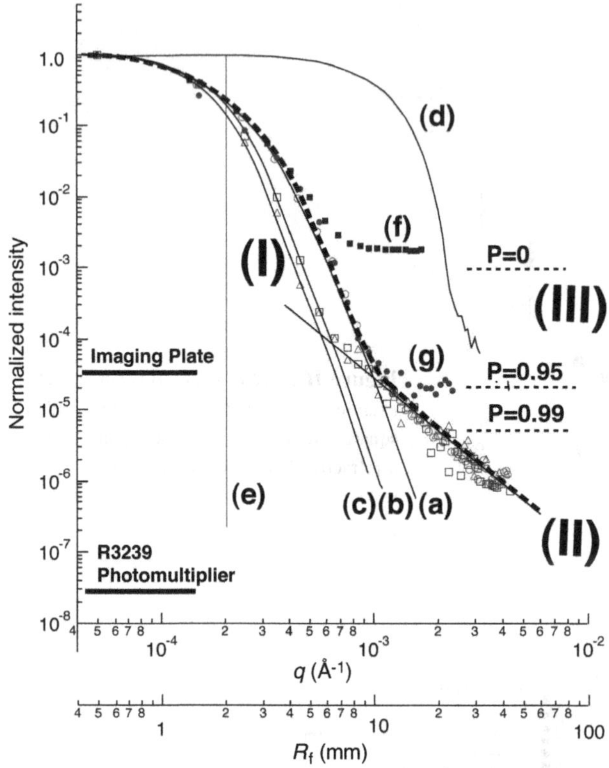


**Figure II.1.3.2.8** Focused beam profiles obtained by refractive lens ((a)–(c)) and magnetic lens ((f) and (g)). Profile (d) is for conventional pinhole SANS. Profile (e) is schematically drawn for focused beam with $\Delta\lambda = 0$.

transmission is constant for all beam paths. However, in the case of polarization, at least half of the total neutrons disappear.

The beam profile obtained with $\lambda_C$ was examined (Figure II.1.3.2.8). Profiles (a)–(c), which were obtained using the refractive lens ($MgF_2$ and $N = 70$), were successfully narrowed in comparison with profile (d) for the conventional SANS collimation ($S_1^p$ and $S_2^p$ = 20 and 8 mm$\phi$, respectively). To detect this narrowed beam ((a)–(c)), the high-resolution detector (0.5 mm resolution) was necessary (R3932 photomultiplier with ZnS/$^6$LiF scintillator for *SANS-J-II*). Profiles (a)–(c) were obtained for different $\Delta\lambda$ when $\Delta\lambda/\lambda = 0.13$, 0.1, and 0.08, respectively. Profiles (a)–(c) are composed of two components, as indicated by $R_{(I)}$ and $R_{(II)}$ in Figure II.1.3.2.7. The steep decay from the top to the –5th order of magnitude (marked by $R_{(I)}$) changes narrowly as $\Delta\lambda/\lambda$ becomes smaller. The decay tail $R_{(I)}$ is attributed to the chromatic aberration due to the wavelength distribution, according to the second term of eq. (II.1.3.2.19). If we employ monochromatic neutrons ($\Delta\lambda/\lambda = 0$), the initial decay should be rectangular, as determined by the first term of eq. (II.1.3.2.19) and as demonstrated by profile (e).

At intensities less than that of the –5th order, profiles (a)–(c) starts to decay gradually according to $q^{-3}$ (denoted by $R_{(II)}$). Component (II) is attributed to *parasitic scattering* from the slit, the air, or the aluminum windows along the path of the beam, and to imperfections in the lens. Even with a perfectly fabricated lens, thermal vibration from the crystal ($MgF_2$) may produce the component (II). The cooling effect of the lens was quantitatively discussed elsewhere (Frielinghaus et al., 2009). Thus we obtain,

$$R_{\mathrm{f}} = R_{(\mathrm{I})} + R_{(\mathrm{II})} \quad \text{(for refractive lens)}, \qquad (\mathrm{II}.1.3.2.21)$$

where $R_{(\mathrm{I})}$ is equivalent to eq. (II.1.3.2.19) and $R_{(\mathrm{II})}$ is due to the parasitic scattering. $q_{\mathrm{min}}$ is experimentally given by $R_{\mathrm{f}}$ in eq. (II.1.3.2.2). As discussed later, this gradual decay $R_{(\mathrm{II})}$ crucially limits $q_{\mathrm{min}}$, especially for the weak scattering power of $\mathrm{d}\Sigma/\mathrm{d}\Omega$ $(q=0) < 100\,\mathrm{cm}^{-1}$.

Polarized neutrons (polarization $P=0.95$ and $\Delta\lambda/\lambda=0.13$) are also focused by the magnetic lens (profile (g)). Similar to profiles (a)–(c), profile (g) initially decays due to the effects of chromatic aberration. However, in the case of the magnetic lens, a nondecaying plateau appears at the intensity of the –5th order. This plateau (third component $R_{(\mathrm{III})}$) originates from the defocusing beam of antiparallel spin. If the unpolarized neutrons ($P=0$) are polarized by the magnetic lens, this plateau appears at the –3rd order of magnitude (profile (e)).

$$R_{\mathrm{f}} = R_{(\mathrm{I})} + R_{(\mathrm{II})} + R_{(\mathrm{III})} \quad \text{(for magnetic lens)}, \qquad (\mathrm{II}.1.3.2.22)$$

where $R_{(\mathrm{I})}$ and $R_{(\mathrm{II})}$ for refractive and magnetic lenses are identical to each other.

The plateau from defocusing the beam is analytically estimated as $9 \times 10^{-4}$ and $2.3 \times 10^{-5}$ for $P=0$ and 0.95, respectively, which depends on the capability of the lens (magnetic field gradient $G$, length $Z_{\mathrm{m}}$, aperture size) and $L_2$. To reduce the defocusing plateau to the –6th order as negligibly small ($R_{(\mathrm{III})} \cong 0$), we need to induce polarization higher than 0.99 (as indicated in Figure II.1.3.2.6. Almost perfect polarization ($P=0.99$) has been achieved with a quadrate magnet polarizer (Oku et al., 2007a).

The R3932 photomultiplier with $\mathrm{ZnS}/^{6}\mathrm{LiF}$ scintillator can discriminate gamma rays, and thus the noise at the detector with the beam shutter closed is minimized at –8th order magnitude (solid line in Figure II.1.3.2.8). The imaging plate with Pb shielding (provided by FUJI Film Co., Ltd.; spatial resolution < 0.1 mm) exhibits noise at the –4th order due to gamma radiation, which is not sufficient for USANS measurements.

$R_{\mathrm{f}}$ (the focused beam size) was examined for changing $S_1^{\mathrm{f}}$ (Figure II.1.3.2.9). For $S_1^{\mathrm{f}} = 1$–$4\,\mathrm{mm}\phi$ ($S_2^{\mathrm{f}} = 20\,\mathrm{mm}\phi$), $R_{\mathrm{f}}$ remains constant and only intensity increases (change (I)), whereas $S_1^{\mathrm{f}}$ is from 5 to $12\,\mathrm{mm}\phi$, $R_{\mathrm{f}}$ starts to increase and intensity remains constant (change (II)). $R_{\mathrm{f}}$ in change (I) observed $S_1$ from 1 to 4 $\mathrm{mm}\phi$ is attributed mainly to the second term of chromatic aberration with $\Delta\lambda/\lambda=0.13$, whereas $R_{\mathrm{f}}$ in change (II) from $S_1=5\,\mathrm{mm}\phi$ to $12\,\mathrm{mm}\phi$ is dominated by $S_1$. For $S_1=2.5\,\mathrm{mm}\phi$, the chromatic aberration (second term in eq. (II.1.3.2.19) with $\Delta\lambda/\lambda=0.13$) is matched with the first aperture size $S_1^{\mathrm{f}}$ (first term in eq. (II.1.3.2.19)). If increased to $S_1^{\mathrm{f}} = 20\,\mathrm{mm}\phi$ following change (II), $R_{\mathrm{f}}$ and, therefore, $q_{\mathrm{min}}$ can reach values obtained by conventional pinhole SANS collimation (profile (d) in Figure II.1.3.2.8). By using these conditions to change $S_1^{\mathrm{f}}$, we are able to increase the direct beam intensity at the sample position. With the refractive lens ($N=40$) with $S_1^{\mathrm{f}} = S_2^{\mathrm{f}} = 20\,\mathrm{mm}\phi$ that was optimized for total lens transmission, in

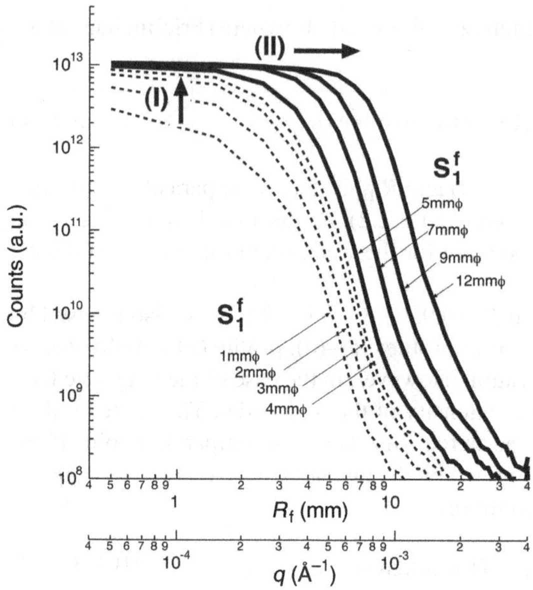


**Figure II.1.3.2.9** Focused beam profiles examined by changing first aperture size $S_1$. Chang (I) and (II) are shown by broken and solid lines, respectively.

comparison with the pinhole collimations, we obtained intensity that was increased by a factor of 3 (Iwase and Koizumi, 2010).

## II.1.3.2.6 DEMONSTRATIONS OF FOCUSING ULTRASMALL-ANGLE SCATTERING

$\sigma_g$ is estimated by curve-fitting a Gaussian function to the focused beam profiles (a) and (d) in Figure II.1.3.2.9. This is then converted to $\sigma_{q,g}$ according to eq. (II.1.3.2.14). Based on $\sigma_{q,g}/q$ shown in Figure II.1.3.2.10, we discuss the advantages of the focusing collimation. The first advantage is minimizing $q_{min}$ to $10^{-4}$ Å$^{-1}$ for medium

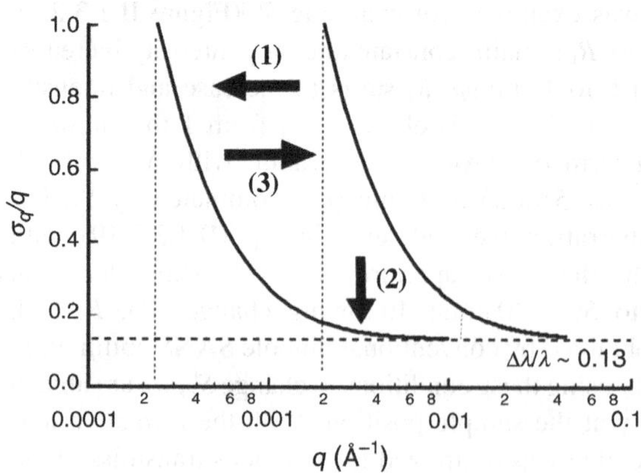


**Figure II.1.3.2.10** Resolution $\sigma_q/q$ as a function of $q$ evaluated from the focused beam in Figure 10. Arrows (1)–(3) indicate the cases of reduction in $q_{min}$, improvement of $\sigma_q$ and intensity gain.

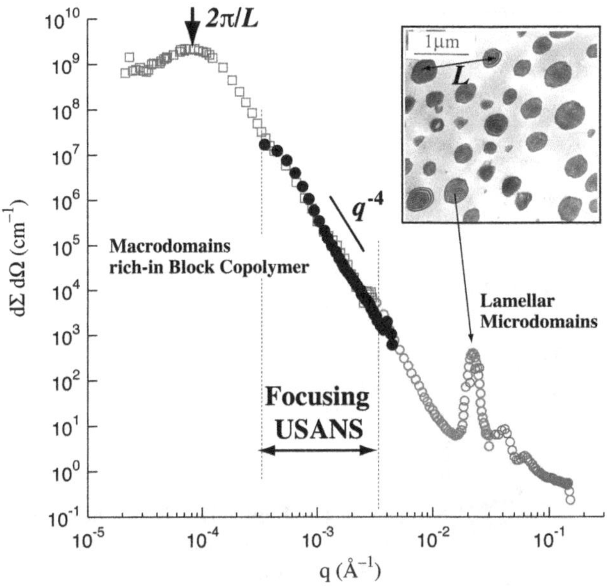


**Figure II.1.3.2.11** Scattering curve obtained for a cast film of block copolymer and homopolymer mixture. Open squares and open circles indicate those by double crystal USANS and pinhole SANS, respectively. Focusing USANS (filled circles) covers medium USANS from $q = 3 \times 10^{-4}$ to $3 \times 10^{-3}$ Å$^{-1}$. An inset indicates an image of phase-separated structure by transmission electron microscopy.

USANS (arrow (1) in Figure II.1.3.2.10). Figure II.1.3.2.11 shows the $q$-profile obtained for a solvent-cast film of the binary mixture of poly(styrene-$b$-isoprene) block copolymer and polystyrene homopolymer. In the film, macrophase separation between the block copolymer and the homopolymer, and microphase separation within the block copolymer are both present, as shown in the inset. Double-crystal USANS covers a $q$-region from $10^{-5} < q < 10^{-4}$ Å$^{-1}$, where we observe a scattering maximum due to interdomain interference from the macrodomains, which is followed by an asymptotic decrease according to $q^{-4}$ (Porod's law). The conventional pinhole SANS ($S_1^f$ and $S_2^f$ are 20 and 8 mm$\phi$, respectively) covers a $q$-region of $q > 3 \times 10^{-3}$ Å$^{-1}$, where scattering maxima due to microdomains appear. Focusing USANS using the refractive lens successfully covers the medium USANS region from $4 \times 10^{-4}$ Å$^{-1}$ to $3 \times 10^{-4}$ Å$^{-1}$, which corresponds to the gap between double-crystal USANS and conventional pinhole SANS.

The tails of the focused beam ($R_{(I)}$ or $R_{(II)}$) limit $q_{min}$ for focusing USANS. This effect strongly depends on the scattering power of a sample specimen ($\sim$d$\Sigma$/d$\Omega$($q = 0$)). Figure II.1.3.2.12 shows the $q$-profiles on PNIPAm gel in $D_2O$, detected by focusing USANS and conventional SANS. As the temperature changes around $T_V = 34$°C, the scattering power increases and $q_{min}$ shifts toward lower values and finally reaches $q_{min} = 4 \times 10^{-4}$ Å$^{-1}$.

Using a circularly symmetric direct beam, the focusing USANS method is superior for observation of anisotropic USANS. Bonse–Hart double-crystal USANS,

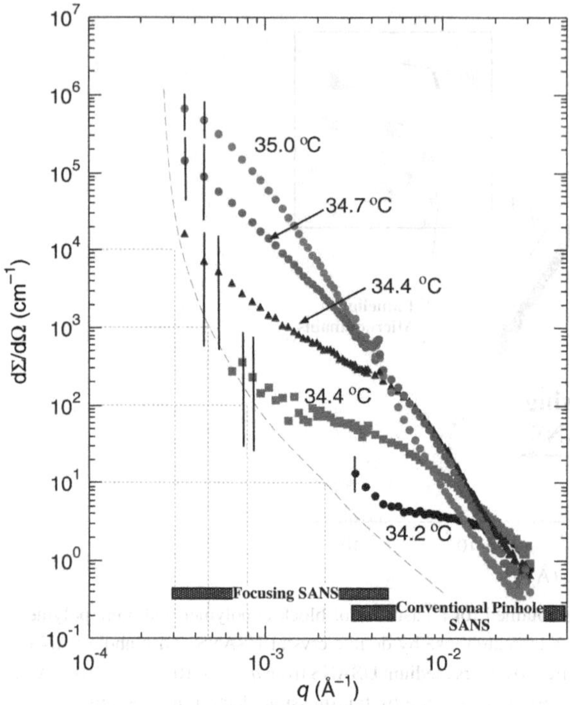


**Figure II.1.3.2.12** Scattering curves obtained for a PNIPA gel swollen with $D_2O$. Focusing USANS covers from $q = 3 \times 10^{-4}$ to $3 \times 10^{-3}$ Å$^{-1}$, whereas pinhole SANS covers for $q > 3 \times 10^{-3}$ Å$^{-1}$.

on the other hand, cannot analyze such anisotropic USANS because of the smearing effect caused by its linear primary beam. Figure II.1.3.2.13 shows the so-called *butterfly scattering* that was obtained for a film of polystyrene/poly(vinylmethylether) (80/20 wt/wt) (Koizumi and Suzuki, 2006). After shearing at 50°C, the

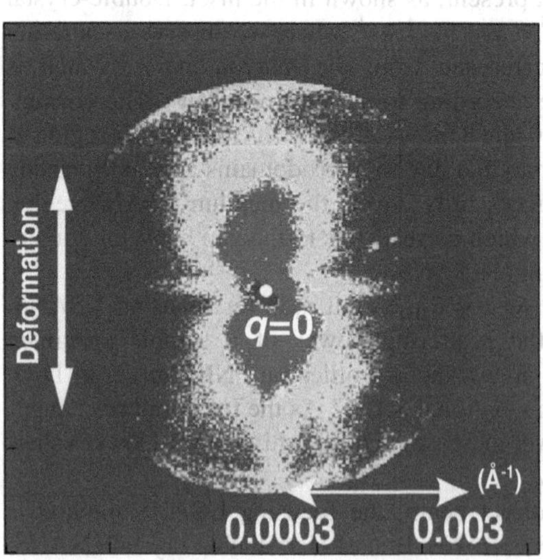


**Figure II.1.3.2.13** Butterfly scattering pattern in medium USANS $q$-region, obtained for a sheared polystyrene/poly (vinylmethylether) film.

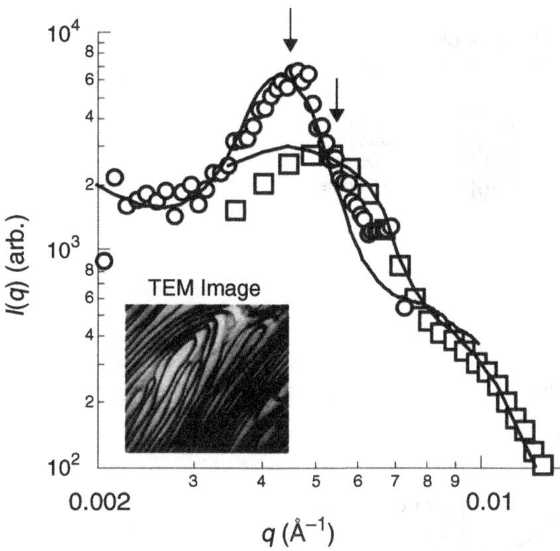


**Figure II.1.3.2.14** Scattering curves for polystyrene-*b*-polyisoprene, detected by focusing and pinhole SANS collimations (circles and squares, respectively). The inset shows TEM image of lamellar microdomain, examined by focusing USANS.

butterfly scattering due to shear-induced phase separation is successfully detected in the USANS $q$-region $0.0004 < q < 0.001$ Å$^{-1}$. If the film specimen is rotated in the primary beam, three-dimensional USANS can be performed; this has already been performed by conventional SANS (Koizumi and Suzuki, 2006).

The second advantage of the technique is the improved $q$-resolution $\sigma_q$ in the conventional SANS $q$-region from 0.002 to 0.01 Å$^{-1}$ (arrow (2) in Figure II.1.3.2.10). In Figure II.1.3.2.10, for $q > 0.004$, $\sigma_q$ of the focusing collimation decreases to the level of $\Delta\lambda/\lambda$. By using the focusing collimation, the scattering maximum due to a lamellar microdomain was observed (Figure II.1.3.2.14) and was sharper than that detected by pinhole SANS. The third advantage is the "intensity gain" in the $q$-region of conventional SANS (arrow (3) in Figure II.1.3.2.10), which was discussed in Section II.1.3.2.4. A $\sim$ 1 means reciprocal angstrome as in line 5 of this page.

## II.1.3.2.7 SUMMARY

The combined SANS technique, that is, the combination of double-crystal USANS and pinhole SANS reinforced by focusing USANS, is a powerful method for investigating the hierarchical structures present in a rich variety of the materials (Motokawa et al. 2007, Yamaguchi et al., 2008; Koizumi et al., 2008; Koga et al., 2008). Pinhole SANS spectrometers at research reactors around the world (HFIR and NIST in the US, FRM-II in Germany, HANARO in Korea), are currently under development, which are based on the similar focusing technique.

Using this method, we expect that a new scientific field of "neutron cell biology" can be explored by elucidating hierarchical structures that can function in a living cell. Figure II.1.3.2.15 shows the results obtained for a model solution of actin cytoskeleton, that is, a mixture of filamentous actin (F-actin) and synthesized cationic polymer of poly-*N*-[3-(dimethylamino)propyl] acrylamide methyl chloride quaternary (PDMAPAA-Q) (Masui and Koizumi, 2009). In this solution, F-actins and PDMAPAA-Q are *finite-sized and hierarchically* condensed as a result of the balance of electrostatic repulsive and attractive interaction. The

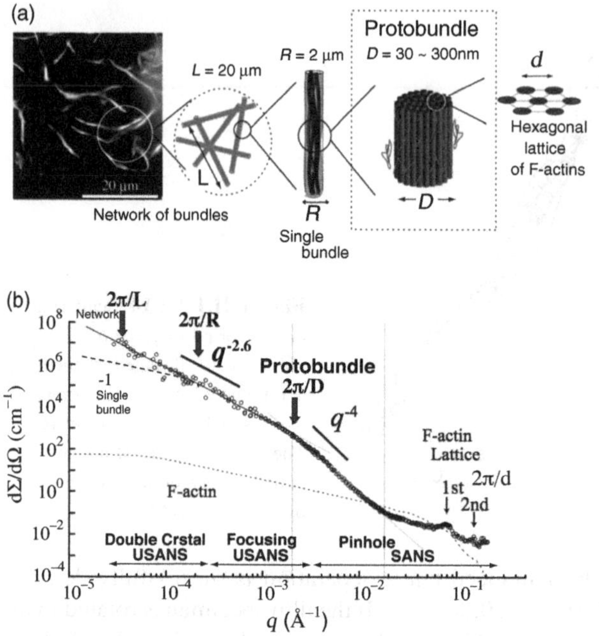


**Figure II.1.3.2.15** (a) Photograph of stretched bundle structure by fluorescence optical micrograph on length scales of 10 μm and schematic illustrations showing protobundle and actin filaments. (b) Scattering curve obtained for a actin solution, covering a wide $q$-region indicating network and stretched bundle (1 ~ 10 μm), a protobundle (10 ~ 100 nm), and actin filament (nm). The thin broken line indicates a scattering curve obtained for the solution of F-actin filaments without bundle formation.

combined SANS method is further reinforced by fluorescence microscopy (FM), which can cover length scales larger than micrometers. FM determined that the complex, referred to as a *superbundle*, appears with a finite size (L = 20 μm and R = 2 μm) and changes in shape from globular to stretched (image in Figure II.1.3.2.15a) by increasing the salt (KCl) concentration $C_{KCl}$ from 0.01 to 0.3 M. From the $q$-behavior of the scattering obtained by the combined SANS method (Figure II.1.3.2.15b), it was found that the superbundle is hierarchically structured, consisting of units of *protobundles*, which are surrounded by polycations and bind F-actins (see Figure II.1.3.2.15a). We also found that the dimensions of a protobundle are strongly influenced by $C_{KCl}$; its diameter (D) increases from 40 to 300 nm, for a $C_{KCl}$ change from 0.01 to 0.3 M. The superbundle morphology is controlled by the bending rigidity of protobundles and increases significantly as the protobundle diameter increases. The formation of stretched bundles, induced by changing the salt concentration, may play a role in cell movement (protrusion of a plasma membrane).

## ACKNOWLEDGMENTS

The author would like to thank Drs. D.F.R.Mildner and B. Hammouda for helpful comments. He also thank Drs. J. Suzuki, T. Oku, and H. Iwase for collaborations on SANS instrument.

# REFERENCES


Alefeld, B., Hayes, C., Mezai, F., Richter, D., and Springer, T. *Physica B* **1997**, *234–236*, 1052.

Born, M. and Wolf, E. Principles of Optics, Pergamon, Oxford, 1975.

Bouwman, W.G., Plomp, J., de Haan, V.O., Kraan, W.H., Well, A.A.v., Habicht, K., Keller, T., and Rekveldt, M.T. *Nucl. Instrum. Methods Phys. Res. A* **2008**, *586*, 9, and references cited therein.

Brûlet, A., Thevenot, V., Lairez, D., Lecommandoux, S., Agut, W., Armes, S.P., Du, J., and Desert, S. *J. Appl. Crystallogr.* **2008**, *41*, 161.

Choi, S.-M., Barker, J.G., Glinka, C.J., Cheng, Y.T., and Gammel, P.L. *J. Appl. Crystallogr.* **2000**, *33*, 793.

Crawford, R.K. and Carpenter, J.M. *J. Appl. Crysallogr.* **1988**, *21*, 589.

Eskildsen, M.R., Gammel, P.L., Issacs, E.D., Detlefs, C., Mortensen, K., and Bishop, D.J. *Nature* **1998**, *391*, 563.

Estermann, I. and Stern, O. *Z. Phys.* **1933**, *85*, 17.

Frielinghaus, H., Pipich, V., Radulescu, A., Heiderich, M., Hanslik, R., Dahlhoff, K., Iwase, H., Koizumi, S., and Schwahn. D. *J. Appl. Crystallogr.* **2009**, *42*, 681–690.

Gähler, R., Kalus, J., and Mampe, W. *J. Phys. E Sci. Instrum.* **1980**, *13*, 546.

Hammouda, B. and Mildner, D.F.R. *J. Appl. Crystallogr.* **2007**, *40*, 250.

Hirota, K., Shinohara, T., Ikeda, K., Mishima, K., Adachi, T., Morishima, T., Sato, S., Oku, T., Yamada, S., Sasao, H., Suzuki, J., and Shimizu, H.M., *Phys. Chem. Chem. Phys.* **2005**, *7*, 1836.

Iwase, H. and Koizumi, S. *J. Appl. Crystallogr.* (in preparation) 2009.

Koga, T., Tanaka, F., Motokawa, R., Koizumi, S., and Winnik, F. *Mocromolecules* **2008**, *41*, 9413.

Koizumi, S. and Suzuki, J. *J. Appl. Crystallogr.* **2006**, *39*, 878.

Koizumi, S., Monkenbusch, M., Richter, D., Schwahn, D., and Farago, B. *J. Chem. Phys.* **2004**, *121*, 12721.

Koizumi, S., Iwase, H., Suzuki, J., Oku, T., Motokawa, R., Sasao, H., Tanaka, H., Yamaguchi, D., Shimizu, H.M., and Hashimoto, T. *Physica B* **2006**, *385–386*, 1000.

Koizumi, S., Iwase, H., Suzuki, J., Oku, T., Motokawa, R., Sasao, H., Tanaka, H., Yamaguchi, D., Shimizu, H.M., and Hashimoto, T. *J. Appl. Crystallogr.* **2007**, *40*, s474.

Koizumi, S., Zhao, Y., Tomita, Y., Kondo, T., Iwase, H., Yamaguchi, D., and Hashimoto, T. *Eur. Phys. J. E* **2008**, *26*, 137.

Lartigue, C., Copley, J.R.D., Mezai, F., and Springer, T. *J. Neut. Res.* **1995**, *5*, 71.

Maier-Leibnitz, H. and Springer, T. *React. Sci. Technol.* **1963**, *17*, 217.

Masui, T., Koizumi, S., Hashimoto, T., Shikinaka, K., Kakugo, A., and Gong, J.P, *Soft Matter.* **2009**, *6*, 2021–2030.

Mikula, P., Lukos, P., and Eichorn, F. *J. Appl. Crystallogr.* **1988**, *21*, 33.

Mildner, D.F.R. and Carpenter, J.M. *J. Appl. Crystallogr.* **1984**, *17*, 249.

Mildner, D.F.R., Hammouda, B., and Kline, S.R. *J. Appl. Crystallogr.* **2005**, *38*, 979.

Motokawa, R., Iida, Y., Zhao,, Y., Hashimoto, T., and Koizumi, S., *Polym. J.* **2007**, *39*, 1312–1318.

Oku, T., et al. *Physica B* **2007a**, *397*, 188.

Oku, T., Iwase, H., Shinohara, T., Yamada, S., Hirota, K., Koizumi, S., Suzuki, J., Hashimoto, T., and Shimizu, H.M. *J. Appl. Crystallogr.* **2007b**, *40*, s408.

Schmatz, W. and Springer, T., Shelton, J., and Ibel, K. *J. Appl. Crystallogr.* **1974**, *7*, 96.

Sears, V.F. Neutron Optics, Oxford University Press, New York, 1989.

Shimizu, HM, Suda, Y., Oku, T., Nakagawa, H., Kato, H., Kamiyama, T., Otani, C., and Sato, H. *Nucl. Instrum. Methods Phys. Res.* **1999**, *A430*, 423.

Shwahn, D., Miksovsky, A., Rauch, H., Seidl, E., and Zugarek, G. *Nucl. Instrum. Methods Phys. Res.* **1985**, *A239*, 229.

Stern, O. *Z. Phys.* **1934**, *89*, 665.

Strobl, M., Treimer, W., Ritoulis, C., Wagh, A.G., Abbas, S., and Manke, I. *J. Appl. Crystallogr.* **2007**, *40*, s463.

Strunz, P., Saroun, J., Mikula, P., Lukas, P., and Eichhorn, F. *J. Appl. Crystallogr.* **1997**, *30*, 844.

Williams, W.G. Polarized Neutron, Clarendon Press, Oxford, 1988.

Yamaguchi, D., Mayama, H., Koizumi, S., Tsujii, K., and Hashimoto, T. *Eur. Phys. J. B.* **2008**, *63*, 153.

# II Instrumentation

## II.2 Neutron Reflectometry

*Naoya Torikai*

### II.2.1 INTRODUCTION

In recent years, material surfaces, interfaces, and thin films have attracted much attention from viewpoints of industrial applications as well as academic researches in a vast field of materials. A material at interface often exhibits peculiar structure and behavior, which are not observed in a bulk, such as lower surface glass transition temperature, dewetting, and preferential surface segregation for soft matters, since it encounters and interacts directly with different materials or phases through a very narrow space of interface. Many practical phenomena relating to our daily life such as adhesion, painting, coating, lubrication, and friction are also largely attributed to interfacial properties of materials. Further, due to a recent large progress in material science and technology, a hybridization of different materials has been much promoted, and a practical size or thickness of materials and devices in their use becomes smaller and thinner. These lead to the situation that a material always possesses interfaces with different component materials, and those interfacial properties affect and sometimes govern the performance of materials and devices with increasing a spatial ratio of the interfacial region occupied in them. Therefore, it is significant to understand interfacial properties and structures of materials, although the detailed investigation on material interfaces brings difficulties to detect them because an interface is generally very thin space and is deeply buried inside materials in most cases.

Neutron reflects and refracts at optically flat interface between two media with different refractive indexes, such as light at water surface, when it impinges onto the interface with a small grazing incident angle. Neutron reflectometry is quite powerful and essential to investigate the structures of material interfaces and thin films due to its high spatial resolution of a subnanometer scale in the depth direction by utilizing the optical reflection property of neutron at material interfaces. This technique can be categorized into an elastic scattering method along with small-angle neutron scattering (SANS) to probe static structures of materials. The use of neutron is

*Neutrons in Soft Matter* Edited by Toyoko Imae, Toshiji Kanaya, Michihiro Furusaka, and Naoya Torikai

advantageous for soft matter researches in several aspects, compared with similar techniques: X-ray reflectometry and ellipsometry, using an electromagnetic wave such as X-ray and light. First, the neutron exhibits a unique scattering ability for elements irrespective of their atomic number through nuclear interaction inherent in each nucleus, compared with an electromagnetic wave that the scattering ability for elements increases with their atomic number, that is, the number of electrons, due to its interaction with electrons. Thus, the neutron shows a relatively high scattering ability for light atoms such as hydrogen, carbon, and oxygen, which are major elements composed of soft matters. It also possesses different scattering ability among isotopes, and especially there exists a large difference between hydrogen (H, coherent scattering length $b = -0.374 \times 10^{-12}$ cm) and deuterium (D, $b = +0.667 \times 10^{-12}$ cm). This is one of the biggest advantages to make use of neutron reflectometry for structural analysis on soft matters, making it possible to enhance a contrast for neutron in samples without changing their physical properties much by replacing H atoms with D atoms (a deuterium labeling method). The deuterium labeling is applied to not only a full and a selective labeling of samples but also a contrast matching or a contrast variation method in more sophisticated manners. In principle, the neutron data on a deuterium labeled sample bring complementary structural information to the other experimental techniques such as X-ray reflectometry and ellipsometry. Another advantage is that the neutron has a high transmissivity and a less absorption to materials because it has no charge and interacts weakly with elements, so that the neutron reflectometry can probe deeply buried interfaces such as solid/liquid interfaces in a nondestructive way, and also can make *in situ* measurements under various sample environments, such as high temperature or high pressure, which need a window to seal. Due to these advantages described above, neutron reflectometry has been widely utilized for structural studies on a variety of material interfaces and thin films for soft matters: a polymer thin film, a Gibbs or a Langmuir monolayer of amphiphilic molecules such as surfactants, lipids, block copolymers, and proteins on water surface, a polymer brush chemically or physically adsorbed on a solid substrate, a Langmuir–Brodgett (LB) film, and so on. Nowadays, almost all neutron facilities possess a few neutron reflectometers for developing studies on material interfaces and thin films irrespective of a type of neutron sources: a reactor and a spallation neutron source. Moreover, the principles of neutron reflection are utilized in neutron optical devices such as a neutron guide, a neutron bender, and a polarizer with metal multilayer supermirrors to guide, to change a direction of, to focus, and to polarize a neutron beam.

A schematic geometry for neutron reflectometry is drawn in Figure II.2.1. Conventionally, in a reflectivity measurement specular reflection, in which the incident angle, $\theta_{in}$, of neutron relative to sample surface is equal to the reflected angle, $\theta_{out}$, has been observed to probe the structures along the $z$-direction perpendicular to the sample surface. On the other hand, at the angular position, in which $\theta_{in} \neq \theta_{out}$, around specular reflection position weak off-specular reflection is observed originating from in-plane structures of the sample in $x$-direction.

In this chapter, the principles of a conventional specular reflectivity measurement are explained using a few model systems with a single interface, a single- and

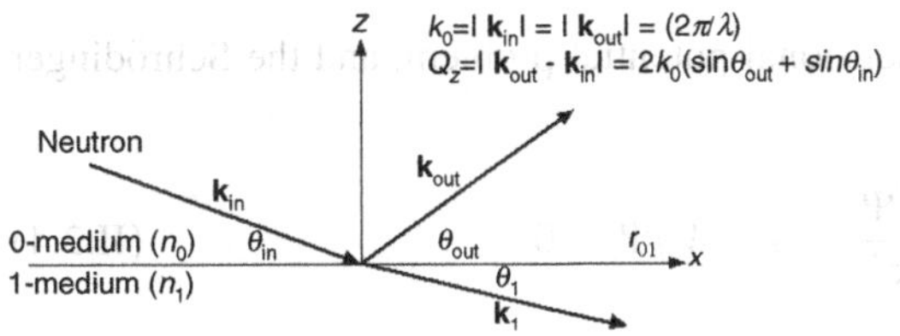


**Figure II.2.1** A schematic illustration of a geometrical configuration for neutron reflection at a single ideal interface between two media with different refractive indexes.

a multilayer film in sequence. Also, a practical experimental procedure and an instrument setup of neutron reflectometer are described, and then a few typical examples on soft matter researches are demonstrated. Finally, a grazing incidence small-angle scattering (GISAS) as well as an off-specular reflection method is briefly explained to explore in-plane structures, and the chapter is summarized with future prospects of neutron reflectometry.

## II.2.2 PRINCIPLES OF NEUTRON REFLECTION

### II.2.2.1 Refractive Index

In this chapter, for simplicity, it is supposed that two media ($i = 0, 1$) with refractive indexes $n_i$ are separated by an ideally flat interface with no roughness, as drawn in Figure II.2.1. The neutron with wave vector $\mathbf{k}_0$ ($k_0 = |\mathbf{k}_0| = 2\pi/\lambda$, where $\lambda$ is the wavelength of neutron) impinges onto the interface between two media from the side of 0-medium (air or vacuum) at small glancing angle $\theta_{in}$. The neutron energy $E$ in air or vacuum is expressed as

$$E = \frac{\hbar^2 k_0^2}{2m} = \frac{h^2}{2m\lambda^2} \qquad \text{(II.2.1)}$$

where $h$ ($= 2\pi\hbar$) is the Plank's constant ($= 6.63 \times 10^{-34}$ J·s) and $m$ is the neutron mass ($= 1.67 \times 10^{-27}$ kg). On the other hand, the neutron in a medium feels a potential attributed to nuclear interaction because soft matters are normally nonmagnetic. Ignoring neutron absorption, the potential $V$ is described by the integral of the Fermi pseudopotential

$$V = \left(\frac{1}{v_s}\right) \int V(\mathbf{r}) d^3\mathbf{r} = \frac{2\pi\hbar^2 \rho \Sigma b}{m}, \qquad \text{(II.2.2)}$$

where

$$V(\mathbf{r}) = \left(\frac{2\pi\hbar^2}{m}\right) \Sigma b \delta(\mathbf{r}), \qquad \text{(II.2.3)}$$

where $v_s$ is the volume of a system, $\Sigma b$ a sum of coherent scattering length of bound atoms and $\rho$ is the number density for materials, and $\delta(r)$ the delta function. Supposing that neutron is regarded as a plane wave, its reflection behavior can be described as a solution of one-dimensional potential problem, and the Schrödinger equation is expressed as

$$\frac{\hbar^2}{2m}\frac{d^2\Psi}{dz^2} + (E - V)\Psi = 0 \qquad \text{(II.2.4)}$$

or

$$\frac{d^2\Psi}{dz^2} + k_1^2\Psi = 0 \qquad \text{(II.2.5)}$$

$$k_1^2 = \frac{2m(E-V)}{\hbar^2} \qquad \text{(II.2.6)}$$

in which $\Psi$ is a wave function of neutron. A refractive index $n_i$ of material is defined as

$$n_1^2 = \frac{k_1^2}{k_0^2} \qquad \text{(II.2.7)}$$

and is described with scattering length density $\rho\Sigma b$ of the material and $\lambda$ of neutron

$$n_1^2 = 1 - \frac{V}{E} = 1 - \frac{\lambda^2\rho\Sigma b}{\pi}. \qquad \text{(II.2.8)}$$

The $\rho\Sigma b$ value for polymer can be calculated as

$$\rho\Sigma b = \frac{dN_A\Sigma b}{M}, \qquad \text{(II.2.9)}$$

where $d$ is the density (g/cm$^3$) of polymer, $M$ is the molecular weight of monomer unit, and $N_A$ is the Avogadro's number ($= 6.02 \times 10^{23}$ mol$^{-1}$). The values of $\rho\Sigma b$ for typical materials are tabulated in Table II.2.1. Most of the materials in the table possess positive values for $\rho\Sigma b$, so that their $n_i$ is slightly smaller than 1, except for light water and hydrogenous polyethylene with negative $\rho\Sigma b$ values. The effect of neutron absorption can be ignored for most of soft matters because of their small absorption cross section, although the absorption correction needs to be considered for X-ray.

### II.2.2.2 Snell's Refraction Law

The incident ("in"), reflected ("out"), and refracted ("1") components of neutron beam illustrated in Figure II.2.1 can be regarded as a plane wave so that their wave function can be expressed as

$$\Psi_j = a_j \exp(ik_j z) \qquad \text{(II.2.10)}$$

where $j =$ "in," "out," or "1".

**Table II.2.1** Scattering Length Density $\rho\Sigma b$ and the Critical $Q_z$ Value for Total Reflection of Typical Materials

| Material | $\rho\Sigma b$ (nm$^{-2}$) | $Q_{z,c}$ (nm$^{-1}$) |
|---|---|---|
| Si | $2.07 \times 10^{-4}$ | 0.10 |
| $SiO_2$ | $3.47 \times 10^{-4}$ | 0.13 |
| Ni | $9.21 \times 10^{-4}$ | 0.22 |
| $H_2O$ | $-0.56 \times 10^{-4}$ | – |
| $D_2O$ | $6.35 \times 10^{-4}$ | 0.18 |
| *h*PE $(C_2H_4)_n$ | $-0.34 \times 10^{-4}$ | – |
| *d*PE $(C_2D_4)_n$ | $7.1 \times 10^{-4}$ | 0.19 |
| *h*PS $(C_8H_8)_n$ | $1.41 \times 10^{-4}$ | 0.08 |
| *d*PS $(C_8D_8)_n$ | $6.47 \times 10^{-4}$ | 0.18 |
| P2VP $(C_7NH_7)_n$ | $1.95 \times 10^{-4}$ | 0.099 |

*h*PE: poly(ethylene-$h_4$), *d*PE: poly(ethylene-$d_4$), *h*PS: poly(styrene-$h_8$), *d*PS: poly(styrene $d_8$), and P2VP: poly(2-vinylpyridine).

By considering the continuity of $\Psi$ and $\nabla\Psi$ at the interface, the following relations are derived

$$a_1 = a_{\rm in} + a_{\rm out} \tag{II.2.11}$$

$$a_1\mathbf{k}_1 = a_{\rm in}\mathbf{k}_{\rm in} + a_{\rm out}\mathbf{k}_{\rm out}. \tag{II.2.12}$$

Eq. (II.2.12) can be separately written for the components parallel ($x-$) and perpendicular ($z-$) to the interface

$$a_1k_1\cos\theta_1 = a_{\rm in}k_{\rm in}\cos\theta_{\rm in} + a_{\rm out}k_{\rm out}\cos\theta_{\rm out} \tag{II.2.13}$$

$$-a_1k_1\sin\theta_1 = -a_{\rm in}k_{\rm in}\sin\theta_{\rm in} + a_{\rm out}k_{\rm out}\sin\theta_{\rm out}. \tag{II.2.14}$$

Then, the Snell's refractive law, which is the one well known as the property of light, is obtained by substituting $\mathbf{k}_0 \equiv \mathbf{k}_{\rm in} = \mathbf{k}_{\rm out}$, and eqs. (II.2.7) and (II.2.11) into eq. (II.2.13):

$$n_0\cos\theta_0 = n_1\cos\theta_1. \tag{II.2.15}$$

Now assuming that a 0-medium is air or vacuum, $n_0 = 1$ and then eq. (II.2.15) can be redescribed as

$$n_1 = \frac{\cos\theta_0}{\cos\theta_1}. \tag{II.2.16}$$

In general, most of materials possess refractive index $n_i$ slightly smaller than 1, so that they show total reflection with reflectivity of unity when neutrons are impinged on sample surface with the incident angle smaller than the critical angle $\theta_c$. The critical angle $\theta_c$ for total reflection is given by substituting $\theta_1 = 0$ into eq. (II.2.16)

$$\cos\theta_c = n_1 \tag{II.2.17}$$

and also is related with scattering length density $\rho\Sigma b$ of material

$$\theta_c = \lambda\left(\frac{\rho\Sigma b}{\pi}\right)^{1/2}. \tag{II.2.18}$$

Then, the critical value of momentum transfer $Q_z$ ($= |\mathbf{k}_{out} - \mathbf{k}_{in}| = 2k_0 \sin\theta_0 = (4\pi/\lambda)\sin\theta_0$) along $z$-direction for total reflection can be expressed as

$$Q_{z,c} = \left(\frac{4\pi}{\lambda}\right)\sin\theta_c = 4(\pi\rho\Sigma b)^{1/2}. \tag{II.2.19}$$

The $Q_{z,c}$ value for typical materials are also tabulated in Table II.2.1. Because the materials such as light water ($H_2O$) and hydrogenated polyethylene (*h*PE) exhibit negative values for $\rho\Sigma b$, they do not have $Q_{z,c}$, that is, no total reflection is presented.

### II.2.2.3 Reflection for a Single Ideal Interface

To simplify a system, let us suppose a reflection of neutron from a single and ideally flat interface without no roughness first. For this case, eq. (II.2.14) for $z$-direction is rewritten using eqs. (II.2.7) and (II.2.11),

$$(a_{in} + a_{out})n_1 \sin\theta_1 = (a_{in} - a_{out})n_0 \sin\theta_0. \tag{II.2.20}$$

Again, suppose that a 0-medium is air or vacuum, $n_0 = 1$ and the reflection coefficient $r_{0,1}$ for an ideal interface is given as

$$r_{0,1} = \frac{a_{out}}{a_{in}} = \frac{k_0 - k_1}{k_0 + k_1}, \tag{II.2.21}$$

then its reflectivity $R$ is derived as

$$R = r_{0,1}^2 = \left(\frac{k_0 - k_1}{k_0 + k_1}\right)^2. \tag{II.2.22}$$

The $r_{0,1}$ and $R$ for an ideal interface are called as the Fresnel reflection coefficient and the Fresnel reflectivity, respectively. Figure II.2.2 shows a specular reflectivity profile as a function of $Q_z$ for a silicon (Si) substrate with an ideal surface ($\sigma = 0$ nm). The parameter $\sigma$ used as an index of surface and interfacial roughness will be explained later. The reflectivity profiles maintain total reflection region up to $Q_{z,c}$ around 0.1 nm$^{-1}$ for a silicon substrate irrespective of $\sigma$ values, and then rapidly

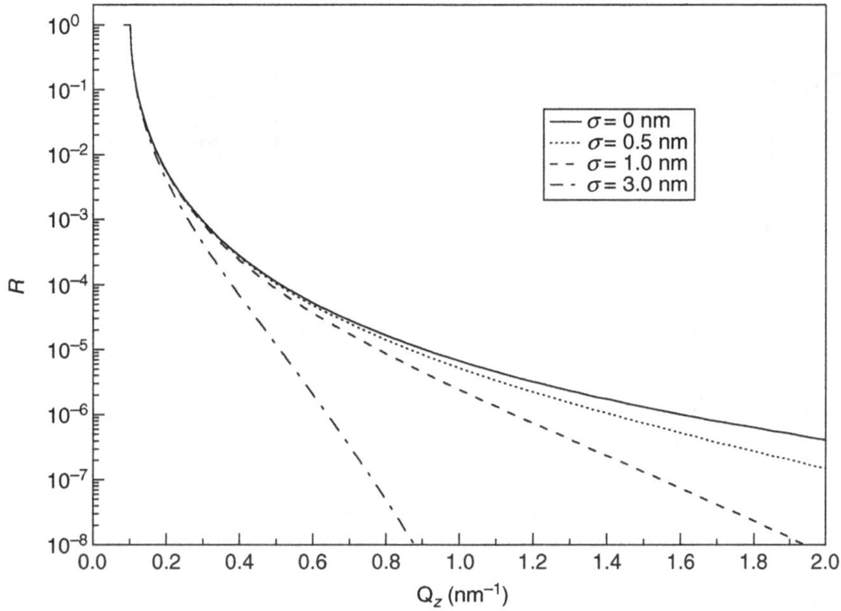


**Figure II.2.2** Specular neutron reflectivity profiles as a function of neutron momentum transfer $Q_z$ along $z$-direction for Si substrates with different surface roughness $\sigma$.

decreases with increasing $Q_z$. The Fresnel reflectivity $R_F$ for an ideal interface shows the dependence of $Q_z^{-4}$, which is well known as the Porod law for small-angle scattering for a flat interface. Therefore, as shown in Figure II.2.3, the reflectivity $R$ multiplied by $Q_z^4$ approaches to the asymptotic constant value at the high-$Q_z$ region.

#### II.2.2.4 Influence of Interfacial Roughness

Although an ideal interface has been discussed up to here, a real material interface is uneven and has more or less roughness. The interfacial roughness influences on reflectivity in different ways, depending on its size relative to neutron coherence length, $l_c$. The $l_c$ value for neutron was experimentally estimated to be approximately a few tens of μm (Richardson et al., 1997; Sferrazza et al., 1997). The interfacial roughness or waviness over a long-range distance larger than $l_c$ causes similar effects to a beam divergence of incident neutron on a flat sample surface. On the other hand, the interfacial roughness over a much shorter-range than $l_c$ is expressed by a smooth interfacial profile and reduces reflectivity very rapidly. Generally, a sample possessing a roughened surface such as a mechanically polished sample or a crystalline polymer film is not adequate for a reflectivity measurement since it loses much neutron intensity by surface roughness.

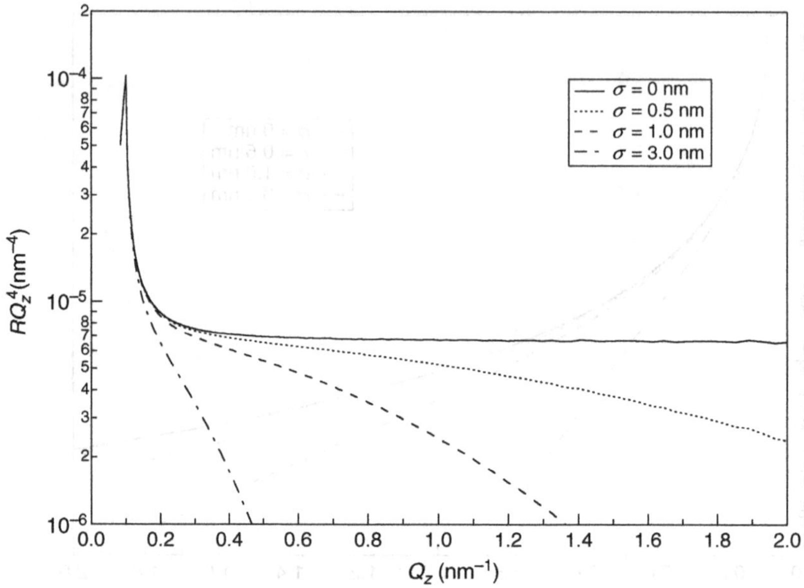


**Figure II.2.3** A plot of $RQ_z^4$ against $Q_z$ for the same specular neutron reflectivities shown in Figure II.2.2.

The Born approximation, in which the effects of multiple scattering are ignored, is adopted to examine the effects of interfacial roughness on neutron reflectivity. Under this approximation, the reflectivity can be described as

$$R(Q_z) = R_{\mathrm{F}}(Q_z)\left|\int\left(\frac{\mathrm{d}P(z)}{\mathrm{d}z}\right)\mathrm{e}^{iQ_z z}\mathrm{d}z\right|^2 \tag{II.2.23}$$

The $P(z)$ is a density profile, so-called interfacial profile, across an interface, obtained by projecting the interface onto the $z$-axis, as illustrated in Figure II.2.4. When the interface possesses roughness, an error function is frequently used as $P(z)$, which is given as

$$P(z) = erf\left(\frac{z}{\sqrt{2}\sigma}\right), \tag{II.2.24}$$

then

$$\frac{\mathrm{d}P(z)}{\mathrm{d}z} = \left(\frac{1}{\sqrt{2\pi\sigma^2}}\right)\exp\left[\frac{-z^2}{2\sigma^2}\right], \tag{II.2.25}$$

which is a form of the Gaussian function with a standard deviation $\sigma$. Here, interfacial thickness $t_{\mathrm{I}}$ is defined as the following

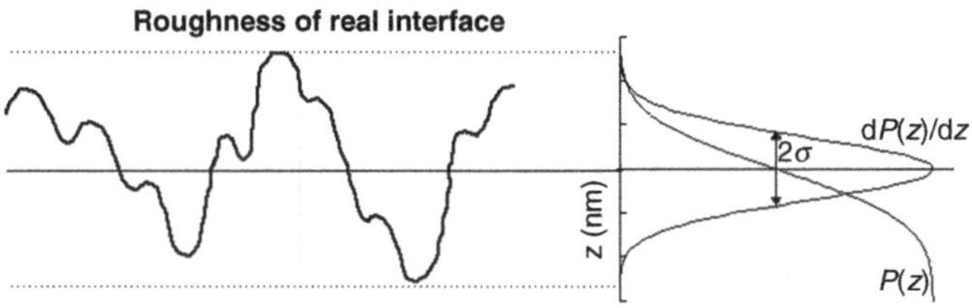


**Figure II.2.4** A schematic illustration of roughness of a real interface and its interfacial profile $P(z)$ and the derivative $\mathrm{d}P(z)/\mathrm{d}z$ with $\sigma$.

$$t_I = \frac{1}{\mathrm{d}P(z)/\mathrm{d}z} \text{ at } P = 0.5. \qquad \text{(II.2.26)}$$

By substituting eq. (II.2.25) into eq. (II.2.26),

$$t_\mathrm{I} = (2\pi)^{1/2}\sigma. \qquad \text{(II.2.27)}$$

On the other hand, by substituting eq. (II.2.25) into eq. (II.2.23), the reflectivity can be expressed as

$$R(Q_z) = R_\mathrm{F}(Q_z)\exp(-Q_z^2\sigma^2). \qquad \text{(II.2.28)}$$

The reflectivity profiles for a silicon substrate with different $\sigma$ values are shown and compared with that for an ideal surface, that is $\sigma = 0$ nm, in Figures II.2.2 and II.2.3. It is noted that the reflectivity profile shows the larger $Q_z$-dependence than the power of -4 at the high-$Q_z$ side, and decreases faster with increasing the value of $\sigma$.

However, eq. (II.2.28) thus derived is not rigorously adequate for describing total reflection region, which is affected much by multiple scattering, because of the presupposition for adopting the Born approximation here. Therefore, in many cases, the interfacial region is regarded as a stacking of numerous layers with an ideal interface between them, of which interfacial profile is approximated to be a step function as illustrated in Figure II.2.5. To calculate reflectivity for this interfacial region, a procedure for a multilayer film, indicated later in this chapter, is adopted.

### II.2.2.5 For a Single-Layer Film

Next, let us consider a single-layer film on a substrate as illustrated in Figure II.2.6. For this case, the thin film has two interfaces: with air (0-medium) and with substrate (2-medium). When neutrons impinge on the film surface from the side of air, interference occurs by a phase difference, corresponding to film thickness $d_\mathrm{i}$, between the neutrons reflected at the film surface and at the interface with the substrate. Parratt (Parratt, 1954) derived the reflection coefficient $r'_{0,1}$ for the film surface considering

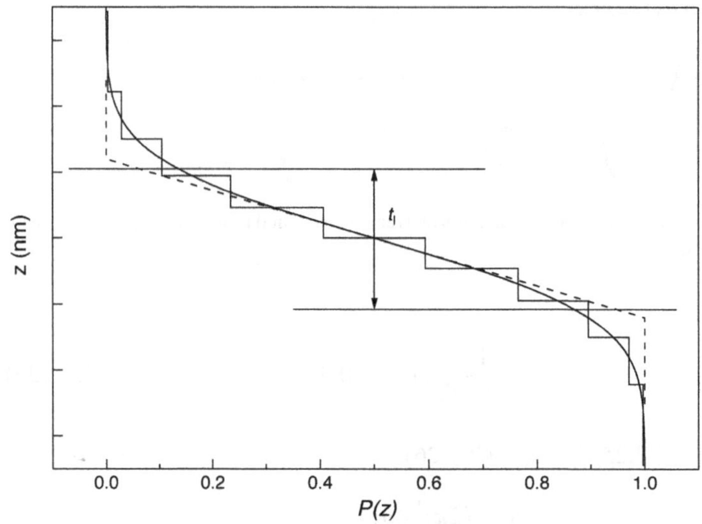


**Figure II.2.5** A drawing of an interfacial profile $P(z)$, its approximate step function expressing several small layers without any roughness between them, and the definition of interfacial thickness $t_I$ given by eq. (II.2.26).

multiple reflections in the film for X-ray:

$$r'_{0,1} = \frac{r_{0,1} + r_{1,2}\exp(2ik_1d_1)}{1 + r_{0,1}r_{1,2}\exp(2ik_1d_1)}, \tag{II.2.29}$$

where $r_{j,j+1}$ expresses the Fresnel reflection coefficient for an ideal interface. The reflectivity $R$ for a single-layer film is derived as

$$R = r'^2_{0,1} = \frac{r^2_{0,1} + r^2_{1,2} + 2r_{0,1}r_{1,2}\cos(2k_1d_1)}{1 + r^2_{0,1}r^2_{1,2} + 2r_{0,1}r_{1,2}\cos(2k_1d_1)}. \tag{II.2.30}$$

The same equation is held for the case of neutron. Figure II.2.7 compares specular reflectivity profiles calculated for a bare silicon substrate, and thin films of deuterated polystyrene ($d$PS, $\rho\Sigma b = 6.47 \times 10^{-4}\,\text{nm}^{-2}$) and hydrogenated

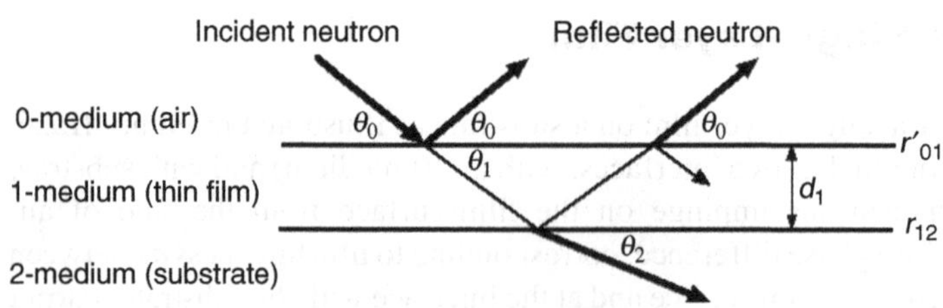


**Figure II.2.6** A schematic drawing of a single-layer film with a thickness $d_i$.

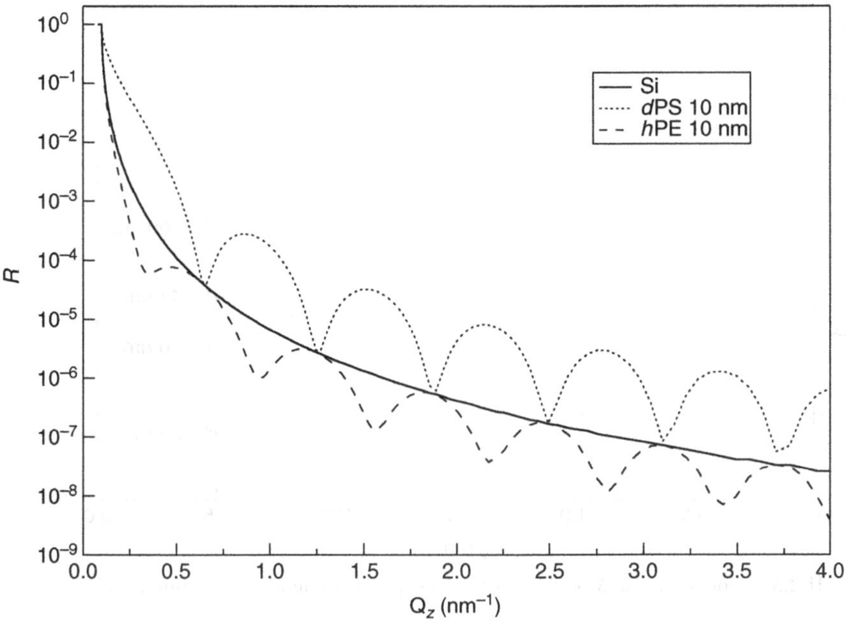


**Figure II.2.7** A profile of $R$ against $Q_z$ for thin films of different polymer with the same film thickness along with the case of a bare Si wafer.

polyethylene (*h*PE, $\rho\Sigma b = -0.34 \times 10^{-4}\,\mathrm{nm}^{-2}$), with the same film thickness of 10 nm, prepared on a Si substrate. Here, the roughness of surface and the interface is not considered into the calculation of reflectivity. The profiles for the two polymer thin films exhibit regular oscillation, the so-called Kiessig fringes, with a constant period corresponding to the film thickness, $d_i$. The film thickness can be evaluated by measuring the frequency of the fringes $\Delta Q_{z,\mathrm{f}}$ precisely as

$$d_i = \frac{2\pi}{\Delta Q_{z,f}}. \qquad \text{(II.2.31)}$$

It is also noted that the fringes for the *d*PS film with a positive $\rho\Sigma b$ value appear in the upper side of the reflectivity profile for the bare Si substrate, while the ones for the *h*PE with a negative value of $\rho\Sigma b$ do in the lower side. Figure II.2.8 also shows specular reflectivity profiles calculated for thin films of the same material with different $d_i$. It is apparent that $\Delta Q_{z,\mathrm{f}}$ of the Kiessig fringes is shortened with increasing $d_i$ according to eq. (II.2.31). It is noted that a measurement resolution in $Q_z$, $\Delta Q_z/Q_z$, defined later in this chapter, is not considered into these calculations of reflectivity. The measurable limit in a small $\Delta Q_{z,\mathrm{f}}$ is determined by the resolution $\Delta Q_z/Q_z$, and poor resolution makes the fringe unclear. The film thickness can be measured up to approximately a few hundreds nm at most with a typical value of $\Delta Q_z/Q_z$ around a few percents.

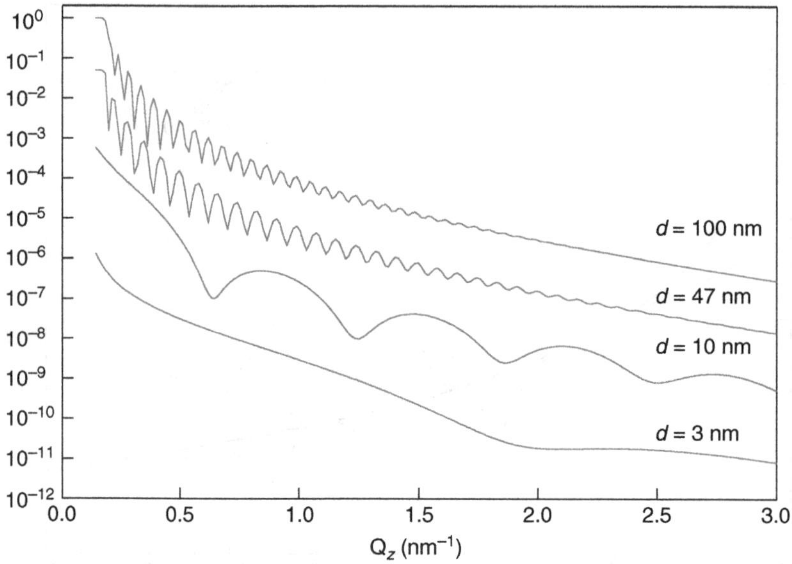


**Figure II.2.8** A comparison in $R$ profiles for the same polymer thin films with different $d_i$.

### II.2.2.6 For a Multilayer Film

For the case of a multilayer film as illustrated in Figure II.2.9, the reflection coefficient $r'_{j,j+1}$ for a arbitral interface dividing two layers in the film is given as

$$r'_{j,j+1} = \frac{r_{j,j+1} + r'_{j+1,j+2}\exp(2ik_{j+1}d_{j+1})}{1 + r_{j,j+1}r'_{j+1,j+2}\exp(2ik_{j+1}d_{j+1})} \tag{II.2.32}$$

considering the contributions of reflection from its lower layers. Therefore, as explained for the case of a single-layer film, the reflection coefficient is calculated from the lowermost interface $r_{n,n+1}$, that is, the interface between the layers of $n$ and $(n+1)$ (a substrate) in Figure II.2.9, to air surface of the film $r'_{0,1}$ in order according to eq. (II.2.32) (a Parratt's recursion algorithm (Parratt, 1954)). Finally, the reflectivity $R$ for a multilayer film is calculated as $r'^2_{0,1}$.

## II.2.3 REFLECTIVITY MEASUREMENT

### II.2.3.1 Instrument

A number of neutron reflectometers are currently working at both types of neutron sources: a reactor and a spallation neutron sources. It is no exaggeration to say that every neutron facility holds at least one reflectometer, and develops structural studies on interfaces and thin films for a variety of materials. The conventional reflectometer has a relatively simple configuration, in which basically a pair of incident slits,

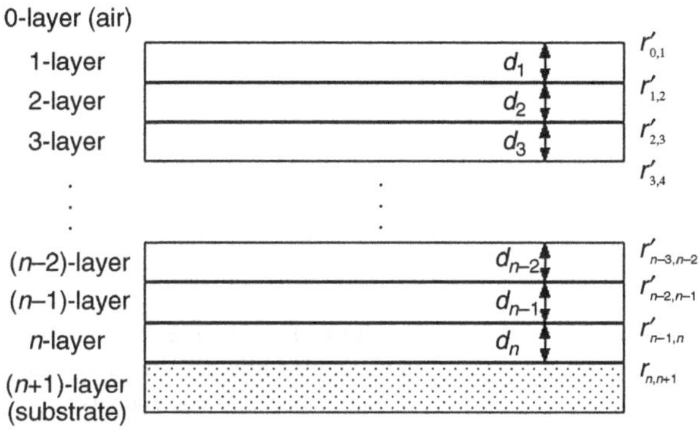


**Figure II.2.9** A schematics of a multilayer sample.

a sample stage and a neutron detector are arranged in order from the upper stream along a neutron beam line irrespective of the type of neutron sources. The slit blades are made of a neutron-absorbing material such as sintered boron carbide ($B_4C$) or cadmium, and the two incident slits with four blades each are separated each other by approximately a few meters to produce a well-collimated beam with a sub-millimeter width in one direction, that is, $z$-direction illustrated in Figure II.2.1. The sample and the detector are mounted on goniometers or translation tables to make them arrange precisely at proper angular position relative to the incident beam. As a detector, a $^3$He proportional counter, or a one- or a two-dimensional position-sensitive detector is mounted at the position approximately a few meters apart from the sample position.

In a conventional specular reflectivity measurement, a variation of neutron intensity reflected from a sample is counted as a function of neutron momentum transfer $Q_z$, defined as $(4\pi/\lambda)\sin\theta$, along the $z$-direction. There are two variables, $\lambda$ and $\theta$, to change the magnitude of $Q_z$. At a reactor source, most reflectometers utilize a constant wavelength ($\lambda$) neutron monochromated by a single crystal such as pyrolytic graphite or beryllium. A so-called $\theta - 2\theta$ scan is made, in which $\theta$ and $2\theta$ are defined as the angles of a sample and a neutron detector, respectively, relative to the incident beam axis to observe specular reflection. The $\theta$ and $2\theta$ are changed together by step keeping a specular angular condition. The reflected neutron intensity measured by a $\theta$–$2\theta$ scan is shown as a function of $\theta$ for a *d*PS thin film in Figure II.2.10. It should be noted that the reflected neutron intensity in total reflection region, that is, in the region of $\theta$ below 0.5°, decreases with decreasing $\theta$ due to a geometrical effect with a very small grazing incidence angle of neutron beam. A few relatively new reflectometers such as D-17 (Institut Laue Langevin (ILL), France) (Cubitt and Fragnet, 2002) and PLATYPUS (Australian Nuclear Science and Technology Organization (ANSTO), Australia) (James et al., 2006) at high-intensity reactor sources can be operated in a time-of-flight (TOF) mode by using a pair of choppers to define the origin of TOF for neutron. On the other hand, the reflect-ometers installed at a spallation neutron source naturally adopt a TOF mode. The pulsed white neutrons with a wide $\lambda$ band, produced at a neutron target by a high-intensity accelerated proton beam, are tightly collimated with a pair of incident slits just before the sample position and then are impinged onto the sample surface at

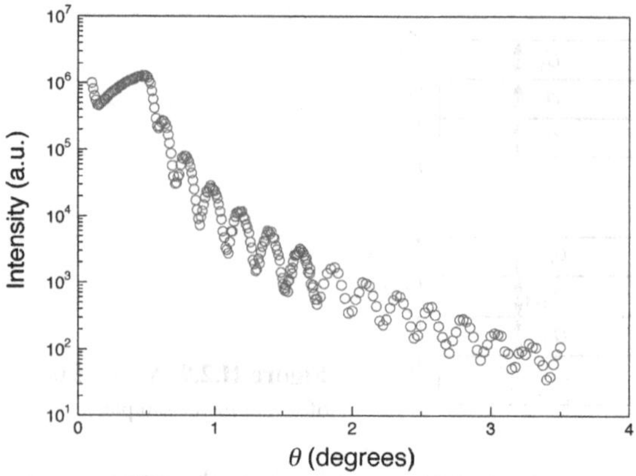


**Figure II.2.10** The variation of neutron intensity reflected from a thin film specimen of homopolymer obtained by a $\theta$–$2\theta$ scan with a constant $\lambda$ mode.

a fixed incident angle. The wavelength of neutrons is determined by measuring a time of flight $t$ when the neutrons produced at the center of neutron target at $t = 0$ s reach at a detector located at the distance $L$ far from the center of neutron target, with the following equation

$$\lambda = \frac{h}{mv} = \frac{ht}{mL} \tag{II.2.33}$$

where $m$ and $v$ are mass and velocity of neutron, respectively. Figure II.2.11 shows incident and reflected neutron spectra for a nickel film with a 50 nm thickness as a function of $\lambda$. The incident neutron spectrum exhibits a peak around $\lambda$ of 0.1 nm. The measurable $Q_z$ range at one incident angle depends on available $\lambda$ range, which is not suffered from flame overlapping that fast neutrons overpass slow ones produced in the precedent flame. The maximum $\lambda$ is provided with a frequency $f$ (Hz) of accelerator operation

$$\lambda_{\max} = \frac{h}{fmL}. \tag{II.2.34}$$

The wider $\lambda$ range of neutrons is available at the longer $f$ of the facility operation. The reflectivity profile is obtained by normalizing the reflected neutron intensity to the incident one, and is plotted together in Figure II.2.11. In an actual measurement at a TOF mode, a few reflectivity profiles are taken at different $\theta$ keeping an angular resolution ($\Delta\theta/\theta$) and an area illuminated on the sample surface unchanged, and then are combined into one profile.

The reflectometers are also categorized by their sample geometry: vertical or horizontal types. On a vertical-type instrument, a sample is vertically mounted, in other words, a reflection plane of the sample is set horizontally. This type of sample geometry is not adequate for the samples with a free interface such as a liquid surface or a liquid/liquid interface, since free interfaces are never hold vertically. Solid samples and solid/liquid interfaces created with a dedicated special cell, except for free interfaces, are measurable on this vertical-type instrument. On the

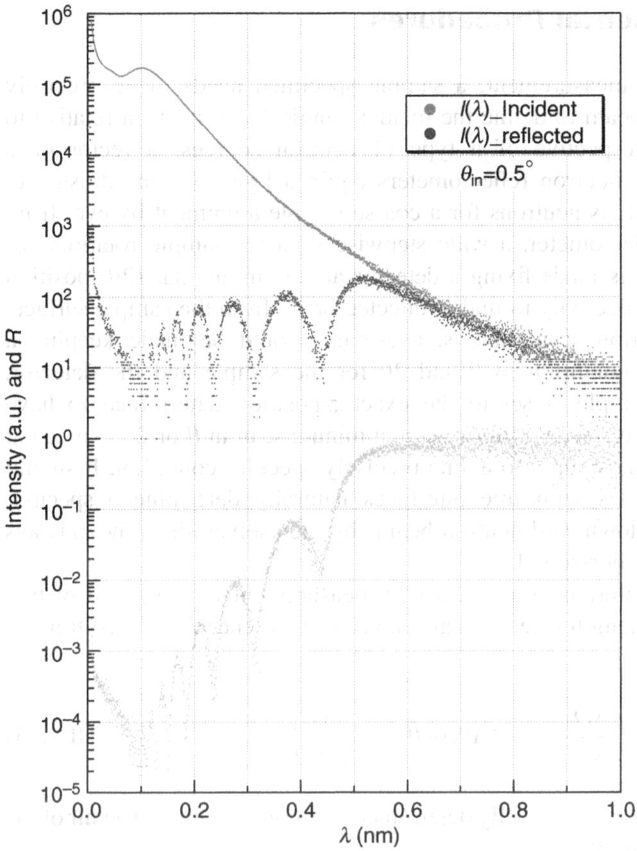


**Figure II.2.11** The spectra of incident as well as reflected neutrons measured for a thin nickel (Ni) film with a TOF mode, and its reflectivity profile as a function of $\lambda$.

other hand, a horizontal-type instrument places a sample just horizontally on its sample stage, that is, the reflection plane is vertical, so that it is suitable for a measurement on free interfaces. However, the beamline design required for the free interface measurement becomes relatively complicated comparing with the vertical type, since the neutron beam has to be guided downward onto the samples with free interfaces with different incident angles. On this type instrument combined with the TOF mode neutron reflection can be measured in a wide range of $Q_z$ at one incident angle without moving a sample at all according to a wide $\lambda$ band of the incident neutrons. The horizontal-type instrument is applicable for almost all types of samples including free interfaces.

Some reflectometers: CRISP (ISIS Facility, UK) (Thomas and Penfold, 1990) at a spallation source, AND/R (NIST, USA) (Dura et al., 2006) at a reactor, and so on are able to produce a polarized neutron beam by using a neutron spin selection device such as a magnetic supermirror, and utilize it for developing the studies on magnetic thin films or artificial lattices as well as soft matters or biomaterials.

### II.2.3.2 Experimental Procedures

For starting reflectivity measurement, a sample specimen needs to be precisely aligned with a neutron beam to define the incident angle $\theta_{in}$ of neutron relative to the sample surface, irrespective of a type of neutron sources: a rector or a spallation source. Some neutron reflectometers equip a laser system adjusted to have the same beam path as neutrons for a coarse sample alignment by eye. If no laser system on the reflectometer, a wide stepwise scan for sample rotation, so-called a $\theta$- or a $\omega$-scan, is made fixing a detector at certain angular ($2\theta$) position having relatively high reflectivity to find a reflected beam from the sample surface. Next, a sample translation scan, that is, a $z$-scan, should be made keeping a predetermined specular condition in $\theta$ and $2\theta$ for the sample and the detector, respectively, and the sample is set to the exact $z$-position determined to have maximum intensity in the $z$-scan. After $z$-scan, a minute scan in $\theta$ or $2\theta$ is made at the fixed $z$-position of the sample to define precisely specular condition. It should be noted that the samples with free interfaces naturally determine a specular angular condition with downward neutron beams by themselves since they always keep the free interfaces horizontal.

Generally, a $Q_z$ resolution in a reflectivity measurement is composed of two contributions from wavelength spread $\Delta\lambda$ and angular divergence $\Delta\theta$ of a collimated neutron beam,

$$\left(\frac{\Delta Q_z}{Q_z}\right)^2 = (\Delta\theta \cot\theta)^2 + \left(\frac{\Delta\lambda}{\lambda}\right)^2. \tag{II.2.35}$$

$\Delta\theta$ of the neutron beam is geometrically determined by a configuration of a pair of the incident slits before a sample,

$$\Delta\theta = \tan^{-1}\left(\frac{W_{S1} + W_{S2}}{2L_1}\right) \tag{II.2.36}$$

where $W_{S1}$ and $W_{S2}$ are opening width of respective incident slits and $L_1$ is the separation distance between them. On the other hand, $\Delta\lambda$ is normally fixed at constant value approximately a few percents in $\Delta\lambda/\lambda$ by a monochrometer, unless a reflectometer adopts a TOF mode at a reactor source. While a reflectometer with a TOF mode, TOF data are acquired with a small time $t$ channel in μs, and rebinned to be constant in $\Delta t/t$, that is, $\Delta\lambda/\lambda$, comparable to the angular resolution.

A reflected neutron intensity is observed as a function of $\theta_{in}$ or $\lambda$ keeping a specular angular condition in $\theta$ and $2\theta$ determined by a combination of prescans of a sample, that is, a $\theta$- or a $2\theta$-scan and a $z$-scan. The incident slit condition is normally adjusted so as to keep an illuminating area of neutron on a sample surface as well as $\Delta\theta$ constant for different $\theta_{in}$. Background intensity is normally measured by placing a sample or a detector at an offset position relative to the specular angular condition, and then subtracted from the reflected neutron counts.

### II.2.3.3 Data Analysis

There exist a few variations in data analysis to obtain a scattering length density $\rho\Sigma b$ distribution along the direction perpendicular to a sample surface from a measured reflectivity profile. However, it should be noted that a "phase" problem missing "phase" information in the measured data is inevitable for reflectivity data analysis as for normal scattering experiments. Also, a limited $Q_z$-range as well as a limited dynamic range for a reflectivity measurement causes ambiguity in the analysis results. Thus, it is experimentally favorable that a reflectivity profile should be measured in the $Q_z$-range and in the range of reflectivity as wide as possible. The knowledge obtained for a sample by the other experimental techniques is quite helpful for the data analysis of reflectivity.

The most general analysis method in reflectometry is a model fitting method. The sample is supposed to consist of sequential stacking of multilayers between which interfaces are an ideal, that is, they are infinitely sharp and have no interfacial thickness. The reflectivity is calculated from the $\rho\Sigma b$ distribution for that multilayer model using an optical matrix method (Born and Wolf, 1970) or a recursion algorithm (Heavens, 1965; Parratt, 1954) given by eq. (II.2.32). In the optical matrix method, a characteristic $2 \times 2$ matrix $M_i$ is defined for the $i$th layer by applying the Maxwell's equations at each interface

$$M_i = \begin{bmatrix} \cos(k_i d_i) & -(i/k_i)\sin(k_i d_i) \\ -ik_i \sin(k_i d_i) & \cos(k_i d_i) \end{bmatrix}, \tag{II.2.37}$$

where $k_i$ is the wave vector of neutron along $z$-direction in the $i$th layer with thickness $d_i$. The reflectivity is expressed from the product of these matrices

$$M = [M_1][M_2][M_3] \ldots [M_n]. \tag{II.2.38}$$

The resultant matrix is written as

$$M = \prod_{i=0}^{i} M_i = \begin{bmatrix} M_{11} & M_{12} \\ M_{21} & M_{22} \end{bmatrix} \tag{II.2.39}$$

and the reflectivity is given by

$$R = \left| \frac{(M_{11} + M_{12}k_{n+1})k_0 - (M_{21} + M_{22})k_{n+1}}{(M_{11} + M_{12}k_{n+1})k_0 + (M_{21} + M_{22})k_{n+1}} \right|^2 \tag{II.2.40}$$

where the subscripts of "$n + 1$" and "0" correspond to a substrate and air, respectively. The reflectivity profile thus calculated and corrected for instrument resolution given as eq. (II.2.35) is fitted to the experimental data through a nonlinear least squares method until a minimum value is obtained for $\chi^2$, that is, a measure of fitting reliability, defined as

$$\chi^2 = \frac{\Sigma(R_{\mathrm{M},i} - R_{\mathrm{C},i})^2}{\sigma_i} \qquad \text{(II.2.41)}$$

where $R_{\mathrm{M},i}$ and $R_{\mathrm{C},i}$ are the measured data with a standard error $\sigma_i$ and calculated reflectivity, respectively, at the $i$th data point.

Another method is a kinematic approximation with a partial structure factor (Lu et al., 1992, 1996, 2000; Lee and Milnes, 1995). This method has been successfully applied to an amphiphilic molecular layer adsorbed at liquid surface. In the kinematic approximation, which is applicable for weak scattering far from total reflection region, specular reflectivity is written as

$$R(Q_z) = \frac{16\pi^2}{Q_z^2} |\rho\Sigma b(Q_z)|^2 \qquad \text{(II.2.42)}$$

where $\rho\Sigma b(Q_z)$ is the one-dimensional Fourier transform of scattering length density distribution, $\rho\Sigma b(z)$, along the $z$-direction perpendicular to a sample surface

$$\rho\Sigma b(Q_z) = \int_{-\infty}^{\infty} \rho\Sigma b(Z) \exp(-iQ_z Z) \mathrm{d}z \qquad \text{(II.2.43)}$$

Using partial structure factors, the contributions from separate parts are described as

$$R(Q_z) = \frac{16\pi^2}{Q_z^2} \left( \Sigma b_i^2 h_{ii}(Q_z) + \Sigma\Sigma 2 b_i b_j h_{ij}(Q_z) \right) \qquad \text{(II.2.44)}$$

where $h_{ii}$ and $h_{ij}$ are the self- and the cross-partial structure factors, respectively, given as

$$h_{ii}(Q_z) = |\rho_i(Q_z)|^2 \qquad \text{(II.2.45)}$$

$$h_{ij}(Q_z) = \mathrm{Re}\{\rho_i(Q_z)\rho_i^*(Q_z)\}^2 \qquad \text{(II.2.46)}$$

and $\rho_i(Q_z)$ is the one-dimensional Fourier transform of number density, $\rho_i(z)$ of the $i$-component along the direction normal to the sample surface. If the adsorbed layer on liquid surface is made up of three components, for instance, a head and a tail group of surfactant, and water, six reflectivity profiles are at least required by utilizing different contrasts to obtain all the partial structure factors. In general, for a sample system with $n$ components, $(1/2)n(n+1)$ reflectivity profiles, that is, different contrast samples including the difference in the contrast between neutron and X-ray, are needed at the minimum.

Moreover, another attempts have been made to analyze specular reflectivity data. Sivia et al. applied maximum entropy and Bayesian spectral analysis (Sivia, 1991a, 1991b; Sivia and Webster 1998) in a model-independent way, and Singh et al. utilized a Fourier-series basis of sine and cosine forms within kinematical theory especially for block copolymer thin films with lamellar microdomains (Singh et al., 1993).

Moreover, Pedersen et al. and Berk et al. independently developed to utilize nonparametric cubic and parametric B-spline curves, respectively (Pedersen, 1992; Pedersen and Hamley 1994; Berk and Majkrzak, 1995). On the other hand, Majkrzak et al. proposed and demonstrated phase determination in different ways: a known reference layer with three different $\rho\Sigma b$, a single buried ferromagnetic layer with a polarized neutron beam, $\rho\Sigma b$ variation of the incident and/or substrate medium for gas/liquid or solid/liquid interfaces (Majkrzak and Berk, 1995, 1998; Majkrzak et al., 2003; Schreyer et al., 1999].

## II.2.4 TYPICAL EXAMPLES

Here, a few typical examples on soft matter researches using conventional specular neutron reflectometry are demonstrated.

The first case is solid thin films of soft matters: a homopolymer, a polymer blend, a block copolymer, a lipid multilayer film, a Langmuir–Blodgett film, and so on, prepared on a solid substrate such as silicon (Torikai et al., 2007) using a spin-coating or a Langmuir–Blodget method. Figure II.2.12 shows a specular neutron reflectivity profile, together with X-ray data, for a microphase-separated diblock copolymer film, composed of deuterated polystyrene (*d*PS) and poly(2-vinylpyridine) (P2VP), which are an incompatible polymer pair showing upper critical solution temperature, prepared by spin-coating from dilute solution on a silicon

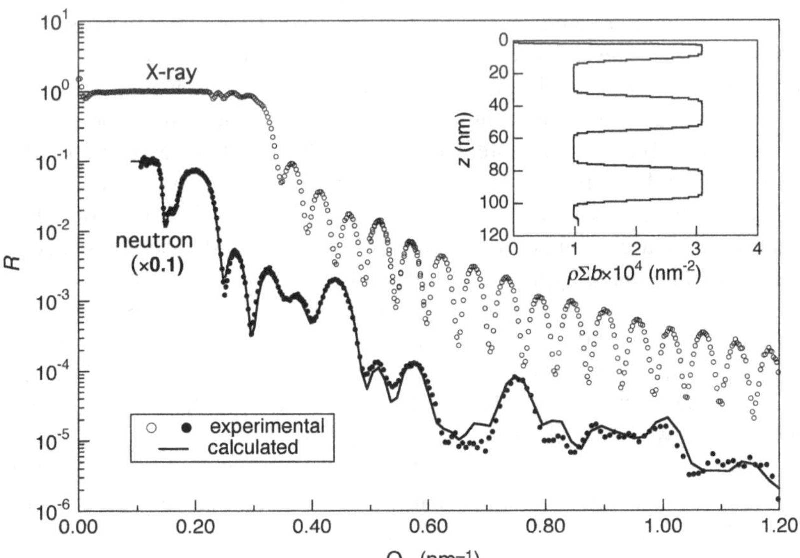


**Figure II.2.12** The specular neutron reflectivity profile along with X-ray one for a phase-separated block copolymer thin film. The inset shows the $\rho\Sigma b$ variation, obtained for the neutron data by a model fitting, along the depth direction.

substrate (Torikai et al., 2000, 2001). The whole PS block chain in a molecule is deuterated to enhance a contrast for neutron between PS and P2VP microdomains. The $\rho\Sigma b$ values inherent for *d*PS and normal hydrogenous one are compared: $6.47 \times 10^{-4}\,\mathrm{nm}^{-2}$ for *d*PS and $1.41 \times 10^{-4}\,\mathrm{nm}^{-2}$ for *h*PS, relative to $1.95 \times 10^{-4}$ $\mathrm{nm}^{-2}$ for P2VP, as tabulated in Table II.2.1. The X-ray profile reveals the Kiessig fringes with a regular frequency like a reflectivity profile for a uniform single-layer film, indicating that the difference in electron density between *d*PS and P2VP is not enough for X-ray to distinguish them. On the other hand, for neutron due to the enhancement in $\rho\Sigma b$ by deuterium labeling of the PS microdomains the profile shows a few distinct Bragg peaks reflecting the formation of periodical lamellar microphase-separated structure preferentially oriented along the direction parallel to the film surface. The solid line on the neutron data (closed circles) is the best-fitted reflectivity profile calculated by the Parratt's recursion algorithm with a depth $\rho\Sigma b$ variation shown in the inset, where 2.5 alternative stacking of *d*PS and P2VP lamellar microdomains is formed with a repeating distance of 43 nm in a 110 nm thick film. The interfacial thickness, $t_\mathrm{I}$, between the microdomains, defined in eq. (II.2.26), was evaluated to be about $3.3 \pm 0.3$ nm. The evaluated value of $t_\mathrm{I}$ was much larger than the predicted interfacial thickness by a mean-field theory, and this discrepancy was interpreted quantitatively by considering an existence of thermal fluctuations on the phase-separated interface. And the same kinds of experiments for block copolymer interface were performed and compared interfacial structures among the block copolymers with different molecular architectures (Torikai et al., 2000, 2001).

Next, an example for the studies on free interfaces, that is, air/liquid, or liquid/liquid interface, is demonstrated. A trough usually coated with Teflon is directly mounted on a sample stage of a horizontal neutron reflectometer, and an *in situ* reflectivity measurement is made on an adsorbed Gibbs monolayer or a spread Langmuir monolayer on air/water interface controlling temperature, or surface pressure with a movable barrier system (Lu et al., 2000). Here, earlier specular neutron reflection studies are shown on one of the most successfully examined surfactant systems, hexadecyltrimethylammonium bromide (C16TAB), adsorbed on air/water interface by a combination of deuterium labeling and a partial structure factor method in a kinematic approximation (Lu et al., 1994, 1995a). To determine the relative position and the distribution width of an alkyl chain (C16), a head group of trimethylammonium (TAB) and water in the adsorbed layer, the following six different deuterium-labeled samples were prepared: *d*C16*h*TAB in null reflecting water (NRW) and $D_2O$, "0"C16*h*TAB in $D_2O$, *d*C16*d*TAB in NRW and $D_2O$, and "0"C16*d*TAB in NRW, where the prefixes *h*, *d*, and "0" indicate that each part is composed of hydrogenous, fully deuterated species and the blend of *h*- and *d*-species in a appropriate ratio so as to make its average scattering length zero, respectively. Also, NRW is a roughly 1 : 9 mixture of $D_2O$ and $H_2O$, of which average $\rho\Sigma b$ value is comparable to air, that is, $0\,\mathrm{nm}^{-2}$. Under a kinematic approximation, the reflectivity $R(Q_z)$ is given by the contribution from each portion in the adsorbed layer, that is, an alkyl chain (C) and a head group (H) of a surfactant, and water (S), as

$$R(Q_z) = \frac{16\pi^2}{Q_z^2}(b_C^2 h_{CC}(Q_z) + b_H^2 h_{HH}(Q_z) + b_S^2 h_{SS}(Q_z)$$
$$+ 2b_C b_H h_{CH}(Q_z) + 2b_C b_S h_{CS}(Q_z) + 2b_H b_S h_{HS}(Q_z)), \quad \text{(II.2.47)}$$

which is composed of the self-terms $h_{ii}(Q_z)$ and the cross-terms $h_{ij}(Q_z)$. The self-partial structure factors $h_{ii}(Q_z)$ are described by the width, $\sigma_i$, and amplitude, $n_{i0}$, of the distribution $n_i(z)$ supposing the Gaussian distribution for surfactant fragments

$$n_i(z) = n_{i0} \exp\left(\frac{-4z^2}{\sigma^2}\right), \quad \text{(II.2.48)}$$

which gives

$$h_{ii}(Q_z) = |n_i(Q_z)|^2 = \left(\frac{\pi\sigma_i^2 n_{i0}^2}{4}\right)\exp\left(\frac{-Q_z^2\sigma_i^2}{8}\right). \quad \text{(II.2.49)}$$

$n_i(Q_z)$ is the one-dimensional Fourier transform of $n_i(z)$. The value of $n_{i0}$ is related to the surface excess $\Gamma_i$ as

$$\Gamma_i = \frac{1}{A_i} = \frac{\sigma_i n_{i0}\pi^{1/2}}{2} \quad \text{(II.2.50)}$$

where $A_i$ is the occupied area for a surfactant molecule in the adsorbed layer. For solvent, the distribution $n_S(z)$ is expressed as

$$n_S(z) = n_{S0}\left(\frac{1}{2} + \frac{1}{2}\left[\tanh\left(\frac{z}{\zeta}\right)\right]\right) \quad \text{(II.2.51)}$$

where $n_{S0}$ and $\zeta$ are a bulk number density of solution and a width parameter of solvent distribution at air/water interface, respectively. The corresponding partial structure factor is

$$h_{SS}(Q_z) = n_{S0}^2\left(\frac{\zeta\pi}{2}\right)^2 \operatorname{cosech}^2\left(\frac{\zeta\pi Q_z}{2}\right). \quad \text{(II.2.52)}$$

On the other hand, cross-partial structure factors $h_{ij}(Q_z)$ are given as

$$h_{CS}(Q_z) = \pm[h_{CC}(Q_z)h_{SS}(Q_z)]^{1/2}\sin(Q_z\delta_{CS}) \quad \text{(II.2.53)}$$

$$h_{CH}(Q_z) = \pm[h_{CC}(Q_z)h_{HH}(Q_z)]^{1/2}\cos(Q_z\delta_{CH}) \quad \text{(II.2.54)}$$

derived from

$$h_{ij}(Q_z) = Re[n_i(Q_z)n_j{}^*(Q_z)\exp\{-iQ_z(\delta_i - \delta_j)\}] \qquad \text{(II.2.55)}$$

between the two distributions centered at $\delta_i$ and $\delta_j$, when $n_C(z)$ and $n_H(z)$ are exactly described by functions symmetrical about their centers while $n_S(z)$ is exactly an odd function (Simister et al., 1992; Lu et al., 1992). Figure II.2.13a and b demonstrates self-partial structure factors, $h_{CC}(Q_z)$ and $h_{HH}(Q_z)$, of the alkyl chain and the head group, respectively, at three concentrations of C16TAB less than its critical micelle concentration (CMC) $\sim 9 \times 10^{-4}$ M, along with the fitting results using the Gaussian distributions eq. (II.2.49). It was clarified that the width of both alkyl chain and head group distributions distinctly decreases, with decreasing surface coverage, that is, with increasing $A_i$. On the other hand, Figure II.2.13c shows cross-partial structure factors $h_{CH}(Q_z)$ with the fits of

$$h_{CH}(Q_z) = \frac{1}{A_i^2}\exp\left[-Q_z^2\left(\frac{\sigma_C^2 + \sigma_H^2}{16}\right)\right]\cos(Q_z\delta_{CH}) \qquad \text{(II.2.56)}$$

by substituting the Gaussian distributions obtained above as fitting results for $h_{CC}(Q_z)$ and $h_{HH}(Q_z)$ into eq. (II.2.54). A similar equation can be derived for $h_{iS}(Q_z)$ as

$$h_{iS}(Q_z) = \left(\frac{n_0}{A_i}\right)\exp\left[\frac{-Q_z^2\sigma_i^2}{16}\right]\left(\frac{\zeta\pi_z}{2}\right)\text{cosech}\left(\frac{\zeta\pi Q_z}{2}\right)\sin(Q_z\delta_{iS}). \qquad \text{(II.2.57)}$$

The values of $A_i$, $\sigma_C$, and $\sigma_H$ in eq. (II.2.56) are already known from the self-partial structure factors $h_{ii}(Q_z)$, so that the $\delta_{CH}$ value can be determined by fitting eq. (II.2.56) to the $h_{CH}(Q_z)$ profiles as shown in Figure II.2.13c. The obtained $\delta_{CH}$ values vary from 0.8 to 0.5 nm in the concentration range studied. Further, the detailed structural analysis was made using the C16TAB with a partially deuterated alkyl chain, that is, *d*C6*h*C10TAB and *h*C10*d*C6TAB. For this, the following five additional samples were prepared: *d*C6"0"C10*h*TAB in NRW and $D_2O$, "0"C10*d*C6*h*TAB in NRW and $D_2O$ and a 1:1 mixture of the two partially deuterated chain compounds in NRW. The obtained volume fraction profiles are shown for the different parts of C16TAB layer adsorbed on air/water interface in Figure II.2.14, and the properties of the adsorbed C16TAB layer, that is, capillary wave, tilt angle of and gauche defects in the alkyl chain, were discussed. Later, further detailed experiments were conducted with the C16TAB with a partially deuterated alkyl chain of more subdivided four atoms ($-C_4D_8-$ or $-C_4D_9$) and two atoms ($-C_2D_4-$ or $-C_2D_5$) (Lu et al., 1995b).

In a measurement on a solid/liquid interface, neutron reflection is observed through relatively transparent materials for neutron such as silicon or quartz by guiding a neutron beam onto the objective interface from the side face of thick ($\sim$10 mm) block of Si or $SiO_2$. The transmission of neutron through Si or $SiO_2$ with 60 mm path length is estimated to be about 0.5 or 0.2, respectively. This geometry of a solid–liquid interface is utilized to prepare various sample environments: a high-

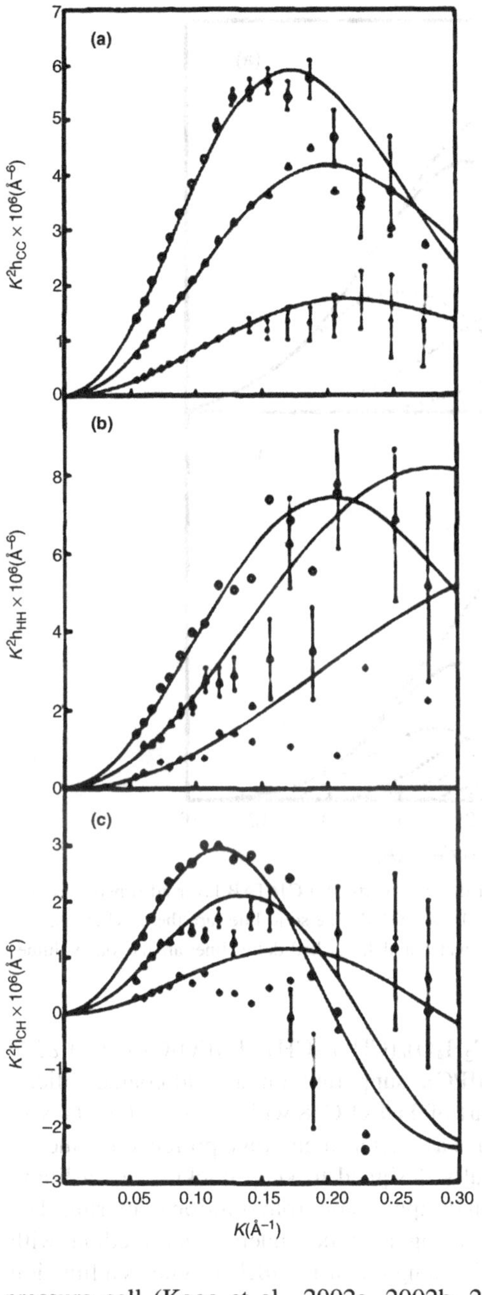


**Figure II.2.13** The partial structure factors of two self-terms (a) $h_{CC}$ and (b) $h_{HH}$, and a cross-term (c) $h_{CH}$ as a function of $\kappa$ ($=Q_z$) for C16TAB at three concentrations, that is, ○ (circle): $9.1 \times 10^{-4}$ M, △: $2.75 \times 10^{-4}$ M, +: $0.7 \times 10^{-4}$ M. The continuous lines in (a) and (b) are a fit using the Gaussian distribution with $\sigma$, while the one in (c) is corresponding to a fit using eq. (II.2.56) (Lu et al., 1994).

pressure cell (Koga et al., 2002a, 2002b, 2003), a flow cell (Baker et al., 1994; Hamilton et al., 1994, 1999), an electrochemical cell as a working electrode (Cooper et al., 2004), and so on. As an example concerning the solid/liquid interface, *in situ* neutron reflectometry is described on a lipid membrane in a wet condition with water. Krueger et al. (2001) demonstrated that a hybrid bilayer membrane (HBM) of

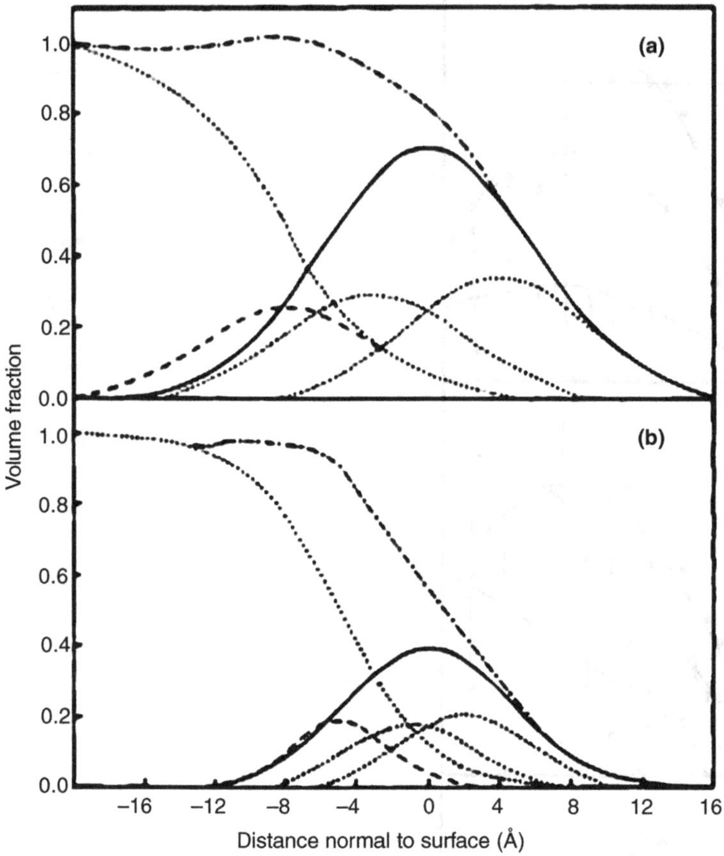


**Figure II.2.14** The volume fraction profiles of different parts of a C16TAB layer adsorbed at water surface at the concentrations of (a) CMC and (b) $0.7 \times 10^{-4}$ M. The solid lines are the alkyl chain, the dashed lines are the head group, the dotted lines are water, and the dashed-dotted lines are the total volume fraction (Lu et al., 1994).

thiahexa(ethylene oxide) alkane ($HS(C_2H_4O)_6(CH_2)_{17}CH_3$, THEO-C18) and *d*54-dimyristoyl phosphatidylcholine (*d*DMPC), supported on a gold-coated silicon substrate through interaction of thiol part of THEO-C18 with a gold, interacts with melittin as a model peptide for antibiotics as well as membrane proteins in aqueous solution. A solid/liquid cell was carefully designed to keep thickness of a liquid reservoir as thin as possible for background suppression from solution scattering. The reflectivity profiles were analyzed by using a model-independent method with parametric B-splines to describe scattering length density, $\rho\Sigma b$, profile as a function of $z$. The obtained families of $\rho\Sigma b$ profiles, which are equally good fits to the reflectivity profiles down to $10^{-8}$ for reflectivity in a $Q_z$ range up to 7 nm$^{-1}$ shown in the inset, for the HBMs with and without melittin in $D_2O$ solution are compared in Figure II.2.15. This clearly shows that the presence of melittin affects the $\rho\Sigma b$ profile for the HBMs: the displacement of $D_2O$ from a lipid head group region to

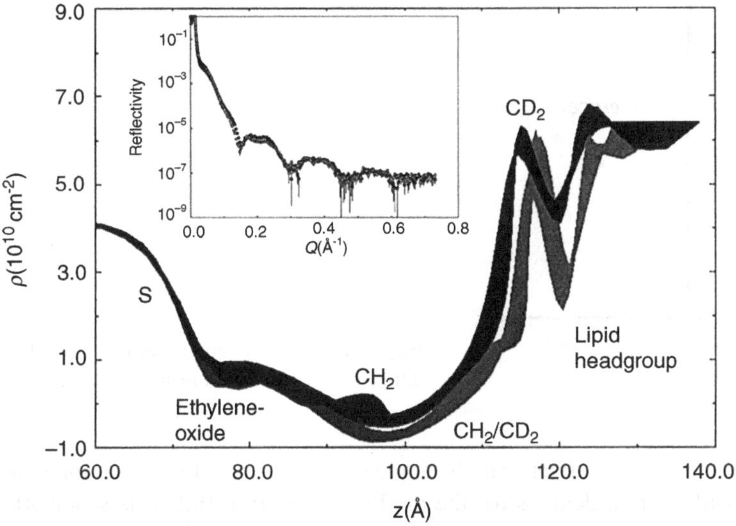


**Figure II.2.15** The families of scattering length density $\rho$ profile obtained for a hybrid bilayer membrane of THEO-C18 and *d*DMPC with (light gray shaded) and without (dark gray shaded) the presence of melittin in $D_2O$ by a model-free method with parametric B-splines. The inset shows specular neutron reflectivity profiles for the ones with (□) and without (•) melittin (Krueger et al., 2001).

accommodate melittin molecule lowered the corresponding $\rho\Sigma b$ value significantly, and the change in alkane chain length shifted the position of interface region with the lipid acyl chain (–$CD_2$–) to higher $z$ value by as much as 0.3 nm.

## II.2.5 OFF-SPECULAR REFLECTION AND GRAZING INCIDENCE SMALL-ANGLE SCATTERING

The preceding sections were focused on a conventional specular reflection technique, which is a powerful method to probe a scattering length density, $\rho\Sigma b$, variation along the direction perpendicular to interface, but is not necessarily sensitive to its in-plane or lateral structure of the sample. Off-specular reflection is observed at the reflection angle $\theta_{out}$, which is not equal to the incident angle $\theta_{in}$ of neutron as shown in Figure II.2.1, according to surface and interfacial roughness or in-plane structures of samples. However, the off-specular reflection possesses generally quite weak intensity, so that it has not been much performed for neutron compared with X-ray due to a limitation in incident neutron flux resulting in poor data statistics so far. For the reflectometer using the neutron with constant $\lambda$ at a reactor source, the off-specular reflection is measured by making a so-called rocking scan, a detector scan or offset scan. In the rocking scan, the angle of a sample relative to the incident neutron is changed step by step keeping the detector position ($2\theta$) fixed, while in the detector scan the detector angle ($2\theta$) is changed around a specular position or a one-dimensional position-sensitive detector (PSD) is used keeping the sample angle

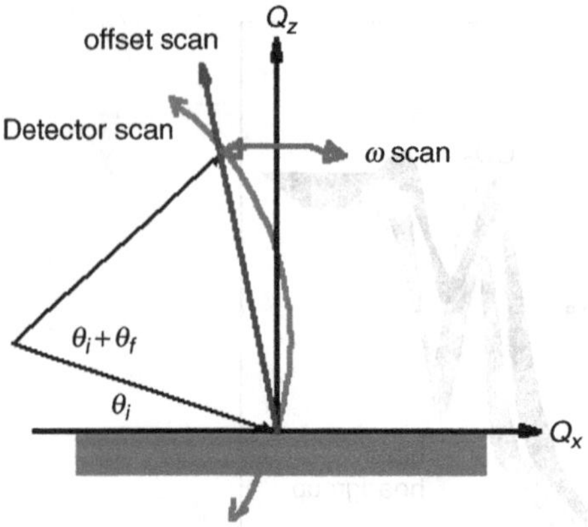


**Figure II.2.16** A schematic drawing of loci for different scan modes in a reciprocal space.

($\theta$ or $\omega$) fixed. Further in the offset scan the angle of a sample relative to the incident beam is offset with certain degrees for the $\theta-2\theta$ scan. Figure II.2.16 is schematic representation of a locus for the off-specular scans in a reciprocal space, where $Q_x$ along the $x$-direction, that is, along the direction parallel to sample surface, is geometrically defined as

$$Q_x = \left(\frac{2\pi}{\lambda}\right)(\cos\theta_{\text{out}} - \cos\theta_{\text{in}}). \tag{II.2.58}$$

On the other hand, a time-of-flight reflectometer using neutrons with a wide $\lambda$ band measures off-specular reflection simultaneously with specular reflection using a one-dimensional PSD. In Figure II.2.17a, a map of reflected neutron intensity is shown for a microphase-separated block copolymer thin film with lamellar microdomains using a one-dimensional PSD combined with a TOF mode at the fixed incident angle $\theta_{\text{in}}$ of 0.6° (Torikai et al., 2007). This map is converted to the one in a reciprocal space of $Q_x$–$Q_z$ as shown in Figure II.2.17b. A few Bragg peak spots are observed along the specular reflection ridge at $Q_x = 0\,\text{nm}^{-1}$, along which normally a specular reflectivity profile is observed. In this case, weak streaks are crossing the specular reflection ridge at the position of the Bragg peaks, implying that microphase-separated interfaces are roughened with some spatial correlations between adjacent interfaces (conformal roughness). The off-specular reflection is quantitatively analyzed based on the concept of distorted wave Born approximation (DWBA). The DWBA is applicable for a region near total reflection, in which Born approximation is not hold due to severe effects of multiple scattering. The formalisms have been extensively developed for off-specular reflection on rough surfaces and roughness correlations in multilayer systems (Sinha et al., 1988; Pynn, 1992; Dietrich and Haase, 1995; Ljungdahl and Lovesey, 1996).

Recently, a GISAS method attracts many interests to explore *nano*scale non-layered structures at interfaces or in thin films. The GISAS geometry is schematically

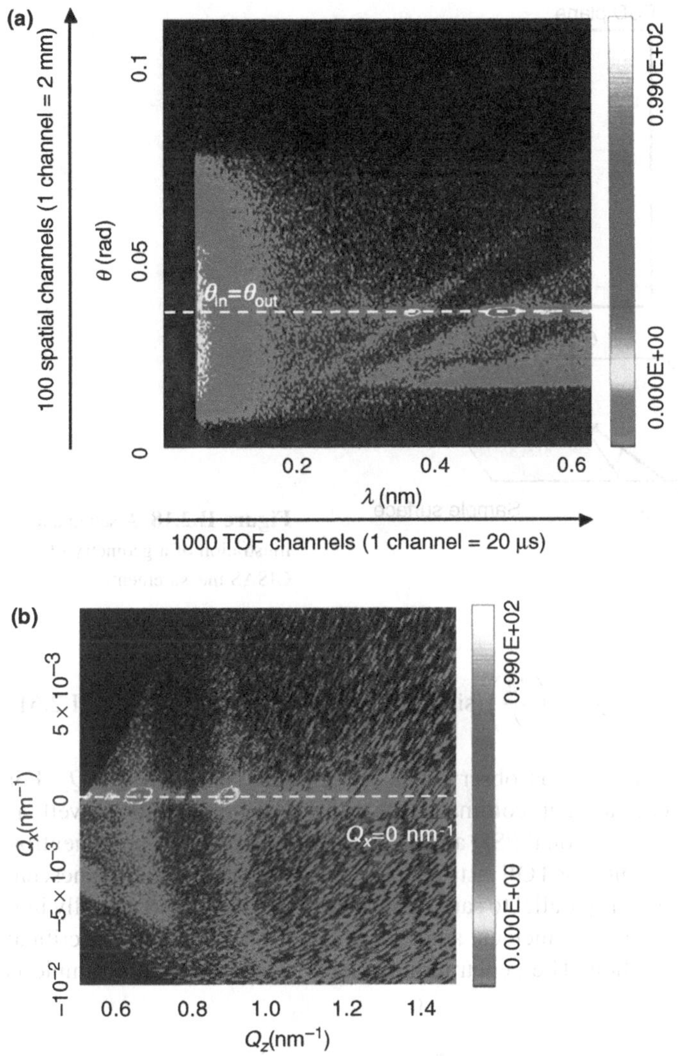


**Figure II.2.17** The neutron intensity map of off-specular reflection from a phase-separated block copolymer thin film in (a) $\theta_{out} - \lambda$ and (b) $Q_x - Q_z$ spaces. The broken lines in these figures are corresponding to a specular reflection ridge, that is, $\theta_{in} = \theta_{out}$ or $Q_x = 0\,\text{nm}^{-1}$.

drawn in Figure II.2.18, where each component of scattering vector is defined as

$$Q_x = \left(\frac{2\pi}{\lambda}\right)(\cos\theta_{out} - \cos\theta_{in}\cos\Psi), \tag{II.2.59}$$

$$Q_y = \left(\frac{2\pi}{\lambda}\right)\cos\theta_{in}\cos\Psi, \tag{II.2.60}$$

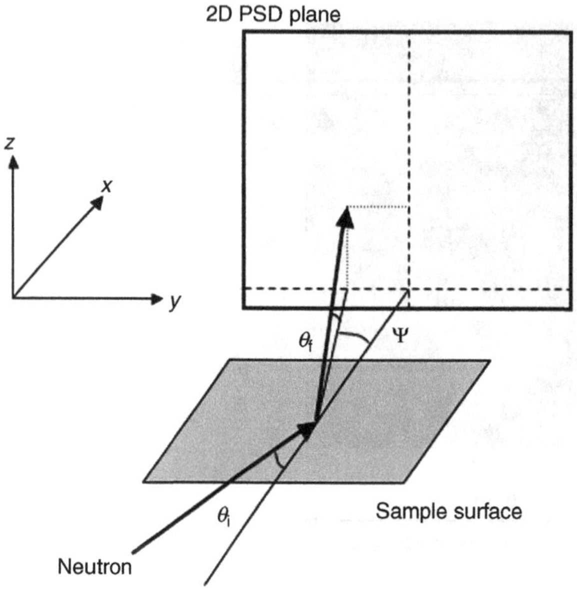


**Figure II.2.18** A schematic illustration of a geometry of GISAS measurement.

$$Q_z = \left(\frac{2\pi}{\lambda}\right)(\sin\theta_{\text{out}} + \sin\theta_{\text{in}}), \tag{II.2.61}$$

and the small-angle scattering is observed in a reciprocal plane of $Q_y$–$Q_z$. For a GISAS measurement, a tight collimated beam along $y$-direction as well as $z$-direction, and a two-dimensional PSD are needed irrespective of a type of neutron instrument, that is, constant-$\lambda$ or TOF instrument. This method utilizes an evanescent neutron wave, which travels parallel to sample surface and decays exponentially into the sample on condition that the incident angle, $\theta_{\text{in}}$, of neutron is less than the critical angle, $\theta_c$, for total reflection. The penetration depth $\Lambda$ of neutron into a sample is given as

$$\Lambda = \frac{\lambda}{\sqrt{2}\pi}\left\{\sqrt{(\theta_0^2-\theta_c^2)^2+4\beta^2}-(\theta_0^2-\theta_c^2)\right\}^{-1/2}, \tag{II.2.62}$$

where $\beta$ is absorption term, which is not considered into eq. (II.2.8), in the definition of refractive index $n$

$$\beta = \frac{\lambda}{4\pi}\rho(\sigma_{\text{inc}} + \sigma_{\text{a}}). \tag{II.2.63}$$

where $\sigma_{\text{inc}}$ and $\sigma_{\text{a}}$ are incoherent and absorption cross sections, respectively. Figure II.2.19 shows the $\Lambda$ variation calculated for $D_2O$ with different $\lambda$ values, 0.25 and 0.9 nm, of neutron as a function of $\theta_{\text{in}}$ normalized to $\theta_c$. This demonstrates that $\Lambda$ of neutron varies with changing $\theta_0$, that is, an exploring depth is controllable in

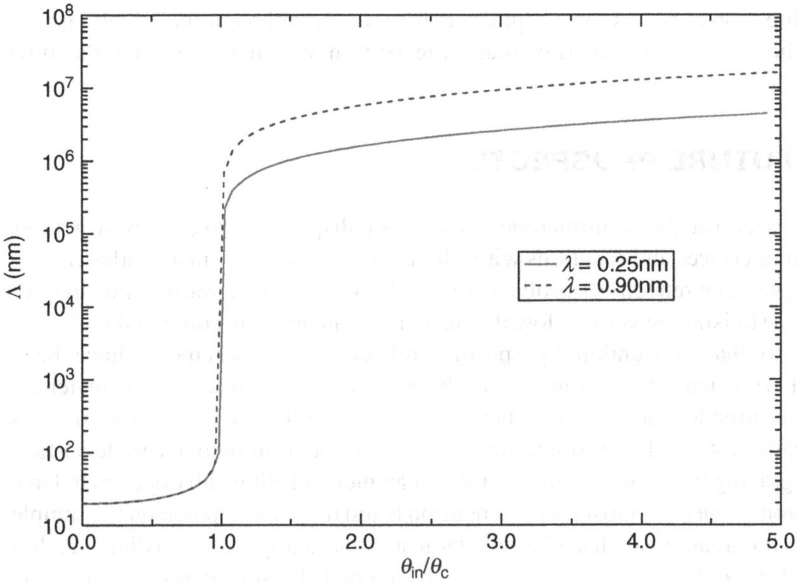


**Figure II.2.19** A penetration depth Λ of neutron with different $\lambda$ for $D_2O$ as a function of the incident angle $\theta_0$ normalized to the critical one $\theta_c$ for total reflection.

a GISAS measurement. One of the earliest pioneering studies (Hamilton et al., 1994, 1999) using a GISANS (or a near-surface small-angle neutron scattering (NS-SANS)) technique investigated the decay kinetics of a near-surface hexagonal phase for a thread-like micelle under Poiseuille shearing flow at the interface between quartz and solution. The scattering patterns at different time after flow cessation are compared in Figure II.2.20. The clear hexagonal ordering of micelles aligned along

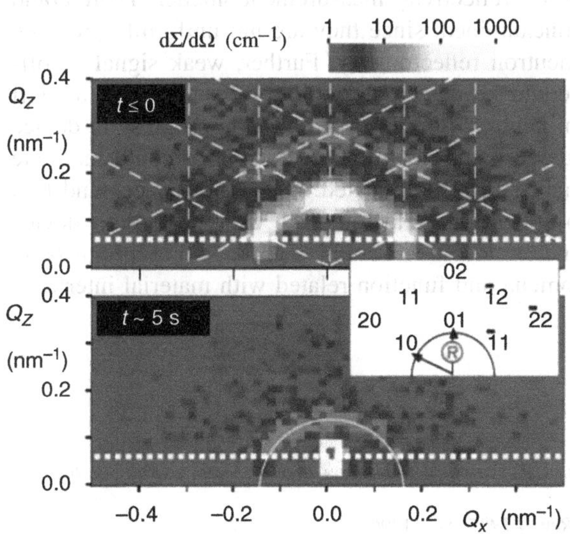


**Figure II.2.20** The two-dimensional GISAS patterns of a thread-like micelle for the fully aligned ($t < 0$ s) and partially relaxed ($t \sim 5$ s) near surface after a cessation of Poiseuille flow (Hamilton et al., 1999).

the flow direction at $t < 0$ s loses rapidly its local order, although the micelle is still aligned, showing a diffusive ring scattering pattern within $t \sim 5$ s after the flow cessation.

## II.2.6 FUTURE PROSPECTS

It is widely accepted that neutron reflectometry is indispensable to explore structures of material interfaces and thin films with a high spatial resolution in a nondestructive way. However, one reflectivity profile takes a relatively long measuring time to have enough data statistics because of low flux in an incident neutron, compared with light or X-ray, so that conventionally specular reflectivity measurements have been conducted with neutron to examine equilibrium or frozen structures of materials in a depth direction so far. The beam size for a reflectivity measurement is geometrically restricted by a small incident angle of neutron onto a sample surface due to the grazing incidence geometry. Even if an incident slit is fully opened to have more neutron counts, the most part of a neutron beam does not impinge on the sample surface. So far, a sample with a relatively large area, typically ~5 cm in diameter, has been used for a reflectivity measurement to have good statistics in reflected neutron intensity. Consequently, the use of the higher flux neutron as an incident beam is advantageous for reflectivity measurements. The recent advent of high-power neutron sources: ILL (France), FRM-II (Germany), OPAL (Australia) for a reactor source, and ISIS (UK), SNS (USA), and J-PARC (Japan) for a spallation source brings much gain in a neutron flux. All these neutron sources possess a few dedicated instruments for neutron reflectometry. A time-resolved measurement with a very short time slice could become mainstream in neutron reflectometry to elucidate nonequilibrium or kinetic phenomena such as interfacial diffusion, adsorption, and phase separation changing their structures with a time. The neutron flux gain also makes a sample size required for a reflectivity measurement smaller. There could have remained many samples unexamined, since they are not uniformly prepared over a large sample area for neutron reflectometry. Further, weak signal of off-specular reflection or GISAS could be detectable with good statistics in a short measurement time by a gain in neutron flux combined with a progress in device developments such as optical devices and neutron detectors. That could facilitate three-dimensional structural analysis for complicated material interfaces and thin films. It is expected that those neutron reflectometers with up-to-date device developments at high-power neutron sources could lead to a big progress in understanding of various phenomena and function related with material interfaces and thin films.

## REFERENCES

Baker, S.M., Smith, G., Pynn, R., Butler, P., Hayter, J., Hamilton, and W., Maid, L. *Rev. Sci. Instrum.* **1994**, *65*, 412.

Berk, N.F. and Majkrzak, C.F. *Phys. Rev. B* **1995**, *51*, 11296.

Born, M. and Wolf, E. Principles of Optics, Pergamon Press, Oxford, 1970.

Cooper, J.M., Cubitt, R., Dalgliesh, R.M., Gadegaard, N., Glidle, A., Hillman, A.R., Mortimer, R.J., Ryder, K.S., and Smith, E.L., *J. Am. Chem. Soc.* **2004**, *126*, 15362.

Cubitt, R. and Fragnet, G., *Appl. Phys. A* **2002**, *74* (Suppl.), S329.

Dietrich, S. and Haase, A. *Phys. Rep.* **1995**, *260*, 1.

Dura, J.A., Pierce, D.J., Majkrzak, C.F., Maliszewskyj, N.C., McGillivray, D.J., Losche, M., O'Donovan, K.V., Mihailescu, M., Perez-Salasa, U., Worcester, D.L., and White, S.H. *Rev. Sci. Instrum.* **2006**, *77*, 074301.

Hamilton, W.A., Butler, P.D., Baker, S.M., Smith, G.S., Hayter, J.B., Magid, L.J., and Pynn, R. *Phys. Rev. Lett.* **1994**, *72*, 2219.

Hamilton, W.A., Butler, P.D., Magid, L.J., Han, Z., and Slawecki, T.M. *Phys. Rev. E* **1999**, *60*, R1146.

Heavens, O.S. Optical Properties of Thin Solid Films, Dover Publications, 1965.

James, M., Nelson, A., Brule, A., and Schulz, J.C. *J. Neutron Res.* **2006**, *14*, 91.

Koga, T., Seo, Y.-S., Zhang, Y., Shin, K., Kusano, K., Nishikawa, K., Rafailovich, M.H., Sokolov, J.C., Chu, B., Peiffer, D., Occhiogrosso, R., and Satija, S.K. *Phys. Rev. Lett.* **2002a**, *89*, 125506.

Koga, T., Seo, Y.-S., Hu, X., Shin, K., Zhang, Y., Rafailovich, M.H., Sokolov, J.C., Chu, B., and Satija, S. K. *Europhys. Lett.* **2002b**, *60*, 559.

Koga, T., Seo, Y.-S., Shin, K., Zhang, Y., Rafailovich, M.H., Sokolov, J.C., Chu, B., and Satija, S.K. *Macromolecules* **2003**, *36*, 5236.

Krueger, S., Meuse, C.W., Majkrzak, C.F., Dura, J.A., Berk, N.F., Tarek, M., and Plant, A.L. *Langmuir* **2001**, *17*, 511.

Lee, E.M. and Milnes, J.E. *J. Appl. Crystallogr.* **1995**, *28*, 518.

Ljungdahl, G. and Lovesey, S.W. *Physica Scripta* **1996**, *53*, 734.

Lu, J.R., Simister, E.A., Lee, E.M., Thomas, R.K., Rennie, A.R., and Penfold, J. *Langmuir* **1992** *8*, 1837.

Lu, J.R., Hromadova, M., Simister, E.A., Thomas, R.K., and Penfold, J. *J. Phys. Chem.* **1994**, *98*, 11519.

Lu, J.R., Thomas, R.K., Binks, B.P., Fletcher, P.D.I., and Penfold, J. *J. Phys. Chem.* **1995a**, *99*, 4113.

Lu, J.R., Li, Z.X., Smallwood, J., Thomas, R.K., and Penfold, J., *J. Phys. Chem.* **1995b**, *99*, 8233.

Lu, J.R., Lee, E.M., and Thomas, R.K. *Acta Crystallogr.* **1996**, *A52*, 11.

Lu, J.R., Thomas, R.K., and Penfold, J. *Adv. Colloid Interface Sci.* **2000**, *84*, 143.

Majkrzak, C.F. and Berk, N.F. *Phys. Rev. B* **1995**, *52*, 10827.

Majkrzak, C.F. and Berk, N.F. *Phys. Rev. B* **1998**, *58*, 15416.

Majkrzak, C.F. and Berk, N.F., Perez-Salas, U.A. *Langmuir* **2003**, *19*, 7796.

Parratt, L.G. *Phys. Rev.* **1954**, *95*, 359.

Pedersen, J.S. *J. Appl. Crystallogr.* **1992**, *25*, 129.

Pedersen, J.S. and Hamley, I.W. *J. Appl. Crystallogr.* **1994**, *27*, 36.

Pynn, R. *Phys. Rev. B* **1992**, *45*, 602.

Richardson, R.M., Webster, J.R.P., and Zarbakhsh, A. *J. Appl. Crystallogr.* **1997**, *30*, 943.

Schreyer, A., Majkrzak, C.F., Berk, N.F., Gruell, H., and Han, C. *J. Phys. Chem. Solids* **1999**, *60*, 1045.

Sferrazza, M., Xiao, C., Jones, R.A.L., Bucknall, D.G., Webster, J., and Penfold, J. *Phys. Rev. Lett.* **1997**, *78*, 3693.

Simister, E.A., Lee, E.M., Thomas, R.K., and Penfold, J. *J. Phys. Chem.* **1992**, *92*, 1373.

Singh, N., Tirrell, M., and Bates, F.S. *J. Appl. Crystallogr.* **1993**, *26*, 650.

Sinha, S.K., Sirota, E.B., Garoff, S., and Stanley, H.B. *Phys. Rev. B* **1988**, *38*, 2297.

Sivia, D.S., Hamilton, W.A., and Smith, G.S. *Physica B*, **1991a**, *173*, 121.

Sivia, D.S., Hamilton, W.A., Smith, G. S., Rieker, T.P., and Pynn, R. *J. Appl. Phys.* **1991b**, *70*, 732.

Sivia, D.S. and Webster, J.R.P. *Physica B*, **1998**, *248*, 327.

Thomas, R.K. and Penfold, J. *J. Phys. Condens. Matter* **1990**, *2*, 1369.

Torikai, N., Matsushita, Y., Langridge, S., Bucknall, D., Penfold, J., and Takeda, M. *Physica B* **2000**, *283*, 12.

Torikai, N., Seki, M., Matsushita, Y., Takeda, M., Soyama, K., Metoki, N., Langridge, S., Bucknall, D., and Penfold, J. *J. Phys. Soc. Jpn.* **2001**, *70* (Suppl. A), 344.

Torikai, N., Yamada, N.L., Noro, A., Harada, M., Kawaguchi, D., Takano, A., and Matsushita, Y. *Polym. J.* **2007**, *39*, 1238.

# II Instrumentation

# II.3 Quasielastic and Inelastic Neutron Scattering

## II.3.1 Neutron Spin Echo Spectroscopy

*Michael Monkenbusch and Dieter Richter**

### II.3.1.1 INTRODUCTION

The neutron scattering intensity that is observed in diffraction experiments is a representation of the ensemble-averaged structure at some definite time. In real space, it corresponds to a flash light photograph of the conformation of scattering centers. As long as the centers are immobile and the structure stays the same during time, the thermal or cold neutrons are scattered elastically and do not change their velocity or wavelength during the scattering event. On the other hand, mobile and moving scattering centers induce a velocity change. This inelastic or quasielastic scattering may spread the neutron velocity by amounts ranging from the full velocity of the incoming neutrons down to quasielastic widths virtually approaching zero. The latter may, for example, occur during cooling of a glass-forming liquid, which slows down by many orders of magnitude upon approach of the glass transition temperature. Most other soft matter samples exhibit their structural identity in the small-angle neutron scattering (SANS) regime of diffraction intensity. The typical time of a structural change that significantly changes the phase of scattered radiation is related to the time that structural important items need to acquire a displacement of $x = 2\lambda \sin\theta$, where $\lambda$ is the (neutron) wavelength and $2\theta$ is the scattering angle. A simple example to arrive at a numerical estimate is a mesoscopic sphere, a nanoparticle, diffusing in a liquid with the viscosity of water. The diffusion constant may then be computed using the Einstein–Stokes expression $D = k_B T/6\pi\eta r$. The time this sphere (or a polymer coil of comparable hydrodynamic radius) needs to move its own radius by diffusion is given by $r^2 = \langle x^2 \rangle = 6tD = tk_B T/\pi\eta r$ that

*Corresponding author.

*Neutrons in Soft Matter* Edited by Toyoko Imae, Toshiji Kanaya, Michihiro Furusaka, and Naoya Torikai

yields $t = r^3\pi\eta/k_BT$. Substituting values $r = 10\,\text{nm}$, $\eta = 10^{-3}\,\text{N/ms}$, $T = 300\,\text{K}$ yields $t = 758\,\text{ns}$. Translated to the frequency domain, this corresponds to a quasielastic energy width $\hbar\omega/e \approx 1\,\text{neV}$. This situation is illustrated in Figure II.3.1.1.

The simple example immediately shows that the typical soft matter systems that extend into the mesoscopic length scale of some nm as, for example, polymer coils with molecular weight of several kg/mol or proteins do and that are embedded in an environment as liquid as water require neV resolution in energy or hundreds of ns in the time domain to investigate their large-scale mobility. Internal motions happen on a smaller timescale and may be faster. The example also explains why the resolution requirements increase steeply with the size of the observed object, respectively, the inverse momentum transfer of observation. Large objects have diffusion coefficients that scale inversely with their size, and the displacement that represents a relevant

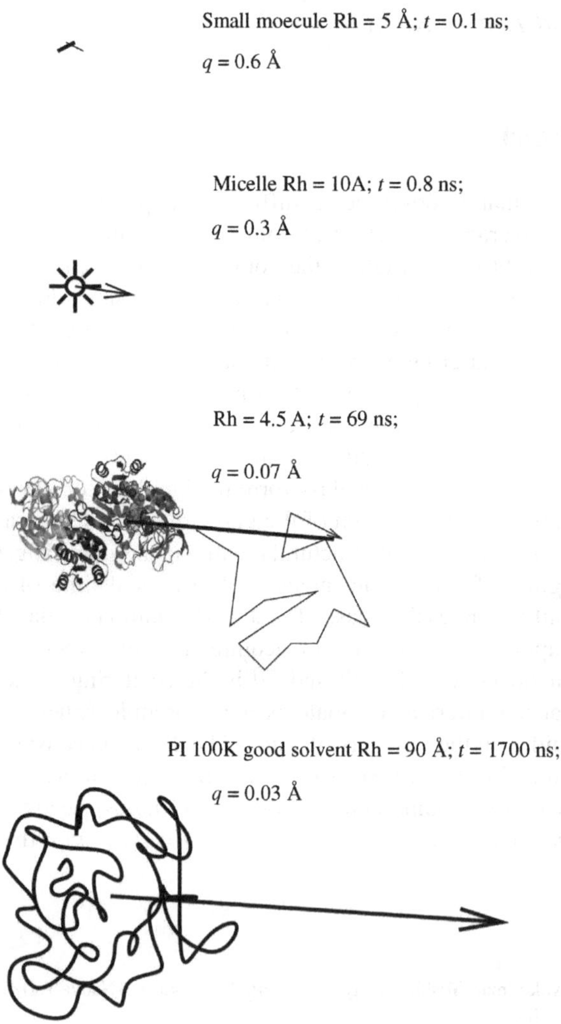


**Figure II.3.1.1** Objects of different size lead to very different diffusion times given by the time it takes for the object to move by one radius. Selecting typically soft matter building blocks (a segment/small molecule, a micelle, a protein, and a macromolecule in the form of a Gaussian coil) the time range an experimental method should cover (here NSE) is inferred.

change (in phase) at the adequate spatial scale of observation (proportional to $1/q$) also scale with size.

With respect to neutron scattering, this means that dynamics investigations of soft matter require much higher resolution than those supplied by conventional time-of-flight (~10 μeV) or backscattering (~1 μeV) spectrometers.

Thus, soft matter research with neutron spectroscopic methods is the realm of neutron spin echo (NSE) spectroscopy, which is able to extend the effective resolution by several orders of magnitude. However, as a Fourier method, NSE yields the intermediate scattering function $S(q, t)$ in the time domain and not the spectral function $S(q, \omega)$. In this respect, it is related to dynamic light scattering (DLS) and photon correlation spectroscopy (XCPS).

The motions that cause the small neutron velocity changes observed in an NSE experiment are thermal equilibrium fluctuations also called Brownian motions. Their driving force stems from stochastic "kicks" from the surrounding molecules that also are subject to Brownian motions. Their combined effect forms the coupling to the "heat bath." Besides this random force, the typical soft matter response is governed by friction and elastic forces. Inertia, which scales with the mass $\sim R^3$ is negligible compared to friction $\sim R$, for smaller (mesoscopic) objects friction dominates. The elasticity in many of the soft matter systems is entirely or to a large part due to entropy change upon deformation or displacement, rubber elasticity is a prominent example. The friction comprises also hydrodynamic coupling effects that yield more complex expressions for the mobility. The primary additional information that can be obtained from NSE experiments pertains the mobility of the constituents of a sample. If frictions are known from other sources also conclusions on elastic forces and their spatial patterns can be extracted. One of the simplest cases, diffusion of very dilute spherical objects, depends on a scalar friction only. As soon as the objects deviate from spherical symmetry (with respect to their scattering length density) rotational diffusion adds additional dynamics to the scattered radiation. At elevated density, diffusion will be modified by hydrodynamic and potential (electrostatic, van der Waals, short-range repulsion, and so on) interactions. If the spherical objects are, for example, droplets in a microemulsion, they have additional internal degrees of freedom. Their surface that is formed by a layer of surfactant molecules may easily be deformed and shows fluctuating deviations from spherical symmetry that also add to the basic diffusion dynamics. Analysis of these systems or other microemulsions with planar membranes (lamellar) or other shapes allows inferring the bending properties of the interfaces. Polymer systems exhibit special dynamical features due to their connectivity. If the segments are connected by flexible joints, the polymer forms a coil with considerable internal flexibility that exhibits the corresponding fluctuations, however, slower, more restricted, and qualitatively different from single-segment diffusion. It is the possibility to vary the scattering contrast within soft matter systems by combining H- with D-containing constituents that renders NSE in the SANS regime so powerful. The objects of interest may be selectively made visible.

Analyzing SANS scattering intensity for dynamical effects implies that the scattering is coherent and normally increases in intensity by several orders of magnitude upon approaching $q = 0$. Spin incoherent scattering, which is the dominant contribution from hydrogen atoms at larger $q$ and which often is in the focus of time-of-flight (TOF) and neutron back scattering spectroscopy (BSS) investigations, poses special difficulties for NSE. The spin-dependent scattering interferes with the spin analysis of the NSE method, but with additional effort it is also possible to measure self-correlation using the proton scattering.

## II.3.1.2 SCATTERING FUNCTIONS

The scattered neutrons are distributed according to the double differential cross section, which for a simple one-component case reads

$$\frac{d\sigma}{d\Omega dE} = \frac{k_i}{k_f} N b^2 S\left(q, \frac{E}{\hbar}\right) \tag{II.3.1.1}$$

where $k_{i,f} = 2\pi/\lambda_{i,f}$ are the incoming and the final neutron wave vectors, $N$ the number of scattering units, and $b$ the scattering length. The scattering function $S(q, \omega)$ contains the information on the structure of the sample and its time variation. In the regime of NSE investigations on soft matter, some approximations may be made: $k_f \cong k_i$ and $E = (\hbar^2 k_{i,f}^2)/2m$, which means that the scattering is nearly elastic (quasielastic with extremely narrow width), in particular is the difference between $k_f$ and $k_i$ much smaller than their typical uncertainty. It is also assumed that a classical description of motions in the sample is adequate, again since at temperatures around ambient $k_B T \gg E$ (26 meV $\gg$ 1 μeV) the detailed balance factor $\exp(E/k_B T)$ (Marshall and Lovesey, 1971) is virtually 1 and therefore $S(q, \omega) \cong S(q, -\omega)$. As will be explained below, the spin manipulations of the NSE spectrometer serve to detect miniscule energy transfers during scattering in a way that finally yields the intermediate scattering function

$$I(q,t) = \hbar \int_{-\infty}^{\infty} e^{i\omega t} S(q,\omega)\, d\omega \overset{\hbar\omega/kT \ll 1}{\widehat{\cong}} \hbar \int_{-\infty}^{\infty} \cos(\omega t) S(q,\omega)\, d\omega$$
$$= \left\langle \frac{1}{N} \sum_{j,l} e^{-i\vec{q}\cdot\{\vec{r}_j(0) - \vec{r}_l(t)\}} \right\rangle \tag{II.3.1.2}$$

rather than $S(q, \omega)$. The pure cosine transform holds for the NSE, soft matter realm. Knowledge of the average of all atomic trajectories would allow computing the right-hand side of eq. (II.3.1.2). Indeed, more recent molecular dynamics calculations reach the required system size and time domain (Narros et al., 2008).

However, in many cases where SANS intensity is analyzed, single-atom coordinates are not adequate, rather a description in terms of a coarse-grained scattering length density contrast $\Delta\rho$ is used:

$$I(q,t) = \frac{1}{V} \int\int e^{iq(r-r')} \langle \Delta\rho(r,t) \Delta\rho(r',0) \rangle d^3r d^3r' \tag{II.3.1.3}$$

with the integration volume $V$. In largely incompressible systems such as complex fluids—virtually all NSE soft matter samples can be considered as incompressible—$\Delta\rho$ results from concentration difference between proton-containing and deuterated molecules or molecules of different chemical nature. The latter, however, usually cannot provide enough scattering contrast compared to that obtained by isotopic labeling. Since the labeled molecules are surrounded by an incompressible liquid even using the coordinate representation for $I(q, t)$ single-atom coordinates are unpractical, rather—as usual in the description of SANS intensities—larger units (segments) are considered and treated as scattering centers at $r_{i,j}$ that differ in the sum of scattering length within their volume from the corresponding sum in the displaced volume of the surrounding liquid.

If the description in terms of a time-dependent density or concentration is less adequate as, for example, for the case of diffusion of segments, molecules, or aggregates a formulation of the scattering function with time-dependent coordinates (of segments, molecules, and so on) may be used. Here, it is often useful to rely on the Gaussian approximation. Its physical significance is that the probability for a certain distance or displacement vector corresponds to a Gaussian at any time, the width of the Gaussian being time dependent. Simple diffusion of a particle fulfils this assumption exactly, many other system do it approximately. Applying this assumption often leads to simple analytic expressions for the intermediate scattering function. The evaluation of the scattering from polymer chains in the Rouse model (see below), for example, relies on the Gaussian approximation. The basic relation is

$$\langle \exp(-i\vec{q}\cdot\{\vec{x}_0-\vec{x}_t\})\rangle = \exp\left[-\frac{1}{6}q^2\langle x^2(t)\rangle\right] \qquad \text{(II.3.1.4)}$$

where the angular brackets represent an ensemble average over a Gaussian distribution and the macroscopic system is isotropic. In cases where the scattering intensity stems from noninterfering centers at $x_i(t)$ eq. (II.3.1.4) directly yields the mean square displacement as:

$$\langle x^2(t)\rangle = -\frac{6}{q^2}\ln\left(\frac{S(q,t)}{S(q)}\right) \qquad \text{(II.3.1.5)}$$

giving direct insight into sub- or superlinear diffusion. An example is shown in Figure II.3.1.7.

## II.3.1.3 INSTRUMENTS AND SIGNALS

The need for extremely high-energy resolution—or better resolution of energy transfer—of more than 1:1000 up to 1:1,000,000 calls for a nonconventional instrument principle. The standard method to prepare an incoming neutron beam with a well-defined incoming energy (velocity, wavelength) and to analyze the energy (velocity, wavelength) distribution after scattering, ends to be feasible at values of about 1:1000 as realized in the backscattering instruments. There, the intensity loss due to the narrow wavelength selection is partly compensated by an increase of detection solid angle. Each detector receives neutrons from square meter sized analyzer mirrors. The associated width in $q$ restricts the use in the SANS regime. Any further narrowing of the bandwidth selection would proportionally reduce the incoming intensity and—in order to match the resolution—also the transmission of the analyzer. The same is true for the standard time-of-flight instrument, improvement of the resolution beyond the normal 1:100 by a factor $f$ would reduce the incoming intensity by the same factor, to match the time resolution the chopper opening time must be shortened accordingly and in addition the time-of-flight uncertainty due to finite sample size and detector thickness must be compensated by the use of smaller samples and thinner detectors (that may be less sensitive). The total intensity reduction therefore significantly exceeds $f^2$.

The NSE method overcomes this limit by eliminating the need to prepare an incoming beam with a wavelength distribution as narrow as the required resolution. What rather counts is the ability to detect energy transfer independent from the initial velocity. The NSE principle (Mezei, 1972, 1980) consists in coding the initial velocity in terms of a spin precession angle individually for each neutron. After scattering at the sample, a symmetric setup decodes the velocity by performing an effectively negative precession. The coding scheme then translates very small differences of the neutron velocity before and after scattering into a spin precession angle with detectable influence on the neutron beam polarization. Thus, broad quasielastic scattering yields a broad distribution of final spin directions and as a consequence a low average polarization. The sensitivity of the coding process on velocity changes may be controlled by an effective magnetic field that determines the precession frequency in the coding and the decoding flight paths or by the average neutron velocity (wavelength) that determines the duration of the influence of the precession field. For a given spectral width of the scattering function, the polarization decreases with increasing coding sensitivity. The latter effect contains the desired information.

Below the result of an exact derivation of the dependence of detected intensity on the instrument parameters and the scattering function is given. But before that, a generic setup is described in order to illustrate the meaning of the different terms in eq. (II.3.1.6), Figure II.3.1.2 shows a schematics of the classical NSE of the IN11 type (Mezei, 1980). Neutrons from a cold source in a neutron guide are coarsely monochromatized by a mechanical velocity selector with $\Delta\lambda/\lambda \cong \Delta v/v \cong 0.1 \ldots 0.2$ the close to triangular transmission function of the velocity selector multiplies to the neutron spectrum. Below, we assume a wavelength distribution $w(\lambda - \lambda_0)$ after the selector, which is approximated by a Gaussian $\Phi_0/(\Lambda\pi^{1/2}) \exp(-[(\lambda - \lambda_0)/2\Lambda]^2)$. With $\lambda_0$, the nominal wavelength, $\Lambda \cong [\Delta\lambda/\lambda]_{\mathrm{FWHM}}/\pi$ and $\Phi_0$ the neutron flux. Before or after the selector, the neutron beam is polarized by separating the spinup and spindown neutrons using reflection from a magnetic multilayer mirror. After polarization, the beam needs a small guide field to avoid uncontrolled spin precession in residual environmental magnetic fields (resulting, for example, from the Earth's magnetic field). Thus, a neutron guide delivers a polarized and moderately monochromatized—and therefore intense—neutron beam to the

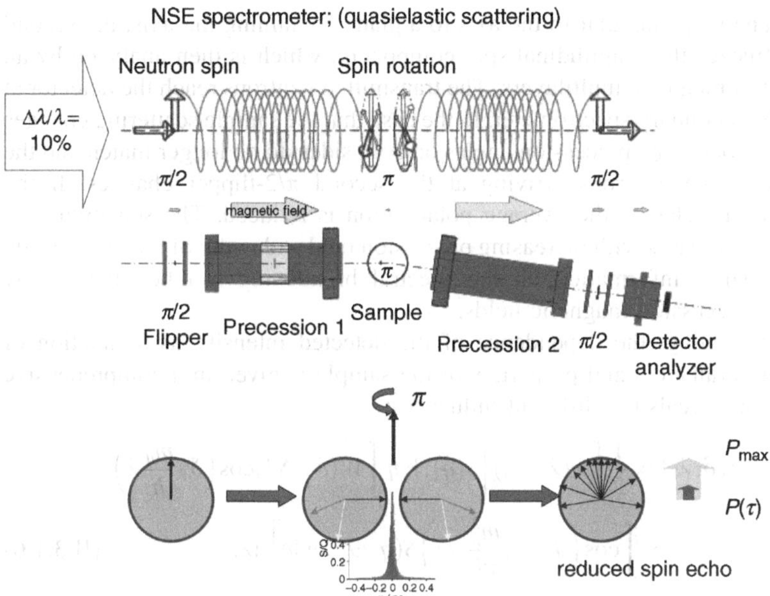


**Figure II.3.1.2** Schematics of a generic spin echo spectrometer (the indicated solenoid configuration corresponds to the JNSE and NIST NG5 spectrometers). See main text for a detailed description.

entry of the proper NSE spectrometer. There the neutrons traverse the first $\pi/2$-flipper, which defines the start of the primary, preparatory coding (precession) region. The $\pi/2$-flipper turns the neutron polarization from longitudinal to perpendicular to the beam axis that is also the direction of the precession field. The subsequent precession is effected by a longitudinal magnetic field created by solenoids enclosing the neutron path. After a few meters flight path, respectively, a few milliseconds in the precession field the neutron leaves the first solenoid set and approaches the sample position. The spin has typically undergone several thousand full rotations around the precession field; the exact final precession angle (modulo $2\pi$) depends sensitively on the exact velocity of the neutron under consideration. The incoming velocity spread causes that the spin directions of the neutrons are equally distributed in a plane perpendicular to the axis. Close to this position, a $\pi$-flipper rotates the spin direction by 180° around a vertical axis, which means that the plane of spin directions is rotated such that a final precession angle of $\Psi$ mod $2\pi = \alpha$ is transformed to $\Psi'$mod $2\pi = -\alpha$. Then the neutron is scattered by the sample and may or may not suffer an energy transfer. In a symmetric decoding precession region, the spin clock of each neutron proceeds in the same way as in the primary coding section. If the scattering was elastic and energy and velocity were unchanged the precession angle at the position of the second $\pi/2$-flipper is ($\Psi' + \Psi$ mod $2\pi = -\alpha + \alpha = 0$). This effect is independent of the starting neutron velocity and holds for all elastically scattered neutrons, which means that the initial polarization of the beam is restored. It is called "spin echo." The name was coined for an analogous effect that was discovered in the early days of nuclear magnetic resonance spectroscopy (Hahn, 1950). Due to the rotation of the spin direction plane perpendicular to the axis to a plane containing the axis, the second $\pi/2$-flipper freezes the longitudinal spin component, which is then analyzed by an arrangement of magnetic multilayers. The transmitted neutrons reach the detector at the end of the secondary spectrometer. In the case that the sample scattering changes the neutron velocity, the precession angles on both sides do no longer match and the polarization of the neutrons arriving at the second $\pi/2$-flipper changes. If the scattering is quasielastic, the average polarization is reduced. The sensitivity for this reduction increases with increasing precession field and width of the quasielastic broadening. Thus, information on the spectral broadening can be obtained by scanning the precession magnetic fields.

In the following, the dependence of the detected intensity as a function of instrumental parameters and properties of the sample is given in a comprehensive expression that reveals the different influences:

$$I_{\mathrm{Det}}[J,\delta,\lambda_0] \propto \frac{1}{2}\int w(\lambda-\lambda_0)\left[S(q) \pm \eta\int W(\delta-\Delta J)\cos\left(\delta\gamma\frac{m_n}{h}\lambda\right)\right.$$
$$\left.\times\int\cos\left(J\lambda^3\gamma\frac{m_n^2}{2\pi h^2}\omega\right)S(q,\omega)\mathrm{d}\omega\mathrm{d}\delta\right]\mathrm{d}\lambda. \qquad \text{(II.3.1.6)}$$

Before discussing the comprehensive formula, we show how the more common simpler expressions are retrieved from it. By replacing $w(\lambda-\lambda_0)$ and $W(\delta-\Delta J)$ by $\delta$-functions and $\eta=1$, we arrive at the expression for an ideal instrument:

$$I_{\mathrm{Det}}[J,\delta,\lambda_0] \propto \frac{1}{2}\left[S(q) \pm \cos\left(\delta\gamma\frac{m_n}{h}\lambda_0\right)\int\cos\left(J\lambda_0^3\gamma\frac{m_n^2}{2\pi h^2}\omega\right)S(q,\omega)\mathrm{d}\omega\right]. \qquad \text{(II.3.1.7)}$$

The parameters field integral $J=\int_{\pi/2\text{-flipper}}^{\pi\text{-flipper}}|B|\mathrm{d}l$, field integral asymmetry $\delta=J_1-J_2$ between first and second precession paths, and the nominal wavelength $\lambda_0$ can be set and varied during an experiment. $\gamma$, $h$, and $m_n$ are the neutron Larmor constant, the Planck's constant, and the neutron mass, respectively. At the symmetry point, $\delta=J_1-J_2=0$ eq. (II.3.1.7) further simplifies as

$$I_{\mathrm{Det}}[J,\delta=0,\lambda_0] \propto \frac{1}{2}\left[S(q) \pm \int\cos\left(\underbrace{J\lambda_0^3\gamma\frac{m_n^2}{2\pi h^2}}_{\psi}\omega\right)S(q,\omega)\mathrm{d}\omega\right]$$
$$=\frac{1}{2}[S(q,0) \pm S(q,t)] \qquad \text{(II.3.1.8)}$$

whereas for a purely elastic scatterer, $S(q,\omega)=S(q)\delta(\omega)$, the simplified expression for the typical echo oscillation is obtained

$$I_{\mathrm{Det}}[\delta,\lambda_0] \propto \frac{1}{2}\left[S(q) \pm S(q)\cos\left(\delta\gamma\frac{m_n}{h}\lambda_0\right)\right]. \qquad \text{(II.3.1.9)}$$

Due to the technical imperfections of the real spectrometer, $\eta$ is introduced in eq. (II.3.1.6) to account for the efficiency of polarizer, analyzer, and flippers and $W(\Delta J) \approx 1/(\Sigma\pi^{1/2})\exp(-[\Delta J/2\Sigma]^2)$ is the distribution (inhomogeneity) of field integrals for different paths within the beam. To derive eq. (II.3.1.6), the precession angle difference was computed using a series expansion to relate $\Delta\lambda$ to $\Delta v$ and an analyzer with transmission $(1 \pm \cos\Delta\Psi)/2$ was assumed. High-resolution NSE spectrometers contain correction elements to compensate for the intrinsic inhomogeneity of the used solenoid and path length differences within the beam. In that case, $\Sigma$ stands for the residual errors after the correction. The Gaussian approximation of the latter is a coarse approximation that enables further analytic treatment:

$$I \propto \frac{1}{2}\int w(\lambda-\lambda_0)\left[S(q) \pm \eta\exp\left(-\Sigma^2\gamma^2\frac{m_n^2}{h^2}\lambda^2\right)\cos\left(\delta\gamma\frac{m_n}{h}\lambda\right)\right.$$
$$\left.\times\int\cos\left(J\lambda^3\gamma\frac{m_n^2}{2\pi h^2}\omega\right)S(q,\omega)\mathrm{d}\omega\right]\mathrm{d}\lambda \qquad \text{(II.3.1.10)}$$

This form for $I$ already reveals that the intensity signal contains a contribution that corresponds to the Fourier transform of the spectral form of the scattering function $I(q,t) = \int\cos(t\omega)S(q,\omega)\mathrm{d}\omega$ where the Fourier time is given by $t = J\lambda^3\gamma(m_n/h)^2/2\pi$ and depends linearly on the field integral and on the cube of the wavelength. The field integral dependence allows variation of $t$ simply by changing the solenoid current. The wavelength dependence is the same as the wavelength dependence of the resolution of any time-of-flight instrument where high resolution here means a large value of $t$. The factor $R = \eta\exp(-\Sigma^2\gamma^2(m_n/h)^2\lambda^2)$ describes the effect of instrumental imperfection by combining polarization efficiency effects with those of field inhomogeneity $\Sigma = \sqrt{\langle\Delta J^2\rangle}$, which is largely proportional to $J$. Taking the Gaussian approximation also into account for the incoming wavelength spectrum and ignoring further terms that stem from the wavelength dependence of $S(q[\lambda],\omega)$ assuming $I(q[\lambda],t[\lambda]) \cong I(q[\lambda_0],t[\lambda_0]) = I(q_0,t)$, we arrive at

$$I \propto \frac{S(q_0)}{2}\left[1 \pm \eta\frac{1}{A}\exp\Bigg(\underbrace{-[\Sigma^2\lambda_0^2+\Lambda^2\delta^2]\gamma^2\frac{m_n^2}{h^2}/A^2}_{-\psi^2}\Bigg)\right.$$
$$\left.\times\cos\left(\delta\gamma\frac{m_n}{h}\lambda_0/A^2\right)\frac{I(q_0,t_0)}{S(q_0)} + \cdots\right] \qquad \text{(II.3.1.11)}$$

where $A = \sqrt{1+4\Sigma^2\Lambda^2\gamma^2\frac{m_n^2}{h^2}}$ is very close to 1 for reasonable parameters $\Sigma$, $\Lambda$ is the value of $g = \gamma(m_n/h) = 4.627\times10^{14}\,\mathrm{T}^{-1}\mathrm{m}^{-2}$. The field integral inhomogeneity $\Sigma$ is virtually proportional to the field integral $\Sigma = \hat{\Sigma}J$ and therefore has an implicit dependence on the Fourier time $t_0$. The diffraction intensity relates to $S(q) = I(q,t=0)$. The cos term in front of $I(q_0,t_0)$ together with the Gaussian describes the echo effect as seen if $\delta$ is scanned; the maximum amplitude of the cos oscillation is called echo amplitude. The goal of a NSE experiment is to extract $I(q,t)$, respectively, $I(q,t)/I(q,t=0)$ from the measured intensities. In Section II.3.1.4, it is explained how this is extracted based on the properties of eq. (II.3.1.11).

Finally, one may wonder how much distortion of the intermediate scattering function is caused by the broad wavelength spread of the incoming neutrons. If one observes the wavelength dependence of the wave vector $q \propto \lambda^{-1}$ and of the Fourier time $t \propto \lambda^3$, higher order corrections enter the expression for the detected intensity:

$$I \propto \frac{1}{2}\left[\overline{S(q)} \pm \eta \frac{1}{A}\exp(-\psi^2)\left\{\cos(\phi)S_0 - \frac{2}{A^2}\frac{\Lambda^2}{\lambda_0^2}\right.\right.$$
$$\left.\left.\times\left(\left[A^2\phi\sin(\phi)+2(\Sigma\lambda_0 g)^2\cos(\phi)\right]S_1 - \frac{1}{2A^2}\cos(\phi)S_2\right) + O\left(\frac{\Lambda^4}{\lambda_0^4}\right)\cdots\right\}\right] \quad \text{(II.3.1.12)}$$

with $S_0 = S(q_0,t_0)$, $S_1 = -q_0(\mathrm{d}/\mathrm{d}q_0)S(q_0,t_0) + 3t_0(\mathrm{d}/\mathrm{d}t_0)S(q_0,t_0)$, and $S_2 = \{q_0^2(\mathrm{d}^2/\mathrm{d}q_0^2) + 2q_0(\mathrm{d}^2/\mathrm{d}q_0) + 9t_0^2(\mathrm{d}^2/\mathrm{d}t_0^2) + 6(\mathrm{d}^2/\mathrm{d}t_0) - 6q_0t_0(\mathrm{d}^2/\mathrm{d}q_0\mathrm{d}t_0)\}S(q_0,t_0)$.

For a 10% FWHM selector, the expansion parameter $(\Lambda/\lambda)^2 = 10^{-3}$, the corresponding coefficients depend on derivatives of the intermediate scattering function. Only scattering functions with steeply varying $S(q)$ have sizeable correction terms. For smoothly varying functions as $S(q,t)$ from polymer melts or solutions, the deviations due to the wavelength spread are less than 1%. This explains why the NSE yields accurate $S(q,t)$ for polymers and similar systems without further correction even if the wavelength spread amounts to 10% ... 20% FWHM. For simple diffusion, the echo and the intermediate scattering function are shown and the effect of a broad wavelength distribution is illustrated in Figure II.3.1.3.

There are a number of variations that lead to different NSE instruments. Starting with the minor conceptual change of replacing the second solenoid by a sector coil. This has been realized as add-on IN11C to the grandfather of all NSE instruments IN11 at the ILL. The sector coil covers a larger range of scattering angles of 30° with one setting (Farago, 1998) yielding a related intensity gain. Due to limitations of the field integral and homogeneity of the sector coil the resolution in terms of maximum Fourier time is limited to 20% of the solenoid version. A more radical approach to simultaneously cover a huge range of scattering angles has been realized in form of the NSE spectrometer SPAN (Pappas et al., 2001, 2002). Here, a completely different geometry of magnet coils is employed (see Figure II.3.1.4). The precession is effected by the radial field in the horizontal midplane (= scattering plane) of a pair of two large current rings below and above that plane. This field is completely symmetric with respect to scattering angle if the sample is positioned in the center point of this arrangement. The field integrals are considerably lower than for the solenoid-type spectrometers. To complete the setup to form a NSE spectrometer further current rings are needed to support the flipper and analyzer functions. Due to the steep $q$-dependence of the typical relaxations in soft matter, however, a full utilization of "0° ... 180°" would lead to a mismatch of resolution between low and high scattering angles that limits the efficiency. In those cases, rather a smaller sector yields usable results at the same time.

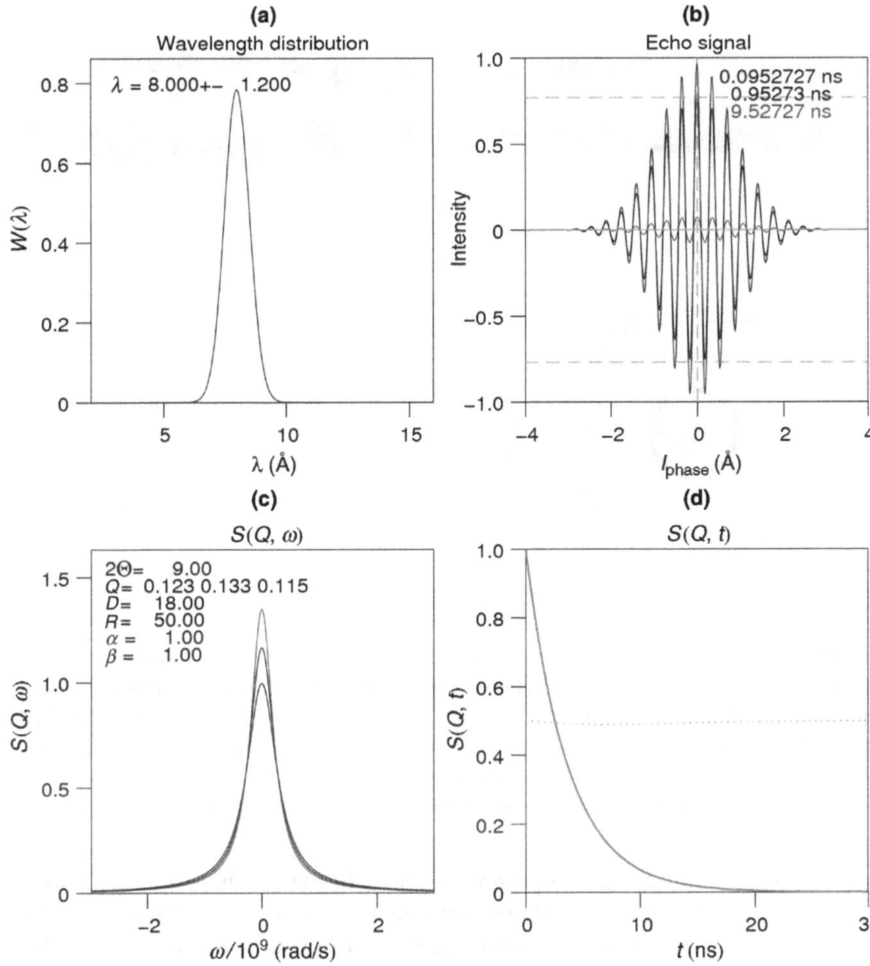


**Figure II.3.1.3** (d) NSE curves (ideal and with finite wavelength spread) as derived from the echo signals shown in (b). The corresponding spectrum (simple diffusion) is shown in (c) for three $q$-values that are within the wavelength band and the corresponding wavelength distribution in the (a). (d) The curves show that the effect of the finite wavelength width is negligible, the dotted curve represents the difference between zero wavelength width and the show finite width multiplied by 10 with an offset of 0.5.

The above variants still rely on the exact spin echo mechanism as explained above. A more different type of NSE spectrometer is the resonance NRSE or zero field spectrometer (Golub and Gahler, 1987). It may very coarsely be imagined as the result of the transformation (of one arm) in a rotating coordinate system such that the main precession field becomes zero. The Bloch equation that governs the dynamics of the expectation value of the spin magnetization in presence of a magnetic field is modified by a transformation into a rotating system such that an extra (pseudo) magnetic field occurs in the direction of the rotation axis and with a strength such that the rotation frequency equals the Larmor precession frequency in that field. This

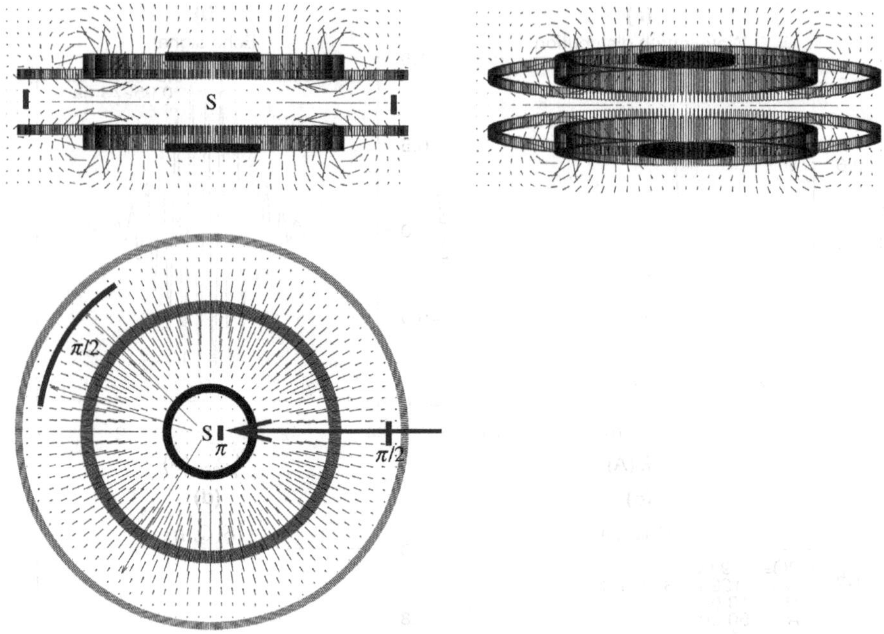


**Figure II.3.1.4** Sketch of the SPAN spectrometer type viewed from the side and from the above. The small arrow indicate the magnetic field, *S* is the sample position in the center. Polarization analysis and detection (not shown here) happens after the second $\pi/2$-flipper. The center ring coils serve to set the field at the sample position, the second rings are the main solenoid that creates the precession field pointing radially toward the sample. The outer rings serve to set a suitable outer field for the $\pi/2$-flippers. For detection, in principle, all positions around the periphery may be used (only one sector is indicated).

transformation eliminates the large precession field and leads to rotating flipper fields, which may be realized by alternating (radio frequency (RF), $\Omega$) currents. The technical realization deviates a bit from the direct result of the rotational transformation; here, the reader is referred to the literature by Longeville (2000) and Haeussler et al. (2007). Realizations of the NRSE principle utilize a doubled RF-flipper at each position (bootstrap) that doubles the rotation effect and reduces the stray field of the flipper. With the translations $4\Omega L = \gamma J, 4(\Omega_1 L_1 - \Omega_2 L_2) = \gamma\delta$ where $L$ is the length of the neutron path between two flippers of one arm and $\Sigma = 4\Omega\sqrt{\langle \Delta L^2 \rangle}$, the NRSE yields the same type of intensity signal as described above for the generic NSE.

A derivation of the NRSE is the so-called MIEZE spectrometer (Köeppe et al., 1999). It may be considered as a time-of-flight instrument with a sinusoidal intensity modulation of the beam in contrast to the short pulses of a conventional TOF spectrometer. Normally, a fast modulation of a neutron beam is only preserved along the path to a distant detector when the flight time difference induced by the initial velocity spread is much smaller than the modulation frequency. The same intensity dilemma as discussed above emerges here when high resolution is required. The spin echo technique as it is employed in the MIEZE type spectrometer is used to prepare a beam with a wide band of wavelength, however, the modulation phase varies with the velocity within that band such that exactly at the location of the detector—and only there—neutrons of all velocities have the same modulation phase. This requires a modulation path that is limited by two RF-flippers running at different frequencies, their difference being the modulation frequency. The equal phase condition at the detector imposes the relation $\Omega_1 L_1 = (\Omega_2 - \Omega_1)L_2$ between the flipper frequencies $\Omega_1$and $\Omega_2$ and the lengths between the flippers $L_1$ and the second flipper and detector $L_2$. The conversion of spin precession to intensity modulation is performed by an analyzer behind the preparation section, but before the sample. After the analyzer, in particular at the sample, the spin state of the neutron does not matter any longer. Therefore, spin incoherent scattering from the sample as well as magnetic fields around or in the sample do not infer with the detection of energy or velocity changes of the neutron. However, for high resolution the flight path length from the second RF-flipper to the sample must be very well defined. For example, if for a neutron of $\lambda = 0.8$ nm with a velocity of 500 m/s an energy transfer of 0.7 μeV corresponding to a Fourier time of ~1 ns shall be detected, velocity differences of less than 0.3 m/s must be detected. Assume $L = 5$ m, then time differences of 6 μs at the detector (166 kHz modulation) and therefore path length differences of 3 mm are relevant. This already imposes severe limits on the sample extension and/or the solid angle. The modulation contrast at the detector is reduced if the scattering is quasielastic and broadens the velocity distribution. It has an analogous role as the echo amplitude for the other spin echo types. Note, however, that until now SPAN is mainly used for the investigation of magnetic fluctuation and soft matter applications of resonance and MIEZE are very sparse. An overview of existing NSE instruments is given in Table II.3.1.1.

## II.3.1.4 CONDUCTING EXPERIMENTS

### II.3.1.4.1 How Intensity Data Relate to *I*(*q*, *t*)/*I*(*q*, 0)

For a given field integral and scattering angle setting, the first parameter to be scanned is the field integral asymmetry $\delta$. This is usually done by varying the current through a phase coil of a few turns wound around one of the main precession solenoids. For NRSE, the translation rules reveal that one may either scan the distance $L$ between RF-flippers or use slightly different RF frequencies $\Omega_{1,2}$ in the first or the second arm. This "phase scan" allows the precise identification of the symmetry point that sensitively depends on small environmental fields. To extract the echo amplitude, it suffices to scan one oscillation. Scanning over a wider range shows the oscillating echo signal modulated by a (pseudo)-Gaussian envelope. This scan allows calibrating the wavelength $\lambda_0$ from the oscillation period and the wavelength distribution width $\Lambda$ from the inverse of the envelope width. Once done, these parameters are fixed inputs to numerical procedures to extract a precise echo amplitude from phase scan intensity data. As eq. (II.3.1.11) shows, the obtained amplitude is still less than the desired (normalized) $I(q,t)$ value but contains the reduction factors $\eta$ and

**Table II.3.1.1** List of Currently Existing Neutron Spin Echo Instruments

| Instrument | Type | Location | Effective field integral (Tm) | Link | References |
|---|---|---|---|---|---|
| IN11 | Generic | ILL, Grenoble | 0.27 | www.ill.eu | Farago (1998) |
| IN11C | Extended angle | ILL | 0.06 | www.ill.eu | Farago (1999) |
| IN15 | Generic | ILL | 0.27 | www.ill.eu | Schleger et al. (1997) |
| J-NSE | Generic | JCNS, Juelich/Garching | 0.5 | www.jcns.de | Monkenbusch et al. (1997) |
| NG5-NSE | Generic | NIST Center for Neutron Research, Gaithersburg | 0.5 | www.ncnr.nist.gov | Rosov et al. (2000) |
| C2-3-1 (C2-2) | Generic | ISSP University, Tokyo; JAERI, Tokai | 0.22 | www.issp.u-tokyo.ac.jp | Takeda et al. (1995, 1999) |
| SNS-NSE | Generic | JCNS, SNS Juelich/Oak Ridge | 1.8 | www.jcns.de | Ohl et al. (2003, 2004, 2005) |
| V5-SPAN | SPAN-type (see text) | HZB, Berlin | 0.06 | www.hmi.de/bensc | Pappas et al. (2001) |
| G1-Bis (Muses) | Zero field type | LLB, Saclay | 0.12 | www-llb.cea.fr | Longeville (2000) |
| RESEDA | Zero field type | TUM, FRMII, Garching | 0.07 | www.frm2.tum.de | Hauessler 2008b |

$\exp[-(\Sigma^2\lambda_0^2)\gamma^2(m_n^2/h^2)]$. An estimate for $\eta$ is obtained as measure for the overall polarization analysis efficiency by measuring the difference of detected intensity with all flippers switched off and with only the $\pi$-flipper active. Ideally, one of these intensities should be zero. Their difference is a measure for the maximum echo amplitude to be expected. To account for $\eta$, the normalizing value $S(q_0)$ is taken as the measured difference. The Gaussian part of the prefactor is an estimate of the signal reduction due to inhomogeneity. Besides the maximum field integral, this factor is the ultimate limitation for the resolution of the NSE instrument in terms of large Fourier time. It is determined by a scattering experiment under otherwise identical conditions with a reference sample that scatters only elastically. In the SANS regime, various carbon powder or graphite (grafoil) samples with large surface area are suited, for large angles a random alloy of TiZr may be used. The reference data are treated as those of the sample and the normalized echo amplitude of the sample is divided by the one from the reference sample. This resolution correction by division corresponds to a deconvolution in the frequency domain. Note that the effect of broad wavelength width in combination with the $\lambda$-dependence of the resolution factor may lead to a distortion of the resolution corrected, normalized intermediate scattering function if the sample and the reference have grossly different shapes of $S(q)$. In particular this effect is expected in the very low $q$ regime.

Data for further Fourier times are collected by repeating the above procedure for different currents in the main coil leading to different effective field integral values. $q$-variation beyond the range covered by a position-sensitive detector at one detector arm setting is performed by moving the second arm, the detector carrier to different central scattering angles. Finally, further variations of external parameters as, for example, the sample temperature are typically done in the outermost loop of this nested sequence of countings. However, the exact sequence in which the different intensity values are measured is somewhat arbitrary and depends on various practical considerations as times needed to change parameters, stability of the instrument and of the environmental fields, and so on.

## II.3.1.4.2 Spin Incoherent Scattering

If the scattering intensity only results from the contrast due to different average scattering length at different positions in the sample system, the above derivation holds. This includes isotopic incoherent scattering. However, if the system exhibits intensity components from spin incoherent scattering, the factor $\eta$ in eq. (II.3.1.11) has to be replaced by $-(1/3)\eta$ because the spin incoherence implies that two-thirds of the scattering events cause a spin flip. This renders two-thirds of the echo amplitude into unpolarized background and leaves one-third of spin-flipped neutrons. The spin flip implies the additional minus sign. If the spin incoherent scattering is the dominant intensity—as for thin fully protonated samples—the described evaluation procedure still works since for the normalizing difference between spinup and spindown counts the new $\eta_{\text{eff}} = -(1/3)\eta$ applies. The situation becomes more involved if coherent and spin incoherent contributions both contribute to the intensity.

In the SANS regime, the incoherent level may be read off the high $q$-background. Fortunately, the intensity towards low and intermediate $q$ increases steeply beyond this level. Nevertheless, NSE experiments also explore the spatial range where the incoherent "background" becomes significant, for example, if dilute protein solutions or glassy dynamics at intermolecular distances are investigated.

In these cases, a composite signal is observed that cannot easily be decomposed. With $\Delta_{\mathrm{coh,inc}} = \left[I^{\uparrow} - I^{\downarrow}\right]_{\mathrm{coh,inc}}$, the normalizing difference between spinup and spin-down counts and $f_{\mathrm{coh,inc}}(t)$ the Fourier time-dependent normalized relaxation functions of the coherent and the incoherent scattering, respectively, we arrive at

$$F_{\mathrm{combined}}(t) = \frac{\Delta_{\mathrm{coh}} f_{\mathrm{coh}}(t) - [(1/3)\Delta_{\mathrm{inc}} f_{\mathrm{inc}}(t)]}{\Delta_{\mathrm{coh}} - [(1/3)\Delta_{\mathrm{inc}}]} \tag{II.3.1.13}$$

In cases where coherent scattering and incoherent scattering show the same dynamics, eq. (II.3.1.13) still yields results that represent that dynamics. However, already then the unfavorable situation may occur that $\Delta_{\mathrm{coh}} - [(1/3)\Delta_{\mathrm{inc}}] = 0$ leading to very large errors. The situation when $f_{\mathrm{coh}}(t)$ and $f_{\mathrm{inc}}(t)$ are different and the incoherent dynamics is faster this may lead to the unusual effect that $F_{\mathrm{combined}}(t)$ increases above 1 with increasing $t$ then exhibits a maximum before the monotonous decay starts. In general, a full analysis of the scattering function under such conditions would need an independent experiment on another type of instrument that yields a coherent–incoherent combination with a linear independent combination of refractors. For example, the MIEZE technique or BSS would have the potential to do so with the coefficients (1,1) instead of (1, −1/3). Up to now, however, this possibility has not yet been tested quantitatively.

### II.3.1.4.3 Sample Requirements

To be able to measure the dynamics of structures in a sample, these structures must be responsible for a reasonable fraction of the scattering intensity in the $q$ range of interest. The total scattering intensity into that solid angle element must be sufficiently intense. To be able to extract the echo amplitude with acceptable error from a phase scan for one $q$ and one Fourier time one needs a number $>$10,000 for the total counts on the detector. For example, we take $\Delta\Omega = 4 \times 10^{-3}$ sr, a SANS sample with $d\Sigma/d\Omega = 0.1\ \mathrm{cm}^{-1}$ and 2 cc volume, an overall average transmission and detection efficiency of $(\eta_{\mathrm{det}} = 0.5) \times (t_{\mathrm{corr}} = 0.8)^3 \times (t_{\mathrm{analyzer}} = 0.4) = 0.1$ and $10^7\ \mathrm{n/cm^2s}$ neutron flux at the sample. That yields 800 cps on a multidetector of $25 \times 25\ \mathrm{cm}^2$ in 4 m distance from the sample. At long wavelength, the neutron flux is one to two orders of magnitude less with a corresponding decrease of expected neutron detection rate. There counting times of several minutes per single point are needed.

For the normal SANS regime, a sample should consist of a fully deuterated matrix, which may be a solvent or a polymer melt containing the objects to be observed. The molecular structures that are in the focus of the investigation should be protonated thus yielding a scattering contrast. For concentrations between 20% and

2%, sample thicknesses of 2–4 mm are used. With a typical sample size of 30 × 30 mm, the required volume to be prepared is 2–4 cc. At the measuring temperature, the samples must be liquid or contain liquid parts in order to show mobilities that are detectable by NSE experiments.

## II.3.1.4.4 Background Subtraction

NSE experiments suffer from different sources of background. On the one hand, constant background contributions, as general background radiation and electronic detector noise, do not modify the final signal as it is obtained in terms of a ratio of difference counts. However, excess background adds to the statistical noise. More severe background contributions stem from parasitic scattering from spectrometer components in the primary beam path and from sample environment and cuvettes. Whereas, scattering from the cuvette adds a normally elastic contribution to the sample scattering, the intensity from parts that have some distance from the sample point exhibit an artificial pseudodynamic effect because they have (very) different magnetic path contributions with different resolution and path integral symmetry properties. Finally, the matrix that in principle is a part of the sample, which is considered as a source of background in terms of scattering contributions that are not related to the fluctuations of the molecular items that are made visible by the (isotopic) scattering contrast. For example, a deuterated solvent or melt exhibits a fast scattering contribution below the time window of the NSE experiment that stems from multiple scattering and impurities—only little of it from the compressibility contribution to the SANS intensity. The effect on the normalized NSE signal of this type of background is a fast initial drop before the first time point such that the normalized curve seems to start at a value below one. Despite the fact that this is a part of the scattering function of the sample as a whole, this contribution is considered as undesired background and the usual evaluation procedure tries to eliminate it. If the sample contains immobile impurities such as large aggregates of residual catalyst or other substrates, the scattering will contain a constant part that also might be considered as background. To subtract these contributions, it is necessary to measure a suitable background sample, normally the deuterated matrix in the same type of cuvette at the same temperature. In more complicated cases, the proper background sample choice may be less unique. For all those cases that are not extremely affected by or sensitive to background effects, the following procedure is good practice. The scattering intensity of the deuterated matrix is measured with polarization analysis. In addition, sample transmission and background transmission are needed. If only the fast inelastic scattering component is to be corrected, these values can be used to determine the correction factor to the normalization. If the sample contains also background components that produce an echo signal, a full NSE run has to be performed with the background sample (with a counting time that is roughly reduced by the factor $\sqrt{I_{\mathrm{bgr}}/I_{\mathrm{sample}}}$). The same is true if there is significant contribution from spectrometer parts (windows, flippers, and correction elements). An extra complication, which we will not further discuss, may come due to scattering from parts in the incoming beam before the sample whose scattering intensity reaches the detector without being reduced by sample transmission. Concerning the procedure to perform the full background subtraction, there are two main choices:subtraction on the level of raw data consisting of single counting results; subtraction after echo amplitudes have been determined from the phase scans. Method (a) is simple and straightforward and requires only the transmission ratio as further input, however, it is prone to errors due to phase drifts caused by external fields. Method (b), on the other hand, relies on the ability to extract echo amplitudes from the low-intensity background data, if this is possible it automatically compensates for those phase drifts. As soon as symmetry phases are fixed from the sample or resolution, method (b) is comparable to method (a). Then only the possibility to treat very fast dynamics contaminations differently from very slow (static) contributions is an extra for method (b). The subtraction is then given by

$$F(q,t) = \frac{2\left(A_{\text{sample}}(q,t)/T_{\text{sample}} - \Phi A_{\text{bgr}}(q,t)/T_{\text{bgr}}\right)}{\left(I^{\uparrow}_{\text{sample}}(q,t) - I^{\downarrow}_{\text{sample}}(q,t)\right)/T_{\text{sample}} - \Phi\left(I^{\uparrow}_{\text{bgr}}(q,t) - I^{\downarrow}_{\text{bgr}}(q,t)\right)/T_{\text{bgr}}} \tag{II.3.1.14}$$

where $I^{\uparrow}(q,t)$ stands for the up (down) intensity obtained with the magnetic setting for Fourier time $t$ but inactive $\pi/2$ flippers, $A_{\text{sample/bgr}}$ are the echo amplitudes and $T_{\text{sample/bgr}}$ the respective transmissions. $\Phi$ accounts for different amounts of material (for example, solvent) in the sample and the background samples. Note that the contribution from the sample or the background substances may have different contributions than the container and the window scattering; in those cases, a more complicated formula analog to eq. (II.3.1.14) applies.

#### II.3.1.4.5 Measuring Times

Measuring times are typically in the range between 3 min and 3 h for one $F(q_i,t_j)$ value, at reasonable average 15–20 min. At one temperature, a complete measurement typically consists of about 100 single measurements $j = 1, \ldots, 5, j = 1, \ldots, 5$, which may of course change depending on individual requirements. Given an average of five samples or different temperatures/pressures, a total of three days + 1 day for resolution, transmission, and other service tasks may be considered as minimum time for an experiment.

### II.3.1.5 INTERPRETATION OF EXPERIMENTS

After performing the raw data treatment including resolution correction and background subtraction, tables of the normalized intermediate scattering function $F(q_i, t_j) = I(q_i, t_j)/I(q_i, 0)$ are obtained. The dynamical features of the investigated system must now be inferred from these values. For simple Fickian diffusion with diffusion coefficient $D$, one expects $F(q, t) = \exp(-Dq^2t)$. Any immovable scatterers contribute a $t$-independent constant $c(q)$. Segmental dynamics in polymers usually leads to $F(q, t) \approx \exp(-[\Gamma(q)t]^{\beta})$, with stretching exponent $\beta \cong 0.5, \ldots, 0.85$ depending on the scattering contrast and environment. The $q$-dependent rate increases, which is $\propto q^4$ for Rouse dynamics of polymer melts in single chain contrast and $\propto q^3$ for Zimm dynamics in polymer solutions. See below for more details on polymers. A first classification of the measured spectra can often be obtained by extracting the initial slope or first cumulant from the relaxation curves. That practically means matching

$$F(q,t) \underbrace{\cong}_{t \to 0} \exp(-[\Gamma_q t + c_2 t^2 + \cdots]) \qquad \text{(II.3.1.15)}$$

The $q$-dependence of this rate carries information on the observed processes. In particular, when the relaxation curve is close to a simple exponential, the results can be given in terms of an effective diffusion constant $D_{\text{eff}}(q) = \Gamma_q / q^2$. Simple diffusion would yield a $q$-independent value, a flexible polymer in solution shows $D_{\text{eff}}(q) \propto q$ in the intermediate $q$-regime. Any peculiar nonconstant shape of the $D_{\text{eff}}(q)$ curve indicates additional degrees of freedom that are explored by thermally excited fluctuations. For example, the extra freedom for a polymer in solution comes from its internal conformational flexibility in addition to the simple diffusive motion of the polymer's center of mass.

Probing scaling properties of the data can also serve to classify the physics and to sensitively test for small deviation from a basic underlying model. Mostly, scaling pertains either the identification of a certain combination of $q$ and t as a single independent variable which is sufficient to describe the data or by a time scaling with friction or viscosity, which yields an indirect $t - T$ scaling for a given temperature dependence of the friction. The first type of scaling is obvious for simple diffusion, where $q^2 t$ is the single independent variable. The below discussed Rouse and Zimm models of polymer dynamics in the melt or in solution exhibit $q^4 t$ and $q^3 t$ scaling, respectively. See Section II.3.1.5.1 for examples and more details.

## II.3.1.5.1 Polymers

The standard reference models to describe the molecular dynamics of linear polymers in the melt or in solution have been formulated half a century ago by Rouse (1953) and Zimm (1956), and at that time mainly in order to explain rheological properties. Later, de Gennes and Duboisviolette (de Gennes, 1967; Duboisviolette and de Gennes, 1967) derived expressions for the scattering function for these models. A comprehensive derivation can be found in the book of Doi and Edwards (1994). The basics of these models for flexible linear polymers are the assumption that the chain has the conformation of a Gaussian coil. The interaction between different segments of the chain is dominated by the conformational entropy change upon stretching (that is the usual rubber elasticity). Thus, the polymer is modeled by a spring and bead chain, the springs resulting from the entropy elasticity of an unspecified number of real segments between the virtual beads. For the simpler Rouse model, all interaction between the considered polymer chain and the surrounding chains forming the melt is lumped into a single local friction of the beads. The resulting force is proportional to the bead velocity. The linear chain exhibits relaxation modes $x_p$ such that $x_n(t) = x_0 + 2\sum_p x_p(t)\cos(np\pi/N)$ with $p = 1 \cdots N$ and relaxation rates of the correlation functions $\langle x_p(t)x_p(0)\rangle \propto \exp(-p^2t/\tau_R)$ with relaxation rates $\propto p^2$, $x_0$ is the center-of-mass coordinate. The solution of the corresponding Langevin equation and its translation to the scattering function has been extensively described in the literature by Doi and Edwards (1994); finally,

$$S(q,t) = \exp[-Dq^2t] \times \frac{1}{N}\sum_{n,n=1\ldots N} \exp\left[\left(\frac{-q^2l^2|n-m|}{6} - \frac{q^2 2R_e^2}{3\pi^2}\sum_{p=1\ldots N}\frac{1}{p^2}\cos\left(\frac{np\pi}{N}\right) \times\cos\left(\frac{mp\pi}{N}\right)\left\{1-\exp\left(-\frac{tp^2}{\tau_R}\right)\right\}\right)\right] \tag{II.3.1.16}$$

with the $R_e^2 = Nl^2$ and the Rouse time $\tau_R = \varsigma R_e^2/3\pi^2 k_B T$, the longest relaxation time of the polymer coil depending on its size $R_e$ and the bead friction $\zeta$. The center-of-mass diffusion of the Rouse chain corresponds to the Einstein formula with the added friction of $N$ beads, $D = k_B T/N$. The second part of eq. (II.3.1.16) describes $I(q, t)$ with dynamics due to internal motions only. The limiting values

$$I(q,t=0) = \frac{1}{N}\sum_{n,m=1\ldots N} \exp\left[\frac{-q^2l^2|n-m|}{6}\right] \cong N\text{Debye}\left(q\frac{R_e}{\sqrt{6}}\right) = 2N\left[\exp\left(-q^2\frac{R_e^2}{6}\right) - 1 + q^2\frac{R_e^2}{6}\right]\Big/\left(q^2\frac{R_e^2}{6}\right)^2 \tag{II.3.1.17}$$

is the well known (Doi and Edwards, 1994) form factor of a Gaussian chain. On the other hand, the infinite time limit

$$\hat{I}(q,t=\infty) = \frac{1}{N}\sum_{n,m=1\ldots N} \exp\left(\frac{-q^2l^2|n-m|}{6} - \frac{q^2 2R_e^2}{3\pi^2}\sum_{p=1\ldots N}\frac{1}{p^2}\cos\left(\frac{np\pi}{N}\right)\cos\left(\frac{mp\pi}{N}\right)\right) \tag{II.3.1.18}$$

yields approximately $N\exp(-\frac{1}{3}q^2R_e^2/6)$ the limiting form factor of the time-averaged coil. (Best-fit within $\pm 0.015\,N$ is $N\exp\left(-\frac{1}{3.2}q^2R_e^2/6\right)$. For large chains $(N\to\infty)$, respectively, $qR_e \gg 1$ and $\hat{I}(q,t=\infty) \cong 0$ the internal dynamics can be described by de Gennes expression:

$$I(q,t)/I(q,0) = F\left(x = \frac{k_B T l^2}{12\varsigma}q^4 t\right) = \int_0^\infty \exp\left[-u - \sqrt{\Gamma_q t}\,h(u/\sqrt{\Gamma_q t})\right] du \tag{II.3.1.19}$$

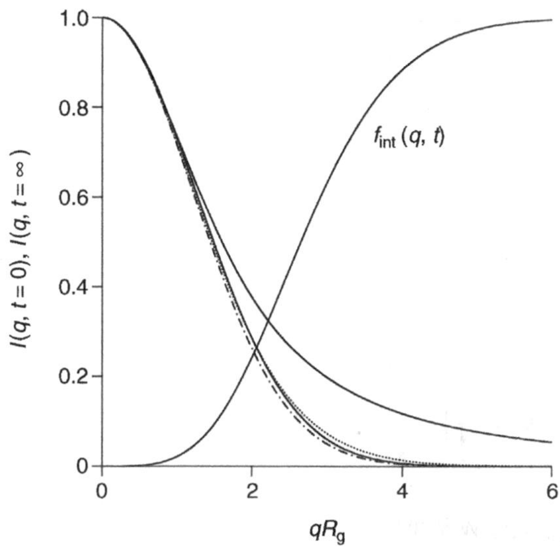


**Figure II.3.1.5** Fraction of internal Rouse $f_{int}$ $(q, t)$ dynamics in $I(q, t)$. The dotted curve corresponds to eq. (II.3.1.18). The dash–dotted curve is a Gaussian that coincides with the Debye curve at $q=0$.

with $h(u) = u\mathrm{erf}(u/2) + 2\exp(-u^2/4)/\sqrt{\pi} - u$ and $\Gamma_q t = (k_B T l^2/12\varsigma)q^4 t$, which immediately implies that the normalized intermediate scattering function depends on $q$ and $t$ only in terms of the scaling variable $x = q^4 t$. The fraction of internal dynamics compared to center-of-mass diffusion is illustrated in Figure II.3.1.5.

The Zimm model differs from the Rouse model due to the consideration of a hydrodynamic coupling between beads within one chain. The coupling is affected by the solvent flow pattern induced by the motion of bead A, which drags on bead B in some distance. Indeed is this coupling the dominant effect in polymer solutions compared to the local bead friction, it leads to a modified mode spectrum $\tau_p = [\eta(\sqrt{N}l)^3/\sqrt{3\pi k_B T}]p^{-3/2}$ and the center-of-mass diffusion constant $D = 8k_B T/3\eta l\sqrt{6\pi^3 N}$, which depend on the solvent viscosity $\eta$. Relpacing $\tau_p$ and $D$ in eq. (II.3.1.16) yields the expression for the intermediate scattering function of the Zimm model. With $x = (k_B T\eta/6\pi)q^3 t$, the scaling form of the Zimm model in the large $q$ limit is given as

$$F(x) = \int_0^\infty \exp\left[-u - x^{2/3}\frac{2}{\pi}\int_0^\infty \frac{\cos(yux^{-2/3})}{y^2}\left\{1-\exp(-y^{2/3}/\sqrt{2})\right\}dy\right]du, \tag{II.3.1.20}$$

which may be approximated by $F(x) \approx \exp\left[-(x/a)^\beta\right]$ with $a \approx 1.354$ and $\beta \approx 0.85$. The differences in scaling behavior and how they can help to classify the data in the one or the other model realm is illustrated in Figure II.3.1.6.

The self-part of the scattering function, which represents the diffusive motion of single segment is

$$\exp(-q^2 Dt) \times \exp\left(-q^2\frac{l^2}{6}\left[\frac{12k_B T}{\pi\zeta l^2}t\right]^{1/2}\right) \tag{II.3.1.21}$$

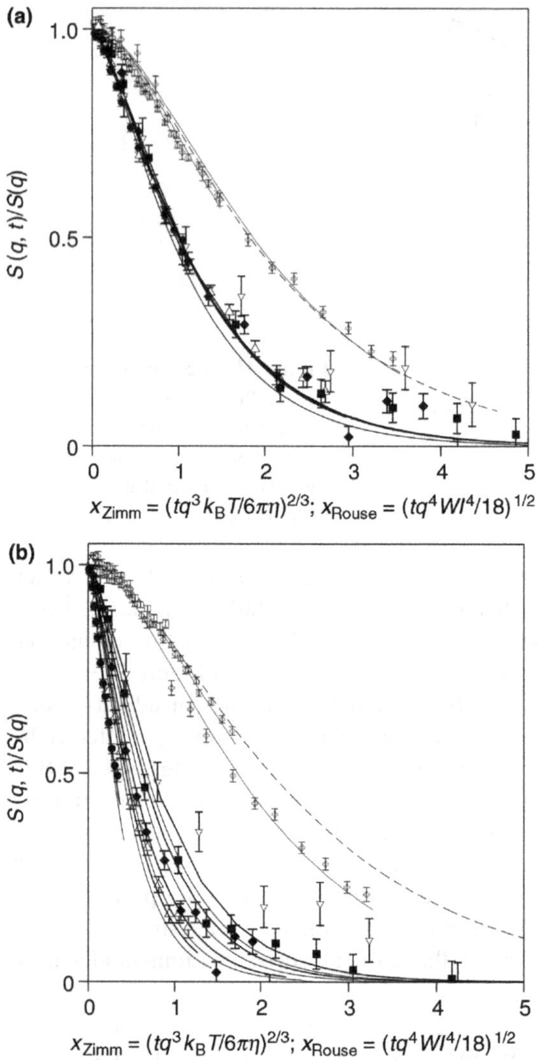


**Figure II.3.1.6** (a) Data from a melt of PEE chains exhibiting Rouse scaling (upper curve) and from a PI solution in THF showing Zimm scaling (lower curves), the covered $q$-range was (0.05–0.18 Å$^{-1}$). (b) The same data, but with exchange scaling schemes, thus demonstrating how sensitive data following one of the paradigms (Rouse or Zimm) may be discriminated.

for Rouse and

$$\exp(-q^2Dt) \times \exp\left(-\frac{q^2}{6} \cdot \frac{lb}{g\pi^{3/2}}\left[\frac{\sqrt{3\pi}k_BT}{\eta}t\right]^{2/3}\right) \tag{II.3.1.22}$$

Zimm dynamics, respectively. Experimentally, these functions are accessible by random segment labeling (Richter et al., 1989) or—in the case of melts—by analyzing the incoherent scattering from a protonated sample (Wischnewski et al., 2003).

With the currently available computing facilities, however, it is also feasible to directly sum eq. (II.3.1.16) and its Zimm equivalent that allows studying the effect of a modified mode spectrum and/or some stretching of the exponential mode relaxations. The latter is suggested by a number of MD calculations that were analyzed in terms of Rouse modes (Smith et al., 2000, 2001).

Below the radius of gyration, both the Rouse and the Zimm models do not have an intrinsic length scale, a fact that is reflected by the scaling property. Of course, the molecular size of a monomer sets a lower length limit to this scale invariance. Beyond the validity of these models, violation of the scaling indicates that the existence of other characteristic lengths. Toward low $q$, this is seen by the transition from (scaling) Rouse dynamics to center-of-mass diffusion. For long chains, the entanglement of chains leads to another scaling violation, which is related to the diameter of the virtual tube that is formed by the entanglements (Edwards, 1967; de Gennes, 1971). NSE offers the unique possibility to investigate the shape of the virtual tube by following the motions of a single labeled chain. After a time, $t > \tau_e$ when all internal modes of the chain have explored their configuration space across the tube but $t > \tau_d$ when the chain leaves the initial tube by curvilinear diffusion motion, the scattering function $S(q, t)$ corresponds to the dynamically accessible virtual tube space. The value of $S(q, t)$ levels off to a plateau, further decay only occurs on timescales of $\tau_d$. The times $\tau_e$ and $\tau_d$ are the "entanglement time," that is, the time a segment may freely diffuse before the tube confinement sets in and the reptation time, respectively. The scattering function for this case may be described by (Schleger et al., 1998; de Gennes, 1981; Wischnewski et al., 2000):

$$\frac{S(q,t)}{S(q)} = \left\{1 - e^{-q^2 d^2/36}\right\} e^{t/\tau_0} erfc\left(\sqrt{t/\tau_0}\right) + e^{-q^2 d^2/36} S_{\text{escape}}(q,t), \qquad \text{(II.3.1.23)}$$

where $\tau_0(q) = 12\varsigma/(k_B T l^2 q^4)$ is the timescale associated with local reptation and $S_{\text{escape}}(q,t)$ for long chains decays on the timescale $\tau_d$ that is much larger than the entanglement time and the characteristic local reptation time in the relevant $q$-range. At short times below $\tau_e$, eq. (II.3.1.23) does not apply. However, there the segment exhibits a free Rouse diffusion until the segments "feel" the tube constraints. This has been shown by following the short-time segment diffusion using the incoherent scattering signal; see Figure II.3.1.7 and Ref. (Wischnewski et al., 2003).

For melts of shorter chains, end effects gain relative importance and thereby reveal the effects of contour length fluctuations (CLFs). Due to fluctuating end motion into the tube and back into some other direction uncorrelated to the previously abandoned initial tube, the correlation decays at the ends at a faster rate than by pure reptation. The CLF effect leads to a faster decay of $S_{\text{escape}}(q, t)$, which can be modeled by an analytic expression (Likhtman and McLeish, 2002; Zamponi et al., 2005). The underlying picture associates this faster decay with decorrelation of the chain ends and their initial tube sections where the fraction of lost initial tube $s(t) = (N_e C_\mu / 2N)[(t/\tau_e)]^{1/4}$ increases with the $t^{1/4}$; $C_\mu \cong 1.5$ is a numerical factor determined by simulation. This

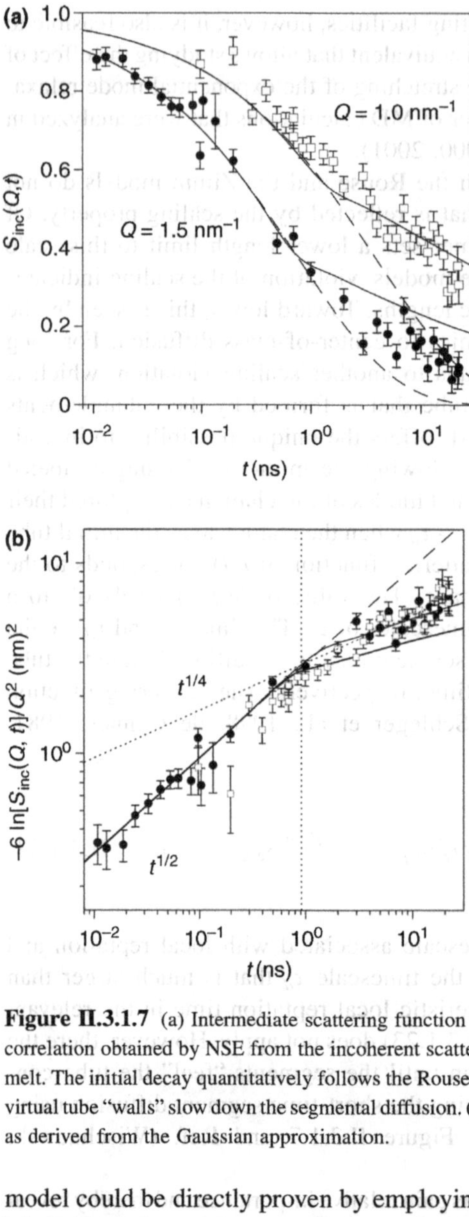


**Figure II.3.1.7** (a) Intermediate scattering function corresponding to the proton (segment) self-correlation obtained by NSE from the incoherent scattering from a high molecular weight polyethylene melt. The initial decay quantitatively follows the Rouse prediction (dashed lines). At times beyond $\tau_e$ the virtual tube "walls" slow down the segmental diffusion. (b) The mean square displacement of the segments as derived from the Gaussian approximation.

model could be directly proven by employing partial labeling. When the ends of the labeled h-chains are deuterated up to a relative end distance $s(t_{max})$, the extra decay due to CLF is no longer observed. The escape term is indistinguishable from the one for very long chains. The length of the labeled section in the experiment had been chosen such that the time window ($t_{min} = 1$ ns, $t_{max} = 160$ ns) of the NSE experiment matches the time the CLF need to reach the h-labeled part (Zamponi et al., 2005). Figure II.3.1.8 compares NSE spectra from a long chain with full tube confinement with a middle labeled short chain and a short chain with ends that undergo visible CLF.

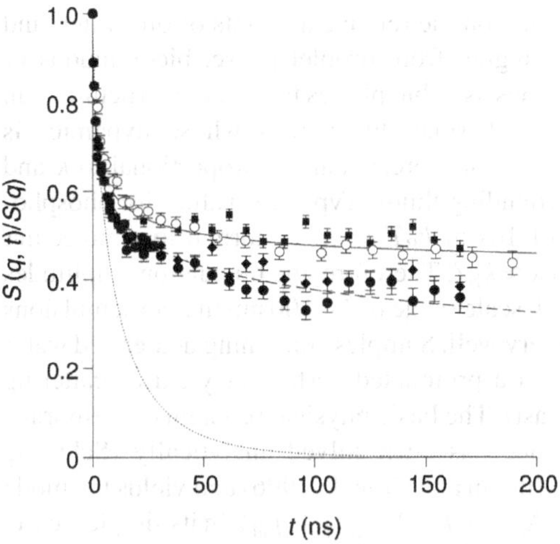


**Figure II.3.1.8** Comparison of the free Rouse chain expectation (dotted line) with experimental data from very long chain polyethylene with al molecular weight of 190 kg/mol (filled squares) and from a shorter chain with d-h-d labeling in a deuterated matrix with 4k-17k-4k g/mol, open circles. Both follow eq. (II.3.1.23) computed with a tube diameter of 49 Å and virtually constant escape term (solid line). The CLF effect is masked because the ends are made invisible by the *d*-labeling. Short *h*-chain (16 and 25 kg/mol) with significant CLF effect drops in the plateau region according to the loss of tube from the ends as described in the text (solid circles; solid diamonds, and dashed line). For all data: $q = 0.115\,\text{Å}^{-1}$ and $T = 509\,\text{K}$.

## II.3.1.5.2 Micelles, Microemulsions, and Membranes

The elementary building block of a microemulsion is the surfactant interface (membrane) that separates regions of two immiscible fluids (oil and water). The surfactant has a hydrophilic and a hydrophobic side one pointing into the water the other into the oil region. If there is only one fluid present, the surfactant molecules aggregate and form micelles. Or double-sided membranes are formed as the phospholipid hull of vesicles and cells. Much of the phase properties are determined by the elastic moduli and the spontaneous curvature of the interface membranes and their interactions. The expression of the membrane free energy per interface area as function of local curvature stems from Helfrich (Helfrich, 1973):

$$F = \frac{1}{2}\kappa\left(\frac{1}{R_1} + \frac{1}{R_2} - \frac{2}{R_0}\right)^2 + \bar{\kappa}\frac{1}{R_1 R_2} \qquad \text{(II.3.1.24)}$$

For closed shapes such as droplets, the second term (saddle-splay bending) contributes a constant to the energy that only depends on the topological connectivity of the shape. Note also that, for $R_0 = 0$, the total energy of a piece of membrane is invariant with respect to a change of length scale.

Depending on the specific values and on the relative amounts of oil, water, and surfactant, a multitude of structures ranging from droplet phase, bicontinuous or sponge phases, as well as ordered varieties as cubic phases to lamellar structures can be obtained. The membranes undergo thermal fluctuations whose dynamics is determined by the balance of elastic restoring forces that are proportional to $\kappa$ and friction due to viscous flow of the surrounding fluids. Typical $\kappa$-values for phospholipid double layers are in the range of 10s of $k_B T$ whereas normal low molecular surfactants form microemulsions with $\kappa \approx k_B T$. Therefore, the fluctuation amplitudes are large and easily detected. The length scale range of 1 – 100 nm in microemulsions matches the SANS and NSE $q$-ranges very well. Samples containing deuterated water and deuterated oil in combination with a protonated surfactant yield a scattering intensity from the interface (film contrast). The basic physical problem of membrane undulation modes on a planar interface has been solved analytically (Schilling et al., 2001; Ramaswamy et al., 1993; Zilman and Granek, 1996) and yields the mode wave vector-dependent relaxation rate $\Lambda_k = \kappa k^3/2(\eta_{\mathrm{water}} + \eta_{\mathrm{oil}})$, in its simplest case. In the lamellar phase the membrane interaction stabilizes the smectic structure and additional contributions to the rate are present. In addition, it is noted here that the large fluctuation amplitudes cause a nonlinear coupling between displacement and the scattering phase that leads to the effect that the mode wave vector $k$ cannot easily be identified with the experimental wave vector $q$. A $k$-mode may contribute to intensity at all $q$-values. Experiments on lamellar phases have some caveats since they tend to partially orient with time or even if they can be reproducibly oriented the effects of bending modulus and membrane interaction combine. Bicontinuous microemulsions enable a somewhat cleaner access to the local bending modulus when measured with NSE (Holderer et al., 2005). At high $q$ values, on a local scale, the membrane dynamics is seen on randomly oriented patches with a size proportional to the correlation size $\zeta$ of the bicontinuous phase (Zilman and Granek, 1996). The normalized intermediate scattering function is predicted to have the form

$$\frac{S(q,t)}{S(q)} = \exp\left[-(\Gamma_q t)^{2/3}\right] \tag{II.3.1.25}$$

where the 3 in the denominator of the stretching exponent relates to the $k^3$ exponent of the mode dispersion in the membranes. The quantitative relation between the rate $\Gamma_q \propto q^3 f(\kappa,\zeta,\eta,T)$ and the bending modulus $\kappa$ also depends on the correlation length $\zeta$ and requires numerical evaluation of $f$ (Holderer et al., 2005). The approximate relation $\Gamma_q = q^3 0.025\gamma_k \sqrt{k_B T/\kappa}(k_B T/\eta)$ where $\gamma_k \cong 1 - 3(k_B T/4\pi\kappa)\ln(q\zeta) \underset{\kappa \gg k_B T}{\longrightarrow} 1$ (Zilman and Granek, 1996) seems to fail for small values of $\kappa \approx k_B T$. The correlation length $\zeta$ may be inferred from SANS data obtained in water versus oil contrast (bulk contrast). The two relevant length scales, namely, domain size $d = 2\pi/q_0$ and correlation length $\zeta$ determine the shape of the SANS curve (Teubner and Strey, 1987):

$$S(q) \propto \left(q^4 - 2\left[q_0^2 - \zeta^{-2}\right]q^2 + \left[q_0^2 + \zeta^{-2}\right]^2\right)^{-1}. \tag{II.3.1.26}$$

These lengths allow for another approach to determine (Gompper et al., 2001):

$$\kappa_{\mathrm{SANS}} = q_0 \zeta \frac{5\sqrt{3} k_{\mathrm{B}} T}{64}. \tag{II.3.1.27}$$

As discussed by Holderer et al. (2005), it is shown that the NSE methods yields the bare bending modulus whereas the SANS method lumps the effect of all short-range fluctuations into a renormalized

$$\kappa_{\mathrm{SANS}} = \kappa + k_{\mathrm{B}} T \left( \frac{\alpha}{4\pi} \right) \ln \Psi \tag{II.3.1.28}$$

where $\alpha = 3$ and $\Psi$ is the volume fraction of the surfactant. Thereby, in the water-C10E4-decane system, the surfactant concentration dependence of $\kappa_{\mathrm{SANS}}$ is explained in reasonable agreement with the bare $\kappa$ from NSE. The obtained bare bending modulus had values close to $k_{\mathrm{B}}T$. Addition of polymeric analogs of C10E4 either as block coplymers or as homopolymers cause the so-called boosting (Jakobs et al., 1999) or antiboosting (Frielinghaus et al., 2004) effect. Boosting means that addition of a small amount of copolymer greatly reduces the total amount of surfactant necessary to hold a given amount of oil and water in the microemulsion phase. The effect is explained by a change of the bending modulus due to polymer addition. The associated changes of a few tenths of $k_{\mathrm{B}}T$ are observed in the NSE and renormalized SANS values (Holderer et al., 2005).

Droplet microemulsions have first been investigated by NSE in the seminal work of Huang et al. (1987). The observed dynamics contains contributions from the center-of-mass diffusion of the droplets and the shape fluctuations of the droplets. The latter depends on the membrane properties and carry information on the bending modulus. The most prominent dynamical shape fluctuation mode corresponds to the spherical harmonics with $l = 2$, an ellipsoidal deviation from the average sphere. Even in film contrast, these (and higher $l$ fluctuations) contributions are masked by the intensity of the average sphere. Only in regions where the form factor $f_0(qR)$ of the sphere is close to zero, the fluctuation scattering yields a detectable contribution (Huang et al., 1987; Farago et al., 1990a and 1990b)

$$S(q,t) \propto \exp\left(-D_{\mathrm{cm}} q^2 t\right) V_{\mathrm{surfactant}}(R) \left[ f_0(qR) + \sum_{l \geq 2} \frac{2l+1}{4\pi} \langle |u_l^2| \rangle f_l(qR) \exp(-\Gamma_l t) \right]. \tag{II.3.1.29}$$

The effective diffusion constant extracted from the initial slope

$$D_{\mathrm{eff}}(q) = -\frac{1}{S(q)} \left( \frac{\mathrm{d}S(q,t)}{\mathrm{d}t} \right) \tag{II.3.1.30}$$

exhibits a maximum where $f_0(qR)$ has a minimum and in particular $\langle |u_2^2| \rangle f_2(qR) \exp(-\Gamma_2 t)$ contributes. Compared to the center-of-mass diffusion, the experiments show a factor of 2 increase of $D_{\text{eff}}(q)$ at $qR = \pi$ where $f_0(qR)$ has its first minimum. In order to extract membrane parameters from the data, it has to be observed that the depth of the minimum depends on the degree of the droplet polydispersity. Thereby, the saddle-splay modulus enters here since it contributes to the total free energy change if the number of droplets changes. The amplitudes, the rates $\left\langle |u_l|^2 \right\rangle$, and $\Gamma_l$ depend on the ratio $\kappa/\kappa$ between the saddle-splay and the bending modulus and $R/R_0$, the ratio between the droplet radius and the spontaneous curvature. The amplitude for $l = 0$ is proportional to the polydispersity index $p$, $\left\langle |u_0|^2 \right\rangle = 4\pi p$. In addition, they contain a concentration-dependent entropy contribution $h(\phi) = -\frac{1}{\phi}(\phi \ln \phi + [1-\phi] \ln[1-\phi])$, for example, see the work by Huang et al. (1987) and Nagao and Seto (2008),

$$\left\langle |u_0|^2 \right\rangle = \frac{k_B T}{\kappa(6-A)}; \quad \left\langle |u_2|^2 \right\rangle = \frac{k_B T}{4\kappa A}; \quad \Gamma_2 = \frac{A}{R^3} \frac{24\kappa}{23\eta_{\text{core}} + 32\eta_{\text{solvent}}}, \tag{II.3.1.31}$$

where $A = \{4(R/R_0) - 3(\bar{\kappa}/\kappa) - [(3k_B T/4\pi\kappa)h(\phi)]\}$. The first term describes the polydispersity, the $\bar{\kappa}$ dependence enters it via $A$. The expressions may be combined to yield (Kawabata et al., 2007)

$$\kappa = \frac{1}{48}\left( \frac{4k_B T}{\left\langle |u_0|^2 \right\rangle} + \Gamma_2 R^3 \frac{23\eta_{\text{core}} + 32\eta_{\text{solvent}}}{3} \right). \tag{II.3.1.32}$$

Note, however, that the initial slope contribution of the $l = 2$ mode is proportional to

$$\left\langle |u_2|^2 \right\rangle \Gamma_2 = \frac{k_B T}{4R^3} \frac{24}{23\eta_{\text{core}} + 32\eta_{\text{solvent}}}, \tag{II.3.1.33}$$

which does not depend on the bending moduli. The latter fact poses a severe difficulty to the precise determination of $\kappa$ from the initial slope. And only the direct observation of the multiexponetial decay as a function of time allows the independent determination of decay rate and amplitude of the deformation mode (Farago et al., 1995).

If one considers the relaxation of the $l = 2$ ellipsoidal distortion besides the shape relaxation toward a sphere, rotational diffusion of the ellipsoid may contribute to the dynamics. In particular, if the viscosity $\eta_{\text{core}}$ of the inner fluid is very high, rotational diffusion could be the dominant relaxation. However, Milner and Safran (1987) estimated the rotation diffusion contribution for equal viscosities as $\Gamma_2^{\text{rot}}/\Gamma_2^{\text{shape}} \approx (k_B T/8\pi\kappa)(24/55)$, which is less than 0.1 for $\kappa \approx k_B T$. Thus, for typical microemulsion, the shape fluctuations dominate. Using this method, the bending moduli of AOT in a water-decane microemulsion has been determined as 3.8 kT and $\bar{\kappa}/\kappa = -1.89$ the effect of adding butanol as cosurfactant was a significant reduction of the bending moduli (Huang et al., 1987). In the aftermath, the topic and method have been resumed with the investigation of the effect of *tert*- and *sec*-butyl alcohol as cosurfactant (Zambrano et al., 2006) reporting similar trends upon alcohol addition. However, the absolute $\kappa$-values are an order of magnitude smaller. Besides, details in the fitting procedure the earlier investigations and the recent ones differ only slightly in the expressions used to connect the membrane properties to the scattering function. The older work neglects the entropic contribution $h(\phi)$, which, however, as the spontaneous curvature cancels out of the result for $\kappa$-value and has the minor influence on the value of $A$. Recently, Nagao and Seto (2008) modified the method by considering the ratio of data from film and bulk contrast. For dense systems, where interference between droplets leads to a structure factor, this method has the advantage that the structure factor divides out of the ratio of the measurement with the two contrasts.

The huge difference in the results on virtually the same system still needs clarification, it seems to be connected to the difficulty to determine $\Gamma_2$ and $\left\langle |u_2|^2 \right\rangle$ independently, which is a badly defined problem if no highly accurate relaxation curves that allow a multiexponetial (explicit or implicit) decomposition are available and treated accordingly. Let us expressed this in a different manner: For these types of problems, it is mandatory to include a shape analysis of the decay curves (especially those obtained around the maximum of $D_{\mathrm{eff}}$) in order to obtain reliable estimates for the membrane parameters.

## II.3.1.5.3 Proteins and Other Biopolymers

Proteins are rather compact molecular items, which serve as biochemical nanomachines to catalyze or control specific reactions in the cell. In the realm of NSE diffusion and large-scale internal domain motions are of particular importance. The most prominent motion of (globular) proteins in aqueous solution close to physiological conditions is the center-of-mass diffusion. On top of this, effects of rotational diffusion and internal domain motions may be present. Under suitable conditions, the contribution due to domain motions can be separated and analyzed with respect to amplitude, mobility (friction), and displacement pattern. If the corresponding scattering can be observed at all, one may immediately infer that the amplitude is in the nanometer range and the mobility is not too far from a freely diffusing domain. NSE observes equilibrium fluctuations and thereby provides information on the mobility, friction, and spring constants.

Thus, the protein nanomachinery is not observed in action, however, additional knowledge on its functioning—beyond that inferred from analyzing the structure—is obtained by watching which part of the "device" moves and when it is shaken. Technically, the effects of center-of-mass diffusion and rotational diffusion have to be accounted for before further internal domain or shape fluctuations can be identified. For this purpose, one needs the $6 \times 6$ hydrodynamic mobility tensor $\overleftrightarrow{D}$ of the (hypothetically) rigid protein. An accurate estimate for $\overleftrightarrow{D}$ can be computed with the program HYDROPRO (de la Torre et al., 2000). With that, the effective diffusion constant as derived from the first cumulant (initial slope) of the intermediate scattering function is

$$D_{\text{eff}}(q) = \frac{\sum_{j,k} \left\langle b_j b_k \exp(i\vec{q}\cdot[\vec{r}_j - \vec{r}_k]) \begin{pmatrix} \vec{q} \\ \vec{q}\times\vec{r}_j \end{pmatrix} \overleftrightarrow{D} \begin{pmatrix} \vec{q} \\ \vec{q}\times\vec{r}_k \end{pmatrix} \right\rangle}{q^2 \sum_{j,k} \langle b_j b_k \exp(i\vec{q}\cdot[\vec{r}_j - \vec{r}_k])\rangle} \qquad \text{(II.3.1.34)}$$

where—depending on the level of coarse graining—$\vec{r}_j$ and $b_j$ stand for the center of scattering and the sum of scattering length of the atoms belonging to group (amino acid ...) $j$ minus the solvent scattering length density contained in a volume of the same size as that of group $j$. Additional contributions $\Delta D_{\text{eff}}(q) = D_{\text{meas}}(q) - D_{\text{eff}}(q)$ are attributed to internal motions. This generic scenario is shown in Figure II.3.1.9. These carry the possible information on the functional domain motions. The feasibility to detect internal domain motions by NSE has been shown using the protein taq-polymerase (Callaway et al., 2005). As a newer and more elaborated example, we discuss results obtained for the protein yeast alcohol dehydrogenase (ADH) (Biehl et al., 2008).

In solution yeast ADH is a tetramer, each subunit has a dumbbell-like structure formed by the globular catalytic domain with two Zn atoms and the globular cofactor binding domain, see Figure II.3.1.10. As cofactor needed for the reduction

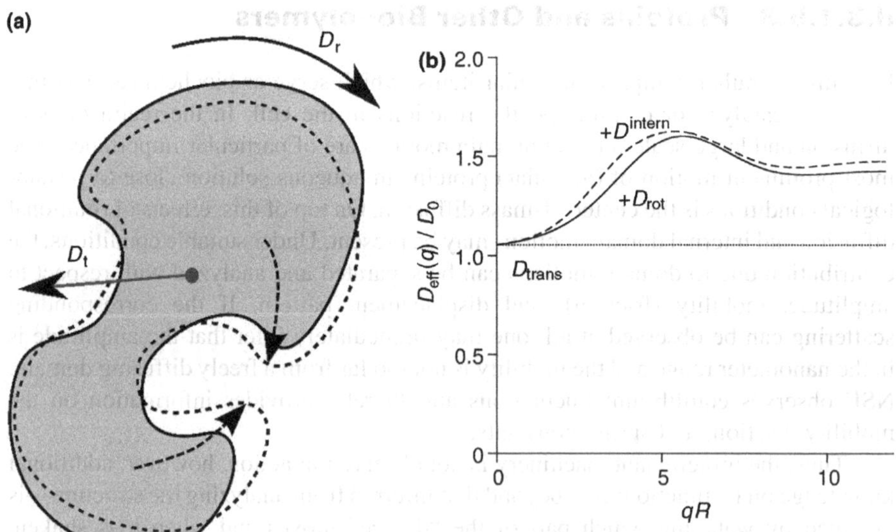


**Figure II.3.1.9** (a) The coherent quasielastic scattering of an aspherical semirigid particle (such as a protein) is dominated by contributions from center-of-mass diffusion, $D_t$, rigid body rotational diffusion, $D_r$, and internal domain (shape) fluctuations. Typical magnitudes of the effect of $D_{\text{eff}}$—as obtained from the initial slope of the NSE curves—for a protein with flexible domains is shown in (b).

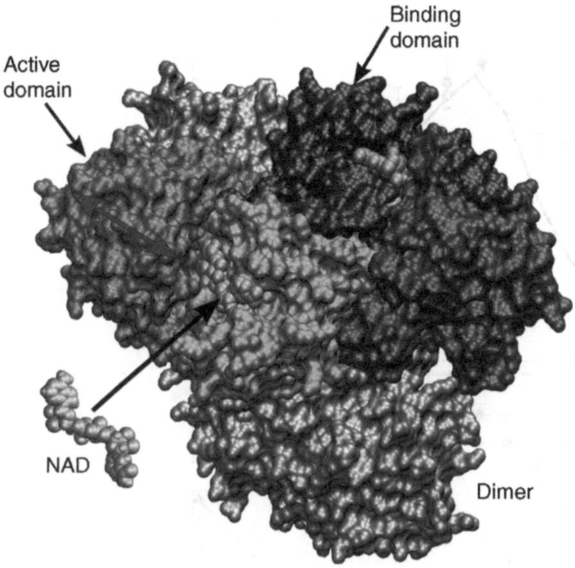


**Figure II.3.1.10** Space filing model of the protein ADH in its form as dimer. The binding of the cofactor NAD is indicated. The NSE experiment was performed on a tetrameric form of ADH, which corresponds to the association of two of these dimers. (See the color version of this figure in Color Plates section.)

(oxidation) reaction of alcohol nicotine adenine dinucleotide (NAD) is needed. The observed internal fluctuations pertains the opening of the cleft between catalytic and binding domain and is thought to assist the incorporation of NAD. Neutron small angle (SANS) experiments showed that ADH tetramers in solution ($D_2O$ + buffer) show a conformation that exactly matches the crystal structure as given in the protein data bank. NSE data in the range from $q = 0.29\,\text{nm}^{-1}$ ($1\,\text{ns} < t < 160\,\text{ns}$) to $q = 0.2\,\text{nm}^{-1}$ ($0.1\,\text{ns} < t < 25\,\text{ns}$) were measured at $T = 5°\text{C}$ and the initial slope was determined by fitting to exponentials. Applying the Einstein–Sokes relation to the limiting value of the diffusion coefficient $D_{\text{meas}}(q \to 0)$ of $23.5 \pm 2\,\mu\text{m}^2/\text{s}$ yields a radius of gyration of a compact sphere of $R_g = 3.5\,\text{nm}$ in good agreement with the value of 3.4 nm from the crystal structure. The effective diffusion exhibits an increase of 30% around $q = 1\,\text{nm}^{-1}$.

Comparing with the results of eq. (II.3.1.34) shows that the low $q$-value can be explained by center-of-mass diffusion and most of the increase at larger $q$ by rotational diffusion. However, in particular, at the leading flank of the maximum in $D_{\text{meas}}(q)$ at $q$ slightly below $1\,\text{nm}^{-1}$, a significant additional contribution $\Delta D_{\text{eff}}(q)$ with a peak at $q = 0.8\,\text{nm}^{-1}$ is observed. The results are summarized in Figure II.3.1.11. Closer inspection of the measured $S(q, t)/S(q)$ curves at that wave vector reveals that the relaxation comprises at least two exponentials. The extra contribution has a relative amplitude of 0.1 and a relaxation time of 35 ns. Further interpretation of these data has been obtained by computing displacement patterns of the molecule in terms of vibrations in a coarse elastic network model. The displacement patterns of low-lying modes have been converted to structure factor contributions in analogy to one-phonon cross sections. The initial slope contribution of the lowest nontrivial mode matches the $q$-dependence of $\Delta D_{\text{eff}}(q)$ quite well; it

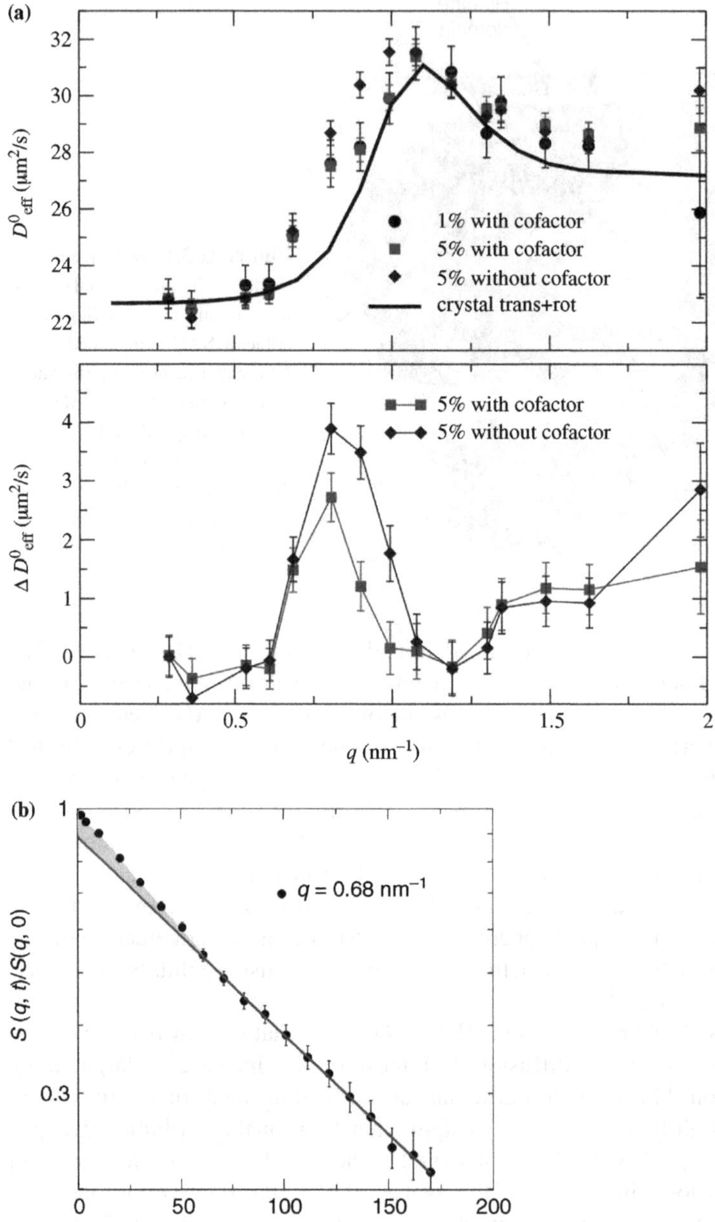


**Figure II.3.1.11** (a) Initial slope derived effective diffusion obtained from NSE data from 5% ADH solutions with and without cofactor in comparison with the rigid body translation and rotation prediction (solid) line. The lower part displays the excess due to internal motion; see the work of Biehl et al. (2008). (b) Direct NSE data from the ADH protein solution at a wave vector where the additional dynamics due to domain fluctuation is strong (shaded region). The line corresponds to the contributions of center-of-mass and rotational diffusion. The effect of the internal modes on the initial slope is in the range of 10%.

corresponds to a cleft opening–closing motion. The addition of NAD reduces the intensity of the measured peak as well as the height of the computed structure factor. These observations strongly suggest that the observed internal dynamics pertains to a cleft opening fluctuation that influences the speed of NAD association/release. A few other low modes may also contribute to some extent.

### II.3.1.5.4 Other Applications

Other applications of NSE on protein solutions focus on the center-of-mass dynamics and its concentration dependence (Longeville and Le Coeur, 2008). This diffusion is influenced by direct (potential) or indirect (hydrodynamics) interactions. At high concentrations, the hydrodynamic interaction leads to an increased effective viscosity and a slowing down of diffusion. The latter effect influences the efficiency of transport in crowded solutions, for example, the oxygen transport by hemoglobin or myoglobin (Longeville and Le Coeur, 2008a). Other proteins such as the spherical ferritin aggregate (Haeussler, 2008a) rather serve as probe particle to probe the theory for colloid systems. In that case, proteins take the role as well-defined small particles. A careful choice can identify proteins that form rigid and virtually spherical objects ideally suited to test the physics of interacting colloidal systems. However, in both cases, the biologically motivated (biophysics) or the colloid physics motivate (biological physics)—the NSE studies aim at the measurement and understanding of the hydrodynamic function $H(q, \phi)$ on the measured diffusion $D(q) = D_0H(q)/S(q)$ where the structure factor $S(q)$ depends on the direct (potential) interactions only and can be inferred from SANS experiments. The $q$-dependence of the hydrodynamic function $H(q)$ resembles $S(q)$ and thereby mitigates modulation of $D(q)$ due to the structure factor (Doster and Longeville, 2007). An important result is that with increasing concentration the diffusion constant (of hemoglobin and myoglobin) goes down exponentially and drops even faster than expected. The additional friction due to protein–protein interaction at $\phi = 0.4$ is nearly 30 times as large as the solvent friction. The unexpected magnitude of the effect is attributed to the water hull of the protein that adds to the volume and thereby rescales the effective volume fraction (Longeville and Le Coeur, 2008).

## II.3.1.6 CONCLUSION

In the very broad field of soft matter systems, the molecular motions that are relevant for the properties and function occur on the scale of their (mesoscopic) molecular building blocks. In this spatial range, of a few nanometers motions have typical times of some nanoseconds or more. Thermal fluctuations or Brownian motions occur in equilibrium and carry information on the mobility of the observed items (molecular groups, protein domains, interfaces, and so on). For spectroscopic methods, the expected timescale translates into a required resolution from μeV to neV. For neutron instruments that offer the unique possibility to modulate visibility of building blocks of the sample by *H*/*D*-labeling, this range is just touched by the spectroscopic backscattering instruments with their resolution of slightly less than 1 μeV. To reach the neV range, neutron spin echo techniques are the only known means to cover that dynamical range. As Fourier method, NSE yields data in the time domain in terms of the intermediate scattering function. The covered time range extends from some ps to μs by combining change of magnetic field integral $BL = 2 \times 10^{-4}$–1 Tm and wavelength $\lambda \cong 0.2$–2 nm to utilize the relation $t \propto BL\lambda^3$. It should be noted that even with this high resolution, the viscosity or equivalent friction on a local scale must not be much higher that of a typical liquid in order to be able to observe the motions.

Thus, with NSE and labeling technique, it is possible to observe the dynamical single-chain structure factor in polymer melts. Thereby, it has been possible to verify and to refine concepts contained in the Rouse model and for longer chains to directly observe their dynamical confinement in a virtual tube due to entanglements. Since the equilibrium conformation of chains in the melt is a Gaussian coil, the tube confinement is only visible in the evolvement of the intermediate scattering function with time. In solutions, the Zimm model takes the role of the Rouse model as prototypical description of the dynamics. All deviations occurring due to restricted flexibility of the polymer chain or mutual interaction a higher concentration reveal themselves first as deviations from the standard Zimm or Rouse behaviors. In microemulsions, the interface between oil and water contains the surfactant. Labeling allows to "stain" just this interface and observe its dynamics. Here, as the friction can largely be inferred from the macroscopic viscosities of oil and water, the dynamics may be used to extract the membrane-bending modules, which determine much of the physics of microemulsions. However, the interpretation of the observed $S(q, t)$ in terms of bending elasticities is involved, depends on the phase and requires sophisticated evaluations. Nevertheless, it has been able to corroborate one of the theoretical expressions for the bending modulus renormalization by comparison of NSE and SANS results. Also the increase of the bending modulus by a few $0.1k_BT$ due to addition of amphiphilic block copolymers and the adverse effect of the corresponding homopolymers could be extracted from NSE results. These findings are in accordance with the theoretical explanations of the boosting or antiboosting effects of polymer addition to microemulsions. Proteins are rather compact macromolecular objects, which can catalyze biochemical reactions in the cell. The question whether they can do this as completely rigid objects or whether some degree of domain motion supports this function has been addressed with NSE. After a careful subtraction of rigid body diffusion effects, a significant signal from mobile domains remains. It could be assigned to motional modes that most probably are needed to give access of the reactants to the binding site in the protein.

The above-mentioned examples are far from complete, but serve as examples what kind of knowledge NSE experiments can supply. There are many additional fields even in soft matter where the method already contributed significantly and will continue to do it in the future. Just some are mentioned here: dynamics of glass-forming systems, gels, rubbers, diffusion in microporous substrates, and more (Richter et al., 2005).

## REFERENCES


Biehl, R., Hoffmann, B., Monkenbusch, M., Falus, P., Prevost, S., Merkel, R., and Richter, D. *Phys. Rev. Lett.* **2008**, *101*, 138102.

Callaway, D.J.E., Bu, Z.M., Biehl, R., Monkenbusch, M., and Richter, D. *Proc. Natl. Acad. Sci. USA* **2005**, *102*, 17646.

De la Torre, J.G., Huertas, M.L., and Carrasco, B. *Biophys. J.* **2000**, *78*, 719–730.

De Gennes, P.G. *Physics* **1967**, *3*, 37.

De Gennes, P.G. *J. Chem. Phys.* **1971**, *55*, 572.

De Gennes P.G. *J. de Physique* **1981**, *42*, 735.

Doi, M. and Edwards, S.F. *International Series of Monographs on Physics*, Series 73, Oxford University Press, Oxford, 1994.

Doster, W. and Longeville, S. *Biophys. J.* **2007**, *93*, 1360.

Duboisviolette, E. and de Gennes, P.G. *Physics* **1967**, *3*, 181.

Edwards S.F. *Proc. Phys Soc.* **1967**, *92*, 9.

Farago, B., Richter, D., Huang, J.S., Safran, S.A., and Milner, S.T. *Prog. Colloid Polym. Sci.* **1990a**, *81*, 60.

Farago, B., Richter, D., Huang, J.S., Safran, S.A., and Milner, S.T. *Phys. Rev. Lett.* **1990b**, *65*, 3348–3351.

Farago, B., Monkenbusch, M., Goecking, K.D., Richter, D., and Huang, J.S. *Physica B* **1995**, *213–214*, 712.

Farago, B. *Physica B* **1998**, *241–243*, 113.

Farago, B. *Physica B* **1999**, *268*, 270.

Frielinghaus, H., Byelov, D., Holderer, O., Allgaier, J., and Richter, D. *Langmuir* **2004**, *20*, 10433.

Gompper, G., Endo, H., Mihailescu, M., Allgaier, J., Monkenbusch, M., Richter, D., Jakobs, B., Sottmann, T., and Strey, R. *Europhys. Lett.* **2001**, *56*, 683.

Golub, R. and Gahler, R. *Phys. Lett. A* **1987**, *123*, 43.

Hahn, E.L. *Phys. Rev. Lett.* **1950**, *80*, 580.

Haeussler, W. *Eur. Biophys. J.* **2008a**, *37*, 563.

Haeussler, W., Streibl, D., and Boeni, P., *Meas. Sci. Technol.* **2008b**, *19*, 034015.

Haeussler, W., Gohla-Neudeckera, B., Schwikowskia, R., Streiblb, D., and Boeni, P. *Physica B* **2007**, *397*, 112.

Helfrich, W. *Z. Naturforsch. C* **1973**, *28*, 693–703.

Holderer, O., Frielinghaus, H., Byelov, D., Monkenbusch, M., Allgaier, J., and Richter, D. *J. Chem. Phys.* **2005**, *122*, 094908.

Huang, J.S., Milner, S.T., Farago, B., Richter, D. *Phys. Rev. Lett.* **1987**, *59*, 2600.

Jakobs, B., Sottmann, T., Strey, R., Allgaier, J., Willner, L., and Richter, D. *Langmuir* **1999**, *15*, 6707.

Kawabata, Y., Seto, H., Nagao, M., and Takeda, T. *J. Chem. Phys.* **2007**, *127*, 044705.

Köppe, M., Bleuel, M., Gähler, R., Golub, R., Hanka, P., Keller, T., Longeville, S., Rauch, U., and Wuttke, J. *Physica B* **1999**, *266*, 75.

Likhtman, A.E. and McLeish T.C.B. *Macromolecules* **2002**, *35*, 6332.

Longeville, S. *J. de Phys. IV* **2000**, *10*, 59.

Longeville, S. and Le Coeur, C. *Chem. Phys.* **2008**, *345*, 298.

Marshall, W. and Lovesey, S.W. *Theory of Thermal Neutron Scattering*, Oxford University Press, London, 1971.

Mezei, F. *Zeitschrift für Physik* **1972**, *255*, 146.

Mezei, F. *Neutron Spin Echo, Lecture Notes in Physics*, Vol. 128. Springer, Berlin, 1980.

Milner, S.T. and Safran, S.A. *Phys. Rev. A* **1987**, *36*, 4371.

Monkenbusch, M. and Richter, D. *Comptes Rendus Physique* **2007**, *8*, 845.

Monkenbusch, M., Schätzler, R. and Richter, D. *Nucl. Instrum. Methods Phys. Res. A* **1997**, *399*, 301–323.

Nagao, M. and Seto, H. *Phys. Rev. E* **2008**, *78*, 011507.

Narros, A., Arbe, A., Alvarez, F., Colmenero, J., and Richter, D. *J. Chem. Phys.* **2008**, *128*, 224905.

OHL, M., MONKENBUSCH, M., and RICHTER, D. *Physica B* **2003**, *335*, 153.

OHL, M. MONKENBUSCH, M., RICHTER, D., PAPPAS, C., LIEUTENANT, K., KRIST, T., ZSIGMOND, G., and MEZEI, F. *Physica B* **2004**, *350*, 147.

OHL, M., MONKENBUSCH, M., KOZIELEWSKI, T., LAATSCH, B., TIEMANN, C., and RICHTER, D. *Physica B* **2005**, *356*, 234.

PAPPAS, C., KISCHNIK, R., and MEZEI, F. *Physica B* **2001**, *297*, 14–17.

PAPPAS, C., TRIOLO, A., KISCHNIK, R., and MEZEI, F. *Appl. Phys. A* 2002, *74* (Suppl.), S286.

RAMASWAMY, S. PROST, J., CAI, W., and LUBENSKY, T.C. *Europhys. Lett.* **1993**, *23*, 271.

RICHTER, D., EWEN, B., FARAGO, B., and WAGNER, T. *Phys. Rev. Lett.* **1989**, *62*, 2140–2143.

RICHTER, D., MONKENBUSCH, M., ARBE, A., and COLMENERO, J. *Advances in Polymer Science*, Vol. 174. Springer, Heidelberg, 2005.

ROSOV, N., RATHGEBER, S., and MONKENBUSCH, M. *Scattering from Polymers*, Series 739, American Chemical Society, Washington, DC, 2000, p. 103.

ROUSE, P.E. *J. Chem. Phys.* **1953**, *21*, 1272.

SCHLEGER, P., ALEFELD, B., BARTHELEMY, J.F., EHLERS, G., FARAGO, B., GIRAUD, P., HAYES, C., KOLLMAR, A., LARTIGUE, C., MEZEI, F., and RICHTER, D. *Physica B* **1997**, *241*, 164.

SCHLEGER, P., FARAGO, B., LARTIGUE, C., KOLLMAR, A. and RICHTER, D. *Phys. Rev. Lett.* **1998**, *81*, 124.

SCHILLING, T., THEISSEN, O., GOMPPER, G. *Eur. Phys. J. E* **2001**, *4*, 103–114.

SMITH, G.D., PAUL, W., MONKENBUSCH, M., and RICHTER, D. *Chem. Phys.* **2000**, *261*, 61.

SMITH, G.D., PAUL, W., MONKENBUSCH, M., and RICHTER, D. *J. Chem. Phys.* **2001**, *114*, 4285.

TAKEDA, T., KOMURA, S., SETO, H., NAGAI, M., KOBAYASHI, H., YOKOI, E., EBISAWA, T., TASAKI, S., ZEYEN, C.M. E., ITO, Y., TAKAHASHI, S., and YOSHIZAWA, H. *Physica B* **1995**, *213*, 863.

TAKEDA, T., SETO, H., KAWABATA, Y., OKUHARA, D., KRIST, T., ZEYEN, C.M.E., ANDERSON, I.E., HOGHOJ, P.E., NAGAO, M., YOSHIZAWA, H., KOMURA, S., EBISAWA, T., TASAKI, S., and MONKENBUSCH, M. *J. Phys. Chem. Solids* **1999**, *60*, 1599–1601.

TEUBNER, M. and STREY, R. *J. Chem. Phys.* **1987**, *87*, 3195.

WISCHNEWSKI, A., RICHTER, D., MONKENBUSCH, M., WILLNER, L., FARAGO, B., EHLERS, G., and SCHLEGER, P. *Physica B* **2000**, *276*, 337.

WISCHNEWSKI, A., MONKENBUSCH, M., WILLNER, L., RICHTER, D., and KALI, G. *Phys. Rev. Lett.* **2003**, *90*, 058302.

ZAMBRANO, E., KOTLARCHYK, M., LANGNER, A., and FARAONE, A. *J. Phys. Condens. Matter* **2006**, *18*, S2451.

ZAMPONI, M., MONKENBUSCH, M., WILLNER, L., WISCHNEWSKI, A., FARAGO, B., and RICHTER, D. *Europhys. Lett.* **2005**, *72*, 1039–1044.

ZILMAN, A., and GRANEK, R. *Phys. Rev. Lett.* **1996**, *77*, 4788.

ZIMM, B.H. *J. Chem. Phys.* **1956**, *24*, 269.

# II Instrumentation

# II.3 Quasielastic and Inelastic Neutron Scattering

## II.3.2 Neutron Backscattering

*Bernhard Frick and Dan Neumann*

### II.3.2.1 INTRODUCTION

Neutron backscattering spectroscopy (BSS) was invented in the 1960s to improve the energy resolution of neutron instrumentation into the sub-μeV region—a range unattainable with conventional triple-axis or time-of-flight spectrometers (Maier-Leibnitz, 1966; Alefeld, 1967, 1969; Alefeld et. al., 1969). In fact, the energy resolution of a triple-axis instrument is limited by the beam divergence, crystal quality, and most importantly the Bragg angle. In fact, a backscattering spectrometer can be considered an extreme case of a triple-axis spectrometer where the resolution limitation imposed by these crystal optics behaviors is overcome. In particular, as its name suggests, "backscattering" refers to neutrons scattered from the monochromator and analyzer crystals through 180°. This allows these instruments to typically provide an energy resolution with a full width at half maximum (FWHM) on the order of 1 μeV, corresponding to a timescale of ns. This range encompasses important dynamics in materials ranging from polymers and biomaterials to porous solids and energy storage materials to magnetic systems. It is important to note that the term backscattering refers only to the neutron optical components that determine the resolution and not to scattering from the sample. Thus, the sample scattering can be investigated through a normal range of scattering angles giving scientists access to a significant range of momentum transfers and therefore length scales of the observed motion.

The energy width provided by Bragg diffraction from a crystal can be obtained simply by differentiating Bragg's law to yield

$$\Delta E = 2E\left(\cot\theta\Delta\theta + \frac{\Delta d}{d}\right), \tag{II.3.2.1}$$

*Neutrons in Soft Matter* Edited by Toyoko Imae, Toshiji Kanaya, Michihiro Furusaka, and Naoya Torikai

where $E$ is the energy of the neutrons, $\theta$ is the Bragg angle, and $d$ is the lattice spacing of the crystal. The $\Delta d/d$ term is usually ignored, but as we will see it is important for backscattering. This equation suggests that the resolution provided by crystal optics can be improved by decreasing $\Delta\theta$ by using perfect crystal monochromators instead of mosaic crystals or by limiting the divergence provided to the crystals using collimators. Either of these measures leads to extremely low count rates that are unsuitable for most experiments. Alternatively, one can minimize $\cot\theta$ by using a Bragg angle of nearly 90° meaning that the neutron will be backscattered from the monochromator. Because the coupling between divergence and final energy becomes negligible in backscattering, this approach also allows one to increase the divergence and thus promises larger count rates. However, it is more difficult to achieve in practice. It was Maier-Leibnitz' group who built the first realistic neutron spectrometer operating at Bragg angles very close to 90° in Garching near Munich (Alefeld, 1967, 1969; Alefeld et. al., 1969; Birr et al., 1971).

To further address the inherently low intensity for high-resolution spectrometers, this group designed and built a large spherical backscattering analyzer thereby increasing the solid angle of analysis (http://www.ill.eu/other_sites/BS-review/index.htm). This relaxes the $Q$-resolution, which is acceptable for scattering laws with weak momentum transfer dependence. Fortunately, this is usually the case for inelastic scattering particularly for polycrystalline or amorphous samples or for incoherent scattering.

The Garching group also devised ways to vary the energy of the neutrons that are incident on the sample without scanning the Bragg angle, which would degrade the energy resolution by moving the monochromator or analyzer away from backscattering. They showed that maintaining the backscattering condition is possible by either rapidly moving the monochromator parallel to the incident beam (a Doppler monochromator) or changing the lattice spacing of the monochromator using thermal expansion via a finely controlled temperature stage. Their first BSS was soon followed by spectrometers in Jülich, Germany (Alefeld, 1972) and at the Institut Laue-Langevin, Grenoble, France (Heidemann, 1978) (http://www.ill.eu/instruments-support/instruments-groups/instruments/in10) (IN10). These cold neutron spectrometers used Doppler drives to change the incident energy. In contrast, a variant of IN10, IN10B (Cook, 1992), and the only thermal neutron BSS IN13 (Heidemann and Buevoz, 1977) (http://www.ill.eu/instruments-support/instruments-groups/instruments/in13) use thermal expansion of a heated monochromator. Both IN10B and IN13 are still in operation. As pulsed neutron sources were commissioned, the primary spectrometers of BSSs were adapted to use the time structure of the source. These TOF-backscattering spectrometers that are inverted geometry time-of-flight spectrometers (Carlile and Adams, 1992) where the energy is analyzed by a large set of analyzer crystals set close to backscattering are discussed later in this chapter.

## II.3.2.2 THE ENERGY RESOLUTION NEAR BACKSCATTERING FROM PERFECT CRYSTALS

Equation (II.3.2.1) showed that the factors contributing to the energy resolution of BSSs can be divided into an angular term and another proportional to $\Delta d$. The angular term can be expanded around a Bragg angle of 90° yielding

$$\frac{\Delta E}{2E} \approx \left(\frac{(\delta\theta + \Delta\theta/2)^2}{2}\right) + \frac{\Delta d}{d}. \tag{II.3.2.2}$$

The first term in eq. (II.3.2.2) is an approximation valid near Bragg angles of 90°, where one can express the deviation from backscattering as contributions from the beam divergence, given by $\Delta\theta$, and deviation from a 90° Bragg angle, given by $\delta\theta$ (see Figure II.3.2.1a). Note that both $\delta\theta$ and $\Delta\theta$ contribute only in second order. Also, in backscattering, the angular deviations $\Delta\theta$ and $\delta\theta$ only provide larger energies (corresponding to shorter wavelengths) compared to the nominal one. By comparing

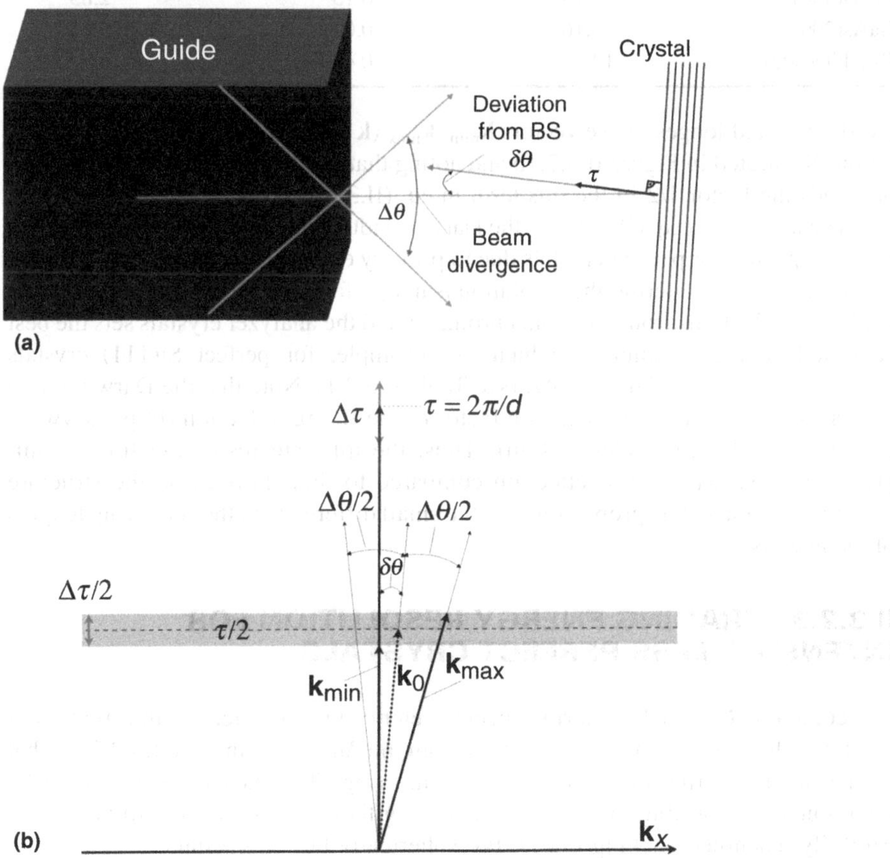


**Figure II.3.2.1** (a) Illustration of beam divergence $\Delta\theta$ and deviation from backscattering $\delta\theta$ for $\delta\theta < \Delta\theta$; the beam divergence is centered around the deviation angle $\delta\theta$. For a neutron guide with $^{nat}Ni$ and for the Si(111) backscattering wavelength 6.27 Å, the maximum deviation from backscattering due to divergence is equal the critical angle, thus $\Delta\theta/2 \sim 0.627°$, therefore, contributing near backscattering about 0.25 μeV to the energy resolution (see eq. (II.3.2.2)). (b) Reciprocal space consideration for estimating the angular deviation contribution to the resolution near backscattering for $\delta\theta < \Delta\theta$. The reciprocal lattice vector $\boldsymbol{\tau}$ has an uncertainty due to primary extinction and lattice strains, which determines which wave vectors (**k**) are accepted out of the divergent beam.

**Table II.3.2.1** Best possible energy resolution for some perfect crystals suitable for backscattering, which in dynamical scattering theory is given by the uncertainty in the reciprocal lattice vector $\tau = 2\pi/d$ corresponding to the reflection used, equivalent to the uncertainty in lattice spacing $d$ (first term of eq. (II.3.2.1). It is crucial to note that the resolution is determined by the structure factor (http://www.ill.eu/other_sites/BS-review/index.htm)

| Crystal Plane | $\Delta\tau/\tau$ ($10^{-5}$) | $\Delta E$ext (μeV) | $\lambda$ (Å) for $\theta = 90°$ |
|---|---|---|---|
| Si(111) | 1.86 | 0.08 | 6.27 |
| Si(311) | 0.51 | 0.08 | 3.27 |
| $CaF_2$(111) | 1.52 | 0.06 | 6.31 |
| $CaF_2$(422) | 0.54 | 0.18 | 2.23 |
| GaAs(400) | 0.75 | 0.15 | 2.83 |
| GaAs(200) | 0.16 | 0.01 | 5.65 |
| Graphite(002) | 12 | 0.44 | 6.7 |

the shortest and longest wave vectors $\mathbf{k}_{min}$, $\mathbf{k}_{max}$ ($\mathbf{k} = 2\pi/\lambda$) scattered under the conditions illustrated in Figure II.3.2.1b and noting that $\Delta\theta$ refers to the full divergence, one finds the factor 1/2 in the this term in eq. (II.3.2.2) (Birr et al., 1971).

The last term in eq. (II.3.2.2) is the Darwin width, which refers to the uncertainty in the $d$-spacing for perfect crystals due to primary extinction. Thus even in the case of "perfect" backscattering the resolution is not perfect. Rather the Darwin width of the Bragg reflections from the monochromator and the analyzer crystals sets the best achievable energy resolution, which, for example, for perfect Si(111) crystals contributes about 0.077 μeV to $\Delta E$ (see Table II.3.2.1). Note that the Darwin width increases linearly with the structure factor of the Bragg reflection (http://www.ill.eu/other_sites/BS-review/index.htm). Thus, the ultimate resolution limit is improved for the GaAs(002) reflection compared to Si(111) because the structure factor for GaAs(002) is proportional to the small difference in the scattering lengths of Ga and As.

## II.3.2.3 TRADING ENERGY RESOLUTION FOR INTENSITY: LESS PERFECT CRYSTALS

The count rate of a backscattering spectrometer can be enhanced with little loss of resolution by matching the contribution from the $\Delta d/d$ term in eq. (II.3.2.2) to that of the angular terms through crystal engineering, This is most often done by intentionally deforming the crystals by bowing them when they are affixed to the carefully machined backing plates. For spherically bowed crystals,

$$\frac{\Delta d}{d} = \left(\frac{\Delta d}{d}\right)_{\text{Darwin}} + \mathrm{P}_{\text{eff}}\left(\frac{t}{R_c}\right) \tag{II.3.2.3}$$

where $(\Delta d/d)_{\text{Darwin}} \approx 1.86 \times 10^{-5}$ is the Darwin peak width, $P_{\text{eff}} = 0.44$ is Poisson's ratio for this crystal orientation, $t$ is the thickness of the crystal (between 250 and 2000 μm) and $R_c$ is the radius of curvature. Thus, by bending the crystals, an increased number of lattice planes contribute to the reflection, sacrificing resolution for intensity. This technique was applied to the early IN10 analyzers, which had a "polished" (unstrained) and "unpolished" analyzer setup. In fact, "unpolished" referred to small hexagonal crystals ($\sim$10 mm side length) with a lapped waver surface, which were glued under deformation onto the spherical support of the analyzers. It is important to note that the measurements indicate the additional strain imposed by bowing is partially relieved due to the finite lateral size of the crystals. Thus, eq. (II.3.2.3) underestimates the required thickness to achieve a desired $\Delta d/d$. This and the more inhomogeneous strain distribution of small crystals led to the use of large deformed wafers (100–120 mm diameter) on all later BSS.

## II.3.2.4 THE ENERGY RESOLUTION OF A COMPLETE BACKSCATTERING SPECTROMETER

The above resolution considerations are sufficient to understand the design of a BSS. In fact, all geometrical contributions to the energy resolution such as beam divergence, beam size, sample size, crystal size, and detector size are taken into account by the angular contributions in eq. (II.3.2.2) and the crystal term by the ($\Delta d/d$) term. The total energy resolution of any BSS is then the convolution of the contributions in eq. (II.3.2.2), arising from the monochromator in the primary spectrometer and from the analyzers in the secondary spectrometer:

$$\delta E = \left(\delta E_{\mathrm{p}}^2 + \delta E_{\mathrm{s}}^2\right)^{1/2}. \tag{II.3.2.4}$$

As we will see later, the resolution of the primary spectrometer is given by a flight time resolution contribution for a TOF-backscattering instrument.

## II.3.2.5 THE FIRST GENERATION OF REACTOR BACKSCATTERING SPECTROMETERS

Figure II.3.2.2 shows a sketch of IN10, which is representative of the first backscattering spectrometers (http://www.ill.eu/instruments-support/instruments-groups/instruments/in10). The arrangement of a perfect backscattering geometry is geometrically difficult, a fact that is reflected in the design of both the primary and the secondary spectrometers. The IN10 monochromator is placed at the end of a natural Ni-guide, which deflects the beam toward a graphite deflector crystal placed above the neutron beam about 6 m upstream from the monochromator. This geometry means that the Bragg angle deviates slightly from backscattering with the angular deviation chosen to be smaller than the critical angle of the neutron guide. The resolution contribution ($\Delta E$) from the monochromator system due to the deviation from backscattering ($\delta\theta$) is estimated to be $\approx 0.1\,\mu$eV and from the guide divergence $(\Delta\theta) \approx 0.25\,\mu$eV. These should be compared to the extinction contribution (from the $\Delta d/d$ term) of 0.08 μeV for Si(111).

The monochromatic beam is sent from the graphite deflector to the sample, passing a chopper with 50% duty cycle. As the detectors must have a direct view of

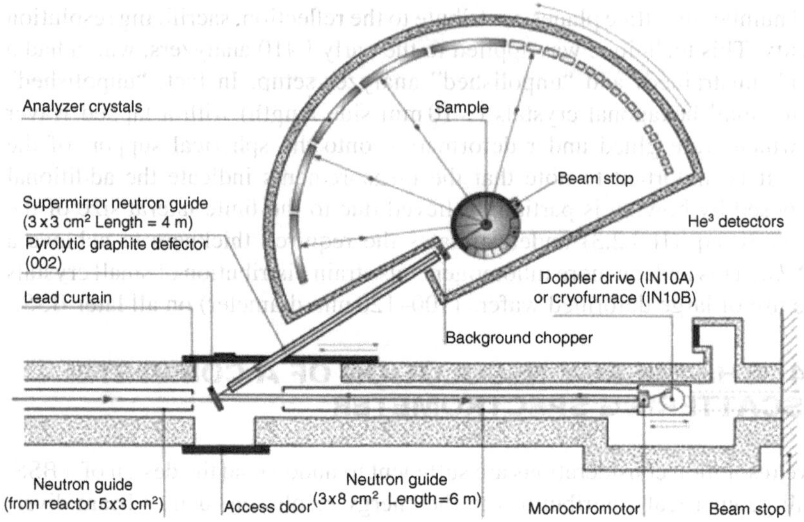


**Figure II.3.2.2** IN10: Representative of a first-generation backscattering instrument (http://www.ill.eu/instruments-support/instruments-groups/instruments/in10).

the sample for the analyzer to be in "exact" backscattering, this chopper is necessary to exclude those neutrons that scatter directly from the sample into the detectors without striking the analyzer. For this purpose, the detector is electronically gated with the chopper phase so that the detectors are inactive when neutrons strike the sample. The neutrons that are scattered from the sample travel to a large, spherically shaped analyzer composed of perfect single crystals, normally of the same kind and the orientation as the monochromator, selects a fixed final energy (2080 μeV for Si (111)). Most importantly, to provide resolution that matches that of the primary spectrometer, the analyzers are aligned in perfect backscattering ($\delta\theta = 0$). This introduces another geometrical difficulty that is solved by requiring that the detected neutrons pass through the sample a second time. Of necessity, this implies the possibility of additional scattering. Fortunately, this induces only a very small correction. This fortuitous situation occurs because most neutron scattering is elastic or nearly so and the distance between the sample and the detectors is small. Thus, there is only a negligible shift in the apparent energy of any twice-scattered neutrons that reach the detectors. This means that the correction primarily appears in the less important $Q$-dependence of the scattering. Furthermore, due to the typical 10% probability of scattering by the sample and the fact that the scattering goes into $4\pi$ sr, the number of these double scattered neutrons seen by the rather limited solid angle of a detector is small.

As we have just seen, the neutrons that scatter from the sample to the analyzer to the detector have a precisely known fixed final energy imposed by the analyzer. All that remains to determine the energy transferred to the sample is to ascertain the initial energy of the detected neutrons. If the incident energy is changed by varying the temperature of the monochromator, with a sufficiently slow temperature variation, the neutron energy can be determined simply by the temperature of the monochromator (and thus the *d*-spacing) at the time the neutrons are detected. For Doppler-equipped spectrometers, the incident neutron energy varies with the instantaneous speed of the monochromator. Thus, the detected neutrons must be related to the Doppler speed at an earlier time. This is possible because both the neutron energy and hence its velocity and the flight distance between the monochromator and the detector are well known. Therefore, it is a simple matter to relate the time at which the neutron is detected to the time it left the monochromator, which is directly related to the initial energy of the neutron. The Doppler frequency is usually lower than the chopper frequency and it is not phase related.

## II.3.2.6 TRADING *Q*-RESOLUTION FOR INTENSITY: FOCUSING

The first reactor BSS example IN10 was placed on a natural Ni-guide and employed a flat monochromator meaning that the guide divergence presented a nonnegligible contribution to the energy resolution of the monochromator system. Moreover, the *Q*-resolution provided by the primary spectrometer was much better than that of the secondary spectrometer. Thus, the developments in neutron optics, which deliver high neutron flux based on focusing and greatly improved guide coatings provided an opportunity to increase the angular divergence, and therefore the flux of neutrons, delivered to the sample. Moreover, the increase in count rate would be at the expense of the rather unimportant *Q*-resolution provided by the primary spectrometer. The use of enhanced focusing optics requires replacing the flat, perfect crystal monochromator with a primary spectrometer having a spherical geometry similar to that developed for the secondary spectrometer of IN10 (Alefeld et al., 1992; Magerl et al., 1992). The first backscattering instrument to profit from these innovations in neutron optics was IN16, a second-generation instrument at the ILL (Frick et al., 1997; Frick and Gonzalez, 2001) (Figure II.3.2.3).

IN16 uses a combination of a vertically focusing, pyrolytic-graphite neutron deflector placed in the primary guide along with a tapered neutron guide to extract a suitable neutron beam and focus it to a small high-intensity beam albeit one with increased angular divergence. This deflector consists of three PG(002) horizontally inclined crystals having a mosaic of 0.5° each, thereby introducing an artificial horizontal mosaic that matches the divergence provided by a $^{58}$Ni -guide while not significantly increasing the vertical divergence. The beam that exits the tapered guide impinges on a second deflector, with the two deflectors arranged in the standard double monochromator focusing geometry. Starting from the second flat PG(002) deflector, the area of the now divergent beam greatly expands before it hits a large spherical monochromator in backscattering geometry. If the focal point created by the optics of deflector–guide combination can be imaged onto the sample by a backscattering monochromator; this design allows the divergence of the incident beam to be quite large without degrading the energy resolution. This means that the beam size at the focal point must be small with respect to the distance between the focal point and the backscattering monochromator. The divergence contribution to

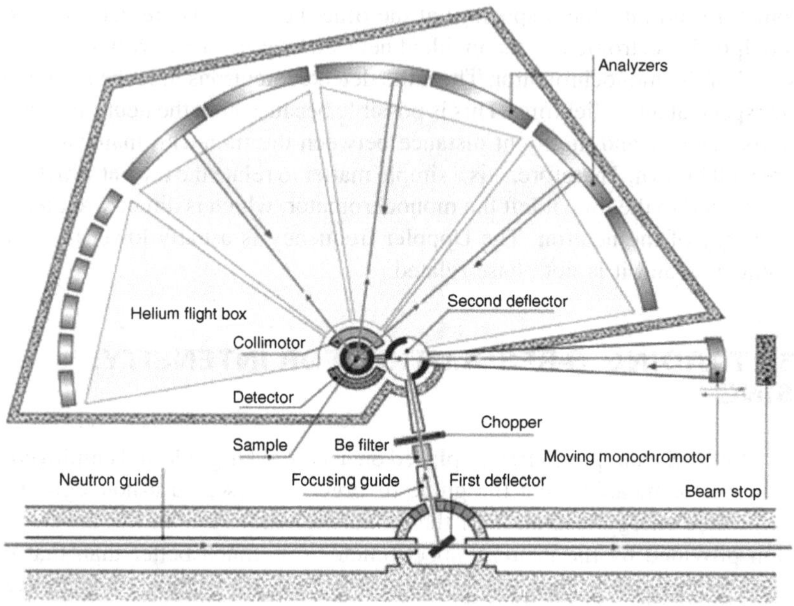


**Figure II.3.2.3** Schematic view of IN16. A double monochromator-guide focusing optics preselects a wavelength band for the backscattering monochromator. The second deflector takes also the role of a chopper (http://www.ill.eu/instruments-support/instruments-groups/instruments/in16).

the energy resolution in such a geometry can be estimated from eq. (II.3.2.2)—for a spot size of $22 \times 22\,\mathrm{mm}^2$ and a distance of 2 m to the spherical monochromator, the divergence contribution to the energy resolution is $\Delta E \approx 0.13\,\mu\mathrm{eV}$, a value reasonably close to the extinction contribution. Unfortunately, deviations from backscattering induced by an ill-defined focal point, monochromator displacement, fixed monochromator radius, and finite crystal size all degrade this "ideal" value.

In order to remain close to backscattering the second deflector on IN16 is mounted on the 50% duty cycle chopper. The chopper speed is matched to the flight time of the neutrons from the chopper to the monochromator and back. A background chopper prevents neutrons from entering the secondary spectrometer during the opening time of the deflector chopper and a cooled Be-filter suppresses the higher order scattering from the first deflector. The secondary spectrometer layout is similar to IN10, but IN16 has a larger analyzer radius (2 m compared to 1.5 m), larger solid angle coverage, a multitube detector assembly, and a diffraction bank below the analyzer area that allows for monitoring the structure of the material being studied. More importantly, IN16 has both a high-energy resolution Si(111) configuration with a spherical array of $4 \times 4\,\mathrm{mm}^2$ crystals ($\Delta E \approx 0.4\,\mu\mathrm{eV}$) and a low-resolution Si(111) configuration of large deformed 0.7 mm thick wafers (($\Delta E \approx 0.85\,\mu\mathrm{eV}$) as well as a Si(311) configuration which allows to access large $Q$-range (3.7 $\mathrm{\AA}^{-1}$ compared to 1.9 $\mathrm{\AA}^{-1}$ for Si(111)). Unfortunately, the chopper is not optimized for the short Si(311) wavelength leading to long counting times. IN16 is a particularly flexible instrument because the secondary spectrometer is under air and the analyzers and the spectrometer are on air pads, which allows relatively rapid configuration changes.

## II.3.2.7 TRADING $Q$-RESOLUTION FOR INTENSITY: PHASE SPACE TRANSFORMATION

With the backscattering spectrometer HFBS at the National Institute of Standards and Technology (NIST) (http://rrdjazz.nist.gov/instruments/hfbs) (Meyer et al., 2003), the recent commissioned SPHERES of the Juelich Center for Neutron Scattering (JCNS) at the Forschungs Reaktor Muenchen (FRM II) (http://www.frm2.tum.de/wissenschaft/spektrometer/spheres/index.html) and the ongoing construction of IN16B a new generation of BSS is introduced. These third-generation spectrometers are similar and they further increase the incident beam divergence using the phase space transformation (PST), first proposed by Schelten and Alefeld (Schelten and Alefeld, 1984). This device requires mounting crystals on a chopper much as is done on the second deflector of IN16, but with a crystal speed at least three times faster. More importantly, it also requires an end-guide position.

As the PST effect has been the subject of many publications (Schelten and Alefeld, 1984; Gehring et al. 1995; Gehring and Neumann, 1997; Kirstein, 1999, 2000; Meyer et al., 2003; Hennig et al., 2009) we will only describe it briefly here. The main purpose of a PST is to offer the optimal phase space to the moving backscattering monochromator. Consider a well-collimated neutron beam such as that provided by a neutron guide. When such a beam diffracts from a stationary mosaic crystal, the phase space element is transformed into the concave-shaped element shown in the top panel of Figure II.3.2.4. Neutron energies are not changed in this process; rather the energies are sorted in angle as indicated by the colors—higher energy neutrons diffracting at lower angles. Note that the energy of many neutrons fall outside the range accepted by the backscattering monochromator, a range indicated by the circular parallel black lines.

The situation is different, however, if the crystal is moving. If the motion is antiparallel to the projection of the direction of the incoming neutrons onto the Bragg planes of the mosaic crystal (i.e., to the left in Figure II.3.2.4), the concave element rotates in phase space. More importantly, the rotation is such that shorter wave vectors become elongated while longer wave vectors are shortened. When the speed is well chosen, this phase space transformation results in the situation shown in the Figure II.3.2.4b. Here, many more of the neutrons have energies lying within the band defined by the parallel solid lines. Moreover, these neutrons subtend a considerably larger angle. Thus, the PST converts a wide wavelength band with lower divergence into a narrower wavelength band with larger divergence—the transformation is from "white to wide" (Schelten and Alefeld, 1984) in agreement with Liouville's theorem. The necessary conditions for employing a PST are (i) the availability of a wide wavelength band (typically this requires an end-guide position), (ii) the mosaic of the PST crystal has to be large enough to accept the wide wavelength band as well as the divergence provided to the PST, and (iii) the deflector (PST) crystal has to move with a speed of 250–300 m/s perpendicular to the scattering plane and the reciprocal lattice vector of the reflection used.

The first spectrometer to utilize a PST chopper was HFBS at NIST (Meyer et al., 2003; Gehring and Neumann, 1997) (http://rrdjazz.nist.gov/instruments/hfbs) (Figure II.3.2.5). (A very similar PST spectrometer, SPHERES, built by JCNS

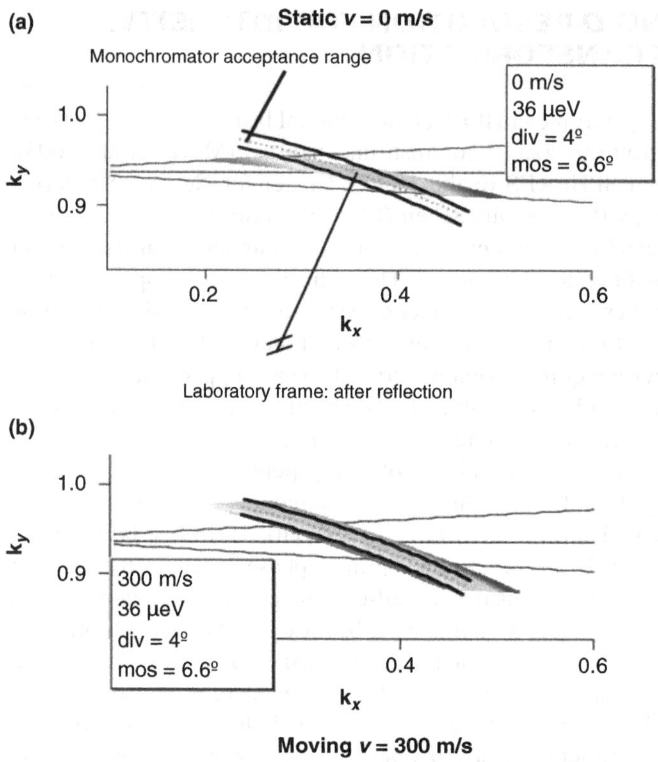


**Figure II.3.2.4** Section of reciprocal space showing the result from phase space calculations of the reflection of a divergent neutron beam (4°) with wide wavelength spread. The colored points refer to neutrons that were reflected by the graphite PG(002) mosaic crystal planes. The circular black parallel curves describe the acceptance of a spherical backscattering monochromator moving at its extreme Doppler velocity corresponding to ±36 μeV. When the area of the colored phase space in the acceptance range of the Doppler monochromator is maximized, the intensity is optimum. (a) The mosaic crystal is at rest. (b) The mosaic crystal is moving in -$\mathbf{k}_x$ direction perpendicular to the reciprocal lattice vector (parallel $\mathbf{k}_y$ direction). (See the color version of this figure in Color Plates section.)

recently began operations at the FRM-II Munich.) The primary spectrometer utilizes the spherical focusing geometry developed for IN16. However, rather than using an in-guide, stationary deflector that would substantially limit the wavelength band available to the PST, the guide transports the neutrons directly to the PST chopper that deflects the neutrons to the monochromator. The neutrons are then backscattered from the monochromator, passing through an opening in the PST chopper to the sample. Note that the linear speed of the crystals that are mounted on the chopper is 250 m/s. Compare this to the approximately 80 m/s crystal speed on the second IN16 deflector and the technical challenge of building such a chopper becomes clear. The additional wavelength band and the much higher crystal speeds combine to provide a measured gain a factor 4 when the crystals are moving at the design speed compared to when they are stationary.

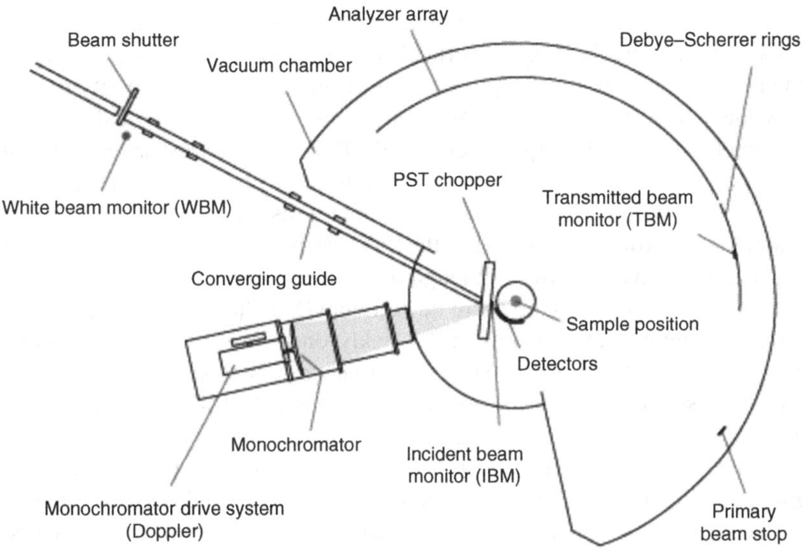


**Figure II.3.2.5** Schematic layout of the HFBS, NIST.

The disk (Figure II.3.2.6) consists of three segments each enclosing 60 graphite crystals having a nominal mosaic of 2.5°. Much like the in-guide deflector on IN16, an artificial mosaic is produced by stacking three PG(002) so that they are horizontally inclined with respect to each other thus introducing an artificial horizontal mosaic of 7.5°. Thus, the beam leaves the chopper with a divergence of 15° seen in the gray scale area of Figure II.3.2.5. Clearly the *Q*-resolution provided by HFBS is quite relaxed.

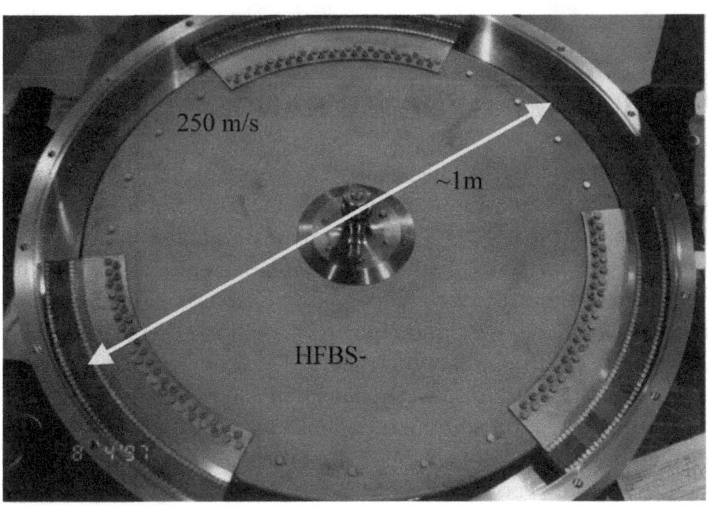


**Figure II.3.2.6** The PST disk and the graphite cassettes at its outer border, packed into cassettes for increasing the mechanical stability.

Due to the stacking arrangement of the graphite crystals in the chopper, the vertical mosaic is only 2.5° thereby limiting the vertical beam divergence and making it easier to focus the neutrons back to the sample position. In addition to the necessity for the crystals to achieve a linear speed of 250 m/s, the rotational frequency (and thus the chopper diameter) are set by the requirement that the PST chopper must have opened in the time it takes neutrons to fly from the chopper to the monochromator and back. The chopper on HFBS is designed so that the 1 m diameter disk turns 180° in this time.

In the similar spectrometer SPHERES, the PST disk has a diameter of about 1.3 m providing a crystal speed of 300 m/s (although it is currently operating in one-third of this design) that was originally believed to be optimum (Schelten and Alefeld, 1984). However, the maximum gain depends only weakly on the speed for a reasonable mosaic of 5–10° and for the divergence that modern neutron optics can provide. More recent simulations suggest that for the geometry of the new IN16B being built at the ILL, the maximum gain is achieved for 250 m/s (Hennig et al., 2009).

A disadvantage of the PST concept is that it is necessary to bring a wide wavelength band (meaning a high flux of neutrons) into the secondary spectrometer. In fact, the beam stop for this is located only about 20–40 cm from the detectors. Therefore, it is difficult to achieve a good signal to background ratio. The HFBS instrument employs a velocity selector that limits the wavelength band and thereby reduces the background by a factor of 8 with only a 15% reduction in the signal. The background is further reduced by reducing the scattering from air in the secondary spectrometer by either building a vacuum chamber (HFBS) or filling the secondary spectrometer with Ar (SPHERES). Finally, like on IN16, one can incorporate a background suppression chopper that prevents neutrons from entering the chamber from the guide when neutrons are striking the sample (IN16B). A combination of measurements and simulations suggests that this device further reduces the background by a factor of 2, albeit with a small decrease in the intensity (Garcia-Sakai et al., 2008).

## II.3.2.8 IMPROVING THE DYNAMIC RANGE

Since the earliest days of backscattering spectrometers, efforts have been made to increase the energy range accessible by these instruments. This desire led to the development of offset monochromators in which the monochromator has a slightly different *d*-spacing from the analyzer. Perhaps the best example of this is the development of SiGe alloys at the ILL for use as a backscattering monochromator (Magerl and Holm, 1990). Found to be of more utility, was adjusting the *d*-spacing with thermal expansion through the use of cryofurnaces. These devices, which are in regular use at IN13 and IN10B, produce a well-controlled and exquisitely uniform temperature environment for the monochromator (Cook et al., 1992). However, the advent of large focusing monochromators made this approach technically daunting. Thus, more recent efforts have concentrated on the development of Doppler monochromators capable of driving larger loads at higher speeds. Because the PST approach presents a larger energy band to the monochromator than is available when using an in-guide deflector, the first significant development along this line was made for HFBS. This spectrometer employs a counterbalanced mechanical device that is capable of providing speeds corresponding to energy transfers of ±50 μeV although in practice vibrational resonances limit the useful range to ±36 μeV. An alternative approach taken at SPHERES is the use of linear motors, moving a carbon fiber piston and monochromator on very thin air cushions (Doppler AEROLAS[1]). The linear motor Doppler drive allows users a great deal of flexibility in choosing a velocity profile and amplitude. Thus, in principle, one can distribute the neutron energies over the range of interest in a way that produces the most information in the shortest amount of time. In order to suppress vibrations, it is mounted on a heavy granite block. The same drive with improved control is now installed at IN16 and IN16B.

## II.3.2.9 ONGOING BACKSCATTERING PROJECTS

At ILL a new BSS, IN16B, is under construction (Figure II.3.2.7) with the aim to combine all progress achieved on IN16, HFBS, and SPHERES (Frick et al., 2006). IN16B will be placed at the end of a very long ballistic cold neutron guide where it profits from modern focusing optics and PST effect, providing a high count rate and a wider dynamic range (Bordallo et al., 2009). In addition, it will be able to sweep to an IN16-like side position, which guarantees good background conditions. Similar to HFBS, IN16B will have vacuum in the secondary spectrometer and in all flight path. Flexibility is maintained for extending to high $Q$ with Si(311) crystals and to high-energy resolution with GaAs(002). Operation as an inverted time-of-flight back-scattering option (BATS) has been designed and is currently under review.

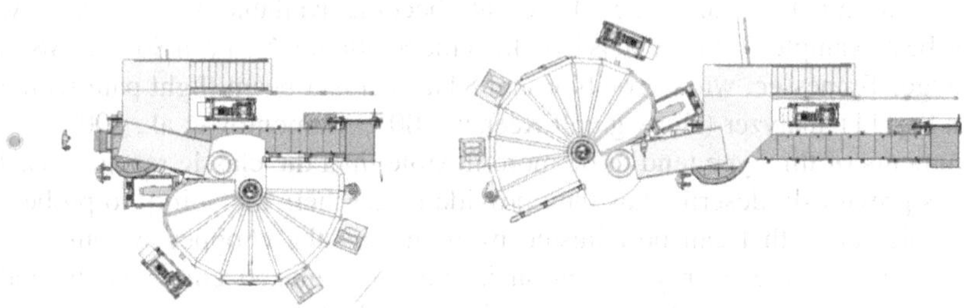


**Figure II.3.2.7** The design of IN16B allows it to sweep between a side position to work in an IN16-like low background mode and a high-flux end-guide position that employs a PST chopper.

[1] Doppler AEROLAS is the linear motor drive that was developed by AEROLAS, Unterhaching, Germany. Commercial equipment, instruments, or materials are identified to foster understanding. Such identification neither imply recommendation or endorsement by the National Institute of Standards and Technology or the Insitut Laue Langevin nor imply that the materials or equipment identified are necessarily the best available for the purpose.

The PST disk of IN16B with a diameter of 66 cm is planned to be more compact than the PST of HFBS and SPHERES. On these spectrometers, three graphite crystals will be assembled in a cassette, with a slight horizontal inclination angle of 2.2° between them to obtain an effective horizontal mosaic of about 6.6°. IN16B assembles single cassettes on a chopper wheel with two (rather than three) windows allowing for a more compact disk. The technical challenge is packaging these rather fragile graphite crystals in a way that allows to rotate them at 243 m/s with a corresponding centrifugal acceleration of $1.8 \times 10^5$ m/s$^2$). Commissioning is envisaged to begin by 2012.

Finally, the Bragg Institute at the Australian Nuclear Science and Technology Organization has announced that it will build a backscattering instrument to be called Emu. While the conceptual design of this instrument has not yet been fixed, commissioning is scheduled to begin by 2013.

## II.3.2.10 BACKSCATTERING AT SPALLATION SOURCES

Inverted geometry time-of-flight (TOF) spectrometers with crystal analyzer systems are a class of instruments that is widely available at pulsed neutron sources (Carlile and Adams, 1992). The basic layout for a TOF-backscattering instrument is shown in Figure II.3.2.8. A pulsed, white neutron beam travels down a long neutron guide before striking the sample. Those scattered neutrons that satisfy the Bragg condition at the analyzer are diffracted to a multidetector array very near the sample where they are recorded as a function of time of flight. As the final energy ($E_S$) is fixed by the analyzer, this allows the determination of the initial velocity of each of the detected neutrons and therefore the energy transfer. However, instruments that routinely operate with the Bragg angle from the crystal analyzer close enough to backscattering to routinely provide resolution better than 10 μeV have only become available over the last few years. The best example of this trend is BASIS at the Spallation Neutron Source (SNS) at Oak Ridge, Tennessee, which combines an 84 m incident beam flight path with a large area Si(111) analyzer (Herwig and Keener, 2002; Mamontov et al., 2008).

Instruments of this type tend to be more flexible than the classic reactor based instruments previously described as they provide researchers the ability to probe a wide dynamic range that can be adjusted by rephasing the chopper system. The disadvantage is that the energy resolution is typically relaxed compared to that provided by the classic reactor-based design. For example, BASIS provides an energy

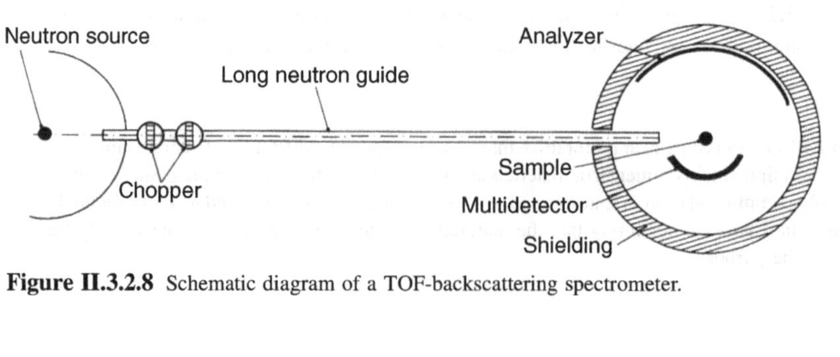


**Figure II.3.2.8** Schematic diagram of a TOF-backscattering spectrometer.

resolution of 3.5–4.0 μeV and a typical dynamic range of ±250 μeV. Alternatively TOF-backscattering spectrometers can employ mica analyzers to improve the resolution. However, this approach decreases the fixed final energy, thereby decreasing the available phase space and greatly reducing the count rate. In most cases, where this compromise has been attempted, it has not proved to be scientifically productive.

An important feature of inverted geometry spectrometers compared to the more typical direct geometry TOF spectrometers described in the next section is that the available energy range in neutron energy loss is typically quite large allowing measurements to be made at very low temperatures. This comes about because direct geometry machines achieve high resolution by reducing the incident neutron energy until only a very narrow energy range is available in neutron energy loss. In addition, inverted geometry instruments do not require high-speed choppers near the sample. This allows more effective use of modern neutron optics. Thus, inverted geometry spectrometers typically enjoy a somewhat higher count rate than their direct geometry counterparts. On the other hand, due to the nearly complete freedom to choose the initial neutron energy, direct geometry instruments provide exceptional experimental flexibility.

## II.3.2.11 ENERGY RESOLUTION FOR A TOF-BACKSCATTERING SPECTROMETER

Taking the approximation that the contributions to the energy resolution ($\delta E$) from the primary TOF spectrometer ($\delta E_p$), which is given by time-of-flight and from the secondary backscattering spectrometer ($\delta E_s$) are independent and that therefore the resolutions add in quadrature (see eq. (II.3.2.3)). For this type of instrument, the contribution to the energy resolution from the primary spectrometer is, to a good approximation, proportional to the ratio of the pulse width ($\Delta t_p$) to the total flight time $t_o$:

$$\delta E_p = 2E_p\left(\frac{\Delta t_p}{t_0}\right) \qquad \text{(II.3.2.5)}$$

Thus, to improve the energy resolution, one must either reduce the width of the pulses or lengthen the instruments. If the pulse width is reduced using a chopper, one loses intensity. Thus, all TOF-backscattering instruments that routinely operate with an energy resolution better than 10 μeV have been on long neutron guides. For example, the source to sample distance on BASIS at the Spallation Neutron Source is 84 m ($t_o \approx 140{,}000$ μs) and the instrument views a poisoned moderator to limit $\Delta t_p$ to 45 μs giving $\delta E_p \approx 1.3$ μeV.

The secondary spectrometer resolution, $\delta E_s$, is given by the same equations as those used to estimate the resolution of the classic, reactor-based backscattering instrument design (eq. (II.3.2.2)). Again the divergence $\Delta\theta$ of the beam when it strikes the analyzer crystals, $\delta\theta$, the average deviation (in the small-angle approximation) from the exact backscattering and the spread in the $d$-spacing of the analyzer crystals due to the Darwin width or to introduced strain are the controlling quantities. As the angular terms appear as the square of the deviation from exact backscattering, obtaining high resolution requires being very near the exact backscattering condition. For example, on BASIS, the nominal value of $\delta\theta$ is 2.1° (0.036 rad) and $\Delta\theta/2$ is 0.35° (0.006 rad) for a sample that completely fills the $3 \times 3\,\text{cm}^2$ beam. Ignoring the (small) contribution from $\Delta d/d$, these two terms combine to give $\delta E_s \approx 3.6\,\mu\text{eV}$. Thus, the resolution of BASIS is dominated by that of the secondary spectrometer. Moreover, this example illustrates the necessity of achieving an analyzer system that operates very close to backscattering ($\delta\theta$ approaching 1°) in order to obtain an energy resolution of about 1 μeV.

An important consideration in all neutron instruments is preventing unwanted neutrons from reaching the detectors. TOF spectrometers can suffer from "frame overlap" that occurs when "fast" neutrons from one pulse catch up to "slow" neutrons in the previous pulse. This problem is particularly acute for the long flight paths needed for high-resolution instruments. To alleviate it, TOF-backscattering instruments need complex chopper systems that limit the bandwidth and thus the dynamic range. Many factors go into the design, with the repetition rate of the source and the length of the instrument crucial. BASIS that operates at 60 Hz with a length of 84 m employs four bandwidth choppers. The number would be greater for longer instruments or for those operating at a higher repetition rate.

Crystal analyzers can also introduce spurious features through higher order reflections that occur at energies $n^2E_s$. Thus, neutrons detected at a given time can be contaminated by a series of other energy transfers. The chopper systems can be used to eliminate all of the neutrons in the beam that have energies equal to $n^2E_s$ for $n \neq 1$—those neutrons that could elastically scatter from the sample and be diffracted by the higher order analyzer reflections. As elastic cross sections are much larger than the inelastic cross sections, these neutrons would cause the most serious problems. However, to completely eliminate spurious scattering from the spectrum, one needs to allow only neutrons with energy $E_s$ to reach the detector. Thus, TOF-backscattering instruments often have a filter (typically Be) to eliminate these neutrons. Finally, the most common analyzer is Si(111) for which all of even $n$ orders of contamination are absent due to the structure factor of Si.

## II.3.2.12 THE FIRST GENERATION OF SPALLATION SOURCE BACKSCATTERING SPECTROMETERS

IRIS, operational at ISIS since 1987, is the prototypical example of the "classic" TOF-backscattering instrument (Carlile and Adams, 1992) (http://wwwisis2.isis.rl.ac.uk/molecularSpectroscopy/iris/index.htm) (Figure II.3.2.9). In its most common mode of operation using a graphite analyzer ((002) reflection, $d = 3.354\,\text{Å}$, $\delta\theta \approx 2.5°$), it provides an energy resolution of 18 μeV for $E_s = 1840\,\mu\text{eV}$. The instrument views a liquid hydrogen cold source via a 34 m long curved neutron guide that terminates with a converging guide. The choppers that reduce frame overlap define the energy range, which is ±400 μeV when centered on the elastic line. In addition, a Be filter (not shown), eliminates higher order reflections from the graphite analyzer.

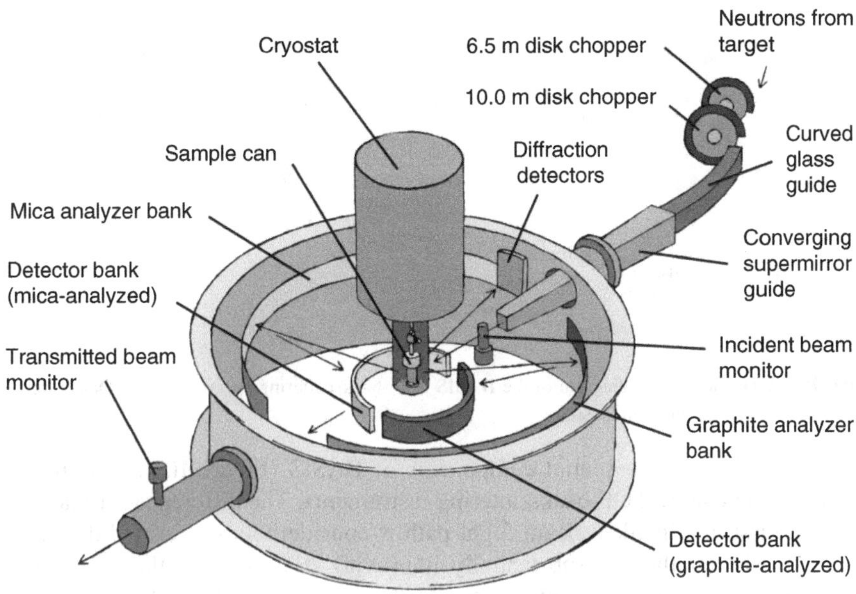


**Figure II.3.2.9** Spallation source backscattering spectrometer IRIS at ISIS.

The scientific necessity to attain higher energy resolution for studies of soft matter and biomolecules keeps driving quasielastic neutron scattering studies to instruments with better resolution. The easiest to attain this on an existing instrument is to effectively lengthen the instrument by decreasing the neutron energy. To accomplish this on IRIS, an analyzer consisting an array of mica crystals that have a considerably larger *d*-spacing provides an energy resolution of 4 μeV for 9.5 Å neutrons.

An interesting aside is that despite considerable attention to eliminating unwanted neutrons, the graphite analyzer on IRIS displayed a very high background. This was eventually traced to thermal diffuse scattering from the graphite analyzer. Cooling the entire analyzer system greatly reduced the problem (Carlile et al., 1994). Another issue for graphite analyzer systems is that graphite is quite expensive. This, along with the necessity of cooling the analyzer to liquid nitrogen temperatures, limits the area subtended by the analyzer, thereby compromising the count rate. For example, on IRIS's sister spectrometer, OSIRIS (Martin et al., 1996), the analyzer covers only 8% of $4\pi$ sr compared to approximately 20% for classic reactor-based instruments (Andersen et al., 2002).

Finally, it is worth noting that the first instrument of this type was actually built at KENS in Japan. This instrument is no longer operational (Inoue et al., 1985).

## II.3.2.13 IMPROVING THE ENERGY RESOLUTION OF TOF-BACKSCATTERING INSTRUMENTS

The advent of the next-generation spallation sources, SNS and J-PARC, has enabled a scientifically important evolution of TOF-backscattering spectrometers to better

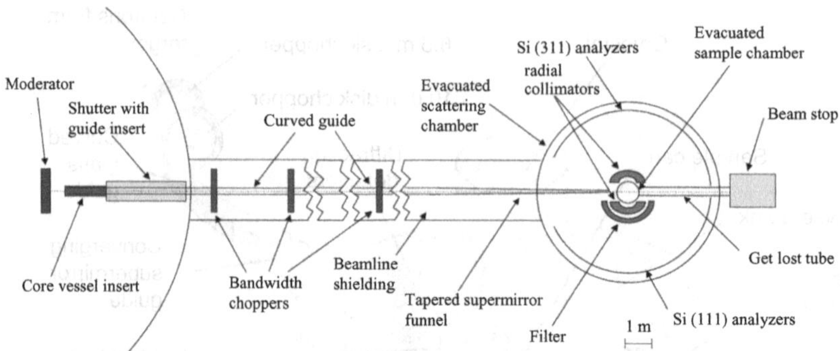


**Figure II.3.2.10** Schematic diagram of the BASIS TOF-backscattering spectrometer at Oak Ridge National Laboratory in the United States.

energy resolution. The essential components of BASIS (Figure II.3.2.10) are the same as for the classic TOF-backscattering instruments. The differences are largely in degree. First, the incident beam flight path is considerably longer and the pulse width ($\Delta t_p$) shorter thereby substantially improving $\delta E_p$. Second, the analyzer is closer to backscattering, which enhances $\delta E_s$. Taken together, these changes yield a substantially improved energy resolution.

The secondary spectrometer provides another distinct difference. The primary analyzer is Si(111) rather than graphite (002). This has many consequences. First, Si possesses much higher energy phonons than graphite. This eliminates the need to cool the analyzer to prevent background from thermal diffuse scattering. Since Si is much cheaper, the analyzer can be significantly bigger (20% of $4\pi$ sr), enhancing the count rate.

The principle difficulty is that Si is not a mosaic crystal. Thus, the crystals must be strained to enhance $\Delta d/d$ and optimize the count rate (see eq. (II.3.2.3). As previously described, this is accomplished by gluing crystals of a chosen thickness $r$ to a carefully machined backing plate. For TOF-backscattering instruments with $\delta\theta \approx 2°$, the wafer thickness $r$ estimated from eq. (II.3.2.2) should be as large as possible so that the crystals can be glued without excessive breakage. For BASIS, $r = 2$ mm.

The DNA spectrometer being built at J-PARC (Arai et al., 2009) seeks to build on these trends. Instead of being placed on a poisoned moderator to shorten the pulse length and enhance the resolution, DNA will be placed on a coupled cold moderator with water premoderator. This moderator displays high peak brightness albeit with a 220 μs pulse length. Thus, DNA will employ a pulse shaping chopper to shorten $\Delta t_p$ to achieve high resolution in the primary spectrometer. The energy resolution of the secondary spectrometer will also be improved compared to BASIS by placing the analyzer closer to exact backscattering (i.e., reducing $\delta\theta$). Due to the high brightness provided by the coupled moderator, it is expected that this instrument will provide count rates comparable to those on BASIS in spite of the resolution being $\approx$2 μeV compared to 3.5 μeV on BASIS. Further improvements in the resolution will likely require reducing $\delta\theta$ below $\approx 1°$. This would necessitate employing a chopper to reduce the duty cycle to <50%. Then, neutrons diffracted from the analyzer could traverse the sample on their way to the detector while neutrons that are directly scattered from the sample to the detector would not be counted. As previously discussed, this approach is employed in all reactor-based backscattering instruments such as IN16 and SPHERES.

There is also a proposal to build a similar instrument with a Si(111) analyzer at ISIS. The instrument, called FIRES, will have an 80 m flight path and much like DNA will view a coupled hydrogen moderator and will generate short pulses using a high-speed chopper (Demmel and Andersen, 2009) (http://wwwisis2.isis.rl.ac.uk/molecularSpectroscopy/fires/index.htm).

## II.3.2.14 OPTIMIZING TOF-BACKSCATTERING INSTRUMENTS FOR NONZERO ENERGY TRANSFER

One problem with TOF-backscattering instruments is that the resolution is typically matched at the elastic line. Thus, as one re-phases the choppers to access large energy transfers, the energy resolution of the primary instrument is degraded compared to that provided by the crystal analyzer. This results in a loss of intensity compared to what would be available if the two halves of the instrument were matched. The MARS spectrometer at the Paul Scherrer Institute (Tregenna et al., 2008) seeks to address this issue by placing the analyzer and detector system on movable supports so that it can be adjusted to increase $\delta\theta$ when large energy transfers are desired. Unfortunately, due to the small mosaic of the mica crystals that are employed to achieve μeV-level resolution at the elastic line, the reflectivity of the crystals is limited as one goes away from backscattering. Thus, the intensity gain is less than would be achieved if the mosaic were larger. Still this ground-breaking instrument provides unique scientific opportunities when high resolution is required at energy transfers significantly away from the elastic line. And further developments to control the mosaic or $\Delta d/d$ of the analyzer crystals could yield significant improvements for high-resolution measurements of dispersionless excitations in materials.

## II.3.2.15 CONCLUSIONS AND OUTLOOK

The development of neutron backscattering spectroscopy, which began about 40 years ago, is in a period of rapid development. This is providing unprecedented scientific opportunities ranging from basic research to provide better understanding of physical phenomena in condensed matter to the development of future technologies such as improved materials for energy conversion and storage to better structural materials including plastics and concrete, to entirely new classes of functional nanomaterials. Because many of these opportunities require understanding the dynamics over wide range of timescales, the scientific community is taking full advantage of the progress in backscattering instrumentation achieved over the last 15 years in terms of increased count rates and improved dynamic ranges. In addition, the higher count rates due to the technical advances described in this chapter may make rarely used options such as Si(311) monochromators and analyzers more attractive and may also provide the opportunity to achieve higher energy resolution using, for example, GaAs(002) crystals. Thus, we eagerly look forward to the developments in the application of neutron backscattering spectroscopy in the coming years.

## REFERENCES

ALEFELD, B. *Mathematisch Naturwissenschaftliche Klasse* **1967**, *11*, 109.

ALEFELD, B. *Z. Phys.* **1969**, *222*, 155.

ALEFELD, B. *Kerntechnik* **1972**, *14*, 15.

ALEFELD, B., BIRR, M., and HEIDEMANN, A. *Naturwissenschaften* **1969**, *56*, 410.

ALEFELD, B., SPRINGER, T., and HEIDEMANN, A. *Nucl. Sci. Eng.* **1992**, *110*, 84.

ANDERSEN, K., MARTIN, Y., MARERO, D., and BARLOW, M.J. *Appl. Phys. A* **2002**, *74* (Suppl.), S237.

ARAI, M., AIZAWA, K., NAKAJIAMA, K., SHIBATA, K., and TAKAHASHI, N. Internal report: JAEA-Review 2009-014, 2009.

BIRR, M., HEIDEMANN, A., and ALEFELD, B. *Nucl. Instrum. Methods Phys. Res.* **1971**, *95*, 435.

BORDALLO, H., FRICK, B., SCHOBER, H., SEYDEL, T. *J. Neutron Res.* **2009**, *16*, 39.

CARLILE, C.J. and ADAMS, M.A. *Physica B* **1992**, *182*, 431.

CARLILE, C.J., ADAMS, M.A., KRISHNA, P.S.R., PRAGER, M., SHIBATA, K., and WESTERHUIJS, P. *Nucl. Instrum. Methods Phys. Res. A* **1994**, *338*, 78.

COOK, J.C., PETRY, W., HEIDEMANN, A., and BARTHELEMY, J.F. *Nucl. Instrum. Methods Phy. Res. A* **1992**, *312*, 553.

DEMMEL, F. and ANDERSEN, K. *Meas. Sci. Technol.* **2009**, *19*, 034021.

FRICK, B., GONZALEZ, M. *Physica B* **2001**, *301*, 8.

FRICK, B., MAGERL, A., BLANC, Y., REBESCO, R. *Physica B* **1997**, *234*, 117.

FRICK, B., BORDALLO, H.N., SEYDEL, T., BARTHéLEMY, J.-F., THOMAS, M., BAZZOLI, D., and SCHOBER, H. *Physica B* **2006**, *385–386*, 1101.

GARCIA-SAKAI, V., MAMONTOV, E., and NEUMANN, D.A. *J. Neutron Res.* **2008**, *16*, 65.

GEHRING, P.M. and NEUMANN, D.A. *Physica B* **1997**, *241–243*, 64.

GEHRING, P.M., BROCKER, C.W., and NEUMANN, D.A. *Mater. Res. Soc. Symp. Proc.* **1995**, *376*, 113.

HEIDEMANN, A. Internal ILL report: ILL78HE144T, 1978.

HEIDEMANN, A. and BUEVOZ, A. IN13: A high resolution spectrometer for short wavelengths, ILL internal Report: ILL77HE24T, 1977.

HERWIG, K.W. and KEENER, W.S. *Appl. Phys. A* **2002**, *74* (Suppl.), S1592.

HENNIG, M., FRICK, B., and SEYDEL, T. Phase Space Transformation, ILL Internal Report 2005, 2009.

INOUE, K., ISHIKAWA, Y., WATANABE, N., KAJI, K., KIYANAGI, Y., IWASA, H., and KOHGI, M. *Nucl. Instrum. Methods Phys. Res. A* **1985**, *238*, 401.

KIRSTEIN, O., GRIMM, H., PRAGER, M., and RICHTER, D. *J. Neutron Res.* **1999**, *8*, 119.

KIRSTEIN, O., PRAGER, M., KOZIELEWSKI, T., and RICHTER, D. *Physica B* **2000**, *283*, 361.

MAGERL, A. and HOLM, C. *Nucl. Instrum. Methods phys. Res. A* **1990**, *290*, 414.

MAGERL, A., FRICK, B., and LISS, K.D. *Proc. SPIE*, **1992**, *1738*, 360.

MAIER-LEIBNITZ, H. *Nukleonik* **1966**, *8*, 61.

MARTIN, D., CAMPBELL, S., and CARLILE, C.J. *J. Phys. Soc. Jpn.* **1996**, *65* (Suppl. A), 245.

MEYER, A., DIMEO, R.M., GEHRING, P.M., and NEUMANN, D.A. *Rev. Sci. Instrum.* **2003**, *74*, 2759.

MAMONTOV, E., ZAMPONI, M., HAMMONS, S., KEENER, W.S., HAGEN, M., HERWIG, K.W. *Neutron News* **2008**, *19*, 22.

TREGENNA-PIGGOT, P.L.W., JURANYI, F., and ALLENSPACH, P. *J. Neutron Res.* **2008**, *16*, 1.

SCHELTEN, J. and ALEFELD, B. Proceedings of Workshop on Neutron Scattering Instrumentation for SNQ; Report Jül-1954, 1984.

# II Instrumentation

# II.3 Quasielastic and Inelastic Neutron Scattering

## II.3.3 Time-of-Flight Spectrometry

*Ruep E. Lechner*

### II.3.3.1 INTRODUCTION

The ensemble of low-energy transfer inelastic and quasielastic (incoherent) neutron scattering (IINS and QENS, respectively) techniques employing time-of-flight (TOF) methods are well suited particularly for the investigation of the dynamic structure of soft matter and biological systems. This will be discussed in detail in this chapter.

Soft matter, polymers, biological macromolecules, and biological systems in general are constructed according to well-defined building schemes exhibiting a certain degree of long-range or intermediate-range order. This order is restricted, implying also an appreciable amount of disorder, for various reasons. One reason is that high structural symmetry—except for some rare cases—is generally absent in native samples. Another is that the order is limited to certain parts of the macromolecules and to part of the degrees of freedom. Furthermore, the ubiquitous presence of solvents in many cases, and in particular of water in biological systems plays an important role. In the latter case, this is generally a prerequisite for the unrestrained performance of biological function. The interaction of water molecules with biological surfaces, their diffusion close to and within the hydration layers covering biological macromolecules, provides the latter with the indispensable additional space for the conformational degrees of freedom required for function. The presence of mobile water molecules not only allows or induces additional short-range translational and rotational diffusive motion of parts of biological macromolecules but also causes damping of low-frequency vibrations in the macromolecules. All these motions, which are believed to be essential for biological function, are an important part of the dynamical characteristics of biological matter. Analogous statements can generally be made in the case of soft matter regarding the relevance of

*Neutrons in Soft Matter* Edited by Toyoko Imae, Toshiji Kanaya, Michihiro Furusaka, and Naoya Torikai

the dynamic structure with or without the presence of molecules of water or of other solvents near the surfaces of macromolecules of functional materials, and with respect to their already known or potential specific technical applications.

In this chapter, the basic principles of QENS and IINS techniques are outlined. Obviously, we need to discuss not only neutron scattering instruments but also theory and methods of analysis, with an emphasis on application to both soft matter and biological problems. The method of QENS focuses on scattering processes involving small amounts of energy exchange, with spectral distributions peaked at zero energy transfer. IINS spectra extend to somewhat higher energies, but are, by principle, also fully overlapping with the QENS energy region. Both together allow us to study dynamical phenomena in the Fourier *time* region from $10^{-13}$ to $10^{-7}$ s. Atomic and molecular motions are explored in *space*, on length scales comparable with the wavelengths of the neutrons used in the scattering experiments. Typical spatial parameters, such as vibrational displacements, jump distances, diffusion paths, and correlation lengths, are amenable to evaluation in the range from $10^{-9}$ to $10^{-6}$ cm. Quasielastic and inelastic neutron scattering experiments on such dynamic processes lead to spectra as a function of energy transfer $\hbar\omega$, with

$$\hbar\omega = E - E_0, \qquad \text{(II.3.3.1)}$$

in a range from 10 to $10^{-5}$ meV, where $E_0$ and $E$ are the neutron energies before and after scattering, respectively. The corresponding momentum transfer $\hbar\mathbf{Q}$ in such a process is proportional to the scattering vector

$$\mathbf{Q} = \mathbf{k} - \mathbf{k}_0, \qquad \text{(II.3.3.2)}$$

where $\mathbf{k}_0$ and $\mathbf{k}$ are the neutron wave vectors before and after scattering, respectively. These vectors have lengths given by the wave numbers $k_0 = |\mathbf{k}_0| = 2\pi/\lambda_0$ and $k = |\mathbf{k}| = 2\pi/\lambda$. The wave vector transfer values for elastic scattering, $Q = |\mathbf{Q}| = (4\pi/\lambda_0)\sin(\varphi/2)$, are typically in the region from 0.1 to 5 Å$^{-1}$ ($\lambda_0$, $\lambda$ = incident and scattered neutron wavelengths, $\varphi$ = scattering angle, that is, the angle between the vectors $\mathbf{k}_0$ and $\mathbf{k}$), such that $2\pi/Q$ ranges from the order of magnitude of interatomic distances to that of diameters of (e.g., biological) macromolecules. The neutron scattering intensity in such a process is proportional to the so-called *scattering function* or *dynamic structure factor* $S(\mathbf{Q}, \omega)$, which can be calculated for typical dynamical processes; the calculation and the determination of this function are the subject of the following paragraphs.

The purpose of QENS experiments is mainly the study of the details of quasielastic and low-energy inelastic spectra that are mostly due to some kind of diffusive and/or damped vibrational atomic and molecular motions.

The interpretation of the scattering function $S(\mathbf{Q}, \omega)$ in terms of diffusive and/or vibrational processes is relatively simple if

- such motions can (at sufficiently high temperatures) be described by classical physics, that is, when quantum effects can be completely or almost completely neglected and

- as in most practical cases—the scattering can be treated in first Born approximation.

This allows an evaluation and an interpretation of $S(\mathbf{Q}, \omega)$ by pair correlation functions $G(\mathbf{r}, t)$ for the scattering nuclei in space and time.

This chapter is organized in several sections dealing with various topics in the following order: Basic theory of neutron scattering; Van Hove correlation functions; incoherent structure factors and dynamical-independence approximation; experimental resolution and observation time; and TOF spectrometers for QENS and IINS spectroscopy. After these basic discussions, a number of specific semiphenomenological models separating motions of vibrational, diffusive reorientational, and translational character from each other will be presented together with pertinent applications, as ingredients for describing the dynamic structure relevant in the context of soft matter. This is done with a perspective toward elucidation of possible correlations between dynamic structure and biological function. We will follow a track leading from indirect studies of function-relevant dynamics via investigations of the dependence on external variables toward a more direct observation using as an example recent energy- and real-time-resolved QENS-TOF measurements on a biological "machine" in operation.

## II.3.3.2 SOME ESSENTIAL ELEMENTS OF BASIC NEUTRON SCATTERING THEORY

Information on the dynamic structure of condensed matter is obtained by analyzing the intensity of neutrons, for example, from a monochromatic unpolarized neutron beam scattered by a sample into a solid angle element $d\Omega$ and an energy interval $d(\hbar\omega)$. This is proportional to the double-differential scattering cross section $d^2\sigma/d\Omega d\omega$:

$$\frac{d^2\sigma}{d\Omega d\omega} = \frac{k}{k_0} S(\mathbf{Q}, \omega). \tag{II.3.3.3}$$

That is, the product of the ratio of the neutron wave numbers and Van Hove's neutron scattering function $S(\mathbf{Q}, \omega)$ (Van Hove, 1954). The latter is related with Van Hove's correlation function $G(\mathbf{r}, t)$ and with the intermediate scattering function $I(\mathbf{Q}, t)$ by the following two Fourier transforms (see Chapter I):

$$S(\mathbf{Q}, \omega) = \frac{1}{2\pi} \int_{-\infty}^{\infty} e^{-i\omega t} I(\mathbf{Q}, t) dt \tag{II.3.3.4}$$

$$I(\mathbf{Q}, t) = \int_{-\infty}^{\infty} e^{i\mathbf{Q}\mathbf{r}} G(\mathbf{r}, t) d\mathbf{r} \tag{II.3.3.5}$$

It should be noted that the calculation of eq. (II.3.3.3) for the double-differential scattering cross section in slow-neutron scattering is based on first-order perturbation theory. As already pointed out by Fermi (1936), the effective interaction potential between neutron and scattering system can be treated as a small perturbation, when the nuclear scattering lengths are known experimentally. This requirement is very well fulfilled for all elements today (Koester et al., 1991; Sears, 1982,1984). The result of the calculation is known as the first Born approximation to the cross section. This approximation is rather satisfactory (Blatt and Weisskopf, 1952), if the probability is negligible that, after having been scattered by one nucleus of the target system, and before escaping from the sample, the neutron collides subsequently with another or several nuclei. In other words, the expressions given in this chapter—as is customary in neutron scattering literature—are valid for single scattering only. It is of course true that multiple-scattering (MSC) contributions to the measured intensity cannot be avoided completely. In experiments, it is therefore necessary to start by minimizing MSC components by an optimization of experiment parameters such as amount and geometry of sample material in the incident neutron beam. Furthermore, it is possible and eventually required to apply a numerical MSC correction to the measured spectra. Analytical methods (Sears, 1975) directly applied to the theoretical models, and model-independent Monte Carlo techniques (see, for example, Lechner et al., 1980; Russina et al., 2000) are employed for this purpose. In the work of Bée (1988), MSC is discussed extensively and in depth regarding both analytical calculations and numerical simulations, including an approximation for the special case of QENS and applications. Multiple scattering will not be discussed further in this chapter. All theoretical expressions of scattering functions dealt with in the following will concern single scattering only.

The most general (quantum mechanical) form (for unpolarized neutrons) of the scattering function $S(\mathbf{Q}, \omega)$ per atom for a scattering system consisting of an ensemble of $N$ atoms with nuclei generally having different scattering lengths $b_i$ reads explicitly:

$$S(\mathbf{Q},\omega) = \frac{1}{N}\sum_{i,j}^{N}\sum_{\psi 0} p(\psi_0) \sum_{\psi 1} \langle \psi_0 | \mathrm{b_j} \exp(-\mathbf{iQr}_j) | \psi_1 \rangle$$
$$\langle \psi_1 | \mathrm{b_i} \exp(\mathbf{iQr}_i) | \psi_0 \rangle * \delta\left[(\omega_{\psi 1} - \omega_{\psi 0}) - \omega\right] \quad \text{(II.3.3.6)}$$

In this expression, we average over all pairs $(i, j)$ of atoms with the bound scattering lengths $(b_i, b_j)$ and the positions $(\mathbf{r}_i, \mathbf{r}_j)$, and sum over all the sample states ($\langle \psi_0 |$ before and $| \psi_1 \rangle$ after the scattering process) characterized by the wave functions $\psi$; $p(\psi_0)$ is the statistical weight of the initial state $\psi_0$, that is, the corresponding Boltzmann distribution. The $\delta$-function expresses the conservation of the total energy of scattered neutron plus scattering system. The wave function $\psi$ and the energy eigenvalues $\omega_\psi$ are calculated from the Schrödinger equation.

Eq. (II.3.3.6) can be appreciably simplified if we limit the discussion to systems consisting of only one type of atom and if the isotopes and nuclear spin orientations are statistically distributed. Then, there is no correlation between the values of the scattering lengths $b_j$ and the atomic positions $\mathbf{r}_j$. Therefore, the averages over isotope and spin orientation distributions and over the initial states of the system can be calculated independently of each other.

As a consequence, the expression for the scattering function $S(\mathbf{Q}, \omega)$ now contains the averages of scattering length products

$$\langle b_j b_i \rangle = \langle b \rangle^2 + \delta_{ij}(\langle b^2 \rangle - \langle b \rangle^2) = b^2_{\text{coh}} + \delta_{ij} b^2_{\text{inc}} = \frac{\sigma_{\text{coh}}}{4\pi} + \frac{\delta_{ij}\sigma_{\text{inc}}}{4\pi} \qquad \text{(II.3.3.7)}$$

permitting the definition of the coherent ($\sigma_{\text{coh}}$) and the incoherent ($\sigma_{\text{inc}}$) neutron-scattering cross sections for a rigidly bound nucleus ($\delta_{ij}$ = Kronecker symbol), which will—in this type of a "monatomic approximation"—allow us to separate the double-differential scattering cross section into a coherent and an incoherent part:

$$\frac{d^2\sigma}{d\Omega d\omega} = \frac{k}{k_0}\left[\left(\frac{\sigma_{\text{coh}}}{4\pi}\right) S_{\text{coh}}(\mathbf{Q},\omega) + \left(\frac{\sigma_{\text{inc}}}{4\pi}\right) S_{\text{inc}}(\mathbf{Q},\omega)\right], \qquad \text{(II.3.3.8)}$$

where $S_{\text{coh}}(\mathbf{Q}, \omega)$ and $S_{\text{inc}}(\mathbf{Q}, \omega)$ are the coherent and the incoherent scattering functions, respectively.

Apart from the relatively simple case of monatomic systems that are not the topic of this book, the "monatomic approximation" is a very useful tool for the study of the dynamics of the organic material encountered in the fields of soft matter and biology because of the rather high concentration of hydrogen atoms in organic matter, as will be explained below.

Eq. (II.3.3.8) consists of two terms each factorized in three independent components: (i) The ratio of the wave numbers $k$ and $k_0$ characterizing the scattering process, (ii) the scattering cross sections for a rigidly bound nucleus, and (iii) the scattering functions. The latter depend on the scattering vector **Q** and the energy transfer $\hbar\omega$ as defined by eqs. (II.3.3.1) and (II.3.3.2). The coherent scattering function, $S_{\text{coh}}(\mathbf{Q}, \omega)$ in the first term, is due to the atom–atom pair correlations, whereas the incoherent scattering function, $S_{\text{inc}}(Q,\omega)$, in the second term, merely conveys self-correlations of single atoms and, as a consequence, only intensities (and not amplitudes) from scattering by different nuclei have to be added.

It must be emphasized that the structural and the dynamical properties of the scattering sample are fully described by $S_{\text{coh}}(\mathbf{Q}, \omega)$ and $S_{\text{inc}}(\mathbf{Q}, \omega)$ which, *in this approximation*, do not depend on neutron–nuclear interaction, that is, on the nuclear cross sections. The scattering functions and their relation with several other functions important for the description of scattering experiments will be discussed below; a number of details of the derivations can also be found in text books (Turchin, 1965; Gurevich and Tarasov, 1968; Squires, 1978; Lovesey, 1984; Bée, 1988).

In the following sections, we will mainly discuss the *incoherent*, and occasionally the *coherent*, scattering function. The reason is that the neutron scattering from soft matter systems and from native biological material is in a majority of cases largely dominated by that of hydrogen atoms that are usually present in large numbers in organic molecules. For the hydrogen nucleus (the proton), the incoherent scattering cross section is between 10 and 20 times larger than the other scattering cross sections, such that the scattering by hydrogen atoms mirrors the macromolecular dynamics to a large extent, and the separation of the incoherent and coherent scattering functions in relevant ($\mathbf{Q}$, $\omega$) domains is often easy, especially when partial deuteration is also employed. In the case of scattering–density fluctuations with sizeable correlation lengths, the coherent scattering contribution becomes significantly higher than the incoherent one in the small-angle scattering range, where the contrast between *the coherent scattering length densities* of different molecules or molecular subunits contributes as the square of the number of scattering centers. The total coherent and incoherent (bound) scattering cross sections are empirically known and can be found in tables (Koester et al., 1991; Sears, 1984). The fundamental aspects of neutron–nucleus scattering are treated in an excellent review (Sears, 1982); see also Chapter I of this volume.

## II.3.3.3 VAN HOVE CORRELATION FUNCTIONS AND THE CLASSICAL APPROXIMATION

Let us now consider the connection between coherent and incoherent scattering functions to be measured and correlation functions to be calculated in more detail. Starting again with the double-differential scattering cross section, but now in the "monatomic approximation," we have

$$\frac{\mathrm{d}^2\sigma}{\mathrm{d}\Omega\mathrm{d}\omega} = \left(\frac{\mathrm{d}^2\sigma}{\mathrm{d}\Omega\mathrm{d}\omega}\right)_{\mathrm{coh}} + \left(\frac{\mathrm{d}^2\sigma}{\mathrm{d}\Omega\mathrm{d}\omega}\right)_{\mathrm{inc}} = \frac{k}{k_0}\frac{(\sigma_{\mathrm{coh}}+\sigma_{\mathrm{inc}})}{4\pi}S(\mathbf{Q},\omega), \qquad \text{(II.3.3.9)}$$

where

$$S(\mathbf{Q},\omega) = \frac{\sigma_{\mathrm{coh}}}{(\sigma_{\mathrm{coh}}+\sigma_{\mathrm{inc}})}S_{\mathrm{coh}}(\mathbf{Q},\omega) + \frac{\sigma_{\mathrm{inc}}}{(\sigma_{\mathrm{coh}}+\sigma_{\mathrm{inc}})}S_{\mathrm{inc}}(\mathbf{Q},\omega). \qquad \text{(II.3.3.10)}$$

Combining eqs. (II.3.3.9) and (II.3.3.10), one obtains eq. (II.3.3.8).

Apart from the fact that the neutron–nuclear interaction has been extracted, the Fourier transformations establishing the pertinent relations between neutron scattering functions, intermediate scattering functions, and Van Hove's correlation functions are analogous to eqs. (II.3.3.4) and (II.3.3.5), as follows:

$$S_{\mathrm{coh}}(\mathbf{Q},\omega) = \frac{1}{2\pi}\int_{-\infty}^{\infty} \mathrm{e}^{-i\omega t} I_{\mathrm{coh}}(\mathbf{Q},t)\mathrm{d}t \qquad \text{(II.3.3.11)}$$

$$I_{\mathrm{coh}}(\mathbf{Q},t) = \int_{-\infty}^{\infty} \mathrm{e}^{i\mathbf{Q}\mathbf{r}} G_{\mathrm{p}}(\mathbf{r},t)\mathrm{d}\mathbf{r} \qquad \text{(II.3.3.12)}$$

$$S_{\mathrm{inc}}(\mathbf{Q},\omega) = \frac{1}{2\pi}\int_{-\infty}^{\infty} \mathrm{e}^{-i\omega t} I_{\mathrm{inc}}(\mathbf{Q},t)\mathrm{d}t \qquad \text{(II.3.3.13)}$$

$$I_{\text{inc}}(\mathbf{Q},t) = \int_{-\infty}^{\infty} e^{i\mathbf{Qr}} G_s(\mathbf{r},t)d\mathbf{r}. \tag{II.3.3.14}$$

Furthermore, we have

$$G_p(\mathbf{r},t) = \frac{1}{(2\pi)^3} \int_{-\infty}^{\infty} e^{i\omega t} \int_{-\infty}^{\infty} e^{-i\mathbf{Qr}} S_{\text{coh}}(\mathbf{Q},\omega) d\,\mathbf{Q}\, d\omega \tag{II.3.3.15}$$

$$G_s(\mathbf{r},t) = \frac{1}{(2\pi)^3} \int_{-\infty}^{\infty} e^{i\omega t} \int_{-\infty}^{\infty} e^{-i\mathbf{Qr}} S_{\text{inc}}(\mathbf{Q},\omega) d\,\mathbf{Q}\, d\omega. \tag{II.3.3.16}$$

Here, Van Hove's pair correlation and self-correlation functions are labeled with the subscripts p and s, respectively.

The interpretation of eqs. (II.3.3.11)–(II.3.3.16) is straightforward in the classical approximation. While the coherent scattering function $S_{\text{coh}}(\mathbf{Q}, \omega)$ is due to the interference of neutron waves scattered by *pairs of different atoms* at positions $\mathbf{r}_i$ and $\mathbf{r}_j$ at different times 0 and $t$, respectively, the incoherent scattering function $S_{\text{inc}}(\mathbf{Q}, \omega)$ is only due to the interference of waves scattered by the *same* nucleus that is in general located at different positions for different times. The classical meaning of the Van Hove correlation functions can therefore be described as follows: $G_p(\mathbf{r}_j - \mathbf{r}_i, t)$ is the conditional probability per unit volume to find an atom (nucleus) at a position $\mathbf{r}_j$ at time $t$, if this or another atom has been at a position $\mathbf{r}_i$, with a distance vector $\mathbf{r} = \mathbf{r}_j - \mathbf{r}_i$, at a previous time $t = 0$. Analogously, the self-correlation function, $G_s(\mathbf{r}, t)$, is the conditional probability per unit volume to find an atom at $\mathbf{r}(t)$ at time $t$, if the same atom has been at the origin $\mathbf{r} = 0$ at $t = 0$.

The conditions for the validity of the classical approximation are that the amounts of energy and momentum exchanged in the scattering process must remain much smaller than the thermal energy, that is, we must have

$$|\hbar\omega| \ll \frac{k_B T}{2}, \tag{II.3.3.17}$$

$$\frac{(\hbar Q)^2}{2M} \ll \frac{k_B T}{2}, \tag{II.3.3.18}$$

where $k_B$ = Boltzmann constant, $T$ = temperature, and $M$ = atomic mass. Therefore, quantum effects are expected for large $\mathbf{Q}$ and large $\omega$ (or for small $\mathbf{r}$ and small $t$). In the realm of quasielastic neutron scattering concerned with ranges of fairly small $\mathbf{Q}$ and $\omega$, the scattering functions may nevertheless be calculated classically, if they are then corrected by the so-called detailed-balance factor (DBF), $\exp[-\hbar\omega/(2k_B T)]$. Because of the energy dependence of level occupation according to the Boltzmann distribution, the exact (quantum mechanical) scattering functions for energy gain and energy loss are always related in the following way:

$$S(-\mathbf{Q},-\omega)=\exp\left(\frac{\hbar\omega}{k_{\mathrm{B}}T}\right)S(\mathbf{Q},\omega). \qquad \text{(II.3.3.19)}$$

This asymmetry with respect to $\hbar\omega=0$ distinguishes $S(\mathbf{Q},\omega)$ from a classically calculated function, $S^{\mathrm{cl}}(\mathbf{Q},\omega)$, which is symmetric in $\hbar\omega$. The symmetric function obtained, if both sides of eq. (II.3.3.19) are multiplied by the detailed-balance factor, is a very good approximation of $S^{\mathrm{cl}}(\mathbf{Q},\omega)$; based on this fact, we can obtain the true $S(\mathbf{Q},\omega)$ as follows:

$$\mathrm{S}(\mathbf{Q},\omega)=\exp\left(\frac{-\hbar\omega}{2k_{\mathrm{B}}T}\right)S^{\mathrm{cl}}(\mathbf{Q},\omega). \qquad \text{(II.3.3.20)}$$

In the "monatomic approximation," analogously, we have

$$S_{\mathrm{coh}}(\mathbf{Q},\omega)=\exp\left(\frac{-\hbar\omega}{2k_{\mathrm{B}}T}\right)S^{\mathrm{cl}}_{\mathrm{coh}}(\mathbf{Q},\omega) \qquad \text{(II.3.3.21)}$$

and for incoherent scattering, we have

$$S_{\mathrm{inc}}(\mathbf{Q},\omega)=\exp\left(\frac{-\hbar\omega}{2k_{\mathrm{B}}T}\right)S^{\mathrm{cl}}_{\mathrm{inc}}(\mathbf{Q},\omega). \qquad \text{(II.3.3.22)}$$

In the subsequent sections of this chapter, all classical scattering functions will be labeled with the superscript cl.

## II.3.3.4 INCOHERENT STRUCTURE FACTORS AND DYNAMICAL-INDEPENDENCE APPROXIMATION

The behavior of the correlation functions at times that are long as compared to the experimental observation time defined by the energy resolution (see next section) is one of the important features to be analyzed in order to understand the observed dynamical structure, especially in the case of incoherent neutron scattering. For example, if an atom is diffusing in a space that is very large as compared to the interatomic distances, the self-correlation function $G_{\mathrm{s}}(\mathbf{r},t)$ vanishes, if $t$ goes to infinity, whereas, for an atom *bound to a finite volume* (e.g., as part of a vibrating or rotating molecule fixed in a crystal), $G_{\mathrm{s}}(\mathbf{r},t)$ approaches a finite value $G_{\mathrm{s}}(\mathbf{r},\infty)$ for $\mathbf{r}$ varying in the interior of this volume. In fact, generally, the self-correlation function can be split into its asymptotic value in the long-time limit and a time-dependent term $\mathrm{G}'_s(\mathbf{r},t)$, according to

$$G_{\mathrm{s}}(\mathbf{r},t)=G_{\mathrm{s}}(\mathbf{r},\infty)+G'_{\mathrm{s}}(\mathbf{r},t). \qquad \text{(II.3.3.23)}$$

The Fourier transform of this expression over space and time according to eqs. (II.3.3.13) and (II.3.3.14) yields

$$S_{\text{inc}}(\mathbf{Q}, w) = {}^{\text{el}}S_{\text{inc}}(\mathbf{Q})\delta(\omega) + {}^{\text{ne}}S_{\text{inc}}(\mathbf{Q}, \omega). \tag{II.3.3.24}$$

Apparently, the incoherent scattering function for a spatially restricted motion is decomposed into a purely elastic line, ${}^{\text{el}}S_{\text{inc}}(\mathbf{Q})\delta(\omega)$, with the integrated intensity ${}^{\text{el}}S_{\text{inc}}(\mathbf{Q})$, and a nonelastic component, ${}^{\text{ne}}S_{\text{inc}}(\mathbf{Q}, \omega)$. The first term is the result of diffraction of the neutron on the "infinite time" *distribution in space of a single nucleus spread over a finite volume by its motion*, as already pointed out by Stiller (Stiller, 1965). Therefore, we can derive information about the structure in a very direct way from *incoherent* scattering (Sköld, 1968). In order to systematically exploit the theoretical fact expressed by eq. (II.3.3.24) in neutron scattering experiments, the concept of the elastic incoherent structure factor (EISF) was formulated (Lechner, 1971). The EISF concept provides a method, which permits the extraction of structural information on all kinds of localized single-particle motion by the determination of the elastic fraction of the measured spectral intensity, as a function of $\mathbf{Q}$. The idea is simple: First, by employing a suitable (e.g., sufficiently high) energy resolution, the measured integrals of elastic (${}^{\text{el}}\text{Int}\ (\mathbf{Q})$) and nonelastic (${}^{\text{ne}}\text{Int}\ (\mathbf{Q})$) components of the scattering function in eq. (II.3.3.24) are determined separately. Then, an intensity ratio involving the two integrals can be defined as

$$\text{EISF} = \frac{{}^{\text{el}}\text{Int}\ (\mathbf{Q})}{{}^{\text{el}}\text{Int}\ (\mathbf{Q}) + {}^{\text{ne}}\text{Int}\ (\mathbf{Q})} = \frac{A\ {}^{\text{el}}S_{\text{inc}}(\mathbf{Q})}{A \int_{-\infty}^{\infty} S_{\text{inc}}(\mathbf{Q}, \omega)d\omega} \tag{II.3.3.25}$$

where A is an experimental normalization factor. This demonstrates that the difficulty of an absolute intensity calibration is avoided in the determination of the EISF by eq. (II.3.3.25). Because the factor $A$ cancels and the integral of the incoherent scattering function is equal to unity by definition, we simply have $\text{EISF} = {}^{\text{el}}S_{\text{inc}}(\mathbf{Q})$.

The coefficient ${}^{\text{el}}S_{\text{inc}}(\mathbf{Q})$ in the last two equations is the EISF in its most general form, since it is the result of the combined action of *all the motions* of the scattering atom. However, the determination of this "global" EISF is not necessarily the immediate or exclusive aim of an experiment. If we are particularly interested in a specific type of motion, it is—under certain conditions—often possible to isolate the effect of this specific motion. The conditions are (i) that the energy range of the latter is reasonably well separated from the rest of the motions, (ii) that a suitable energy resolution can be employed (see the corresponding discussion in Section II.3.3.5), and last but not least (iii) that the dynamics of the motion of interest can be considered as approximately independent of the other relevant modes of the system (see eqs. (II.3.3.26) and (II.3.3.27)). In this case, the EISF concept is applied directly to the elastic component of the specific type of motion, for example, to stochastic molecular rotation, rather than to the combined effect of all atomic motions.

Let us briefly discuss the dynamical-independence approximation (for more details, see the work by Leadbetter and Lechner (1979), pp. 289–290; Lechner (1983), pp. 186–188; Lechner and Longeville (2006b), pp. 356–357; and Bée (1988), pp. 66–67): The weakness or absence of coupling between the different kinds of motion of the *same atom* allows an independent theoretical treatment of individual types of motion. This procedure, which is obviously much simpler than the full treatment taking account of all kinds of coupling, permits a modular construction of the Van Hove functions $I_{inc}(\mathbf{Q}, t)$ and $S_{inc}(\mathbf{Q}, \omega)$ that can be obtained by combining together the scattering functions of individual motions. For example, the calculation is greatly facilitated if, for example, vibrational, rotational, and translational motions are assumed to be (dynamically) independent. The intermediate scattering function is then a product, while the scattering function is a convolution of the pertinent modular functions:

$$I_{inc}(\mathbf{Q}, t) = I_{vib}(\mathbf{Q}, t) I_{rot}(\mathbf{Q}, t) I_{trans}(\mathbf{Q}, t), \qquad \text{(II.3.3.26)}$$

$$S_{inc}(\mathbf{Q}, \omega) = S_{vib}(\mathbf{Q}, \omega) \otimes S_{rot}(\mathbf{Q}, \omega) \otimes S_{trans}(\mathbf{Q}, \omega). \qquad \text{(II.3.3.27)}$$

Here, $I_{vib}(\mathbf{Q}, t)$, $I_{rot}(\mathbf{Q}, t)$, $I_{trans}(\mathbf{Q}, t)$ are the incoherent intermediate and $S_{vib}(\mathbf{Q}, \omega)$, $S_{rot}(\mathbf{Q}, \omega)$, $S_{trans}(\mathbf{Q}, \omega)$ the incoherent Van Hove scattering functions of the three individual types of motion; the symbol $\otimes$ stands for the convolution in energy transfer $\hbar\omega$.

The expression to be used for the analysis of such specific motions is formally the same as eq. (II.3.3.25). But now we have to replace the nonelastic integral (${}^{ne}\text{Int}\ (\mathbf{Q})$) by the corresponding quasielastic integral (${}^{qe}\text{Int}\ (\mathbf{Q})$), and keep in mind that the incoherent scattering function, $S_{inc}(\mathbf{Q}, \omega)$, must now be replaced by a partial incoherent scattering function, ${}^{qe}S_{inc}(\mathbf{Q}, \omega)$, consisting merely of an elastic term (measured integral: ${}^{el}\text{Int}\ (\mathbf{Q})$) and a quasielastic term (measured integral: ${}^{qe}\text{Int}\ (\mathbf{Q})$) that correspond to the specific motion under study. The EISF is then given as

$$\text{EISF} = \frac{{}^{el}\text{Int}(\mathbf{Q})}{{}^{el}\text{Int}(\mathbf{Q}) + {}^{qe}\text{Int}(\mathbf{Q})} = \frac{A\ {}^{el}S_{inc}(\mathbf{Q})}{A \int_{-\infty}^{\infty} S_{inc}(\mathbf{Q}, \omega)\mathrm{d}\omega} \qquad \text{(II.3.3.28)}$$

Here, the effect of faster motions has been subtracted as a flat "inelastic" background and what remains of the inelastic effect is only an attenuating Debye–Waller, factor bound to be cancelled, because here it is included in the normalization factor $A$. Note that ${}^{qe}S_{inc}(\mathbf{Q}, \omega)$ by definition has the same normalization as $S_{inc}(\mathbf{Q}, \omega)$.

It is the possibility of isolating the EISF of *specific modes of motions*, which has proved of most practical importance for the application of the technique. This isolation, of course, means separating the elastic from the quasielastic component of the QENS spectrum. Due to an important sum rule concerning the incoherent scattering function, namely, the property that its integral is equal to unity, we have the incoherent structure factor sum rule,

$$\text{EISF}(\mathbf{Q}) + \text{QISF}(\mathbf{Q}) = 1, \qquad \text{(II.3.3.29)}$$

where QISF(**Q**) = $^{qe}$Int (**Q**)/A is the quasielastic incoherent structure factor. The latter is the **Q**-dependent spectral weight of the quasielastic component. It obviously contains the same structural information as the EISF, and is sometimes used instead of the latter, for technical reasons (e.g., in case of Bragg contamination of the EISF). From the shape of the EISF as a function of the dimensionless parameter $QR$ ($R$ = radius of rotation or radius of a spherical volume of diffusion, for example), the mechanism of the concerned motion can be recognized. The EISF method can therefore be used as a strategy for finding the appropriate differential equations and their boundary conditions for the dynamical mechanisms of localized atomic motions in condensed matter. For more details about derivation, properties and application of incoherent structure factors, the reader is referred to the review articles by Leadbetter and Lechner (1979), Lechner (1983), Lechner and Longeville (2006b), and to the book by Bée (1988, 2003).

## II.3.3.5 EXPERIMENTAL ENERGY RESOLUTION AND OBSERVATION TIME

Modern neutron time-of-flight spectrometers offer the very important capability to vary the energy resolution $\Delta(\hbar\omega)$ continuously over three orders of magnitude on one single instrument (Lechner, 1985, 1991). This range can be extended to five orders of magnitude if other high-resolution techniques such as backscattering and neutron spin echo methods are combined with the TOF spectrometers. The variable neutron energy resolution $\Delta(\hbar\omega)$, and the implied variable observation time $\Delta t_{\mathrm{H}}$, which is the decay time on the Fourier timescale of the Fourier transformed resolution function, are connected by the following exact uncertainty relation (see the works of Lechner, 1983, 1994, 2001b)

$$H_{\mathrm{RES}} \cdot (\Delta t)_{\mathrm{H}} = c\hbar, \qquad \text{(II.3.3.30)}$$

where, for practical reasons, $H_{\mathrm{RES}} = \Delta(\hbar\omega)$ has been defined as the half width at half maximum (HWHM) of the resolution function. The constant coefficient $c$ is equal to 1, in the case of a (hypothetical) Lorentzian resolution function, whereas in the more realistic case of a Gaussian resolution function, $c = 2\sqrt{\ln 2}$, with definitions of energy width and decay time analogous to those used in eq. (II.3.3.30).

The necessity to discuss resolution functions reflects the fact that the scattering functions $S_{\mathrm{coh}}$ and $S_{\mathrm{inc}}$ defined in eqs. (II.3.3.11)–(II.3.3.14), as well as the corresponding correlation functions $G_{\mathrm{p}}$, $G_{\mathrm{s}}$, and the intermediate scattering functions $I_{\mathrm{coh}}$, $I_{\mathrm{inc}}$ cannot be determined experimentally in their pure forms. The reason is, that, for example, the scattering functions are in principle broadened by the experimental resolution functions $R$(**Q**, $\omega$) in the four-dimensional (**Q**, $\omega$)-space. In the case of incoherent scattering, the **Q**-spread of the resolution can often be neglected, when the studied functions are only rather slowly varying with **Q**. In this chapter, we will focus on this case. It is sufficient to compare the measured resolution-broadened "scattering function," [$S_{\mathrm{inc}}$(**Q**, $\omega$)]$_{\mathrm{meas}}$ to a convolution of the measured energy-resolution function $R(\omega)$ with a theoretical function $[S_{inc}(\mathbf{Q}, \omega)]_{theo}$ being developed, in order to simulate $S_{inc}(\mathbf{Q}, \omega)$:

$$[S_{inc}(\mathbf{Q}, \omega)]_{meas} = \int_{-\infty}^{\infty} [S_{inc}(\mathbf{Q}, \omega')]_{theo} R(\omega - \omega') d\omega' \qquad \text{(II.3.3.31)}$$

While the resolution function $R(\omega)$ has the effect of broadening the neutron scattering function along the energy transfer coordinate of the experiment, the observation function $R^*(t)$ is a factor attenuating the Van Hove correlation function with increasing Fourier time: $R^*(t)$ is the Fourier transform of $R(\omega)$ and is therefore in most practical cases a function essentially decaying with increasing time. The net effect is that the correlation functions are observed in a Fourier time window, with an upper limit controlled by the decay time constant $\Delta t_H$ of the observation function. The low-time limit of this window has a different origin: For instruments working in $(\mathbf{Q}, \omega)$-space, it is mainly a consequence of the (always limited) statistical accuracy of the measurement, because quasielastic intensities typically decrease with increasing energy transfer, and therefore counting statistics become the poorer the larger the energy transfers are as compared to the energy-resolution width.

For comparison, we note that in spin echo experiments, where *intermediate* scattering functions are measured directly as a function of Fourier time, the energy-resolution problem requires a different treatment. Here, instead of the necessity to fold scattering functions with energy-resolution functions, the correction for this resolution effect essentially reduces to dividing the measured spectra by the experimental observation function $R^*(t)$. One should, however, not forget that in addition to the energy resolution problem, in principle, because measurements are carried out as a function of time, there is also a time-resolution broadening to be expected. The latter is mainly caused by the finite width of the wavelength band selected by the velocity selector of the instrument (see the study by Lechner and Longeville, 2006a, pp. 347–348).

Regarding the capability of a TOF spectrometer to vary the energy resolution $\Delta(\hbar\omega)$ continuously over several orders of magnitude, it is obvious that this is very useful when, as usual, the scattering function is a superposition of several or even a large number of dynamical modes with different frequency ranges, which is certainly true in the complex systems encountered in the fields of soft matter and biology. In order to be able to extract information on all relevant motional components, one needs to carry out several measurements with different resolutions. For example, quasielastic neutron scattering spectra obtained with one single energy resolution only, usually furnish incomplete information. To avoid wrong conclusions, it is typically necessary to employ at least three—if not many—different energy resolutions in the study of a given problem. Figure II.3.3.1 shows as an example three spectra from a study of purple membrane (see Section II.3.3.8.1) for a description of this biological system), which demonstrate the qualitative similarity, but a striking quantitative difference of the spectra in such a series of measurements (J. Fitter and R.E. Lechner, 1997, unpublished data (with the TOF spectrometer NEAT); see the work by Rufflé et al., 2000).

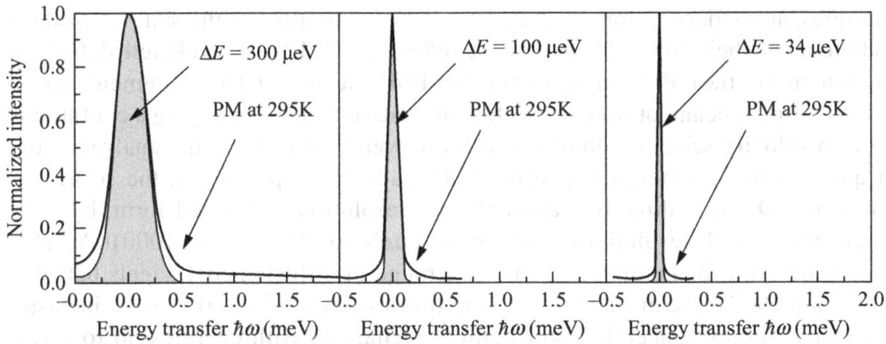


**Figure II.3.3.1** Purple membrane spectra measured with the TOF spectrometer NEAT at BENSC in Berlin, with three different energy resolutions (300, 100, and 34 μeV (FWHM)) using incident wavelengths $\lambda_0 = 4.54, 5.1$, and 6.2 Å, respectively. The shaded spectra represent the resolution function obtained from a vanadium standard sample. In spite of identical dynamics the three PM spectra show strong quasielastic components with rather different apparent linewidths. This is of course due to the different observation times (i.e., different energy resolutions) employed, emphasizing dynamical aspects on different timescales of the system under study. Sample: stacks of purple membrane equilibrated at 98% relative humidity ($D_2O$), at room temperature. Illuminated sample size: $30 \times 60\,mm^2$. Measurement times: 3 h ($\Delta E = 300\,\mu eV$), 7 h ($\Delta E = 100\,\mu eV$) and 14.4 h ($\Delta E = 34\,\mu eV$), respectively. Spectra measured by J. Fitter and R.E. Lechner with the multichopper time-of-flight spectrometer NEAT (see Section II.3.3.6): Unpublished data taken by J. Fitter and R. E. Lechner around 1997; see Rufflé et al., (2000).

The principle of experimental observation time, energy, and Fourier time windows in quasielastic neutron scattering, and their relevance for the determination of dynamic structure, and especially in problems concerning diffusive atomic and molecular motions in condensed matter, has been discussed extensively in the works of Lechner (1983, 1994, 2001b), Lechner and Longeville (2006a, 2006b), Springer and Lechner, (2005). For further detailed fundamental discussions of the Van Hove theory, quasielastic and inelastic neutron scattering, we refer (in chronological order) to the reviews, monographs, and books especially devoted to this topic (Turchin, 1965; Gurevich and Tarasov, 1968; Lovesey and Springer, 1977; Squires, 1978; Windsor, 1981; Lechner and Riekel, 1983; Richter, 1983; Lovesey, 1984; Price and Sköld, 1987; Bée, 1988; Springer and Lechner, 2005; Lechner and Longeville, 2006a, 2006b). General reviews of localized and long-range diffusion studies employing quasielastic neutron scattering techniques are given in several references (Bée, 2003; Springer and Lechner, 2005).

## II.3.3.6 TOF SPECTROMETERS FOR QUASIELASTIC AND INELASTIC NEUTRON SCATTERING

Time-of-flight (TOF) neutron scattering spectrometers record neutron intensity as a function of momentum $\hbar\mathbf{Q}$ and energy $\hbar\omega$ exchanged with a specimen of interest. The wavelength, velocity, or energy of neutrons, or their distribution as a function of these variables, are defined before, and analyzed after scattering by the sample. Several techniques are being used: *Crystal Bragg reflection* (XTL) for wavelength definition as a function of the reflection angle, neutron TOF selection and measurement using a pulsed incident beam, or initial wavelength band selection employing a continuous neutron velocity selector combined with a wavelength filter on the analyzer side. In quasielastic and inelastic neutron TOF scattering experiments, the scattering function $S(\mathbf{Q}, \omega)$ is directly measured, in a resolution-broadened form, by TOF spectrometry with resolutions from about 1 μeV to about a few 1000 μeV. This allows to cover, for example, by QENS, a range of diffusion coefficients between $10^{-12}$ and $10^{-8}$ m$^2$/s, or of characteristic times from $10^{-9}$ to $10^{-13}$ s. In inelastic scattering studies concerning soft matter, excitations with energies up to a few hundred meV can be investigated. A number of TOF methods actually relevant for soft matter studies will be discussed in the following.

## II.3.3.6.1 Multichopper TOF–TOF Spectrometers

In a TOF–TOF spectrometer, the wavelengths of incident and scattered neutrons, $\lambda_0$ and $\lambda$, are both measured by a time-of-flight method. A chopper system located in front of the sample defines $\lambda_0$ using at least two choppers with a mutual phase shift. The scattered neutron wavelength $\lambda$ is then determined by measuring the neutron flight time from the sample to the detector. For thermal neutrons, Fermi-choppers are often used, whereas disk-choppers are preferentially employed in the case of cold neutrons. A multidisk chopper time-of-flight (MTOF) instrument is illustrated schematically in Figure II.3.3.2 (Lechner, 2001a). The two principal choppers, CH1 and CH2, the sample S and the detectors D are separated by the distances $L_{12}$, $L_{2S}$, and $L_{SD}$, respectively, which have values of the order of several meters. CH1 and CH2 create neutron pulses with widths $\tau_1$ and $\tau_2$, and define the bandwidth of the incident neutron wavelength $\lambda_0$. More precisely, the phase difference between CH1 and CH2 allows the latter to select the "monochromatic" wavelength of the experiment. The scattering processes in the sample then cause neutron wavelength shifts to smaller or larger values of $\lambda$.

In Figure II.3.3.3, a neutron flight-path diagram is shown. It explains the method in more detail and demonstrates the filter action of the various disks of the chopper cascade. While the neutron time-of-flight is displayed along the horizontal axis, the flight-path between the different elements of the chopper cascade is represented on the vertical axis. The filter choppers, CHP and CHR perform premonochromatization and pulse-frequency reduction of the beam, respectively, in order to avoid frame-overlap at the monochromator disk CH2 and at the detectors. The spectra shown schematically on the top of the figure, correspond to a study of the rotational motion of $OH^-$ ions (Smit et al., 1979), carried out using 4 Å neutrons in 1975 with the MTOF spectrometer IN5 at ILL. The diagram also demonstrates the periodicity of the data acquisition procedure, inherent in a pulsed experiment. Each TOF period $P_{spec}$ in principle contains one spectrum. But the duration of the measurement must cover a large number of such periods, the spectra of which (about $10^6$ for a measurement time

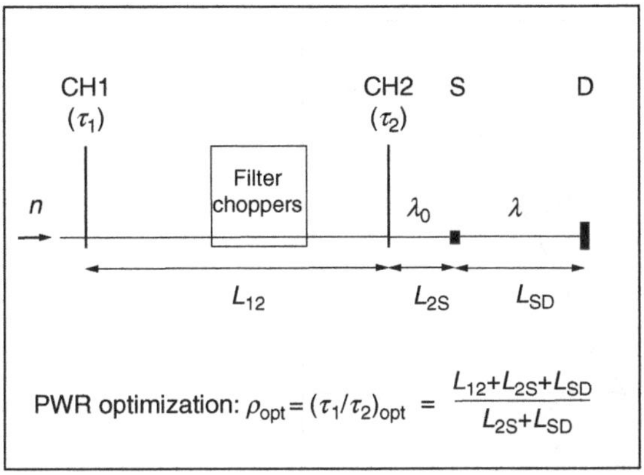


**Figure II.3.3.2** Schematic sketch of an MTOF spectrometer: CH1 and CH2 are the two principal choppers defining the monochromatic neutron pulse and its wavelength bandwidth; S = sample, D = detectors; $L_{12}$, $L_{2S}$, and $L_{SD}$ are the distances between these elements of the instrument; $\tau_1$ and $\tau_2$ are the widths of the pulses created by CH1 and CH2, $\lambda_0$ and $\lambda$ the incident and the scattered neutron wavelengths (Lechner, 2001a). Inset: the pulse-width ratio (PWR) optimization formula for elastic and quasielastic scattering (Lechner, 1985, 1991). Typical instruments of this type in operation at continuous neutron sources are IN5 (upgraded version Ollivier et al., 2004) at ILL in Grenoble, MIBEMOL (Hautecler et al., 1985) at LLB in Saclay, both France; NEAT (Lechner, 1991, 1992, 1996, 2001a; Lechner et al. 1996; Rufflé et al., 2000), at HZB in Berlin, Germany; and DCS (Copley, 1990) at NIST in Gaithersburg, USA. A number of new MTOF spectrometers are under construction at new reactor neutron sources, for example, at the HANARO facility in Daejeon, South Korea, and at the new high-flux spallation neutron sources SNS in Oak Ridge, USA, and at J-PARC in Tokai, Japan.

of 3 h) are added together, in order to obtain sufficient statistical accuracy. The pulse repetition rate $(P_{spec})^{-1}$ of the experiment is limited by the necessity of avoiding frame overlap corresponding to the superposition of the fastest neutrons (scattered with energy gain) within a time-of-flight period and the slowest neutrons (scattered with energy loss) from the previous period. This requires

$$P_{spec}(\mu s) = 252.78 C \lambda_0(\text{Å}) L_{SD}(\text{m}), \qquad \text{(II.3.3.32)}$$

where practical experimental units are indicated. The dimensionless constant $C$ has to be chosen depending on the width of the quasielastic spectrum and the corresponding decay of its intensity on the low-energy side. $C$ is usually in the range of $1.2 \leq C \leq 1.8$. For a given incident neutron wavelength, the total intensity at the detectors is essentially governed by the factor $(\tau_1\tau_2)$, that is, by the product of the two chopper opening times (Lechner, 1985, 1991). The latter also control the resolution, and thus intensity and resolution are connected through these parameters. The optimization formula for elastic scattering is shown as an inset in Figure II.3.3.2. Explicit expressions of $\Delta(\hbar\omega)$ for both the elastic and the inelastic energy resolution widths (HWHM) at the detector are discussed in the studies by Lechner (1985, 1991). The continuous variation of the energy resolution is achieved by varying the chopper pulse widths $\tau_1$ and $\tau_2$ (which, for example, in the case of the MTOF spectrometer, NEAT is possible by a factor between 1 and 40), and by choosing

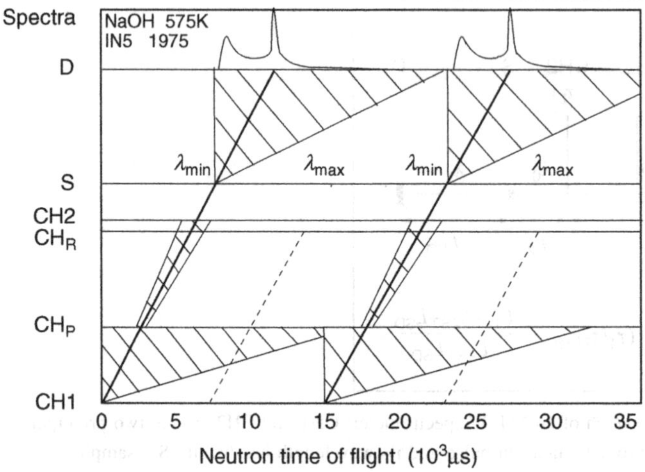


**Figure II.3.3.3** Neutron flight-path diagram: It demonstrates the filter action of the various disks of the chopper cascade. The vertical axis represents the flight-path between the different elements of the chopper cascade. CH1 defines the initial time distribution of the neutron pulse; CHP is the premonochromator; CHR is employed for pulse frequency reduction, in order to avoid frame-overlap at the detectors; finally, CH2 selects the "monochromatic" wavelength band for the experiment. The TOF spectra shown schematically on the top of the figure, correspond to a study of the rotational motion of $OH^-$ ions (Smit et al., 1979) with the MTOF spectrometer IN5 at ILL, carried out in 1975 using 4 Å neutrons. Each spectrum covers one TOF period $P_{spec}$; see text for more details (Lechner, 2001a).

the incident wavelength (yielding another factor of up to about 30 for the wavelength range from 4 to 12 Å). A number of applications of this technique using NEAT at HMI in Berlin are described in the work of Lechner and Longeville (2006b). The most important and the unique feature of this type of instrument is this capability of varying the energy resolution continuously over several orders of magnitude (see Section II.3.3.5). More detailed discussions can be found in the study by Lechner and Longeville (2006a, 2006b).

In Figure II.3.3.4, a schematic sketch of NEAT is shown as an example of a typical conventional TOF–TOF spectrometer that was built in the years 1988 to 1995 at the continuous reactor neutron source BER-II in Berlin.

#### II.3.3.6.2 "Inverted-Geometry" TOF–XTL Spectrometers

Another type of spectrometer uses the TOF–XTL technique. This type of hybrid instrument employs a pulsed polychromatic ("white") incident beam and single crystals as analyzing filters. It is well adapted to the time-structure of spallation neutron sources. The energies of the incident neutrons are measured with TOF techniques, while the energy of scattered neutrons is fixed by the analyzers. For high energy-resolution, the crystals are used in backscattering (BSC) or near-BSC

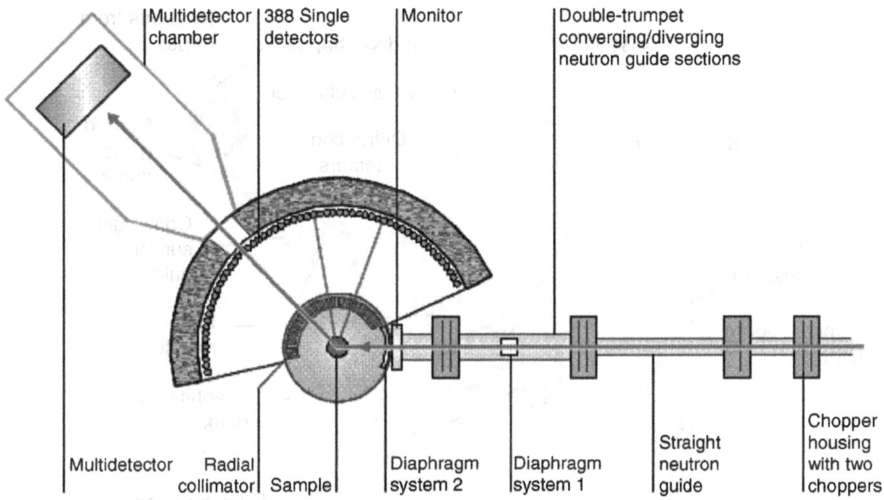


**Figure II.3.3.4** Schematic sketch of the TOF–TOF spectrometer NEAT at HZB. This spectrometer has seven synchronized disk-choppers, rotation speed up to 20,000 rpm (distance between first and last chopper: 12 m), 388 detectors at 2.5 m distance from the sample axis, detector coverage from 14° to 137° in the horizontal plane. In addition, for higher energy resolution and better angular resolution, a two-dimensional multidetector can be placed at any scattering angle within the 14° to 137° range (Lechner, 1991, 1992, 1996,, 2001a; Lechner et al. 1996; Rufflé et al., 2000).

geometry ($\theta \cong \pi/2$), whereas for more moderate resolution, $\theta < \pi/2$ is chosen. Since the quality requirements for the crystals depend on these configurations, and for other practical reasons, dedicated instruments have been built for each case.

A typical example with analyzer in near-BSC geometry is represented by the inverted geometry instrument IRIS (Carlile and Adams, 1992) at ISIS in Chilton, which is depicted schematically in Figure II.3.3.5.

Depending on the type of crystals used, several discrete values of energy resolution in the range from about 1 μeV to 55 μeV are achieved. The energy resolution function of such a spectrometer is essentially given by the convolution of four contributions arising from (i) the finite pulse width produced by the cold moderator of the spallation source or by the pulse-shaping chopper, (ii) the uncertainty of the analyzer crystals' lattice spacing, (iii) the beam divergence and crystal mosaicity, and finally (iv) the time-spread caused by sample and detector thickness. More information can be found in the study by Lechner and Longeville (2006a).

Another remarkable, similar inverted spectrometer at ISIS is OSIRIS, with the following spectrometer specifications:

Energy resolution (FWHM): 25 μeV (PG002); 99 μeV (PG004)

Momentum transfers: $0.3\,\text{Å}^{-1} < Q < 1.8\,\text{Å}^{-1}$ (PG002); $0.37\,\text{Å}^{-1} < Q < 3.6\,\text{Å}^{-1}$ (PG004)

Scattering angle range: $11° < 2\theta < 155°$

Solid angle: 1.09 sr

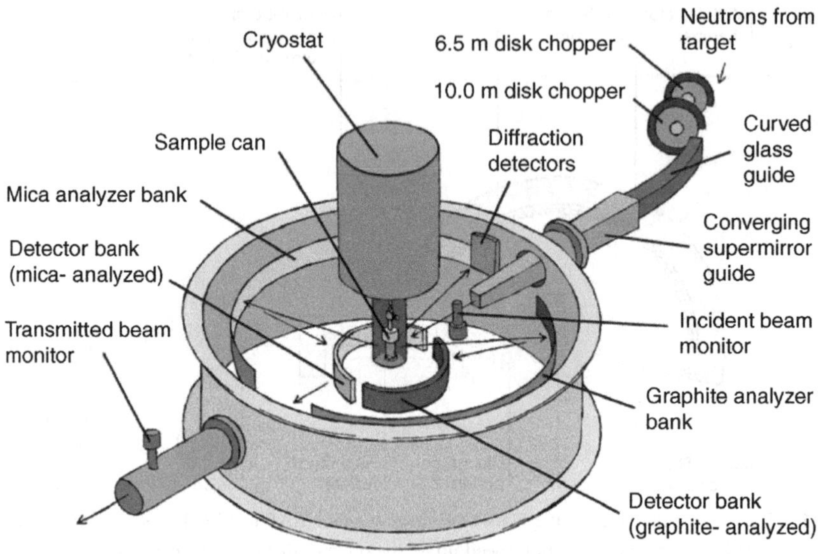


**Figure II.3.3.5** Design of the high-resolution inverted-geometry backscattering spectrometer IRIS (Carlile and Adams, 1992) on the ISIS pulsed source. The sample is located in the center of the analyzer vacuum vessel, at a distance of 36.5 m from the cold hydrogen moderator. The incident "white" beam reaches the sample through a converging supermirror guide. Graphite (resolutions 17.5 and 55 μeV) and mica (resolutions 1, 4.2, and 11 μeV) analyzer banks are arranged laterally on opposite sides. The detectors are mounted around the sample, in a plane slightly below the latter.

## II.3.3.6.3 "Direct-Geometry" XTL–TOF Spectrometers

XTL–TOF spectrometers (Windsor, 1981) basically employ a crystal monochromator to create a continuous monochromatic beam, that is, the incident neutron beam is monochromatized by Bragg reflection. For a given Bragg angle $\Theta$, and a corresponding reciprocal lattice vector **G**, the monochromator selects a certain neutron wavelength $\lambda_0$. The monochromatic beam is then periodically chopped by a disk- or Fermi-chopper, before it hits the sample. The energy distribution of the scattered neutrons is obtained by measuring their time of flight from the sample to the detectors that cover a large range of solid angle and consequently of $Q$-values. A more sophisticated version uses several crystal monochromators (located at slightly different positions and with slightly different orientations on the neutron guide) that reflect several different wavelengths selected. Due to the correlation between the monochromator reflection angle and the wavelength of reflected neutrons (and their velocities), the Fermi-chopper consecutively transmits neutron pulses from these different parts of the incident neutron beam, so that all these neutrons arrive on the detector at the same time. This is the so-called time-focusing principle. The prototype of this spectrometer is IN6 (Scherm et al., 1976) at the ILL. A more recent version of this instrument type is the spectrometer

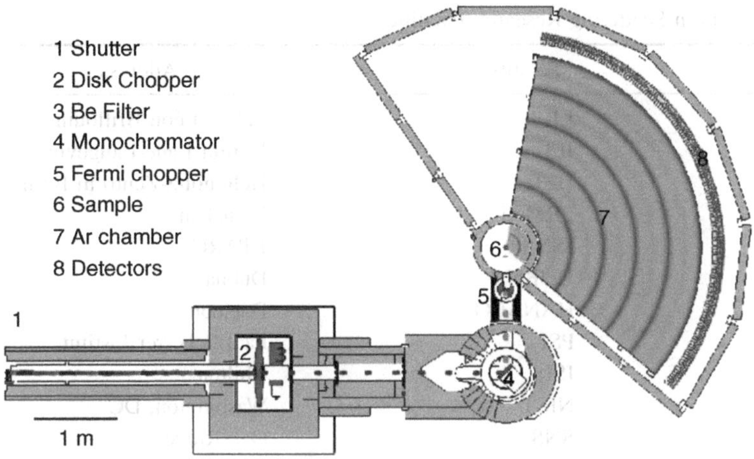


**Figure II.3.3.6** FOCUS spectrometer at Paul-Scherrer Institut (PSI) (Mesot et al., 1996); FOCUS is a typical XTL–TOF spectrometer, that is, a time-of-flight instrument with a Bragg monochromator and a TOF analyzer. While the monochromator selects the incident neutron energy $E_0$, the energy of the scattered neutrons $E$ is determined by measuring the neutron flight time.

FOCUS at PSI (Mesot et al., 1996) shown by the schematic representation in Figure II.3.3.6. This spectrometer uses a monochromator covering a large beam area, which is composed of several tens of crystal pieces with horizontal and vertical focusing (variable radius of curvature). The distance between the guide exit and the monochromator can be varied in order to achieve either a high-intensity, but low-resolution mode or one with lower intensity, but higher resolution. This type of time-of-flight instrument is characterized by five main parameters (Mutka, 1994): the lattice spacing $d$ of the crystals, the monochromator Bragg angle $\theta_M$, the width $W$ of the monochromator, the distances from the guide exit to the monochromator and from the monochromator to the sample, $d_{GM}$ and $d_{MS}$, respectively. The width of the wavelength distribution obtained by this setup is essentially given by

$$\Delta\lambda_0 = d\sin(\theta_M)W\cos(\theta_M)\left|\frac{1}{d_{GM}} - \frac{1}{d_{MS}}\right| \qquad \text{(II.3.3.33)}$$

Further contributions to the energy resolution, that is, the energy transfer uncertainty at the detector, are the mosaic spread $\Delta\theta_M$ of the monochromator, the incident beam divergence, and the sample–detector time-of-flight spread due to finite thicknesses of sample and detectors. If these contributions are independent of each other, they may be added quadratically. When $d_{GM} \cong d_{MS}$, the term given by eq. (II.3.3.33) vanishes, and the highest possible resolution is achieved, but at the expense of beam intensity. On the other hand, the neutron intensity is maximized, when $d_{GM}$ is chosen as small as possible, the resolution is then lower.

**Table II.3.3.1** Neutron Scattering Research Facilities

| Country | Facility | Address |
|---|---|---|
| France | LLB | Labor. Léon-Brillouin |
| France | ILL | Institut Laue-Langevin |
| Germany | HZB | Helmholtz-Zentrum Berlin |
| Germany | FRM-II | München |
| Japan | JSNS | J-PARC |
| Russia | DUBNA | Dubna |
| South Korea | HANARO | Daejeon |
| Switzerland | PSI | Paul-Scherrer Institut |
| UK | ISIS | Chilton |
| USA | NIST | Washington, DC |
| USA | SNS | Oak Ridge |

#### II.3.3.6.4 Worldwide Inventory of TOF Spectrometers Suitable for Quasielastic and Inelastic Soft Matter and Biological Applications at Neutron Scattering Research Facilities

A worldwide inventory of TOF spectrometers suitable for quasielastic and inelastic soft matter and biological applications could easily be set up. It would contain a number of at least 20 powerful neutron scattering instruments. Since, however, the up-to-date technical information relevant for the user can best be found in the Web sites of the neutron scattering research facilities, we will limit this topic here to giving a list of those institutions (see Table II.3.3.1):

### II.3.3.7 SPECIFIC SEMIPHENOMENOLOGICAL MODELS AS INGREDIENTS FOR DESCRIBING DYNAMIC STRUCTURES RELEVANT IN THE CONTEXT OF SOFT MATTER

In this section, we will consider examples of a relatively complex dynamical structure, but for the relatively simple case of incoherent scattering. In order to develop a theoretical scattering function $[S_{\text{inc}}(\mathbf{Q}, \omega)]_{\text{theo}}$ for a comparison with measured incoherent neutron scattering spectra, we generally start using the right-hand side of eq. (II.3.3.27).

$$[S_{\text{inc}}(\mathbf{Q}, \omega)]_{\text{theo}} = S_{\text{vib}}(\mathbf{Q}, \omega) \otimes S_{\text{rot}}(\mathbf{Q}, \omega) \otimes S_{\text{trans}}(\mathbf{Q}, \omega). \qquad \text{(II.3.3.34)}$$

The three individual scattering functions, representing, for example, vibrational, stochastic rotational, and diffusive translational motions (of the same atom), can or will often be calculated classically for each type of motion. In that case, they will be symmetrical in energy transfer and therefore have to be corrected by multiplication with the DBF (see eq. (II.3.3.22)), which conveniently appears in front of the expression. Then, the scattering function $[S_{\text{inc}}(\mathbf{Q},\omega)]_{\text{theo}}$ has the correct asymmetry:

$$[S_{\text{inc}}(\mathbf{Q},\omega)]_{\text{theo}} = \exp\left(\frac{-\hbar\omega}{2k_{\text{B}}T}\right) S^{\text{cl}}_{\text{vib}}(\mathbf{Q},\omega) \otimes S^{\text{cl}}_{\text{rot}}(\mathbf{Q},\omega) \otimes S^{\text{cl}}_{\text{trans}}(\mathbf{Q},\omega). \quad \text{(II.3.3.35)}$$

This equation, which is based on the dynamical-independence assumption in the monatomic approximation, may describe the motional behavior of one single type of atom within a specific dynamical environment as part of the molecular structure. In the case of soft matter and biological macromolecules, even if the observed spectral intensity is essentially only due to scattering by hydrogen atoms, the latter can be in a variety of different bonding situations and may be located in many different molecular subunits (e.g., polar groups, side groups of polypeptide chains, and phospholipids). Therefore, a dynamical model scattering function that takes into account, for example, the $m$ different possible dynamical environments of hydrogen atoms ("protons") and their consequently different motional behavior, will have to be a finite sum of the corresponding m individual scattering functions:

$$[S_{\text{inc}}(\mathbf{Q},\omega)]_{\text{theo}} = \exp\left(\frac{-\hbar\omega}{2k_{\text{B}}T}\right) \sum_{i=1}^{m} F_i [S^{\text{cl}}_{\text{vib}}(\mathbf{Q},\omega) \otimes S^{\text{cl}}_{\text{rot}}(\mathbf{Q},\omega) \otimes S^{\text{cl}}_{\text{trans}}(\mathbf{Q},\omega)]_i. \quad \text{(II.3.3.36)}$$

where the weight factors $F_i$ are the corresponding fractions of the total number of protons, for the $m$ different hydrogen atom types, so that $\sum_{i=1}^{m} F_i = 1$.

When $[S_{\text{inc}}(\mathbf{Q},\omega)]_{\text{theo}}$ is to be compared with an experimental scattering function $[S_{\text{inc}}(\mathbf{Q},\omega)]_{\text{meas}}$, we write

$$[S_{\text{inc}}(\mathbf{Q},\omega)]_{\text{meas}} = F_N [S_{\text{inc}}(\mathbf{Q},\omega)_{\text{theo}}] \otimes R(\omega), \quad \text{(II.3.3.37)}$$

which is composed of an experimental normalization factor $F_{\text{N}}$, and the convolution of the theoretical model function $[S_{\text{inc}}(\mathbf{Q},\omega)]_{\text{theo}}$ describing the dynamics of the sample system, with the experimentally obtained resolution function $R(\omega)$. The first step of the analysis will often be a merely qualitative evaluation of spectral shapes by a purely phenomenological study employing an expression of the simpler form (II.3.3.35), which may be further simplified by minimizing the number of terms and fit parameters, in order to obtain a phenomenological characterization of the results. From this, a viable procedure is then developed for applying more specific expressions of the type given by eq. (II.3.3.36), and using physically meaningful parameters. Let us now consider the individual scattering functions of eqs. (II.3.3.35) and (II.3.3.36) explicitly.

### II.3.3.7.1 Long-Range Translational Diffusion

The function $S^{cl}_{trans}(\mathbf{Q}, \omega)$ stands for translational diffusion, for instance, in the bulk of a solvent, e.g., water, on a macromolecular surface or on the surface of a membrane.

Scattering functions for translational diffusion are often derived from the Chudley–Elliott (CE) jump diffusion model. In its simplest form, this theory starts from the description of a random walk of the diffusing particle on a Bravais lattice of sites available for diffusion (Chudley and Elliott, 1961) assuming equivalence of all sites, negligible jump time, and absence of correlations between different diffusing particles and between successive jumps. For three-dimensional quasiisotropic systems, such as polycrystalline powders or noncrystalline biological matter, the result obtained after orientational averaging is a single Lorentzian,

$$S^{cl}_{trans}(\mathrm{Q}, \omega) = H_{trans}(\mathrm{Q})/\{[H_{trans}(\mathrm{Q})]^2 + \omega^2\}/\pi \tag{II.3.3.38}$$

with the HWHM,

$$H_{trans}(\mathrm{Q}) = \{1-[\sin(\mathrm{Q}d)]/(\mathrm{Q}d)\}/\tau \tag{II.3.3.39}$$

where $d$ is the jump-distance of the diffusing particle, $1/\tau = 6D_s/d^2$, and $D_s$ is the self-diffusion coefficient in three dimensions.

In the case of isotropic two-dimensional diffusion in a plane, such as the surface of a biological membrane (see below), the orientation of this plane with respect to the scattering vector **Q** has to be taken into account and the corresponding expression for the width is

$$H_{trans}(\mathrm{Q}) = \{1-[\sin(\mathrm{Q_p}d)]/(\mathrm{Q_p}d)\}/\tau, \tag{II.3.3.40}$$

where

$$\mathrm{Q_p} = \mathrm{Q}\sin|(\varphi/2) \pm 45°| \tag{II.3.3.41}$$

are the components parallel to the membrane plane of the scattering vector **Q** for the different sample orientations, $\alpha = 45°$ (minus sign) and $\alpha = 135°$ (plus sign), respectively; the jump rate is $1/\tau = 4D_s/d^2$; $D_s$ = self-diffusion coefficient in two dimensions; $\varphi$ = scattering angle. As usual, in the continuous diffusion (low **Q**) limit,

$$H_{trans}(\mathrm{Q}) = D_s\mathrm{Q}^2. \tag{II.3.3.42}$$

As an example, the two-dimensional diffusion of water on the planar surface of purple membrane (see Section II.3.3.8.1) for a description of this biological system) has been studied with quasielastic neutron scattering (QENS) and pulsed field gradient nuclear magnetic resonance (PFG-NMR) methods (Lechner, 1994; Lechner et al., 1994a,b). This work concerns the transport of protons at room temperature, with $H_2O$ molecules as intermittent vehicles diffusing along the surface of purple membrane (PM).

PM stacks (see sample description in Section II.3.3.8 and Figure II.3.3.11) had been oriented by drying at 86% relative humidity (r.h.) on aluminium foils, followed by equilibration at 100% r.h. for a short time, until an intermembrane water layer thickness of about 10 Å was achieved. Subsequently, the stacks were sealed in circular slab-shaped aluminium containers and studied as a function of sample orientation. This procedure was carried out with two different kinds of sample with identical weights of PM, one hydrated with $D_2O$, and the other hydrated with $H_2O$. A subtraction of spectra measured with $D_2O$ from those measured with $H_2O$ gives to a first approximation data, which mainly represent scattering from the solvent $H_2O$. The following results were obtained:

**(i)** Translational diffusion of water molecules at room temperature within the thin water layers on the PM surface is fast, although about five times slower than in bulk water; translational diffusion coefficient: $D_s = 4.4 \times 10^{-6}\,\mathrm{cm^2/s}$. The QENS measurements are in good agreement with the PFG-NMR results. It should be noted that this relatively fast two-dimensional diffusion process is likely to occur—as much as water molecules are involved as proton vehicles—mainly in the central water layer in between adjacent membranes, that is, at an average distance of about 5 Å from the two membrane surfaces. This central layer corresponds to the third (roughly) monomolecular water layer with respect to each of the two surfaces, where the first layer (of hydration water) is known to be rather tightly bound to the membranes.

**(ii)** Fast diffusion was observed only parallel, but not perpendicular to the PM planes, that is, this diffusion process is clearly two-dimensional on the timescale of the experiments.

**(iii)** From the QENS linewidth data plotted for the two different sample orientations of the experiment as a function of $Q^2$ in Figure II.3.3.7, the translational jump distance was found: $d = 4.1$ Å. This is three times larger than that of excess water (Teixeira et al., 1985) at the same temperature. The value of $d$ was obtained by a comparison of the linewidth data to the theoretical width curve $H_{\mathrm{trans}}(\mathbf{Q})$ corresponding to an isotropic approximation of the Chudley–Elliott jump diffusion model in two dimensions. It was calculated using eq. (II.3.3.40). Such a large value of the jump distance is plausible, if this corresponds to a distance between potential minima relevant for the diffusive motion on this timescale (i.e., sites of transient binding for water molecules on the membrane surface).

**(iv)** The diffusion is much slower at 86% r.h.; this observation may be related to the slowing down of biological function of PM (and in particular the proton pumping mechanism of bacteriorhodopsin) with decreasing hydration level.

**(v)** The translational diffusion is accompanied by local diffusive proton motion, presumably related to the rotation of water molecules. This local rotational motion was found to be about six times slower than the corresponding type of motion in (pure) excess water (Teixeira et al., 1985). Qualitatively, the measured data further imply that rotational motions parallel to the membrane plane (i.e., rotation axis perpendicular to the membrane surface) have a higher rate than motions perpendicular to the membrane plane.

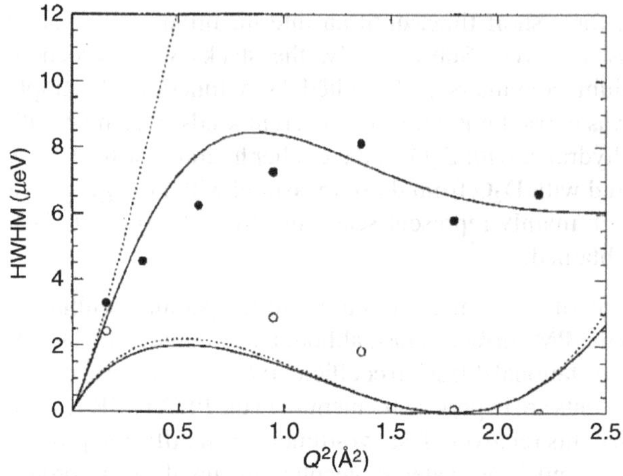


**Figure II.3.3.7** 2D-translational diffusion linewidths of water protons in the interbilayer hydration sheets of purple membrane stacks: linewidths plotted as a function of $Q^2$. The experimental values are compared to the theoretical width curve $H_{trans}(\mathbf{Q})$ corresponding to an isotropic approximation of the Chudley–Elliott jump diffusion model in two dimensions, calculated using eq. (II.3.3.40) for the two different sample orientations of the experiment ($\alpha = 45°$, with **Q** perpendicular to the surface: open circles; $\alpha = 135°$, with **Q** parallel to the membrane surface: full circles). The curves were obtained for a 2-dimensional jump distance of 4.1 Å. Dotted straight line: behavior of the width according to expression (II.3.3.42), that is, the continuous diffusion model valid only in the low-$Q$ limit (Figure from: Lechner et al., 1994b).

Another biological membrane, fully hydrated porcine stratum corneum, was studied recently with the combination of QENS, PFG-NMR, and water sorption techniques (Pieper et al., 2003). Although skin is providing the body with a protective barrier, since it shows a significant resistance to the adsorption and penetration of chemicals, transdermal administration is nowadays considered to be a viable route of great importance for noninvasive drug delivery, and therefore significant efforts have been devoted to the development of relevant therapeutic approaches. Stratum corneum (SC) is the superficial layer of epidermis, to which the principal barrier function of the skin is attributed (Menon, 2002). SC is a heterogeneous membrane, which under physiological conditions has a typical composition of 20% lipids, 40% protein, and 15–20% water. The presence of three water phases was deduced from water sorption isotherm data: bound (water phase A), weakly bound (water phase B), and bulk water (water phase C).

Water phase A accounts for about 6% of the weight of dry SC and is assumed to be bound on polar lipid heads. Therefore, these water molecules are almost immobile with respect to the motion of translational diffusion. Based on the QENS results, their microscopic dynamics has been characterized by the simple approximation of a spherical rotation with a rotational diffusion coefficient of 0.025 meV.

The weakly bound water, that is, water phase B, also corresponds to about 6% of the weight of dry SC and forms additional layers between those of water phase A. For this component, self-diffusion coefficients for motions parallel and perpendicular to the membrane plane of $D_{\parallel} = 3.30 \times 10^{-10}\, m^2/s$ and $D_{\perp} = 1.56 \times 10^{-10}\, m^2/s$, respectively, were determined by PFG-NMR. From QENS, its rotational diffusion coefficient was found to be 0.030 meV. While $D_{\parallel}$ is quite typical for translational water diffusion parallel to the planes of hydrated membranes, the value of $D_{\perp}$ appears to be too high to be solely attributed to a structural heterogeneity and/or a mosaicity present in SC samples. Thus, it can be concluded that pathways for water transport perpendicular to the membrane planes of SC lipid bilayers are present.

Finally, the QENS data prove the presence of bulk water (water phase C) in fully hydrated SC samples, which can be described by a 3D translational diffusion with $D = 2.36 \times 10^{-9}$ m$^2$/s and a rotational diffusion coefficient of 0.0943 meV. This bulk water might be located in pools in the corneocytes as well as within discontinuities of the "perfect" lipid bilayer arrangement.

For more information regarding these and other neutron scattering investigations of diffusion problems, the reader is referred to the works of Lechner (1983, 1994, 1995), Richter (1983), as well as Springer and Lechner (2005). In the following, we will, however, drop the function $S^{\text{cl}}_{\text{trans}}(\mathbf{Q}, \omega)$ in eqs. (II.3.3.35) and (II.3.3.36) because the system to be considered below essentially does not exhibit a translational diffusion broadening measurable with the energy resolution employed in that case.

### II.3.3.7.2 Confined or Localized Diffusive Atomic and Molecular Motions

The second type of scattering function in eqs. (II.3.3.35) and (II.3.3.36), $S^{\text{cl}}_{\text{rot}}(\mathbf{Q}, \omega)$, corresponding to localized relaxations, can often be represented by an expression of the following general form:

$$S^{\text{cl}}_{\text{rot}}(\mathbf{Q}, \omega) = A_0(\mathbf{Q})\delta(\omega) + \sum_{j=1}^{n} A_j(\mathbf{Q}) L_j(H_j, \omega). \qquad \text{(II.3.3.43)}$$

This quasielastic term comprises a finite or infinite series of Lorentzian components $L_n(H_n, \omega)$ that are peaked at zero energy. It stands for the correlation functions due to localized diffusive (i.e., confined stochastic) motions often exhibiting a multi-exponential decay in time, which are considered to be an important part of the internal motions of soft matter and biological macromolecules. This type of decay is well known for specific situations such as confined atomic or molecular diffusion, for example, between a finite number of discrete sites (see, for example, the study by Lechner, 1977 and Lechner and Riekel, 1983, pp. 52–55; see also below). Other often employed models leading to scattering functions with infinite series of Lorentzian components $L_n(H_n, \omega)$ comprise diffusion on a spherical surface (Sears, 1967) or in a spherical cavity (Volino and Dianoux, 1980). Although a large number of models of localized stochastic motions have been developed, corresponding to the solutions of diffusion equations with different boundary conditions, we will limit further citations to only three more articles of pertinent interest (Dianoux and Volino, 1977; Volino and Dianoux, 1978; Volino et al., 2006). While the first concerns, the random motion of uniaxial rotators, such as long, rod-shaped molecules, in an $N$-fold cosine potential, the second deals with diffusion in a one-dimensional cosine potential, whereas in the third case the confinement of one-, two-, and three-dimensional localized diffusive translational motion is defined by the soft, more realistic boundaries of Gaussian wells.

The fractional intensities of the elastic and the quasielastic contributions in eq. (II.3.3.43) are given by the EISF and QISF. Here, $A_0(\mathbf{Q})$ is the EISF of the localized diffusive motion and the functions $A_j(\mathbf{Q})$ are the $n$ corresponding QISFs of the $n$ Lorentzian components, $L_j(H_j, \omega)$, with half widths (HWHM) $H_j$, and with the integrated quasielastic structure factor fulfilling the sum rule,

$$\mathrm{QISF} = \sum_{j=1}^{n} \mathrm{QISF_j}(\mathbf{Q}) = \sum_{j=1}^{n} A_j(\mathbf{Q}) = 1 - \mathrm{EISF}. \qquad \text{(II.3.3.44)}$$

A very clear demonstration of the effect on neutron spectra of confined diffusive molecular motion between a finite number of discrete orientations (effectively resulting in hydrogen atom exchange between discrete sites) can be found in a study of the dynamic equilibrium between the various rotationally isomeric conformations of the molecules in the plastic phase of succinonitrile (Lechner et al., 1977), which is also briefly described in the work of Lechner and Riekel (1983, pp. 52–55).

Although in soft matter and in biological macromolecules, because of the structural complexity, it is more difficult to apply this method, it has been successfully used, for instance, to identify and to classify different kinds of localized diffusive motions of molecular subunits in PM. These motions occur at physiological conditions in PM containing the integral membrane–protein bacteriorhodopsin (BR), which is a light-driven proton pump in *Halobacterium salinarum* (Fitter et al., 1996a). In the analysis of the data taken with two different energy resolutions, an expression of the type given by eq. (II.3.3.36) has been employed specifying suitable model functions for the expected types of localized motions of different classes of hydrogen atoms belonging to different subunits of the macromolecular system. The coefficients $F_i$ were chosen according to the fraction each class represents in the total number of hydrogen atoms. From the results on the hydration dependence of these dynamical effects and of their correlation with the proton pump activity, it has been concluded that the observed stochastic reorientations of methyl groups and of other side groups of the polypeptide chain are relevant for the biological function of this protein.

Another, relatively simple approach would consist in replacing the multiexponential decay model by a Kohlrausch–Williams–Watt (KWW) function (Williams and Watts, 1969) yielding a stretched exponential decay. Furthermore, the fractional Fokker–Planck equation has been proposed recently (Metzler and Klafter, 2000) for constructing nonexponentially decaying solutions (Kneller, 2005), and—last not least—the model of coupled Langevin oscillators describing multiscale relaxation dynamics has been applied recently to a protein (Hinsen et al., 2000). Analogous methods of calculation could in principle be applied to the much more complex geometries encountered in the reality of soft matter and biology. Given, however, the size of the systems to be studied, and given the related mathematical complexity, any of these approaches is likely beyond the scope of the immediate analysis of an experiment. The second, equally relevant reason for refraining from such implied theoretical complexity is that in an initial stage of experiments to be analyzed, these have often been carried out with a single or only a few values of the energy resolution. As a consequence, the observation time windows are rather restricted and do not allow to distinguish more than a fairly small number of different Lorentzian components. For the same reason, rather than aiming immediately at a characterization of the sample dynamics on timescales extending over several orders of magnitude, a beginning analysis has to be restricted to the available observation time windows, so that information can be obtained on the dynamical behavior within these time intervals, generally as a function of certain external parameters, such as temperature. For this purpose, it is often advisable and sufficient to employ only a few phenomenological Lorentzians. Many model functions of the type given by eq. (II.3.3.43) are described explicitly in the literature; for example, see the work of Leadbetter and Lechner (1979); Bée (1988); Lechner (1994); Springer and Lechner (2005).

## II.3.3.7.3 Harmonic Vibrations and Vibrational Density of States

Differently from QENS spectra, inelastic incoherent neutron scattering (IINS) spectra due to harmonic vibrations with little or no damping in soft matter and biological macromolecules such as proteins at relatively low temperatures, are not peaked at zero energy transfer, but extend over large energy ranges up to a few hundred meV, when internal molecular modes are included. In addition, they generally display a broad low-frequency peak centered at energy transfers somewhere in the range from 2 to 5 meV, representing an excess of vibrational modes, as compared to the low-frequency region of a Debye-like density of states. The origin of this so-called "Boson peak" known to be characteristic for both, proteins and glassy systems, is not yet fully understood, but it is believed to be related to the disordered nature of these systems.

An approximate, "effective" vibrational density of states for the wave-like propagating harmonic vibrations can be derived from such IINS spectra by applying the cubic phonon-expansion (Sjölander, 1958) (see also the review by Lechner and Riekel, 1983, pp. 26–27). At sufficiently low temperatures, multiphonon contributions can be neglected, so that only the single-phonon term of the phonon-expansion must be retained. Practical procedures of application are described in more detail in the review by Lechner and Longeville (2006b, pp. 361–367). Let us just mention as an example an investigation concerning the density of vibrational states of the light-harvesting complex of photosystem II of green plants (Pieper et al., 2002, 2004).

Photosynthesis is a fundamental physiological process in nature representing the transformation of solar radiation into storable chemical energy. The photosynthetic apparatus of all oxygen-evolving photoautotrophic organisms such as cyanobacteria, algae, and green plants generally consists of two types of highly specialized pigment–protein complexes referred to as "antennae" and "reaction centers," respectively. The major antenna complex of green plants is the light-harvesting complex of photosystem II (LHC II), which binds more than 65% of the total chlorophyll associated with photosystem II (for a review, see the work by Paulsen, 1995).

The primary steps of photosynthesis comprise the generation of electronically excited states by light absorption in pigment–protein complexes referred to as "antennae" as well as ultrafast excitation energy transfer (EET) to reaction center complexes (for a review, see the work by Renger, 1999). A detailed understanding of this rapid dynamics requires information on pigment–pigment and pigment–protein interactions as well as a thorough investigation of the coupling of the purely electronic transitions of a pigment to low-frequency vibrational modes of the protein matrix (electron–phonon coupling).

IINS is a technique of choice for studying the vibrational density of states of photosynthetic pigment–protein complexes. This experimental approach can yield very useful information independent from and complementary to data from optical spectroscopy, which often are inherently difficult to interpret.

In this chapter, we briefly report on results from IINS experiments on the photosynthetic antenna complex LHC II from spinach, using the time-of-flight spectrometer NEAT. A special preparation protocol was developed, which ensured that trimeric LHC II was properly solubilized in a $D_2O$ containing buffer solution and that the solvent scattering was significantly reduced.

Three important features were found to characterize the dynamics of LHC II at temperatures of 5–100 K:(a) a linear $Q^2$-dependence of the elastic intensity, (b) a linear $Q^2$-dependence of the inelastic peak intensity divided by the Debye–Waller factor, and (c) a linear temperature dependence of the average atomic mean square displacement $\langle u^2 \rangle$. Based on these findings, it can be concluded that both, the elastic and the inelastic contributions of the IINS data obtained, follow the $Q$-dependence of the single-phonon approximation of the phonon-expansion, valid at low temperatures, as mentioned above. Thus, the IINS data display the signature of essentially undamped harmonic vibrational protein dynamics of LHC II in the investigated temperature range. The effective density of vibrational states was derived from the experimental data.

## II.3.3.7.4 Damped Harmonic Oscillators at Low Energies

For the low-energy region of vibrational motions in spectra measured near room temperature (i.e., under physiological conditions, in the case of biological specimens) or above, it is usually indicated to introduce a damped-harmonic oscillator (DHO) model, since the inherent disorder in soft matter and biological systems and the presence of solvent molecules often causes damping of low-frequency vibrations in the macromolecules, which in these systems seems to be at the origin of the ubiquitous "Boson peak" component. The well-known DHO frequency distribution derived in the framework of linear response theory, and using the fluctuation–dissipation theorem (see, for example, the book by Lovesey, 1984, pp. 296–301), is—at least phenomenologically—a rather realistic model. It has been applied for describing inelastic low-energy spectral features in many areas in the past, for example, in cases as different as the central peak phenomenon and phonon softening near structural phase transitions of $SrTiO_3$ and $KMnF_3$ (Shapiro et al., 1972), the spectral line shapes near the roton minimum of superfluid $^4$He (Tarvin and Passell, 1979), the dynamics of light and heavy water at room temperature (Teixeira et al., 1985; Longeville and Lechner, 2000), as well as our work concerning the dynamical structure of a biological membrane containing the proton pump bR (Pieper et al., 2008, 2009). In the latter case, in order to meet the requirements of the sum rules and to obtain the correct dimensionality, the DHO frequency distribution (Lovesey, 1984, pp. 296–301) has been implemented with proper normalization, including an appropriate $Q$-dependence, into the total incoherent scattering function (Lechner et al., 2007). With this normalization, the damped harmonic oscillator scattering function reads,

$$S_{\text{DHO}}(\mathrm{Q},\omega) = \exp(-\langle u^2\rangle Q^2) * \{\delta(\omega) + [\exp(\langle u^2\rangle Q^2)-1]\} * \text{DHO}\} \quad \text{(II.3.3.45)}$$

where

$$\text{DHO} = n(\omega,T) * \frac{g(\omega,\Omega,\Gamma)}{(k_{\text{B}}T)\pi}. \quad \text{(II.3.3.46)}$$

This is a product, essentially of the Bose phonon-population factor,

$$n(\omega,T) = \left[\exp\left(\frac{\hbar\omega}{k_{\text{B}}T}\right)-1\right]^{-1}, \quad \text{(II.3.3.47)}$$

multiplied with the dimensionless vibrational density of states

$$g(\omega,\Omega,\Gamma) = \frac{\hbar\omega * \Omega^2 * \Gamma}{[(\hbar\omega)^2-\Omega^2]^2 + \Gamma^2 * (\hbar\omega)^2} \quad \text{(II.3.3.48)}$$

and divided by the thermal energy $k_{\text{B}}T$; $\Omega$ is the DHO frequency; and $\Gamma$ is the damping constant. Please note that the term $(k_{\text{B}}T)\pi$ in the denominator of eq. (II.3.3.46) represents the $\omega$-integral of the DHO function in the high-temperature limit[1].

The function DHO implicitly contains the detailed-balance factor. Since, as explained in the text following eq. (II.3.3.34), it is convenient to write the DBF as a factor in front of the classical scattering function, we require the classically symmetrical version of $S_{\text{DHO}}(\mathbf{Q},\omega)$. In that case the DBF must be extracted from the DHO expression (II.3.3.45). The term DHO in eq. (II.3.3.45) is then replaced by its symmetrized form $\text{DHO}^{\text{cl}}$ obtained when dividing DHO by the DBF, so that

$$\text{DHO}^{\text{cl}} = \frac{\text{DHO}}{\text{DBF}} = \frac{n(\omega,T) * g(\omega,\Omega,\Gamma)}{\text{DBF} * k_{\text{B}}T * \pi} \quad \text{(II.3.3.49)}$$

[1] With this approximative normalization factor employed for simplicity, the integral over $\omega$ of the DHO function defined by eq. (II.3.3.46) is equal to unity as required, to a very good approximation near room temperature and above. At temperatures lower than 295K (not considered in the example of application discussed in Section 8), the integral slowly increases, but the error starts to exceed 1% only below 250 K.

where

$$\frac{n(\omega,T)}{\text{DBF}} = \frac{\exp(+\hbar\omega/2k_{\text{B}}T)}{\exp(\hbar\omega/k_{\text{B}}T)-1}. \tag{II.3.3.50}$$

As a consequence, let us now rewrite eq. (II.3.3.45):

$$S^{\text{cl}}_{\text{DHO}}(\text{Q},\omega) = \{\exp(-\langle u^2\rangle Q^2) * \delta(\omega) + [1-\exp(-\langle u2\rangle Q^2)] * \text{DHO}^{\text{cl}}\}. \tag{II.3.3.51}$$

The $Q$-dependent weight, $\exp(-\langle u^2\rangle Q^2)$, of the $\delta(\omega)$-function is the Debye–Waller factor characterized by the "global" vibrational mean square displacement $\langle u^2\rangle$. We are employing $u = u_Q$ throughout this chapter, which is the projection of the atomic displacement onto the wave vector **Q**, since only the components of displacements along the direction of **Q** are observed in neutron scattering experiments (see the review by Leadbetter and Lechner, 1979, p. 290); however, please note that, instead, some authors write $\exp[-(\langle u^2\rangle/3)Q^2]$ for the Debye–Waller factor, implying a different but equally correct definition of the mean square displacement, where $\langle u^2\rangle = 3\langle {u_Q}^2\rangle$ is the mean square displacement averaged over all spatial directions, in three dimensions. Together with the $Q$-dependent factor in front of the second term, and the constant factors in eqs. (II.3.3.46) and (II.3.3.48), the Debye–Waller factor provides the required normalization of the scattering function $S^{\text{cl}}_{\text{DHO}}(\mathbf{Q},\omega)$. It should be noted that we are employing expression (II.3.3.51) as a semiphenomenological or even purely phenomenological model. Although it explicitly contains only one single DHO frequency $\Omega$, an appropriate choice of the damping factor $\Gamma$ provides a broad and a relatively flat frequency distribution that in many cases is able to describe very well the low-energy vibrational part (often called "Boson peak") of the observed neutron spectra and permits the separation of the latter from the quasielastic components which are due to confined stochastic motions mainly of small side groups in the soft matter or biological macromolecules.

If one DHO function turns out to be insufficient, it is also possible to introduce a second or more such functions ($m$). Instead of eq. (II.3.3.51), we should then write,

$$S^{\text{cl}}_{\text{DHO}}(\text{Q},\omega) = \{\exp(-\langle u^2\rangle Q^2) * \delta(\omega) + [1-\exp(-\langle u^2\rangle Q^2)] * \sum_{i=1}^{m} w_i * \text{DHO}^{\text{cl}}_i\} \tag{II.3.3.52}$$

where the weight factors are normalized by $\Sigma^m_{i=1} w_i = 1$.

The complete resolution-broadened scattering function including vibrations and localized diffusive motions is given as (see also eq. (II.3.3.31)):

$$[\text{S}_{\text{inc}}(\mathbf{Q},\omega)]_{\text{meas}} = \{\exp(-\hbar\omega/2k_{\text{B}}T) * [S^{\text{cl}}_{\text{DHO}}(\text{Q},\omega) \otimes S^{\text{cl}}_{\text{rot}}(\mathbf{Q},\omega)]\} \otimes R(\omega) \tag{II.3.3.53}$$

This very useful expression has been successfully applied in an energy- *and* real-time-resolved QENS-TOF investigation (Pieper et al., 2008, 2009) that will be described in Section II.3.3.8. Many examples of complex dynamical structure can be treated in an analogous way by implementing suitable model functions for the relevant motional components into the framework of the above-described equations.

## II.3.3.8 THE BIOLOGICAL FUNCTION OF A MACROMOLECULAR SYSTEM AND ITS DYNAMICAL STRUCTURE

We will now demonstrate the application of the above-discussed theoretical framework in the analysis of experiments carried out in the period from 1994 to 2009 with QENS using time-of-flight techniques. The basic topic of these studies is the investigation of the relationship between the biological function of macromolecular systems and their dynamical structure. The experiments were performed primarily, but not exclusively, with the multichopper TOF spectrometer NEAT (see Section II.3.3.6) at the BER II reactor operated then by the Hahn-Meitner-Institut (HMI), now by the Helmholtz-Zentrum Berlin (HZB). Regarding the considered biological system, we will mainly focus on purple membrane with its integral protein bacteriorhodopsin. We will also mention a number of other experiments carried out by various authors in different neutron scattering centers, as much as this is required for a reasonable degree of completeness.

### II.3.3.8.1 The Biological Function of Bacteriorhodopsin Embedded in the Purple Membrane

The PM of *Halobacterium salinarum* is a lipid bilayer system containing a two-dimensional hexagonal arrangement of trimers of the integral protein BR (Oesterhelt and Stoeckenius, 1971). The biological function of BR is to ensure proton transport across the membrane. This provides the proton "current" required to drive, for instance, the ATP production facility called ATP synthase, a proton complex embedded in another part of the membrane (see Figure II.3.3.8), as well as the movement of the flagellae employed by the halobacteria dwelling in salt lakes to allow them moving toward or away from sunlight.

BR is a well-known proton pump and the prototype of a membrane protein. The absorption of a light quantum by light-adapted BR leads to the conformational transition from the all-*trans* to the 13-*cis* isomer of the retinal chromophore and initiates a sequence of interconversions between a number of spectroscopic intermediate states (" intermediates ") usually designated by the letters K, L, M, N, and O, finally leading back to the initial state of BR. We note that the M state is further subdivided into two spectroscopically indistinguishable states M1 and M2. A thermodynamic analysis (Váró and Lanyi, 1991a) has suggested that the free energy initially retained in the chromophore is transferred to the protein in an irreversible transition occurring between the states M1 and M2 of the photocycle. Large decreases in enthalpy and entropy are associated with this. The drop in enthalpy drives a large conformational change involving the whole protein molecule, which occurs about 1 ms after the absorption of the light quantum and is connected itself

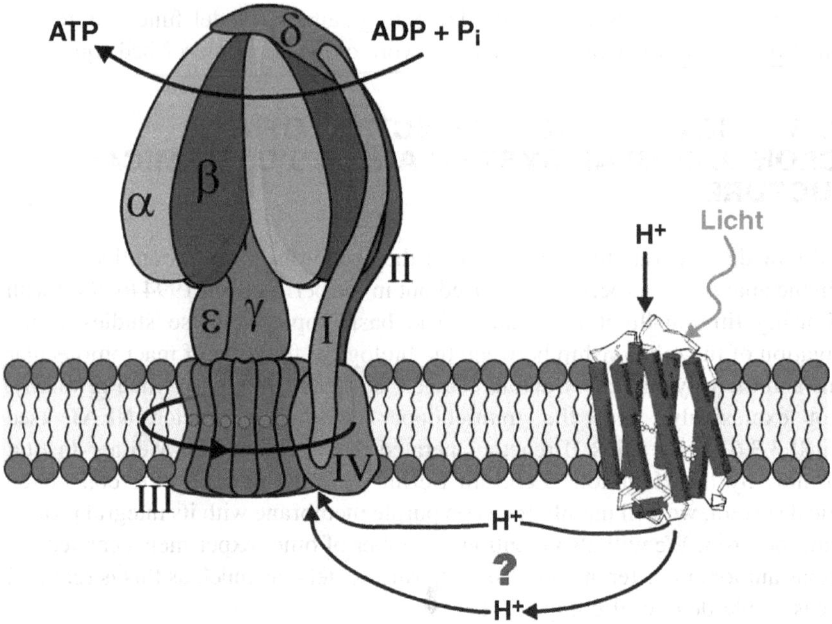


**Figure II.3.3.8** Schematic drawing of an energy transducing biological membrane. The light-driven proton pump BR, embedded (right side of the figure) in the 5 nm thick lipid bilayer called "purple membrane", converts sunlight into an electrochemical proton gradient, which in turn powers ATP synthesis via a stepmotor-like machinery rotating in 120° steps to reconstitute one ATP-molecule per step in the protein ATP synthase (left side). This figure (from the literature by Dencher et al., 1999); courtesy of H. Seelert and N. A. Dencher, 2009) illustrates an experiment, where the optical proton sensors fluorescein (F) and pyranine (P) are bound to specific sites on both proteins or reside in the aqueous bulk phase, respectively, to temporally and spatially monitor proton transfer processes. The Roman and Greek letters designate the subunits of the protein complex.

with a drop in entropy. The conformational change has been observed (i) in neutron diffraction on the ground-state $BR_{568}$ and on the $M_{412}$ intermediate, both trapped at 93 K (Dencher et al., 1989), (ii) in time-resolved synchrotron radiation diffraction (Koch et al., 1991), (iii) by infrared spectroscopy experiments (Ormos, 1991), and has been confirmed by an electron diffraction analysis of the M intermediate trapped by rapid freezing (Subramaniam et al., 1993). This kind of trapping is possible, because the thermal decay of the M state is inhibited below about 220 K (Tokunaga and Iwasa, 1982).

Under physiological conditions, as a result of the reactions occurring during this photocycle (Figure II.3.3.9), a proton is released from the retinal's Schiff base and pumped to the exterior surface of BR during the first half of the cycle, while another proton is taken up on the cytoplasmic side in the entropy-driven second half to replace the first proton at its original binding site, as the protein relaxes to its ground state (for a review, see the work by Oesterhelt and Tittor, 1989).

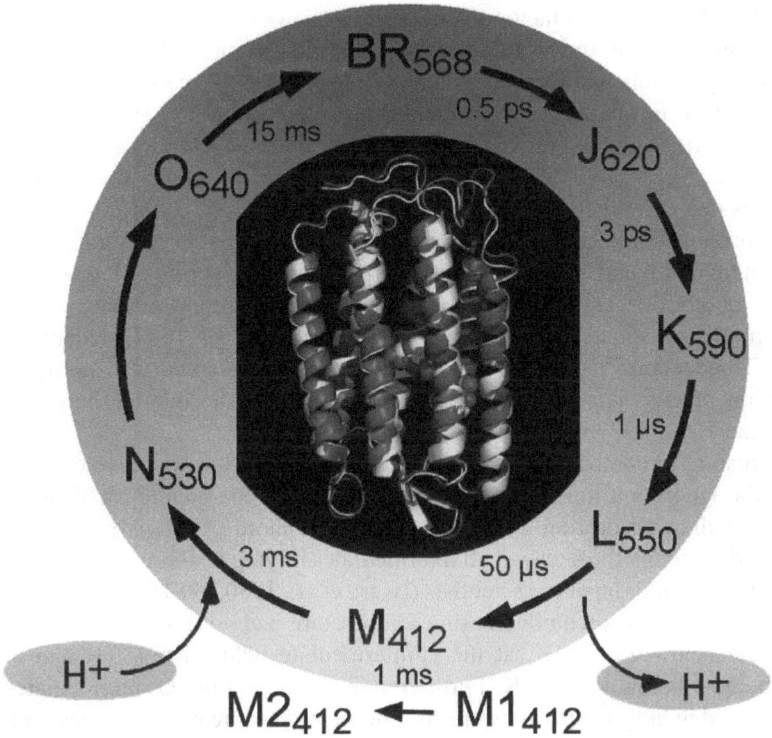


**Figure II.3.3.9** Photocycle of BR at room temperature. The ground-state and the intermediate states (J, K, L, M, N, and O) are characterized by their absorption maxima (subscripts indicate the corresponding wavelengths in nanometers) and decay times; the latter are noted near the arrows connecting consecutive intermediates. The intermediate state $M_{412}$ is further subdivided into two spectroscopically indistinguishable states $M1_{412}$ and $M2_{412}$. They differ due to a conformational change involving the whole protein molecule, which occurs about 1 ms after the protein has absorbed the light quantum in its ground state, $BR_{568}$. The inset shows the structures of a BR monomer in the ground-state $BR_{568}$ (purple) and in the $M_{412}$-intermediate (yellow), respectively, according to the study by Sass et al. (2000). The release and uptake of $H^+$ are also indicated by arrows. Figure taken from the study by Pieper et al. (2009), modified by courtesy of Thomas Hauß. (See the color version of this figure in Color Plates section.)

## II.3.3.8.2 Dependence of the Biological Function on External Variables

It is known that the amount of water associated with BR is one of the most important factors controlling the kinetics of the photocycle. Experimental observations have indicated that the gradual removal of water from PM and thus from the protein modifies the photocycle and has specific consequences on the rates of individual transitions between intermediates. While the processes leading to M formation are only moderately affected by changes in the amount of water present, the relaxation rate of the M decay strongly depends on the level of hydration. Between 90% and

75% r.h., the rate decreases by a factor of ten and slows down by another two orders of magnitude, when the hydration water is removed almost completely by drying at 7% r.h. or less (Korenstein and Hess, 1977a, 1982). More recent experimental results (Hauß et al., 1997), in qualitative agreement with (Korenstein and Hess, 1977a), suggest that from 100% to 85% r.h. the M decay rate decreases by a moderate factor of 2, whereas a much stronger decrease of the M decay rate essentially starts, when the humidity is decreased below from about 85% to 80% r.h.

With decreasing humidity less and less light-induced charge translocation occurs across the protein (Váró and Keszthelyi, 1993). Upon further lowering of the relative humidity, the photocycle and the actual vectorial proton transport activity of the proton pump are drastically modified (Váró and Lanyi, 1991b; Thiedemann et al., 1992; Thiedemann, 1994). Furthermore, it has been observed that the light-dark adaptation process is strongly affected by changes of the relative humidity (Korenstein and Hess, 1977b; Kouyama et al., 1985).

It is interesting to compare the hydration dependence of the photocycle to the variation of the latter as a function of temperature. While the quantum efficiency relevant for the light absorption at the retinal chromophore of BR is practically temperature independent, the decay of all intermediates relying on thermal activation obviously is not. In 1980, Japanese scientists (Iwasa et al., 1980, 1981) observed that the photocycle slows down with decreasing temperature and appears to stop at the M intermediate below $T \approx 220$ K, at the L-intermediate below $T \approx 180$ K and at the K-intermediate below $T \approx 150$ K. Furthermore, the N and the O intermediates were not observed by low-temperature spectrophotometry. More recent studies with higher precision have pushed the limits of observability of intermediates down to lower temperatures. According to these results, the photocycle is, from a practical point of view, "frozen in" at intermediate states, so that BR is "captured" in the M2-state at 260 K (Ormos, 1991), in the M1-state near 230 K, in the L-state near 155 K, and in the K-state near 90 K (Dencher et al., 2000).

### II.3.3.8.3 The Dynamical Transition

In 1989, a paper "Dynamical transition of myoglobin revealed by inelastic neutron scattering" was published (Doster et al., 1989), in which the authors presented an analysis of low-energy neutron scattering spectra obtained from hydrated myoglobin in the large temperature range from 4 to 350 K, and on the experimental timescale from 0.1 to 100 ps (observation time, see Section II.3.3.5). The new and rather interesting observation reported was that at temperatures below 180 K, the globular oxygen storage protein myoglobin in muscle tissue behaves like a harmonic solid: It exhibits, on this timescale, only vibrational motions with a linear temperature dependence of the atomic mean square displacement. Above 180 K, a striking "dynamical transition" was discovered, arising from the additional excitation of nonvibrational motions. The corresponding quasielastic spectral component was interpreted by the authors using a workable, but not unique, phenomenological model corresponding to two-site "torsional" jumps of hydrogen bearing parts of the protein molecule between states of different energy. The extra mobility is reflected in a strong deviation at higher temperatures from the linear *T*-dependence of the mean square displacement, a feature already previously observed for the heme iron by Mößbauer spectroscopy (Parak et al., 1980, 1982; Keller and Debrunner, 1980; Bauminger et al., 1983). Further information on this neutron scattering study can be found in the work of Doster et al. (1990) and Cusack and Doster (1990).

It is interesting to note that, as monitored by flash photolysis (Austin et al., 1975), the dynamical transition effect correlates with the temperature dependence of ligand binding rates at the heme iron, that is, with the biological function of myoglobin. The local molecular flexibility gradually introduced, when the system coming from lower temperature arrives at the dynamical transition, seems to be a prerequisite of function, since the diffusion of $O_2$ through the close-packed protein matrix and the binding to the heme iron require the assistance of density fluctuations.

It has soon turned out that the dynamical phenomena described by Doster et al. (1989) are not only a general property of globular proteins but also beyond this appear to be present in all hydrated biological matter. For example, in PM features of a dynamical transition were observed above 230 K by Ferrand et al. (Ferrand, 1993a,1993b), provided the hydration of the membranes was sufficient. Starting at this temperature, the isotropically averaged "mean square displacement" obtained applying a Gaussian model for the *Q*-dependence of the elastic intensity of hydrogens in PM, deviates from linearity and rises steeply with *T*. This observation is clearly important. The qualitative examination of the vibrational and quasielastic parts of the spectra obtained with the time-of-flight spectrometer IN6 (energy resolution about 100 μeV) at the ILL in Grenoble, yielded two observations: (i) a broad asymmetric inelastic peak with a high-frequency tail and a maximum close to 2.5 meV. This peak shifts significantly and particularly in the wet sample, to lower frequencies as the temperature rises, indicating a "softening" of the vibrational spectrum; (ii) quasielastic broadening appears mainly for the wet sample at and above 240 K.

Light activation of BR taking place in normal, i.e. physiological conditions regarding temperature and humidity, also functions below 230 K, when the membrane-protein complex is vibrating harmonically. The ability of the protein to functionally relax and complete the photocycle initiated by the absorption of a photon, however, is strongly correlated with the onset of low-frequency, large-amplitude nonharmonic atomic motions in the membrane. For a normally hydrated sample, this appeared to occur near 230 K, where the dynamical transition from a low-temperature harmonic to a nonharmonic regime was reported to have been observed by Ferrand et al. (1993a,1993b). In moderately dry samples, on the other hand, in which the photocycle is slowed down by several orders of magnitude, no transition occurs and protein motions remain approximately harmonic up to room temperature. The authors concluded, in agreement with a previously made hypothesis (Zaccai, 1987), "that BR was functional in warm, moist conditions, because they created a "soft" environment within the membrane, in which large-amplitude (anharmonic) atomic motions are possible.

### II.3.3.8.4 Purple Membrane Samples for Neutron Scattering Experiments

It should be possible to *indirectly* elucidate important aspects of biological function by comparing its variation as a function of external variables to the accompanying change in the biological system's static and dynamic structure. Below, we will discuss a series of neutron scattering investigations aiming at the discovery of correlations between the known dependence of the photocycle on external variables such as temperature ($T$), hydration ($h$), lipidation ($l$), and corresponding $T$-, $h$- and $l$-dependent effects observed in the dynamic structure of multilayered stacks of PM.

Let us first describe the type of samples used in the neutron scattering experiments. A single purple membrane patch (thickness: about 5 nm; diameter: about 0.5–1 μm) is a rather perfect two-dimensional hexagonal single crystal constituted by

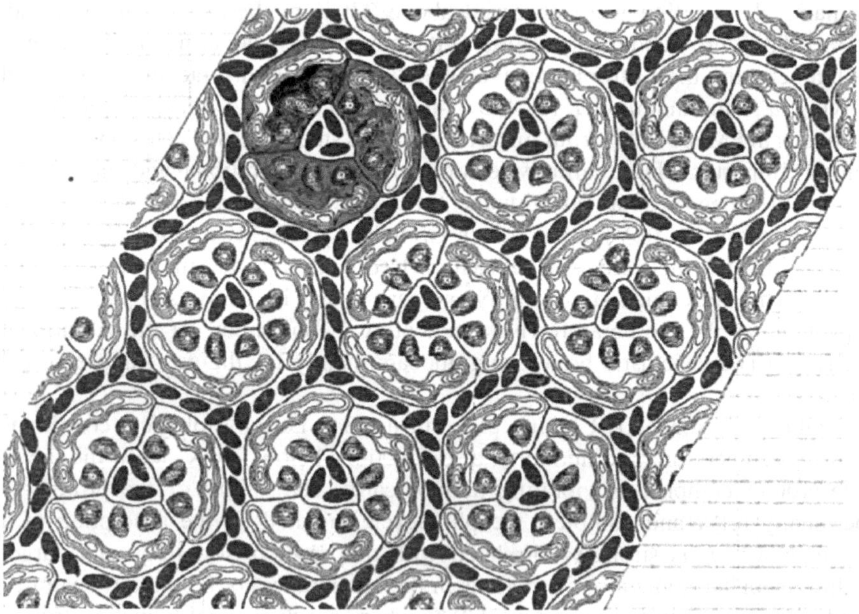

**Figure II.3.3.10** 2D-hexagonal structure of BR trimers in purple membrane (projection perpendicular to the membrane plane): Projected density map of native purple membrane obtained by direct imaging in the electron microscope. The probable molecular envelope of single protein molecules is shown to outline the spaces available to lipid molecules. The possible distribution of lipid molecules in between the BR trimers is indicated schematically (for clarity, one of the trimers is additionally shaded). The lipid molecules are assumed to contain parallel dihydrophytoyl chains, modeled here as ovals with a minor axis of 5 Å and a major axis of 10 Å. Figure reconstituted by R.E. Lechner after the literature by Glaeser et al. (1985).

trimers of the protein bacteriorhodopsin. Figure II.3.3.10 shows a density map of native purple membrane obtained by direct imaging in the electron microscope. The figure represents a projection perpendicular to the membrane plane of the 2D-hexagonal arrangement of BR trimers (reconstituted by R. E. Lechner after the study by Glaeser et al., 1985).

It is essential that for an experiment the membranes need to be available in amounts of a few hundred milligrams. Therefore, a large number of PM patches are required to form the sample.

The preparation of samples starts with the isolation of purple membranes from *H. salinarum*. The PM suspension is deposited on aluminium foils and dried under controlled conditions for several days. Oriented films of PM stacks are obtained with membrane plane quasiparallel to the surface of the aluminium foil. Samples are rehydrated with $D_2O$ and equilibrated at different hydration levels using vapor exchange over pure $D_2O$ (e.g., high hydration level, "wet" $h = 0.28$ g of $D_2O$ per gram of PM, corresponding to 340 water molecules/BR molecule) and over saturated salt solution, respectively (KCL solution, for example, yields a low hydration level, "dry" $h = 0.03$ g of $D_2O$ per gram of PM, corresponding to 40 water molecules/BR molecule).

Membrane samples for neutron scattering experiments can be obtained as hydrated multilamellar stacks of about $2 \times 10^4$ aligned layers made of purple membrane patches, with membranes quasiparallel to the surface of a sample support, mosaic spread 12° (FWHM). Theses samples consist of alternating layers of membrane sheets and of (heavy) water (Figure II.3.3.11). Depending on the degree of hydration, the multilayer stacks have a lamellar lattice constant between about 53 and 68 Å. Finally, the samples are sealed in a circular slab-shaped aluminium container and positioned in the neutron beam. In Figure II.3.3.11, a schematic sketch of a purple membrane sample is shown as an "in-plane" projection. To make it easier to understand, the picture is simplified: only two membrane layers with one water layer in between are shown, and instead of BR-trimers embedded between small areas of lipid bilayers, only one integral bacteriorhodopsin molecule with its proton conducting (proton pump) channel is depicted in each of the two membranes. The proton channel is represented by a computer graphics model of the proton pathway across BR (Dencher et al., 1992a,1992b).

With respect to our purpose, this type of sample reveals several advantages: (i) Lamellar membrane stacks represent a sample that shares important properties with "natural membranes" (i.e., membranes in water solution), even if hydrated at low levels down to only two layers of water between two adjacent membrane sheets. One is dealing in many cases with fully functional membrane proteins, and there are no artificial protein–protein contacts (which often occur in hydrated powders of globular proteins). (ii) Lamellar membrane stacks can be used to study the function of bacteriorhodopsin at different hydration levels (for example, by time-resolved optical absorption spectroscopy (Thiedemann et al., 1992)), as well as structural and dynamical properties of the lipids, of the membrane proteins, and of the aqueous solvent molecules. (iii) Lamellar membrane stacks represent a partly oriented sample. Therefore, one can examine many properties as a function of orientation

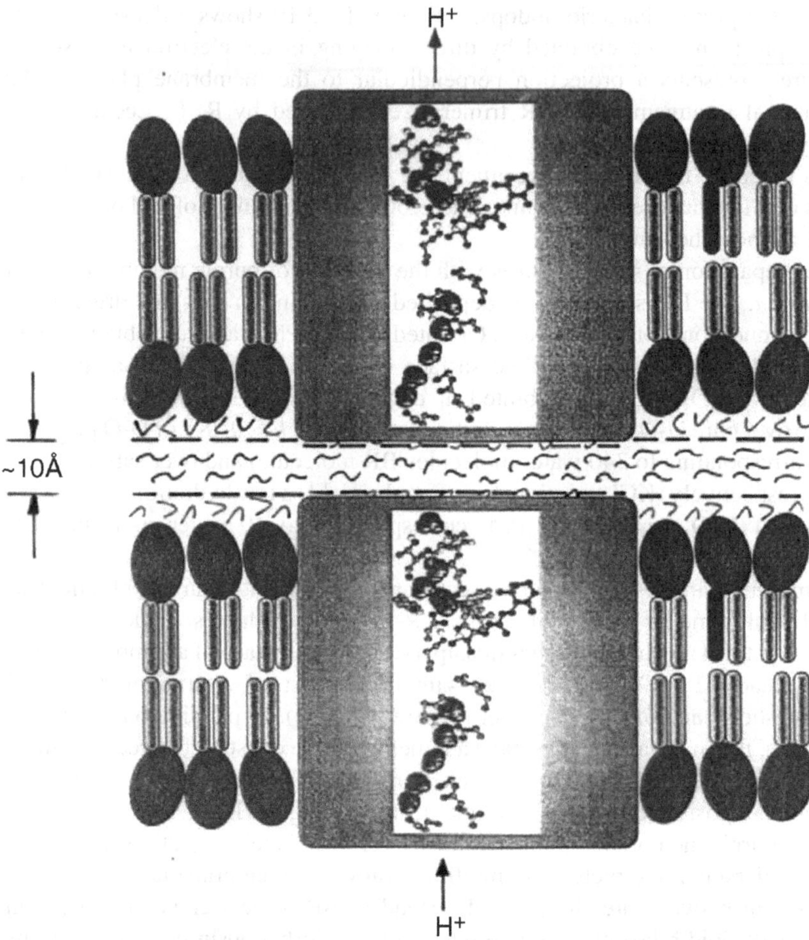


**Figure II.3.3.11** Schematic sketch of a purple membrane sample (projection "in-plane"): Samples are produced as multilayer stacks of about $2 \times 10^4$ aligned quasiparallel purple membrane layers, mosaic spread: 12° (FWHM); exemplary lamellar lattice constant: 60 Å. To make the figure more intelligible, the picture is simplified: only two membrane layers with one water layer (short "wavy" lines) in between are shown, and instead of BR-trimers embedded between small areas of lipid bilayers, only one integral bacteriorhodopsin molecule is depicted in each of the two membranes. The proton conducting (proton pump) channel is represented by a computer graphics model of the proton pathway across BR, emphasizing two chains of hydrogen-bonded water molecules: one for proton release (upper part), one for proton uptake (lower part). The positions of water molecules reflect the van der Waals contact radii. The upper end of the channel is located at the extracellular surface of BR, toward which the proton is ejected (Dencher et al., 1992a,1992b).

(e.g., diffusion of solvent molecules parallel and perpendicular to the membrane surface, see Section II.3.3.7). All neutron scattering experiments on purple membrane, which are described in the following Sections, have been performed with this kind of hydrated multilamellar membrane sheets.

## II.3.3.8.5 Quantitative Analysis of QENS Spectra from Purple Membrane: The Dependence on Temperature

After the qualitative analysis in terms of an isotropically averaged *T*-dependent "mean square displacement" obtained from the quasielastic PM spectra (Ferrand et al., 1993a,1993b), it seemed necessary to extract more quantitative details from such data by examining and characterizing the spectral shapes. It should be remembered that the above-mentioned observation does not concern a true mean square displacement in the Gaussian sense because it was obtained from data containing large non-Gaussian effects due to stochastic localized molecular motions. Figure II.3.3.12 shows an example of QENS measurements taken for this purpose with the same instrument (IN6) at nine different temperatures from 100 to 297 K (Fitter et al., 1996b). The elastic energy resolution was 81 μeV (FWHM). In this first trial, a fit was performed aiming at the experimental separation of the spectra into three components, an elastic, a quasielastic, and an inelastic one in order to distinguish contributions due to stochastic fluctuations from those caused by

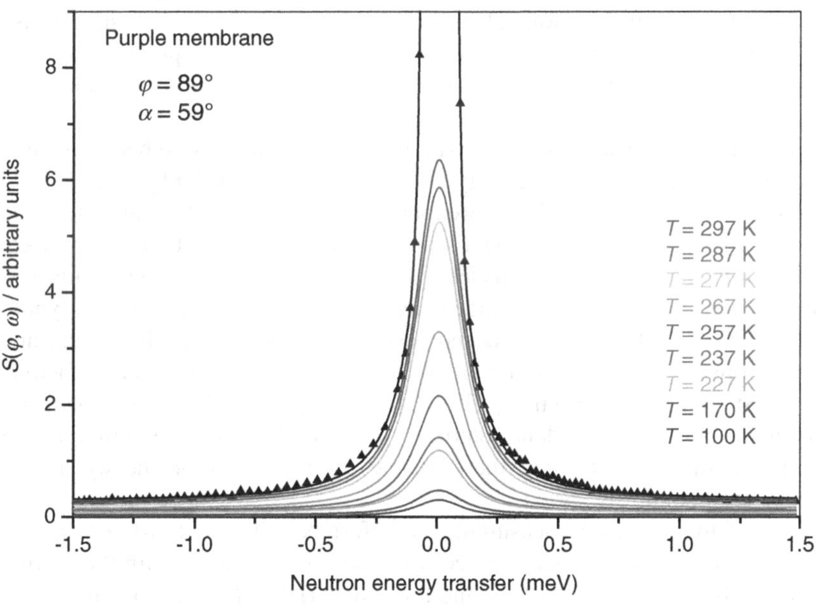


**Figure II.3.3.12** Result of a phenomenological single-Lorentzian fit (see text for details) applied in the energy window ±1.5 meV to spectra measured at different temperatures with the focusing time-of-flight spectrometer IN6 (ILL, Grenoble), with an elastic energy resolution of 81 μeV (FWHM). Experimental points (solid triangles) and total scattering function (solid line fitting the experimental points) composed of a Lorentzian plus an elastic resolution component are shown for a spectrum measured at scattering angle $\varphi = 89°$. The additional solid lines are the Lorentzians representing the quasielastic components obtained from spectra measured at nine different temperatures from 100 K (bottom) to 297 K (top). The level of the flat inelastic "background" (for clarity not shown in the figure) is approximately reached by the values of the Lorentzian functions at the borders of the energy window. The orientation of the slab-shaped sample with respect to the incident beam was $\alpha = 59°$ (see the work of Fitter et al., 1996b).

vibrational atomic or molecular motions. The method used goes back to early work on localized diffusive atomic motions in crystals, described in several reviews (see, for example, the study by Leadbetter and Lechner, 1979, pp. 289–290 and the study by Lechner and Riekel, 1983, pp. 26–27 and by Lechner, 1994, pp. 65–67. The purely phenomenological fit function is composed of a measured elastic resolution component, a single Lorentzian with energy width and relative intensity as parameters, plus a flat inelastic term. This is the simplest possible form of the scattering function for localized diffusive molecular motions given in Section II.3.3.7, eq. (II.3.3.43), which already permits to quantify the signature of the dynamical transition in terms of the spectral change occurring as a function of temperature. Some authors considered it as remarkable that the quasielastic linewidth found with this simple method of analysis was approximately *T*- and *Q*-independent. The *Q*-independence indicates that the major part of the scattering is due to localized diffusive motions of hydrogen atoms within restricted volumes (with a size of a few Å). Regarding the apparent *T*-independence, it must be said that this result should not be overestimated: As already mentioned in Section II.3.3.5, one usually needs to carry out several measurements with different energy resolutions, in order to be able to extract more detailed information on the molecular dynamics of a complex system, because QENS spectra taken with one single energy resolution (corresponding to only one single observation time window), will generally not furnish information as complete as possible and desirable.

For this reason, the *T*-dependent measurements on PM have soon been extended employing three different elastic energy resolutions (FWHM): 100 μeV (NEAT at HMI, Berlin), 80 μeV (IN6 at ILL, Grenoble), and 1.8 μeV (IN10 at ILL). The measured spectra were analyzed using a scattering function of the type given by eq. (II.3.3.43), multiplied by a vibrational Debye–Waller factor, $\exp(-\langle u^2\rangle Q^2)$. This type of expression permits a model-independent separation of the scattered intensity into its elastic and quasielastic components and the determination of their linewidths and integral fractions, that is, their elastic and quasielastic incoherent structure factors, EISF and $\mathrm{QISF}_j$; $\mathrm{QISF}_j = A_j(\mathbf{Q})$ in eq. (II.3.3.43). The *Q*-dependent functions $A_j(\mathbf{Q})$ represent an average *T*-dependent internal molecular flexibility within the PM protein–lipid complex. They can therefore be used to monitor the dynamical transition as temperature is varied.

Combining the data from measurements performed with three different energy resolutions, three different quasielastic components were isolated, with their structure factors $\mathrm{QISF}_1$, $\mathrm{QISF}_2$, and $\mathrm{QISF}_3$, and linewidths (HWHM) $H_1 = 4.0$ meV (IN6 and NEAT), $H_2 = 110$ and 120 μeV (IN6 and NEAT, respectively), $H_3 = 5.5$ μeV (IN10). The corresponding correlation times are $\tau_1 = 0.16$ ps (IN6 and NEAT), $\tau_2 = 6$ and 5.5 ps (IN6 and NEAT, respectively), and $\tau_3 = 120$ ps (IN10).

The fact that we obtain different linewidths, even for only slightly different energy resolutions ($H_{\mathrm{RES}} = 100$ and 80 μeV, respectively; see also Section II.3.3.5), is a strong indication that the quasielastic spectrum of purple membranes is characterized by a much larger number (larger than two) of Lorentzians showing a broad and more or less continuous distribution of linewidths. The energy linewidths obtained in the phenomenological fit at specific energy resolutions must be understood as mean values of a distribution of linewidths accessible to measurement for the employed observation-time window (see Section II.3.3.5 and the work of Lechner, 2001b), and the related motions are characterized by the corresponding range of different correlation times.

The temperature dependence of the structure factors $QISF_1$, $QISF_2$, and $QISF_3$ is plotted in Figure II.3.3.13 for the selected value of $Q = 1.7\,\text{Å}^{-1}$. These results (Fitter et al., 1997b) obtained from different observation time windows, clearly suggest that in purple membrane the dynamical transition from purely harmonic vibrations to an onset of additional stochastic molecular fluctuations takes place at a temperature slightly below, but not higher than 200 K, for localized diffusive motions on three different timescales, namely, around 0.1, 6, and 120 ps. For the slowest motions (correlation time 120 ps) studied with the highest resolution (IN10), we were not able to measure enough spectra in the neighborhood of 200 K to permit deriving such a conclusion from the spectra alone, but an elastic-window scan seemed to reveal that the linear range in the temperature dependence of the "elastic" intensity also ends near 200 K. It must, however, be remembered that the decrease of statistical accuracy

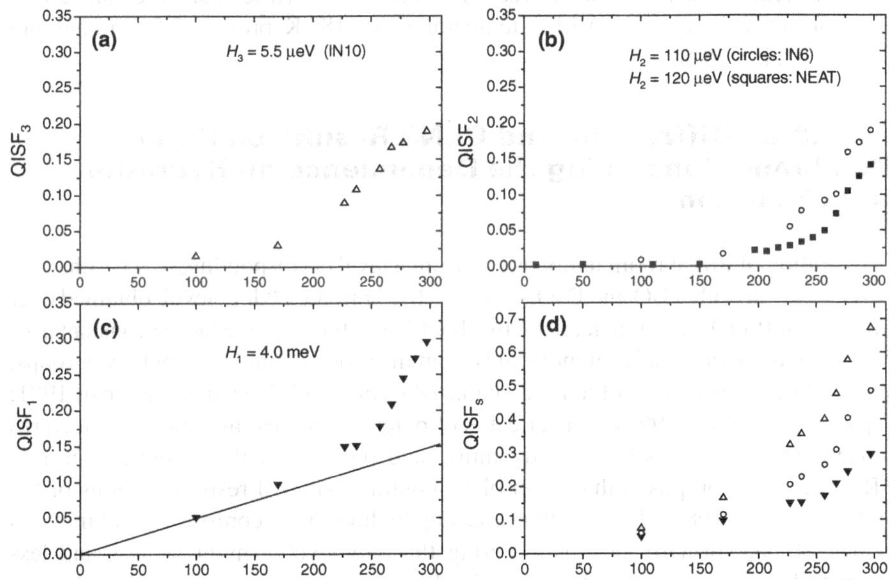


**Figure II.3.3.13** *T*-dependent incoherent quasielastic structure factors ($QISF_j$) of PM, for the selected value of $Q = 1.7\,\text{Å}^{-1}$, were determined in measurements employing three different elastic energy resolutions (FWHM): 100 μeV (NEAT at HMI, Berlin), 80 μeV (IN6 at ILL, Grenoble), and 1.8 μeV (IN10 at ILL). The measured values of $QISF_1$, $QISF_2$, and $QISF_3$, corresponding to the linewidths (HWHM) (a) $H_3 = 5.5$ μeV (IN10), (b) $H_2 = 110$ and 120 μeV (IN6 and NEAT, respectively), and (c) $H_1 = 4.0$ meV (IN6 and NEAT), are plotted as a function of *T*. (d) shows the plots of the sums $QISF_s$ of individual $QISF_j$ curves, namely the 3 curves $QISF_1$, [$QISF_1 + QISF_2$ (IN6)], and [$QISF_1 + QISF_2$ (IN6) + $QISF_3$]. These results (Fitter et al., 1997b) obtained from different observation time windows, clearly suggest that in purple membrane the dynamical transition from purely harmonic vibrations to an onset of additional stochastic molecular fluctuations takes place near 200 K (see text for more information).

in the determination of QISFs, when the temperature is lowered toward the dynamical transition, does not allow to determine to a high precision, where exactly on the $T$-scale the stochastic molecular motions start to become significant. Below that temperature, the intensity fraction $QISF_1$ of the term with the largest width ($H_1$) reveals by its linear $T$-behavior that at these low temperatures it corresponds to vibrational motions alone, while above it also has a sizeable stochastic component. In the case of the other two terms ($QISF_2$ and $QISF_3$), the vibrational contributions are apparently insignificant at any temperature. We furthermore note that an additional steep increase in the quasielastic intensity is observed at temperatures above 250 K. In Section II.3.3.8.6, we will see that this is connected with a complex process of water melting and rehydration of the membrane layers. Further discussions on these experimental results can be found in a brief review (Fitter et al., 1999a). In this context, it is interesting to note that very recently we have succeeded in discovering dynamic transition related effects in purple membrane to a high precision for much longer times using dielectric spectroscopy in the frequency range from $10^{-3}$ to $10^6$ Hz. These data show corresponding discontinuities of the dielectric permittivity $\varepsilon'$ in the temperature region between 185 and 195 K (Buchsteiner et al., 2007) (compare to a dynamic transition temperature of 180 K proposed for myoglobin (Doster et al., 1989)).

### II.3.3.8.6 Diffraction and QENS Results on Purple Membrane Concerning the Dependence on Hydration and Lipidation

Dehydration of PM at room temperature, from a level corresponding to 93% r.h. (i.e., "wet" level) in both, $H_2O$ and $D_2O$, down to "0%" r.h. (i.e., "dry" level, obtained over silica gel or $P_2O_5$), causes a decrease by about 1 Å of the in-plane lattice constant $d_i$ of PM due to removal of solvent molecules from the space around the lipid headgroups, leading to condensation of the lipid chains (Zaccai, 1987; Rogan and Zaccai, 1981; Papadopoulos et al., 1990). This effect is of potential interest in view of conjectures concerning correlations between dynamics and function of the membrane protein BR. It has been proposed that an in-plane contraction[2] will restrict motions of the protein (Zaccai, 1987). Thus, with decreasing hydration the conformational motions of the retinal, normally occurring during the photocycle, might be less and less compatible with a correspondingly diminished capability for conformation changes of the protein. This conjecture is supported by QENS studies on the dynamical structure of natural PM composed of BR and lipids with a weight ratio of 3:1 in comparison to delipidated PM containing only 5 wt% of lipids. The results yield strong evidence that it is mainly the presence of hydration water attached to lipid headgroups at high hydration levels, which increases the internal flexibility of the protein–lipid complex (Fitter et al., 1997a, 1999a).

[2] The observed $d_i$-values were (Zaccai, 1987) 62.5 Å at 93 % r.h. and 61.4 Å at 0% r.h.; in another experiment (Papadopoulos et al., 1990) the values of 62.1 and 60.7 Å have been found for 100 and 0% r.h., respectively.

In order to study the influence of lipids on the dynamical behavior of BR, measurements were performed with natural PM composed of 75% BR (w/w) and 25% lipid (w/w) as well as delipidated PM having only 5% lipid (w/w) (Fitter et al., 1998a). Both types of membranes had been hydrated at 75% r.h. and at 98% r.h. (in $D_2O$ atmosphere). The measured spectra were again analyzed using a scattering function of the type given by eq. (II.3.3.43), multiplied by a vibrational Debye–Waller factor, $\exp(-\langle u^2\rangle Q^2)$. A simple model was used, which should reproduce some general properties of the complex dynamical behavior. It describes continuous diffusion inside the volume of a sphere with the radius $r$, where the EISF is given by (Volino and Dianoux, 1980)

$$A_0(Q) = \left[3\frac{\sin(Qr)-(Qr)\cdot\cos(Qr)}{(Qr)^3}\right]^2. \tag{II.3.3.54}$$

Because the data are limited in momentum transfer $Q$ (with $Q_{\max} = 2.25\,\text{Å}^{-1}$) and the analysis was restricted to an energy transfer range with $|\hbar\omega| \le 0.8$ meV, only the first quasielastic Lorentzian component of the infinite series given by the complete model (Volino and Dianoux, 1980) was considered. Therefore, the theoretical scattering function is parameterized only by the radius $r$ (determining $A_0(Q)$ and $A_1(Q) = 1 - A_0(Q)$ and the width of the quasielastic contribution $H_1 = (\tau_1)^{-1}$. These parameters must be understood as values averaged over all the motions observable in the time window of the experiments (a few picoseconds). The radii $r$, determining the spherical volume of stochastic localized motions, can be interpreted as "amplitudes" of these motions. These "large amplitude" motions give the main contribution to what we call "internal flexibility."

The "mean square displacements" $\langle u^2\rangle$ represent amplitudes of mainly vibrational motions. All spectra show a $Q$-independent linewidth of $H = 150\,\mu$eV, which corresponds to a correlation time of $\tau_1 = 4.4$ ps. The comparison of internal molecular motions occurring in natural PM and in delipidated PM samples revealed the following results (see Table II.3.3.2):

1. We find more internal flexibility, related to more quasielastic scattering with larger radii $r$ and to larger $\langle u2\rangle$ values, in natural PM as compared to delipidated PM (see Table II.3.3.2). A steeper decrease of $A_0$ with $Q$ is related to larger "amplitudes" $r$ and therefore to a larger internal flexibility. The difference in flexibility between natural and delipidated PM is relatively weak in the case of "dry" (75% r.h.) samples and more pronounced in the case of "wet" (98% r.h.) samples.

**Table II.3.3.2** Parameter as Obtained from the Model Fit (Fitter et al., 1998b)

| | $PM_{delip}$ (Wet) | $PM_{nat}$ (Wet) | $PM_{delip}$ (Dry) | $PM_{nat}$ (Dry) |
|---|---|---|---|---|
| Radius (Å) | 0.57 | 0.77 | 0.45 | 0.52 |
| $\langle u^2\rangle$ (Å$^2$) | 0.052 | 0.096 | 0.076 | 0.130 |

2. With respect to the obtained radii *r*, and as already known from previous studies, "wet" samples exhibit more internal flexibility than "dry" samples. This effect is more pronounced in the case of natural PM samples as compared to delipidated PM samples. Due to damping of the vibrational motions in wet samples, the mean square displacements $\langle u2 \rangle$ are smaller as compared to dry samples.

These results indicate that either the lipids themselves are more flexible than the membrane protein BR (a high lipid content leads to an increased "overall" internal flexibility in the PM as compared to a low lipid content), or that more lipids increase the flexibility of the membrane proteins. A further, and most probable, possibility is a combination of both influences. It is reasonable to assume that in the tightly packed protein–lipid complex, the dynamics of BR is coupled with the dynamics of the lipids. The fact that the differences in the flexibility between natural and delipidated PM samples are much more pronounced in "wet" samples, gives strong evidence that mainly the presence of hydration water, which is attached to the polar lipid head groups at high hydration levels, increases the internal flexibility of the protein–lipid complex (Fitter et al., 1997a, 1998a). This hypothesis is supported by diffraction experiments performed with natural PM, where at high hydration levels a larger in-plane cell parameter was found (62.4 Å) as compared to dry samples (61.4 Å). The reduction of the in-plane cell dimension is related to a removal of solvent molecules that were around the lipid head groups (Zaccai, 1987; Zaccai and Gilmore, 1979). Although, on the basis of the present data, we are not able to distinguish between dynamics of the lipids and dynamics of the BR, we can draw the conclusion that the amount of lipids and their ability to attract solvent molecules is strongly related to "large amplitude" motions of the whole purple membrane (Fitter et al., 1997a). These "stochastic large amplitude" motions seem to be essential for the light-induced proton pumping in bacteriorhodopsin. Time constants of the spectroscopic M intermediate, which are related to the M decay, are significantly larger in dry and/or delipidated purple membranes as compared to wet and natural membranes. Less efficient proton pumping (indicated by a prolonged M decay) was found to be accompanied by a reduced internal flexibility of the purple membrane (Fitter et al., 1998a).

In order to elucidate the influence of the hydration, the dynamical properties have been studied as a function of hydration level by measuring natural PM samples hydrated at levels *h* ranging from 0.05 to 0.56 g $D_2O$/g PM. Similar to the investigations on the temperature dependence, the data (measured with NEAT, $\lambda_0 = 5.1$ Å, $H_{RES} = 100\,\mu eV$) were fitted with two quasielastic components. The effect of hydration on the dynamical behavior was also analyzed concerning the *h*-dependent incoherent quasielastic structure factors. The main features are the following: (i) Faster stochastic motions are not very much influenced by the hydration level. In contrast to this, "slower" motions are much more pronounced at high hydration levels as compared to low hydration levels. (ii) Even at rather low hydration levels, an appreciable part of slow stochastic motions (~7% of the total scattering) is still present. In contrast to the temperature dependence, where it is possible to freeze

out these motions at low temperatures, it seems to be not possible to dry out these motions at the lowest hydration levels of the range given above.
Recently, it has indeed been found that those tertiary structural changes believed to be essential for proton pumping, and accompanying the photocycle of BR under physiological conditions, that is, close to room temperature and at high hydration (Dencher et al., 1989; Koch et al., 1991; Hauß et al., 1994; Subramaniam et al., 1993), are hydration dependent, decrease with relative humidity and do no longer occur at 57% r.h. and below (Sass et al., 1997).

Phenomenologically even more impressive than the effect of dehydration on the in-plane lattice constant is its influence on the lamellar structure of PM multilayer systems. The lamellar spacing $d_l$ decreases drastically upon removal of water. For instance, when "wet" PM was controlled by equilibration at 93% r.h. (in the presence of a saturated $Na_2SO_4$ solution at 20°C) a value of ≈55 Å was found for $d_l$ at room temperature; drying PM over $P_2O_5$, on the other hand, gave $d_l \approx 49$ Å (Zaccai, 1987). Usually, such specimens are prepared starting from suspensions of the membrane sheets in distilled water. It is, however, interesting to note that the lamellar spacing of dried PM depends appreciably on the procedure of preparation. Thus, when samples were dried over concentrated sulfuric acid *in vacuo* from low ionic strength suspensions of different pH values, $d_l$ varied from 47.8 to 53.0 Å with pH increasing from 2 to 10 (Henderson, 1975). It is likely that also in the "wet" 'case spacing values will to some extent be function of the method of preparation. In the following Section, we discuss observations showing how closely the dependences of PM multilayer systems on temperature and hydration can be coupled.

## II.3.3.8.7 Dehydration–Rehydration Transition of Purple Membrane as a Function of Temperature

After completion of the PFG–NMR study on two-dimensional proton diffusion in thin water layers intercalated between membranes within PM multilayer stacks (Lechner 1994; Lechner et al., 1994b), see Section II.3.3.7, we decided to use the same technique for looking also at the temperature dependence of this phenomenon. Almost immediately it was discovered—in the first instant indeed surprising—that during a cooling phase below 263 K the diffusion coefficient $D_s$ suddenly "disappeared," that is, became unmeasurably small, while in the subsequent heating period, $D_s$ reappeared near 263 K, but was increasing along a different curve. A supercooling and hysteresis effect had clearly been observed. This observation together with the temperature behavior above 250 K of the quasielastic intensity of $D_2O$-hydrated purple membranes (steep increase with $T$; see Figure II.3.3.13) prompted us to carry out a neutron diffraction investigation of the lamellar structure of PM multilayer stacks as a function of temperature. Although a large number of studies by different authors, concerning the temperature variation of biologically relevant properties of PM, had been carried out on PM multilayer systems before, the $T$-dependence of the lamellar lattice constant of these systems does not seem to have been studied previously as a function of temperature.

The experiment was conducted using samples that had been equilibrated at room temperature in the presence of water vapor at a number of different relative humidities. They had weights between 200 and 300 mg, which were hermetically sealed in circular slab-shaped aluminum containers of 50 mm inner diameter and oriented in a cryostat with respect to the neutron beam. Most of the diffraction measurements were carried out on $D_2O$-hydrated specimens because of the much better contrast.

The measurements were carried out with the neutron diffractometer V1 at the cold-neutron guide NL1A of the Berlin Neutron Scattering Center (BENSC). Data were taken in the temperature range $295 \geq T \geq 180$ K by performing cooling and heating cycles. The wavelength values of the monochromatic neutron beam used in different experiment periods were $\lambda = 5.37, 5.42$, and 5.56 Å. The lamellar spacing $d_l$, which represents the thickness of the purple membrane including the water layer located in between two adjacent membranes, was determined using first- and second-order reflections, requiring small scattering angles $2\theta$ ranging from 3° to 12°. The formation of ice crystals at low temperatures was monitored on the same samples by the measurement of Bragg peaks in the neighborhood of $2\theta = 90°$.

In the diffraction experiment as a function of temperature, the following qualitative behavior of $d_l$ was observed (Lechner et al., 1998), see Figure II.3.3.14: During the cooling cycle of samples equilibrated at 100% r.h., starting at 295 K, $d_l$ stays approximately constant down to a few degrees below the freezing point of bulk water ($T_f = 273.15$ K for pure $H_2O$ and 276.97 K for pure $D_2O$). Then, at a temperature $T_{fh}$, a discontinuity occurs, which is connected with a large decrease in $d_l$ by a step of the order of 20–30%. The step size, as well as the room temperature value of $d_l$, increases with the length of time, during which the sample had been equilibrated at 100% r.h.. This demonstrates the difficulty of reaching true equilibrium in the case of 100% r.h. in view of the usual divergence-like behavior of sorption isotherms in the neighborhood of the 100% r.h. limit. Below $T_{fh}$, further cooling is accompanied by a continuous decrease of $d_l$. But the slope of this decrease diminishes upon cooling and becomes very small below about 240 K. Upon reheating the sample, $d_l$ follows a continuous curve, starts to increase noticeably near a temperature $T^*_{mh} \approx 255$ K and exhibits a hysteresis-type behavior above this temperature. It does not show a step at the temperature $T_{fh}$, but a discontinuity now occurs at a somewhat higher temperature denoted by $T_{mh}$ (but still below the melting point of bulk water). At $T_{mh}$, the lamellar lattice constant reaches a value close to the level originally observed at room temperature. Beyond that point, it stays roughly constant. This behavior is reproducible, as was shown by several consecutive cooling/heating cycles. In fact, the reproducibility is rather perfect for heating curves, and equally for cooling curves, except for the exact value of $T_{fh}$ in the latter case, which appears to be reproducible within 1 K. This is truly remarkable, because the unambiguous relation between temperature and hydration in the closed multilayer membrane system makes it possible to realize hydration level scans of the membranes in a reproducible way by varying the temperature over a large range below room temperature, including the region of this dehydration/rehydration transition.

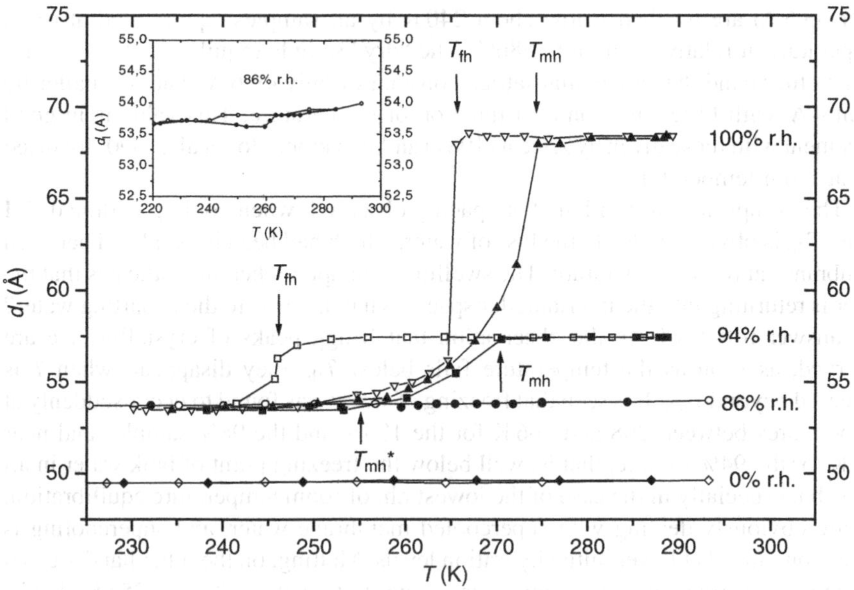


**Figure II.3.3.14** Lamellar lattice constant $d_l$ as a function of temperature, observed in cooling/heating cycles for four different samples, that had been equilibrated at room temperature at different relative humidities ($D_2O$ vapor): H2 (100% r.h.), HN1 (94% r.h.), RL2 (86% r.h.), and H2 (0% r.h.). The freezing and melting discontinuities, $T_{fh}$ and $T_{mh}$, of the membrane water are indicated, using the same symbols for 100% r.h. and for 94% r.h.; $T_{mh}*$ is the approximate starting point of the membrane rehydration region. The inset shows the RL2 (86% r.h.) results on an expanded scale, in order to make the very small freezing and melting discontinuities visible also in this case. Open symbols: cooling branches of the measurements; full symbols: heating branches of the measurements; and lines have been drawn as guides through the experimental points (Figure from: Lechner et al., 1998).

For samples hydrated at 94% r.h. and less, much smaller steps in $d_l$ are observed upon cooling, and at much lower temperatures. When reheating, the lattice spacing follows a curve very similar to the 100% r.h. case, until the original layer thickness is again recovered. This happens at somewhat lower temperature than in the case of higher hydration. Figure II.3.3.14 shows as examples the temperature dependences found for four different samples, which had been equilibrated at room temperature at 100%, 94%, 86%, and 0% r.h. ($D_2O$ vapor), respectively. The "94%" sample, for example, starting with $d_l = 57.5$ Å at room temperature, exhibits its "freezing" discontinuity at $T_{fh} \approx 246$ K, approaches $d_l \approx 53.5$ Å at lower temperatures and shows its melting discontinuity at $T_{mh} \approx 270$ K. No drastic change in $d_l$ was seen upon temperature variation, when the humidity of equilibration at room temperature had been 86%. Close inspection of the data nevertheless revealed a tiny freezing discontinuity at $T_{fh} \approx 243$ K and a small melting discontinuity at $T_{mh} \approx 263$ K (see inset of Figure II.3.3.14). The "86%" sample started with $d_l \approx 54.0$ Å near room temperature, and slowly approached $d_l \approx 53.7$ Å below 243 K. It is interesting to note that the same or very closely the same low-temperature limiting values of $d_l$

(near 53.5 Å) are reached below about 240 K by all samples equilibrated at room temperature at relative humidity ≥86%. The "dry" sample (equilibrated at 0% r.h.), on the other hand, has a lamellar lattice constant of only 49.6 Å, that is, smaller by about 4 Å, with little variation as a function of temperature. This value is in good agreement with those given by Zaccai (1987) and Papadopoulos et al. (1990) for dried PM at room temperature.

The abrupt decrease in lamellar spacing observed, when cooling hydrated PM below $T_{fh}$, is obviously due to the loss of water, which had been intercalated between membranes at room temperature. The swelling seen upon reheating indicates that the water is returning into the interlamellar space. What happens to the departing water? The answer follows from the observation that Bragg peaks of crystalline ice are observed, as soon as the temperature falls below $T_{fh}$; they disappear, when $T$ is increased beyond $T_{mh}$. It is seen that freezing of water was found to occur suddenly at temperatures between 268 and 266 K for the 100% and the 98% samples and near 245 K for the 94% sample, that is, well below the freezing point of bulk water in all cases, but especially in the case of the lowest r.h. of room temperature equilibration. We are obviously dealing with supercooled membrane water, and supercooling is more pronounced at lower initial hydration levels. Melting, on the other hand, occurs as a continuous transition in a temperature range starting at $T = T_{mh}{}^{*} \approx 255$ K, that is, already below the membrane water "freezing point $T_{fh}$," and extending up to the membrane water "melting point $T_{mh}$," which is above $T_{fh}$.

First, it is interesting to note that below 240 K, in the samples equilibrated at a humidity of 86% r.h. or more, there is still water in the interbilayer space, since the lamellar spacing in this case is larger by about 4 Å than that of vacuum dried samples. The practical absence of Bragg reflections (Lechner et al., 1998) suggests that this water is essentially noncrystalline. Because of the ruggedness of the membrane surfaces, it does not seem very likely that these water sheets with an average thickness of only about 4 Å are single crystalline, whereas poly-crystalline ice layers with these dimensions would have been seen in the experiment. Second, at temperatures below about 240 K, the same lamellar spacing value is obtained for samples equilibrated at all relative humidities ≥86%. Since the next nearest neighbor distance of water molecules in liquid water near room temperature and in hexagonal ice is about 4.5 Å, and in view of the ruggedness of the membrane surface, this water layer, with a minimum thickness of about 4 Å, consists of approximately two continuous monolayers of water, one belonging to each membrane surface. This layer, plus the water in shallow surface depressions in the lipid areas of the membrane, presumed to be bordered by protruding protein loops of up to about 4 Å height (Jaffe and Glaeser, 1987), the water penetrating into the membrane around lipid headgroups (Zaccai and Gilmore, 1979) and the few water molecules bound in the interior of the protein (Hauß et al., 1997), represent the so-called "nonfreezing water" component much discussed in the literature (Blaurock, 1975; Blaurock and Stoeckenius, 1971). By comparison with lamellar lattice spacings and weights determined at room temperature, we have deduced a value of $h = (0.24 \pm 0.02)$ [g $D_2O$/g BR] for this "nonfreezing" hydration limit. Such a level would be obtained by equilibration at about 84% r.h. at room temperature. It is lower than corresponding limiting values in water-soluble proteins, which are of the order of $h = 0.3$–$0.4$ [g $H_2O$/g BR] (Kuntz and Kauzmann, 1974; Rupley and Careri, 1991). The present results do not tell us unambigously, to what extent this water is still in the liquid or rather in a glassy state ("solid amorphous water" or "amorphous ice" (Doster et al., 1986) below 240 K. All we can say from these results is that this component does not spontaneously transform into the crystalline state. The most likely reason for this is that the vapor pressure of this water component at $T \leq 240$ K is practically equal to that of ice at the same temperatures. Future very-high-resolution quasielastic neutron scattering experiments and pulsed field gradient NMR measurements might be able to yield more information on this question.

Concluding this topic, regarding our above-described observations, we note that similar ones have been made in multilayer samples of disk membranes (containing the integral membrane photoreceptor protein rhodopsin) (Fitter et al., 1998b), and in lamellar suspensions of lipids (Gleeson et al., 1994). Characteristic features due to dehydration and rehydration of PM multilayers have also been seen in $T$-dependent studies of the dielectric permittivity $\varepsilon'$ with dielectric spectroscopy (Buchsteiner et al., 2007). The qualitative features of the observed dehydration/rehydration phenomena appear to be a general property of lipid–protein membranes. They demonstrate the phenomenon of the temperature-dependent dehydration and rehydration of such membranes, which is caused by the presence of hydration forces and by the specific temperature dependence of the chemical potential of water. These forces cause a local freezing point depression and are the main reason why nucleation of ice crystals first occurs outside of the space between bilayers. The temperature variation of the chemical potentials leads to a difference in vapor pressure, causing hydration water to be extracted from the interbilayer space in the presence of these ice crystals. Vice versa, the water originating from the melting of the same ice crystals, within the closed system, is "sucked" back into the interbilayer space. This effect is due to the temperature-dependent change in the balance between the hydration forces and the chemical force resulting from ice formation. It is important to note that this leads to what amounts to a spatial separation of the crystallized water from the biological surface. Multilayers present the advantage, that the extent of this separation can be studied in a very direct way by the measurement of the lamellar lattice spacing, when the water is literally displaced and leaves the interlamellar space.

It should be emphasized, that the dynamical consequences of the dynamical transition are not limited to a small temperature region. Depending on the timescales studied in the above-mentioned experiments, diffusive motions with different correlation times (note that jump rates are essentially reciprocal correlation times) not only start to become observable between 190 and 230 K but also, as the temperature is increased, their amplitudes, expressed in terms of QISFs, increase continuously in a rather drastic way at least up to room temperature. This means, that more and more conformational degrees of freedom of the macromolecular biological system become available for molecular motions.

What we have learned from the present study of the lamellar spacing helps to understand a good part of the temperature-dependent dynamic structure observations above the "dynamical transition": Their specific characteristics are obviously strongly correlated with the dehydration/rehydration behavior of the purple membrane as a function of temperature. As discussed above, the temperature-induced rehydration of PM starts at $T_{mh}^* \approx 255$ K, that is, halfway between 200 K and room temperature, when most of the increase with temperature of $\langle u^2 \rangle$ and the QISFs has yet to come. Indeed, at the same point $T_{mh}^*$, with the precision to which such a temperature can be given in the absence of a pronounced discontinuity, high-resolution QENS results (Fitter et al., 1996b, 1997b) have revealed a significant steepening of the slopes of QISF curves, which means an acceleration of the increase of molecular flexibility, as the membrane hydration starts to increase. This effect, which had not been interpreted in detail before, is clearly due to the rehydration of the membrane. Somewhat later, qualitatively similar effects were observed in $\langle u^2 \rangle$ curves (Réat et al., 1998; Lehnert et al., 2002; Kühn et al., 2005). The papers of (Réat et al., 1998; Lehnert et al., 2002) report on results obtained with fully deuterated purple membranes containing BR with the three H-labelled groups H-retinal, H-tryptophan, and H-methionine (in the center of the protein). These results show that the labelled parts in the core region of the protein are more rigid than the membrane globally. A relevance of this observation for the valve function of the retinal in the proton channel of BR is suggested. In the study by Kühn et al. (2005), the existence of a correlation between the temperature dependence of $\langle u^2 \rangle$ and that of a functionally important charge transfer efficiency in photosytem II is demonstrated.

In fact, we may say that the harmonic-to-diffusive "dynamical" transition in multilayer membrane systems is followed at somewhat higher temperatures by a second dynamical transition caused by the rehydration of the membranes. This transition is also continuous, extends over the temperature region, where rehydration takes place, that is, from $T_{mh}^* \approx 255$ K to the melting discontinuity $T_{mh}$, and appreciably enhances the amplitudes of spatially restricted diffusive molecular motions. As a consequence, in the presence of a sufficient amount of hydration water, and under conditions allowing rapid diffusion of water molecules on the membrane surface (Lechner, 1994; Lechner et al., 1994a; Lechner et al., 1994b), the bacteriorhodopsin molecule can deploy the full hierarchy of those molecular motions required for its biological function. This suggests the extent to which the very presence of liquid water itself and the molecular mobility garantied by it are essential for the functioning of the biological system.

### II.3.3.8.8 Lamellar Lattice Constant of PM Multilayer and QISF: Structure–Dynamics–Function Correlation of *T*-Dependences

We have already described above the benefit one gains from working with PM multilayer stacks in the form of hermetically sealed (closed) systems in temperature-dependent experiments. This has permitted to carry out isochorous studies of the *T*-dependent lamellar structure, as well as of the dynamics of purple membrane, on the same samples and under the exactly identical conditions of hydration and

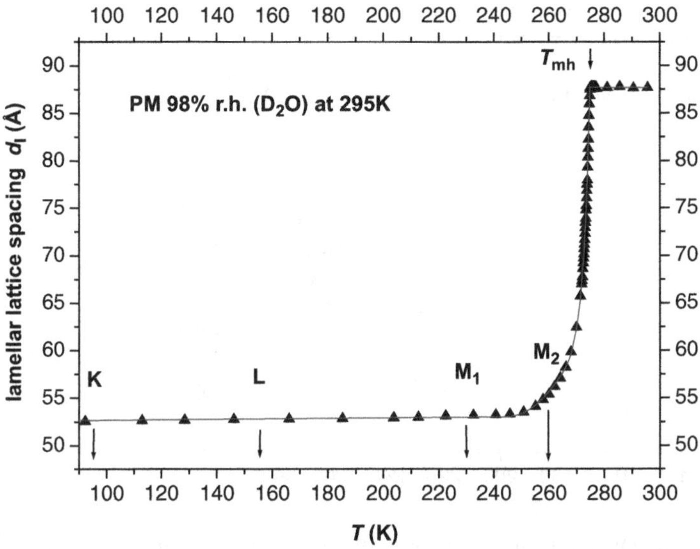


**Figure II.3.3.15** Lamellar lattice constant of a PM multilayer stack as a function of *T* in a heating cycle; the sample (labeled H1) had been equilibrated in $D_2O$ vapor at room temperature, at 98% r.h. The experimental values (triangles) obtained with the membrane-diffractometer V2 at BENSC in Berlin, are shown together with a theoretical model curve (see below). The vertical arrows indicate the approximate limiting temperatures, where—in the course of the photocycle—BR is captured in the intermediates K, L, M1, and M2, respectively (Dencher et al., 2000). $T_{mh}$ = melting discontinuity point of membrane water (Lechner et al., 1998).

temperature. In such experiments, one uses the remarkably strict coupling between temperature and hydration at constant volume, which prevails during the continuous, perfectly reproducible, dynamical and dehydration–rehydration transitions (see Sections II.3.3.8.3, II.3.3.8.5, and II.3.3.8.7) and the study by Lechner et al. (1998).

The results concerning the static lamellar structure were obtained with the membrane-diffractometer V2 at BENSC in Berlin and are shown in Figure II.3.3.15. The measured lamellar lattice constant (triangles) of the PM multilayer stack is shown, together with a theoretical model curve, as a function of *T* in a heating scan. The vertical arrows labeled with letters, indicate the approximate limiting temperatures, where—in the course of the photocycle—BR is captured in the intermediates K, L, M1, and M2, respectively.

Figure II.3.3.16 displays effects caused by both the dynamical and the rehydration transition of PM. The data shown were derived from QENS spectra (resolution: 93 μeV) of the same PM sample as used in the experiment of Figure II.3.3.15. These results concerning the dynamics were taken with the direct-geometry TOF spectrometer IN6 at ILL. The parameters directly obtained from the measurements are the fraction of quasielastic scattering (i.e., the QISF) and its energy width, both as a function of *Q*. The QISF permits a weighted average of the amplitudes of localized diffusive motions of the hydrogen atoms to be calculated, which are bound in the macromolecules. This is obtained in the present example by

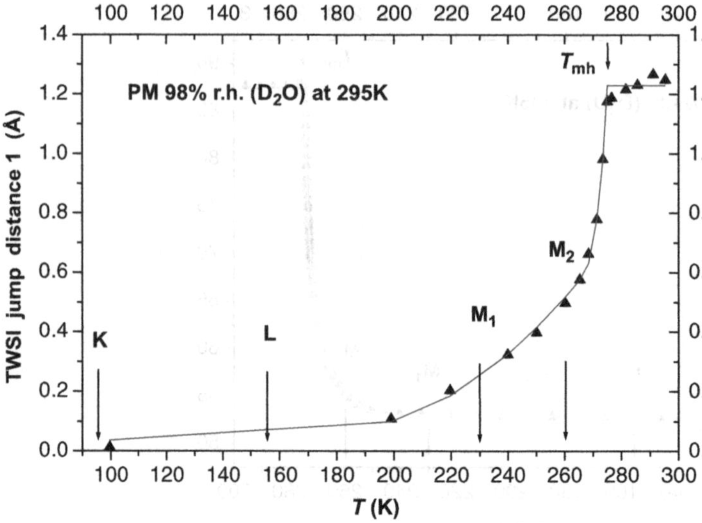


**Figure II.3.3.16** Two-site (TWSI) jump distance $d_{2S}$ of the PM sample H1 (same as in Figure II.3.3.15) as a function of $T$ in a heating cycle, as determined from the experimental QISF. This phenomenological parameter, derived from QENS spectra (resolution: 93 ≈ μeV) measured with the direct-geometry TOF spectrometer IN6 at ILL, stands for the molecular flexibility. It measures a weighted average of the amplitudes of stochastic (localized diffusive) motions of the hydrogen atoms bound in the macromolecules. The experimental values (triangles) are shown together with a theoretical model curve (see below). Vertical arrows: same meaning as in Figure II.3.3.15 (Figure from: Lechner et al., 2006).

calculating this function for orientationally averaged two-site jumps, with a jump distance $d_{2S}$ as parameter of the fit[3]. The values of $d_{2S}$, which would yield the measured QISF, were determined as a function of T in a heating scan (triangles in Figure II.3.3.16). They are shown together with a theoretical model curve (Lechner et al., 2006). The model that includes the consideration of hydration forces and of the $T$-dependent competition between the chemical potentials of water and ice will be explained later in a full paper. The vertical arrows indicate the same limiting temperatures as in Figure II.3.3.15.

The following interesting observations are made: Figure II.3.3.15 shows that in the $T$-region below $T \approx 250$ K, except for a small thermal-expansion effect, the value of $d_l$ is practically constant. This implies that the thickness of the water layer in the space between consecutive membranes of the stack essentially does not vary.

Above $T \approx 250$ K, the lattice spacing $d_l$ starts to increase in a continuous fashion. It increases drastically with rising $T$, until the final value is reached at $T_{mh}$. It is

[3] It should be noted that this method is similar to the one described in Section II.3.3.8.6, where the radius of a spherical volume is determined instead of a two-site jump-distance. Although the model of spherical volume diffusion and that of stochastic, two-site jumps are physically very different; their use for the present purpose is equally justified and, because of the orientational averaging of diffusion and jump vectors in the case of quasi-isotropic samples, leads to similar qualitative results, as long as one does not have to consider very specific anisotropic diffusion geometries.

important to note that the capturing temperatures of the intermediate states K, L, and M1 fall into the region, where the (approximately monomolecular) hydration–water layer on each membrane surface has constant thickness, whereas the location of M2 on the temperature scale corresponds to an increase of about 3.5 Å in the lamellar lattice spacing, and therefore also in the thickness of the intermembrane water layer. This indicates the presence of an approximately complete second monomolecular water layer on each surface of the membrane, at this temperature. It suggests that in the course of the photocycle, the intermediate M2 can only be reached, if the membrane is covered with at least two monomolecular layers of water, which is in agreement with the qualitative observation made at room temperature, that below 80% r.h., the M decay rate decreases strongly, that is, that more than 80% r.h. is required for the photocycle to proceed (at a reasonable speed) beyond the M2 state (Lechner et al., 1998).

The $T$-variation of the jump distance $d_{2S}$ (Figure II.3.3.16) below $T \approx 250$ K is fundamentally different from that of the lamellar lattice constant. The value of $d_{2S}$ stays close to zero below 200 K, but grows continuously and significantly beyond this point. Up to $T \approx 250$ K, this is clearly a pure effect of thermal activation, since the amount of water present in the membrane multilayer is constant in this $T$-region. Therefore, this phenomenon must be attributed to the intrinsic dynamical transition of the PM complex including the monomolecular water layer bound to the membrane surface. Above $T \approx 250$ K, however, the addition of liquid interbilayer water visibly accelerates the increase in flexibility, as demonstrated by a drastic increase of the jump distance.

This method has allowed us, for the first time, to separate the pure effect of thermal activation involving the plasticizing action of hydration water, from effects due to the variation of the hydration level of PM with its integral protein, the proton pump BR, and to correlate this with the $T$-dependent functional activity of BR. In detail, our observations suggest the following interpretation of the capturing phenomenon of intermediate states (Ormos, 1991; Dencher et al., 2000): The monomolecular hydration layer exclusively present at low $T$ starts to melt (probably at first only with respect to degrees of freedom of localized diffusive motions) near 220 K. As a consequence, close to $T \approx 230$ K, the PM hydration–shell complex has acquired sufficient flexibility for being able to reach the M1-state of the photocycle. At somewhat higher temperatures, near $T \approx 260$ K, the photocycle succeeds in proceeding further, so as to reach the M2-state. This is, because the required additional flexibility, not present at lower temperatures, is now provided, when the hydration–shell at each of two adjacent membrane surfaces has grown in thickness to contain at least two monomolecular liquid–water layers.

This clearly suggests that the photocycle, in order to be complete—and thus the proton pumping function of bacteriorhodopsin to be operational—requires stochastic molecular dynamics at least in the ps range and is turned off when, because of the missing plasticizing action of liquid water, the corresponding molecular motions are frozen or dried out. Finally, it should be remembered that the molecular mobility ("flexibility") observed in the present work refers to the ps timescale corresponding to the 93 μeV resolution of this experiment.

## II.3.3.8.9 A Real-Time-Resolved QENS Study of BR with Functional Activity Induced by a Pulsed Laser Synchronized with a Pulsed Neutron Beam Supports Dynamics–Function Relationship of the Proton Pump

With the idea that knowledge about the dynamical properties of a protein are of essential importance for understanding the structure–dynamics–function relationship at the atomic level, a number of neutron scattering efforts have been made in this direction. All the results concerning the successful search for correlations between internal protein dynamics and functionality, which have been discussed in several of the preceding sections, however, have only been obtained *indirectly* in steady-state experiments by variation of external parameters such as temperature hydration and lipidaton. A previous attempt to measure changes in protein dynamics during the photocycle of BR by QENS was also carried out as a steady-state experiment using the D96N-mutant of BR (Fitter et al., 1999b). In this approach, the M intermediate was trapped at low temperature, where diffusive protein motions are widely suppressed, so that no alteration of protein dynamics could be observed. Just recently, however, we have reported the first *direct* experimental observation of modulated protein dynamics during the photocycle of BR using a novel type of laser-neutron pump–probe experiment (Pieper et al., 2008, 2009). This real-time approach combines *in situ* pulsed optical activation of the BR photocycle (see Section II.3.3.8.1)) with a time-dependent search for light-induced alterations of protein dynamics using an (intrinsically) pulsed TOF-QENS spectrometer. As a first successful application, we present data obtained selectively in the ground state and in the M intermediate of BR.

In this chapter, we have to appreciably abbreviate a detailed description (given in the study by Pieper et al., 2009) of the extensive series of optical preparation experiments necessary to ensure that the suitable timescales and the optimum excitation conditions for the QENS experiment are known. Thin films of wild-type PM hydrated at 98% r.h. were used as sample systems, which exhibit an internal protein flexibility above the threshold for fully functional proton pumping with a concomitantly large quasielastic contribution to the QENS spectra (Fitter et al., 1999b) at a defined, reproducible hydration level. In PM, about 75% of the total incoherent scattering is due to BR (Fitter et al., 1999b) so that the QENS data are dominated by the BR contribution.

First, the kinetics of the light-induced formation and decay of the M intermediate of BR have been carefully studied by laser intensity-dependent absorption difference spectroscopy directly for the special case of $D_2O$-hydrated PM films. The absorption changes at 412 nm are indicative for both the kinetics and the time-dependent extent of the formation of the M intermediate in the experiment (Figure II.3.3.17).

The timescale of the laser-excited QENS experiment critically depends on the laser pulse intensity employed. A threshold value of 10 mJ/cm$^2$ has been determined, below which the photocycle of BR can be activated under nearly physiological conditions.

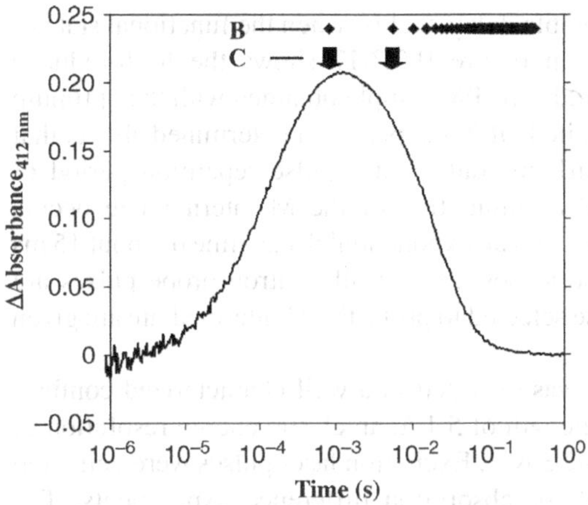


**Figure II.3.3.17** Transient absorbance changes of BR at 412 nm at room temperature after laser excitation at 532 nm and $t = 0$ obtained using a laser pulse energy of 8 mJ/cm$^2$ and a laser pulse repetition time of 400 ms, that is, exactly those conditions used for the light-excited QENS experiments. The abscissa indicates the real-time scale, where laser excitation happens at the origin. Black diamonds and arrows indicate the arrival times of the neutron pulses at the sample relative to the laser pulse in two different measurement modes, that is, experiments without (B, black diamonds) and with time selection (C, arrows), respectively. Here, the letters B and C correspond to the subscripts of the fit functions to the respective QENS spectra shown in Figure II.3.3.18 (Figure from: Pieper et al., 2009).

Another important factor in a pulsed, repetitive QENS experiment is the stability of the PM sample during the time of a typical QENS experiment of 6 h.

Since the kinetics of the BR photocycle is also strongly dependent on temperature, it had to be proved that the excitation conditions chosen do not lead to a temporary temperature increase during the experiment. Finally, the fraction of BR molecules activated by the actinic laser pulse has been determined by comparison of the absorption of the ground-state $BR_{568}$ with the maximum absorbance difference obtained at 568 nm. According to this ratio, approximately 20% of the BR molecules undergo the photocycle simultaneously for a ground-state absorption adjusted to a maximum of 3 at 568 nm. Obviously, the ratio of "laser-excited" to "dark" sample becomes worse with increasing sample thickness and critically determines the maximum permissible amount of sample for laser-excited QENS experiments, in general. In the present case, the ground-state absorption of 3 corresponds to a PM sample with a total mass of only $\approx$30 mg, which is about a factor of 10 smaller than the typical mass of protein samples for QENS experiments. This implies the need for significantly longer measurement times in laser-excited QENS measurements when aiming for the same data statistics as in conventional experiments.

The laser-excited QENS experiment presented here follows the principle of a pump–probe experiment, where the functional state of the PM sample is first conditioned by the laser pump pulse and then subsequently tested by neutron probe pulses. Thus, the laser activation of the sample has to be synchronized with the generally pulsed QENS measurement and the neutron pulses have to probe the sample's protein dynamics after a suitable delay, that is, when the functional state of interest is present. The black trace in Figure II.3.3.17 shows the flash-induced absorption changes at 412 nm for a wild-type PM sample obtained with the optimum excitation conditions for the laser-excited QENS experiment determined above, that is, using laser pulse energies of 8 mJ/cm$^2$ only and a pulse repetition period of ≈400 ms. In this case, the maximal accumulation of the M intermediate occurs around 1 ms after the actinic laser flash, and an exponential decay time of about 15 ms is observed for this state. The temporal positions of all neutron probe pulses are shown by black diamonds, while those selected to probe the M intermediate are given by black arrows in Figure II.3.3.17.

The time-of-flight spectrometer was operated in a well-characterized configuration with an incident neutron wavelength of 5.1 Å, an elastic energy resolution of 93 μeV and an elastic $Q$-range of 0.3–2.3 Å$^{-1}$. Excitation laser pulses were delivered by the same NdYAG laser as used for absorption difference experiments. The technical details of the laser setup on the spectrometer can be found in the study by Pieper et al. (2009).

#### *II.3.3.8.9.1 Experiments and Analysis*

QENS spectra of PM hydrated at 98% r.h. were first measured in the dark (see black squares in Frame I of Figure II.3.3.18). The data are described using a phenomenological model function taking into account both localized diffusive and vibrational protein motions. From the literature by Pieper et al. (2009), we have

$$[S_{\rm inc}(\mathbf{Q},\omega)]_{\rm theo} = \mathrm{e}^{-\langle u^2\rangle Q^2}\left\{A_0(\mathbf{Q})\delta(\omega) + \sum_n A_n(\mathbf{Q})L_n(H_n,\omega)\right\} + S_{\rm in}(\mathrm{Q},\omega). \tag{II.3.3.55}$$

This expression is derived from eq. (II.3.3.53). That is, for the formulation of the vibrational term, the inelastic component of a DHO scattering function, $S_{\rm DHO}(\mathrm{Q},\omega)$, was used, that is, the inelastic term in the right-hand side expression of eq. (II.3.3.45) including the DBF implicitly (see eqs. (II.3.3.45)–(II.3.3.51)). Here, the DBF must equally be included implicitly as a factor in the numerator of the Lorentzian functions $L_n(H_n,\omega)$ in the localized diffusion term of eq. (II.3.3.55). Because the DHO function represents a rather broad energy distribution as compared to the quasielastic and elastic spectral components, it was possible to simplify the calculation by setting the Debye–Waller factor in front of the expression and implementing the inelastic part as an additive term denoted $S_{\rm in}(\mathrm{Q},\omega)$, instead of performing the energy convolution with the localized diffusion component.

It turns out that a satisfactory fit of the data generally requires two quasielastic components with mean relaxation times $\tau_1 = 12.2$ ps (quasielastic structure factor, QISF$_1$) and $\tau_2 = 1$ ps (quasielastic structure factor, QISF$_2$), respectively (see $S_{\rm A}(\mathbf{Q},\omega)$ in Frame I of Figure II.3.3.18). These quasielastic components represent localized diffusive protein motions. The slower relaxation time (12.2 ps) dominates the

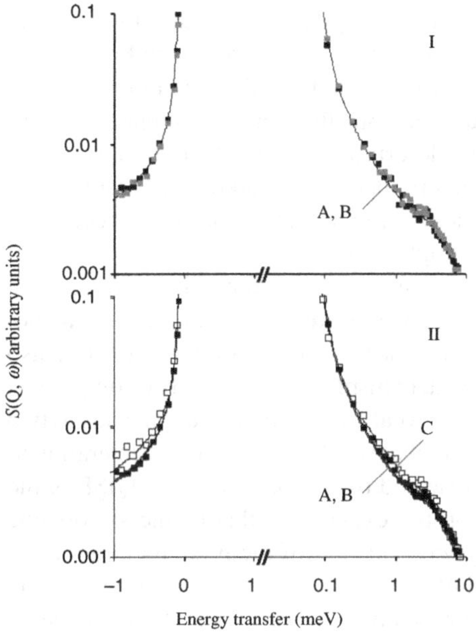


**Figure II.3.3.18** QENS spectra of wild-type PM hydrated in $D_2O$ vapor of 98% r.h. (squares) measured with NEAT, with an incident neutron wavelength of 5.1 Å and an elastic energy resolution of 93 μeV at room temperature (295 K). QENS data are averaged over all scattering angles so that for elastic scattering a mean value of $Q = 1.51\ \text{Å}^{-1}$ was obtained. The energy transfer axis is composed of a linear (left) and a logarithmic part (right side). Light-induced QENS spectra were obtained using repetitive laser excitation with a pulse energy of 8 mJ/cm$^2$ and a laser pulse repetition time of 400 ms. Frame I shows a comparison of the PM QENS spectra obtained in the dark (black squares) and upon laser illumination without time selectivity (grey squares). The full line is a fit of the data. The labels A and B correspond to the subscripts of the respective fit functions $S_A(\mathbf{Q}, \omega)$ and $S_B(\mathbf{Q}, \omega)$, respectively. Frame II shows the PM QENS spectrum obtained upon laser illumination without time selectivity (black squares; please note that these are the same data as the grey squares of Frame I) together with a QENS spectrum measured selectively during the M intermediate of BR (open squares). The full lines are fits of the data labeled with the subscripts of the fit functions $S_A(\mathbf{Q}, \omega)$, $S_B(\mathbf{Q}, \omega)$, and $S_C(\mathbf{Q}, \omega)$, respectively (Figure after Pieper et al., 2009).

spectrum. The vibrational motions are modeled by a separate DHO lineshape function (see above), which mainly accounts for the broad distribution of protein vibrations resembling the so-called Boson peak. In the following analysis, the light-induced changes in the protein flexibility of PM are described by variation of only the amplitude factors of the two quasielastic components in the case of diffusive protein motions (referred to as "quasielastic structure factors" (QISFs)) and of the vibrational mean square displacement $\langle u^2 \rangle$ in the case of protein vibrations.

The effect of laser excitation on BR protein dynamics was probed by measuring a QENS spectrum of PM selectively during the presence of the M intermediate of BR (see open squares in Frame II of Figure II.3.3.18). Here, the QENS spectra from only those neutron pulses are added up, which probe the sample at time delays of 0.8 and

5.6 ms after each laser flash. The position of the probe pulses is indicated by arrows in Figure II.3.3.17. A comparison to the fit of the dark PM spectrum (see Frame II of Figure II.3.3.18) reveals that there is a statistically significant deviation in the energy transfer range between –1 and + 4 meV, that is, in both, the quasi- and inelastic part, of the QENS spectrum. To achieve a satisfactory description of these data (see $S_C(\mathbf{Q}, \omega)$ in Frame II of Figure II.3.3.18), the vibrational term determined above was first fitted to the high-frequency part of the Boson peak, then the remaining deviation was accounted for by adjusting the EISF and QISFs accordingly. Note, however, that due to a well-known sum rule (see eq. (II.3.3.44)) only two of the three amplitude factors can be varied independently. Furthermore, a change of only the EISF and QISFs was not sufficient to account for the observed changes in the QENS spectrum. A proper fit also requires the change of the vibrational mean square displacement $\langle u^2 \rangle$. The comparison of the values resulting for the free parameters reveals that light excitation leads to two effects on the protein dynamics in the M intermediate of BR: (a) a strong acceleration of the localized diffusive protein motions represented by an increase of the QISF of the fast quasielastic component (i.e., $QISF_2$) on the expense of that of the slower one ($QISF_1$) and (b) a larger mean square displacement $\langle u^2 \rangle$ of the protein vibrations.

The reversibility of the light-induced effect discussed above can be tested by adding up the QENS spectra of all neutron probe pulses regardless of the actual delay relative to the laser excitation pulse. The temporal position of these probe pulses is indicated by black diamonds in Figure II.3.3.17. It is obvious that most of these probe pulses test the sample in the ground state of BR. The integral contribution of the M intermediate to this spectrum can be estimated to be only ~1.5%. This data set without time selectivity is represented by grey squares in Frame I, and by black squares in Frame II of Figure II.3.3.18 and can be described to a very good approximation by the same fit used to describe the QENS spectrum of PM in the dark so that $S_A(\mathbf{Q}, \omega) = S_B(\mathbf{Q}, \omega)$. In this regard, note that all data sets described by $S_A(\mathbf{Q}, \omega)$, $S_B(\mathbf{Q}, \omega)$, and $S_C(\mathbf{Q}, \omega)$, respectively, were obtained using the same sample sets with and without time selectivity for the M intermediate so that the applied standard QENS data correction procedures are exactly the same. Therefore, the changes induced by laser excitation appear to be completely reversible for the excitation conditions applied in the present experiment.

As mentioned above, a homogeneous illumination of the sample across its finite thickness cannot be achieved. Thus, the light-induced QENS spectrum shown in Figure II.3.3.18 cannot be directly identified with that of the M intermediate of BR. In order to isolate the spectral contribution due to the M state, a decomposition of the QENS spectrum obtained selectively during the presence of the M intermediate can, in first approximation, be based on the fact that we know the experimentally determined fraction of 20% of the BR molecules to be in the M intermediate and on the assumption that most of the residual 80% of the BR molecules remain in the ground state. Therefore, in eq. (II.3.3.56), we write the sum of the QENS spectra measured selectively at delay times $t = 0.8$ and 5.6 ms, $S_M$ $(\mathbf{Q}, \omega)$, as a fraction of excited molecules with $x = 0.2$. This assumption is justified for those BR molecules, which do not absorb a photon of the excitation laser pulse, but neglects two other small fractions of BR: (i) molecules remaining in the L-intermediate and

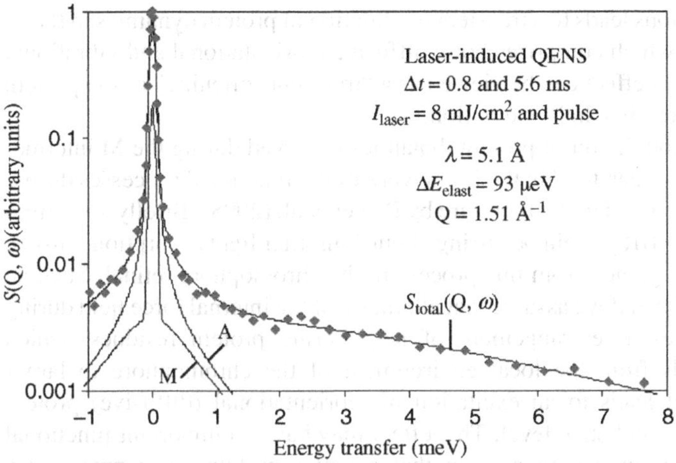


**Figure II.3.3.19** QENS spectrum of wild-type PM obtained during the M intermediate of BR (lozenge-shaped "diamonds") using an incident neutron wavelength of 5.1 Å and an elastic energy resolution of 93 μeV at room temperature (295 K). QENS data are averaged over all angles so that $Q = 1.51$ Å$^{-1}$ for elastic scattering. The fit function $S_{\text{total}}(\mathbf{Q}, \omega)$ of the light-induced PM QENS spectrum obtained during the M intermediate is composed of two components: (a) a fraction of 80% of the BR molecules remaining in the dark (0.8 $S_A$ $(\mathbf{Q}, \omega)$) and (b) an experimentally determined fraction of 20% of the BR molecules in the M intermediate after absorption of a light quantum (0.2 $S_M(Q, \omega)$) (Figure from: Pieper et al., 2009).

(ii) molecules having advanced to the N-intermediate during the arrival time interval of the neutron pulse. The remaining fraction $(1 - x)$ of BR molecules remaining in the dark is represented by the function $S_A(\mathbf{Q}, \omega)$.

$$S_{\text{total}}(\mathbf{Q}, \omega, t = 0.8 \text{ and } 5.6\,\text{ms}) = (1-x)S_A(\mathbf{Q}, \omega) + xS_M(\mathbf{Q}, w). \qquad \text{(II.3.3.56)}$$

A fit using this model function is shown in Figure II.3.3.19 together with its two contributions $S_A(\mathbf{Q}, \omega)$ and $S_M(\mathbf{Q}, \omega)$, respectively. As for the fit shown in Frame II of Figure II.3.3.18, a change of both the QISFs and the vibrational mean square displacement is required to account for the observed differences in the two QENS spectra. However, these changes can now directly be related to the fraction of BR molecules undergoing the photocycle and an approximate QENS spectrum of the M intermediate, $S_M(\mathbf{Q}, \omega)$, is obtained. This separation reveals that the light-induced effect on the protein dynamics is much more pronounced than indicated by a direct fit of the selectively obtained QENS spectrum of PM represented by $S_C(\mathbf{Q}, \omega)$ in Figure II.3.3.18. The parameters gathered by the spectral separation are indicative of a significant acceleration of the diffusive protein motions. It is worth to note that—contrary to the situation in the case of the ground state—only the fast picosecond-component is necessary to describe the quasielastic part of the QENS spectrum of the M intermediate. In addition, there is also a considerable increase of the vibrational mean square displacement $\langle u^2 \rangle$. As argued above, a comparison to the fits shown in Frame I of Figure II.3.3.18 proves that these changes are completely reversible. Therefore, laser excitation of wild-type PM at room temperature under nearly physiological conditions leads to a transient modulation of protein dynamics in the M intermediate of BR, which comprises both, diffusive reorientational and vibrational motions. The observed effect can be viewed as a "transient softening" of the protein in the M intermediate after light excitation.

This transient modulation of protein dynamics observed during the M intermediate of BR has to be understood in the framework of the functional processes during the photocycle as discussed in the literature by Pieper et al. (2008). Briefly, the origin of the softening of the BR protein occurring about 1 ms after light absorption must be the excitation energy gained from this process in the chromophore retinal. Such an effect can be rationalized if we assume that alteration of the internal force field during the photocycle drives a rearrangement of neighboring protein residues, which subsequently extends from the local environment of the chromophore to larger protein domains and leads to an excitation of reorientational (diffusive) protein motions above the ground-state level. This effect may have an important functional role because it is reasonable to assume that a softer protein environment is a prerequisite to overcome potential barriers during the large-scale structural changes in the M intermediate of wild-type BR. Once the temporary softening decays, the structural transition becomes irreversible ensuring an unidirectional proton transport, before a new proton arrives to reprotonate the Schiff's base. This implies that the functional importance of protein dynamics goes beyond that of a "lubricating grease," which accomodates functionally important conformational changes. Rather, protein mobility may play a very specific role in physiological processes requiring large-scale structural changes as demonstrated here for the case of the BR photocycle. In conclusion, the transient alteration of protein dynamics within a functional process of a membrane protein is a direct proof for the functional significance of protein conformational flexibility. An interesting discussion of the dynamics–function relationship has also been given recently by Sakai and Arbe (2009).

## II.3.3.9 SUMMARY AND CONCLUSION

This chapter begins with a description of the theoretical tools employed in the analysis of the experimental results obtained with QENS and IINS techniques. The latter were also described briefly. The role of experimental energy resolution and observation time as variables in modern neutron scattering experiments, rather than just as parameter values, was emphasized particularly. After these more basic sections, a number of specific semiphenomenological models permitting the separation of motions with vibrational, diffusive reorientational, or translational character from each other were presented together with pertinent applications. Such models can be considered as modular ingredients for describing dynamic structures relevant in the context of soft matter. This presentation was focusing particularly on the elucidation of possible correlations between dynamic structure and biological function, that is, of the dynamics–function relationship of macromolecular systems in biology. Starting from indirect studies of function-relevant dynamics, the evolution of pertinent neutron scattering studies with this perspective was followed over the past 20 years via investigations of the dependence of dynamic structure and biological function on external variables toward a more direct observation of such correlations using recent energy- *and* real-time-resolved QENS-TOF measurements on a biological "machine" in operation as an example.

Regarding the considered biological systems, we focused mainly on the results concerning PM of *H. salinarum* with its integral protein, the proton pump BR, but not without also mentioning some other interesting examples.

The important role of the presence of solvents in many cases, and in particular of water in biological systems, was discussed. In the latter case, this is generally a prerequisite for the unrestrained performance of biological function. Mobile water molecules not only allow or induce additional short-range translational and rotational diffusive motion of parts of biological macromolecules but also cause damping of low-frequency vibrations in the macromolecules.

The interaction between water and surfaces it hydrates is reflected, for instance, in the characteristic modifications of the parameters of long-range translational and of localized rotational diffusion. In particular, two-dimensional diffusion of protons carried along, at least intermittently, by water molecules as vehicles in thin water layers parallel to PM surfaces, was found to be five times slower than diffusion of water molecules in the bulk. It is important to remember that this diffusion process has direct functional relevance, since the proton current sustained by the proton pumping protein BR is energizing the ATP-producing protein–complex ATP-synthase. Regarding diffusion parameters, similar results were obtained for two-dimensional water diffusion close to another biological membrane, fully hydrated porcine stratum corneum, the superficial layer of epidermis.

The translational diffusion is accompanied by local diffusive proton motion, presumably related to the rotation of water molecules. This local rotational motion was found to be about six times slower than the corresponding type of motion in (pure) excess water.

In the context of the temperature-dependent dynamical transition and the dehydration/rehydration (freezing/melting) transition, the results of studies of confined or localized diffusive atomic and molecular motions occurring in biological systems were discussed. The QENS technique has been successfully used, for example, to identify and to classify different kinds of localized diffusive motions of molecular subunits in PM. From the results on the temperature and hydration dependence of these dynamical effects and of their correlation with the activity of the proton pump BR, it has been concluded that the observed stochastic reorientations of methyl groups and of other side groups of the polypeptide chain are relevant for the biological function of this protein.

We have also discussed examples of extracting from IINS spectra an approximate, "effective" vibrational density of states for the wave-like propagating harmonic vibrations, for example, in the photosynthetic antenna complex LHC II from spinach. At low temperatures (here, 5–100 K), such IINS data display the signature of essentially undamped harmonic vibrational protein dynamics. At much higher temperatures, under physiological conditions (e.g., room temperature), it is indicated to introduce a DHO model, since the inherent disorder in soft matter and biological systems and the presence of solvent molecules often cause damping of low-frequency vibrations in the macromolecules, which in these systems seems to be at the origin of the ubiquitous "Boson peak" component. In the presence of the stochastic molecular motions, which have been recognized as so essential for biological functionality, it is very important in the analysis of neutron spectra to dispose of such a suitable model, in order to enable a clean isolation of the stochastic effects.

We have reported on the benefit one gains from working with PM multilayer stacks in the form of hermetically sealed (closed) systems in temperature-dependent experiments. This has permitted to carry out isochorous studies of the *T*-dependent lamellar structure, as well as of the dynamics of purple membrane, on the same samples and under exactly identical conditions of hydration and temperature. In such experiments, one uses the remarkably strict coupling between temperature and hydration at constant volume, which prevails during the continuous, perfectly reproducible, dynamical, and dehydration–rehydration transitions.

This method has allowed us, for the first time, to separate the pure effect of thermal activation involving the plasticizing action of hydration water, from effects due to the variation of the hydration level of PM with its integral protein, the proton pump BR, and to correlate this with the *T*-dependent functional activity of BR. In detail, our observations suggest the following interpretation of the capturing phenomenon of intermediate states: The monomolecular hydration layer exclusively present at low *T*, starts to melt (probably at first only with respect to degrees of freedom of localized diffusive motions) near 220 K. As a consequence, close to $T \approx 230$ K, the PM hydration–shell complex has acquired sufficient flexibility for being able to reach the M1-state of the photocycle. At higher temperatures, near $T \approx 260$ K, the photocycle succeeds in proceeding further, so as to reach the M2-state. This is, because the required additional flexibility, not present at lower temperatures, is now provided, when the hydration–shell at each of two adjacent membrane surfaces has grown in thickness to contain at least two monomolecular liquid–water layers.

This clearly suggests that the photocycle, in order to be complete—and thus the proton pumping function of bacteriorhodopsin to be operational—requires stochastic molecular dynamics at least in the ps range and is turned off when, because of the missing plasticizing action of liquid water, the corresponding molecular motions are frozen or dried out.

All in all, the results obtained from neutron scattering experiments confirm an important structure–dynamics–function correlation: the molecular "flexibility" mediated through internal stochastic atomic and molecular motions in biological systems is a precondition of biological function. It is this internal molecular mobility, which is essential for the relaxational part of biological functionality and is thus one of the main basic biophysical ingredients of life itself. Its ubiquitous existence in biology appears to be not only a consequence of but also a prerequisite for Darwin's evolution of living organisms.

Finally, it should be emphasized that we are only at the beginning of potentially much more extensive investigations on the structure–dynamics–function relationship in many very important biological macromolecules, which can be carried out in the future using the above-described methods plus new developments, and employing the upcoming new high-flux neutron sources in the USA, Japan, and Europe.

## ACKNOWLEDGMENT

I am obliged to Jacques Ollivier for having carried out (in December 1999) the numerical verification of the temperature-dependent normalization of the DHO function.

## REFERENCES

Austin, R.H., Beeson, K.W., Eisenstein, L., Frauenfelder, H., and Gunsales, I.C. *Biochemistry* **1975**, *14*, 5355.

Bauminger, E.R., Cohen, S.G., Nowik, I., Ofer, S., and Yariv, J. *Proc. Natl. Acad. Sci. USA* **1983**, *80*, 736.

Bée, M. *Quasielastic Neutron Scattering: Principles and Applications in Solid State Chemistry, Biology and Materials Science*, Adam Hilger, Bristol, 1988, pp. 107–147.

Bée, M. *Chem. Phys.* **2003**, *292*, 121. Special issue: *Quasielastic Neutron Scattering of Structural Dynamics in Condensed Matter,* Lechner, R.E (Guest Editor).

Blatt, J.M. and Weisskopf, V.F. *Theoretical Nuclear Physics*, John Wiley, 1952.

Blaurock, A.E. *J. Mol. Biol.* **1975**, *93*, 139.

Blaurock, A.E. and Stoeckenius, W. *Nat. New Biol.* **1971**, *233*, 152.

Buchsteiner, A., Lechner, R.E., Hauß, Th., and Dencher, N.A. *J. Mol. Biol.* **2007**, *371*, 914.

Carlile, C.J. and Adams, M.A. *Physica B* **1992**, *82*, 431.

Chudley, C.T. and Elliott, R.J. *Proc. Phys. Soc. (London)* **1961**, *77*, 353.

Copley, J.R.D. *Nucl. Instrum. Methods Phy. Res. A* **1990**, *291*, 519.

Cusack, S. and Doster, W. *Biophys. J.* **1990**, *58*, 243.

Dencher, N.A., Dresselhaus, D., Zaccai, G., and Büldt, G. *Proc. Natl. Acad. Sci. USA* **1989**, *86*, 7876.

Dencher, N.A., Büldt, G., Heberle, J., Höltje, H.-D., and Höltje, M. In: Bountis, T. (editor). *Proton Transfer in Hydrogen-Bonded Systems, NATO ASI Series B Physics*, Plenum Press, New York, Vol. *291*, 1992a, p. 171.

Dencher, N.A., Heberle, J., Büldt, G., Höltje, H.-D., and Höltje, M. In: Pullman, et al. (editors). *Membrane Proteins: Structures, Interactions and Models*, Kluwer Academic Publishers, Dordrecht, The Netherlands, 1992b, p. 69.

Dencher, N.A., Fitter, J., and Lechner, R.E. In: Bellissent-Funel, M.-C. (editors). *Hydration Processes in Biology, Series A: Life Sciences*, IOS Press, 1999, p. 195.

Dencher, N.A., Sass, H.J., and Büldt, G. *Biochim. Biophys. Acta* **2000**, *1460*, 192.

Dianoux, A.J. and Volino, F. *Mol. Phys.* **1977**, *34*, 1263.

Doster, W., Bachleitner, A., Dunau, R., Hiebl, M., and Lüscher, E. *Biophys. J.* **1986**, *50*, 213.

Doster, W., Cusack, S., and Petry, W. *Nature* **1989**, *337*, 754.

Doster, W., Cusack, S., and Petry, W. *Phys. Rev Lett.* **1990**, *65*, 1080.

Fermi, E. *Ric. Sci.* **1936**, *1*, 13.

Ferrand, M., Dianoux, A.J., Petry, W., and Zaccai, G. *Proc. Natl. Acad. Sci. USA* **1993a**, *90*, 9668.

Ferrand, M., Petry, W., Dianoux, A.J., and Zaccai, G. *Physica A* **1993b**, *201*, 425.

Fitter, J., Lechner, R.E., Büldt, G., and Dencher, N.A. *Proc. Natl. Acad. Sci. USA* **1996a**, *93*, 7600.

Fitter, J., Lechner, R.E., Büldt, G., and Dencher, N.A. *Physica B* **1996b**, *226*, 61.

Fitter, J., Lechner, R.E., and Dencher, N.A. In: Cusack, S., Büttner, H., Ferrand, M., Langan, P., and Timmins, P. (editors). *Biological Macromolecular Dynamics*, Adenine Press, Schenectady, NY, 1997a, p. 123.

Fitter, J., Lechner, R.E., and Dencher, N.A. *Biophys. J.* **1997b**, *73*, 2126.

Fitter, J., Verclas, S.A.W., Lechner, R.E., Seelert, H., and Dencher, N.A. *FEBS Lett.* **1998a**, *433*, 321.

Fitter, J., Ernst, O.P., Hauß, T., Lechner, R.E., Hofmann, K.P., and Dencher, N.A. *Eur. Biophys. J.* **1998b**, *27*, 638.

Fitter, J., Lechner, R.E., and Dencher, N.A. *J. Phys. Chem. B* **1999a**, *103*, 8036.

Fitter, J., Verclas, S.A.W., Lechner, R.E., Büldt, G., Ernst, O.P., Hofmann, K.P., and Dencher, N.A. *Physica B* **1999b**, *266*, 35.

Glaeser, R.M., Jubb, J.S., and Henderson, R. *Biophys. J.* **1985**, *48*, 775.

Gleeson, J.T., Erramilli, S., and Gruner, S.M. *Biophys. J.* **1994**, *67*, 706.

Gurevich, I.I. and Tarasov, L.V. *Low-Energy Neutron Physics*, North-Holland Publishing Company, Amsterdam, 1968.

Hauß, T., Büldt, G., Heyn, M.P., and Dencher, N.A. *Proc. Natl. Acad. Sci. USA* **1994**, *91*, 11854.

Hauß, T., Papadopoulos, G., Verclas, S.A.W., Büldt, G., and Dencher, N.A. *Physica B* **1997**, *234–236*, 217.

Hautecler, S., Legrand, E., Vansteelandt, L., d'Hooghe, P., Rooms, G., Seeger, A., Schalt, W., and Gobert, G. *Neutron Scattering in the 'Nineties*, IAEA, Vienna, 1985, p. 211.

Henderson, R. *J. Mol. Biol.* **1975**, *93*, 123.

Hinsen, K., Petrescu, A.-J., Dellerue, S., Bellissent-Funel, M., and Kneller, G. *Chem. Phys.* **2000**, *261*, 25.

Iwasa, T., Tokunaga, F., and Yoshizawa, T. *Biophys. Struct. Mech.* **1980**, *6*, 253.

Iwasa, T., Tokunaga, F., and Yoshizawa, T. *Photochem. Photobiol.* **1981**, *33*, 539.

Jaffe, J.S. and Glaeser, R.M. *Ultramicroscopy* **1987**, *23*, 17.

Keller, H. and Debrunner, P. *Phys. Rev. Lett.* **1980**, *45*, 68.

Kneller, G.R. *Phys. Chem. Chem. Phys.* **2005**, *7*, 2641.

Koch, M.H.J., Dencher, N.A., Oesterhelt, D., Plöhn, H.J., Rapp, G., and Büldt, G. *EMBO J.* **1991**, *10*, 521.

Koester, L., Rauch, H., and Seymann, E. *At. Data Nucl. Data Tables* **1991**, *49*, 65.

Korenstein, R., and Hess, B. *Nature* **1977a**, *270*, 184.

Korenstein, R. and Hess, B. *FEBS Lett.* **1977b**, *82*, 7.

Korenstein, R. and Hess, B. *Methods Enzymol.* **1982**, *88*, 180.

Kouyama, T., Bogomolni, R.A., and Stoeckenius, W. *Biophys. J.* **1985**, *48*, 201.

Kühn, P., Pieper, J., Kaminskaya, O., Eckert, H.-J., Lechner, R.E., Shuvalov, V., and Renger, G. *Photosyn. Res.* **2005**, *84*, 317.

Kuntz Jr, I.D., and Kauzmann, W. Hydration of proteins and polypeptides. In: Anfinsen, C.B., Edsall, J.T., and Richards, F.M. (editors). *Advances in Protein Chemistry*, Academic Press, New York, London, Vol. *28*, 1974, p. 239.

Leadbetter, A.J. and Lechner, R.E. In: Sherwood, J.N. (editor). *The Plastically Crystalline State*, John Wiley & Sons, New York, 1979, p. 285.

Lechner, R.E., Study of molecular rotation by quasielastic neutron scattering, Lecture presented at the College 4 Meeting, Institut Laue-Langevin, Grenoble, France, Nov. 18 1971, 18, (unpublished).

Lechner, R.E. In: Benière, F. and Catlow, C.R.A. (editors). *Mass Transport in Solids*, NATO ASI Series B: Physics, Vol. 97, Plenum Press, New York, 1983, p. 169.

Lechner, R.E. *Neutron Scattering in the 'Nineties*, IAEA, Vienna, 1985, p. 401.

Lechner, R.E. *Proceedings of the International Collaboration on Advanced Neutron Sources (ICANS-XI)*, Tsukuba, Japan, 1990; KEK Report 90-25 National Laboratory for High-Energy Physics, Tsukuba, 1991, p. 717.

Lechner, R.E. *Physica B* 1992, 180–181 973.

Lechner, R.E. Quasielastic neutron scattering. In: Colmenero, J., Alegría, A., and Bermejo, F.J. (editors). *Proceedings of the Quasielastic Neutron Scattering Workshop QENS'93*, San Sebastián, Spain, 1993; World Scientific, Singapore, 1994, p. 62.

Lechner, R.E. *Solid State Ionics* **1995**, *77*, 280.

Lechner, R.E. *Neutron News* **1996**, *7*, 9.

Lechner, R.E. *Proceedings of ICANS-XV Conference*, Tsukuba, 2000 In: Suzuki, J.,and Itoh, S. (editors). JAERI, Tokai-mura, Japan, 2001a, 357.

Lechner, R.E. *Physica B* **2001b**, *301*, 83.

LECHNER, R.E. and LONGEVILLE, S. Neutron scattering in biology. In: FITTER, J., GUTBERLET, T., and KATSARAS, J. (editors) *Quasielastic Neutron Scattering in Biology, Part I: Methods*, Springer-Verlag, Berlin, 2006a, p. 309, Chapter 15.

LECHNER, R.E. and LONGEVILLE, S. Neutron scattering in biology. In: FITTER, J., GUTBERLET, T., and KATSARAS, J. (editors). *Quasielastic Neutron Scattering in Biology, Part II: Applications*, Springer-Verlag, Berlin, 2006b, p. 355, Chapter 16.

LECHNER, R.E. and RIEKEL, C. In: HÖHLER, G. (editor) *Neutron Scattering and Muon Spin Rotation*, Springer Tracts in Modern Physics, Springer-Verlag, Berlin, **1983**, *101*, p. 1.

LECHNER, R.E., AMOUREUX, J.P., BÉE, M., and FOURET, R. *Commun. Phys.* **1977**, *2*, 207.

LECHNER, R.E., BADUREK, G., DIANOUX, A.J., HERVET, H., and VOLINO, F. *J. Chem. Phys.* **1980**, *73*, 934.

LECHNER, R.E., DENCHER, N.A., FITTER, J., and BÜLDT, G., and BELUSHKIN, A.V. *Biophys. Chem.* **1994a**, *49*, 91.

LECHNER, R.E., DENCHER, N.A., FITTER, J., and DIPPEL, Th. *Solid State Ionics* **1994b**, *70–71*, 296.

LECHNER, R.E., MELZER, R., and FITTER, J. *Physica B*, **1996**, *226*, 86.

LECHNER, R.E., FITTER, J., DENCHER, N.A., and HAUß, T. *J. Mol. Biol.* **1998**, *277*, 593.

LECHNER, R.E., FITTER, J., DENCHER, N.A., and HAUß, T. *Physica B: Physics of Condensed Matter* **2006,** 385–386, 835.

LECHNER, R.E., PIEPER, J., BUCHSTEINER, A., HAUß, T., and DENCHER, N.A. In: SOKOL, P.E., KAISER, H., BAXTER, D., PYNN, R., BOSSEV, D., and LEUSCHNER, M. (editors). Quasi-Elastic Neutron Scattering Conference 2006 (QENS2006) Materials Research Society, Warrendale, PA, 2007, 63, postprint version (a pdf file may be obtained by email request from ruep.lechner@gmail.com).

LEHNERT, U., RÉAT, V., KEßLER, B., OESTERHELT, D., and ZACCAI, G. *Appl. Phys. A* **2002**, *74*, (Suppl.), S1287.

LONGEVILLE, G. and LECHNER, R.E. *Physica B* **2000**, *276*, 183.

LOVESEY, S.W. and SPRINGER, T. *Dynamics of Solids and Liquids by Neutron Scattering, Topics in Current Physics*, Springer-Verlag, *Berlin*, 1977, p. 3.

LOVESEY, S.W. *Theory of Neutron Scattering from Condensed Matter*, Clarendon Press, Oxford, 1984.

MENON, G.K. *Adv. Drug Del. Rev.* **2002**, *54*, (Suppl. 1), S3.

MESOT, J., JANSSEN, S., HOLITZNER, L., and HEMPELMANN, R. *J. Neutron Res.* **1996**, *3*, 293.

METZLER, R. and KLAFTER, J. *Phys. Rep.* **2000**, *339*, 1.

MUTKA, H. *Nucl. Instrum. Methods Phys. Res. A* **1994**, *338*, 144.

OESTERHELT, D. and STOECKENIUS, W. *Nat. New Biol.* **1971**, *233*, 149.

OESTERHELT, D. and TITTOR, J. *Trends Biochem. Sci.* **1989**, *14*, 57.

OLLIVIER, J., PLAZANET, M., SCHOBER, H. and COOK, J.C. *Physica B* **2004**, *350*, 173.

ORMOS, P. *Proc. Natl. Acad. Sci. USA* **1991**, *88*, 473.

PAPADOPOULOS, G., DENCHER, N.A., ZACCAI, G., and BÜLDT, G. *J. Mol. Biol.* **1990**, *214*, 15.

PARAK, F., FROLOV, E.N., KONONENKO, A.A., MÖßBAUER, R.L., and GOLDANSKII, V.I., et al. *FEBS Lett.* **1980**, *117*, 368.

PARAK, F., KNAPP, E.W., and KUCHEIDA, D. *J. Mol. Biol.* **1982**, *161*, 177.

PAULSEN, H. *Photochem. Photobiol.* **1995**, *62*, 367.

PIEPER, J., IRRGANG, K.-D., RENGER, G., and LECHNER, R.E. *Appl. Phys. A* **2002**, *74*, (Suppl.), S1264.

PIEPER, J., CHARALAMBOPOULOU, G., STERIOTIS, T., VASENKOV, S., DESMEDT, A., and LECHNER, R.E. *Chem. Phys.* **2003**, *292*, 465. Special Issue.

PIEPER, J., IRRGANG, K.-D., RENGER, G., and LECHNER, R.E. *J. Phys. Chem. B* **2004**, *108*, 10556.

PIEPER, J., BUCHSTEINER, A., DENCHER, N.A., LECHNER, R.E., and HAUß, Th. *Phys. Rev. Lett.* **2008**, *100*, 228103–228101.

PIEPER, J., BUCHSTEINER, A., DENCHER, N.A., LECHNER, R.E., and HAUß, Th. *Photochem. Photobiol.* **2009**, *85*, 590.

PRICE, D.L. and SKÖLD, K. *Neutron Scattering, Methods of Experimental Physics*. Academic Press, London, 1987, p. 23, Part B, Chapters 8–11.

RÉAT, V., PATZELT, H., FERRAND, M., PFISTER, C., OESTERHELT, D., and ZACCAI, G. *Proc. Natl. Acad. Sci. USA* **1998**, *95*, 4970.

RENGER, G. In: SINGHAL, G.S. RENGER, G. SOPORY, K., IRRGANGM, K.-D., and GOVINDJEE, J.A. (editors.) *Concepts in Photobiology and Photomorphogenesis*, Narosa Publishing House, New Delhi, India, 1999, p. 52.

RICHTER, D. *Neutron Scattering and Muon Spin Rotation, Springer Tracts in Modern Physics*, Springer-Verlag, *Berlin*, **1983**, Vol. *101*, p. 85.

ROGAN, P.K. and ZACCAI, G. *J. Mol. Biol.* **1981**, *145*, 281.

RUFFLÉ, B., OLLIVIER, J., LONGEVILLE, S., and LECHNER, R.E. *Nucl. Instrum. Methods Phys. Res. A* **2000**, *449*, 322.

RUPLEY, J.A. and CARERI, G. Protein hydration and function. In: ANFINSEN, C.B. EDSALL, J.T. RICHARDS, F.M. and EISENBERG, D.S. (editors). *Advances in Protein Chemistry*, Vol. *41*, Academic Press, New York, 1991, p. 37.

RUSSINA, M., MEZEI, F., LECHNER, R.E., LONGEVILLE, S., and URBAN, B. *Phys. Rev. Lett.* **2000**, *84*, 3630; see also Russina, M., Mezei, F., Lechner, R.E., Longeville, S., and Urban, B. *Phys. Rev. Lett.* **2000**, *85*, 5670.

SAKAI, V.G. and ARBE, A. *Curr. Opin. Colloid Interface Sci.* **2009**, *14*, 381.

SASS, H.J., SCHACHOWA, I.W., RAPP, G., KOCH, M.H.J., OESTERHELT, D., DENCHER, N.A., and BÜLDT, G. *EMBO J.* **1997**, *16*, 1484.

SASS, H.J., BÜLDT, G., GESSENICH, R., HEHN, D., NEFF, D., SCHLESINGER, R., BERENDZEN, J., and ORMOS, P. *Nature* **2000**, *406*, 649.

SCHERM, R., CARLILE, C., DIANOUX, A.J., SUCK, J., and WHITE, J. *ILL Int. Rep.* **1976**, *76*, S235.

SEARS, V.F. *Adv. Phys.* **1975**, *24*, 1.

SEARS, V.F. *Can. J. Phys.* **1967**, *45*, 237.

SEARS, V.F. *Phys. Rep.* **1982**, *82*, 1.

SEARS, V.F., Atomic Energy of Canada Ltd. Report AECL 8490, Chalk River, Ontario, June 1984.

SHAPIRO, S.M., AXE, J.D., and SHIRANE, G. *Phys. Rev. B* **1972**, *6*, 4332.

SJÖLANDER, A. *Arkiv för Fysik* **1958**, *14*, 315.

SKÖLD, K. *J. Chem. Phys.* **1968**, *49*, 2443.

SMIT, J.G., DACHS, H., and LECHNER, R.E. *Solid State Commun.* **1979**, *29*, 219.

SPRINGER, T. and LECHNER, R.E. In: HEITJANS, P. and KÄRGER, J. (editors). *Diffusion in Condensed Matter*, Springer-Verlag, Berlin, 2005, p. 93.

SQUIRES, G.L. *Introduction to the Theory of Thermal Neutron Scattering*, University Press, Cambridge, 1978.

STILLER, H. *Inelastic Scattering of Neutrons*, Vol. 2, IAEA, Vienna, 1965, p. 179.

SUBRAMANIAM, S., GERSTEIN, M., OESTERHELT, D., and HENDERSON, R. *EMBO J.* **1993**, *12*, 1.

TARVIN, J.A. and PASSELL, L. *Phys. Rev. B* **1979**, *19*, 1458.

TEIXEIRA, J., BELLISSENT-FUNEL, M.-C., CHEN, S.H., and DORNER, B. *Phys. Rev. Lett.* **1985**, *54*, 2681.

THIEDEMANN, G.U., HEBERLE, J., and DENCHER, N.A. In: RIGAUD, J.L. (editor) *Structures and Functions of Retinal Proteins*, Vol. 221, Colloque INSERM, John Libbey Eurotext, London, 1992, p. 217.

THIEDEMANN, G.U. Ph.D. thesis, Fachbereich Physik, Freie Universität Berlin 1994.

TOKUNAGA, F. and IWASA, T. *Methods Enzymol.* **1982**, *88*, 163.

TURCHIN, V.F. *Slow Neutrons, Israel Program for Scientific Translations*, Sivan Press, Jerusalem, 1965; Russian original: Moscow, 1963.

Van HOVE, L. *Phys. Rev.* **1954**, *95*, 249.

VÁRÓ, G. and KESZTHELYI, L. *Biophys. J.* **1993**, *43*, 47.

VÁRÓ, G. and LANYI, J.K. *Biochemistry* **1991a**, *30*, 5016.

VÁRÓ, G. and LANYI, J.K. *Biophys. J* **1991b**, *59*, 313.

VOLINO, F. and DIANOUX, A.J. *Mol. Phys.* **1978**, *36*, 389.

VOLINO, F. and DIANOUX, A. *Mol. Phys.* **1980**, *41*, 271.

VOLINO, F., PERRIN, J.-C., and LYONNARD, S. *J. Phys. Chem. B* **2006**, *110*, 11217.

WILLIAMS, G. and WATTS, D. *Trans. Faraday. Soc.* **1969**, *66*, 80.

WINDSOR, C.G. *Pulsed Neutron Scattering*, Taylor & Francis, London, 1981.

ZACCAI, G. *J. Mol. Biol.* **1987**, *194*, 569.

ZACCAI, G. and GILMORE, D.J. *J. Mol. Biol.* **1979**, *132*, 181.

# II Instrumentation

## II.4 Neutron Imaging

*Nobuyuki Takenaka*

### II.4.1 INTRODUCTION

An image can be constructed with a matrix of two-dimensional data by plotting the values of the data two-dimensionally as the brightness. A black and white image can be obtained if the data size is, for example, more than 500 × 500 and the data dynamic range is more than 8 bits. Every two-dimensional measurement of the physical quantities concerning neutrons may be called the neutron imaging in this meaning. However, the neutron imaging in this chapter is limited to the measurement of two- or three-dimensional spatial distribution of the physical values obtained after interactions between an object and the neutrons. The coordinates of the image are the same as or transformable to those of the object.

Neutron radiography is a conventional and typical neutron imaging method. An optical image is obtained by the neutrons passing through an object in the neutron radiography system. It is just the Roentgen using neutron rays instead of X-rays. Two-dimensional still and dynamic images and three-dimensional images by the computer tomography (CT) reconstruction can be obtained in the same way as the X-rays radiography. Continuous thermal or cold neutron rays from a reactor or an accelerator are used for the neutron radiography. The energy of the neutron rays has the Gaussian distribution and is not monochromatic. Recently, new neutron imaging methods different from the traditional neutron radiography are being developed using a pulsed neutron source.

In this chapter, principles, facilities, and examples of application of the neutron radiography are described and some new imaging methods are introduced.

### II.4.2 NEUTRON RADIOGRAPHY

The neutron radiography was carried out and applied to a patent at first by H. Kallman and E. Kuhn in Germany by using neutrons generated by an accelerator in 1935, just 3 years after the discovery of neutron by J. Chadwic. The first paper was published by

*Neutrons in Soft Matter* Edited by Toyoko Imae, Toshiji Kanaya, Michihiro Furusaka, and Naoya Torikai

O. Peter in 1946 (Peter, 1946). Applications of the neutron radiography had been conducted since 1950s and the techniques were summarized as a book at first by H. Berger in 1965 (Berger, 1965). The first international conference on the neutron radiography was held in 1981. International conferences or topical meetings have been held every 2 years. The proceedings and the special issues of them were published where most of activities on the neutron radiography researches and applications can be found (Barton, 1983, 1987, 1994a, 1994b; Fujine, 1990, 2001; MacGillivray, 1994; Kobayashi, 1996; Fischer, 1997; Lehmann, 1999; Brenizer, 2004; Chirco, 2005a, 2005b; Buecherl, 2005; Arif, 2008; Takenaka, 2009).

## II.4.2.1 Principles

Figure II.4.1 shows a schematic diagram of a traditional neutron radiography system. Neutrons from a reactor or an accelerator are collimated and the parallel beam is irradiated to the object. The neutron image after the object is converted to an optical image by the scintillation converter. No scintlillators detect a neutron directly because the neutron is electrically neutral. Secondary charged particles generated by neutron capture scintillate the converter. A few isotopes, $^3$He, $^6$Li, $^{10}$B, Cd, and Gd are used for the neutron capture. An aluminum plate coated by mixture of enriched $^6$LiF and ZnS powders with a binder is often used as the scintillator. The converters named NE426 are commercially sold. The optical image is recorded by the camera in the mirror. A cooled-CCD camera is often used for still images and a CCD or a C-MOS video camera for movies. One or two mirrors are used to keep the camera from damages due to the direct beam irradiation. The converter, the mirrors, and the camera are often equipped in a light-tight camera obscura.

Radioactive rays are attenuated exponentially by absorption and scattering. The total attenuation in an object is expressed as

$$I = I_0 \exp(-\mu t), \tag{II.4.1}$$

where $I_0$, $I$ (cm$^2$s), $t$ (cm), and $\mu$ (cm$^{-1}$) are the initial intensity of the radioactive rays, the intensity after the object, the thickness of the object, and the attenuation coefficient that is the total macroscopic cross section determined by the radioactive rays and materials of the object, respectively. Mass attenuation coefficient $\mu_m$(cm$^2$/g) is defined with the density of the object, $\rho$ (g/cm$^3$), as $\mu_m = \mu/\rho$.

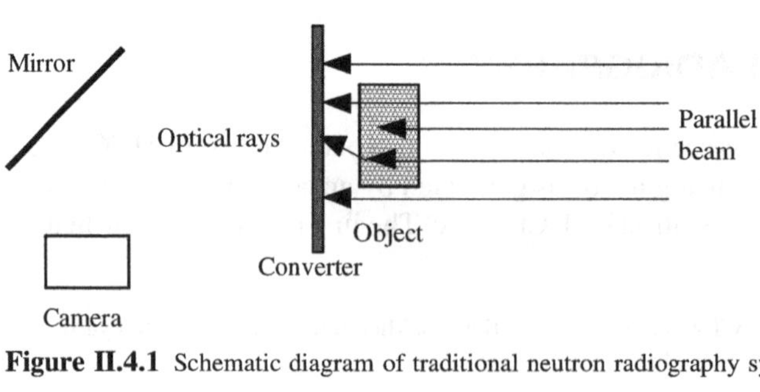


**Figure II.4.1** Schematic diagram of traditional neutron radiography system.

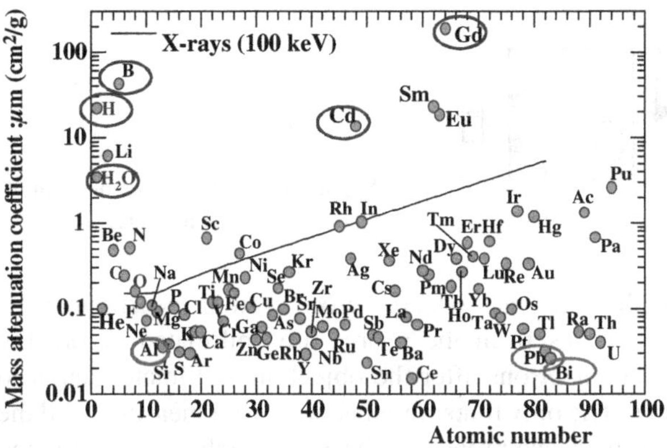


**Figure II.4.2** Mass attenuation coefficients of elements.

Figure II.4.2 shows the mass attenuation coefficients of natural elements for thermal neutron rays by the symbols and X-rays of 100 keV by the solid line against atomic numbers. X-rays around 100 keV are used for the Roentgen. The mass attenuation coefficients of X-rays increase with increasing the atomic number, while those of neutron rays depend on each element. They are high for light elements such as H, Li, and B and some special elements such as Cd and Gd. The scattering is dominant in the attenuation of H and the absorption is dominant in the others. They are low for most of metallic elements including typical heavy metals of Pb and Bi. The attenuation rate of Al is quite low and aluminum tapes, supports, and containers are often used in the neutron radiography experiments.

X-rays radiography is suitable for visualization of high atomic number elements inside of light ones such as the Roentgen for a human body. Neutron radiography is applicable to visualization of objects including hydrogen through metallic bodies. Many liquids are hydride and most of the machine bodies are metallic. Therefore, neutron radiography can be used as Roentgen of machines for industrial applications. A typical comparison of the optical, the X-rays, and the neutron image of a gas lighter is shown in Figure II.4.3a–c (Hiraoka, 1995). Metallic components of the tank, the springs, and the pipes in the lighter can be seen in the X-rays image in Figure II.4.3b. The parts made of plastic resin and the oil in the metallic tank in it are visualized in the neutron image in Figure II.4.3c.

The neutron radiography images are expected qualitatively by the linear correlation of eq. (II.4.1) using the total macroscopic cross section shown in Figure II.4.2. However, brightness distribution of the neutron radiography image is not quantitative due to the effects of scattered neutrons and some noises.

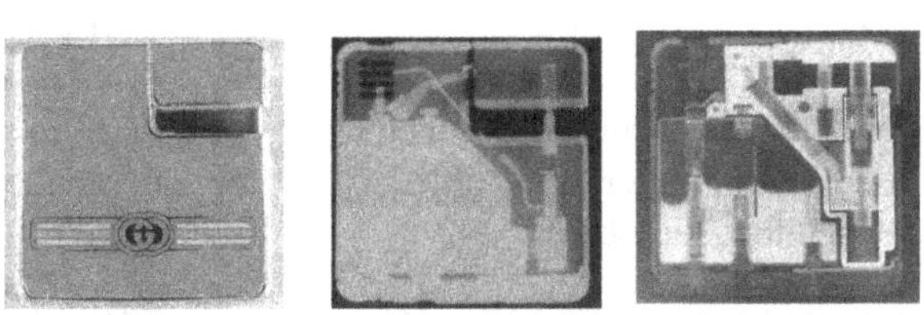

(a) (b) (c)

**Figure II.4.3** Comparison of gas lighter images. (**a**) Optical. (**b**) X-rays. (**c**) Thermal neutron.

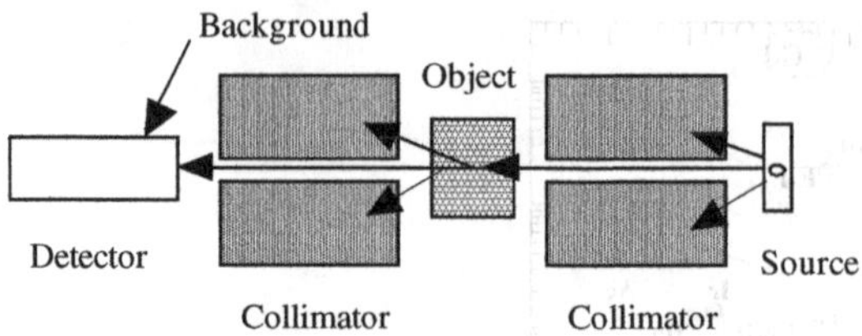


**Figure II.4.4** Schematic diagram of total cross-section measurement.

A schematic diagram of the total macroscopic cross section measurement is illustrated in Figure II.4.4. Neutrons from the source are collimated before the irradiation to obtain a beam. The neutrons after the object are collimated again to eliminate the scattered neutrons. The only transmitted neutrons are measured and the total attenuation rate is determined. Neutrons for the radiography are collimated before the irradiation to obtain a column beam but there is no collimator after the object as shown in Figure II.4.2. The neutrons scattered in the object may reach to the other points of the converter in the image and increase the brightness at the points. The optical ray after the converter may be also scattered in the camera obscura. The scattering effects always make the image brighter.

The optical ray brightness of the conventional converters is proportional to the neutron intensity. The two-dimensional brightness distribution of a neutron radiography image, $S(x, y)$, is expressed as

$$S(x,y) = G(x,y)\exp\left(-\sum_i \rho_i \mu_{m,i} t_i(x,y)\right) + O(x,y) \qquad \text{(II.4.2)}$$

where $t_i(x, y)$ is the effective thickness of the $i$th element along the beam line, $G(x, y)$ is the gain spatially distributed due to spatial nonuniformity of the beam intensity and the converter sensitivity and $O(x, y)$ is the offset. The offset term is divided into two terms. One is due to electrical offset of the camera system, electrical noises, and noises due to radioactive rays especially $\gamma$-rays, which are not much affected by the object for the neutron radiography testing. The other is the offset due to the scattering of neutrons and optical rays in the camera obscure and depends on the object. If the latter offset is negligible, the gain and the offset can be determined taking a flat field image without the object,

$$S_1(x,y) = G(x,y) + O(x,y), \qquad \text{(II.4.3)}$$

and an offset image without neutrons but with $\gamma$-rays,

$$S_0(x,y) = O(x,y) \qquad \text{(II.4.4)}$$

by using a shutter that stops the neutron beam and transmits the $\gamma$-rays. A sheet or a plate containing boron is often used as the shutter. If the object is in a container made of the $w$th element and an image without the object

$$S_w(x,y) = G(x,y)\exp(-\rho_w \mu_{m,w} t_w(x,y)) + O(x,y) \qquad \text{(II.4.5)}$$

can be obtained, an image without the container can be vanished and an image of only the object can be image-processed. This method is effective when the object moves or diffuses.

Hydrogen has high scattering cross section. Some special techniques are required when the object contains hydrogen and the scattering effects are negligible. Experimental methods to reduce or compensate the neutron scattering effects are summarized, for example, in a paper (Takenaka, 2001) and a numerical compensation method using a Monte Carlo simulation is found in a study (Hassanein, 2008).

### II.4.2.2 Facilities

Neutron radiography can be conducted when the camera obscure is placed at a neutron beam port. Neutron radiography systems are equipped at many neutron sources. High-performance neutron radiography facilities are listed in Table II.4.1. The performances of neutron radiography are shown by the neutron flux, $\phi$ ($n$/cm$^2$s) and the collimator rate, *L*/*D*. *L* and *D* correspond to the distance between the object and the neutron source and the size of the neutron source, respectively. *L*/*D* is equivalent to an inverse of the tangential of the beam angle width. The image brightness is proportional to the flux and the image blur decreases with increasing the *L*/*D*. The performance of neutron radiography depends on the source and the camera system. Roughly speaking for the traditional system, a clear 16 bits still image can be obtained for a few seconds and a movie of 8 bits and 30 frames/s is available for $\phi = 10^8$. The blur, $\delta$, is determined by the distance, $z$, between the object and the converter as

$$\delta = \frac{z}{L/D}. \tag{II.4.6}$$

*L*/*D* is changeable in some facilities and the flux decreases roughly inverse proportional to the square of *L*/*D*.

The NE426 had been often used for the traditional neutron radiography system described in these facilities. Quantum efficiency of the NE426 is 15–25% and its spatial resolution, that is, the blur due to the converter, is 100–200 µm. Recently, new devices for high-performance neutron radiography are being developed to obtain higher spatial resolution and brighter images. The converters up to a few ten $\mu$m in the spatial resolution are being developed at PSI (Frei, 2008, 2009) and JAEA (Yasuda, 2008). New neutron imaging detectors using a MCP (Micro Channel Plate) containing $^{10}$B are being developed on a target of 10 µm in spatial resolution (Tremsin, 2009). Image intensifiers I.I.s for neutrons are being developed for neutron radiography high-speed imaging up to 1000 frames/s. The quantum efficient is higher than 70% and moreover the image is efficiently intensified electrically (Nittoh, 2009).

### II.4.2.3 Applications

Neutron radiography has been applied to many fields, science, engineering, biology, agriculture, archeology, and so on. Some imaging results are introduced in this section.

**Table II.4.1** High-Performance Neutron Radiography Facilities

| Institute | Neutron Source | Name of Facility | Neutron Flux ($n/cm^2s$) | *L/D* | View Size (mm) | Note |
|---|---|---|---|---|---|---|
| JAEA | JRR-3 | TNRF | $1.2 \times 10^8$ | 185 × 154 | 255 × 305 | Reactor 20 MW |
| ILL | HFR | NEUTROGRAPH | $3.0 \times 10^9$ | 150 | 160 × 200 | Reactor 58 MW |
| HZB | BER II CNS | VONRAD | From $2 \times 10^8$ to $6 \times 10^6$ | 70–500 | From 30 × 50 to 100 × 100 | Reactor 10 MW |
| TUM | FRM-II | ANTARES | From $9.7 \times 10^7$ to $2.5 \times 10^7$ | 402/795 | From 320 × 320 to 360 × 360 | Reactor 20 MW |
| | FRM-II CNS | NECTAR | $4.9 \times 10^6$ | 230 | 300 × 300 | |
| NIST | NBSR | NIF | From$3.0 \times 10^8$ to $2.5 \times 10^5$ | 100–6,000 | 80–260 diameter. | Reactor 20 MW |
| PSI | SINQ | NEUTRA | From $2.8 \times 10^7$ to $4 \times 10^6$ | 200–550 | 150–400 diameter. | Accellerator spollation source |
| | SINQ CNS | ICON | From $2.3 \times 10^7$ to $2.1 \times 10^3$ | 151–24,200 | From 30 × 70 to 120 × 310 | |

JAEA: Japan Atomic Enegy Agency; ILL: Institut Laue-Langevin, HZB: Helmholtz Center Berlin (formerly Hahn-Meitner Institut); TUM: Techniche Universitat Munchen; NIST: National Institute of Standards and Technology; PSI: Paul Sherrer Institut.

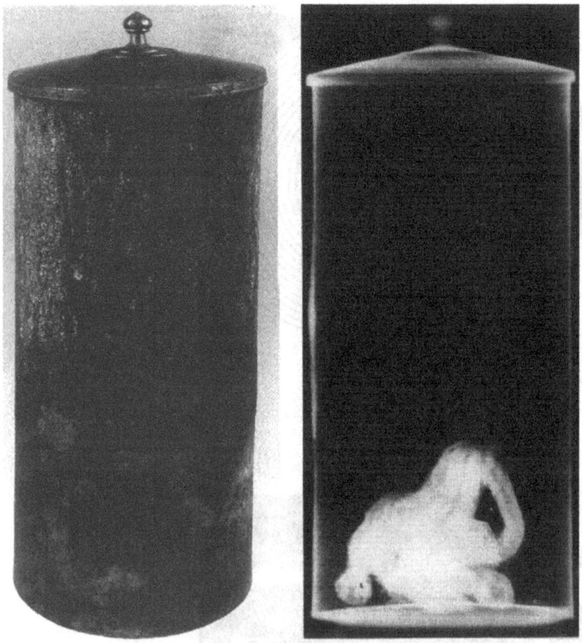

**Figure II.4.5** A sutra in a bronze container.

Figure II.4.5 shows the optical and the neutron image of an archeological object, a roll of paper where Buddhism sutra was written in a bronze container (Hiraoka 1995). A lid of such a container is often fixed to the container cylinder due to rust. An archeologist is often worrying whether he forces to open the lid to see the sutra or leaves it to keep the storage condition for future researches. Nondestructive examination to see the paper roll condition through the metallic container is very attractive to the archeologist. The neutron radiography image in Figure II.4.5 shows that the paper roll containing hydrogen is clearly visualized through the bronze container. It was shown that the sutra was deformed and damaged. The archeologist decided not to open the lid.

The optical and the neutron image of a wire solder rolled spirally are shown in Figure II.4.6 (Hiraoka, 1995). A solder is an alloy of Sn and Pb that are transparent to a neutron beam. The diameter of the spiral wire in the neutron image is much smaller than that in the optical image. The solder wire is cored with paste containing oil. The paste string is visualized by the neutron radiography through the solder. This visualization was applied to the nondestructive testing of the solder manufacturing process.

X-ray and neutron images of bullets of a rifle are shown in Figure II.4.7 (Hiraoka 1995). Gunpowder in the cartridge cases of the bullets can be clearly seen through the metallic shells in the neutron image since the powder contains hydrogen and nitrogen. Nondestructive examinations of solid rocket engines and fuses using the gunpowder are conducted by neutron radiography.

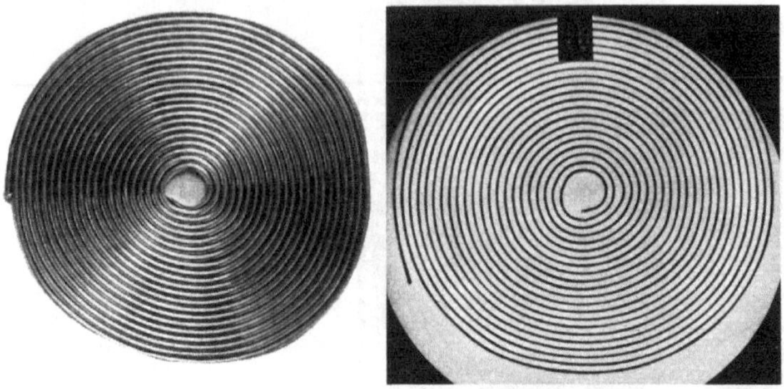

**Figure II.4.6** A wire solder with paste.

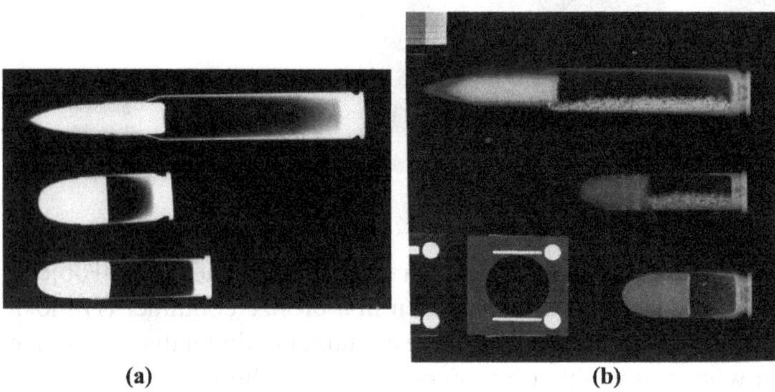

**Figure II.4.7** Nondestructive examination of bullets of a rifle. (**a**) X-ray. (**b**) Neutron.

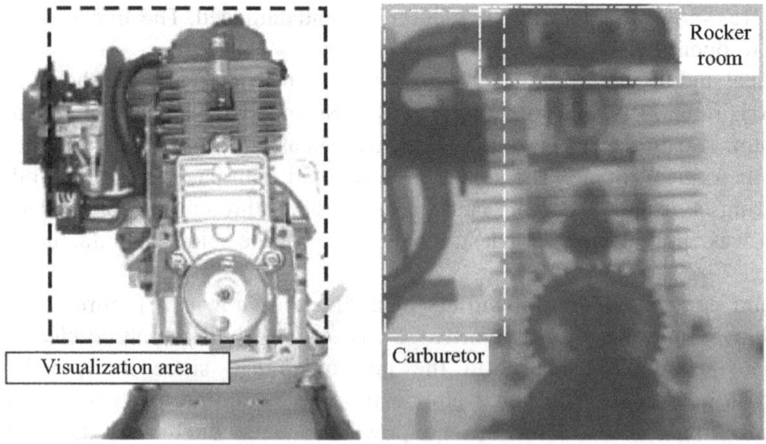


**Figure II.4.8** A small four-cycle engine.

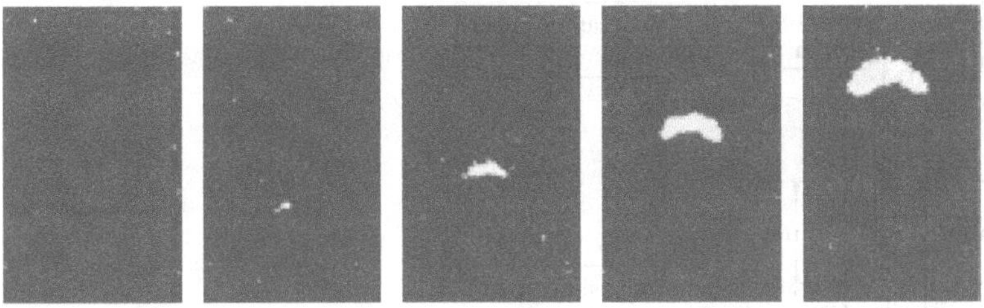

**Figure II.4.9** Visualization of water vapor bubble in Pb–Bi alloy.

A frame of a movie of a small four-cycle engine is shown in Figure II.4.8 (Nakamura, 2009). Most of engine bodies are made of aluminum alloy. The inner components of the engine, the gear, the shafts, the valves made of iron alloys and the lubrication oil can be clearly seen. The frame rate and the shutter speed were 30 frames/s and 1/500 s in this movie by using the neutron I.I. (Nittoh, 2009). Frame rates more than 1000 frames/s are expected for the industrial application.

A water vapor bubble growing and rising in a pool of molten Pb–Bi alloy is visualized in Figure II.4.9 (Saito, 2009). Melting point of the alloy was 393 K. The heavy metal can be seen through by neutron radiography.

Hydrogen distributions in hydrogen storage alloy can be visualized by neutron radiography. The transmission images and the cross section images by the CT reconstruction in a block made of TiCrV alloy in a metallic container are shown in Figure II.4.10 (Matsubayashi, 2008).

Polymer electrolyte fuel cells (PEFCs) are being developed for automobiles and household cogeneration systems. Recently, applications of neutron radiography to the PEFC researches have been used in many facilities. The most recent reports can be found in the special issue (Takenaka, 2009) by using the system at JAEA, PSI, NIST, Pennsylvania State University, and Korean Atomic Energy Research Institute.

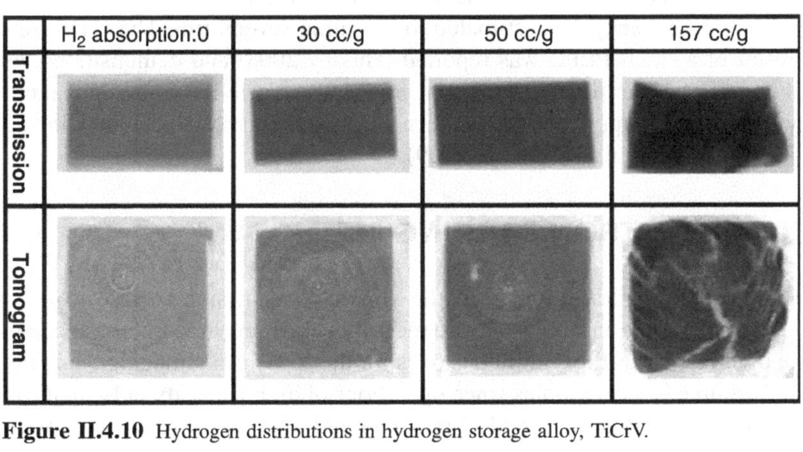


**Figure II.4.10** Hydrogen distributions in hydrogen storage alloy, TiCrV.

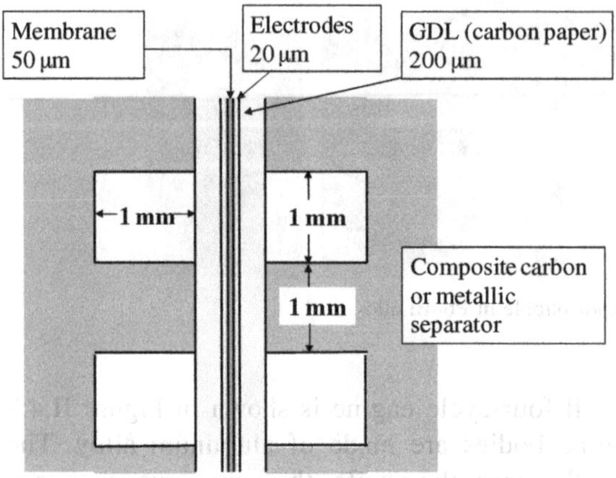


**Figure II.4.11** Schematic diagram of a typical PEFC cross section.

Figure II.4.11 shows a schematic diagram of the typical PEFC cross section. The separators with gas flow channel, the porous gas diffusion layers (GDLs), the membrane and catalyst electrodes assembly (MEA) are placed as shown in Figure II.4.11. Humid hydrogen gas and air are supplied to the PEFC to avoid the membrane from dried. Water is generated in the cathode electrode. Liquid water is condensed and appears in the electrode, the GDL and the channel and may prevent the oxygen from being supplied to the electrode from the channel. Water distributions visualized by irradiating neutrons perpendicular or parallel to the MEA, that is, horizontally or perpendicularly, respectively, in the figure have been conducted. The CT reconstructions were also conducted. High spatial resolution system up to 10 μm is required to discuss the water distribution in the MEA and the GDL and is being developed as described above (Frei, 2009; Yasuda, 2008; Tremsin, 2009). Two-dimensional distribution of water thickness integrated perpendicular to the membrane plane was measured quantitatively by an umbra method (Takenaka, 2001) as shown in Figure II.4.12 (Murakawa, 2009). The effective fuel cell size was 50 mm × 50 mm and the quantitative measurement pixel size was 3 mm × 1 mm. The thickness measurement efficiency was estimated to be within several μm. The CT reconstruction result of a small PEFC was reported (Hussey 2008) and demonstrated, as shown in Figure II.4.13, in http://physics.nist.gov/MajResFac/NIF/tomography.html.

Many other recent applications of the neutron imaging can be found in the special issue of the journal (Takenaka, 2009).

## II.4.3 NEW NEUTRON IMAGING METHODS

New imaging techniques different from the simple transmission methods of the neutron radiography as described above are recently being developed. Pulse neutron sources are used as well as the continuous ones for them. Neutron imaging using the pulse neutron source is recent topics since operations of high-intensity pulse neutron sources base on an accelerator, J-PARC in Japan and SNS in the USA have been

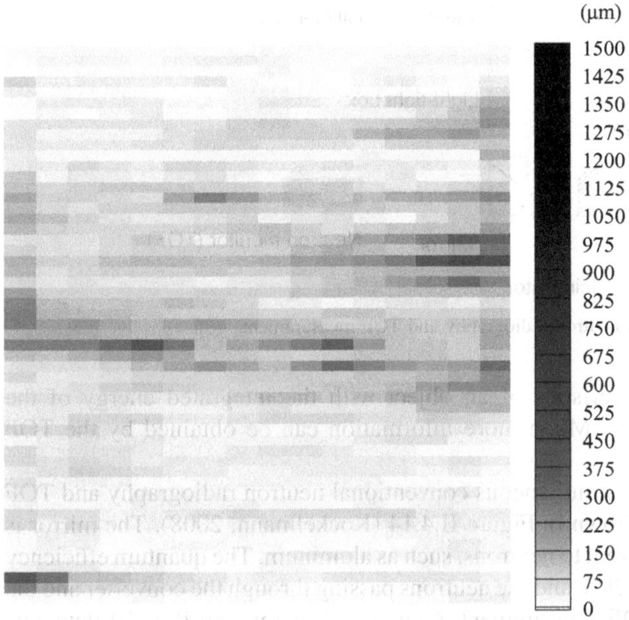


**Figure II.4.12** Quantitative water thickness distribution measurement in PEFC, 3 mm × 1 mm in pixel size.

started. Images by neutrons with various monochromatic energies can be obtained by the TOF (time-of-flight) method by using the pulse neutron source.

The TOF methods are applied to the transmitted neutrons through an object in the neutron imaging. The conventional neutron radiography gives transmitted images

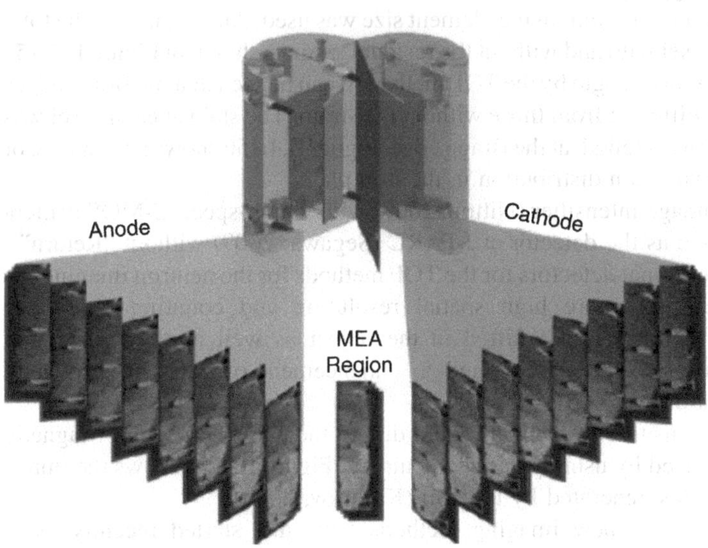


**Figure II.4.13** CT reconstruction of 3D water distribution in a small PEFC.

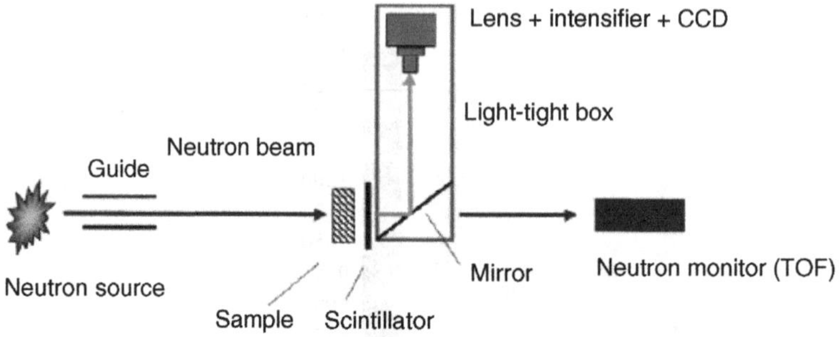


**Figure II.4.14** Simultaneous neutron radiography and TOF measurement.

of the attenuation characteristics of an object with the integrated energy of the continuous neutron source. Much more information can be obtained by the TOF methods.

An illustration of the simultaneous conventional neutron radiography and TOF measurement system is shown in Figure II.4.14 (Kockelmann, 2008). The mirror is made of materials transparent to neutrons, such as aluminum. The quantum efficiency of the converter is around 20% and the neutrons passing through the converter and the mirror are used for the TOF measurement. A neutron detector with 10 × 10 elements, 2 mm × 2 mm in the element size is used for this system. A high spatial resolution radiography image with energy integrated neutrons and energy dependent attenuation characteristics images in 2 mm spatial resolution can be obtained by this method. A steel welding testing for neutron energy around the Bragg edge was reported by using this system (Kockelmann, 2008).

Strain distributions in an iron plate, 100 mm × 40 mm × 5 mm, were measured by determining the Bragg edge wavelength (Iwase, 2009). A neutron detector with 8 × 8 elements and 1.7 mm × 1.7 mm in the element size was used. An example of the total cross section at a pixel with and without the residual strain is shown in Figure II.4.15a against the neutron wavelength by the TOF method. It can be seen that the Bragg edges with the strain are different from those without the strain. The stain at each pixel was measured from the wavelength at the Bragg edge. Figure II.4.15b shows an example of the two-dimensional strain distribution in the iron plate.

The neutron image intensifier (Nittoh, 2009) with a high-speed C-MOS camera was tried to be used as the detector at J-PARC (Segawa, 2009) without "Return".

The two-dimensional detectors for the TOF methods for the neutron imaging are under development for more high spatial resolution and counting ratio with "Return". Phase of neutrons is shifted in the object as well as the intensity is attenuated. The phase shift was applied to enhancement of edges as shown in Figure II.4.16 (Kardjrov, 2005).

Neutrons are sensitive to magnetic field due to their magnetic spin. Magnetic fields can be visualized by using polarized neutrons. Figure II.4.17 shows the image of the magnetic fields generated by the coil (Kardjrov, 2009).

Applications of the new imaging methods have just started recently. New applications in various fields are expected.

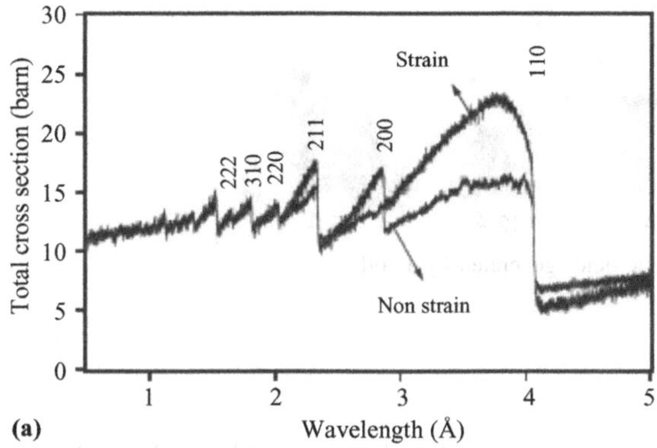


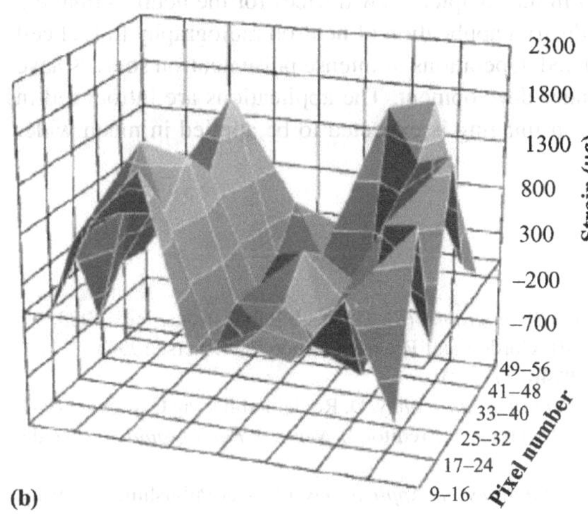


**Figure II.4.15** (**a**) Total cross section of an iron plate with and without stain. (**b**) Two-dimensional residual strain distribution.

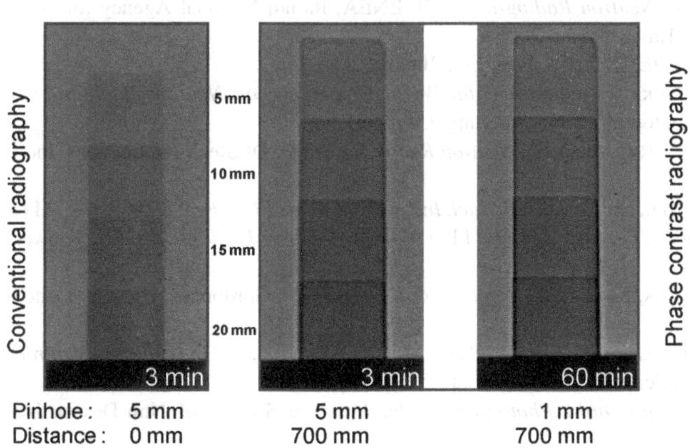


**Figure II.4.16** Edge enhancement of an aluminum step by a phase contrast method.

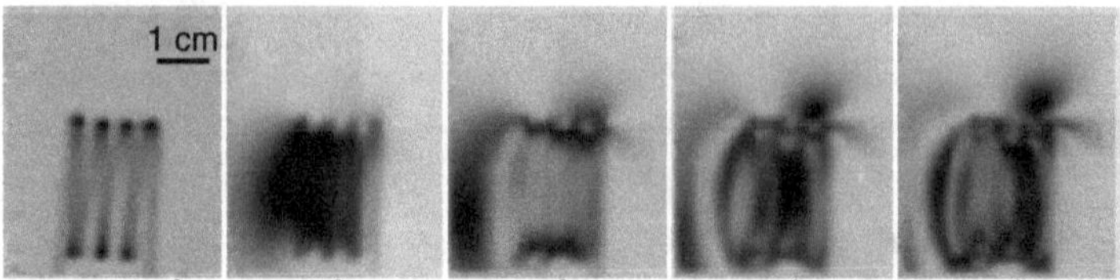


**Figure II.4.17** Visualization of magnetic fields generated by a coil.

## II.4.4 SUMMARY

Principles, facilities, and applications of the conventional neutron radiography and new neutron imaging methods are described in this chapter. New devices for the neutron imaging are being developed. Projects on the industrial application of neutron radiography to fuel cell and engine researches have been conducted. Operations of intense pulse neutron sources have started and the imaging systems are under development. The applications are introduced in limited fields in this chapter. The neutron imaging is expected to be applied in much wider fields.

## REFERENCES


Arif, M. and Downing, R.G. (editors). *Neutron Radiography (8)*, DEStech Publications, Inc., 2008.

Barton, J.P. (editor). *Neutron Radiography (4)*, Gordon and Breach Science Publishers, 1994a.

Barton, J.P. (editor). *Nondestr. Test. Eval.* **1994b**, *11*(2–3).

Barton, J.P. and von der Hardt, P. (editors). *Neutron Radiography*, D. Reidel Publishing Company, 1983.

Barton, J.P., Farny, G., Person, J.-L., and Roettger, H. (editors). *Neutron Radiography*, D. Reidel Publishing Company, 1987.

Berger, H. *Neutron Radiography: Methods, Capabilities and Applications*, Elsevier Publishing Company, 1965, modified photocopy edition, Industrial Quality Inc., 1995.

Brenizer, J. and Tsukimura, R. (editors). *Appl. Radiat. Isot.* **2004**, 61.

Buecherl, T., Schilinger, B., Tuerler, A., and Boeni, P. (editors). *Nucl. Instrum. Methods Phys. Res. A* **2005**, *542*(1–3).

Chirco, P. and Rosa, R. (editors). *Neutron Radiography (7)*, ENEA, Italian National Agency for New Technologies, Energy and the Environment, 2005a.

Chirco, P. and Rosa, R. (editors). *IEEE Trans. Nucl. Sci*, **2005b** *52*(1).

Fischer, C.O., Stade, J., and Bock, W. (editors). *Fifth World Conference on Neutron Radiography*, Deutsche Gesellschaft fuer Zerstoerungsfreie Pruefung e.V., 1997.

Frei, G. In: Arif, M. and Downing, R.G. (editors). *Neutron Radiography (8)*, DEStech Publications, Inc., 2008, pp. 599–607.

Frei, G., Lehmann, E.H., Mannes, D., and Boillat, P. *Nucl. Instrum. Methods Phys. Res. A* **2009**, *605*, 111.

Fujine, S., Kanda, K., Matsumoto, G., and Barton, J.P. (editors). *Neutron Radiography (3)*, Kluwer Academic Publishers, 1990.

Fujine, S., Kobayashi, H., and Kanda, K. (editors). *Neutron Radiography (6)*, Gordon and Breach Science Publishers, 2001.

Hassanein, R., Vonobel, P., and Lehmann, E. In: Arif, M. and Downing, R.G. (editors). *Neutron Radiography (8)*, DEStech Publications, Inc., 2008, pp. 206–216.

Hiraoka, E. (editor), *Neutron Radiography Photography*, The Japanese Society of Non-Destructive Inspection, 1995 (in Japanese).

Hussey, D.S., Owejan, J.P., Jacobson, D.L., Trabold, T.A., Gagliardo, J., Baker, D.R., Caulk, D.A., and Arif, M. In: Arif, M. and Downing, R.G. (editors). *Neutron Radiography (8)*, DEStech Publications, Inc., 2008, pp. 470–476.

Iwase, K., Sakuma, K., Kamyama, T., and Kiyanagi, Y. *Nucl. Instrum. Methods Phys. Res. A* **2009**, *605*, 1.

Kardjrov, N., Lee, S.W., Lehmann, E., Lim, I.C., Sim, C.M., and Vonotober, P. *Nucl. Instrum. Methods Phys. Res. A* **2005**, *542*, 100.

Kardjrov, N., Hilger, A., Manke, I., Strobl, M., Dawson, M., and Bahart, J. *Nucl. Instrum. Methods Phys. Res. A* **2009**, *605*, 13.

Kobayashi, H. and Mochiki, K. (editors). *Nucl. Instrum. Methods Phys. Res. A* **1996**, *377*(1).

Kockellmann, W., Sanisteban, J.R., Vontobel, P., Frei, G., and Hehmann, E.H. In: Arif, M. and Downing, R.G. (editors). *Neutron Radiography (8)*, DEStech Publications, Inc., 2008, pp. 217–226.

Lehmann, E., Pleinert, H., and Koerner, S. (editors). *Nucl. Instrum. Methods Phys. Res. A* **1999**, *424*(1).

MacGillivray, G.M. and Brenizer, J.S. (editors). *Proceedings of the First International Topical Meeting on Neutron Radiography System Design and Characterization*, AECL Research Chalk River Laboratories, Chalk River, Ontario, Canada, 1994.

Matsubayashi, M., Iimura, H., Yasuda, R., Ito, H., Kubo, K., Arashima, H., and Ebisawa, T. In: Arif, M. and Downing, R.G. (editors). *Neutron Radiography (8)*, DEStech Publications, Inc., 2008, pp. 432–439.

Murakawa, H., Ueda, T., Yoshida, T., Sugimoto, K., Asano, H., Takenaka, N., Iikura, H., Yasuda, R., and Matsubayashi, M. *Nucl. Instrum. Methods Phys. Res. A* **2009**, *605*, 127.

Nakamura, M., Sugimoto, K., Asano, H., Murakawa, H., Takenaka, N., and Mochiki, K. *Nucl. Instrum. Methods Phys. Res. A* **2009**, *605*, 204.

Nittoh, K., Konagai, C., Noji, T., and Miyabe, K. *Nucl. Instrum. Methods Phys. Res. A* **2009**, *605*, 107.

Peter, O. *Z. Naturforsch.* **1946**, *1*, 557.

Saito, Y. *Nucl. Instrum. Methods Phys. Res. A* **2009**, *605*, 192.

Segawa, M., Kai, T., Sakai, T., Shinohara, T., Nakamura, T., Matsubayashi, M., Harada, M., Oikawa, K., Maekawa, F., Ooi, M., and Kureta, M. *Proceedings of the 2009 Fall Meeting of the Atomic Energy Society of Japan*, H32, 2002 (in Japanese)

Takenaka, N., Asano, H., Fujii, T., and Matsubayashi, M. *Nondestr. Test. Eval.* **2001**, *16*, 348.

Takenaka, N., Mochiki, K., Kawabata, Y., and Asano, H. (editors). *Nucl. Instrum. Methods Phys. Res. A* **2009**, 605.

Tremsin, A.S., McPhate, J.B., Vallerge, J.V., Siegmund, O.H.W., Hull, J.S., Feller, W.B., and Lehnann, E. *Nucl. Instrum. Methods Phys. Res. A* **2009**, *605*, 103.

Yasuda, R., Hyashida, H., Iikura, H., Sakai, T., and Matsubayashi, M. *Proceedings of the Sixth International Topical Meeting on Neutron Radiography*, Kobe, Japan, September 14–18, 2008, p. 81.

# III Data Treatment and Sample Environment

# III.1 Practical Aspects of SANS Experiments

*George D. Wignall*

## III.1.1 INTRODUCTION

Many of the powerful techniques developed to study materials in the past several decades are rendered intractable by their very power when applied to materials of the complexity of a polymer alloy, colloidal suspension, or microporous medium. They yield information in such detail that the features that make the bulk material interesting are obscured in the mass. What is often required is a probe that is sensitive to possibly new features in the microscopic structure, while remaining insensitive to those aspects of the structure that are essentially common to the class of materials being studied. In considering a blend of two known polymers, for example, a technique such as conventional crystallography provides a wealth of information on the atomic structure of the chemical monomers, containing new information on whether or not the blend is actually thermodynamically stable.

Small–angle neutron scattering (SANS) is a technique in which the microscopic length scale probed in the sample may be chosen to suit the property being studied, while remaining insensitive to much finer or much coarser details. Before the application of SANS to study polymer structure, chain conformation studies were limited to light (LS) and small–angle X–ray scattering (SAXS) techniques, usually conducted in dilute solution due to the difficulties of separating the inter– and intrachain contributions to the structure. The unique role of neutron scattering in soft condensed matter arises from the difference in the coherent scattering length between deuterium ($b_D = 0.67 \times 10^{-12}$ cm) and hydrogen ($b_H = -0.37 \times 10^{-12}$ cm), which results in a marked difference in scattering power (contrast) between molecules synthesized from normal (hydrogenous) and deuterated monomer units. Thus, deuterium labeling techniques may be used to "stain" molecules and make them "visible" in

*Neutrons in Soft Matter* Edited by Toyoko Imae, Toshiji Kanaya, Michihiro Furusaka, and Naoya Torikai

the condensed state and other crowded environments, such as concentrated solutions of overlapping chains. For over three decades, SANS has proven to be a powerful tool for studies of individual molecules in such soft matter systems and made it possible to extract information about their size, shape, and associations.

Since most soft materials are essentially transparent to neutrons of the wavelengths typically used in SANS (~5–20 Å), bulk samples may be studied, and sample environments are easily varied over a wide range of pressures and temperatures. Furthermore, the need to site these instruments at suitable steady-state (reactor) or pulsed sources has offered to SANS users routine access to state-of-the-art centralized facilities such as pressure cells, furnaces, refrigerators, and so on. For these and other reasons, SANS has become very widely used, inter alia, for the study of polymers, biological macromolecules, colloids, gels, micellar, and microemulsion systems, as illustrated in a recent review (Melnichenko and Wignall, 2007).

Thus, scattering techniques have been employed for decades to study soft mater, for example, in the work of Bunn (1935), who used X-rays to determine the crystal structure of polyethylene via Bragg's law

$$\lambda = 2D \sin\theta, \qquad \text{(III.1.1)}$$

where $D$ is the distance between crystallographic planes, $\lambda$ is the wavelength, and $2\theta$ is the angle of scatter. For elastic scattering (where the energies of the incident and scattered neutrons are the same), the intensity is measured as a function of the momentum transfer

$$Q = 4\pi\lambda^{-1} \sin\theta. \qquad \text{(III.1.2)}$$

Combining eqs. (III.1.1) and (III.1.2) gives $D = 2\pi/Q$, and so to study length scales ~10–2000 Å that are important for soft matter (e.g., the size of the polymer "coil"), we need to work at low $Q$ values ($10^{-3}$ to 1 Å$^{-1}$) and collect data at small angles ($\theta < 10°$) using long wavelength ($5 < \lambda < 20$ Å) or "cold' neutrons.

For sample containment, there are several materials (e.g., quartz, single-crystal Si) that have very little absorption or scattering for neutrons. For SAXS, on the other hand, materials that have high absorption (to define a SAXS beam) also have high scattering power, as both parameters are a strong function of the atomic number, and parasitic scattering is usually higher for SAXS. Similarly, it is much harder to contain samples in a SAXS camera as most materials have substantial absorption, which attenuates the beam. Thus, the high penetrating power of neutrons makes it relatively easy to contain samples with a minimum of instrumental backgrounds, and SANS is particularly suited to study the structure of matter under pressure due to the high transmission of many of the materials used in the construction of pressure vessels (Wignall, 1999).

This chapter will therefore discuss practical aspects of the technique and summarize some of the lessons that have been learned in the past three decades, with the aim of aiding potential new users of the technique, who have a general scientific background, but no specialist knowledge of scattering, to understand the potential of the technique and apply it to areas of their own particular research interests.

## III.1.2 SANS INSTRUMENTATION

The first requirement of current and future research on soft matter using SANS is to have access to suitable facilities, many of which are currently located on 20–40-year-old reactors and may become unavailable in future. The first modern instrument (Schelten, 1972, 1981) was built in the late 1960s at the FRJ-2 reactor in Kernforschungsanlage (Forschungszentrum), Jülich, Germany, and pioneered the use of long-wavelength neutrons and large overall instrument length ($\sim$40 m). This was a direct consequence of the low source brilliance of neutron sources, which are orders of magnitude below that of X-ray sources. To compensate for this difference, it is necessary to use large sample areas (1–20 $cm^2$), which means that the size of the instrument must be large in order to maintain resolution in the range of 10–2000 Å.

The FRJ-2 SANS facility was the first to employ a large overall instrument size and to boost the flux of the long-wavelength ($\lambda > 5$ Å) component of the Maxwellian spectrum by moderating the neutrons to a lower temperature by means of a cold source containing a small volume of liquid hydrogen at $T \sim 20$K. This gives flux gains of over an order of magnitude at $\lambda \sim 10$ Å (see below). The D11 facility, built during the 1970s on the high flux reactor at the Institut Laue-Langevin (ILL), Grenoble, France, incorporated many of the features of the FRJ-2 instrument, including a cold source and long ($\sim$ 80 m) dimensions (Ibel, 1976; Lindner et al., 1992).

Subsequently, over 20 other SANS facilities have been constructed worldwide, though several of these instruments are no longer operational, including the pioneering modern SANS facility (Schelten, 1972, 1981). This trend can be expected to continue as many currently available facilities were constructed on reactors built in the 1960s and 1970s, and forward surveys have indicated that over the next two decades, the installed capacity of neutron beams for research could decrease substantially. Fortunately, the decline in the availability of reactor-based SANS instruments, as exemplified by the shutdowns of the Forschungszentrum (Jülich), Brookhaven, and Risø (Copenhagen) reactors, has been offset by two competing trends. First, several new reactors have been constructed (e.g., at the Australian Nuclear Science and Technology Organization High Flux Reactor, Lucas Heights, and at the Forschungsreaktor, München, Germany) along with upgrades to existing sources (e.g., at the ILL in the mid-1990s and Oak Ridge National Laboratory (ORNL) during 2000–2006).

In addition, a range of accelerator-based SANS instruments have been developed over the past 15 years, and in particular, the Spallation Neutron Source (SNS) has been brought on line at Oak Ridge. In Europe, the second target station at the ISIS facility (ISIS-2) and the new facilities at the Munich reactor will do much to offset and even reverse predicted decline in the availability of SANS facilities. Also, reactors have been optimized over the past several decades, and the flux of instruments planned on new or upgraded reactor sources will either be less than or equal to the current state-of-the-art instruments (e.g., at D22 at the ILL). However, this is not the case for pulsed facilities, which have only just begun to reach their full potential, so we can still expect substantial gains over the current facilities via the SNS, ISIS–2, and so on.

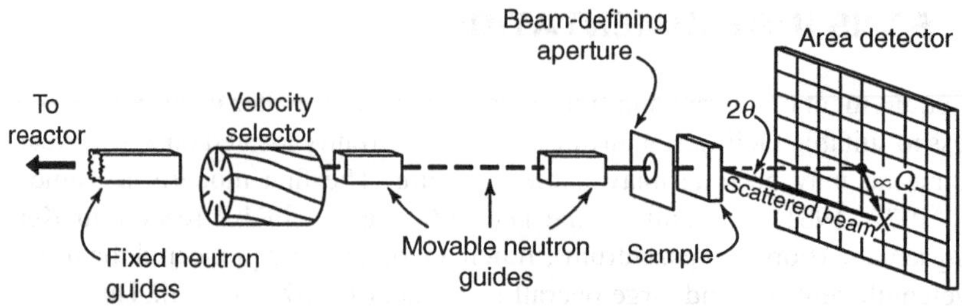


**Figure III.1.1** Schematic outline of a reactor-based facility.

A schematic diagram of a reactor-based facility is shown in Figure III.1.1. Fission neutrons are produced in the core, which is surrounded by a moderator (e.g., $D_2O$, $H_2O$) and a reflector (e.g., Be, graphite) that reduce the neutron energy. A typical moderator/reflector temperature is 310K that produces a Maxwellian spectrum of wavelength that is peaked at $\lambda \sim 1$ Å (thermal neutrons). Because of the $\lambda^{-4}$ factor that enters into the calculation (Schelten, 1981) of the scattering power for a given resolution, it is highly advantageous to use long wavelengths and increase the flux in this region. This may be accomplished by further moderating the neutrons to a lower temperature by means of a cold source containing a small volume of liquid or superfluid hydrogen, placed near the end of the beam tube. Alternative refrigerants include liquid deuterium, and the SANS cameras on the FRJ-2 and ILL reactors were the first to use the combination of a cold source and neutron guide tubes, as proposed by Maier-Leibnitz and Springer (1966). These are often coated by natural Ni or isotopic $Ni^{58}$ and operate by total internal reflection to transport the neutron beam from the cold source to the sample, in a manner analogous to the way light may be transported by fiber optics. The guide system (Figure III.1.1) provides a gap for the insertion of a velocity selector to define the wavelength, typically in the range $5 < \lambda < 20$ Å, and bandwidth ($\Delta A/\lambda \sim 5$–35%) of the neutron beam. In addition to fixed neutron guides, most instruments have translatable guide sections and apertures that may be moved in and out of the neutron beam to define the incident beam collimation. This is followed by an accessible section (1–2 m) at the sample position to accommodate sample changers, pressure cells, furnaces, and so on. Thus, when all the moveable guides are removed from the beam, the source slit is typically ~10–20 m from the sample, and this distance is reduced to 1–2 m when all the guides translated into the beam to increase the sample flux. An area detector (typically a $64 \times 64\,cm^2$ or $100 \times 100\,cm^2$ proportional counter) is often positioned via a motor-driven carrier mounted on rails in the post-sample flight tube, ~1–20 m in length. Like the incident neutron guides, this is normally evacuated to reduce air scatter, which would otherwise be strong, given the overall instrument length ~20–40 m.

Reactor sources also produce appreciable background (e.g., fast neutrons, $\gamma$-rays) that can be recorded by area detectors. By introducing curvature into the guides, it is possible to separate this component, which is not reflected as efficiently as cold neutrons. Alternatively, the beam may be deflected by supermirrors, based on the discrete thin-film multilayer equations of Hayter and Mook (1989), and such mirrors may be designed to reflect up to 3–4 times the critical angle for internal reflection that can be achieved by natural Ni guide coatings ($\theta_c \sim 0.1\lambda$ (Å)).

## III.1.3 SANS BEAM COLLIMATION AND SAMPLE CONTAINMENT

SANS experiments are rather simple, compared to SAXS, because of the high transparency of materials such as quartz and silicon, and so many different sample cells, enclosures, environments have been constructed over the past three decades. The size of the beam at the sample is usually defined by slits (irises) made of neutron absorbing materials (e.g., $Li^6$, cadmium, boron), for which the ratio of scattering to absorption is virtually zero. This has the result that neutron beams can be very well collimated, as indicated in Figure III.1.2, which shows the intensity distribution across a typical cold neutron beam (Schmatz et al., 1974), showing that the ratio of parasitic scattering to the main beam intensity is very small (typically $\leq 10^{-5}$ within ~1 mm from the beam stop). As mentioned previously, for SAXS, materials have high absorption (to define a SAXS beam) and high scattering power, as both parameters are a strong function of the atomic number, and parasitic scattering is usually higher for SAXS (Alexander, 1969). For singly scattered neutrons, the intensity $I(Q)$ is proportional to the sample thickness ($t$) and transmission ($T$) (see below), which is related to the linear attenuation coefficient, $\mu$, via $T = e^{-\mu t}$. Thus, the intensity is maximized for $\mu t = 1$, and typical sample thicknesses are ~1–2 mm for

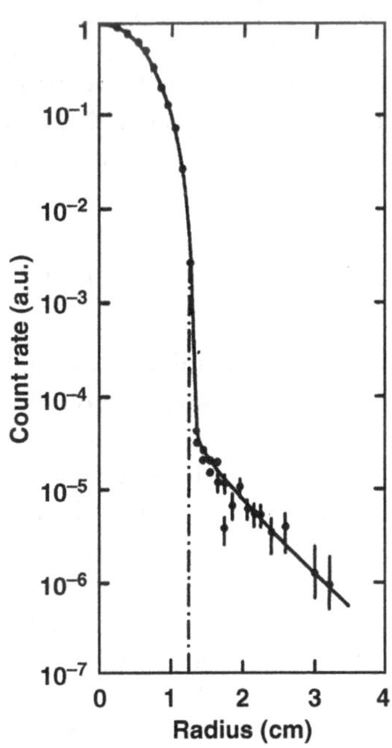


**Figure III.1.2** Intensity distribution across the primary beam. The geometrical boundary of the beam is indicated by the dash-dotted vertical line; 2 cm was equivalent to a $Q$ value of $2.5 \times 10^{-3}$ Å$^{-1}$ (Schmatz et al., 1974).

predominantly hydrogenous samples ($H_2O$, H blanks, etc.) and ~1 cm for predominantly deuterated materials (e.g., $D_2O$ and D blanks). The effect of the thickness on coherent multiple scattering (MS) and incoherent backgrounds is discussed in more detail in Sections III.1.4 and III.1.5.

## III.1.4 DETECTOR SENSITIVITY CALIBRATION AND INCOHERENT BACKGROUND SCATTERING

The majority of area detectors are multiwire proportional counters (Abele et al., 1981; Ibel, 1976), with active areas up to 1 $m^2$ and an element (cell) size ~0.5–1 $cm^2$, which is chosen to be of the same order as the sample size to equalize the various contributions to the instrumental resolution (Schelten, 1981). In general, the detector response function, $R(Q)$, may be approximated by a Gaussian distribution with a full width at half maximum ~0.5–1 cm, and the spatial variation of the detector efficiency ($\varepsilon$) is usually measured via a predominantly incoherent scatterer (light water, vanadium) that has an angle-independent intensity in the $Q$ range measured. Thus, to a first approximation, any variation in the measured signal can be attributed to the detector efficiency and used in the data analysis software to correct this effect along with instrumental backgrounds.

Measurements at higher $Q$ values ($0.1 < Q < 1$ Å$^{-1}$) are particularly sensitive to the incoherent background, which can be of the same order of magnitude as the coherent signal. The absolute cross section, $[(d\Sigma/d\Omega)(Q)]$, is defined (Turchin, 1965) as the ratio of the number of neutrons (neutron/s) scattered per second into unit solid angle divided by the incident neutron flux (neutron/($cm^2$ s)) and thus has the dimension of area ($cm^2$). On normalizing to unit sample volume, $d\Sigma/d\Omega(Q)$ has the unit of $cm^{-1}$ and is equivalent to the Rayleigh ratio used in light scattering (Kirste et al., 1975; Wignall, 1987). From the above definition, the relationship between the cross section and the measured count rate in a detector element with area, $\Delta a$, and counting efficiency, $\varepsilon$, situated normal to the scattered beam at a distance, $r$, from the sample is given by

$$\frac{d\Sigma}{\delta\Omega}(Q) = \frac{I(Q)r^2}{\varepsilon \cdot I_0 \cdot \Delta a \cdot A \cdot t \cdot T'} \qquad \text{(III.1.3)}$$

where $I_0$ is the intensity (count/(s $cm^2$)) on a sample of area $A$, thickness $t$, and irradiated volume ($At$). As mentioned above, the measured transmission $T$ is given by $T = \exp(-\mu t)$ and accounts for the attenuation of the beam passing through the sample. For SANS, it is assumed that the attenuation factor is the same for all scattered neutrons and this approximation is reasonable for $\theta < 10°$. Similarly, eq. (III.1.3) assumes that the solid angle subtended by a detector element is independent of $2\theta$ and this approximation again holds for small angles where cos $2\theta$ is close to unity. Thus, measurements on samples with different dimensions ($t$, $A$) and transmission ($T$) may be normalized to the same volume to give a (coherent) cross section, which is an intensive (material) property, independent of the sample dimensions. This assertion is based on the assumption that neutrons are scattered only once before

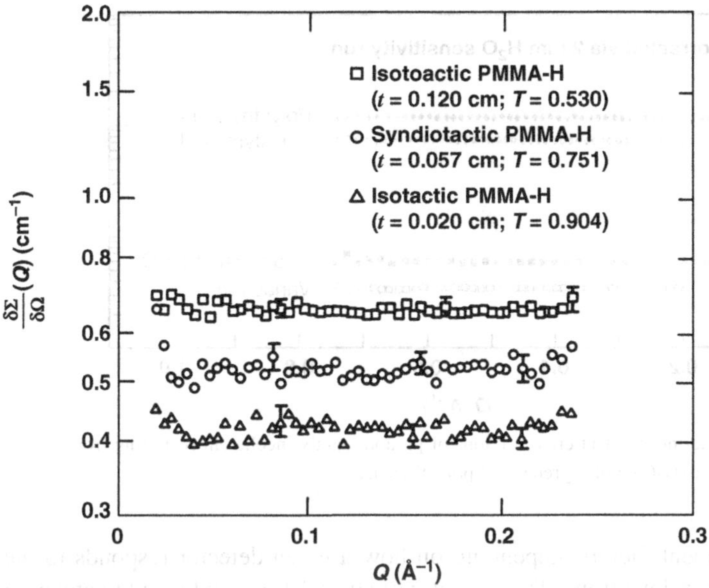


**Figure III.1.3** $d\Sigma/d\Omega(Q)$ versus $Q$ for three fully hydrogenated PMMA-H blanks.

being detected and this has been shown to be a reasonable approximation for coherent SANS from polymers (Goyal et al., 1983) and other materials (Schelten and Schmatz, 1980), as discussed below. For incoherent scattering, however, this assumption is not valid and 1–2 mm samples containing hydrogen ($H_2O$, $H^1$-labeled polymers, etc.) give rise to appreciable multiple scattering. This is illustrated in Figure III.1.3, which shows the apparent cross section (O'Reilly et al., 1985) produced by three hydrogenous (protonated) PMMA-H blanks, after normalizing via eq. (III.1.3). Because the data contain appreciable multiple scattering (which is not proportional to the thickness or transmission), the data cannot be normalized to a true cross section that is independent of the sample dimensions. However, the incoherent background is independent of $Q$, and empirical methods have been developed to subtract this background to a good approximation (Hayashi et al., 1976; Dubner et al., 1990).

Figure III.1.4 compares the angle-independent, predominantly incoherent, cross sections of thin (~1 mm) samples of vanadium, heavy water ($D_2O$), and hydrogenous ($H^1$-labeled) polystyrene (PS) and polyethylene (Wignall, 2004), measured with an incident wavelength of 4.75 Å. The scattering from light water ($H_2O$) is also predominantly incoherent and has much higher intrinsic scattering than vanadium or $D_2O$. Because of this, for 1–2 mm samples, the multiple scattering is much higher (>30%) than for vanadium (~10%) and cannot be calculated to the same degree of accuracy (Wignall, 1987), and an appreciable fraction of neutrons are scattered inelastically (i.e., with a change of energy on scattering). Such effects are very difficult to model (Copley, 1988; Boyer and King, 1988) and, moreover, the detector efficiency is a function of the wavelength, and this introduces sample- and

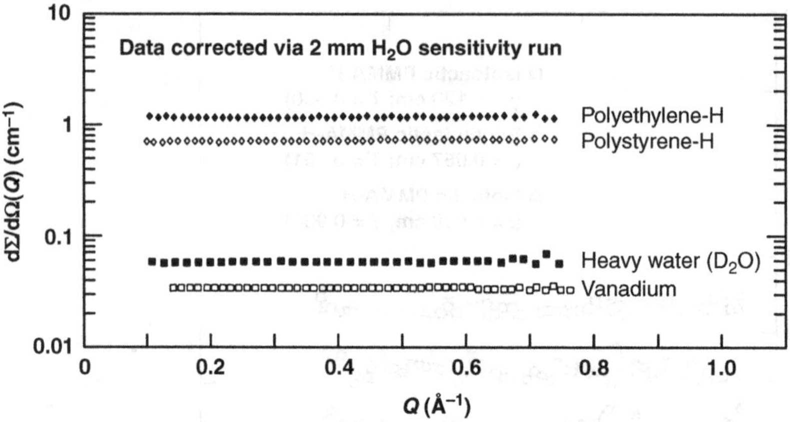


**Figure III.1.4** Angle-independent cross sections of predominantly incoherent scattering from vanadium, heavy water ($D_2O$), polystyrene, and polyethylene.

instrument-dependent factors, depending on how a given detector responds to the inelastically scattered neutrons. Thus, the use of eq. (III.1.3) would lead to apparent cross sections that are functions of wavelength and are also detector dependent (May et al., 1982). Also, because of the strong multiple scattering, the intensity for water or protonated polymer samples is not proportional to the product *tT* (eq. (III.1.3)), and hence it is not possible to define a true cross section that is a material (intensive) property, independent of the sample dimensions. The scattering is a nonlinear function of the sample thickness and this is illustrated in Figure III.1.5, which shows the apparent "cross sections" produced by applying eq. (III.1.3) to water samples. For a sample thickness ~1 mm, the cross section is $\sim 1\,\text{cm}^{-1}$ for $H_2O$ (compared to $\sim 0.06\,\text{cm}^{-1}$ for $D_2O$ as shown in Figure III.1.4). However, due to strong multiple scattering, the apparent "cross section" varies by >1000% as the $H_2O$ thickness increases from 1 to 10 mm.

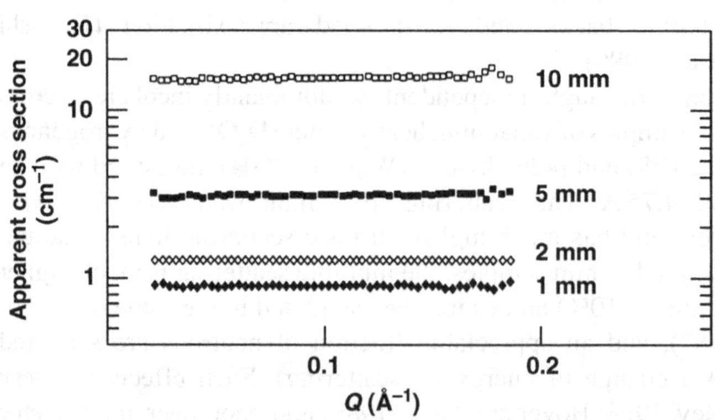


**Figure III.1.5** Apparent cross sections from 1, 2, 5, and 10 mm thick samples of water ($H_2O$).

A further complication is that the cross section of a hydrogen atom is dependent on the incident wavelength and other factors. The cross sections and scattering lengths quoted for hydrogen normally refer to bound protons and neglect inelastic effects arising from interchange of energy with the neutron. For coherent scattering, which is a collective effect arising from the interference of scattered waves over a large correlation volume, this approximation is reasonable, especially at low $Q$ where recoil effects are small. However, for incoherent scattering, which depends on the uncorrelated motion of individual atoms, inelastic effects become increasingly important for longer wavelengths, with the result that the H incoherent cross section is a function of both the incident neutron energy and the sample temperature (Maconnachie, 1984). Thus, the theoretical cross section cannot simply be used to calculate the background because the hydrogen incoherent cross section ($\sigma_{inc} = 79.7 \times 10^{-24}\,cm^2$), although widely quoted in the literature (Bacon, 1975), almost never applies to real polymer systems (Wignall, 1987, 2006).

For example, the effective incoherent cross section $\sigma_{inc}$ changes by about 30% for poly(methyl methacrylate) as $\lambda$ changes from 4.7 to 10 Å (Maconnachie, 1984). In addition, because of inelastic effects due to torsion, rotation, and vibration, the effective incoherent cross section is a function of the particular chemical group (methyl, hydroxyl, etc.) in which the proton is situated (Coyne and Wu, 1989). These complications are illustrated in Figure III.1.6, which shows the total cross section for a water molecule as a function of the incident wavelength. For shorter wavelengths, the energies associated with the vibrational and translational motion of water molecules are negligible and the cross section approaches the sum of the free atom cross sections ($\sigma_{FREE}$). However, as $\lambda$ increases, the scattering does not plateau at the bound atom cross section ($\sigma_{BOUND} \sim 167 \times 10^{-24}\,cm^2$) and varies continuously with energy (Wignall, 1987), changing appreciably over the range of wavelengths normally used in SANS studies ($5 < \lambda < 20$ Å).

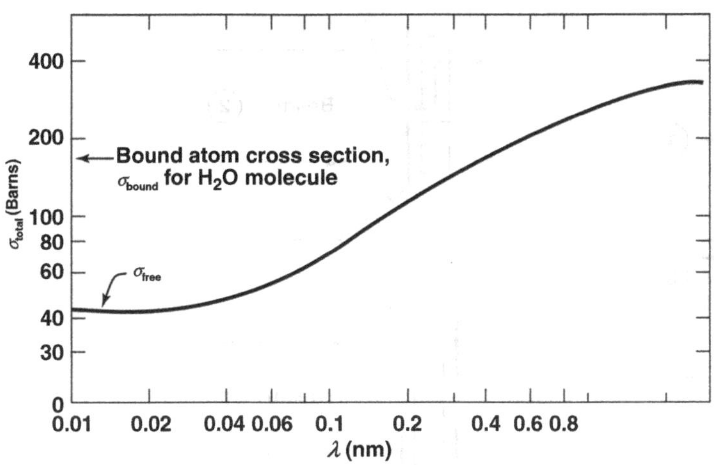


**Figure III.1.6** Total cross section for water molecule ($H_2O$) versus neutron wavelength, $\lambda$ (nm), at $T = 293$K according to Brookhaven National Laboratory Tables (Garber and Kinser, 1976).

As mentioned above, the spatial variation of the detector efficiency (ε) is usually measured via the scattering from $H_2O$ or a protonated polymer, and Figures III.1.4 and III.1.5 show that despite the fact that multiple scattering in such materials is not fully understood, the data for predominantly protonated materials ($H_2O$, polymethyl methacrylate, polystyrene, etc.) are independent of angle. Thus, to a good approximation, the variation in the measured signal is proportional to the detector efficiency and may be used in the data analysis software to correct this effect on a cell-by-cell basis. Second-order corrections representing departures from truly isotropic scattering and unequal path lengths through different regions of the active gas (e.g., $He^3$) are usually wavelength, instrument, and even detector dependent (May et al., 1982), and such adjustments may be customized for a particular facility (e.g., Lindner et al., 1992).

The fact that predominantly hydrogenous materials have such strong multiple scattering means that the actual "incoherent" level measured for such systems depends on the slit configuration. This is illustrated in Figure III.1.7, which shows an incident beam with different slit geometries. Even though the diameter of the

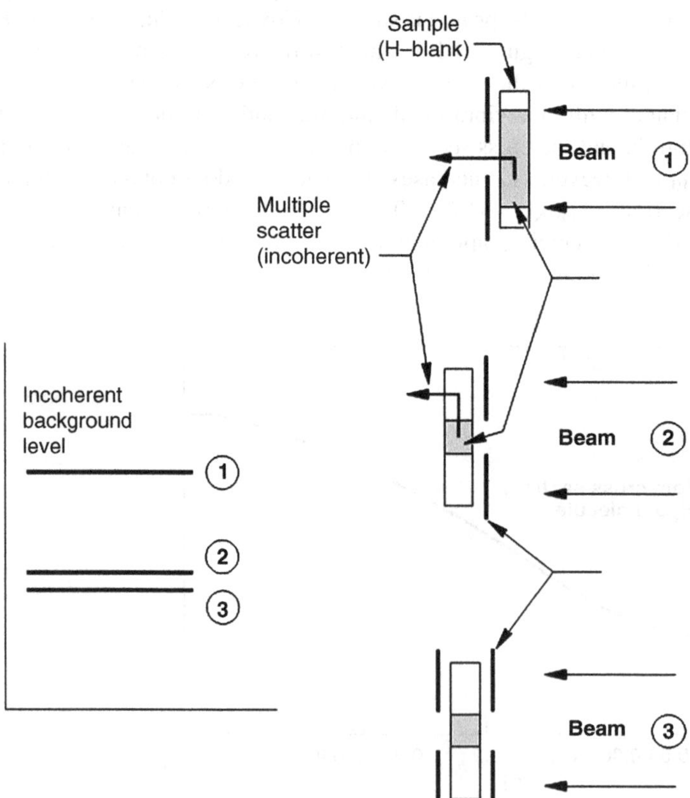


**Figure III.1.7** Effect of slit geometry on incoherent background level for H blanks.

smallest slit is the same for all three slit geometries, different sample volumes are irradiated and produce different levels of flat "incoherent" scattering. In geometry, the incident beam irradiates the largest sample volume (indicated via the shaded portion), and thus multiple scattering from this volume can enter the slit area from volumes larger than the slit dimensions, as indicated. Similarly, for case , single scattering can deflect neutrons to the unirradiated volume as the beam exits and when these neutrons are rescattered, they can also enter the detector as an apparent "incoherent" component. In case , the slits on the input and exit beams minimize the apparent backgrounds produced by multiple scattering, as illustrated schematically on the left-hand side of Figure III.1.7. As mentioned above, measurements at $Q$ values in the range $0.1 < Q < 1\,\text{Å}^{-1}$ are particularly sensitive to the incoherent background, which can be of the same order of magnitude as the coherent signal, so such effects should be considered when choosing the appropriate background to subtract. A more detailed description of the evaluation of incoherent background from soft matter has been given by Shibayama et al. (2005).

## III.1.5 THE EFFECT OF SAMPLE THICKNESS

As indicated in Figure III.1.3, for predominantly incoherent scatterers such as $H^1$-labeled "blanks", the data contain appreciable multiple scattering, and the use of eq. (III.1.3) only leads to a true cross section in the limit of very thin (~0.1 mm) samples. For samples with thicknesses more typical of actual polymer samples, empirical methods must be used to estimate this background (Hayashi et al., 1976; Dubner et al., 1990). Similarly, for partially labeled soft matter samples, containing appreciable amounts of hydrogen, there will be a flat incoherent background that will need to be subtracted to measure the coherent component of the scattering. For example, the scattering from blends of monodisperse deuterated and protonated molecules is well known as a function of the polymerization index $N$ (assumed to be the same for both labeled and unlabeled molecules) and the coherent cross section is given (Wignall, 1987; Wignall and Bates, 1987) by

$$\frac{d\Sigma(Q)}{d\Omega} = V^{-1} N \varphi_D \varphi_H (a_H - a_D)^2 P(Q), \qquad \text{(III.1.4)}$$

where $\varphi_D$ and $\varphi_H$ are the volume fractions of labeled (deuterated) and unlabeled polymer, $a_D$ and $a_H$ are the coherent scattering amplitudes of labeled and unlabeled monomer (repeat) units, $V$ is the segment volume, and $P(Q)$ is the form factor of the polymer molecule. For atactic polystyrene, the Debye Gaussian coil model (Flory, 1949) has been shown to give a good description of the scattering up to $Q \sim 0.4\,\text{Å}^{-1}$:

$$P(Q) = 2(R_g^2 Q^2 + \exp(-R_g^2 Q^2) + \exp(-R_g^2 Q^2) - 1)/(R_g^4 Q^4), \qquad \text{(III.1.5)}$$

where $R_g$ is the radius of gyration of the chain.

Eq. (III.1.5) has been shown to hold for all $X_D$ with a maximum scattered intensity at $\varphi_D = \varphi_H = 0.5$ (Wignall et al., 1981), and at this level of labeling, incoherent background forms only a small perturbation ($d\Sigma/d\Omega \sim 0.5\,cm^{-1}$) to the coherent scattering ($d\Sigma/d\Omega(0) = 74\,cm^{-1}$) for $N \sim 10^3$. For such labeled soft matter samples with $d\Sigma/d\Omega(0) \sim 10^2$–$10^3\,cm^{-1}$, coherent–coherent multiple scattering has been shown to be negligible (Goyal et al., 1983), and similarly for hard matter systems with similar cross section (Schelten and Schmatz, 1980). For example, aluminum single crystals with small percentages (0.4–0.8%) of voids produced by neutron irradiation have been widely used for absolute calibration (Wignall, 1987). At low $Q$ ($Q < 0.014\,Å^{-1}$) the scattering is isotropic and obeys the well-known Guinier formula (Guinier and Fournet, 1955). A typical Guinier plot of $\ln(d\Sigma/d\Omega(Q))$ versus $Q^2$ is shown in Figure III.1.8 with $d\Sigma/d\Omega(0) = 227\,cm^{-1}$ and $R_g = 223$ Å. The incoherent cross section of Al is negligible, so such samples are predominantly coherent scatterers, for which calculations indicate that multiple scattering is negligible (Schelten and Schmatz, 1980). For soft matter samples with similar cross sections, multiple coherent–coherent scattering is also negligible and the scattering is proportional to the thickness and the transmission. Thus, eq. (III.1.3) may be used to collapse data from samples with different thicknesses and transmissions to the same cross section as indicated in Figure III.1.9 for samples of polyvinyl alcohol gels, where the thickness varies by a factor of 4, but both data sets superimpose (Kanaya et al., 1992). For materials with $d\Sigma/d\Omega(0) > 10^4\,cm^{-1}$, however, multiple coherent–coherent scattering can produce quantitative distortions of the data. This is illustrated in Figure III.1.10, which shows the cross section of a porous sedimentary rock, where the pore–rock interface may be described as a surface fractal (fractal dimension, $D_s = 2.82$; upper size limit $\zeta = 1.2\,\mu m$) over 3 orders of magnitude in the length scale and 10 orders of magnitude in intensity. Such a wide range of the

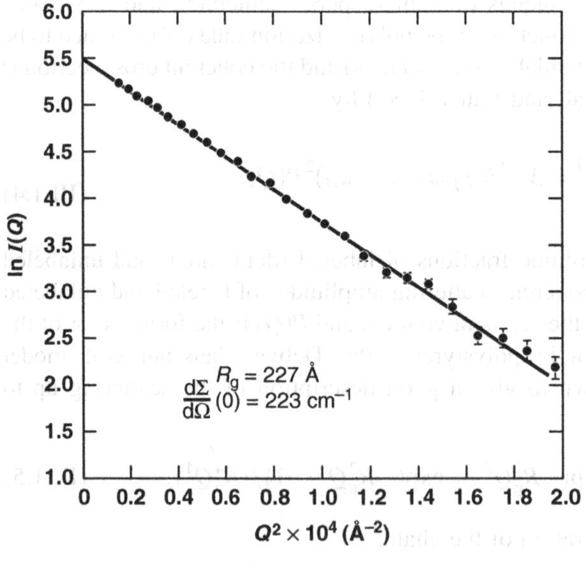


**Figure III.1.8** Guinier plot for irradiated aluminum (Al-8).

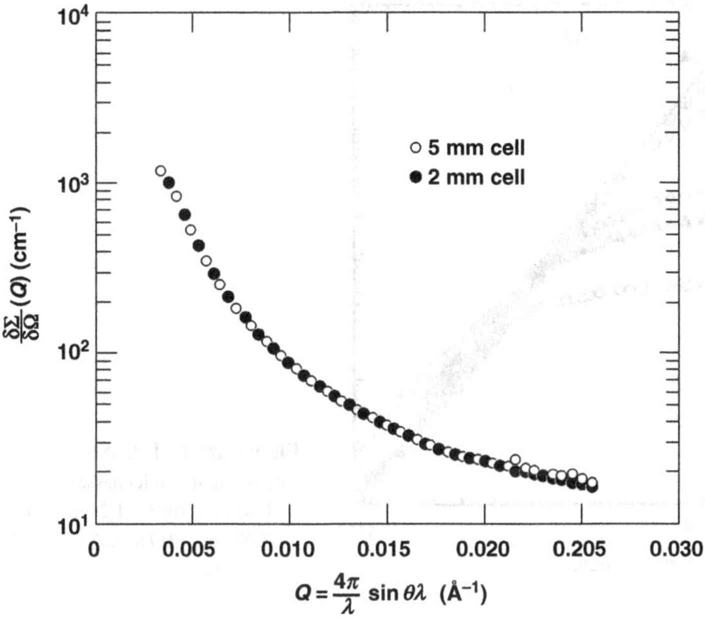


**Figure III.1.9** $d\Sigma/d\Omega(Q)$ versus $Q$ for polyvinyl alcohol gel in cells with different path lengths.

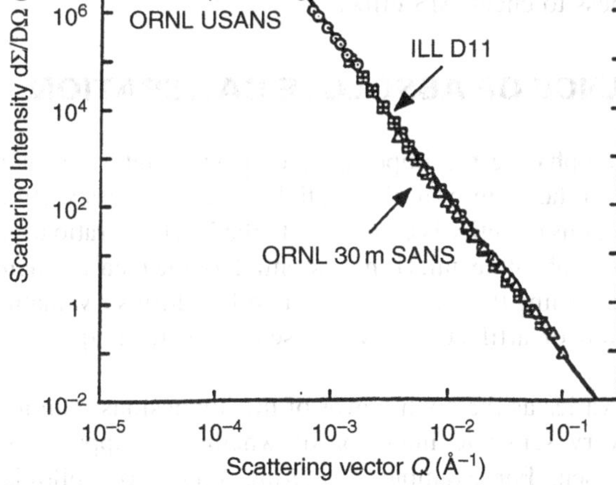


**Figure III.1.10** Overlap of ORNL/ILL data from sedimentary rock showing that the pore–rock interface is a surface fractal ($D_S = 2.82$) over ten orders of magnitude in $d\Sigma/d\Omega$ (Radlinski et al., 1999).

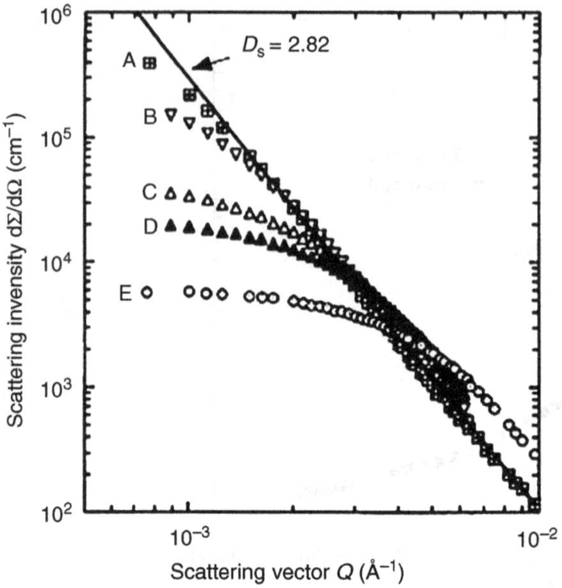


**Figure III.1.11** SANS data for various sample thicknesses: (a) $t = 0.63$ mm, (b) $t = 1.2$ mm, (c) $t = 3.09$ mm, (d) $t = 4.23$ mm, and (e) $t = 7.4$ mm.

scattering vector, $Q$, was achieved by combining data from two different SANS facilities and also an ultrasmall-angle neutron scattering (USANS) instrument (Radlinski et al., 1999). To check possible multiple scattering effects, measurements were performed on samples of several thicknesses (Figure III.1.11). The data exhibit pronounced MS effects that are particularly misleading in the small $Q$ region, where there is a similarity between the saturation caused by MS and the flattening out due to the finite size of fractal scatterers. In this study, it was possible to thin down rock samples until MS became irrelevant, as indicated in Figure III.1.11 showing SANS data for various sample thicknesses at an incident wavelength of $\lambda = 14$ Å. The data indicate that for thicknesses ~0.1 mm, MS effects are absent and thus for studies of samples with similar cross sections, it is recommended that experiments be performed as a function of thickness to check MS effects.

## III.1.6 THE IMPORTANCE OF ABSOLUTE CALIBRATION

The following examples will emphasize the importance of placing intensity data on an absolute scale, typically in the form of a differential scattering cross section $d\Sigma/d\Omega(Q)$. As mentioned above, this quantity is equivalent to the Rayleigh ratio used in light scattering. While the use of absolute units is not essential for the measurement of spatial dimensions (e.g., determining the $R_g$ of a polymer coil), it forms a valuable diagnostic tool for the detection of artifacts, to which scattering techniques are sometimes vulnerable.

Because the cross section varies as the sixth power of the dimensions (Guinier and Fournet, 1955), it is a very sensitive indicator of whether an appropriate structural model has been chosen. For example, scattering studies of colloidal micellar solutions may be modeled by core–shell spherical micelles as a function of a set of parameters describing particle structure and interactions [4]. On an arbitrary intensity scale, Hayter and Penhold (1983) have pointed out that it is possible to produce excellent fits of the particle *shape* that may be in error by as much as 3–4 orders of magnitude in intensity. Thus, absolute calibration allows such artifacts to be recognized, and the model parameters may be restricted to those that reproduce the observed cross section. This behavior illustrates the point referred to above that the intensity is extremely sensitive to the particle or molecular dimensions and even an approximate (~25%) absolute calibration is sufficient to reveal the presence of such artifacts.

Similarly, absolute SANS measurements of melt-crystallized blends of $H^1$- and $D^2$-labeled polyethylene (PEH and PED) showed that the scattering could also exceed the expected intensity for randomly mixed molecules by orders of magnitude. This indicated that some kind of aggregation phenomenon was taking place (Schelten et al., 1977). Figure III.1.12 shows a Zimm plot ($d\Sigma/d\Omega^{-1}$ versus $Q^2$) of the SANS differential scattering cross section for 6.0 wt% of PED in a matrix of PEH after quenching from the melt, and the polymerization index calculated from the extrapolated cross section ($d\Sigma/d\Omega(0) = 28.0 \pm 2\,\mathrm{cm}^{-1}$) is $N = 1600$, which is of the same order as the value from gel permeation chromatography (Wignall, 2004). However, when the same sample is slowly cooled from the melt (Figure III.1.13), the extrapolated cross section increases by over an order of magnitude, along with the "apparent" molecular weight. It is clear that these data do not originate in the scattering from single molecules, and it has been shown that the excess intensity is caused by aggregation or clustering of the labeled molecules (Schelten et al., 1977; Wignall, 2004), though this would not be clear if the data were in arbitrary units. This behavior illustrates the point referred to above that the intensity is extremely sensitive to the particle or molecular dimensions and even an approximate absolute calibration is sufficient to reveal the presence of such artifacts.

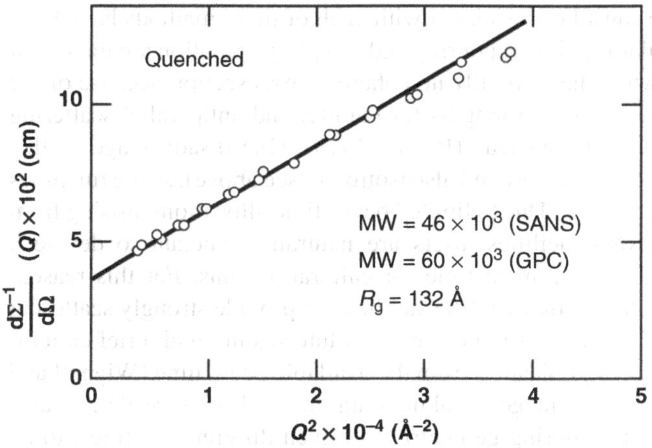


**Figure III.1.12** Zimm plot for 6 wt% PED molecules in PEH matrix *quenched* from the melt.

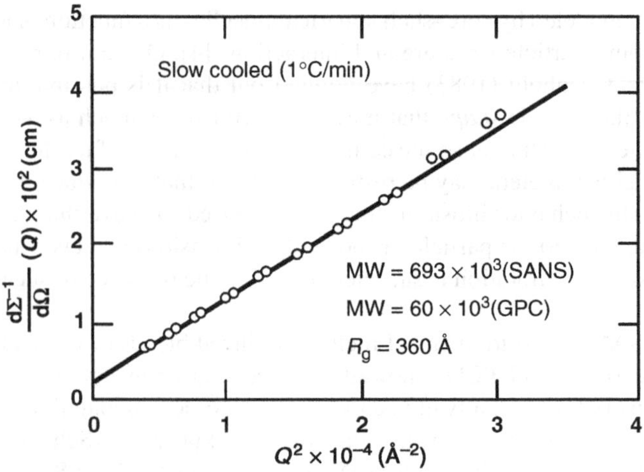


**Figure III.1.13** Zimm plot for 6 wt% PED molecules in PEH matrix *slow cooled* (1°C/min) from the melt.

In view of the maturity of the SANS technique, it is surprising that data are still published in arbitrary units that are functions of the timescale of the experiment and/or the sample thickness. Referring to eq. (III.1.3), absolute calibration reduces to measuring the constant $K_N = \varepsilon I_0 \Delta a$, which may be determined by comparison with a standard of known cross section run in the same scattering geometry for the same time (Wignall and Bates, 1987). If an incident beam intensity monitor is employed, as is normally the case, comparisons are made for the same number of monitor counts, that is, the same number of incident neutrons. Various calibration measurements have been used to measure the calibration constant, including direct measurement of the beam flux, calibration via a predominantly incoherent scattering material (e.g., vanadium or water), monodisperse blends of $H^1$- and $D^2$-labeled homopolymers, and various other standards.

Specific factors that must be considered with each of these methods have been discussed, and in particular multiple scattering and sample preparation are important when using vanadium, which has virtually no coherent cross section because of the fortuitous combination of scattering lengths for parallel and antiparallel scattering with respect to the spin of the nucleus (Bacon, 1975). One disadvantage of this standard is that the cross section is low and also isotropic (see above), so the run times for calibration are relatively long. Due to limited beam-time allocations, arising from the high demand for SANS facilities, users are naturally reluctant to devote a significant fraction of their instrument time for calibration runs. For this reason, it has been a matter of policy at many SANS facilities to provide strongly scattering precalibrated samples to allow users to perform absolute scaling with brief calibration runs that do not detract significantly from the available beam time (Wignall and Bates, 1987). A significant advantage of taking data on an absolute scale is that it allows data from different scattering geometries, or even different facilities, to be superimposed on the same scale (Figure III.1.10).

To the author's knowledge, the use of such precalibrated samples was pioneered by Schelten (personal communication, 1973) at the FRJ-2 SANS facility, where an isotropically scattering polyethylene (Lupolen®) standard was calibrated against vanadium. As described previously, Figure III.1.4 compares the angle-independent, predominantly incoherent, cross sections of thin (~1 mm) samples of vanadium with normal ($H^1$-labeled) polystyrene and polyethylene, both of which have an order of magnitude higher signal, thus shortening the time for calibration runs. However, even with the higher cross section of protonated polymers, such isotropic scatterers cannot be used at low $Q$ values (long sample–detector distances, $r$) as the intensity falls as $1/r^2$, and standardization involves a measurement at a low sample–detector distance, followed by scaling to the $r$ value of the measurement via the inverse square law (Schelten, personal communication, 1973).

The scattering from light water ($H_2O$) is an order of magnitude higher than that of vanadium, and therefore this system has the advantage of higher intrinsic scattering for calibration purposes and hence has lower sensitivity to statistical errors and artifacts than vanadium. One disadvantage is that for 1–2 mm samples, the multiple scattering is much higher (>30%) than that for vanadium (~10%) and cannot be calculated to the same degree of accuracy because an appreciable fraction of neutrons are scattered inelastically. As explained above, such effects are difficult to model and the use of eq. (III.1.3) leads to apparent cross sections that are functions of wavelength and are also detector dependent. In spite of this, such samples may still be used for calibration, provided the thickness is minimized (~1 mm) and they are calibrated against primary standards for a given instrument to take advantage of the intrinsically high signal-to-noise ratio for light water samples (Jacrot, 1976; May et al., 1982; Wignall, 1987; Lindner et al., 2000).

## III.1.7 ISOTOPE EFFECTS

SANS studies of deuterium-labeled polymers were initially based on the assumption that the molecular configurations and interactions are independent of deuteration or, alternatively, that the Flory–Huggins interaction parameter between labeled and unlabeled segments of the same species, $\chi_{HD}$, is zero. However, it was subsequently discovered that isotopic substitution *can* influence polymer thermodynamics, as deuterated and protonated polyethylene exhibit melting temperatures differing by ~6°C, so their mixtures can segregate in the solid state due to differential crystallization effects (Schelten et al., 1977; Bates et al., 1987). Also, the theta temperature ($T_\Theta$) of polystyrene solutions was shown to depend on the isotopic constitution of both polymer and solvent (Strazielle and Benoît, 1975), and the critical temperature of polystyrene–polyvinylmethylether blends depends on the PS isotope employed (Yang et al., 1983). Thus, isotopic labeling may influence phase transitions and Buckingham and Hentschel (1984) suggested that this might arise from a finite interaction parameter between the segments of the $H^1$- and $D^2$-labeled species ($\chi_{HD} \sim 10^{-4} - 10^{-3}$). Subsequently, SANS was used to measure $\chi_{HD}$ for a range of isotopic mixtures (Bates and Wignall, 1986a, 1986b; Bates et al. 1985, 1988a,

1988b; Schwahn et al., 1989; Lapp et al., 1985; Londono et al., 1994). For a blend of two isotopic polymer species (A and B), one of which is deuterium labeled, the coherent cross section (after subtracting the coherent and incoherent backgrounds) is given (Wignall, 1987, 2004) by

$$\frac{d\Sigma(Q)}{d\Omega} = V^{-1}(a_H - a_D)^2 S(Q), \qquad \text{(III.1.6)}$$

where the segment volume ($V$) is assumed to be the same for each isomer. $S(Q)$ is the structure factor, which contains information regarding both molecular architecture and thermodynamic interactions. In the mean field random phase approximation (RPA), $S(Q)$ is given by (deGennes, 1979)

$$S^{-1}(Q) = (\varphi_A N_A P_A(Q, R_{gA}))^{-1} + ((1-\varphi_A) N_B P_B(Q, R_{gB}))^{-1} - 2\chi_{HD}, \qquad \text{(III.1.7)}$$

where $\varphi_A$ is the volume fraction of the A species ($\varphi_B = 1 - \varphi_A$) and $R_{gA}$, $R_{gB}$; $N_A$, $N_B$; $P_A(Q)$, $P_B(Q)$ are, respectively, the radii of gyration, polymerization indices, and single chain form factors of the two species. The intrachain functions, $P_A(Q)$ and $P_B(Q)$, are represented by Debye functions (eq. (III.1.5)), based on the assumption of a Gaussian distribution of chain elements.

Eqs. (III.1.6) and (III.1.7) may be generalized to the case where the species are chemically different (with unequal segment volumes, $V_A$ and $V_B$) and applied to polymer blends (Wignall, 2004). However, when applied to isotopic mixtures, the $H^1$- and $D^2$-labeled molecules may be regarded as different "species" with the same segment volume ($V$) and volume fractions, $\varphi_A = \varphi_H$ and $\varphi_B = \varphi_D$. The RPA (eq. (III.1.7)) may then be fitted to the data with $\chi_{HD}$ as the only adjustable parameter (Bates and Wignall, 1986a, 1986b; Bates et al. 1985, 1986, 1988a, 1988b; Londono et al., 1994). Measurements on polybutadiene (Bates et al., 1985, 1986, 1988a, 1988b), polystyrene (Schwahn et al., 1989), polybutene (Bates et al., 1986a, 1986b), polyethylene (Londono et al., 1994), and polydimethylsiloxane (Lapp et al., 1985) confirm the existence of a universal isotope effect, arising from small differences in volume and polarizability between C–$H^1$ and C–$D^2$ bonds. Table III.1.1 lists typical values of the isotopic interaction parameter in the range $0.2 < \varphi_D < 0.8$, where $\chi_{HD}$ has been shown to be relatively independent of concentration (Londono et al., 1994).

**Table III.1.1** Typical Isotopic Interaction Parameters for Various Polymers (Wignall, 2004)

| Polymer | $T$ (°C) | $\chi_{HD}$ ($\times 10^4$) |
|---|---|---|
| Polystyrene | 160 | 1.8 |
| 1,4-Polybutadiene | 50 | 7.2 |
| 1,2-Polybutadiene (polyvinyl ethylene) | 47 | 6.8 |
| 1,2-Polybutene (polyethyl ethylene) | 47 | 8.8 |
| Polydimethylsiloxane | 296 | 17 |
| Polyethylene | 160 | 4.0 |

The above results raise the important question of how SANS studies are influenced by isotope effects. The initial SANS experiments on polymers relied on analogies with LS, where the limit of zero concentration was required to eliminate interchain scattering. Under such conditions, the isotope effect contributes almost insignificantly to the intensity (Wignall, 2004). Upon recognizing that information on chain dimensions could also be obtained from concentrated isotopic mixtures (e.g., eq. (III.1.4)), many experiments were conducted with higher labeling levels to enhance the intensity, and it is under these conditions that isotope-induced segregation effects are manifested. In the condensed state, many of the systems studied are solids at room temperature and have been exposed for only a limited time in the liquid state, for example, during melt pressing. For polybutadiene, with a glass transition temperature below −90°C, isotopic blends are liquid at room temperature, and this facilitates the attainment of equilibrium. Hence, isotope effects can be particularly dramatic in this system, and Figure III.1.14 shows the scattering cross section of mixtures of deuterated ($N_D = 4600$) and protonated ($N_H = 960$) polybutadienes as a function of temperature (Bates et al., 1985). It can be seen that the extrapolated zero $Q$ cross section exceeds by large factors the value it would have ($\sim 100\,\mathrm{cm}^{-1}$) if the $H^1$–$D^2$ interactions were negligible. The excess scattering is caused by the proximity of a phase boundary, and for sufficiently high molecular weight, this system will even phase separate (Bates and Wignall, 1986a, 1986b), as will other isotopic mixtures such as polyethylene (Londono et al., 1994). Thus, it is prudent to evaluate future experiments, based on measured values of $\chi_{HD}$ (Table III.1.1), and to check for excess scattering. This is best accomplished by calibrating data on an absolute scale and comparing the measured and theoretical intensities. Some examples of how to make such comparisons are given by Wignall (2006).

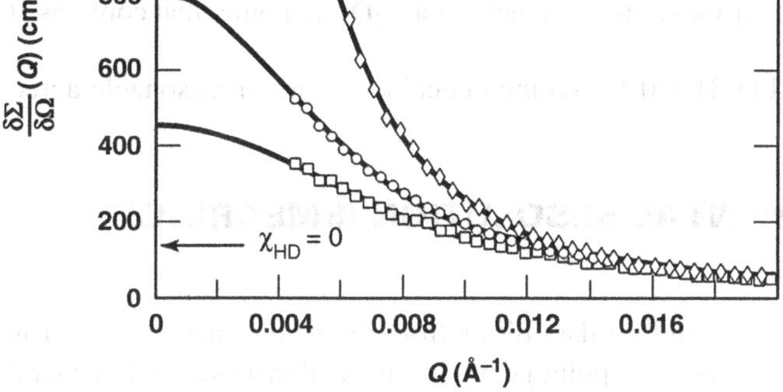


**Figure III.1.14** $d\Sigma/d\Omega(Q)$ versus $Q$ for blend of 69 vol% protonated and 31% deuterated 1,4-polybutadiene at the critical composition (Bates et al., 1985).

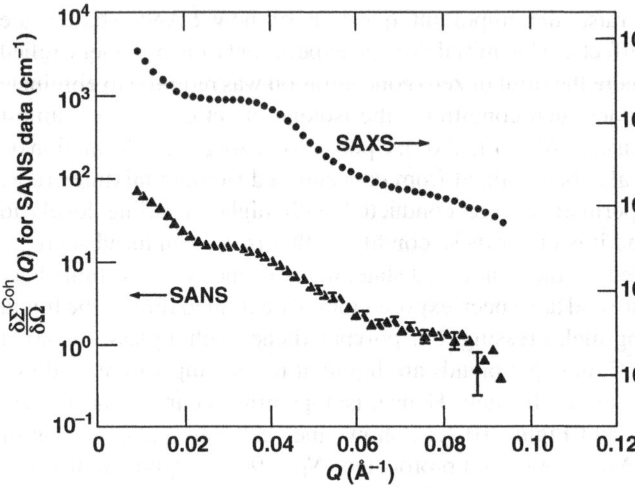


**Figure III.1.15** $d\Sigma/d\Omega(Q)$ versus $Q$ for deuterated polyethylene sample after subtraction of incoherent background.

When both SAXS and SANS data are taken from the same sample, it is sometimes possible to calculate the ratio of the two cross sections, or to scale the scattering with a particular type of radiation (e.g., SAXS) from a measurement with another incident radiation. For example, Figure III.1.15 shows the SAXS and SANS signals from a sample of deuteropolyethylene (Wignall, 2006). For fully deuterated material, the SANS incoherent cross section is small, and the system is essentially a two-phase system that has virtually identical SAXS and SANS cross sections apart from a scale factor. Figure III.1.15 shows absolute SAXS and SANS data for the same sample of PED that should scale as the ratio of the electron density to the scattering length density, and the ratio, $R$, of the two cross sections is given by

$$R = \frac{(0.282 \times 10^{-12} \times 16)^2}{(4.00 \times 10^{-12})^2} = 1.27, \qquad \text{(III.1.8)}$$

where $r_T = 0.282 \times 10^{-12}$ cm is the Thompson scattering factor of one electron, and $4.00 \times 10^{-12}$ cm is the neutron scattering length of a $C_2D_4$ monomer that contains 16 electrons.

Thus, the measured $(1.31 \pm 0.1)$ and theoretical ratios are in reasonable agreement (Wignall, 2006).

## III.1.8 INSTRUMENTAL RESOLUTION (SMEARING) EFFECTS

Experimentally measured scattering data differ from the actual (theoretical) cross sections because of departures from point geometry in a real instrument. In general, instrumental resolution effects are smaller for SANS than for SAXS. This is because most SANS experiments are performed in point geometry, whereas a significant proportion of X-ray experiments have used long slit sources (e.g., Kratky cameras), where smearing effects are larger, particularly at small angles (Schmidt, 1970, 1988; Glatter, 1982; Wignall et al., 1988). Less attention has been paid to resolution effects in SANS experiments, largely because the corrections are in general smaller for point geometry. However, the corrections are not always negligible, particularly for sharply varying scattering patterns and large scattering dimensions.

In a pinhole SANS instrument (Figure III.1.1), there are essentially three contributions to the smearing of an ideal curve: (a) the finite angular divergence of the beam, $\Delta\theta/\theta$, (b) the finite resolution of the detector, $R(Q)$, and (c) the polychromatic nature of the beam, $\Delta\lambda/\lambda$. For many systems, the scattering is azimuthally symmetric about the incident beam; that is, $d\Sigma/d\Omega(Q)$ is a function only of the magnitude of the scattering vector $|\mathbf{Q}| = 4\pi\lambda^{-1}\sin\theta$. In this case, once the instrumental parameters are well characterized, it is possible by numerical techniques not only to smear a given ideal scattering curve, but also to desmear an observed pattern by means of an indirect Fourier transform (IFT) to obtain the actual $Q$ dependence (Glatter, 1977, 1982; Moore, 1980; Ramakrishnan, 1985). Where the assumption of azimuthal symmetry cannot be made, the above smearing and desmearing procedures are not applicable, and alternative procedures based on Monte Carlo (MC) techniques have been developed that simulate the experimental smearing of a given theoretical scattering pattern that can be expressed analytically or numerically (Wignall et al., 1988). This procedure permits the estimation of resolution effects even in anisotropic systems, but cannot facilitate the desmearing of the observed pattern. Taken together, MC and IFT methods permit a realistic evaluation of the circumstances where resolution effects warrant correction. Both procedures have been illustrated via a range of results of experiments that have been performed on a typical pinhole SANS facility (Wignall et al., 1988), where it was shown that for experiments with scattering dimensions <200 Å, smearing effects are small (<5%). For example, the Guinier radius of monodisperse voids in an Al single crystal (Figure III.1.8) may be resolved without correcting for smearing effects when the incident wavelength spread $\Delta\lambda/\lambda < 10\%$ and the sample–detector and sample–source distances are sufficiently large (~10–20 m). Furthermore, dimensions up to ~1000 Å may be accurately resolved after proper evaluation of resolution effects. Smearing effects may be reduced by decreasing the wavelength range ($\Delta\lambda/\lambda$) or the angular spread ($\Delta\theta/\theta$), though the measured intensity is a strong function of the resolution and Schelten (1981) has pointed out that a reduction of a factor of 2 in $\Delta Q/Q$ will reduce the scattered intensity by over three orders of magnitude.

In addition to instrumental resolution effects, SANS data can also be smeared by integrating over a finite range of scattering dimensions in the system under study. For example, polymer latexes are generally spherical, with radii typically in the range $10^2$–$10^3$ Å and have a finite range of particle radii. Some such particle size distributions may be described (Wai et al., 1987; Fisher et al., 1988) by a zero-order logarithmic distribution (ZOLD), where the frequency of particles of radius $R$ is a function of the average size and the standard deviation, $\sigma$. Figure III.1.16 shows desmeared SANS data for a PS–PMMA latex, compared to the solid sphere scattering function (Rayleigh, 1911), using the ZOLD with an average diameter $D = 2R = 1008$ Å and

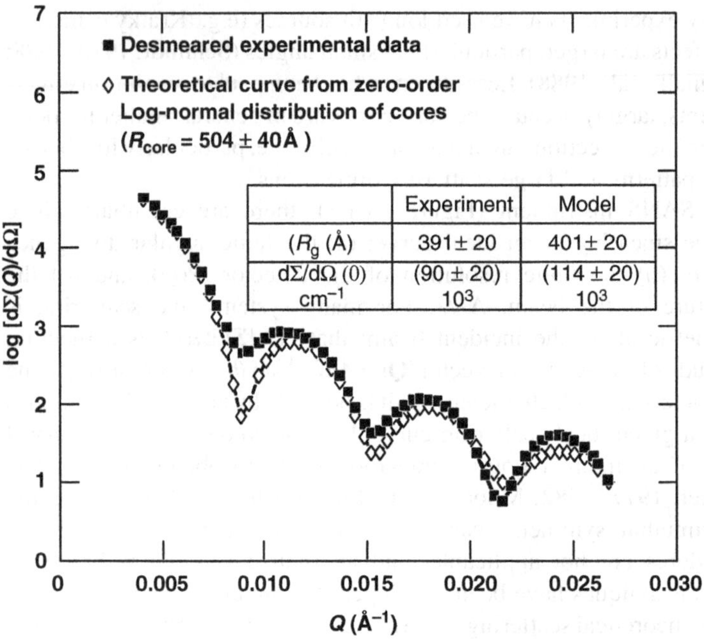


**Figure III.1.16** Comparison of experimental SANS data and theoretical scattering function for PS–PMMA core latexes in $D_2O$.

$\sigma = 92$ Å. In addition to the shape of the scattering envelope, the scattering intensity provides an independent check on the model if the data are measured in absolute units and the concentration of particles and the scattering length density are known. The measured absolute cross section and $R_g$ are shown in Figure III.1.16, along with those calculated from the latex dimensions determined independently by LS and transmission electron microscopy. Both parameters agree with the model within the experimental error, thus supporting the validity of this methodology.

## III.1.9 SOME OTHER POTENTIAL ARTIFACTS IN SANS AND SUGGESTIONS FOR FURTHER READING

Space limitations have precluded discussion of every potential artifact in the SANS technique and the above discussion has focused on the main possible errors. For readers interested in the possibility of exploring the above topics in more detail and investigating some of the subjects that were omitted, the following sources for further reading are suggested. Detailed discussions concerning the design and optimization of SANS instruments are given by Falcao et al. (1994), May (1994), and Schelten (1972). Fuller descriptions of instrumental smearing (resolution) effects are given by Pederson et al. (1990) and Schmidt (1970, 1988), and the effects of isotopic substitution on phase boundaries and melting points have been discussed by Shibayama et al. (1985), Bates et al. (1987), and Wignall et al., 2002 and 2006.

To date, the treatment of SANS data has been based on the assumption that the scattering is predominantly elastic and the data may be integrated over all energies to give the time-averaged structure of the system. However, few studies have been performed to check whether inelastic effects may be safely neglected, as currently assumed, and those that have been undertaken indicate that this may not always be the case. In particular, for soft matter, inelastic effects can be large even at low $Q$ values, as shown by Rennie and Ghosh (1990), who analyzed the energy spectrum of a beam of 12 Å neutrons, scattered from a variety of solvents and polymer solutions. The data indicated that less than half of the neutrons were scattered elastically as assumed by conventional SANS methodology. Strong inelastic effects have also been observed on pulsed facilities (Heenan et al., 1997), where they are particularly important as the assignment of the correct $Q$ values is based on the time of flight, which is strongly affected by changes in the neutron velocity on scattering. Newer detectors currently being installed on several facilities permit time-stamping, which gives the option of restricting the measured data to events corresponding to strictly elastic scattering, and allow systematic studies of inelastic effects in SANS to delineate more precisely the types of experiments where the "standard" methodology is appropriate.

## ACKNOWLEDGMENTS

The author wishes to thank Dr. J. Schelten for permission to use Figure III.1.2 and recognizes his many contributions to the developments of the SANS technique in Europe. Similarly, Dr. W. C. Koehler played an analogous role in the advancement of this technique in the United States, and together they laid much of the foundations for the application of SANS to soft matter.

## REFERENCES

Abele, R.K. Allin, G.W. Clay, W.T. Fowler, C.E., and Kopp, M.K. *IEEE Trans. Nucl. Sci.* **1981**, *NS 28*, 811.

Alexander, L.E. *X-Ray Diffraction Methods in Polymer Science*, Krieger Publishing Co. Inc., London, 1969.

Bacon, G.E. *Neutron Scattering*, Clarendon Press, Oxford, 1975, p. 48.

Bates, F.S. and Wignall, G.D. *Macromolecules* **1986a**, *19*, 932.

Bates, F.S. and Wignall, G.D. *Phys. Rev. Lett.* **1986b**, *57*, 1429.

Bates, F.S., Wignall, G.D., and Koehler, W.C. *Phys. Rev. Lett.* **1985**, *55*, 2425.

Bates, F.S. Dierker, S.B., and Wignall, G.D. *Macromolecules* **1986**, *19*, 1938.

Bates, F.S. Keith, H.D., and McWhan, D.B. *Macromolecules* **1987**, *20*, 3065.

Bates, F.S. Muthukumar, M. Wignall, G.D., and Fetters, L.J. *J. Chem. Phys.* **1988a**, *21*, 535.

Bates, F.S. Muthukumar, M. Wignall, G.D., and Fetters, L.J. *J. Chem. Phys.* **1988b**, *21*, 1086.

Boyer, W. and King, J. S. *J. Appl. Crystallogr.* **1988**, *21*, 818.

Buckingham, A.B. and Hentschel, H.E. *J. Polym. Sci., Polym. Phys. Ed.* **1984**, *18*, 853.

Bunn, C.W. *Trans. Faraday Soc.* **1935**, *35*, 482.

Copley, J.R.D. *J. Appl. Crystallogr.* **1988**, *21*, 639.

Coyne, L.D. and Wu, W.L. *Polym. Commun.* **1989**, *30*, 312.

DeGennes, P.G. *Scaling Concepts in Polymer Physics*, Cornell University Press, Ithaca, NY, 1979, Chapter 5.

DUBNER, W.S., SCHULTZ, J.M., and WIGNALL, G.D. *J. Appl. Crystallogr.* **1990**, *23*, 469.
FALCAO, J. *J. Appl. Crystallogr.* **1994**, *27*, 30.
FISHER, L.W. MELPOLDER, S.M. O'REILLY, J.M. RAMAKRISHNAN, V.R., and WIGNALL, G.D. *J. Colloid Interface Sci.* **1988**, *123*, 24.
FLORY, P.J. *J. Chem. Phys.* **1949**, *17*, 303.
GARBER, GLATTER, D.I. and KINSER, R.R. *Neutron Cross Sections*, BNL 325, 3rd ed., Brookhaven National Laboratory, Upton, NY, 1976.
GLATTER, O. *J. Appl. Crystallogr.* **1977**, *10*, 415.
GLATTER O. In: GLATTER O. and KRATKY O. (editors). *Small-Angle X-Ray Scattering*, Academic Press, London, 1982, p. 131.
GOYAL, P.S. KING, J.S., and SUMMERFIELD, G.C. *Polymer* **1983**, *24*, 131.
GUINIER, A. and FOURNET, G. *Small-Angle Scattering of X-Rays*, Wiley, New York, 1955.
HAYASHI, H. HAMADA, F., and NAKAJIMA, A. *Macromolecules* **1976**, *9*, 543.
HAYTER, J.B. and PENFOLD, J. *Colloid Polym. Sci.* **1983**, *261*, 1022.
HAYTER, J.B. and MOOK, H.A. *J. Appl. Crystallogr.* **1989**, *22*, 35.
HEENAN, R.K. PENFOLD, J., and KING, S.M. *J. Appl. Crystallogr.* **1977**, *30*, 1140.
IBEL, K. *J. Appl. Crystallogr.* **1976**, *9*, 196.
JACROT, B. *Rep. Prog. Phys.* **1976**, *39*, 911.
KANAYA, T. OHKURA, M. KAJI, K. FURUSAKA, M. MIAWA, M. YAMAOKA, H., and WIGNALL, G.D. *Physica* **1992**, *B120*, 549.
KIRSTE, R.G. KRUSE, W.A., and IBEL, K. *Polymer* **1975**, *16*, 120.
LAPP, A. PICOT, C., and BENOÎT, H. *Macromolecules* **1985**, *18*, 2437.
LINDNER, P. MAY, R.P., and TIMMINS, P. *Physica B* **1992**, *B180–181*, 967.
LINDNER, P. LECLERCQ, F., and DAMAY, P. *Physica B* **2000**, *291*, 152.
LONDONO, J.D. NARTEN, A.H. WIGNALL, G.D. HONNELL, K.G. JOHNSON, T.W., and BATES, F.S. *Macromolecules* **1994**, *27*, 2864.
MACONNACHIE, A. *Polymer* **1984**, *25*, 1068.
MAIER-LEIBNITZ, H., and SPRINGER, T. *Annu. Rev. Nucl. Sci.* **1966**, *16*, 207.
MAY, R.P. *J. Appl. Crystallogr.* **1994**, *27*, 298.
MAY, R.P. IBEL, K., and HAAS, J. *J. Appl. Crystallogr.* **1982**, *15*, 15.
MELNICHENKO, Y.B., and WIGNALL, G.D. *J. Appl. Phys.* **2007**, *102*, 021102.
MOORE, P.B. *J. Appl. Crystallogr.* **1980**, *13*, 168.
O'REILLY, J.M. TEEGARDEN, J.M., and WIGNALL, G.D. *Macromolecules* **1985**, *18*, 2747.
PEDERSON, J.S., POSSELT, D., and MORTENSEN, K. *J. Appl. Cryst.,* **1990**, *23*, 321.
RAMAKRISHNAN, V.R. *J. Appl. Phys.* **1985**, *18*, 42.
RADLINSKI, A.P. RADLINSKA, E.Z. AGAMALIAN, M. WIGNALL, G.D. LINDNER, P., and RANDL, O.G. *Phys. Rev. Lett.* **1999**, *82*, 3078.
RAYLEIGH, L. *Proc. R. Soc. Lond. A* **1911**, *84*, 24.
RENNIE, A., and GHOSH, R. *Inst. Phys. Conf. Ser.* **1990**, *107*, 233.
SCHELTEN, J. *Kerntechik* **1972**, *14*, 86.
SCHELTEN, J. In: CHEN, S.H. CHU, B., and NOSSAL, R. (editors). *Scattering Techniques Applied to Supramolecular and Nonequilibrium Systems*, NATO Advanced Study Series, Plenum Press, **1981**, *73*, 35.
SCHELTEN, J. and SCHMATZ, W. *J. Appl. Crystallogr.* **1980**, *13*, 385.
SCHELTEN, J. WIGNALL, G.D. LONGMAN, G.W., and BALLARD, D.G.H. *Polymer* **1977**, *18*, 111.
SCHMATZ, W. SPRINGER, T. SCHELTEN, J., and IBEL, K. *J. Appl. Crystallogr.* **1974**, *7*, 96.
SCHMIDT, P.W. *J. Appl. Crystallogr.* **1970**, *21*, 602.
SCHMIDT, P.W. *J. Appl. Crystallogr.* **1988**, *3*, 137.
SCHWAHN, D., HAHN, K., STREIB, J., and SPRINGER, T. *J. Chem. Phys.* **1989**, *93*, 8383.
SHIBAYAMA, M. YANG, H. STEIN, R.S., and HAN, C.C. *Macromolecules* **1985**, *18*, 2179.
SHIBAYAMA, M. NAGAO, G. OKABE, S., and KARINO, T. *J. Phys. Soc. Jpn.* **2005**, *74*, 2728.
STRAZIELLE, C. and BENOÎT, H. *Macromolecules* **1975**, *8*, 203.
TURCHIN, V.F. *Slow Neutrons*, Sivan Press, Jerusalem, 1965, p. 16.

WAI, M.P. GELMAN, R.A. FATICA, M.J. HOERL, R.H., and WIGNALL, G.D. *Polymer* **1987**, *28*, 918.

WIGNALL, G.D. *Encyclopedia of Polymer Science and Engineering*, Vol. 10, Wiley, New York, 1987, p. 112.

WIGNALL, G.D. *J. Phys. Condens. Matter* **1999**, *11*, R157.

WIGNALL, G.D.In: MARK, J.E. (editor). *Physical Properties of Polymers*, Cambridge University Press, 2004, p. 424, Chapter 7.

WIGNALL, G.D.In: MARK, J.E. (ed.). *Polymer Properties Handbook*, 2nd edition, Springer Verlag, 2006, p. 407.

WIGNALL, G.D. and BATES, F.S. *J. Appl. Crystallogr.* **1987**, *20*, 28.

WIGNALL, G.D. HENDRICKS, R.W. KOEHLER, W.C. LIN, J.S. WAI, M.P. THOMAS, E.L., and STEIN, R.S. *Polymer* **1981**, *22*, 886.

WIGNALL, G.D. CHRISTEN, D.K., and RAMAKRISHNAN, V.R. *J. Appl. Crystallogr.* **1988**, *21*, 438.

WIGNALL, G.D., BENOÎT, H., HASHIMOTO, T., HIGGINS, J.S., KING, S., LODGE, T.P., MORTENSEN, K., and RYAN, A.J. In: KAHOVEC, J. (editor). *Scattering Methods for the Investigation of Polymers*, Vol. 190, Wiley-VCH, Weinheim, 2002, p. 185.

YANG, H. HADZIIOANNOU, G. and STEIN, R.S. *J. Polym. Sci., Polym. Phys. Ed.* **1983**, *21*, 159.

# III Data Treatment and Sample Environment

## III.2 Structure Analysis

*Hideki Seto*

### III.2.1 PRINCIPLES

The basic equation for X-ray or neutron scattering is the Bragg's formula:

$$\sin\theta = \frac{n\lambda}{2d} = \frac{n\lambda}{4\pi}Q, \qquad \text{(III.2.1)}$$

where $\theta$, $\lambda$, $d$, and Q correspond to a scattering angle, wavelength of an incident neutron or X-ray beam, characteristic repeat distance of a structure, and wave number of the periodicity, respectively. In cases of investigating crystal structures, typical value of $d$ is an order of Å. Normally, the wavelength of X-ray for diffraction study is less than 1.6 Å and of neutron is between 1 and 10 Å, and the scattering angle $2\theta$ expands from about 10° to 180°. A diffraction profile from a crystalline structure typically displays a combination of the Bragg peaks distributing in a reciprocal space. Thus, the first mission of the "standard crystallography" is to explain positions and intensities of these Bragg peaks.

On the other hand, in cases of characteristic sizes of structures being between 10 and 1000 Å, Q is in the order of from $10^{-3}$ to $10^{-1}$ Å$^{-1}$ and $\theta$ is less than 1°. These large-scale objects are not always aligned regularly like atoms in crystal structures. A scattering intensity observed by SAS always has a diffusive profile; monotonic decrease of scattering intensity with increasing Q or only a broad peak is observed. In addition, an observed profile usually displays as a powder pattern, that is, no orientational anisotropy. In such cases, an azimuthal average will be calculated from an observed profile, and a one-dimensional scattering profile $I$(Q) should be explained assuming some models.

Because all the systems for small-angle scattering studies include a plenty of atoms in their characteristic repeat distances, materials to be studied by small-angle scattering are considered to be a continuous medium in contrast to the

*Neutrons in Soft Matter* Edited by Toyoko Imae, Toshiji Kanaya, Michihiro Furusaka, and Naoya Torikai

standard way of diffraction from a crystalline structure in which a material is looked upon as a periodic arrangement of atoms. A scattering amplitude density could be defined as

$$\rho(\mathbf{r}) = \frac{1}{v}\sum_{i \in v} b_i \delta(\mathbf{r}-\mathbf{r}_i), \tag{III.2.2}$$

where $v$ is a microscopic volume larger than an atomic volume, and $b_i$ and $\mathbf{r}_i$ are scattering amplitude and position vector of the $i$th atom, respectively. Self-correlation function of the scattering amplitude density $\rho(\mathbf{r})$ is calculated from

$$P_a(\mathbf{r}) = \frac{1}{V}\int \rho(\mathbf{r}')\rho(\mathbf{r}'+\mathbf{r})d\mathbf{r}' \equiv \langle \rho(0)\rho(\mathbf{r})\rangle, \tag{III.2.3}$$

which is the Patterson function. Scattering function can be obtained by a Fourier transform of $P_a(\mathbf{r})$ as

$$I(\mathbf{Q}) = \frac{1}{V}\left|\sum_i f_i \exp(i\mathbf{Q}\cdot\mathbf{r})\right|^2 = \frac{1}{V}\left|\int \rho(\mathbf{r})\exp(i\mathbf{Q}\cdot\mathbf{r})d\mathbf{r}\right|^2 = \int P_a(\mathbf{r})\exp(i\mathbf{Q}\cdot\mathbf{r})d\mathbf{r}, \tag{III.2.4}$$

where $V$ is the whole volume of the sample. In this definition, the form factor, that is, the Fourier transform of the scattering density distribution, is defined by

$$A(\mathbf{Q}) = \sum_i b_i \exp(i\mathbf{Q}\cdot\mathbf{r}) = \int \rho_i(\mathbf{r})\exp(i\mathbf{Q}\cdot\mathbf{r})d\mathbf{r}. \tag{III.2.5}$$

Therefore, the mission of the small-angle scattering data analysis is the discovery of a correct model of $\rho(\mathbf{r})$ to explain the experimentally observed scattering profiles.

## III.2.2 ISOLATED PARTICLE

A disperse system to be studied by small-angle scattering is regarded as a matrix with particles embedded in it. It appears that if all the particles in "solution" are identical, the total scattering intensity is proportional to the scattering intensity of a particle, averaged over all orientations. Thus, let us start from deriving basic relations dealing with small-angle scattering by an isolated particle.

If there are $n$ atoms with scattering length $f_i(\mathbf{Q})$, which may be regarded as constants in the small-angle region, $f_i(\mathbf{Q}) = f_i(0) = f_i$, and coordinates $\mathbf{r}_i$ within a particle, then one can write the averaged intensity as

$$I(\mathrm{Q}) = \sum_{i=1}^{n}\sum_{j=1}^{n} f_i f_j \frac{\sin(\mathrm{Q}r_{ij})}{\mathrm{Q}r_{ij}}, \tag{III.2.6}$$

where $r_{ij} = |\mathbf{r}_i - \mathbf{r}_j|$, while the sum is taken over all the atoms in a particle. This equation was first derived by Debye. Thus, it is called as "Debye equation." If a density distribution $\rho(\mathbf{r})$ is used and not an atomic set to describe a particle, then the Debye equation takes the form

$$I(Q) = \iint_V \rho(\mathbf{r}_1)\rho(\mathbf{r}_2)\frac{\sin(Qr_{12})}{Qr_{12}} d\mathbf{r}_1 d\mathbf{r}_2. \tag{III.2.7}$$

We shall see later that this form is appropriate for some applications.

Equation (III.2.4) for the scattering intensity by a fixed particle can be expressed in the form

$$I(\mathbf{Q}) = \int_v P_a(\mathbf{r})\cos(\mathbf{Qr})d\mathbf{r}, \tag{III.2.8}$$

where even function $P_a(\mathbf{r})$ corresponding to real function $\rho(\mathbf{r})$ was used and the cosine Fourier transform was applied. The small-angle scattering intensity is obtained by averaging $I(\mathbf{Q})$ over the solid angle in reciprocal space:

$$\begin{aligned} I(Q) &= \frac{1}{4\pi}\int_\Omega I(\mathbf{Q})d\Omega \\ &= \frac{1}{4\pi}\int_0^\infty \int_0^{4\pi} r^2 dr d\omega \int_0^{4\pi} P_a(\mathbf{r})\cos(\mathbf{Qr})d\Omega \\ &= \int_0^\infty r^2 dr \int_0^{4\pi} P_a(\mathbf{r})d\omega \frac{1}{4\pi}\int_0^{4\pi} \cos(\mathbf{Qr})d\Omega. \end{aligned} \tag{III.2.9}$$

Averaging $\cos(\mathbf{Qr})$ over all orientation yields

$$\langle\cos(\mathbf{Qr})\rangle = \int_0^\pi \cos[Qr\cos(\varphi)]\frac{\sin(\varphi)}{2}d\varphi = \frac{\sin(Qr)}{Qr}, \tag{III.2.10}$$

where $\varphi$ is the angle between vectors $\mathbf{Q}$ and $\mathbf{r}$, and $\sin(\varphi)d\varphi/2$ represents the probability of the angle lying inside the interval $(\varphi, \varphi + d\varphi)$. Thus, we obtain

$$I(Q) = 4\pi\int_0^\infty \gamma(r)\frac{\sin(Qr)}{Qr} r^2 dr, \tag{III.2.11}$$

where

$$\gamma(r) = \frac{1}{4\pi}\int_0^{4\pi} P_a(\mathbf{r})d\omega = \langle\rho(\mathbf{r}) \times \rho(-\mathbf{r})\rangle \tag{III.2.12}$$

is the correlation function of a particle, in other words, the averaged self-convolution of the density distribution. This function can be written by the inverse transform

$$\gamma(\mathrm{r}) = \frac{1}{2\pi^2}\int_0^\infty I(\mathrm{Q})\frac{\sin(\mathrm{Qr})}{\mathrm{Qr}}\mathrm{Q}^2\mathrm{dQ}. \tag{III.2.13}$$

If $D$ is the largest dimension in a particle, $\gamma(\mathrm{r}) = 0$ for $\mathrm{r} > D$ can be assumed and eq. (III.2.9) may be rewritten as

$$I(\mathrm{Q}) = 4\pi\int_0^D \gamma(\mathrm{r})\frac{\sin(\mathrm{Qr})}{\mathrm{Qr}}\mathrm{r}^2\mathrm{dr}(= |A(\mathrm{Q})|^2). \tag{III.2.14}$$

Both the function $\gamma(\mathrm{r})$ and the function

$$p(\mathrm{r}) = \mathrm{r}^2\gamma(\mathrm{r}) \tag{III.2.15}$$

are the distance distribution function, which are widely used in small-angle scattering. These functions depend both on a particle's geometry, expressing numerically the set of distances joining the volume elements within a particle, and on a particle's inner inhomogeneity distribution.

In case of a spherical particle with radius $R$ and the scattering amplitude $\rho_0$, the form factor $A(\mathrm{Q})$ can be calculated as

$$\begin{aligned}
A(\mathrm{Q}) &= 4\pi\rho_0\int_0^R \frac{\sin(\mathrm{Qr})}{\mathrm{Qr}}\mathrm{r}^2\mathrm{dr} \\
&= 4\pi\rho_0\int_0^{QR} \frac{\sin(x)}{x}\frac{x^2}{\mathrm{Q}^2}\frac{\mathrm{dx}}{\mathrm{Q}} \\
&= \frac{4\pi\rho_0}{\mathrm{Q}^3}\int_0^{QR} x\sin(x)\mathrm{dx} \\
&= \frac{4}{3}\pi R^3\rho_0\left(3\frac{\sin(\mathrm{Q}R)-\mathrm{Q}R\cos(\mathrm{Q}R)}{(\mathrm{Q}R)^3}\right),
\end{aligned} \tag{III.2.16}$$

with substituting $Qr = x$. Thus, the scattering intensity for an isolated spherical particle is given by

$$I(\mathrm{Q}) = \rho_0^2 v^2\left\{3\frac{\sin(\mathrm{Q}R)-\mathrm{Q}R\cos(\mathrm{Q}R)}{(\mathrm{Q}R)^3}\right\}^2, \tag{III.2.17}$$

where $v = (3/4)\pi R^3$ is the particle volume. The shape of this scattering function is shown in Figures III.2.1 (linear scale) and III.2.2 (semilogarithmic scale) as a function of $QR$.

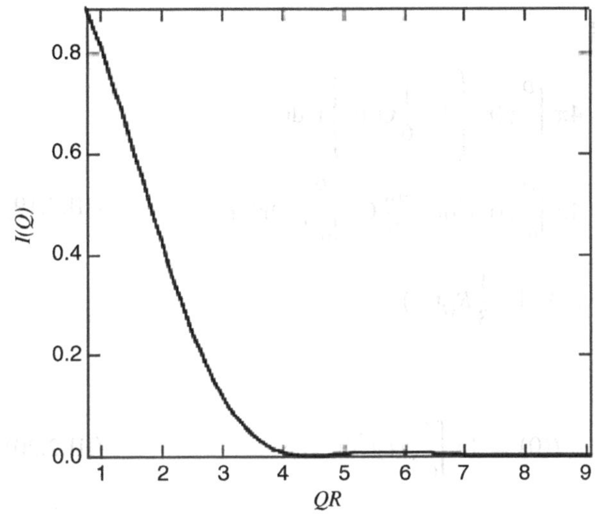


**Figure III.2.1** Small-angle scattering profile for a spherical particle.

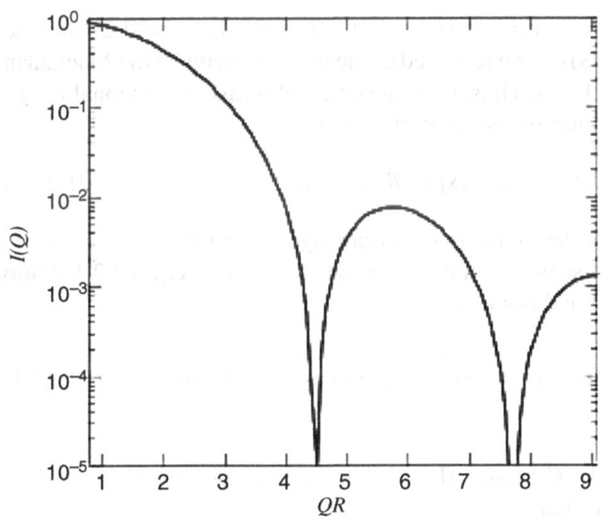


**Figure III.2.2** The same data as of Figure III.2.1 shown in semilogarithmic scale.

## III.2.3 GUINIER APPROXIMATION

To investigate the behavior of $I(Q)$ at very low angle, one can substitute the Maclaurin series,

$$\frac{\sin(Qr)}{Qr} = 1 - \frac{Q^2r^2}{6} + \frac{Q^4r^4}{120} - \ldots, \tag{III.2.18}$$

into eq. (III.2.14), obtaining

$$
\begin{aligned}
I(\mathrm{Q}) &= 4\pi \int_0^D \gamma(\mathrm{r})\left\{1-\frac{1}{6}\mathrm{Q}^2\mathrm{r}^2\right\}\mathrm{r}^2\mathrm{dr} \\
&= 4\pi \int_0^D \gamma(\mathrm{r})\mathrm{r}^2\mathrm{dr}-\frac{4\pi}{6}\mathrm{Q}^2\int_0^D \gamma(\mathrm{r})\mathrm{r}^4\mathrm{dr} \\
&= I(0)(1-\frac{1}{3}R_G^2\mathrm{Q}^2),
\end{aligned}
\tag{III.2.19}
$$

where

$$I(0) = 4\pi \int_0^D \gamma(\mathrm{r})\mathrm{r}^2\mathrm{dr} \tag{III.2.20}$$

and

$$R_\mathrm{G}^2 = \frac{1}{2}\int_0^D \gamma(\mathrm{r})\mathrm{r}^4\mathrm{dr} \Big/ \int_0^D \gamma(\mathrm{r})\mathrm{r}^2\mathrm{dr}, \tag{III.2.21}$$

in the vicinity of $\mathrm{Q}=0$ with restricting to the first two terms in eq. (III.2.18). The right-hand side of eq. (III.2.18) can be regarded as the first two terms of the Maclaurin series of function $\exp(-R_G^2\mathrm{Q}^2/3)$. Thus, to an accuracy of terms proportional to $\mathrm{Q}^4$, one can write for the beginning of the scattering curve

$$I(\mathrm{Q}) = I(0)\exp(-R_\mathrm{G}^2\mathrm{Q}^2/3). \tag{III.2.22}$$

This is the Guinier equation, derived about 70 years ago (Guinier, 1939). To relate parameters $I(0)$ and $R_\mathrm{G}$ to the structure of a particle, one substitutes eq. (III.2.18) into the Debye equation (III.2.7) and obtains

$$I(\mathrm{Q}) = \iint_V \rho(\mathbf{r}_1)\rho(\mathbf{r}_2)\mathbf{dr}_1\mathbf{dr}_2-\frac{\mathrm{Q}^2}{6}\iint_V \rho(\mathbf{r}_1)\rho(\mathbf{r}_2)|\mathbf{r}_1-\mathbf{r}_2|^2\mathbf{dr}_1\mathbf{dr}_2. \tag{III.2.23}$$

Comparing eq. (III.2.23) with eq. (III.2.19), one can deduce from the leading terms on the right-hand side that

$$I(0) = \left|\int \rho(\mathbf{r})\mathbf{dr}\right|^2, \tag{III.2.24}$$

which is just the square of the total particle scattering length. From the comparison of the coefficients of the $\mathrm{Q}^2$ terms, the following relation can be obtained:

$$R_G^2 = \frac{\iint_V \rho(\mathbf{r}_1)\rho(\mathbf{r}_2)|\mathbf{r}_1-\mathbf{r}_2|^2\mathbf{dr}_1\mathbf{dr}_2}{2\iint_V \rho(\mathbf{r}_1)\rho(\mathbf{r}_2)\mathbf{dr}_1\mathbf{dr}_2}. \tag{III.2.25}$$

If one places the origin of coordinates at the point $\mathbf{r}_0$, then simple calculations give

$$R_G^2 = \frac{\int \rho(\mathbf{r}-\mathbf{r}_0)|\mathbf{r}-\mathbf{r}_0|^2 d\mathbf{r}}{\int \rho(\mathbf{r})d\mathbf{r}} - \left[\frac{\rho(\mathbf{r}-\mathbf{r}_0)(\mathbf{r}-\mathbf{r}_0)d\mathbf{r}}{\rho(\mathbf{r})d\mathbf{r}}\right]^2. \quad \text{(III.2.26)}$$

It should be noted that the integrals in eq. (III.2.25) is independent of the actual position of the origin of coordinates, because both integrals are taken over the whole space. Hence, one can match point $\mathbf{r}_0$ with the center of mass of a particle. The second term in eq. (III.2.26) becomes zero, and thus

$$R_G^2 = \int_V \rho(\mathbf{r})r^2 d\mathbf{r} \Big/ \int_V \rho(\mathbf{r})d\mathbf{r}, \quad \text{(III.2.27)}$$

which is the radius of gyration of a particle about its center of mass.

Therefore, we have come to an important conclusion. In the very small-angle region, regardless of the structure of a particle, the scattering curve can be described with only two parameters: $I(0)$, characterizing the total amount of scattering objects, and $R_G$, bearing information on its distribution with respect to the particle's center of mass. For example, the radius of gyration of a uniform spherical particle with radius $R$ can be simply calculated as

$$R_G^2 = \frac{\int_0^R r^4 dr}{\int_0^R r^2 dr} = \frac{3}{5}R^2. \quad \text{(III.2.28)}$$

In Figures III.2.3 and III.2.4, the Guinier function is plotted as a function of $QR$ with the accurate solution for the spherical particle with radius $R$, which is the same curve as shown in Figures III.2.1 and III.2.2. From these figures, one can notice that the Guinier approximation is applicable only at the region of $QR < \pi$.

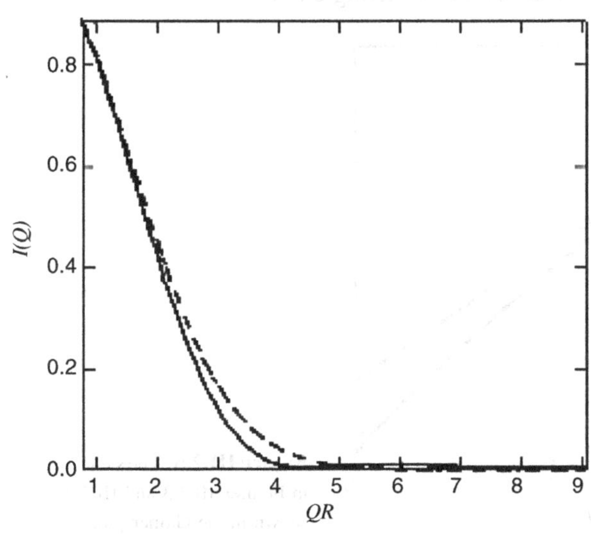


**Figure III.2.3** Guinier approximation (dashed line) is shown by comparing with the SAS scattering from the spherical particle (solid line.)

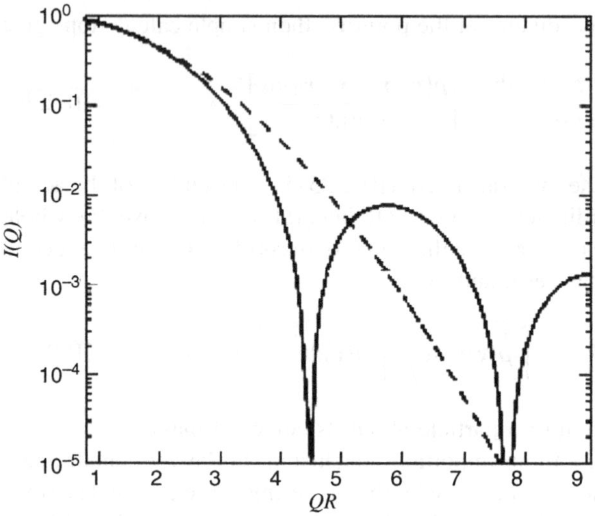


**Figure III.2.4** The Guiner function and the scattering from the spherical particle in the semilogarithmic scale.

By taking a natural logarithm of eq. (III.2.22), one has

$$\ln I(\mathrm{Q}) \approx \ln I(0) - R_{\mathrm{G}}^2 \mathrm{Q}^2/3. \qquad \text{(III.2.29)}$$

Therefore, $R_G$ can be determined from the slope of the linear part of the ln $I$(Q) versus $Q^2$ plot, which is called the "Guinier plot." The same data shown in Figure III.2.3 are indicated in Figure III.2.5. Expressions for the radius of gyration via the dimensions of simple uniform geometrical bodies are summarized in Table III.2.1.

The accuracy of calculation of $R_G$ depends on a number of factors (for details refer to Section 3.3 in Feigin and Svergun (1987)). Therefore, one has to be careful to estimate the Guinier radius from observed scattering curves.

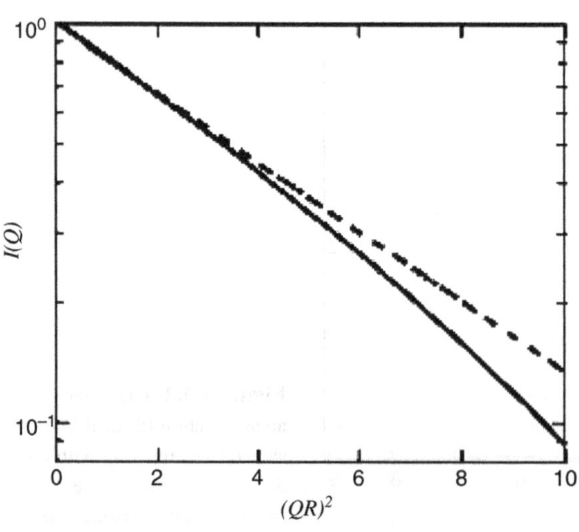


**Figure III.2.5** The same data as in Figures III.2.3 and III.2.4 shown in the Guiner plot.

**Table III.2.1** Radii of Gyration of Some Homogeneous Bodies

| | |
|---|---|
| Sphere of radius $R$ | $R_G^2 = \frac{3}{5}R^2$ |
| Spherical shell with radii $R_1 > R_2$ | $R_G^2 = \frac{3}{5}\frac{R_1^5 - R_2^5}{R_1^3 - R_2^3}$ |
| Ellipse with semiaxes $a$ and $b$ | $R_G^2 = \frac{a^2 + b^2}{4}$ |
| Ellipsoid with semiaxes $a$, $b$, and $c$ | $R_G^2 = \frac{a^2 + b^2 + c^2}{5}$ |
| Prism with edges $A$, $B$, and $C$ | $R_G^2 = \frac{A^2 + B^2 + C^2}{12}$ |
| Elliptical cylinder with semiaxes $a$ and $b$ and height $h$ | $R_G^2 = \frac{a^2 + b^2}{4} + \frac{h^2}{12}$ |
| Hollow circular cylinder with radii $R_1 > R_2$ and height $h$ | $R_G^2 = \frac{R_1{}^2 + R_2{}^2}{2} + \frac{h^2}{12}$ |

## III.2.4 ANISOMETRIC PARTICLES

Scattering by rodlike particles that have dimensions considerably larger along $z$-axis than along other two ($x$ and $y$) can also be approximated. If the scattering density distribution can be written as

$$\rho(x, y, z) = \rho(x, y) \cdot \Pi(z, L), \tag{III.2.30}$$

then the three-dimensional intensity is calculated as

$$I(\mathbf{Q}) = A_0^2(Q_x, Q_y) \cdot L^2 \cdot \delta^2(L, Q_z), \tag{III.2.31}$$

where

$$\Pi(x, a) = \begin{cases} 1, & |x| \leq a/2 \\ 0, & |x| > a/2 \end{cases}, \tag{III.2.32}$$

$$\delta(u, v) = \frac{\sin(\pi u v)}{\pi u v}, \tag{III.2.33}$$

$$A_0(Q_x, Q_y) = \int\int_{S_c} \rho(x, y)\exp[i(xQ_x + yQ_y)]\mathrm{d}x\mathrm{d}y, \tag{III.2.34}$$

while $S_c$ is the cross section of the rodlike particle in the $x$–$y$ plane, and $L$ is the length of the particle. When $L$ is assumed to be large enough, function $\delta(L, Q_z)$ is nearly equal to $\delta(Q_z)$, and most of the scattering intensity is concentrated near the $Q_x$–$Q_y$ plane. From eq. (III.2.33), $\delta(L, Q_z)$ is written as

$$\delta(Q_z, L) = \frac{\sin(\pi L|\mathbf{Q}|t)}{\pi L|\mathbf{Q}|t} = \frac{\sin(LQt/2)}{LQt/2}, \tag{III.2.35}$$

where $t = \cos\alpha$, $\alpha$ being the angle between the scattering vector $\mathbf{Q}$ and the $Q_z$ axis in reciprocal space. The upper limit of integration should be unity; however, $\delta^2(L, Q_z)$ is negligibly small at $t > 1$ if $Q \geq 2\pi L^{-1}$ is satisfied. Therefore, the integration can be performed up to infinity and the following equations

$$\langle L^2\delta^2(L, Q_z)\rangle = L^2\int_0^\infty \frac{\sin(LQt/2)}{LQt/2}\,dt = L\frac{\pi}{Q} \equiv \langle F_L^2\rangle \tag{III.2.36}$$

and

$$I(Q) = \langle F_L^2\rangle\langle A_0^2(Q_x, Q_y)\rangle = L\frac{\pi}{Q}I_c(Q), \tag{III.2.37}$$

where $\langle F_L^2\rangle$ is a "length factor," can be obtained.

When similar calculations as in eqs. (III.2.9)–(III.2.12) are performed and two-dimensional averaging of $\exp[i(xQ_x + yQ_y)]$ resulting in the Bessel function $J_0(Qr)$ is taken into consideration,

$$I_c(Q) = 2\pi\int_0^d \gamma_c(r)J_0(Qr)r\,dr \tag{III.2.38}$$

is obtained, where $d$ is the diameter of the cross section $S_C$ and

$$\gamma_c(r) = \frac{1}{2\pi}\int_0^{2\pi} d\omega\int_{S_c}\rho(\mathbf{u})\rho(\mathbf{u}+\mathbf{r})\,dx\,dy, \tag{III.2.39}$$

where vectors $\mathbf{u}$ and $\mathbf{r}$ correspond to the $x$–$y$ plane. For small values of Q, the expansion of the Bessel function in a Maclaurin series

$$J_0(x) = 1 - \frac{x^2}{4} + \ldots \tag{III.2.40}$$

gives the formula similar to eq. (III.2.22),

$$I_c(Q) = I_c(0)\exp(-R_c^2Q^2/2), \tag{III.2.41}$$

where $I_c(0)$ and $R_c$ denote the amount of scattering matter in the cross section $S_c$ and its radius of gyration, respectively. In case of the uniform rodlike particles with length $L$, radius $d$ ($\ll L$), and scattering density $\rho_0$, this function can be written as

$$I(Q) = \frac{\pi N v^2}{QVL}\rho_0^2\exp(-d^2Q^2/4), \tag{III.2.42}$$

where $v = \pi d^2 L$ is the particle volume, $N$ the number of particles, and $V$ the whole volume of the system. This type of scattering curve is verified by the plot of log[$Q$ $I$(Q)] against $Q^2$, which is called the "first Kratky plot" and the radius $d$ is obtained from the slope.

For the case of disk-like particles, a similar calculation can be performed and the scattering function at small Q region is approximated by

$$I_t(Q) = I_t(0)\exp(-R_t^2 Q^2). \tag{III.2.43}$$

For the scattering function of the uniform disk-like particles with thickness $T$ and the area $S$, one has

$$I(Q) = \frac{2\pi N v^2}{Q^2 VS}\rho_0^2 \exp(-T^2 Q^2/12), \tag{III.2.44}$$

where $v = TS$ is the particle volume. Plotting log[$Q^2 I$(Q)] against $Q^2$ (second Kratky plot) shows whether an observed small-angle scattering is explained by this function or not, and the slope gives the thickness of the disk-like particles.

## III.2.5 POLYDISPERSE SYSTEM

If spherical particles have a size distribution $f(R)$ as shown in Figure III.2.6, eq. (III.2.17) should be integrated over the radius $R$ to give a small-angle scattering intensity

$$I(Q) = \left| \int_0^\infty \rho_0 v \left\{ 3\frac{\sin QR - QR\cos QR}{(QR)^3} \right\} f(R)dR \right|^2. \tag{III.2.45}$$

Kotlarchyk and Chen (1983) derived an analytical form of the scattering function for the polydisperse droplet in cases that the rectangular distribution and the Schultz size distribution

$$f_s(R) = \left(\frac{Z+1}{\bar{R}}\right)^{Z+1} R^Z \exp\left[-\left(\frac{Z+1}{\bar{R}}\right)R\right] \Big/ \Gamma(Z+1), Z > -1, \tag{III.2.46}$$

where $\bar{R}$ is the mean of the distribution, $Z$ a width parameter, and $\Gamma(X)$ the gamma function. The integral in eq. (III.2.45) could be calculated explicitly, and they found

$$I(Q) = 8\pi^2 \rho_0^2 \bar{R}^6 (Z+1)^{-6} \alpha^{Z+7} G_1(Q), \tag{III.2.47}$$

where

$$\begin{aligned} G_1(Q) = {} & \alpha^{-(Z+1)} - (4+\alpha^2)^{-(Z+1)/2} \cos\left[(Z+1)\tan^{-1}\frac{2}{\alpha}\right] \\ & + (Z+2)(Z+1)\left\{\alpha^{-(Z+3)} + (4+\alpha^2)^{-(Z+3)/2}\cos\left[(Z+3)\tan^{-1}\frac{2}{\alpha}\right]\right\}, \\ & -2(Z+1)(4+\alpha^2)^{-(Z+2)/2}\sin\left[(Z+2)\tan^{-1}\frac{2}{\alpha}\right] \end{aligned} \tag{III.2.48}$$

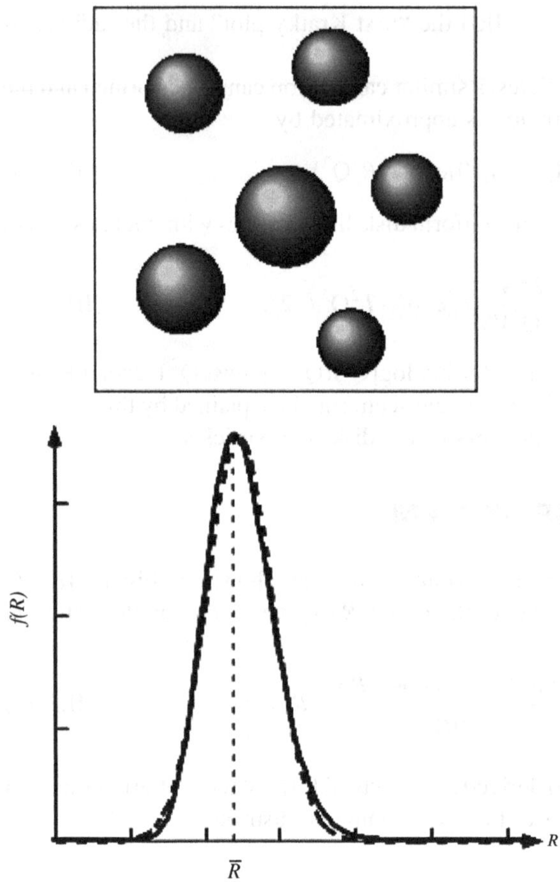


**Figure III.2.6** Size distribution $f(R)$ of the polydisperse particles. The solid line indicates the Schultz distribution and the dashed line the Gaussian distribution.

$$\alpha = (Z+1)/Q\bar{R}. \quad \text{(III.2.49)}$$

Figures III.2.7 and III.2.8 compare the small-angle scattering from the monodisperse particles of radius $R = 45$, 50, and 55 Å with that from polydisperse particles of $\bar{R} = 43$ Å and $Z = 30$.

## III.2.6 POROD LAW

The behavior of $I(Q)$ at Q tending to infinity was investigated by Porod (1951). He showed, in the system with clear interface between particle and surrounding solvent, the asymptotic trend of the scattering intensity for Q tending to infinity proportional to $Q^{-4}$, and it is readily related to the particle surface $S$,

$$I(Q) \approx \frac{2\pi}{Q^4}\rho^2 S \cdot (Q \to \infty), \quad \text{(III.2.50)}$$

which is called the "Porod law."

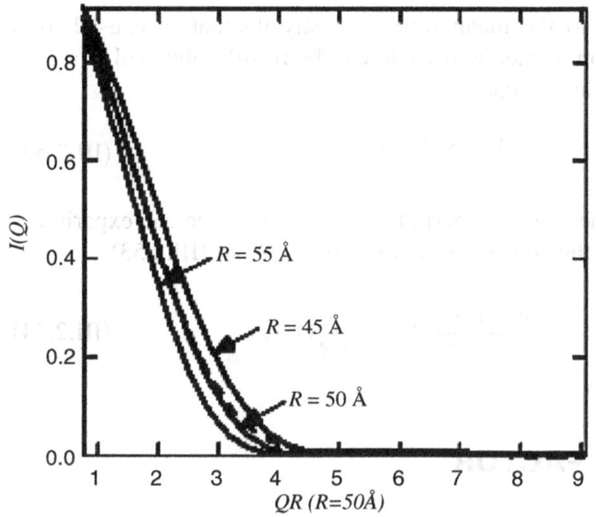


**Figure III.2.7** SAS profiles from spherical particles with various radii. The dashed line indicates the profile from the polydisperse spherical particles with Schultz size distribution.

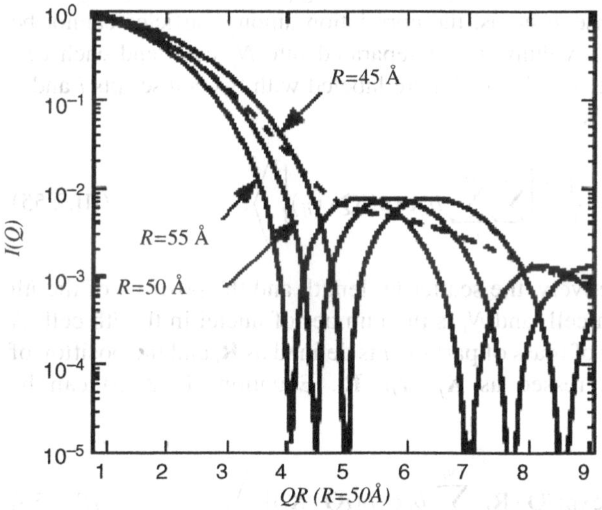


**Figure III.2.8** The same data as of Figure III.2.7 shown in semilogarithmic scale.

The Porod invariant (Porod, 1952) is an important integral characteristic of the scattering intensity:

$$I^{\text{total}} = \int_0^{\infty} I(\text{Q})\text{Q}^2 d\text{Q}. \tag{III.2.51}$$

This invariant is proportional to the total scattered energy because the factor $\text{Q}^2$ is due to the use of polar coordinates. For particle scattering, eqs. (III.2.12) and (III.2.13) give

$$I^{\text{total}} = 2\pi^2\gamma(0) = 2\pi^2 \int_v \rho^2(\mathbf{r})d\mathbf{r}; \tag{III.2.52}$$

that is, $I^{\text{total}}$ is proportional to the mean square density fluctuation caused by a particle. The volume of a homogeneous particle can be readily obtained from the value of the invariant, since in this case

$$I^{\text{total}} = 2\pi^2\rho^2 V. \tag{III.2.53}$$

For the simple case of the spherical particle with radius $R$, one can experimentally obtain the curvature of the sphere from eqs. (III.2.50) and (III.2.53):

$$\frac{I(\mathrm{Q})}{I^{\text{total}}} = \frac{2\pi\rho^2 \cdot 4\pi R^2}{2\pi^2 \frac{4}{3}\pi R^3}\mathrm{Q}^{-4} = \frac{3}{\pi R}\mathrm{Q}^{-4}. \tag{III.2.54}$$

## III.2.7 STRUCTURE FACTOR

In the above section, systems without interference among particles have been treated. However, especially for dense systems, the correlation among particles cannot be ignored. Imagine the sample volume to be separated into $N_{\text{p}}$ cells and each cell contains exactly one particle. All the nuclei are labeled with two subscripts $i$ and $j$, and the scattering intensity is written as

$$I(\mathbf{Q}) = \frac{1}{V}\left\langle \left| \sum_{i=1}^{N_P}\sum_{j=1}^{N_i} b_{ij}\exp(i\mathbf{Q}\cdot\mathbf{r}_{ij}) \right|^2 \right\rangle, \tag{III.2.55}$$

where $b_{ij}$ and $\mathbf{r}_{ij}$ are, respectively, the scattering length and the position of the $j$th nucleus contained in the $i$th cell, and $N_i$ is the number of nuclei in the $i$th cell. A position vector of the center of mass of particle $i$ is defined as $\mathbf{R}_i$ and the position of a nucleus is relatively expressed as $\mathbf{X}_j = \mathbf{r}_{ij} - \mathbf{R}_i$. Equation (III.2.55) can be rewritten as

$$I(\mathbf{Q}) = \frac{1}{V}\left\langle \left| \sum_{i=1}^{N_P}\exp(i\mathbf{Q}\cdot\mathbf{R}_i)\sum_{j=1}^{N_i} b_{ij}\exp(i\mathbf{Q}\cdot\mathbf{X}_j) \right|^2 \right\rangle. \tag{III.2.56}$$

Defining the form factor $A_i(\mathbf{Q})$ of the $i$th particle,

$$A_i(\mathbf{Q}) = \sum_{j=1}^{N_i} b_{ij}\exp(i\mathbf{Q}\cdot\mathbf{X}_j), \tag{III.2.57}$$

the small-angle scattering reads

$$I(\mathbf{Q}) = \frac{1}{V}\left\langle \sum_{i=1}^{N_P}\sum_{i'=1}^{N_P} A_{i'}^*(\mathbf{Q})A_i(\mathbf{Q})\exp[i\mathbf{Q}\cdot(\mathbf{R}_i - \mathbf{R}_{i'})] \right\rangle. \tag{III.2.58}$$

Considering eq. (III.2.5), the form factor is rewritten in terms of scattering length densities and the scattering from the cell $i$ is separated from the particle $i$ and the solvent.

$$A_i(\mathrm{Q}) = \int_{\text{particle } i} [\rho_i(\mathbf{r}) - \rho_s] \exp(i\mathbf{Q}\cdot\mathbf{r}) d\mathbf{r} + \rho_s \int_{\text{cell } i} \exp(i\mathbf{Q}\cdot\mathbf{r}) d\mathbf{r}, \qquad \text{(III.2.59)}$$

where $\rho_s$ is the scattering length density of the solvent. Since the second term is simply a delta function centered at $\mathrm{Q}=0$, one obtains

$$A_i(\mathrm{Q}) = \int_{\text{particle } i} [\rho_i(\mathbf{r}) - \rho_s] \exp(i\mathbf{Q}\cdot\mathbf{r}) d\mathbf{r}, \ (\mathbf{Q} \neq 0). \qquad \text{(III.2.60)}$$

Assuming that the particle size and orientation are uncorrelated with the position of the particles,

$$I(\mathbf{Q}) = \frac{1}{V} \left\langle \sum_{i=1}^{N_p} \sum_{i'=1}^{N_p} \langle A_{i'}^*(\mathbf{Q}) A_i(\mathbf{Q}) \rangle \exp[i\mathbf{Q}\cdot(\mathbf{R}_i - \mathbf{R}_{i'})] \right\rangle, \qquad \text{(III.2.61)}$$

where the inner brackets represent an average weighted by the distribution of particle sizes and orientations. This average can be written as

$$\langle A_{i'}^*(\mathbf{Q}) A_i(\mathbf{Q}) \rangle = [\langle |A(\mathbf{Q})|^2 \rangle - |\langle A(\mathbf{Q}) \rangle|^2] \delta_{ii'} + |\langle A(\mathbf{Q}) \rangle|^2, \qquad \text{(III.2.62)}$$

and the small-angle scattering intensity reads

$$I(\mathbf{Q}) = n_p [\langle |A(\mathbf{Q})|^2 \rangle - |\langle A(\mathbf{Q}) \rangle|^2] + n_p |\langle A(\mathbf{Q}) \rangle|^2 S(\mathbf{Q}), \qquad \text{(III.2.63)}$$

where $n_p = N_p/V$ is the average number density of particles in the sample and $S(\mathbf{Q})$ is the interparticle structure factor defined by

$$S(\mathbf{Q}) = \frac{1}{N_p} \left\langle \sum_{i=1}^{N_p} \sum_{i'=1}^{N_p} \exp[i\mathbf{Q}\cdot(\mathbf{R}_i - \mathbf{R}_{i'})] \right\rangle. \qquad \text{(III.2.64)}$$

If there is no correlation among particles, $S(\mathbf{Q})$ becomes unity and the small-angle scattering is simply written as

$$I(\mathbf{Q}) = n_p P(\mathbf{Q}) = n_p \langle |A(\mathbf{Q})|^2 \rangle. \qquad \text{(III.2.65)}$$

When the particles has the same shape and the size, the form factor $A_i(\mathbf{Q})$ is identical for all the particles and $I(\mathbf{Q})$ is expressed as

$$I(\mathbf{Q}) = n_p P(\mathbf{Q}) S(\mathbf{Q}) \qquad \text{(III.2.66)}$$

because

$$\langle |A(\mathbf{Q})|^2 \rangle = |\langle A(\mathbf{Q}) \rangle|^2 \tag{III.2.67}$$

is fulfilled.

More general case was examined by Kotlarchyk and Chen (1983). If the particle size and/or the orientation are not the same for each particle, the form factor $A_i(\mathbf{Q})$ varies from particle to particle, and it is better to rewrite eq. (III.2.66) as

$$I(\mathbf{Q}) = n_p P(\mathbf{Q}) S'(\mathbf{Q}), \tag{III.2.68}$$

where

$$S'(\mathbf{Q}) = 1 + \beta(\mathbf{Q})[S(\mathbf{Q}) - 1], \tag{III.2.69}$$

$$\beta(\mathbf{Q}) = \langle |A(\mathbf{Q})|^2 \rangle / |\langle A(\mathbf{Q}) \rangle|^2. \tag{III.2.70}$$

The factor $\beta(\mathbf{Q})$ varies between zero and one and suppresses the oscillations of the structure factor $S(\mathbf{Q})$ in the observed scattering from a polydisperse or nonspherical system of particles. They calculated this term explicitly for the cases of the polydisperse spherical particles. For example, in the case of the Schultz size distribution, they evaluated

$$\beta(\mathbf{Q}) = 2\alpha^{Z+1}(1+\alpha^2)^{-(Z+1)} G_2^2(\mathbf{Q}) / G_1(\mathbf{Q}), \tag{III.2.71}$$

using eq. (III.2.48) and

$$G_2(\mathbf{Q}) = \sin\left[(Z+1)\tan^{-1}\frac{1}{\alpha}\right] - (Z+1)(1+\alpha^2)^{-1/2}\cos\left[(Z+2)\tan^{-1}\frac{1}{\alpha}\right]. \tag{III.2.72}$$

In case of critical phenomena associated with a drastic increase of a particle density fluctuation, the interparticle structure factor is characterized by the correlation length and the isothermal compressibility.

## III.2.8 RELATIVE FORM FACTOR

Neglecting the fluctuations due to the form factor, a coherent small-angle scattering intensity is written as

$$I(\mathrm{Q}) = n_p P(\mathrm{Q}) S(\mathrm{Q}). \tag{III.2.73}$$

In principle, this form is applicable in the monodisperse case. In the case of small-angle neutron scattering (SANS), a form factor depends on the contrast between a scatterer and a background, $\Delta\rho$, because $A(\mathrm{Q})$ is the Fourier transform of the scattering amplitude density difference between them, which can be given as follows:

$$A(\mathrm{Q}) = \int \Delta\rho(\mathrm{r})\exp(-i\mathrm{Qr})d\mathrm{r}. \tag{III.2.74}$$

On the other hand, $S(\mathrm{Q})$ does not depend on the contrast because it indicates the time-averaged correlation of the center of mass of objects, and the properties of molecules and their assemblies do not depend on the contrast.

By using the SANS technique, it is easy to obtain different scattering contrast conditions by selective deuteration of the ingredients. If one changes only the scattering contrast, keeping unchanged the sample composition and the external conditions, for example, temperature, pressure, and so on, a relative form factor, $R(\mathrm{Q})$, can be introduced as the ratio of the scattering intensities from the deuterated sample, $I^{\mathrm{d}}(\mathrm{Q})$, and the protonated sample, $I^{\mathrm{p}}(\mathrm{Q})$, as follows:

$$R(\mathrm{Q}) = \frac{I^p(\mathrm{Q})}{I^d(\mathrm{Q})} = \frac{P^p(\mathrm{Q})}{P^d(\mathrm{Q})}; \tag{III.2.75}$$

that is, the ratio of the scattering intensities of each contrast is identical to the ratio of the form factors of each contrast. This relation is independent of $S(\mathrm{Q})$, and therefore one can evaluate $P(\mathrm{Q})$ of the system without the influence of $S(\mathrm{Q})$.

In reality, the systems measured by SANS are not always the monodisperse systems, although the system has the size distribution and/or the directional inhomogeneity. For polydisperse systems, eq. (III.2.73) can be written as eq. (III.2.68). For example, in the case of the polydisperse spheres, $P(\mathrm{Q})$ and $|\langle A(\mathrm{Q})\rangle|^2$ are defined as

$$P(\mathrm{Q}) = \left\langle |A(\mathrm{Q})|^2 \right\rangle = \int |A(\mathrm{Q},\mathrm{r})|^2 h(\mathrm{r})\mathrm{dr} \tag{III.2.76}$$

and

$$|\langle A(\mathrm{Q})\rangle|^2 = \left| \int A(\mathrm{Q},\mathrm{r})h(\mathrm{r})\mathrm{dr} \right|^2, \tag{III.2.77}$$

where r is the radius of the spheres, $h(\mathrm{r})$ the distribution function of r, and $A(\mathrm{Q}, \mathrm{r})$ the form factor of the spheres with radius r. From these relations, $\beta(\mathrm{Q})$ and thus $S'(\mathrm{Q})$ do not depend on the scattering contrast.

The polydispersity of the system is low enough, the first term in eq. (III.2.63) is negligibly small, and $\beta(\mathrm{Q})$ and $S'(\mathrm{Q})$ are considered to be independent of the scattering contrast. In such case, eq. (III.2.75) can be written as follows:

$$R(\mathrm{Q}) = \frac{I^p(\mathrm{Q})}{I^d(\mathrm{Q})} \cong \frac{P^p(\mathrm{Q})}{P^d(\mathrm{Q})}. \tag{III.2.78}$$

Therefore, the relative form factor method is valid for the small polydispersity cases. This method was proposed by independent groups (Rollet et al., 2002; Nagao et al., 2003) and applied successfully to the water-in-oil droplet microemulsion system (Nagao et al., 2007).

## REFERENCES


Feigin, L. A. and Svergun, D. I. *Structure Analysis by Small-Angle X-Ray and Neutron Scattering*, Plenum Press, New York, 1987. This textbook can be freely downloaded from http://www.embl-hamburg.de/ExternalInfo/Research/Sax/reprints/feigin_svergun_1987.pdf

Guinier, A. *Ann. Phys.* **1939**, *12*, 161.

Kotlarchyk, M. and Chen, S.-H. *J. Chem. Phys.* **1983**, *79*, 2461.

Nagao, M. Seto, H. Shibayama M. and Yamada, N.L. *J. Appl. Crystallogr.* **2003**, *36*, 602.

Nagao, M., Seto H., and Yamada, N.L. *Phys. Rev. E* **2007**, *75*, 061401.

Porod, G. *Kolloid-Z.* **1951**, *124*, 83.

Porod, G. *Kolloid-Z.* 1952, *125*, *51*, 109.

Rollet, A. L., Diat O., and Gebel, G. *J. Phys. Chem. B* **2002**, *106*, 3033.

# III Data Treatment and Sample Environment

# III.3 Calculation of Real Space Parameters and *Ab Initio* Models from Isotropic Elastic SANS Patterns

*Peter. V. Konarev and Dmitri I. Svergun**

## III.3.1 INTRODUCTION

This chapter describes the basic equations and algorithms for the conversion of information from neutron scattering patterns into the structural parameters of the object. We shall restrict ourselves to coherent and elastic small-angle neutron scattering (SANS), which is employed in structural studies of soft condensed matter. Furthermore, the isotropic scattering case will be considered, when the object under study is a disordered system of particles floating in solution or embedded in a matrix. The data analysis approaches used in SANS are similar to (and partially come from) small-angle X-ray scattering (SAXS). Perhaps the most significant difference (and advantage of SANS) is the possibility of isotopic contrast variation by hydrogen/deuterium exchange, which yields unique information about the internal structure of multicomponent particles. Contrast variation will be described in detail in Chapter III.4.; here, we shall present the enhanced possibilities of this technique for *ab initio* low-resolution structure analysis.

After a short reminder of the main theoretical aspects of elastic coherent SANS from systems of chaotically distributed particles, we shall present basic data processing procedures to compute overall parameters and characteristic functions. This will be followed by a section devoted to advanced *ab initio* approaches for 3D model building. Finally, a few examples will be given illustrating the applications of the presented methods to the analysis of solutions of biological macromolecules.

*Corresponding author

*Neutrons in Soft Matter* Edited by Toyoko Imae, Toshiji Kanaya, Michihiro Furusaka, and Naoya Torikai

## III.3.2 FROM THE SMALL-ANGLE SCATTERING CURVE TO OVERALL STRUCTURE PARAMETERS

The physical basis and theoretical aspects of neutron scattering are described in Chapter I. Below, we shall recapitulate for convenience of the reader the most important concepts relevant to the structural analysis of isotropic systems. For a full description of the theoretical principles, mathematical apparatus, and the relations between SANS and SAXS, the reader is referred to textbooks (Feigin and Svergun, 1987) or to recent reviews (Koch et al., 2003; Svergun and Koch, 2003).

### III.3.2.1 Basics of X-Ray and Neutron Scattering

The scattering phenomenon emerges due to interactions of incoming neutrons with nuclei in the sample. The strength of interaction is proportional to the scattering cross section, that is, the number of neutrons per solid angle scattered by the nuclei. For structural studies, only elastic scattering effects are relevant, where there is no energy exchange between the neutron and the nucleus and the scattering length of a nucleus is a real number. The total scattering length consists of a spin-dependent component and a contribution from the nuclear potential, $\sigma_{tot} = \sigma_{spin} + \sigma_{pot}$. The potential component is always structurally related, whereas the spin scattering contains for randomly oriented spins no structural information and produces a flat background only. Spin contrast variation techniques using polarized neutrons and oriented spins in strong magnetic fields can give further advantages for structural analysis (Stuhrmann et al., 1986; Willumeit et al., 1996) and will be considered in Chapter III.4.

When the sample is illuminated by a monochromatic neutron beam with de Broglie wavelength $\lambda$, that is, with wave vector $k_0 = |\mathbf{k}_0| = 2\pi/\lambda$, the neutrons interact with the nuclei within the object, the latter becoming sources of scattered neutrons. For elastic scattering, the wavelength $\lambda$ of the scattered neutrons is equal to that of the incoming beam and the modulus of the scattered wave $k_1 = |\mathbf{k}_1|$ is equal to $k_0$. The amplitude of the wave scattered by each atom/nucleus is described by its scattering length, $f$ (the name reflects the fact that this characteristic has the units of length, that is, centimeter). X-rays interact with electrons, and the atomic scattering length $f_x$ is proportional to the number of electrons. The interaction of neutrons with the nuclear potential is described by the nuclear scattering length $f_p$, which does not display systematic dependence on the atomic number, being rather sensitive to the isotopic content. The case of hydrogen and deuterium deserves special attention. Whereas for X-rays the two isotopes are indistinguishable (both have one electron), a large difference is observed between H and D atoms in the case of neutron scattering. This difference provides the basis for the use of selective deuteration and measurements in water/heavy water mixtures for the analysis of the internal structure of complex particles.

The above considerations give a simplified picture of interactions of X-rays and neutrons with matter, leaving aside the phenomena of anomalous (wavelength-dependent) X-ray scattering and spin-dependent neutron scattering. The reader is referred to original papers and reviews describing application of the advanced techniques making use of these effects in SAXS (Stuhrmann and Notbohm, 1981; Stuhrmann 1991) and SANS (Stuhrmann et al., 1986; Stuhrmann, 1991).

The scattering process involves a transformation ("conversion") from the "real" space (coordinates **r** of the object) to the "reciprocal" space, that is, coordinates of scattering vectors $\mathbf{Q}=(s, \Omega)=\mathbf{k}_1-\mathbf{k}_0$. Here, the momentum transfer $Q=4\pi\lambda^{-1}\sin(\theta)$, where $2\theta$ is the scattering angle and $\Omega$ is the direction of the scattering vector. The transformation is mathematically described by the Fourier operator

$$A(\mathbf{Q}) = \Im[\rho(\mathbf{r})] = \int \rho(\mathbf{r})\exp(i\mathbf{Qr})d\mathbf{r}, \qquad \text{(III.3.1)}$$

where $\rho(\mathbf{r})$ is the scattering length density distribution, $A(\mathbf{Q})$ is the scattering amplitude, and the integration is performed over the illuminated sample volume.

Following the reciprocal properties of Fourier transformation in eq. (III.3.1), where a product of **Qr** is entered, the larger the size in real space, the shorter the corresponding reciprocal space vector. The real space resolution of the scattering data can be estimated as $d=2\pi/Q$, which is a counterpart of the well-known Bragg equation in crystallography. The scattering at small angles, that is, at small $Q$, provides therefore information about large distances (much larger than the wavelength), that is, about overall structure at a low (typically, 2–3 nm) resolution.

To describe the scattering from assemblies of atoms, the scattering length density distribution $\rho(\mathbf{r})$ can be represented as the total scattering length of the nuclei per unit volume. Considering the scattering from systems containing individual particles and assuming that the matrix bulk (e.g., solvent in protein solutions) has a constant scattering density $\rho_s$, the difference scattering amplitude from a single particle relative to that of the equivalent matrix (solvent) volume is defined by the Fourier transform of the excess scattering length density $\Delta\rho(\mathbf{r})=\rho(\mathbf{r})-\rho_s$,

$$A(\mathbf{Q}) = [\Delta\rho(\mathbf{r})] = \Im\int_V \Delta\rho(\mathbf{r})\exp(i\mathbf{Qr})d\mathbf{r}, \qquad \text{(III.3.2)}$$

where the integration is performed over the particle volume $V$, which represents a coherently scattering volume (positions of atoms are correlated). Experimentally one measures the scattering intensity, that is, the number of neutrons scattered in the given direction **Q**, which is proportional to the squared amplitude $I(\mathbf{Q})=[A(\mathbf{Q})]^2$.

If one considers an ensemble of identical particles, the total scattering will depend on the distribution of these particles and two major limiting cases should be considered. In the case of an ideal single crystal, all particles in the sample have defined correlated orientations and are regularly distributed in space, so that scattering amplitudes of individual particles have to be summed up accounting for all interparticle interferences. As a result, the total scattered intensity is redistributed along specific directions defined by the reciprocal lattice and the discrete three-dimensional function $I(\mathbf{Q}_{hkl})$ measured correspond to the density distribution in a single unit cell of the crystal. If the particles are randomly distributed and their positions and orientations are uncorrelated, their scattering intensities rather than their amplitudes are summed (no interference). Accordingly, the intensity from the entire ensemble is a continuous isotropic function proportional to the scattering from a single particle averaged over all orientations in reciprocal space $I(Q) = \langle I(\mathbf{Q}) \rangle_{\Omega}$. Dilute solutions of monodisperse noninteracting biological macromolecules often correspond to this second limiting case. The main structural task in this case is to reconstruct the particle structure (i.e., its excess scattering length density distribution $\Delta\rho(\mathbf{r})$) at low resolution from the scattering data. The scattering intensity from such systems is proportional to the squared contrast $(\Delta\rho)^2$, where $\Delta\rho$ is the average excess scattering length density. The X-ray and neutron scattering densities and matching points of biological macromolecules in $H_2O/D_2O$ solutions are listed in Table III.3.1.

In practice, one often has to deal with nonideal cases when the particles differ in size and/or shape, and/or interparticle interactions cannot be neglected. In particular, polydisperse systems, such as micelles, microemulsions, block copolymers, or metal nanoparticles, often consist of particles with similar shapes that differ only in size. Such systems are conveniently described by the volume distribution function $D(R) = N(R)V(R)$, where $N(R)$ is the number of particles with characteristic size $R$ and $V(R)$ is the volume of the particles of this size. Assuming the contrast $\Delta\rho$ to be the same for all particles, the scattering intensity is given by the integral

$$I(Q) = (\Delta\rho)^2 \int_{R_{\min}}^{R_{\max}} D(R)V(R)i_0(QR)\mathrm{d}R, \qquad \text{(III.3.3)}$$

where $i_0(QR)$ is the normalized scattering intensity of the particle ($i_0(0) = 1$), and $R_{\min}$ and $R_{\max}$ are the minimum and maximum particle sizes, respectively.

**Table III.3.1** X-Ray and Neutron Scattering Densities of Components of Biological Complexes

| | X-rays | | Neutrons | | |
|---|---|---|---|---|---|
| Component | $\rho$ ($10^{10}$ cm$^{-2}$) | Matching solvent | $\rho$ in $H_2O$ ($10^{10}$ cm$^{-2}$) | $\rho$ in $D_2O$ ($10^{10}$ cm$^{-2}$) | Matching % $D_2O$ |
| $H_2O$ | 9.42 | – | −0.6 | – | – |
| $D_2O$ | 9.42 | – | 6.4 | – | – |
| Lipids | 8.46 | – | 0.3 | −6.0 | ≈15% |
| Proteins | 11.8 | 65% sucrose | 1.8 | 3.1 | ≈40% |
| D proteins | 11.8 | 65% sucrose | 6.6 | 8.0 | – |
| Nucleic acids | 15.5 | – | 3.7 | 4.8 | ≈70% |
| D nucleic acids | 15.5 | – | 6.6 | 7.7 | – |

For X-rays, the scattering length density is often expressed in terms of electron density (the number of electrons/Å$^3$; 1 electron/Å$^3$ = $2.82 \times 10^{11}$ cm). The increase in the scattering length density of the same component in $D_2O$ compared to $H_2O$ is due to the H/D substitution of exchangeable protons.

In the next section, the main SAS equations is briefly presented, allowing one to compute the structural parameters for monodisperse and polydisperse systems from the experimental data.

### III.3.2.2 Overall Structural Parameters

Some integral parameters (called invariants) can be computed from the SAS data without model assumptions, and in monodisperse systems they are directly related to the weight and geometrical characteristics of the particles. The best known parameter is the radius of gyration, $R_g$, derived from the so-called Guinier plot (Guinier, 1939):

$$I(Q) \cong I(0)\exp\left(-\frac{1}{3}R_g^2Q^2\right). \tag{III.3.4}$$

This approximation is valid for very small angles ($Q < 1.3/R_g$) and the Guinier plot is very useful at the first stages of data analysis. For ideal monodisperse systems, the plot ($\ln(I(Q))$ versus $Q^2$) should be a linear function, the intercept of whose gives the forward scattering $I(0)$. The slope of the plot yields the $R_g$. Deviations from linearity may point to interparticle interactions or polydispersity of the sample.

The calculation of $R_g$ using the Guinier plot, although typically the first and obligatory step used in data analysis, is usually quite a subjective procedure. It often requires extensive user intervention to find an appropriate range yielding the linear fit. Given the modern trend toward high-throughput scattering studies, automation of this procedure is important. Recently, a program AUTORG for a fully automated determination of $R_g$ for monodisperse systems was developed (Petoukhov et al., 2007). First, the initial portion of the data is analyzed and the range showing unreasonable upward or downward trends (e.g. caused by the beam stop or strong background near the primary beam) is discarded. Then the data range where the scattering intensity decays by an order of magnitude is taken and all possible intervals for Guinier plots are analyzed. For each interval ($Q_{min}$, $Q_{max}$) containing a minimum of five experimental points, a weighted linear fit in eq. (III.3.4) is calculated by least squares and $R_g$ is computed. Each interval fulfilling the conditions $Q_{min}R_g < 1$ and $Q_{max}R_g < 1.3$ and showing no major systematic deviations in the fit is considered consistent. If no consistent intervals are found, the program tries to find intervals with weakened $QR_g$ conditions, but simultaneously reduces the estimate of the data quality. Each consistent interval is rated according to its length (number of points fitted) and discrepancy (root mean square deviation of the fit), and the interval with the best rating is selected. The accuracy of $R_g$ is estimated by taking into account not only the error propagation in the selected fit as usual, but also the deviation of $R_g$ values calculated from other consistent intervals. This accounts to some extent for systematic errors in the $R_g$ determination. An estimate of the overall data quality is then expressed using several criteria: (i) how many consistent intervals were found, (ii) whether the $QR_g$ conditions were weakened or not, (iii) how many starting points were discarded, (iv) whether there is an indication of effects like aggregation, and (v) how accurate is the value of $R_g$. This procedure allows a reliable automated estimate of $R_g$ for monodisperse systems with particles of sufficiently high contrast $\Delta\rho$ (e.g., proteins or nucleoprotein complexes in $H_2O$ or $D_2O$).

The extrapolation to zero scattering angle using eq. (III.3.2) yields the forward scattering $I(0) = (\Delta\rho)^2 V$, where $V$ is the excluded particle volume. As the contrast is defined by the chemical composition of the particle and of the matrix (solvent), the $I(0)$ value permits one to evaluate the molecular mass (MM) of the solute from a comparison with the scattering of a reference sample or a secondary standard in SAXS (Kratky and Pilz, 1972), or incoherent scattering from water in SANS (Zaccai and Jacrot, 1983).

The $R_g$ value characterizes the overall particle size and anisometry. Similar equations hold for very elongated and very flat particles

$$QI(Q) \cong I_C(0)\exp\left(-\frac{1}{2}R_c^2Q^2\right); \qquad Q^2I(Q) \cong I_T(0)\exp(-R_t^2Q^2) \qquad \text{(III.3.5)}$$

to compute the radii of gyration of the cross section $R_c$ and of the thickness $R_t$, respectively.

Another important overall parameter is the so-called Porod invariant (Porod, 1982):

$$Q_P = \int_0^\infty Q^2 I(Q)\mathrm{d}Q = 2\pi^2 \int_V (\Delta\rho(\mathbf{r}))^2 \mathbf{d}r. \qquad \text{(III.3.6)}$$

For homogeneous particles, $Q_P = 2\pi^2(\Delta\rho)^2V$, and taking into account that $I(0) = (\Delta\rho)^2V$, the excluded particle volume is $V = 2\pi^2 I(0)Q_P^{-1}$. Furthermore, for particles with sharp boundaries, $I(Q)$ should decay as $Q^{-4}(\pi S/Q_P V)$, where $S$ is the particle surface that allows one to estimate the specific surface of the particle, $S/V$. These equations hold for homogeneous particles. To reduce the influence of the scattering from inhomogeneities, an appropriate constant is subtracted from the experimental data using the approximation

$$Q^4 I(Q) \approx BQ^4 + A.$$

All the above parameters can also be computed for polydisperse systems, but one must keep in mind that in contrast to monodisperse systems, the experimentally obtained values are not related to single particles but are rather averages over the ensembles.

### III.3.2.3 Characteristic Functions

The Fourier transformation of the scattering intensity from monodisperse systems yields the distance distribution function of the particle

$$p(r) = \frac{r^2}{2\pi^2}\int_0^\infty Q^2 I(Q)\frac{\sin Qr}{Qr}\mathrm{d}r, \qquad \text{(III.3.7)}$$

which is a spherically averaged autocorrelation function of the excess scattering density $\Delta\rho(r)$. This function is equal to zero for the distances $r$ exceeding the maximum particle diameter $D_{max}$, which allows one to estimate this parameter. Furthermore, the appearance of $p(r)$ provides visual information about the particle shape as illustrated in Figure III.3.1, presenting typical scattering patterns and distance distribution functions of geometrical bodies with the same maximum size. Globular particles (curve 1) display bell-shaped $p(r)$ functions with a maximum at about $D_{max}/2$. Elongated particles have skewed distributions with a clear maximum at small distances corresponding to the radius of the cross section (curve 2). Flattened particles display a rather broad maximum (curve 3), also shifted to distances smaller than $D_{max}/2$. A maximum shifted toward distances larger than $D_{max}/2$ is usually indicative of a hollow particle (curve 4). Particles consisting of two separated subunits (e.g., dumbbells) may display two maxima: the first corresponding to the intrasubunit distances and the second yielding the separation between the subunits (curve 5).

Even for simple geometrical bodies, there are only a few cases where $I(s)$ and/or $p(r)$ functions can be expressed analytically. The best known are the expressions for a solid sphere of radius $R$: $I(Q) = A(Q)^2$, $A(Q) = (4\pi R^3/3)[\sin(x) - x\cos(x)]/x^3$, where $x = QR$, and $p(r) = (4\pi R^3/3)r^2(1 - 3t/4 + t^3/16)$, where $t = r/R$. Semianalytical equations for the intensities of ellipsoids, cylinders, and prisms were derived by Mittelbach and Porod in the 1960s, and later analytical formulas for the $p(r)$ function were published for some shapes (e.g., cube (Goodisman, 1980) and tetrahedron

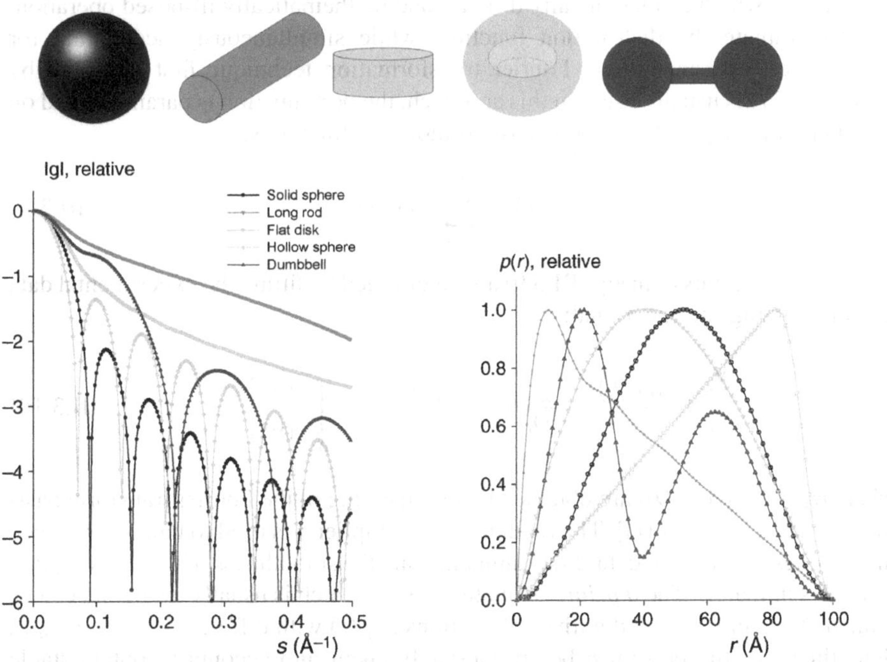


**Figure III.3.1** Scattering intensities (left) and distance distribution functions (right) from different geometrical bodies with the same maximum size.

(Gille, 2003)). Several analytical and semianalytical equations for $I(Q)$ from geometrical bodies can be found in Feigin and Svergun (1987).

Reliable computation of $p(r)$ is a necessary step in the analysis of SANS data, but direct implementation of eq. (III.3.7) is usually difficult as the experimental data $I(Q)$ is measured only in a limited angular interval $[Q_{min}, Q_{max}]$ rather than $[0,\infty]$. Moreover, the neutron scattering data usually contain smearing effects due to the finite beam size, divergence, and/or polychromaticity. These effects are expressed as

$$J(Q) = W[I(Q)] = \int_{-\infty}^{\infty}\int_{-\infty}^{\infty}\int_{0}^{\infty} W_{\mathrm{w}}(u)W_{\mathrm{l}}(t)W_{\lambda}(\lambda)I\left\{\left[(Q-u)^2+t^2\right]^{1/2}\lambda^{-1}\right\}\mathrm{d}\lambda\,\mathrm{d}t\,\mathrm{d}u, \tag{III.3.8}$$

where $J(Q)$ is the experimental scattering intensity for a sample, $W_{\mathrm{w}}$ (slit-width smearing) and $W_{\mathrm{t}}$ (slit-height smearing") are the weighting functions determined by the instrument geometry, and $W_{\lambda}$ is the weight function of the wavelength distribution.

The scattering pattern $J(Q)$ measured in a SANS experiment may thus differ significantly from the ideal curve $I(Q)$. In particular, the smearing effects are rather pronounced for ultrasmall-angle neutron scattering (USANS) experiments with Bonse–Hart collimation (Agamalian et al., 1997). Different approaches have been proposed in the past to reconstruct $I(Q)$ prior to computation of $p(r)$ (Heine and Roppert, 1962; Lake, 1967), but this operation requires the inversion of the integral eq. (III.3.8), which is a technically difficult and mathematically ill-posed operation.

To evaluate the distribution function, while simultaneously accounting for smearing effects, an indirect Fourier transformation technique first proposed by Glatter (1977) is usually used. In this approach, the $p(r)$ function is parameterized on the interval $[0, D_{max}]$ by a set of basic orthogonal functions:

$$p(r) = \sum_{k=1}^{K} c_k \varphi_k(r). \tag{III.3.9}$$

The coefficients $c_k$ in eq. (III.3.9) are determined by fitting the experimental data and minimizing the functional:

$$\Phi_{\alpha} = \sum_{i=1}^{N}\left[\frac{J_{\mathrm{exp}}(Q_i)-\sum_{k=1}^{K} c_k\psi_k(Q_i)}{\sigma(Q_i)}\right]^2 + \alpha\int_{0}^{D_{\max}}\left[\frac{\mathrm{d}p}{\mathrm{d}r}\right]^2\mathrm{d}r, \tag{III.3.10}$$

where $\psi_k(Q) = W[\xi_k(Q)]$ are obtained by the smearing of the Fourier transform basis functions $\xi_k(Q) = \Im[\varphi_k(r)]$. The regularizing multiplier $\alpha$ is used to balance between the goodness of fit to the data (first summand) and the smoothness of the $p(r)$ function (second summand). The *a priori* estimate of $D_{max}$, which is usually available, can be refined at a later stage by iterative calculations of $p(r)$ with different values of $D_{max}$. Smoothness of the $p(r)$ must be appropriately taken into account to obtain stable solutions, but this approach does not require extrapolation of the data beyond the experimentally measured interval and allows one to account for instrumental smearing. Different implementations of the indirect transformation method in computer programs are available (Glatter, 1977; Moore, 1980; Svergun, 1992; Svergun et al. 1988).

The main problem when using the indirect transform technique is to select the proper value of the regularizing multiplier $\alpha$. With too small values, the solutions are unstable to experimental errors, whereas with too large values, the solutions display systematic deviations from the experimental data. In the GNOM program (Semenyuk and Svergun, 1991; Svergun, 1992), a set of perceptual criteria describes the quality of the solution to guide the user in the choice of $\alpha$. The program either finds the optimal solution automatically or detects dubious assumptions about the system (e.g., the value of $D_{max}$). The calculated $p(r)$ is used to obtain the radius of gyration of the particle $R_g$ as

$$R_g^2 = \frac{\int r^2 p(r) \mathrm{d}r}{2 \int p(r) \mathrm{d}r}. \qquad \text{(III.3.11)}$$

As the entire scattering curve is used in the evaluation of these two parameters, rather than a limited angular range at low angles, the values are more reliable and less sensitive to residual interaction effects or low levels of aggregation than those obtained from the Guinier approximation.

To automatically calculate the distance distribution function of the particle and determine its maximum size, an automated version of the program GNOM (called AUTOGNOM) was developed (Petoukhov et al., 2007). In the original version of GNOM, the maximum particle size $D_{max}$ is a user-defined parameter and successive calculations with different $D_{max}$ values are sometimes required to select its optimum value. This optimum $D_{max}$ should provide a smooth real space distance distribution function $p(r)$ such that $p(D_{max})$ and its first derivative $p'(D_{max})$ approach zero, and the back-transformed intensity from the $p(r)$ fits the experimental data. In the program AUTOGNOM, multiple GNOM runs are automatically performed to find the optimum $D_{max}$ and $p(r)$ function. The $D_{max}$ values ranging from $2R_g$ to $4R_g$ are scanned with a step of $0.1R_g$, where $R_g$ is the radius of gyration provided by AUTORG (Petoukhov et al., 2007). The calculated $p(r)$ functions for different $D_{max}$ and corresponding fits to the experimental curves are compared using the perceptual criteria of GNOM, where the smoothness of $p(r)$, absence of systematic deviations in the fit, and other quantities characterizing the solution are merged into a total quality estimate. Moreover, the appropriately normalized value of $p'(D_{max})$ is added to the estimate to ensure that the $p(r)$ function goes smoothly to zero. The best solution according to AUTOGNOM is selected and the function $p(r)$ together with calculated overall parameters is stored.

It is also possible to apply the indirect transformation methods for polydisperse systems where particles have similar shapes and differ only in size (eq. (III.3.3)). In most practical cases, one assumes that the particle form factor is known (in particular, for isotropic systems, the particles are often considered spherical) and eq. (III.3.3) is employed to obtain the volume distribution function $D(R)$. This can be done with the help of the indirect transformation method described above (the function $D(R)$ is expanded into orthogonal functions as in eq. (III.3.9) on the interval [$R_{min}$, $R_{max}$]). The structural parameters of polydisperse systems do not correspond to those of a single particle but are obtained by averaging over the ensemble. Thus, for a polydisperse system of solid spheres, the average sphere radius is expressed as

$$\langle R^2 \rangle_z = \int_{R_{min}}^{R_{max}} R^5 D(R) \mathrm{d}R \left[ \int_{R_{min}}^{R_{max}} R^3 D(R) \mathrm{d}R \right]^{-1}. \quad \text{(III.3.12)}$$

An example indicating the importance of smearing effects in SANS experiments is shown in Figure III.3.2 on the data from bacteriophage T7 in $D_2O$ buffer, collected

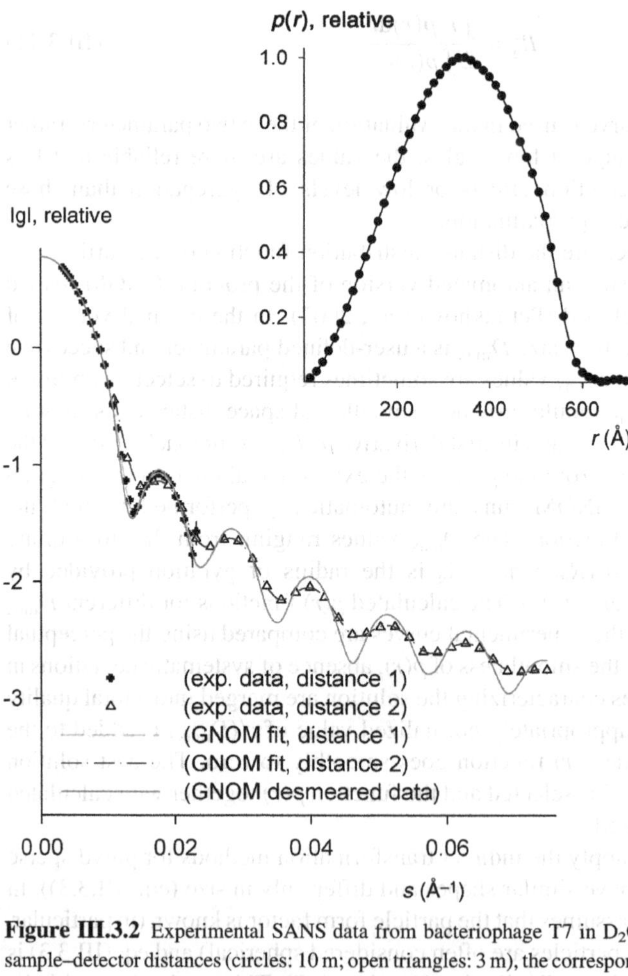


**Figure III.3.2** Experimental SANS data from bacteriophage T7 in $D_2O$ buffer measured at two sample–detector distances (circles: 10 m; open triangles: 3 m), the corresponding GNOM fits (dashed lines and blue dashed lines), and the desmeared data obtained by back-transforming the distance distribution function p(r) (solid line). The latter function is displayed in the insert.

at the Institute Laue-Langevin SANS beamline D11 (Grenoble, France). Bacteriophage T7 is a large bacterial virus with MM of 56 MDa consisting of an icosahedral protein capsid (diameter of about 600 Å) that contains a double-stranded DNA molecule. Two different sample–detector distances were used to record the data in different angular ranges. Figure III.3.2 shows two raw experimental data curves (circles and open triangles) together with the fits from GNOM (dashed line curves) and the desmeared data obtained by GNOM (solid line curve). As one can clearly see for the data measured at lower scattering angles, the difference between experimental smeared data and the desmeared curve by GNOM is not very significant, whereas for the data measured at higher angular range (shorter sample–detector distance), the difference between the two curves is rather pronounced. The skewed shape of distance distribution function $p(r)$ with the maximum shifted to larger intraparticle distances is typical for hollow particles (Figure III.3.2, insert). This profile is in agreement with a core–shell-like structure of the virus, whereby the DNA having a lower contrast in $D_2O$ than the protein (see Table III.3.1) is located inside the protein capsid of the phage.

## III.3.3 FROM THE SMALL-ANGLE SCATTERING CURVE TO 3D *AB INITIO* LOW-RESOLUTION SHAPE

The parameters of the particles directly determined from the experimental data as described in the previous section provide limited information about the overall particle structure. In the past, analysis in terms of three-dimensional models was limited to simple geometrical bodies (e.g., ellipsoids, cylinders, etc.) or was performed by a trial-and-error modeling (Feigin and Svergun, 1987; Glatter and Kratky, 1982) using information from other methods such as electron microscopy as a constraint (Pilz et al., 1972; Tardieu and Vachette, 1982). The recent decade brought a breakthrough in SAS data analysis methods, which tremendously improved the resolution and reliability of the models constructed from the experimental data. In particular, it has become possible to reconstruct three-dimensional low-resolution shape and domain structure of particles *ab initio*. These novel data analysis methods mainly developed for the interpretation of scattering patterns from macromolecular solutions in terms of three-dimensional structures, also applicable to other systems, will be presented here. These new methods are equally well suited for SAXS and SANS data but are especially powerful in combination with SANS contrast variation, as it will be shown later.

### III.3.3.1 *Ab Initio* Shape Determination in the Case of One and Multicomponent Systems

Reconstruction of 3D structure from isotropic scattering data *ab initio* is possible only for monodisperse ensembles of particles without interactions, and only at a low resolution (1–2 nm). At this resolution, the search is usually limited to homogeneous models (shape determination). Given the loss of information due to the spherical average of the scattering data, unambiguous shape reconstruction is fundamentally impossible and care must be taken not to overinterpret the data. In the past, shape modeling was done on a trial-and-error basis by computing scattering patterns from different shapes and comparing them with the experimental data. The first and very elegant *ab initio* approach was proposed by Stuhrmann (1970b), where the shape was described by an angular envelope function expressed using spherical harmonics. This approach was further developed (Svergun and Stuhrmann, 1991) and the first publicly available program SASHA was written (Svergun et al., 1997). It was demonstrated (Svergun et al., 1996) that under certain circumstances a unique 3D shape can be extracted from the SAS data (up to an enantiomorphic shape that always gives the same scattering curve; this ambiguity also holds for all *ab initio* methods described below).

The rapidly growing number crunching capacity of modern computers made it possible to use Monte Carlo-type searches for *ab initio* analysis, which permits one to construct yet more detailed models. The idea of this search was first proposed by Chacon (1998, 2000) and implemented in a program DALAI_GA. Here, a sphere with diameter $D_{max}$ is filled with a large number $M \gg 1$ of densely packed beads. Each bead belongs either to the particle (index $= 1$) or to the solvent (index $= 0$), and the shape is thus described by a binary string of length $M$. Starting from a random string, a genetic algorithm searches for a model that fits the data. Below, we briefly describe a more general approach (Svergun, 1999), which is also suitable for the analysis of contrast variation data from multicomponent complexes (*ab initio* shape determination will be a particular case of this procedure).

Let us consider a particle consisting of $K$ components with distinctly different scattering length densities (e.g., for a binary nucleoprotein complex, $K = 2$ and these components would be the protein and nucleic acid moieties). A volume sufficiently large to enclose this particle (e.g., a sphere of radius $R = D_{max}/2$) is filled with packed small spheres of radius $r_0 \ll R$ (called beads or "dummy atoms"). Each bead is assigned an index $X_j$ indicating the component ("phase") to which it belongs ($X_j$ ranges from 0 (solvent) to $K$). As the bead positions are fixed, the structure of such a model is completely described by a phase assignment (configuration) vector $X$ with $M \approx (R/r_0)^3$ components. If the beads of the $k$th phase have contrast $\Delta\rho_k$, the scattering intensity is

$$I(Q) = \left\langle \sum_{k=1}^{K} \Delta\rho_k A_k^2(\mathbf{Q}) \right\rangle_{\Omega} = \sum_{k=1}^{K} \Delta\rho_k \mathrm{I}_k(Q) + 2 \sum_{n>k} \Delta\rho_k \Delta\rho_n I_{kn}(Q), \quad \text{(III.3.13)}$$

where $A_k(\mathbf{Q})$ and $I_k(Q)$ are the scattering amplitude and intensity, respectively, from the volume occupied by the $k$th phase, and $I_{kn}(Q)$ are the cross-terms. To rapidly evaluate the scattering from such a model, spherical harmonics are employed. Each three-dimensional scattering amplitude from the individual phase is represented as a series

$$A(\mathbf{Q}) = \sum_{l=0}^{\infty} \sum_{m=-l}^{l} A_{lm}(Q) Y_{lm}(\Omega). \qquad \text{(III.3.14)}$$

Here, the spherical harmonics are angle-dependent functions $Y_{lm}(\Omega)$ defined on a surface of the unit sphere, and $A_{lm}(Q)$ are radially dependent functions. This representation, first introduced in SAS by Harrison (1969) and Stuhrmann (1970a), is extremely useful to describe the isotropic scattering and will be widely employed in the algorithms described below. Most important, the use of spherical harmonics allows one to derive a closed expression for the spherically averaged intensity that, for the multiphase model, is written as

$$I(Q) = 2\pi^2 \sum_{l=0}^{\infty} \sum_{m=-l}^{l} \left\{ \sum_{k=1}^{K} \left[ \Delta\rho_k A_{lm}^{(k)}(Q) \right]^2 + 2 \sum_{n>k} \Delta\rho_k A_{lm}^{(k)}(Q) \Delta\rho_n \left[ A_{lm}^{(n)}(Q) \right]^* \right\}, \qquad \text{(III.3.15)}$$

where the partial amplitudes are

$$A_{lm}^{(k)}(Q) = i^l \sqrt{2/\pi} v_a \sum_j j_l(Qr_j) Y_{lm}^*(\omega_j). \qquad \text{(III.3.16)}$$

Here, the sum runs over the beads of the $k$th phase, $(r_j \omega_j) = \mathbf{r}_j$ are their polar coordinates, $v_a = (4\pi r_0^3/3)/0.74$ is the displaced volume per bead.

If a set of $N_C \geq 1$ contrast variation curves $I^{(j)}{}_{\exp}(s), j = 1, \ldots, N_C$ is available, one can search for a configuration $X$ fitting the multiple curves simultaneously, that is, minimizing the overall discrepancy

$$\chi_{\text{ov}}^2(X) = \sum_{j=1}^{N_C} \chi_j^2 \qquad \text{(III.3.17)}$$

where the individual discrepancies are between the experimental and calculated curves for the given contrast.

The bead models contain thousands of elements, and many configurations may be found compatible with the experimental data. To constrain the solution, a penalty term $P(X)$ is introduced ensuring compactness and connectivity of the individual components in the resulting model. The goal function to be minimized takes the form $f(X) = \chi^2 + \alpha P(X)$, where $\alpha > 0$ is the penalty weight ensuring proper account of the constraint.

Given the large number of variables, the minimization can only be done using a Monte Carlo-type search. One of the most suitable methods is simulated annealing

(SA) (Kirkpatrick et al., 1983). The idea in this method is, having started from a random vector $X$, to perform random modifications of this vector $X$, always moving to configurations that decrease $f(X)$) but sometimes also to those that increase $f(X)$. The probability of accepting this last type of move decreases in the course of the minimization (the system is "cooled"). Initially, the temperature is high and the changes almost random, whereas at the end a configuration corresponding (nearly) to the minimum of the goal function is reached.

In the multiphase *ab initio* analysis program MONSA (Petoukhov and Svergun, 2006; Svergun and Nierhaus, 2000), assignment of a single bead is changed at each move and the amplitudes in eq. (III.3.13.14) are only updated but not recalculated. This, together with the use of spherical harmonics, accelerates the computations significantly and permits to run SA procedures requiring millions of function evaluations in reasonable times (depending on the task and on the computer, a few hours to a few days). A full-scale application of this method to analyze the contrast variation data from Met–InlB complex is given in Section III.3.3.2. For a particular case of a single-component particle ($K = 1$), this general approach reduces to the *ab initio* shape determination procedure implemented in the program DAMMIN (Svergun, 1999). This program allows one to account for *a priori* information about the particle (e.g., its anisometry and symmetry). Recently, an enhanced version of DAMMIN (called DAMMIF) was implemented (Franke and Svergun, 2009), which is faster by a factor of 25–40 and also avoids limitations of the finite search volume. Other Monte Carlo-based *ab initio* approaches are also available, including the original genetic algorithm programs DALAI_GA (Chacon 1998, 2000), GA_STRUCT (Heller et al., 2002), and SAXS3D (Bada et al., 2000).

Running any Monte Carlo shape determination program several times from random starts usually produces somewhat corresponding to nearly identical scattering patterns. The models obtained in independent runs can be superimposed and averaged to obtain both the most probable model and an averaged model. This analysis is done automatically in the program package DAMAVER (Volkov and Svergun, 2003) employing the program SUPCOMB (Kozin and Svergun, 2001) to align and superimpose two arbitrary low- or high-resolution models represented by ensembles of points. This procedure also allows one to assess the uniqueness of the solution (Volkov and Svergun, 2003).

### III.3.3.2 Examples of Practical Applications

In this section, a few practical examples are considered, illustrating the efficiency of the data analysis methods introduced above for the structural retrieval of information from macromolecular solutions using SANS. An example of shape determination from the SANS data is given by the analysis of the experimental scattering pattern from *E. coli* thioredoxin reductase (TR). The high-resolution crystal structure of this dimeric protein (MM of a monomer 34 kDa) has been solved and is available from the Protein Data Bank (PDB code: 1TDE) (Waksman et al., 1994). TR catalyzes the reduction by nicotinamide adenine dinucleotide phosphate (NADPH) of a redox-active disulfide bond in the small protein thioredoxin. The enzyme mechanism involves the transfer of reducing equivalents for reduced NAPDH to a disulfide bond in the enzyme via a flavin adenine dinucleotide (FAD). Figure III.3.3a and b displays the neutron scattering pattern of TR in $D_2O$ buffer and the distance distribution function *p*(*r*) calculated from the SANS data (Svergun et al., 1998). The linear appearance of the Guinier plot (Figure III.3.3a, insert) is typical for monodisperse systems. The curve computed from the high-resolution TR model by the program CRYSON (Svergun et al., 1998) coincides well with the experimental data (with the discrepancy factor of $\chi = 2.0$) (Figure III.3.3a, dashed curve), confirming that the dimeric crystal structure of TR is not significantly changed in solution. The low-resolution shape of TR was reconstructed *ab initio* by DAMMIN. The scattering computed from a typical DAMMIN model without symmetry restrictions (Figure III.3.3a, solid line) provides good agreement with the experimental data ($\chi = 1.5$). As one can see from the superimposition in Figure III.3.3c and d, the *ab initio* DAMMIN model of TR (obtained without knowledge of the crystallographic structure) appears fully compatible with the high-resolution crystal structure of the homodimeric TR. This demonstrates the practical usefulness of *ab initio* shape determination from SANS (and SAXS) data, which is now a routine tool employed by numerous groups (Baldock et al., 2006; Durand et al., 2006; Gherardi et al., 2006; Nagar et al., 2006; Smolle et al., 2006; Whitten et al., 2007).

Another example is devoted to the SANS analysis of a protein–protein complex and illustrates the power of specific deuteration and contrast variation using $H_2O/D_2O$ mixtures (Niemann et al., 2008). The *Listeria monocytogenes* surface protein InlB binds to the extracellular domain (ectodomain) of the human receptor tyrosine kinase Met, the product of the c-*met* proto-oncogene. InlB binding activates the Met receptor, leading to the uptake of *Listeria* into normally nonphagocytic host cells. The 32 kDa N-terminal half of InlB ($InlB_{321}$) is sufficient for Met binding and activation. The Met extracellular region consists of a large 60 kDa globular domain (containing an N-terminal Sema domain, and a cysteine-rich PSI domain) followed by four immunoglobulin-like Ig domains (about 10 kDa each). The analysis of the complexes between this Met-binding domain of InlB and various constructs of the Met ectodomain showed that $InlB_{321}$ consistently binds the Met ectodomain with a 1:1 stoichiometry. The scattering experiments were performed to elucidate the quaternary structure of the complex and, in particular, to identify how $InlB_{321}$ binds to the Met ectodomain.

SAXS experiments on the complex, although they are able to provide its overall shape, cannot distinguish between the moieties belonging to $InlB_{321}$ and Met, which makes it difficult to characterize the binding interface. Neutron contrast variation on the samples with selectively deuterated $InlB_{321}$ was therefore performed by Niemann et al. (2008), who collected a total of 7 SAXS curves and 35 contrast variation SANS curves from $InlB_{321}$, constructs of Met (containing variable number of Ig domains) and of their complexes. Rigid body modeling using the program SASREF (Petoukhov and Svergun, 2006) was employed utilizing the high-resolution or homology models of individual domains to find their configuration, simultaneously fitting the available experimental data (part of the data is displayed in Figure III.3.4a ). The Met

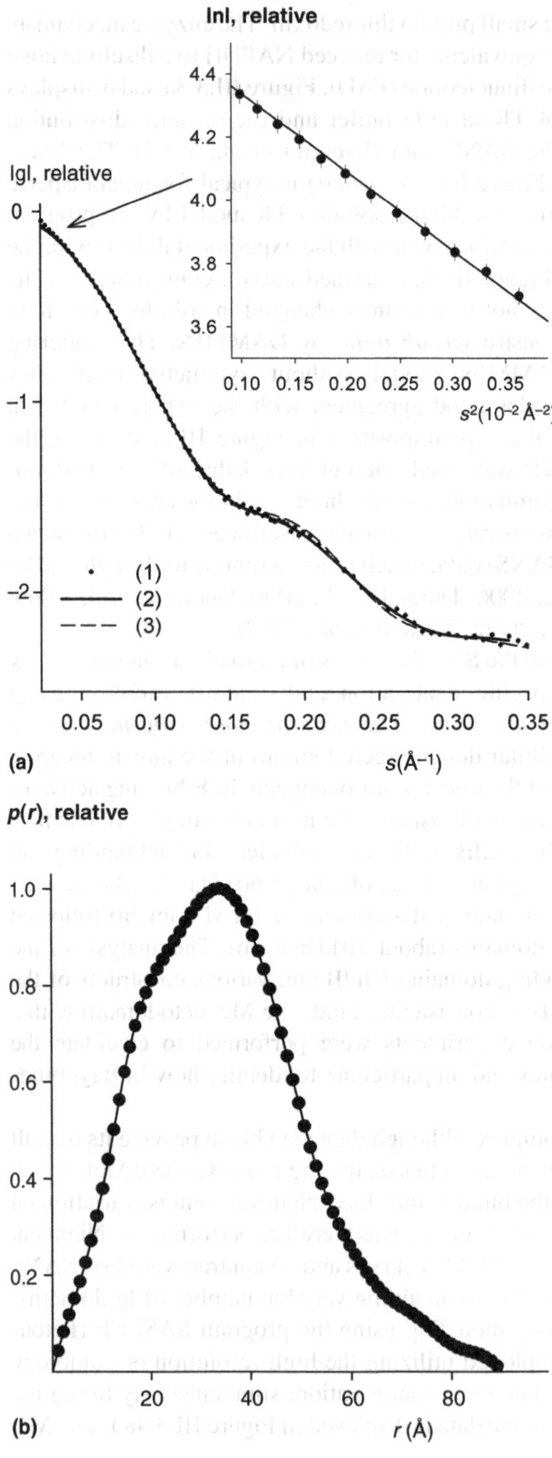


**Figure III.3.3** (a) Experimental SANS data from TR solution in $D_2O$ at 10 mg/mL: (1) experimental scattering data, (2) *ab initio* fit from the program DAMMIN (Svergun, 1999), and (3) scattering from the crystallographic homodimer calculated by the program CRYSON (Svergun et al., 1998). Guinier plot is shown in the insert. (b) The distance distribution function *p*(*r*) computed from the SANS data by GNOM. (c and d) *Ab initio* bead model obtained by DAMMIN (gray semitransparent spheres) superimposed with the crystallographic model of TR dimmer (PDB entry: 1TDE) (Waksman et al., 1994). The two monomers are displayed as $C_\alpha$ traces. Panel (d) is rotated 90° clockwise around the vertical axis.

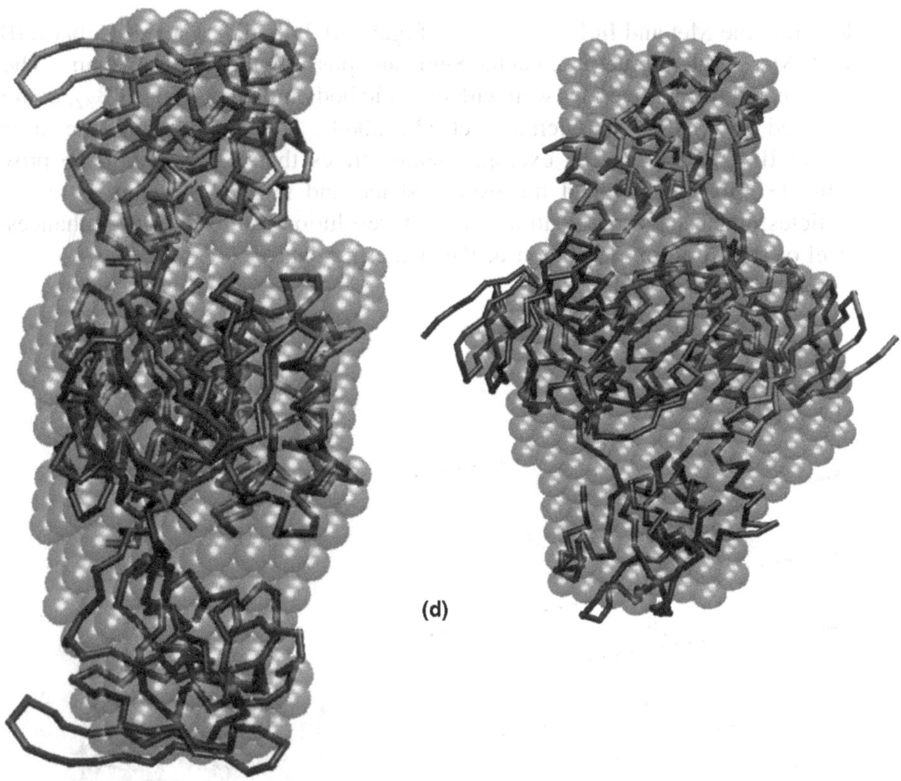


**Figure III.3.3** *(Continued)*

receptor was found to have an elongated shape, whereby the four Ig domains form a bent, rather than a fully extended, conformation. In the complex, $InlB_{321}$ binds to Sema and the first Ig domain of Met (Figure III.3.4b and c). This model is in agreement with the recent crystal structure of the shorter Met fragment in complex with $InlB_{321}$ (Niemann et al., 2007).

Here, we shall demonstrate that the overall structure of the complex and the way of $InlB_{321}$ binding can be obtained from SANS data *ab initio* without any information about the high-resolution structures of the domains. For this, we took a partial data set containing 10 SANS curves, six from the complex of full-length Met with fully (100%) deuterated $InlB_{321}$ measured in solutions with 0%, 35%, 50%, 60%, 81%, and 100% $D_2O$ and four from the complex of full-length Met with partially (50%) deuterated $InlB_{321}$ at 0%, 50%, 81%, and 100% of $D_2O$. In this case, Met–$InlB_{321}$ can be treated as two-component system for *ab initio* shape reconstruction with MONSA, which was measured with different contrast conditions. Indeed, specific deuteration of $InlB_{321}$ makes its contrast significantly different from that of Met, such that MONSA can be employed (phase 1 = Met, phase 2 = $InlB_{321}$). Several runs of MONSA allowed us to obtain reproducible *ab initio* models, neatly fitting simultaneously all 10 scattering curves (Figure III.3.4a). Moreover, the restored shape depicting the Met and $InlB_{321}$ moieties (Figure III.3.4b and c) suggests that $InlB_{321}$ binds Met in the middle (i.e., via the Sema and possibly the first Ig domain). The *ab initio* model further overlaps well with the rigid body model of Met–$InlB_{321}$ complex obtained by SASREF (Niemann et al., 2008), as evident from the overlap Figure III.3.4b and c. The example demonstrates that SANS is able to provide valuable information about the overall shape and internal structure of complex particles, and also its combination with high-resolution methods further enhances the level of structural interpretation of the scattering data.

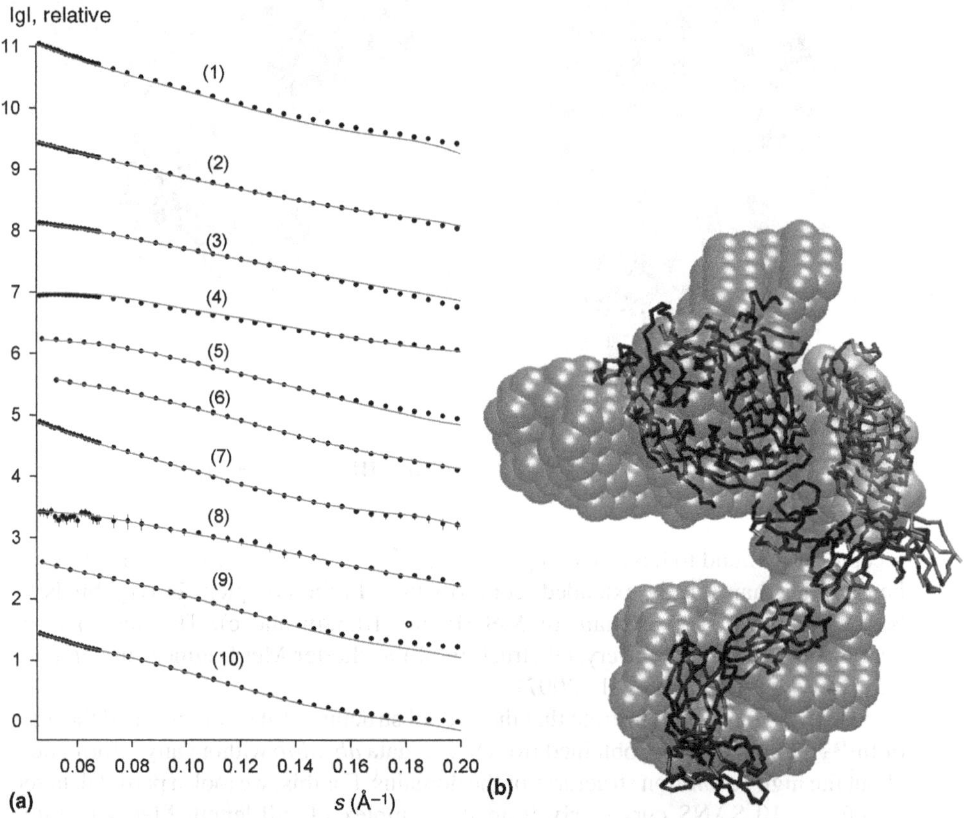


**Figure III.3.4** (a) SANS contrast variation of $InlB_{321}$:Met complex. Experimental data are denoted by dots; the fits by MONSA are presented as red solid lines. Curves (1–6) correspond to 100% deuterated $InlB_{321}$ in complex with Met measured at 0%, 35%, 50%, 60%, 81%, and 100% of $D_2O$, respectively. Curves (7–10) correspond to 50% deuterated $InlB_{321}$ in complex with Met measured at 0%, 50%, 81%, and 100% of $D_2O$, respectively. (b and c) *Ab initio* bead model of Met/Inlb complex obtained by MONSA (gray semitransparent spheres correspond to Met and orange semitransparent spheres depict $InlB_{321}$) superimposed with the rigid body model of the complex constructed by SASREF (Niemann et al., 2008). The model of full-length Met is displayed as blue $C_\alpha$ traces (top: Sema domain, bottom: Ig domains) and the $InlB_{321}$ molecule as red $C_\alpha$ traces. Panel (c) is rotated counterclockwise around the vertical axis. (See the color version of this figure in Color Plate section.)

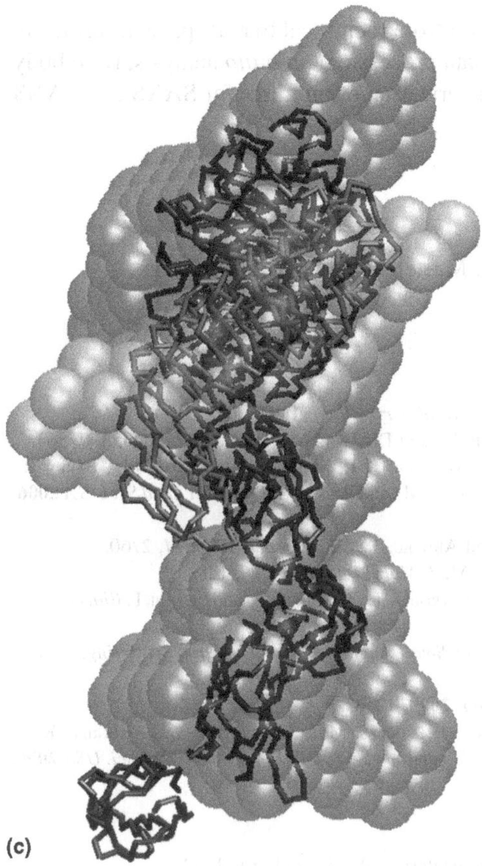

(c) Figure III.3.4 (*Continued*)

## III.3.4 CONCLUSION

The potential of small-angle scattering as a structural technique has recently been boosted from the point of view of both experimental facilities and analysis methods. In SAXS, the most notable is the advent of high brilliance third-generation synchrotron radiation sources; for SANS, further development of the detectors, neutron guides, and specific deuteration facilities are the major factors. This experimental progress is accompanied by the development of novel data analysis algorithms, which made it possible to improve the resolution and reliability of the models constructed from the SAS data. In many practical cases, the fitting of SANS contrast variation data can be combined with the simultaneous fitting of SAXS data, resulting in more reliable and stable solutions.

Many advanced modeling programs are publicly accessible on the Web. In particular, a comprehensive program suite ATSAS (Konarev et al., 2006) is available from www.embl-hamburg.de/ExternalInfo/Research/Sax/. This suite covers the major steps for converting the information from reciprocal to real space described in this section and includes programs for data processing, *ab initio* analysis, rigid body refinement, 3D visualization, and characterization of mixtures for SAXS and SANS from isotropic systems.

## ACKNOWLEDGMENT

The authors are very grateful to Dr. H. Mertens for helpful discussions.

## REFERENCES

Agamalian, M., Wignall, G.D., and Triolo, R. *J. Appl. Crystallogr.* **1997**, *30*, 345.

Bada, M., Walther, D., Arcangioli, B., Doniach, S., and Delarue, M. *J. Mol. Biol.* **2000**, *300*, 563.

Baldock, C., Siegler, V., Bax, D.V., Cain, S.A., Mellody, K.T., Marson, A., Haston, J.L., Berry, R., Wang, M.C., Grossmann, J.G., Roessle, M., Kielty, C.M., and Wess, T.J. *Proc Natl Acad Sci U S A* **2006**, *103*, 11922.

Chacon, P., Moran, F., Diaz, J.F., Pantos, E., and Andreu, J.M. *Biophys. J.* **1998**, *74*, 2760.

Chacon, P., Diaz, J.F., Moran, F., and Andreu, J.M. *J. Mol. Biol.* **2000**, *299*, 1289.

Durand, D., Cannella, D., Dubosclard, V., Pebay-Peyroula, E., Vachette, P., and Fieschi, F. *Biochemistry* **2006**, *45*, 7185.

Feigin, L.A. and Svergun, D.I. *Structure Analysis by Small-Angle X-Ray and Neutron Scattering*, Plenum Press, 1987.

Franke, D. and Svergun, D.I. *J. Appl. Crystallogr.* **2009**, *42*, 342.

Gherardi, E., Sandin, S., Petoukhov, M.V., Finch, J., Youles, M.E., Ofverstedt, L.G., Miguel, R.N., Blundell, T.L., Vande Woude, G.F., Skoglund, U., and Svergun, D.I. *Proc. Natl. Acad. Sci. USA* **2006**, *103*, 4046.

Gille, W. *J. Appl. Crystallogr.* **2003**, *36*, 850.

Glatter, O. *J. Appl. Crystallogr.* **1977**, *10*, 415.

Glatter, O. and Kratky, O. Small Angle X-Ray Scattering, Academic Press, 1982.

Goodisman, J. *J. Appl. Crystallogr.* **1980**, *13*, 132.

Grossmann, J.G. Roessle, M. Kielty, C.M. Wess, T.J. *Proc. Natl. Acad. Sci.* USA **2006**, *103*, 11922.

Guinier, A. *Ann. Phys. (Paris)* **1939**, *12*, 161.

Harrison, S.C. *J. Mol. Biol.* **1969**, *42*, 457.

Heine, S. and Roppert, J. *Acta Phys. Austriaca* **1962**, *15*, 148.

Heller, W.T., Abusamhadneh, E., Finley, N., Rosevear, P.R., and Trewhella, J. *Biochemistry* **2002**, *41*, 15654.

Kirkpatrick, S., Gelatt, C.D. Jr., and Vecci, M.P. *Science* **1983**, *220*, 671.

Koch, M.H., Vachette, P., and Svergun, D.I. *Q Rev. Biophys.* **2003**, *36*, 147.

Konarev, P.V., Petoukhov, M.V., Volkov, V.V., and Svergun, D.I. *J. Appl. Crystallogr.* **2006**, *39*, 277.

Kozin, M.B. and Svergun, D.I. *J. Appl. Crystallogr.* **2001**, *34*, 33.

Kratky, O. and Pilz, I. *Q Rev, Biophys*, **1972**, *5*, 481.

Lake, J.A. *Acta Crystallogr.* **1967**, *23*, 191.

Moore, P.B. *J. Appl. Crystallogr.* **1980**, *13*, 168.

Nagar, B., Hantschel, O., Seeliger, M., Davies, J.M., Weis, W.I., Superti-Furga, G., and Kuriyan, J. *Mol Cell* **2006**, *21*, 787.

Niemann, H.H., Jager, V., Butler, P.J., van den Heuvel, J., Schmidt, S., Ferraris, D., Gherardi, E., and Heinz, D.W. *Cell* **2007**, *130*, 235.

Niemann, H.H., Petoukhov, M.V., Hartlein, M., Moulin, M., Gherardi, E., Timmins, P., Heinz, D.W., and Svergun, D.I. *J. Mol. Biol.* **2008**, *377*, 489.

Petoukhov, M.V. and Svergun, D.I. *Eur. Biophys. J.* **2006**, *35*, 567.

Petoukhov, M.V., Konarev, P.V., Kikhney, A.G., and Svergun, D.I. *J. Appl. Crystallogr.* **2007**, *40*, s223.

Pilz, I., Glatter, O. and Kratky, O. *Z. Naturforsch. B* **1972**, *27*, 518.

Porod, G. General theory. In: Glatter, O. and Kratky, O. (editors). *Small-Angle X-ray Scattering*, Academic Press, 1982, p. 17.

Semenyuk, A.V. and Svergun, D.I. *J. Appl. Crystallogr.* **1991**, *24*, 537.

Smolle, M., Prior, A.E., Brown, A.E., Cooper, A., Byron, O., and Lindsay, J.G. *J. Biol. Chem.* **2006**, *281*, 19772.

Stuhrmann, H.B. *Acta. Crystallogr.* **1970a**, *A26*, 297.

Stuhrmann, H.B. *Z. Phys. Chem.* **1970b**, *72*, 177.

Stuhrmann, H.B. *Biochimie* **1991**, *73*, 899.

Stuhrmann, H.B. and Notbohm, H. *Proc. Natl. Acad. Sci. USA* **1981**, *78*, 6216.

Stuhrmann, H.B., Scharpf, O., Krumpolc, M., Niinikoski, T.O., Rieubland, M., and Rijllart, A. *Eur. Biophys. J.* **1986**, *14*, 1.

Stuhrmann, H.B., Goerigk, G., and Munk, B. Anomalous X-ray scattering.In: Ebashi, S., Koch, M., and Rubenstein, E. (editors). Handbook on Synchrotron Radiation, Elsevier Science Publishers, 1991.

Svergun, D.I. *J. Appl. Crystallogr.* **1992**, *25*, 495.

Svergun, D.I. *Biophys. J.* **1999**, *76*, 2879.

Svergun, D.I. and Koch, M.H.J. *Rep. Prog. Phys.* **2003**, *66*, 1735.

Svergun, D.I. and Nierhaus, K.H. *J. Biol. Chem.* **2000**, *275*, 14432.

Svergun, D.I. and Stuhrmann, H.B. *Acta Crystallogr.* **1991**, *A47*, 736.

Svergun, D.I., Volkov, V.V., Kozin, M.B., and Stuhrmann, H.B. *Acta Crystallogr.* **1996**, *A52*, 419.

Svergun, D.I., Volkov, V.V., Kozin, M.B., Stuhrmann, H.B., Barberato, C., and Koch, M.H.J. *J. Appl. Crystallogr.* **1997**, *30*, 798.

Svergun, D.I., Semenyuk, A.V., and Feigin, L.A. *Acta Crystallgr.* **1988**, *A44*, 244.

Svergun, D.I., Richard, S., Koch, M.H.J. Sayers, Z., Kuprin, S., and Zaccai, G. *Proc. Natl. Acad. Sci. USA* **1998**, *95*, 2267.

Tardieu, A. and Vachette, P. *EMBO J.* **1982**, *1*, 35.

Volkov, V.V. and Svergun, D.I. *J. Appl. Crystallogr.* **2003**, *36*, 860.

Waksman, G., Krishna, T.S.R., Williams, C.H. Jr., and Kuriyan, J. *J. Mol. Biol.* **1994**, *236*, 800.

Whitten, A.E., Jacques, D.A., Hammouda, B., Hanley, T., King, G.F., Guss, J.M., Trewhella, J., and Langley, D.B. *J. Mol. Biol.* **2007**, *368*, 407.

Willumeit, R., Burkhardt, N., Diedrich, G., Zhao, J., Nierhaus, K.H., and Stuhrmann, H.B. *J. Mol. Struct.* **1996**, *383*, 201.

Zaccai, G. and Jacrot, B. *Annu. Rev. Biophys. Bioeng.* **1983**, *12*, 139.

# III Data Treatment and Sample Environment

## III.4 Contrast Variation

*Mitsuhiro Hirai*

### III.4.1 INTRODUCTION

Small-angle neutron scattering (SANS) and small-angle X-ray scattering (SAXS) techniques provide structural information of various materials with a spatial distance ranging from $\sim$1 to $\sim 10^2$ nm (for ultrasmall-angle scattering (SAS), up to $\sim 10^3$ nm) that would correspond to an intermediate range between the regions covered by crystal structure analysis and light scattering methods. As the window of selectable thermal neutron wavelengths ($\sim$0.1–1 nm) is rather wide, neutrons would be particularly suitable for observation of structures and some heterogeneities in materials in a wider space range than with X-ray. Therefore, from the beginning of the full-scale use of SANS in 1970s to the present, SANS has been intensively used for studies such as superstructures of biological substances and phase separation phenomena of polymers and alloy systems. Due to coarse graining of objects (as they appear at low resolution) in small-angle scattering, large structures appear to be amazingly impressive at high contrast such as a ridge and scenery under the dim light of the sundown. In SAS studies of soft matters, the concept of "contrast" is of fundamental importance. Various types of contrast variation methods have been developed, especially in neutron scattering. All of them aim at the determination of structures by changing phases of scattered neutrons in various ways. This chapter briefly summarizes some representative contrast variation methods by focusing on the underlying physical nature of contrast. Although the following sections pick up a few conveniently chosen applications or simulations of contrast variation methods of biological systems, it would be needless to say that these methods are also feasible for the study of polymers and solid alloy systems.

*Neutrons in Soft Matter* Edited by Toyoko Imae, Toshiji Kanaya, Michihiro Furusaka, and Naoya Torikai

## III.4.2 CONCEPT OF CONTRAST IN SMALL-ANGLE SCATTERING AND BASIC SCATTERING FUNCTIONS

Commonly, an identifiable spatial distance $d$ resolved by some kind of experimental observation is called a structural resolution. In small-angle scattering methods, it is convenient to define two different kind of resolutions, both of which start from the modulus $q$ of the scattering vector, where $q = (4\pi \sin \theta)/\lambda$ ($2\theta$, scattering angle; $\lambda$, wavelength). As $q = 2\pi/d$, a higher structural resolution $d$ is achieved by extending the measurement to larger $q$, that is, to larger scattering angles and/or smaller wavelengths. The study of larger structures requires scattering experiments at lower $q$. It has therefore become practice to define a low-$q$ limit, which corresponds to the maximum particle size accessible to a scattering experiment. Using wavelengths smaller than 1 nm, as this is the case with X-rays or thermal neutrons, the scattering angles encountered in low-$q$ scattering are small, whence the term small-angle scattering. Structures at low resolution are described by a scattering density distribution. The scattering density at some point then is the average length $b$ of all atoms in a small cube the size of which compares to the structural resolution at the high-$q$ limit of SAS, which rarely drops below 1 nm. A volume element $\Delta v = 1\ \text{nm}^3$ contains typically 150 atoms, which is largely sufficient to give a fairly accurate value of the scattering density $\bar{\rho}$.

$$\bar{\rho} = \frac{\sum_i b_i}{\Delta v}, \tag{III.4.1}$$

where $b_i$ is the scattering amplitude of each atom in the object. The scattering amplitude of an atom (called atomic form factor) is given by Fourier transform of the scattering potential. For X-rays, it will depend on the number of the atomic electrons and their resonance state. For neutrons, it is the (spin dependent) nuclear potential and magnetic moment of the atomic electrons that define the scattering length. Note, that the variation of the atomic scattering factor of X-ray scattering is insignificant at small scattering angles and that the scattering length of neutron scattering does not vary at all with the scattering angle.

Generalizing the above concept, the huge number of atoms of the sample then is described by a scattering density distribution $\rho(\mathbf{r})$ the features of which match the structural resolution imposed the upper $q$ limit. Very often the sample contains particles embedded in a matrix. The scattering density distribution $\rho(\mathbf{r})$ giving rise to the scattering intensity is the difference between the scattering density distribution of an object (a solute particle), $\rho_s(\mathbf{r})$, and that of a matrix (a solvent or a medium where the objects are embedded), $\rho_m(\mathbf{r})$.

$$\rho(\mathbf{r}) = \rho_s(\mathbf{r}) - \rho_m(\mathbf{r}). \tag{III.4.2}$$

Furthermore, the scattering density of the solute, $\rho_s(\mathbf{r})$, can be expressed as the sum of its average scattering density $\bar{\rho}_s$, which exists the particle shape $\rho_v(\mathbf{r})$, and the residual internal scattering density function $\rho_f(\mathbf{r})$. If the scattering density distribution of the matrix is regarded to be homogeneous within an upper $q$ resolution limit and has the average value of $\bar{\rho}_m$, $\rho(\mathbf{r})$ is given as

$$\begin{aligned}\rho(\mathbf{r}) &= \{\bar{\rho}_s\rho_v(\mathbf{r}) + \rho_f(\mathbf{r})\} - \bar{\rho}_m\rho_v(\mathbf{r}) \\ &= \Delta\rho\rho_v(\mathbf{r}) + \rho_f(\mathbf{r})\end{aligned}, \qquad \text{(III.4.3)}$$

where

$$\rho_v(\mathbf{r}) = \begin{cases} 1 \text{ (inside of scatterer)} \\ 0 \text{ (outside of scatterer)} \end{cases}, \int_V \rho_f(\mathbf{r})d\mathbf{r} = 0, \ \bar{\rho}_s = \frac{1}{V}\int_V \rho_s(\mathbf{r})dv. \qquad \text{(III.4.4)}$$

($V$ is the volume of the scatterer)

$\Delta\rho = \bar{\rho}_s - \bar{\rho}_m$ is the average excess scattering density between the object and the matrix, which is called contrast (Stuhrmann and Kirste, 1965). Therefore, the observable scattering function, $I(\mathbf{q})$, is given by the Fourier transform (in following, expressed by the symbol, $F$) of the autocorrelation function of $\rho(\mathbf{r})^*\rho(-\mathbf{r})$ (*, convolution integral; $R_e$, real part) as follows:

$$\begin{aligned}I(\mathbf{q}) &= F\{\rho(\mathbf{r})^*\rho(-\mathbf{r})\} \\ &= (\Delta\rho)^2|A_v(\mathbf{q})|^2 + 2\Delta\rho R_e\{A_v(\mathbf{q})A_f^*(\mathbf{q})\} + |A_f(\mathbf{q})|^2, \\ &= (\Delta\rho)^2 I_v(\mathbf{q}) + \Delta\rho I_{vf}(\mathbf{q}) + I_f(\mathbf{q})\end{aligned} \qquad \text{(III.4.5)}$$

where $A_v(\mathbf{q})$, $A_f(\mathbf{q})$, $I_v(\mathbf{q})$, and $I_f(\mathbf{q})$ are the scattering amplitudes and scattering functions of the shape function $\rho_v(\mathbf{r})$ and the scattering density fluctuation function $\rho_f(\mathbf{r})$, respectively. $I_{vf}(\mathbf{q})$ is the scattering function of the interference term (convolution term) between the shape and fluctuation functions. $I_v(\mathbf{q})$, $I_f(\mathbf{q})$, and $I_{vf}(\mathbf{q})$ are called the basic scattering functions of the object. Figure III.4.1 shows the concept of contrast and basic scattering functions based on the separation of the scattering density function of the object into its shape and internal scattering density fluctuation functions. Clearly, we can obtain each scattering function by solving a simultaneous equation if we observe $I(\mathbf{q})$ at not less than three different contrasts. If the dissolved

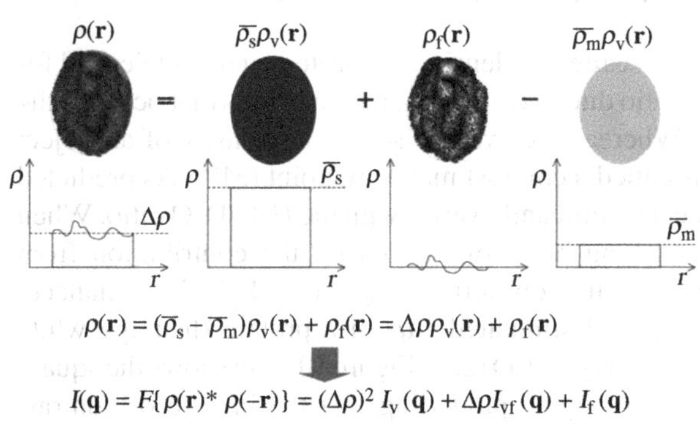


**Figure III.4.1** Concept of contrast and basic scattering functions. Separation of the scattering density function of the object into its shape and internal scattering density fluctuation functions.

particles are randomly dispersed in such as solutions, the spatial average of $I(\mathbf{q})$ is observed.

$$I(q) = \langle I(\mathbf{q}) \rangle = (\Delta\rho)^2 I_v(q) + \Delta\rho I_{vf}(q) + I_f(q). \qquad \text{(III.4.6)}$$

## III.4.3 SOLVENT CONTRAST VARIATION METHOD

When the objects disperse in a solvent, the contrast of the object can be varied by changing the average scattering density $\bar{\rho}_m$ of a solvent, which is called "solvent contrast variation method" (Stuhrmann, 1974; Ibel and Stuhrmann, 1975). In the case of SAXS, the $\bar{\rho}_m$ is varied by adding small molecules (Tardieu, 1976), such as salts, sucrose, and glycerol. However, except for the use of a quite high concentration (more than 4 M) of salt (such as NaBr), even at high concentration of small molecules it would be hard to exceed the $\bar{\rho}_s$ of an object in many cases of soft matters. In addition, some effect on the structure and interaction of the objects by addition of amounts of small molecules cannot be avoided through a change in charge and hydration on the surface of the object, or through an increase of osmotic pressure (Leneveu et al., 1977; Onai and Hirai, 2007). While, in the case of SANS it is easily varied by changing the molar ratio between the nondeuterated and deuterated solvents for SANS based on the isotope dependence of scattering amplitude of elements. The mixing of nondeuterated and deuterated solvents is expected not to alter solute structures. Thus, the solvent contrast variation method of SANS is more feasible, and has been applied to both aqueous and organic solutions. The solvent contrast variation method of SANS is based on the premise that basic scattering functions are invariant against a change of a solvent as for its use. In the case of aqueous solutions, the mixture of light and heavy water ($H_2O$ and $D_2O$) is used as a solvent. The $H_2O$–$D_2O$ mixing method is quite simple and important especially for studies of biological materials because the average scattering densities of all biological constituents (proteins, nucleic acids, lipids, oligosaccharides) for neutron are located in between those of $H_2O$ and $D_2O$ (Stuhrmann and Miller, 1978). Figure III.4.2a shows the $H_2O/D_2O$ dependence of the average scattering densities of some of biological constituents, where the scattering amplitudes and volumes of amino acids, nucleotides, and lipids given by Jacrot (1976) were used for the calculation.

In Figure III.4.2a, the average scattering densities of protein and nucleic acid for neutron vary with the $H_2O/D_2O$ ratio due to H–D exchange reactions that occur in the dissociation moieties of them. Whereas the average scattering density of an object accords with that of a solvent is called a contrast matching point (MP). As predicted from eq. (III.4.6), $I(q)$ of an object significantly varies with the $H_2O/D_2O$ ratio. When approaching to the contrast matching point of the object, the contribution from the internal scattering density fluctuation term, $I_f(q)$, is relatively enhanced. Figure III.4.3a shows the example of simulated $I(q)$ of a protein (hen-egg-white lysozyme, HEWL) depending on the $H_2O/D_2O$ ratio. Figure III.4.3b shows the square of the radius of gyration ($R_g$) of lysozyme plotted against the inverse of contrast

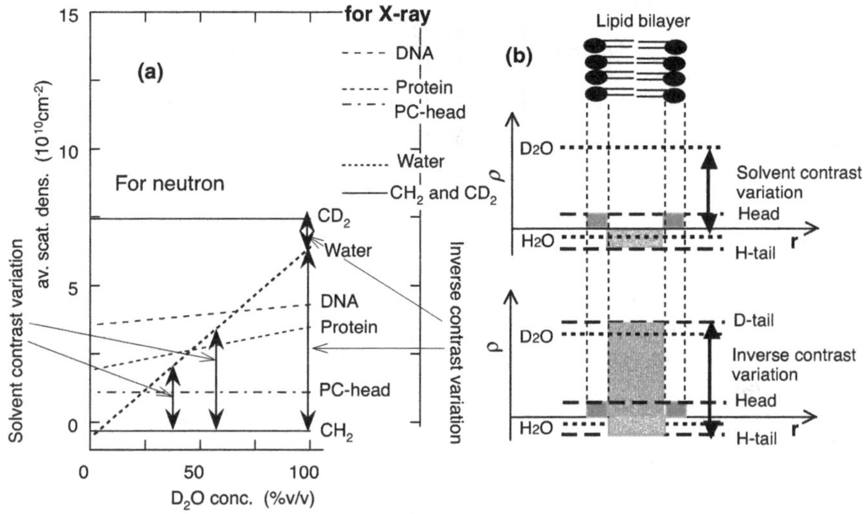


**Figure III.4.2** (a) $H_2O/D_2O$-dependence of average scattering densities of biological constituents (protein, nucleic acid, alkyl chain, and phosphocholine head of lipid) for neutron in comparison with those for X-ray, where the thick arrows indicate how to change the contrast of alkyl chain ($CH_2$) of fatty acid in solvent contrast variation and label contrast variation methods and (b) schematic difference between these methods for a model structure of lipid bilayer composed of deuterated and nondeuterated alkyl chains (also see Section III.4.4).

(called Stuhrmann plot). The dependence of $R_g$ on the contrast is quite important in SAS studies. According to the definition of $R_g$ (Guinier and Fournet, 1955, pp. 149) and eq. (III.4.3), the $R_g$ depending on the contrast was derived as follows (Stuhrmann and Kirste, 1967):

$$R_g^2 = R_{g0}^2 + \frac{\alpha}{\Delta\rho} - \frac{\beta}{(\Delta\rho)^2}, \tag{III.4.7}$$

where $R_{g0}$ (the mechanical radius of gyration of an object), $\alpha$, and $\beta$ are defined as

$$\begin{aligned} R_{g0}^2 &= \frac{1}{V}\int_V r^2 \rho_v(\mathbf{r}) d^3r, \\ \alpha &= \frac{1}{V}\int_V r^2 \rho_f(\mathbf{r}) d^3r, \\ \beta &= \frac{1}{V^2}\left\{\int_V r\rho_f(\mathbf{r}) d^3r\right\}^2. \end{aligned} \tag{III.4.8}$$

In eq. (III.4.7), the $R_{g0}^2$ is an invariant for the change of contrast. The $\alpha$ factor indicates the relative arrangement of higher and lower density regions within the object with

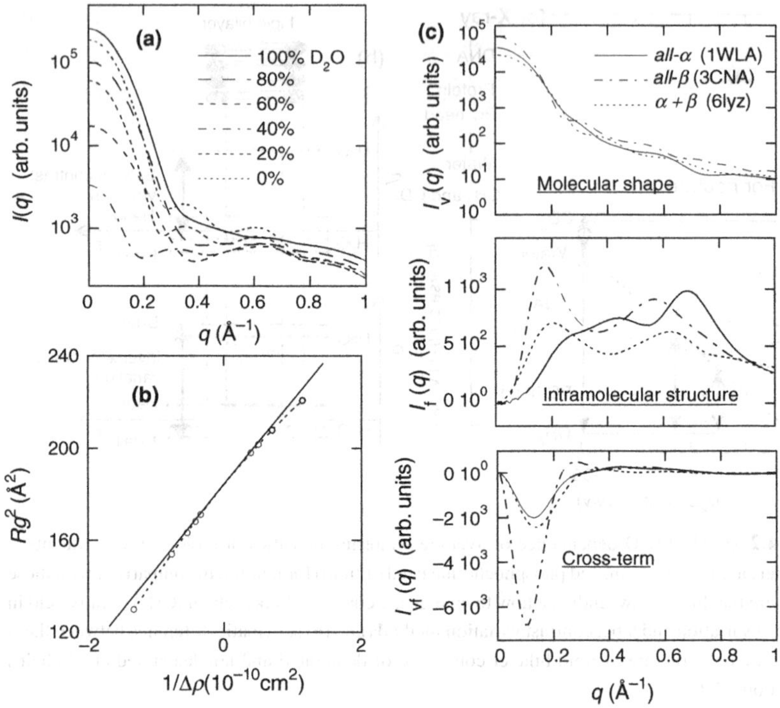


**Figure III.4.3** Simulated scattering functions and basic scattering functions of protein in solution. (a) Simulated scattering function of a protein (hen-egg-white lysozyme) depending on $H_2O/D_2O$ ratio using the program of CRYSON based on its PDB atomic coordinates (6lyz). (b) Square of radius of gyration of lysozyme depending on the inverse of contrast (called Stuhrmann plot) obtained from figure (a), reflecting internal scattering density distribution. (c) Basic scattering functions (from molecular shape, internal scattering density fluctuation of molecule, interference term between them) of three typical proteins (myoglobin, 1wla; concanavalin A, 3cna, lysozyme, 6lyz) classified in different structural categories (*all-α*, *all-β*, $\alpha + \beta$) by SCOP database.

respect to the center of gravity of its volume. If the outside region has a higher (lower) density, $\alpha$ takes a positive (negative) value. The $\beta$ factor indicates the deviation of the center of gravity of the scattering density distribution of the object from the center of gravity of the volume. So-called, the Stuhrmann plot ($R_g^2$ versus $1/\Delta\rho$) is very useful to understand an internal scattering distribution of an object consisting of different components. In Figure III.4.3b, the slope of the Stuhrmann plot, $\alpha$, is positive and the $\beta$ factor is not zero, reflecting the internal scattering density distribution of lysozyme that has already found experimentally (Stuhrmann and Fuess, 1976). The basic scattering functions of some proteins obtained from the simulated scattering functions are shown in Figure III.4.3c, where those proteins are classified as *all-α*, *all-β*, $\alpha + \beta$ proteins, respectively, according to the Structure Classification of Proteins called SCOP (Murzin et al., 1995). The theoretical SANS functions of the proteins in solutions were calculated by using a program called CRYSON (Svergun et al., 1995).

This program using a spherical harmonics expansion method (Stuhrmann, 1970) takes into account of the $H_2O/D_2O$ ratio of a solvent and of the existence of hydration shell surrounding a protein (Svergun et al., 1998) based on its atomic coordinates registered in Protein Data Bank (PDB). In Figure III.4.3 the PDB file codes of the proteins used for the theoretical calculations are given. It should be mentioned that the concept of "contrast" stands on the physical basis that solvent molecules are regarded to give homogeneous background scattering, therefore, this concept no longer holds at a high-angle scattering limit due to the finite volume of a solvent molecule, as shown by wide-angle X-ray scattering (WAXS) results (Hirai et al., 2002). In Figure III.4.3c, $I_v(q)$ reflecting the molecular shape decreases monotonously after showing a significant decrease below $q = {\sim}0.2\,\text{Å}^{-1}$. Whereas $I_f(q)$ caused by the internal scattering density distribution fluctuates up to $q = {\sim}0.8\,\text{Å}^{-1}$, which reflects well the structural characteristic of the protein inside. From the early stage of SANS studies, the separations of the basic scattering functions were performed experimentally for a few of proteins (Stuhrmann, 1973; Stuhrmann and Fuess, 1976; Stuhrmann et al., 1976a,b).

The above simulation agrees with the results of protein structures in solutions using the WAXS method with a high resolution at a third-generation synchrotron source (SR) (Hirai et al., 2002, 2004). As shown clearly, the observed WAXS function of a protein in a solution reflects well the characteristics of hierarchical structural levels within the protein. Figure III.4.4 shows the experimental and theoretical WAXS functions, where the scattering function in the regions of A ($q < {\sim}0.2\,\text{Å}^{-1}$), B (${\sim}0.25\,\text{Å}^{-1} < q < {\sim}0.5\,\text{Å}^{-1}$), C (${\sim}0.5\,\text{Å}^{-1} < q < {\sim}0.8\,\text{Å}^{-1}$), and D (${\sim}1.1\,\text{Å}^{-1} < q < {\sim}1.9\,\text{Å}^{-1}$) mostly correspond to its different hierarchical structure levels, that is, to the quaternary and tertiary structures, the interdomain correlation and intradomain structure, and the secondary structures including the closely packed side chains, respectively. A theoretical study, applying a principal components analysis to WAXS functions computed from the atomic coordinates of a set of 498 protein domains representing all of known fold space, shows that the scattering functions of proteins in solutions not only contain substantial information about those three-dimensional structures but also can be used as a powerful constraint in homology modeling of protein structures (Makowski et al., 2008). Thus, internal structures of proteins in solutions deeply relate to those functions and structural stabilities, and it should be emphasized that to get full regional information of hierarchical structures of proteins by using SAS methods important at all. Even for the recent development of SR-WAXS, the use of neutron would be more advantageous than that of X-ray for the determination of internal structures of proteins principally, since the $H_2O$–$D_2O$ mixing method used for neutron able to separate the observed data into each basic scattering function experimentally. Furthermore, a new generation high-intensity pulse neutron source is well suited for constructing a wide-angle neutron scattering (WANS) instrument that can provide scattering data with high statistics up to high-$q$ region (${\sim}5\,\text{Å}^{-1}$) by use of wide-band neutrons with time-of-flight (TOF) method, which is enough for analyzing hierarchal structures of soft matters. A pulse-WANS method will advance studies of internal structures of proteins in solutions at those functional states.

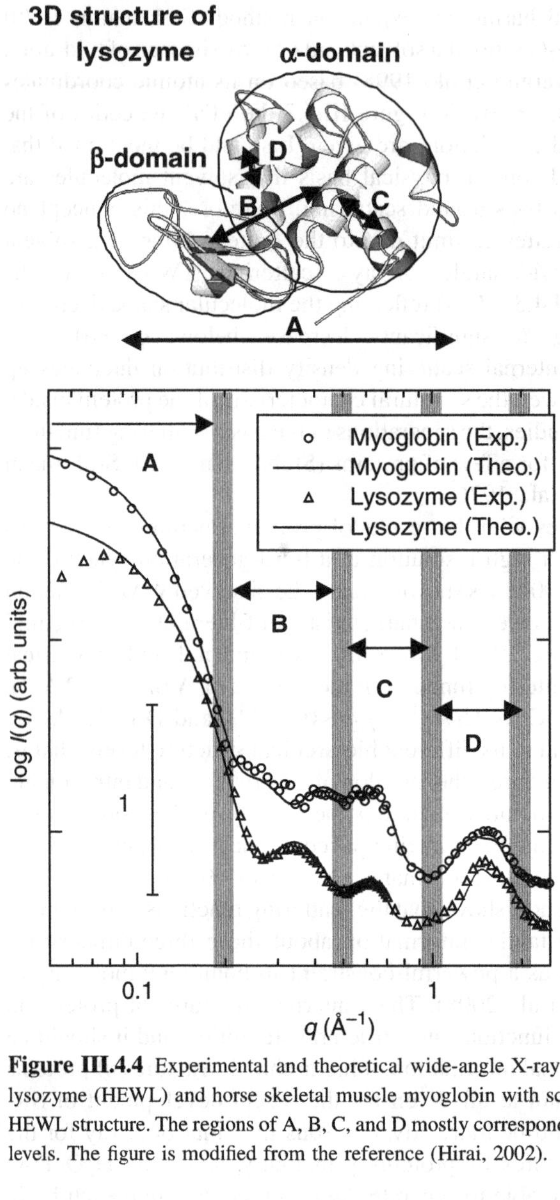


**Figure III.4.4** Experimental and theoretical wide-angle X-ray scattering curves of hen-egg-white lysozyme (HEWL) and horse skeletal muscle myoglobin with schematic image of three-dimensional HEWL structure. The regions of A, B, C, and D mostly correspond to the different hierarchical structural levels. The figure is modified from the reference (Hirai, 2002).

As the average scattering densities of proteins, nucleic acids, and lipids are greatly different, there are many cases to measure a complex consisting of these ingredients at the matching point of each ingredient. For example, in the case of biological substances, such as protein–lipid or protein–nucleic acid complex (including lipoproteins and nucleoproteins) we can roughly determine the structure of one of the ingredients in the complex by matching out the other ingredients when the internal scattering fluctuation of each ingredient is negligibly small compared with the difference in the average scattering densities of the ingredients. More precisely, in analyses using the solvent contrast variations we shall call attention to the following basic thing. The scattering density distribution function of an object consisting of $n$ units (ingredients) has to be expressed as eq. (III.4.9).

$$\rho(\mathbf{r}) = \sum_{i=1}^{n} \{\Delta\rho_i \rho_{v_i}(\mathbf{r}) + \rho_{f_i}(\mathbf{r})\}^* \delta(\mathbf{r} - \mathbf{r}_i), \tag{III.4.9}$$

where $\mathbf{r}_i$ is the position coordinate of the ingredient $i$. Then, the scattering function of the object consisting of two ingredients (ingredient 1, 2) is as follows:

$$\begin{aligned} I(q) &= \langle I(\mathbf{q}) \rangle \\ &= \{(\Delta\rho_1)^2 I_{v_1}(q) + \Delta\rho_1 I_{v_1 f_1}(q) + I_{f_1}(q)\} \\ &\quad + \{(\Delta\rho_2)^2 I_{v_2}(q) + \Delta\rho_2 I_{v_2 f_2}(q) + I_{f2}(q)\} \\ &\quad + \langle \{\Delta\rho_1 \Delta\rho_2 I_{v_1 v_2}(\mathbf{q}) \exp(i\mathbf{q} \cdot \mathbf{r}_{12})\} \rangle \\ &\quad + \langle \{\Delta\rho_1 I_{v_1 f_2}(\mathbf{q}) \exp(i\mathbf{q} \cdot \mathbf{r}_{12})\} \rangle \\ &\quad + \langle \{\Delta\rho_2 I_{f_1 v_2}(\mathbf{q}) \exp(i\mathbf{q} \cdot {}_{12})\} \rangle \end{aligned}, \tag{III.4.10}$$

where $\mathbf{r}_{12} = \mathbf{r}_1 - \mathbf{r}_2$; $I_{v_1 v_2}(\mathbf{q})$, $I_{v_1 f_2}(\mathbf{q})$, and $I_{f_1 v_2}(\mathbf{q})$ are the scattering functions of the convolution terms of shape–shape and shape–internal fluctuation of different ingredients, respectively. Cleary, even if we perform the measurement at the MP of ingredient-1 ($\Delta\rho_1 = 0$), the contribution of the scattering density fluctuation of the ingredient-1 cannot be removed principally. Therefore, for detailed discussion about only structure of each ingredient, it is necessary to measure $I(q)$ of the object at several contrasts, which is also important to avoid another artifact caused by some heterogeneity in the distribution of dissociation moieties and in those H–D exchange reactions.

In spite of unique and prominent characteristics of neutrons probing structure and dynamics of soft materials, it often becomes the problem that statistical accuracy of data are decreased by incoherent scattering from hydrogen atoms in materials, especially with the method of solvent contrast variation. Therefore, it would be useful to show how to estimate an approximate value of statistics precision of scattering data before performing SANS experiments by taking up an example of proteins in aqueous solutions. The zero-angle coherent scattering intensity from an object, $I_{\mathrm{coh}}(0)$ is given as

$$I^{\mathrm{s}}_{\mathrm{coh}}(0) = N_{\mathrm{s}} \sigma^{\mathrm{s}}_{\mathrm{coh}} = \frac{\mathrm{N_a} c d}{M_{\mathrm{s}}} \left\{ 4\pi (\Delta\rho V_{\mathrm{s}})^2 \right\}, \tag{III.4.11}$$

where $N_{\mathrm{s}}$, $\sigma^{\mathrm{s}}_{\mathrm{coh}}$, $M_{\mathrm{s}}$, $V_{\mathrm{s}}$, and $c$ are the number in unit volume, the coherent scattering cross section, the molecular volume, and the concentration of the object; $d$ and $N_{\mathrm{a}}$ are the density of the solution and the Avogadro's number, respectively. If we neglect the contribution of hydrogen atoms that the solute particles bring in a diluted solution, the incoherent scattering $I_{inc}$ from the solvent is given as follows:

$$I_{inc} = N_w \sigma_{inc}^{water} = \frac{N_a(1-cv_a)}{(1-\alpha)M_{H_2O}+\alpha M_{D_2O}}\{(1-\alpha)\sigma_{inc}^{H}+\alpha\sigma_{inc}^{D}\}, \quad \text{(III.4.12)}$$

where $N_w$ and $\sigma_{inc}^{water}$ are the number concentration and incoherent scattering cross section of water; $M_{H_2O}$, $M_{D_2O}$, and $\alpha$ are the molecular weights of $H_2O$ and $D_2O$, and those molar ratios in the solvent; $\sigma_{inc}^{H}$ ($79.91 \times 10^{-24}$ cm$^2$) and $\sigma_{inc}^{D}$ ($2.04 \times 10^{-24}$ cm$^2$) are the incoherent scattering cross sections of hydrogen and deuterium; $v_a$ is the partial specific volume of the object in the solvent. Then, the signal-to-noise (S/N) ratio of the scattering data at zero-angle can be defined as $I_{coh}^{s}(0)/I_{inc}$.

$$I_{coh}^{s}(0)/I_{inc} = \frac{cd\{(1-\alpha)M_{H_2O}+\alpha M_{D_2O}\}\{4\pi(\Delta\rho V_s)^2\}}{(1-cv_a)M_s\{(1-\alpha)\sigma_{inc}^{H}+\alpha\sigma_{inc}^{D}\}}. \quad \text{(III.4.13)}$$

In the case of lysozyme solution ($M_s = 14{,}300$ Da, $V_s = 1.77 \times 10^{-20}$ cm$^3$, $c = 0.01$ g/mL, $v_a = 0.73$, $d = \sim 1$ g/mL), the average scattering densities of lysozyme and solvent depending on $H_2O/D_2O$ ratio are given as $\bar{\rho}_s = \{1.91(1-\alpha)+3.48\alpha\} \times 10^{10}$ cm$^{-2}$ and $\bar{\rho}_{water} = \{-0.562(1-\alpha)+6.404\alpha\} \times 10^{10}$ cm$^{-2}$, respectively, for the neutron wavelength of 1 Å. As the contrast is defined by $\Delta\rho = \bar{\rho}_s - \bar{\rho}_{water}$, the $I_{coh}^{s}(0)/I_{inc}$ value is, for example, 0.38 at $\alpha = 0$, 0.07 at $\alpha = 0.3$, and 23 at $\alpha = 1$ (at the highest contrast condition of lysozyme in an aqueous solvent). Even such a simple calculation helps one to evaluate experimental measurement times. Of course, the coherent scattering from an object, $I_{coh}^{s}(q)$, significantly decreases depending on $q$ value (as shown in Figure III.4.3), whereas the incoherent scattering from a solvent, $I_{inc}$, is essentially isotropic and constant in whole $q$ range. Thus, the value of $I_{coh}^{s}(0)/I_{inc}$ decreases much more at larger $q$. In addition, the scattering function, $I(q)$, of the object is obtained from the difference between the observed scattering intensities of the solution and the solvent, then, the experimental error of $I(q)$ will be increased in about two times. Thus, it is important to minimize $I_{inc}$ in the experiments using the solvent contrast variation method. Furthermore, by considering a measurement condition (wavelength of neutron, incident beam intensity, size of an object), we can also predict the absolute value of the coherent scattering intensity from the object as shown below (Guinier and Fournet, 1955).

$$I_s(\lambda) = \varepsilon T I_0(\lambda)\lambda^2(\Delta\rho)^2 \bar{l} v_f d_s S, \quad \text{(III.4.14)}$$

where $I_s(\lambda)$ and $I_0(\lambda)$ are the absolute scattering intensity from one object and the incident beam intensity; $\varepsilon$ is the detection efficiency; $T$ is the neutron transmission of the sample; $\bar{l}$ and $v_f$ are the mean chord length and the volume fraction of the object (for a spherical object with radius $R_0$, $\bar{l} = 3R_0/2$) (Guinier and Fournet, 1955, pp. 18); $d_s$ and $S$ are the thickness and the cross section of the incident beam on the sample, respectively. In particular, the factor of $\lambda^2(\Delta\rho)^2\bar{l}v_f d_s \equiv W$ means the ratio of neutrons scattered. In the case of lysozyme of 1% w/v in $D_2O$ ($\lambda = 7 \times 10^{-8}$ cm, $d_s = 0.2$ cm,

$\bar{l} = \sim 2.5 \times 10^{-7}$ cm, and $v_f \approx 0.01 v_a = 0.0073$), the value of $W$ is ~0.0015. Then the scattered neutrons at $I_0(\lambda) = 10^5$ neutron/s/cm$^2$ is ~70 neutron/s ($T = 0.95$, $\varepsilon = 0.5$, $S = 1$ cm$^2$). By using eqs. (III.4.13) and (III.4.14) together, one can approximately predict an expected value of the statistics precision of data.

## III.4.4 LABEL TRIANGULATION METHOD

The label triangulation method uses an isotopic substitution of hydrogen and deuterium atoms in an object based on a triangular surveying method (Engelmann and Moore, 1972). It was developed at the early stage of SANS studies, almost simultaneously with the solvent contrast variation method, which is suitable for determining a spatial arrangement of the ingredients or subunits constituting a supermolecule. The label triangulation method was applied in a long-term study (for more than 20 years) to map the spatial arrangement of 21 subunit-proteins constituting the 30S ribosomal subunit of *E. coli* including 16S RNA (Capel et al., 1987). The principle is the following. For example, to determine the spatial distance between a pair of subunits (1 and 2) within the object composed of $n$ subunits, we prepare samples in four ways that the subunit-1, or the subunit-2, or both are selectively deuterated or nondeuterated. In the following, these samples are expressed with $1_H\&2_H$ (both nondeuterated), $1_D\&2_D$ (both deuterated), $1_D\&2_H$ (deuterated subunit-1), and $1_H\&2_D$ (deuterated subunit-2), respectively. Then, according to eqs. (III.4.6) and (III.4.7), the difference between the scattering functions of two solutions (equimolar mixed solutions of $[1_H\&2_H] + [1_D\&2_D]$ and $[1_D\&2_H] + [1_H\&2_D]$ is

$$\begin{aligned}\Delta I(q) &= (\Delta\rho_1^D - \Delta\rho_1^H)(\Delta\rho_2^D - \Delta\rho_2^H)\langle\Delta I(\mathbf{q})\rangle \\ &= (\Delta\rho_1^D - \Delta\rho_1^H)(\Delta\rho_2^D - \Delta\rho_2^H)\langle I_{v_1v_2}(\mathbf{q})\exp(i\mathbf{q}\cdot\mathbf{r}_{12})\rangle \\ &\approx (\Delta\rho_1^D - \Delta\rho_1^H)(\Delta\rho_2^D - \Delta\rho_2^H) I_{v_1v_2}(q)\frac{\sin(qr_{12})}{qr_{12}}\end{aligned} \qquad (III.4.15)$$

where $I_{v_1v_2}(q)$ is the scattering function of the convolution term of the shape functions of the subunit-1 and the subunit-2. Figure III.4.5 shows the schematic picture of the label triangulation method. An advantage of this method is that the contribution from the structures of other subunits (except 1 and 2 subunits) and those interferences with 1 or 2 subunit are eliminated entirely. That from the internal scattering density fluctuations of 1 and 2 subunits are also eliminated. Thus, we can take out only the distance correlation between the subunits notified. In this way, if the measurement about another pair of subunits is performed sequentially, the spatial arrangement of all constituents of a supermolecule can be determined. In addition, by the combination of radii of gyration of different pairs of subunits, $R_{g_{ij}}$ for the pair of $i$ and $j$ subunits, the radius of gyration of each subunit is also estimated.

$$R_{g_{ij}}^2 = R_{g_i}^2 + R_{g_j}^2 + r_{ij}^2, \qquad (III.4.16)$$

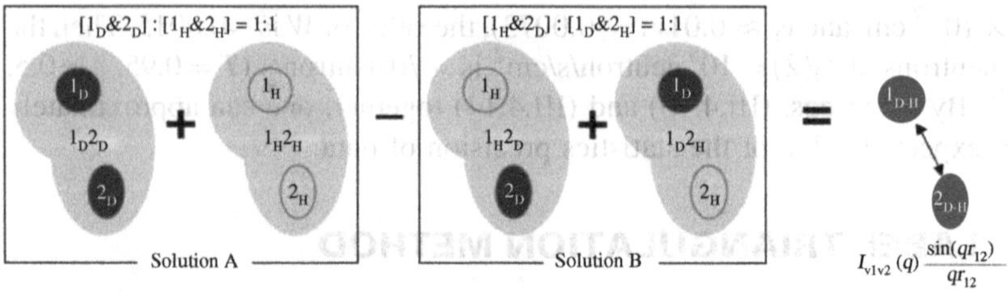


**Figure III.4.5** A schematic picture of label triangulation method. The distance correlation of the specific ingredient (1, 2) in a particle consisting of many other ingredients can be determined. Two different solutions have to be prepared. Solution A includes a particle consisting of both nondeuterated ingredients of 1 and 2 ($1_H\&2_H$) and that consisting of both deuterated ones ($1_D\&2_D$) in same amounts; solution B includes particles consisting either of 1 and 2 deuterated $1_D\&2_H$ (deuterated subunit-1), and $1_H\&2_D$ (deuterated subunit-2) in same amounts. The difference between the observed scattering functions of A and B gives the distance correlation between the ingredients of 1 and 2.

where $R_{g_{ij}}$ is obtained from the distance distribution function as follows:

$$R_{g_{ij}}^2 = \int_0^{\infty} p_{ij}(r)r^2\,\mathrm{d}r \Big/ \int_0^{\infty} p_{ij}(r)\,\mathrm{d}r$$

$$p_{ij}(r) = \int_0^{D_{\max}} qr\,\Delta I(q)\sin(qr)\,\mathrm{d}r, \quad D_{\max}: \text{maximumdimension of a system(apair).} \tag{III.4.17}$$

If one can regard each subunit as a hard sphere, its radius, $R$, is given by $R = \sqrt{3/5}R_g$. The study of 30S ribosomal subunit using the label triangulation method continued for around 20 years, and succeeded the mapping of all subunits (Capel et al., 1987). It should be noted that the label triangulation method is restricted to the determination of intersubunit distances. By definition, it does not supply any reference to the total structure. Therefore, in order to determine the total arrangement (structure) of all subunits in the supramolecule would need further structural information using such as electron microscopy as in the case of the 30S ribosome study.

## III.4.5 LABEL (INVERSE) CONTRAST VARIATION METHOD

This method is also a kind of isotopic substitution method, which changes the contrast of an object, not by varying the $H_2O/D_2O$ ratio, but by the deuteration ratio of the object. This is also called "inverse contrast variation" (Knoll and Schmidt, 1985). The basic equations are the same as in eqs. (III.4.5) and (III.4.6), and eqs. (III.4.9) and (III.4.10), whereas $\Delta\rho$ depends on the deuteration ratio. In experiments using the label contrast variation method, we usually fix the scattering density of a solvent to the average scattering density of other ingredients constituting the object (whence, the contrast of them is set to be nearly 0 in eq. (III.4.10)), and perform the SANS measurements by changing the deuteration ratio of a particular ingredient. As similar as in the case of the solvent contrast variation method, the basic scattering function of the particular ingredient can be derived from the observed scattering functions for at least three differently deuterated samples. As an extension of the label contrast variation method, there is a study of 50S ribosomal subunit of *E. coli* (constituting 5S RNA, 23S RNA, and 34 proteins) where the average scattering densities of other ingredients (proteins and rRNA) except a particular protein were skillfully deuterated (84%-deuterated proteins, 76%-deuterated rRNA) to match the average scattering density of $D_2O$ solvent, named "transparent method," and subjected to label contrast variation and label triangulation methods together (Nierhaus et al., 1983). They determined not only *in situ* structures of particular proteins but also the distance correlation between them. The advantage of the label contrast variation method is to remarkably reduce a fall of the statistics of scattering data by the incoherent scattering background from $H_2O$ and also to avoid an influence on $I_f(\mathbf{q})$ (in eq. (III.4.5)) caused by some heterogeneity in the H–D exchange reaction between the solute and the solvent. Therefore, even for simple systems such as lipid mixtures, a detailed structure analysis by using a model is possible so that scattering data with high statistics precision can be attained at several contrasts.

There is an example applying the label contrast variation method with a modeling analysis to a binary lipid system (Hirai et al., 2003). It was performed to clarify an asymmetric internal structure of lipid bilayer of a small unilamellar vesicle (SUV) composed of phospholipid (dipalmitoylphosphatidylcholine, DPPC) and glycosphingolipid (monosialoganglioside, $G_{M1}$) as a model of so-called "lipid raft." The strategy used in this study is shown in Figure III.4.2b. Lipid rafts, or called glycosphingolipids (GSLs) signaling microdomains, in plasma membrane have been proposed to have functions as platforms of membrane-associated events such as signal transduction, cell adhesion, lipid/protein sorting, and so on (Simons and Ikonen, 1997, 2000). Gangliosides are major components of GSLs and rich in central nervous systems. The functions of lipid rafts have been assumed to closely relate to the peculiar features of GSLs. Figure III.4.6a shows the SANS profile of the SUV of $G_{M1}$–DPPC mixture (molar ratio of [nondeuterated $G_{M1}$]/[DPPC] $= 0.1/1$) depending on the change of the inverse contrast, where the contrast was varied by changing the molar ratio between deuterated and nondeuterated DPPC as [d-DPPC]/[h-DPPC] $= 1/0$, 0.7/0.3, 0.3/0.7, 0/1. In spite of 1% w/v solute in $D_2O$ solvent and of the variation of contrast in a wide range (shown in Figure III.4.6a), the statistics of all scattering data are good enough. As shown in the inset of Figure III.4.6a, the above four samples with different contrasts for neutron give

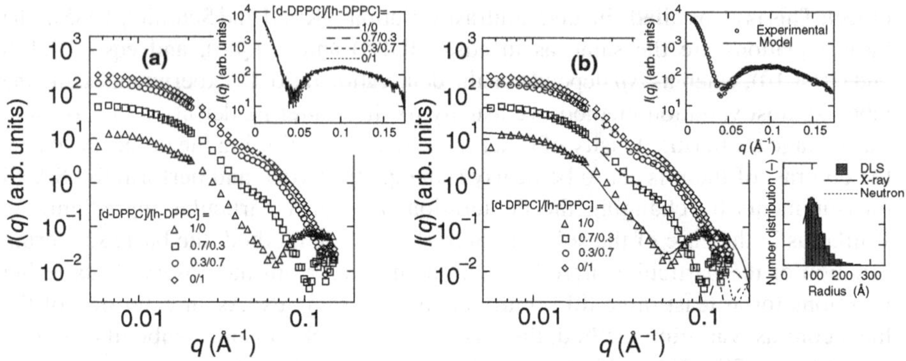


**Figure III.4.6** (a) Experimental SANS profile of $[G_{M1}]/[DPPC] = 0.1/1$ SUV depending on label contrast variation, namely on the change of the molar ratio between deuterated and nondeuterated DPPC. [d-DPPC]/[h-DPPC] = 1/0, 0.7/0.3, 0.3/0.7, 0/1, inset: experimental SAXS profiles of the above four samples and (b) best-fitted theoretical scattering functions superposed the experimental data in figure (a), where the upper inset shows the theoretical and experimental X-ray ones. The size distribution functions obtained by the DLS measurement and the modeling analysis are also shown in the lower inset. These figures are modified from the reference (Hirai et al., 2003).

the same SAXS profile, indicating the completeness of the isotopic substitution of d-DPPC and h-DPPC. In other words, Figure III.4.6a shows the scattering profiles of the same SUV at five different phases (one for X-ray, four for neutron). All experimental SANS and SAXS profiles and other parameters obtained from them (the size distribution, the contrast dependence of $I^{s}_{coh}(0)$ and $R_g$) were explained well by a unique model with a detail internal structure of bilayer, as shown in Figure III.4.6b. They used the following scattering function $I(q)$.

$$I(q) \propto \int_{R_{\min}}^{\infty} I_s(q,R)D(R)dR, \qquad \text{(III.4.18)}$$

where $D(R)$ is the size distribution function of SUVs, $I_s(q,R)$ is the form factor of the SUV with radius $R$, $R_{\min}$ is a minimum particle radius defined by the bilayer thickness. The form factor $I_s(q,R)$ of the SUV was simplified by a particle consisting of five spherical shells with different scattering densities, given by

$$I_s(q,R) = 9\{\bar{\rho}_1 V_1 j_1(qR_1)/(qR_1) + \sum_{i=2}^{n} (\bar{\rho}_i - \bar{\rho}_{i-1}) V_i j_1(qR_i)/(qR_i)\}^2, \qquad \text{(III.4.19)}$$

where $\bar{\rho}_i$ is the contrast of $i$th shell, $j_1$ is the spherical Bessel function of the first rank. Based on the above equations, all SANS and SAXS profiles in Figure III.4.6a are fitted well by a unique model as shown in Figure III.4.6b. The size distribution obtained by the fitting is also in good agreement with that

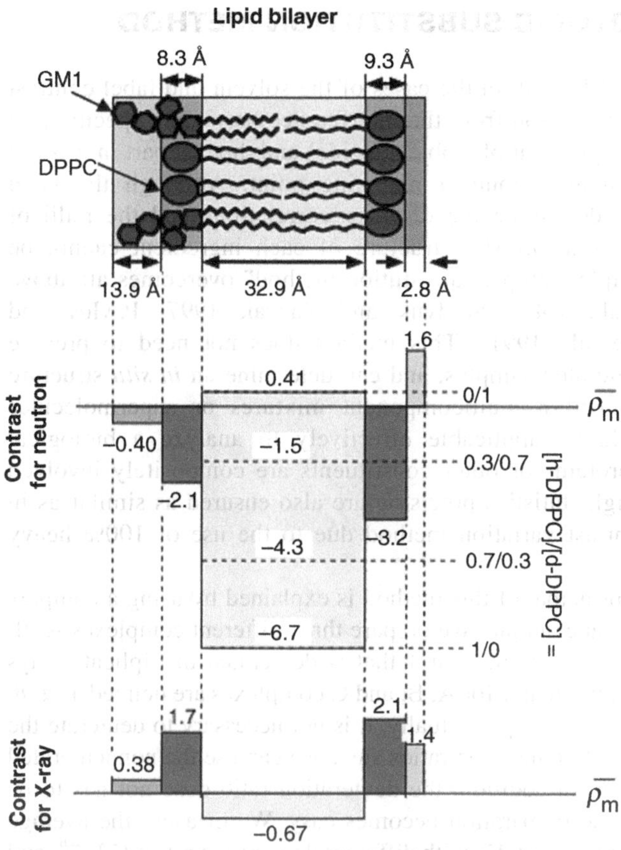


**Figure III.4.7** Asymmetric bilayer structure determined by the modeling analysis shown in Figure III.4.6b, where the obtained structural parameters within the bilayer (thickness and relative contrast of each shell) are given. The figures are modified from the reference (Hirai et al., 2003).

observed by a dynamic light scattering (DLS) measurement (the inset of Figure III.4.6b). The final model optimized by the least square fitting, shown in Figure III.4.7, clearly shows an asymmetric bilayer structure in which $G_{M1}$ molecules preferentially locate at the outer leaflet of the bilayer. The reason why they reasonably succeeded in the modeling is apparently ascribed to that all scattering data obtained at different contrasts have similar statistic precisions due to the use of the label contrast variation. As shown in the above, the method using different sources such as X-ray and neutron for changing a contrast is called "source" contrast variation method where we can use a common word, not electron or scattering length density, but scattering amplitude simply. In addition, it should be mentioned that the combination of use of neutron and X-ray is quite useful especially for SAS studies.

## III.4.6 TRIPLE ISOTOPIC SUBSTITUTION METHOD

As clearly seen from eq. (III.4.10), in the cases of the solvent and label contrast variation methods, the contribution from the interference between a specific part in a system (or a specific ingredient of a object inside) and the rest part in it is not completely removed even at a contrast matching point. Although the label triangulation method can determine the distance correlations and the radii of gyration of all ingredients, an *in situ* structure of each ingredient cannot be determined. So-called "triple isotopic substitution method" overcomes all above problems (Serdyuk et al., 1989; Serdyuk and Zaccai, 1997; Pavlov and Serdyuk, 1987; Pavlov et al., 1991). This method does not need to prepare highly or specifically deuterated samples, and can determine an *in situ* structure of a specific ingredient within multicomponent mixtures or supermolecules (assembles), which would be applicable effectively to analyze a biological reaction where various proteins or other constituents are compositely involved in. In addition, data of high statistics precision are also ensured as similar as in the case of the label contrast variation method due to the use of 100% heavy water as a solvent.

In the following, the principal of this method is explained by using a complex consisting of $n$ subunits as an example. We prepare three different complexes A, B, and C that contain a specific ingredient of 1 that is deuterated in triplicate ways (deuteration ratios of the ingredient-1 for A, B, and C complexes are defined as $\alpha$, $\beta$, and $\gamma$, respectively, where $\alpha < \beta < \gamma$). Actually, it is not necessary to deuterate the ingredient-1 at three different deuteration ratios since we can use the nondeuterated ingredient-1 as the rest one. In addition, the deuteration ratio dose not has to be defined, therefore the sample preparation becomes easy. We measure the average scattering densities of the ingredient-1's with different deuteration ratios ($\bar{\rho}_1^\alpha$, $\bar{\rho}_1^\beta$, and $\bar{\rho}_1^\gamma$), and determine $\delta$ satisfying $(1-\delta)\bar{\rho}_1^\alpha + \delta\bar{\rho}_1^\gamma = \bar{\rho}_1^\beta$. The above scattering densities would be determined not only by SANS measurements, but also by nuclear magnetic resonance measurements since the scattering density of a deuterated sample is proportional to its deuteration ratio, namely, $\delta = (\bar{\rho}_1^\beta - \bar{\rho}_1^\alpha)/(\bar{\rho}_1^\gamma - \bar{\rho}_1^\alpha) = (\beta-\alpha)/(\gamma-\alpha)$.

According to eq. (III.4.9), the scattering density distribution function of each complex ($\rho_A(\mathbf{r})$, $\rho_B(\mathbf{r})$, $\rho_C(\mathbf{r})$) is given as

$$\rho_A(\mathbf{r}) = \rho_1^\alpha(\mathbf{r})*\delta(\mathbf{r}-\mathbf{r}_1) + \sum_{i\neq 1}^{n} \rho_i(\mathbf{r})*\delta(\mathbf{r}-\mathbf{r}_i),$$

$$\rho_B(\mathbf{r}) = \rho_1^\beta(\mathbf{r})*\delta(\mathbf{r}-\mathbf{r}_1) + \sum_{i\neq 1}^{n} \rho_i(\mathbf{r})*\delta(\mathbf{r}-\mathbf{r}_i), \qquad \text{(III.4.20)}$$

$$\rho_C(\mathbf{r}) = \rho_1^\gamma(\mathbf{r})*\delta(\mathbf{r}-\mathbf{r}_1) + \sum_{i\neq 1}^{n} \rho_i(\mathbf{r})*\delta(\mathbf{r}-\mathbf{r}_i),$$

where the shape function ($\rho_v(\mathbf{r})$) and the internal scattering density fluctuation function ($\rho_f(\mathbf{r})$) of each ingredient are omitted in order to obtain a less complicated result. Here we prepare two different samples; (A + C solution) including complex A and complex C in the molar ratio $(1-\delta) : \delta$ ($\delta \leq 1$), and (B solution) including only complex B with the same molar concentration of the A + C solution. The scattering functions of these solutions are

$$\begin{aligned}
I_{\mathrm{A+C}}(q) &= (1-\delta)I_1^\alpha(q) + \delta I_1^\gamma(q) \\
&\quad + 2(1-\delta)\left\langle R_e\left\{A_1^\alpha(\mathbf{q})\sum_{i\neq 1}^{n} A_i^*(\mathbf{q})\exp(i\mathbf{q}\cdot\mathbf{r}_{1i}\right\}\right\rangle \\
&\quad + 2\delta\left\langle R_e\left\{A_1^\gamma(\mathbf{q})\sum_{i\neq 1}^{n} A_i^*(\mathbf{q})\exp(i\mathbf{q}\cdot\mathbf{r}_{1i})\right\}\right\rangle \\
&\quad + \left\langle\left\{\sum_{i\neq 1}^{n}\sum_{j\neq 1}^{n} A_i(\mathbf{q})A_j^*(\mathbf{q})\exp(i\mathbf{q}\cdot\mathbf{r}_{ij})\right\}\right\rangle, \qquad \text{(III.4.21)} \\
I_{\mathrm{B}}(q) &= I_1^\beta(q) \\
&\quad + 2\left\langle R_e\left\{A_1^\beta(\mathbf{q})\sum_{i\neq 1}^{n} A_i^*(\mathbf{q})\exp(i\mathbf{q}\cdot\mathbf{r}_{1i}\right\}\right\rangle \\
&\quad + \left\langle\left\{\sum_{i\neq 1}^{n}\sum_{j\neq 1}^{n} A_i(\mathbf{q})A_j^*(\mathbf{q})\exp(i\mathbf{q}\cdot\mathbf{r}_{ij})\right\}\right\rangle.
\end{aligned}$$

In addition, the scattering density distribution functions of the ingredient-1 deuterated differently satisfy the relation of $(1-\delta)\rho_1^\alpha(\mathbf{r}) + \delta\rho_1^\gamma(\mathbf{r}) = \rho_1^\beta(\mathbf{r})$, then $(1-\delta)A_1^\alpha(\mathbf{q}) + \delta A_1^\gamma(\mathbf{q}) = A_1^\beta(\mathbf{q})$ is given. Finally, we can obtain the difference between the scattering functions of A + C and B solutions as follows:

$$I_{\mathrm{A+C}}(q) - I_{\mathrm{B}}(q) = \delta(1-\delta)\left\langle\left|A_1^\gamma(\mathbf{q}) - A_1^\alpha(\mathbf{q})\right|^2\right\rangle \equiv \delta(1-\delta)I_1^{\gamma-\alpha}(q). \qquad \text{(III.4.22)}$$

The term of $I_1^{\gamma-\alpha}(q)$ indicates the scattering function of the ingredient-1 with the average of $(\bar{\rho}_1^\gamma - \bar{\rho}_1^\alpha)$. Thus, except for the scattering function of the ingredient-1, all other interference terms are completely removed.

Although in the above explanation we took up the case of a supramolecule, clearly, the triple isotopic substitution method is also applicable to a complex reaction system where various ingredients are involved in (Figure III.4.7a ). One should note that this method is to be feasible experimentally in neutron scattering measurements of various kinds of soft matter samples since one does not have to prepare a sample with a special deuteration ratio and can use a conventional instrument, such as SANS and neutron spin echo, if the $\delta$ value is obtained correctly.

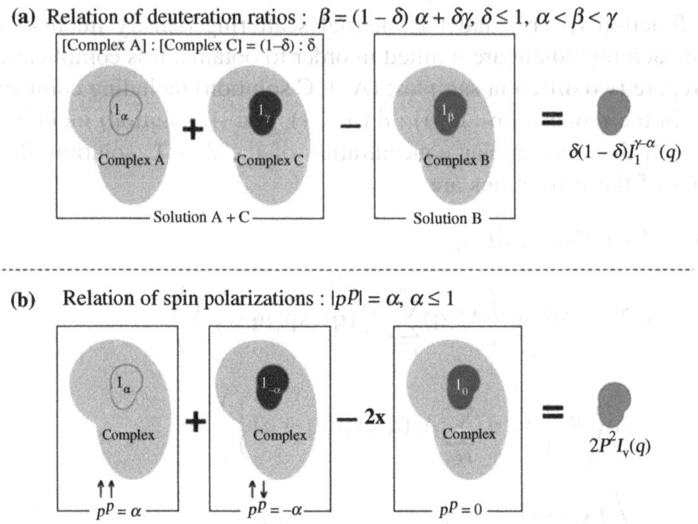


**Figure III.4.8** A schematic picture of separation of the scattering function of a specific ingredient in a multicomponent system by using triple isotopic substitution method (a) and by using spin contrast variation method (b). In (b), ↑↑ and ↑↓ show the spin states of incident neutrons and nuclei. In comparison with the label triangulation method, both methods can determine an *in situ* structure of the ingredient constituting a complex system (not only a particle system but also a reaction system).

## III.4.7 SPIN CONTRAST VARIATION METHOD

The spin contrast variation method in SANS is based on the mechanism of spin-dependent neutron scattering of polarized nuclei (Abragam et al., 1982), namely, which uses the dependency of scattering amplitude of an atom on the polarizations of nuclear spin and neutron spin. The polarizations of hydrogen, deuterium, and carbon-13 nuclear spins themselves are achieved mainly by using the dynamic nuclear spin polarization (DNP) method (Boer et al., 1974). The phenomena of nuclear spin-dependent neutron scattering were observed first from an inorganic crystal with polarized neutron diffraction (Hayter et al., 1974). The spin contrast variation method in SANS using DNP, proposed by Stuhrmann et al. (Knop et al., 1986), can be regarded as a kind of isotopic labeling methods, which has both characteristics of the label contrast variation method and the triple isomorphic substitution method. This method has been applied intensively to a series of studies on 50S ribosomal subunit and on 70S ribosome attached with tRNA and mRNA for around 10 years (Knop et al., 1992; Nierhaus et al., 1998; Junemann et al., 1998; Willumeit et al., 2001). The reason why the DNP–SANS method has been used is that with this method the scattering density of a small molecule embedded in a huge molecule is greatly enhanced and thus allows its structure determination *in situ*. For example, although mRNA fragment comprises only about 0.6% of the total mass of the 70S ribosome, Junemann et al. have succeeded the mapping of this fragment. The DNP–SANS method was also applied to polymers from the earlier stage

(Glättli et al., 1989). As the method of nuclear spin contrast variation is still less known among researchers in small-angle scattering, a short introduction into this new field might be useful.

### III.4.7.1 Fundamentals of Nuclear Spin-Dependent Scattering Amplitudes

The interaction of neutron with a nonmagnetic atom is known to be given by the Fermi scattering amplitude operator as follows (see, e.g., Lovesey, 1984).

$$\mathbf{A} = b + B\mathbf{I}\cdot\mathbf{s}, \tag{III.4.23}$$

where **I** and **s** are the spin operator of the nucleus with spin $I$ and that of the neutron with spin 1/2 ($\mathbf{s} = \mathbf{s}/2$ for neutron; **s** is the Pauli spin operator). $b$, $B$, and the incoherent cross section of the atom are expressed by the scattering amplitudes, $b^{(\pm)}$, corresponding to the eigenvalues for the two possible states of the total spin of the neutron–atom system ($I + 1/2$ and $I - 1/2$) as follows:1

$$b = \frac{(I+1)b^{(+)} + Ib^{(-)}}{2I+1}, \quad B = \frac{b^{(+)} - b^{(-)}}{2I+1}, \sigma_{\mathrm{inc}} = 4\pi\frac{(I+1)I}{(2I+1)^2}(b^{(+)} - b^{(-)})^2. \tag{III.4.24}$$

When the directions of the spins of the neutrons and the nuclei are completely random and the average of the spins of the neutrons and of the nuclei are independent, we get $\bar{\mathbf{A}} = b$. As $b^{(+)} = 1.085 \times 10^{-12}$ cm and $b^{(-)} = -4.74 \times 10^{-12}$ cm, for example, the values of $b$, $B$, and $\sigma_{\mathrm{inc}}$ of hydrogen with $I = 1/2$ are obtained by eq. (III.4.24) to be $-0.375 \times 10^{-12}$ cm, $2.92 \times 10^{-12}$ cm, and $79.9 \times 10^{-24}$ cm$^2$, respectively. Table III.4.1 lists the values of $b$, $B$, and $\sigma_{\mathrm{inc}}$ of other nuclei relating to biological systems in comparison with those for X-ray. It should be mentioned that the contrast variation methods in the above sections only use the first term in eq. (III.4.23) that is independent on the spins of nuclei. In the case of the spin contrast variation method, we definitely use the scattering amplitude $B$ for the nuclei with spin, such as

**Table III.4.1** Neutron and X-Ray Scattering Amplitudes of Some Relevant Atoms

| Nucleus | Spin $I$ | $b$ ($10^{-12}$cm) | $B$ ($10^{-12}$cm) | $\sigma_{\mathrm{inc}}$ ($10^{-24}$cm$^2$) | $f_{\text{X-ray}}$ ($10^{-12}$cm) |
|---|---|---|---|---|---|
| $^1$H | 1/2 | −0.374 | 2.912 | 79.9 | 0.28 |
| $^2$H (D) | 1 | 0.667 | 0.285 | 2.04 | 0.28 |
| $^{12}$C | 0 | 0.665 | 0 | 0 | 1.69 |
| $^{14}$N | 1 | 0.937 | 0.14 | 0.49 | 1.97 |
| $^{16}$O | 0 | 0.580 | 0 | 0 | 2.25 |
| $^{31}$P | 1/2 | 0.517 | 0.026 | 0.006 | 4.23 |
| $^{32}$S | 0 | 0.285 | 0 | 0 | 4.5 |

$b$ and $B$ are spin-independent and spin-dependent coherent scattering lengths of neutron, $\sigma_{\mathrm{inc}}$ is the incoherent neutron scattering cross section, and $f_{\text{X-ray}}$ is the scattering amplitude for X-ray (1 Å).

hydrogen, then, we are able to vary the "contrast" by using the second term (the spin-dependent scattering amplitude) in eq. (III.4.23) and to get a large gain of measurable scattering intensities.

As shown in Table III.4.1, the $B$ values of hydrogen and deuterium nuclei with $I \neq 0$ are relatively large in comparison with those of other nuclei. Therefore, the spin contrast variation is effectively applicable to materials containing amounts of hydrogen atoms (also deuterium atoms, by deuteration), such as polymers and biological materials. For neutrons with polarization $p = 1$, the spin-dependent scattering amplitudes ($A_{\mathrm{H}}$, $A_{\mathrm{D}}$) of hydrogen and deuterium nuclei with polarization $P_{\mathrm{H}}$ and $P_{\mathrm{D}}$ are given as follows:

$$\begin{aligned} A_{\mathrm{H}} &= (-0.374 + 1.456P_{\mathrm{H}}) \times 10^{-12}\ \mathrm{cm}, \\ A_{\mathrm{D}} &= (0.667 + 0.27P_{\mathrm{D}}) \times 10^{-12}\ \mathrm{cm}. \end{aligned} \tag{III.4.25}$$

Thus, by varying simply the product of the spin polarizations of neutron and hydrogen (or deuterium) nuclei in the range of $-1 \leq P_{\mathrm{H}}$ (or $P_{\mathrm{D}}$) $\leq 1$, the scattering amplitude and the scattering cross section of hydrogen nucleus can be reinforced in ~5 and ~25 times at $P_{\mathrm{H}} = -1$, respectively.

### III.4.7.2 Definition of Basic Scattering Functions and Spin Contrast

We consider a situation that neutrons with polarization $p$ impinge on a cluster composed of $M$ spinless nuclei ($I = 0$) and $N$ nuclei with spin ($I \neq 0$) and that there is no correlation among the relative orientations of those spins (both spins of neutron and nuclei are polarized along the same axis). We separately describe the structure of the cluster with two different scattering density distribution functions, namely, the polarization-independent function $\rho_{\mathrm{U}}(\mathbf{r})$ and the polarization-dependent function $\rho_{\mathrm{V}}(\mathbf{r})$.

$$\begin{aligned} \rho_{\mathrm{U}}(\mathbf{r}) &= \sum_{i=1}^{M+N} b_i \delta(\mathbf{r}-\mathbf{r}_i), \\ \rho_{\mathrm{V}}(\mathbf{r}) &= \sum_{i=1}^{N} B_i I_i P_i \delta(\mathbf{r}-\mathbf{r}_i), \end{aligned} \tag{III.4.26}$$

where $\mathbf{r}_i$ is the position of the $i$th nucleus; that has the $B_i$, $I_i$, and $P_i$ are the spin-dependent amplitude, the nuclear spin, and the polarization of $i$th nucleus. Then the two different scattering amplitudes are obtained by the Fourier transform of $\rho_{\mathrm{U}}(\mathbf{r})$ and $\rho_{\mathrm{V}}(\mathbf{r})$.

$$\begin{aligned} U(\mathbf{q}) = F\{\rho_{\mathrm{U}}(\mathbf{r})\} &= \sum_{i=1}^{M+N} b_i \exp(-i\mathbf{q}\cdot\mathbf{r}_i), \\ V(\mathbf{q}) = F\{\rho_{\mathrm{V}}(\mathbf{r})\} &= \sum_{i=1}^{N} P_i B_i I_i \exp(-i\mathbf{q}\cdot\mathbf{r}_i), \end{aligned} \tag{III.4.27}$$

where $U(\mathbf{q})$ is the invariant amplitude against the spin polarization and $V(\mathbf{q})$ is the polarization-dependent one. In the case that the clusters are orientated randomly and contain only one nuclear species with nonzero spin (spin $I$, polarization $P$, spin-dependent amplitude $B$), the coherent scattering function and the incoherent scattering cross section are simplified as (Abragam et al., 1982)

$$\begin{aligned} I_{\rm coh}(q) &= \left\langle |U(\mathbf{q})|^2 \right\rangle + 2p\langle R_e\{U(\mathbf{q})V^*(\mathbf{q})\}\rangle + \left\langle |V(\mathbf{q})|^2 \right\rangle, \\ &= I_{\rm U}(q) + pPI_{\rm UV}(q) + P^2 I_{\rm V}(q) \end{aligned} \tag{III.4.28}$$

$$\begin{aligned} \sigma_{\rm inc} &= 4\pi \sum_{i=1}^{N} B_i^2 \{I_i(I_i+1) - pP_i I_i - P_i^2 I_i^2\}. \\ &= 4\pi N B^2 \{I(I+1) - pPI - P^2 I^2\} \end{aligned} \tag{III.4.29}$$

In eq. (III.4.28), the terms of $I_{\rm U}(q)$, $I_{\rm UV}(q)$, and $I_{\rm V}(q)$ are the basic scattering functions in the spin contrast variation method. $I_{\rm U}(q)$ is equivalent to $I(q)$ in eq. (III.4.6) that is independent on spin polarization. According to eq. (III.4.28), clearly each basic scattering function can be separated by a series of measurements of $I_{\rm coh}(q)$ at different combinations of the nuclear and neutron spin polarizations, in a way which is quite similar to that of eq. (III.4.6). Experimentally we measure $I_{\rm coh}(q)$ at $P = 0$, $pP = \alpha$, and $pP = -\alpha$ since the direction of the polarization of the incident neutron (where, $|p|$ = constant) can be turned over promptly by using a neutron spin flipper. $I_{\rm coh}(q)$ at each spin combination is

$$\begin{aligned} I_{\rm coh}^{P=0}(q) &= I_{\rm U}(q) \\ I_{\rm coh}^{pP=\alpha}(q) &= I_{\rm U}(q) + \alpha I_{\rm UV}(q) + P^2 I_{\rm V}(q). \\ I_{\rm coh}^{pP=-\alpha}(q) &= I_{\rm U}(q) - \alpha I_{\rm UV}(q) + P^2 I_{\rm V}(q) \end{aligned} \tag{III.4.30}$$

Then,

$$\begin{aligned} I_{\rm coh}^{P=0}(q) &= I_{\rm U}(q) \\ I_{\rm UV}(q) &= [I_{\rm coh}^{pP=\alpha}(q) - I_{\rm coh}^{pP=-\alpha}(q)]/(2\alpha) \quad . \\ I_{\rm V}(q) &= [I_{\rm coh}^{pP=\alpha}(q) + I_{\rm coh}^{pP=-\alpha}(q) - 2I_{\rm coh}^{P=0}(q)]/(2P^2) \end{aligned} \tag{III.4.31}$$

It is important that the derivation process and meaning of $I_{\rm V}(q)$ are as same as those of $I_1^{\alpha-\gamma}(q)$ in eq. (III.4.22), suggesting that the $I_{\rm V}(q)$ obtained reflects only a spin-labeled component in the cluster without any interference from other parts. Thus, the spin contrast variation method realizes the same advantage of the label contrast variation method (shown in Figure III.4.8b ), and, in addition, would overcome its some difficulty in the sample preparation based on a correct $\delta$ value in eq. (III.4.22).

It is useful to introduce a definition of contrast in this technique as in the same manner of eqs. (III.4.1)–(III.4.4). The zero-angle scattering intensity of $I_{\text{coh}}(q)$ in eq. (III.4.28) is given by

$$I_{\text{coh}}(0) = \left[\sum_{i=1}^{M+N} b_i\right]^2 + 2NpPBI \sum_{i=1}^{M+N} b_i + (NPBI)^2. \tag{III.4.32}$$

When $|p| \approx 1$ that is mostly available for cold neutrons, the above equation can be rewritten as

$$I_{\text{coh}}(0) = \left[\sum_{i=1}^{M+N} b_i \pm NPBI\right]^2. \tag{III.4.33}$$

Then we can define the polarization-independent and polarization-dependent average scattering densities of the cluster with volume $V$.

$$\bar{\rho}_{\text{U}} = \sum_{i=1}^{M+N} b_i/V, \quad \bar{\rho}_{\text{V}}(P) = NBIP/V. \tag{III.4.34}$$

In the case of the object clusters dispersed (or embedded) in a matrix, the two different contrasts of the cluster, $\Delta\rho_{\text{U}}$ and $\Delta\rho_{\text{V}}(P)$ corresponding respectively to the polarization-independent and polarization-dependent portions, are also defined by

$$\Delta\rho_{\text{U}} = \bar{\rho}_{\text{U}} - \bar{\rho}_{\text{m}}^{U}, \quad \Delta\rho_{\text{V}}(P) = \bar{\rho}_{\text{V}}(P) - \bar{\rho}_{\text{m}}^{V}(P) \equiv \Delta\rho'_{\text{V}}(P), \tag{III.4.35}$$

where $\bar{\rho}_{\text{m}}^{\text{U}}$ and $\bar{\rho}_{\text{m}}^{\text{V}}(P)$ are the polarization-independent and polarization-dependent average scattering densities of the matrix that would also contain both nuclei with and without spins. The term $\Delta\rho'_{\text{V}}(P)$, called spin contrast, is a function of $P$ in the case that the matrix contains the same one isotopic species with nonzero spin of the cluster. Therefore, the whole contrast of the cluster can be defined by $\Delta\rho \equiv \Delta\rho_{\text{U}} \pm \Delta\rho'_{\text{V}}(P)$.

### III.4.7.3 Dynamic Nuclear Spin Polarization SANS and Polarization Dependence of Scattering Densities of Biological Components

The theoretical and experimental details of DNP are explained by Abragam and Goldman (1982). By the presence of paramagnetic electrons (unpaired electrons) (with spin **s**) in compounds (called paramagnetic centers), the DNP method can align highly the nuclear spins **I** (Larmor frequency $\nu_{\text{I}}$) of hydrogen and deuterium in polymers and biological macromolecules (called polarized targets) along the direction of a high external magnetic field below 1K via microwave irradiation at a slightly off electron paramagnetic resonant (EPR) frequency of paramagnetic centers that permits simultaneous reversals of **s** and **I** (flip–flops or flip–flips) since the strong dipolar interaction between these spins scrambles the electronic and nuclear spin states. One of the spins of hydrogen and deuterium can be also depolarized selectively by irradiation at a nuclear magnetic resonant (NMR) frequency of one of them (called selective depolarization method). Under a magnetic field ($H_0$, T) the polarizations of hydrogen and deuterium nuclei and of paramagnetic electron in thermal equilibrium at temperature $T$ (by assuming all spin systems have the same spin temperature of $T$) are given by using the Brillouin functions of $I = 1/2$ for hydrogen, $I = 1$ for deuterium, and $s = 1/2$ for electron, respectively.

$$\begin{aligned} P_{\mathrm{H}} &= \tanh(h\nu_{\mathrm{H}}/\tau) \\ P_{\mathrm{D}} &= 4\tanh(h\nu_{\mathrm{D}}/\tau)/\{3+\tanh^2(h\nu_{\mathrm{D}}/\tau)\}, \\ P_{\mathrm{s}} &= \tanh(h\nu_{\mathrm{e}}/\tau) \end{aligned} \tag{III.4.36}$$

where $\tau = 2k_{\mathrm{B}}T$ ($k_{\mathrm{B}}$, Boltzmann constant); $\nu_{\mathrm{H}}$, $\nu_{\mathrm{D}}$, and $\nu_{\mathrm{s}}$ are the Larmor frequencies of hydrogen, deuterium, and electron, respectively. These values are simply given as follows (Kittel, 1974):

$$\nu_{\mathrm{H}} = 42.6H_0(\mathrm{MHz}),\ \nu_{\mathrm{D}} = 6.54H_0(\mathrm{MHz}),\quad \nu_{\mathrm{s}} = 70H_0(\mathrm{GHz}). \tag{III.4.37}$$

For example, under $H_0 = 2.5$ T, the polarizations can be calculated.

| $T$ (K) | $P_{\mathrm{e}}$ | $P_{\mathrm{H}}$ | $P_{\mathrm{D}}$ |
|---|---|---|---|
| 4.2 | 3.801e – 01 | 6.088e – 04 | 1.246e – 04 |
| 1.0 | 9.329e – 01 | 2.557e – 03 | 5.234e – 04 |
| 0.5 | 9.976e – 01 | 5.113e – 03 | 1.047e – 03 |
| 0.05 | 1.000e + 00 | 5.109e–02 | 1.047e – 02 |

Thus, below 0.5 K, the polarization of paramagnetic electron is nearly 100%. In addition, at low temperature, the spin-lattice relaxation rate of nuclei (typically $1/T_{\mathrm{I}} = \sim 10^{-3}\,\mathrm{s}^{-1}$) is much smaller (slower) than that of electrons (typically $1/T_{\mathrm{s}} = \sim 10^{3}\,\mathrm{s}^{-1}$). Then, an external microwave energy at $h(\nu_{\mathrm{s}} \pm \nu_{\mathrm{I}})$ can induce simultaneous reversals of **s** and **I**, namely, either flip–flops ($h(\nu_s - \nu_{\mathrm{I}})$ or flip–flips ($h(\nu_{\mathrm{s}} + \nu_{\mathrm{I}})$), and the cycle (thermal equilibrium position of **s**—flip–flops (or flip–flips) of **s** and **I**—flip back into thermal equilibrium position of **s**) will store either "up" or "down" state of **I** ($|P_{\mathrm{I}}| \leq |P_{\mathrm{s}}|$). In other words, the DNP method can transfer a high polarization of paramagnetic electrons to that of nuclear spins by dipolar interaction between them. The condition to perform this cycle effectively is given by

$$f = \frac{N_{\mathrm{I}}N_{\mathrm{s}}}{T_{\mathrm{I}}T_{\mathrm{s}}} \ll 1, \tag{III.4.38}$$

where $N_{\mathrm{I}}$ and $N_{\mathrm{s}}$ are the numbers of nucleus and paramagnetic electron. For this purpose the concentration of paramagnetic centers should be controlled to reduce a dipolar interaction between them but to have an appropriate dipolar interaction between nucleus and paramagnetic center, simultaneously.

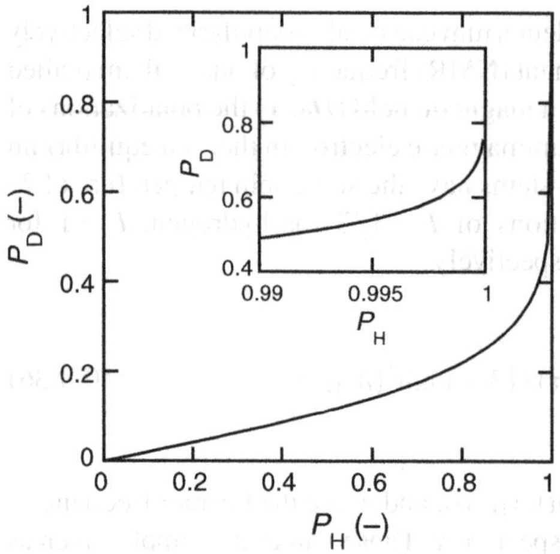


**Figure III.4.9** Relation between deuterium spin polarization $P_D$ and proton spin polarization $P_H$ at the same spin temperature ($T_H = T_D$).

In many cases of the use of DNP–SANS, samples contain both hydrogen and deuterium atoms due to partial deuteration or protiation. Figure III.4.9 shows the relation between the polarizations of hydrogen atom (proton) and deuterium one under the condition of equal spin temperature $T$ by thermal contact ($T_H = T_D = T$ in eq. (III.4.36). Figures III.4.10 and III.4.11 show the cross sections and coherent scattering amplitudes of these isotopes depending on $P_H$, respectively, where $p = 1$ in eqs. (III.4.25) and (III.4.29). Figure III.4.12 shows the simulated values of the average scattering densities of representative biological constituents depending on proton spin polarization $P_H$, where $p = 1$, and the parameters of other nuclei are in Table III.4.1, where the volumes of these components used were those given by Jacrot (1976). Under a measurement condition of DNP–SANS for aqueous solutions

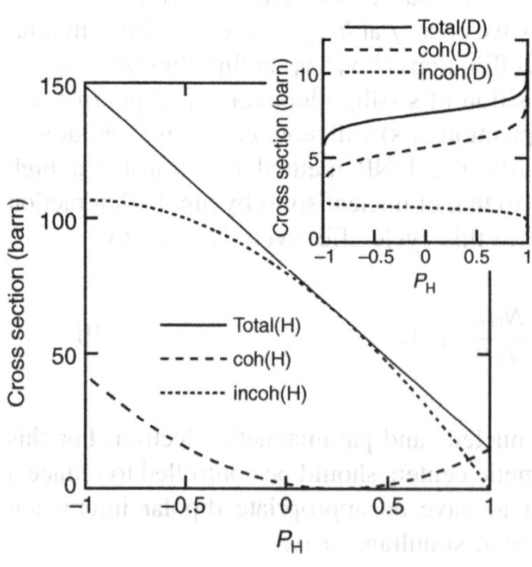


**Figure III.4.10** Coherent and incoherent cross sections of hydrogen and deuterium atoms depending on proton spin polarization $P_H$ at the same spin temperature ($T_H = T_D$), where neutron spin polarization $p = 1$.

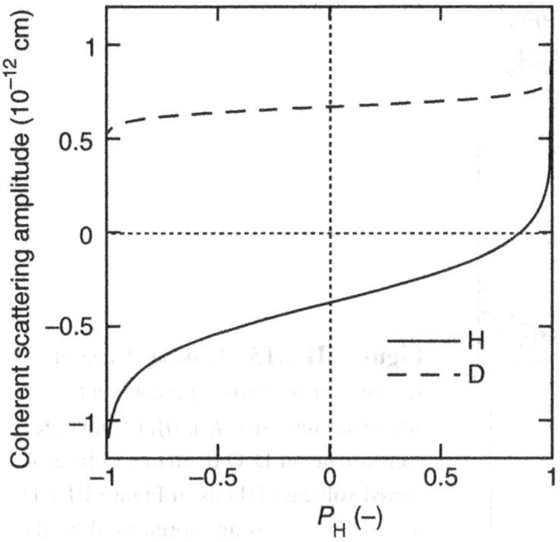


**Figure III.4.11** Coherent scattering amplitudes of hydrogen and deuterium atoms depending on proton spin polarization $P_H$ at the same spin temperature ($T_H = T_D$) and at $p = 1$.

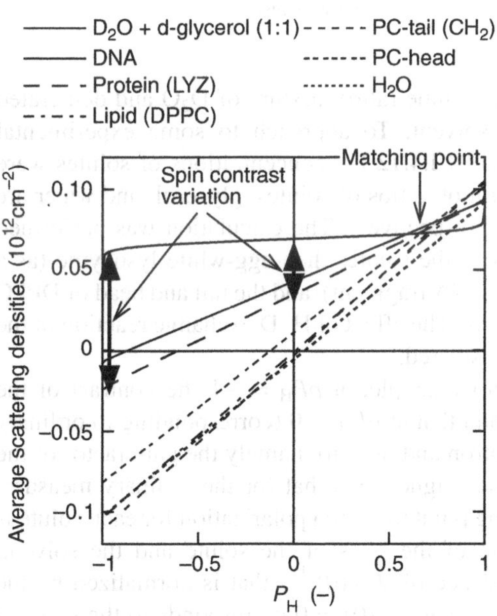


**Figure III.4.12** Average scattering densities of some relevant constituents of biological materials (protein, DNA, and the tail and head portions of phospholipid) and $D_2O$-deuterated glycerol mixed solvent (1:1) depending on proton spin polarization $P_H$ at same spin temperature ($T_H = T_D$), where neutron spin polarization $p = 1$. The arrow at $P_H = 0$ indicates the contrast at $pP_H = 0$ corresponding to the ordinary SANS measurements; the arrow at $P_H = -1$, the contrast at $pP_H = -1$ corresponding to the DNP–SANS measurements that the spin directions of neutron and proton are antiparallel. In "spin contrast variation method," the contrast is varied by proton spin polarization using a dynamic nuclear spin polarization technique.

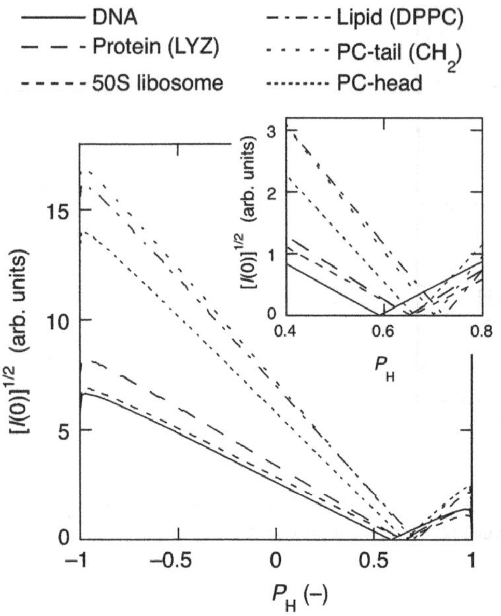


**Figure III.4.13** $P_H$ dependence of square root of zero-angle coherent scattering intensity, $I_{coh}(0)$, of biological constituents in $D_2O$-deuterated glycerol mixed solvent (1:1) as in Figure III.4.11. $[I_{coh}(0)]^{1/2}$ values are normalized by the molecular volumes. The $P_H$ value at $I_{coh}(0) = 0$ corresponds to the contrast matching point of each solute. The solutes concentrations are 0.02 g/mL for all constituents.

below 1K, we ordinary use a 1:1 (in volume ratio) mixture of $D_2O$ and deuterated glycerol (antifreezing liquid) as a solvent. To approach to some experimental conditions, in the simulation of Figure III.4.12 the concentrations of solutes were fixed to be 0.02 g/mL, and the deuteration ratios of solutes, glycerol, and water are placed to be 0%, 98.5%, and 99.5%, respectively. The calculation was performed based on the chemical compositions of the solutes: hen-egg-white lysozyme (as a protein), nucleotide of A-T-G-C (as a DNA fragment), and the tail and head of DPPC (as a glycerophospholipid), respectively. The effect of H–D exchange reaction of the solutes with the solvent was also considered.

As seen from Figure III.4.12, for example, at $pP_H = -1$ the contrast of the protein is about three times higher than that at $pP_H = 0$ (corresponding to ordinary measurements using unpolarized neutron and proton), namely the gain factor of the scattering intensity is around 10 times higher than that for the ordinary measurements. In addition, a contrast matching point to proton polarization for each solute is found from the point of intersection of the lines of the solute and the solvent. Figure III.4.13 shows the $P_H$ dependence of $[I_{coh}(0)]^{1/2}$ that is normalized by the volume of each molecule. The $P_H$ value at $I_{coh}(0) = 0$ corresponds to the contrast matching point of each solute. These values calculated are ~0.67 for PC-head, ~0.7 for PC-tail ($CH_2$), ~0.72 for DPPC (due to the presence of $CH_3$ at the end of the tail portion), ~0.65 for protein (lysozyme), and ~0.59 for DNA. The above values, especially for proteins and nucleic acids, shift depending on the compositions of amino acids and nucleotides. Then, for example, in the case of protein–nucleic acid complex, such as 50S libosome and nucleosome, one can evaluate the matching point of the complex to take a middle value from ~0.59 to ~0.65.

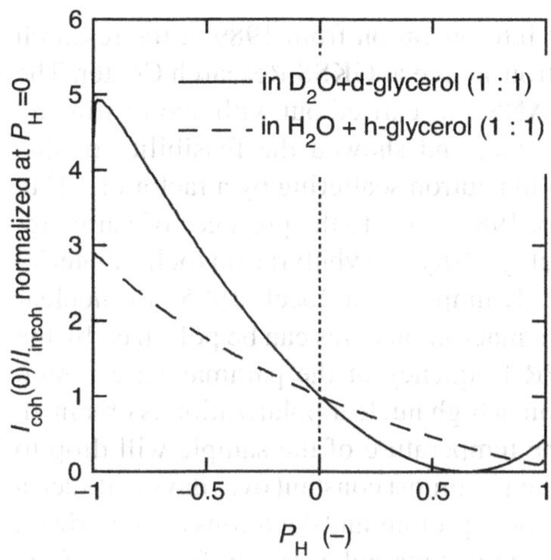


**Figure III.4.14** Signal-to-noise ratio (defined as $I_{coh}(0)/I_{inc}$) of the spin contrast variation method at $p = 1$ for 0.02 g/mL lysozyme in (99.5% $D_2O$ + 98.5% deuterated glycerol (1:1)) and (100% $H_2O$ + nondeuterated glycerol (1:1)). $I_{coh}(0)/I_{inc}$ at $P_H \neq 0$ is normalized by that at $P_H = 0$.

As same as in Section III.4.3, it would be also useful to consider the S/N ratio (defined as $I_{coh}(0)/I_{inc}$) in the case of the spin contrast variation method at $p = 1$. To compare with the case of the solvent contrast variation method using unpolarized neutrons, in Figure III.4.14 the calculated $I_{coh}(0)/I_{inc}$ ratio at $P_H \neq 0$ is normalized by that at $P_H = 0$. The calculation was performed on 0.02 g/mL lysozyme in two different water–glycerol (1:1) mixed solvents, one is 99.5% $D_2O$ + 98.5% deuterated glycerol, and other is 100% $H_2O$ + nondeuterated glycerol. At $P_H \geq 0$, the S/N ratio becomes smaller by getting closer to the contrast matching point of lysozyme ($P_H = \sim 0.65$), however, one should mention that in the solvent contrast variation method, the decrease of S/N approaching to the matching point is significantly lager than the above case. At $P_H \leq 0$, in spite of the increase of the incoherent scattering (see Figure III.4.10), the S/N ratio becomes lager in several times than that at $P_H = 0$ due to the fast increase of the contrast. Anyway, one can expect an improvement of experimental error of observed data. As in ordinal SANS measurements, the $P_H$ dependence of the scattering and transmission from a solvent, a sample cell, and so on, has to be measured and used for the background correction of scattering data from samples.

### III.4.7.4 Experimental of Dynamic Nuclear Spin Polarization SANS

The spin contrast variation method needs to use polarized neutron scattering instrument and DNP techniques. The first measurements of the scattering of polarized proton spin clusters of various proteins in deuterated medium confirmed the predicted concept of the nuclear spin contrast variation method in macromolecular structure research (Knop et al., 1986). The first SANS instrument dedicated mainly to the use of nuclear spin contrast variation method in the fields of molecular biology and polymer researches was put into operation from 1989 at the research reactor FRG-1 equipped with a cold neutron source at GKSS Research Center. The first full-scale experiment using DNP–SANS was carried out with apoferritin and deuterated 50S subunit of *E. coli* ribosome, and showed the feasibility of this method owing to the increase of the gain in neutron scattering by a factor of $10^3$ in comparison to earlier experiments (Knop, 1989). Due to the presence of paramagnetic centers, such as sodium bis(2-ethyl-2-hydroxybutyrato)oxochromate(V) monohydrate (abbreviated EHBA-Cr(V): (Krumpolc and Rocek, 1985), the nuclear spins of hydrogen and deuterium atoms in macromolecules can be polarized by the microwave irradiation at around the EPR frequency of the paramagnetic center. After some hours of microwave irradiation, a high nuclear polarization is obtained. The microwaves are switched off and the temperature of the sample will drop to about 100 mK. The nuclear polarization then is almost constant over days and such a sample is called a "frozen spin" target. As both protons and deuterons are polarized, it might become necessary to have only one isotope polarized. In fact, one of the spins can be saturated (depolarized) by supplying the radiowave at the NMR frequency of hydrogen or deuterium. At a temperature of 100 mK, the system with one isotope strongly polarized and the other isotope completely depolarized is still remarkably stable.

The main devices of a polarized target station are shown schematically in Figure III.4.15. A sample is immersed in a liquid $^4He$ bath transparent to incident

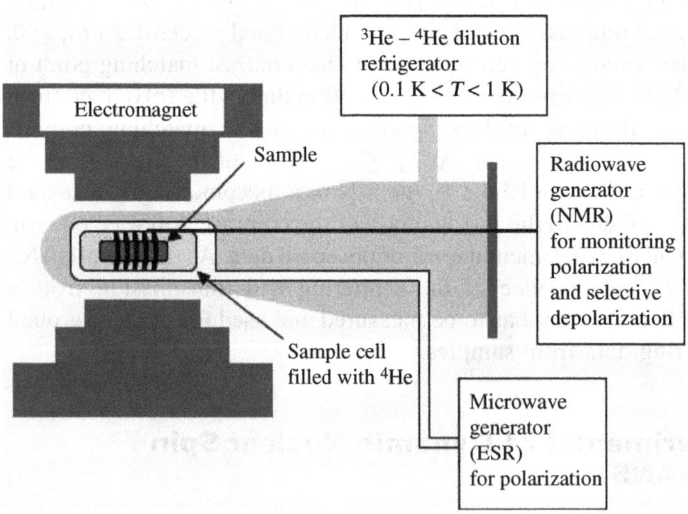


**Figure III.4.15** Schematic picture of a polarized target station used for DNP–SANS measurements. Nuclear spins of hydrogen and deuterium atoms are polarized by the microwave irradiation at around electron paramagnetic resonant (EPR) frequency of the paramagnetic center. The polarization of the target is monitored by a continues-wave NMR with a Q-meter circuit. This NMR coil is also used for selective depolarization of one of the spins by supplying the radiowave at a NMR frequency of hydrogen or deuterium after high steady states of these nuclear spin polarizations are attained.

neutrons, which is cooled by an efficient thermal coupling to the mixing chamber of a $^{3}$He –$^{4}$He dilution refrigerator that is needed to cope with the high-power input from the microwave generator. The liquid sample, for example, a solution of a protein in a glycerol–water mixture doped with a small amount of paramagnetic substance, is rapidly quenched to liquid nitrogen temperature, and is transferred into a sample holder that is connected to the top of the refrigerator. The polarization of the target is monitored by a continues-wave NMR with a Q-meter circuit that is also used for the above depolarization. Under 2.5 T magnetic field, the maximum polarization of hydrogen nuclei of water–glycerol mixed solvent was attained ~95% (Knop, 1989). The inversion of the combination of the spin directions of nuclei and neutron ($pP = \pm\alpha$) is attained immediately by using a spin flipper such as Mezei-type flat coil.

The characteristics of the spin contrast variation methods are summarized as follows:

1. The contrast of an object can be changed by using a single sample (approximately three times for proteins at $pP = -1$).
2. The coherent scattering from an ingredient labeled by hydrogen or deuterium in a multicomponent system (or complex) can be significantly reinforced selectively, especially for the case that a small special ingredient labeled by hydrogen is embedded in the large deuterated system (selective depolarization of deuterium nuclei).
3. Although a special ingredient deuterated is assumed to still contain protons, about 1–2% in regions where they are not expected to be, the *in situ* structure of the special ingredient in a multicomponent system can be determined approximately.
4. In comparison with the conventional contrast variation methods, the number of the samples used for an experiment decreases, and some labor about the specimen preparation will be reduced.

One of the problems of this method may be the doping of paramagnetic centers into samples as impurities and also the development of stable paramagnetic centers suitable for samples not to affect those properties. Another problem may be some heterogeneity or time dependency of polarization and relaxation process of nuclear spins around paramagnetic centers in a polarized target due to spin diffusion mechanism. However, in this decade, the DNP–SANS method has been progressed continuously, which proposes the use of native radical centers (Stuhrmann, 2007a, 2007b), such as radicals in proteins, or the method of the creation of paramagnetic centers at a specific part in a sample by radiation. In addition, a further advanced method using a two-channel DNP (van den Brandt et al., 2003, 2004) and its application studies (van den Brandt et al., 2006, 2007) indicate that time-resolved DNP–SANS can make use of spatial polarization gradients created around paramagnetic centers at the onset of dynamic nuclear and that such a phenomenon would be used for analyzing a local structure near paramagnetic centers in a target sample (Stuhrmann, 2008). Further details on DNP-SANS is given by Stuhrmann (2004).

## III.4.8 SUMMARY

In the above we have described about the outline of the principles of different types of contrast variation methods in neutron scattering. We did not explain SAS data analysis and treatment since there exist excellent books (Glatter and Kratky, 1982; Feigin and Svergun, 1987; the latter treats both SAXS and SANS). Needless to say, readers are recommended to refer the original papers of contrast variation methods. Which method is better for use depends not only on properties of samples but also on experimental environments to prepare them. Each method has both advantage and disadvantage. Therefore, a combination of different contrast variation methods would provide more fruitful information of a target material. Isotopic labeling, especially H–D substitution of an object in deuterated matrix, is essential important for the use for neutron to reduce incoherent backgrounds, even in the case of DNP–SANS. According to recent appreciable progresses in NMR techniques and in those studies of proteins in solutions, collaborations with NMR and neutron scientists seem to advance the use of H–D isotopic labeling for structure analyses of soft matters in the both fields. Although the DNP–SANS method needs polarized neutron scattering optics with dynamic nuclear spin polarization techniques, this method will become a key method at a new neutron source to analyze complex systems composed of various elements such as protein, lipid, and so on.

## REFERENCES

ABRAGAM, A. and GOLDMAN, M. In: KRUMHANSL, J.A., MARSHAL, W., and WILKINSON, D.A. (editors), *Nuclear Magnetism: Order and Disorder*, Clarendon Press, Oxford, 1982.

ABRAGAM, A., BACCHELLA, G.L., COUSTHAM, J., GLÄTTLI, H., FOURMOND, M., MALINOVSKI, A., MERIEL, P., PINOT, M., and ROUBEAU, P. The interest of spin dependent neutron nuclear-scattering amplitudes. *J. de Phys.* **1982**, *43*, 373.

BOER, W.D., BORGHINI, M., MORIMOTO, K., NIINIKOSKI, T.O., and UDO, F. Dynamic polarization of protons, deuterons, and C-13 nuclei - Thermal contact between nuclear spins and an electron spin-spin interaction reservoir. *J. Low Temp. Phys.* **1974**, *15*, 249.

CAPEL, M.S., ENGELMAN, D.M., FREEBORN, B.R., KJELDGAARD, M., LANGER, J.A., RAMAKRISHNAN, V., SCHINDLER, D.G., SCHNEIDER, D.K., SCHOENBORN, B.P., SILLERSIY, I.Y., YABUKI, S., and MOORE, P.B. A complete mapping of the proteins in the small ribosomal-subunit of Escherichia-coli. *Science* **1987**, *238*, 1403.

ENGELMANN, D.M. and MOORE, P.B. New method for determination of biological quaternary structure by neutron-scattering. *Proc. Natl. Acad. Sci. USA* **1972**, *69*, 1997.

FEIGIN, L.A. and SVERGUN, D.I. In: TAYLOR, G.W. (editor), *Structure Analysis by Small-Angle X-Ray and Neutron Scattering*, Plenum Press, 1987.

GLATTER, O.and KRATKY, O. (editor). *Small Angle X-Ray Scattering*, Academic Press, 1982.

GLÄTTLI, H., FERMON, C., EISENKREMER, M., and PINOT, M. Small-angle neutron-scattering with nuclear-polarization of polymers. *J. Phys. France* **1989**, *50*, 2375.

GUINIER, A. and FOURNET, G. *Small-angle scattering of X-rays*, John Wiley & Sons, Inc., New York, 1955.

HAYTER, J.B., JENKIN, G.T., and WHITE, J.W. Polarized-neutron diffraction from spin-polarized protons—tool in structure determination. *Phys. Rev. Lett.* **1974**, *33*, 696.

HIRAI, M., IWASE, H., HAYAKAWA, T., MIURA, K., and INOUE, K. Structural hierarchy of several proteins observed by wide-angle solution scattering. *J. Syn. Rad.* **2002**, *9*, 202.

HIRAI, M., IWASE, H., HAYAKAWA, T., and TAKAHASHI, H. Determination of asymmetric structure of ganglioside-DPPC mixed vesicle using SANS, SAXS, and DLS. *Biophys. J.* **2003**, *85*, 1600.

HIRAI, M., KOIZUMI, M., HAYAKAWA, T., TAKAHASHI, H., ABE, S., HIRAI, H., MIURA, K., and INOUE, K. Hierarchical map of protein unfolding and refolding at thermal equilibrium revealed by wide-angle X-ray scattering. *Biochemistry* **2004**, *43*, 9036.

IBEL, K.H. and STUHRMANN, B. Comparison of neutron and X-ray-scattering of dilute myoglobin solutions. *J. Mol. Biol.* **1975**, *93*, 255.

JACROT, B. The study of biological structures by neutron scattering from solution. *Rep. Prog. Phys.* **1976**, *39*, 911.

JUNEMANN, R., BURKHARDT, N., WADZACK, J., SCHMITT, M., WILLUMEIT, R., STUHRMANN, H.B., and NIERHAUS, K.H. Small angle scattering in ribosomal structure research: Localization of the messenger RNA within ribosomal elongation states. *Biol. Chem.* **1998**, *379*, 807.

KITTEL, C. *Introduction to Solid State Physics*, John Wiley & Sons, 1974.

KNOLL, W. and SCHMIDT, G. The inverse contrast variation in small-angle neutron-scattering - A sensitive technique for the evaluation of lipid phase-diagrams. *J. Appl. Cryst.* **1985**, *18*, 65.

KNOP, W., NIERHAUS, K.H., NOVOTNY, V., NIINIKOSKI, O., KRUMPOLC, M., RIEUBLAND, J. M., RIJLLART, A., SCHÄRPF, O., SCHINK, H.J., STUHRMANN, H.B., and WAGNER, R. Polarized neutron-scattering from dynamic polarized targets of biological origin. *Helv. Phys. Acta.* **1986**, *50*, 741.

KNOP, W., SCHINK, H. J., H. B. STUHRMANN, WAGNER, R., WENKOWESSOUNI, M., SCHÄRPF, O., KRUMPOLC, M., NIINIKOSKI, T. O., RIEUBLAND, M., and RIJLLART, A. Ploarized neutron-scattering by polarized protons of bovine serum-albumin in deuterated solvent. *J. Appl. Cryst.* **1989**, *22*, 352.

KNOP, W., HIRAI, M., SCHINK, H.J., STUHRMANN, H.B., WAGNER, R., ZHAO, J., SCHÄRPF, O., CRICHTON, R.P., KRUMPOLC, M., NIERHAUS, K.H., NIINIKOSKI, T.O., and RIJLLART, A. A new polarized target for neutron-scattering studies on biomolecules - 1st results from apoferritin and the deuterated 50s subunit of ribosomes. *J. Appl. Cryst.* **1992**, *25*, 155.

KRUMPOLC, M. and ROCEK, J. Chromium (V) oxidations of organic-compounds. *Inorg. Chem.* **1985**, *24*, 617.

LENEVEU, D.M., RAND, R.P., and PARSEGIAN, V.A. Measurement and modification of forces between lecitin bilayers. *Biophs. J.* **1977**, *18*, 209.

LOVESEY, S.W. *Theory of Neutron Scattering from Condensed Matter*, Vol. 1, Oxford Science Publications, 1984.

MAKOWSKI, L., RODI, D.J., MANDAVA, S., DEVARAPALLI, S., and FISCHETTI, R.F. Characterization of protein fold by wide-angle X-ray solution scattering. *J. Mol. Biol.* **2008**, *383*, 731.

MURZIN, A.G., BRENNER, S.E., HUBBARD, T., and CHOTHIA, C. SOPC - A structural classification of proteins for the investigation of sequences and structures. *J. Mol. Biol.* **1995**, *247*, 536.

NIERHAUS, K.H., LIETZKE, R., MAY, R.P., NOWOTNY, V., SCHULZE, H., SIMPSON, K., WURMBACH, P., and STUHRMANN, H.B. Shape determinations of ribosomal-proteins insitu. *Proc. Natl. Acad. Sci. USA* **1983**, *80*, 2889.

NIERHAUS, K.H., WADZACK, J. BURKHARDT, N., JÜNEMANN, R., MEERWINCK, W., WILLUMEIT, R., and STUHRMANN, H.B. Structure of the elongating ribosome: Arrangement of the two tRNAs before and after translocation. *Proc. Natl. Acad. Sci. USA* **1998**, *95*, 945.

ONAI, T. and HIRAI, M. Effect of osmotic pressure on ganglioside-cholesterol-DOPC lipid mixture. *J. Phys.: Conf. Series* **2007**, *83*, 012016.

PAVLOV, M.Y. and SERDYUK, I.N. 3-isotopic-substitutions method in small-angle neutron-scattering. *J. Appl. Cryst.* **1987**, *20*, 105.

PAVLOV, M.Y., RUBLEVSKAYA, I.N., SERDYUK, I.N., ZACCAI, G., LEBERMAN, R., and OSTANEVICH, Y.M. Experimental-verification of the triple isotopic-substitution method in small-angle neutron-scattering. *J. Appl. Cryst.* **1991**, *24*, 243.

SERDYUK, I.N. and ZACCAI, G. Triple isotopic substitution method in small-angle neutron scattering: application to some problems of structural biology. *J. Appl. Cryst.* **1997**, *30*, 787.

SERDYUK, I.N., PAVLOV, M.Y., RUBLEVSKAYA, I.N., and ZACCAI, G. A method of triple isotopic substitutions in SANS - Theory and 1st experimental results. *Physica B* **1989**, *156–157*, 461.

SIMONS, K. and IKONEN, E. Functional rafts in cell membranes. *Nature* **1997**, *387*, 569.

SIMONS, K. and IKONEN, E. Cell biology - How cells handle cholesterol. *Science* **2000** *290*, 1721.

STUHRMANN, H.B. Interpretation of small-angle scattering functions of dilute solutions and gases - A representation of structures related to a one-particle-scattering. *Acta Cryst.* **1970**, *A26*, 297.

STUHRMANN, H.B. Comparison of 3 basic scattering functions of myoglobin in solution with those from known structure in crystalline state. *J. Mol. Biol.* **1973**, *77*, 363.

Stuhrmann, H.B. Neutron small-angle scattering of biological macromolecules in solution. *J. Appl. Cryst.* **1974**, *7*, 173.

Stuhrmann, H.B. and Fuess, H. Neutron small-angle scattering study of hen egg-white lysozyme. *Acta Cryst.* **1976**, *A32*, 67.

Stuhrmann, H.B. Unique aspects of neutron scattering for the study of biological systems. *Rep. Prog. Phys.* **2004**, *67*, 1073.

Stuhrmann, H.B. Contrast variation in X-ray and neutron scattering. *J. Appl. Cryst.* **2007a**, *40*, s23.

Stuhrmann, H.B. The electron-spin-nuclear-spin interaction studied by polarized neutron scattering. *Acta Cryst.* **2007b**, *A63*, 455.

Stuhrmann, H.B. Small-angle scattering and its interplay with crystallography, contrast variation in SAXS and SANS. *Acta Cryst.* **2008**, *A64*, 181.

Stuhrmann, H.B. and Kirste, R.G. Elimination der intrapartikularen untergrundstreuung bei der rontgenkleinwinkelstreuung an kompakten teilchen (proteinen). *Z. Phys. Chem. Frankfurt* **1965**, *46*, 247.

Stuhrmann, H.B. and Kirste, R.G. Elimination of intraparticular base scattering in X-ray small angle scattering on compact particles. 2. *Z. Phys. Chem. Frankfurt* **1967**, *5–6*, 334.

Stuhrmann, H.B. and Miller, A. Small-angle scattering of biological structures. *J. Appl. Cryst.* **1978**, *11*, 325.

Stuhrmann, H.B., Haas, J., and Ibel, K. Low-angle neutron-scattering of ferritin studied by contrast variation. *J. Mol. Biol.* **1976a**, *100*, 399.

Stuhrmann, H.B., Haas, J., Ibel, K., Dewolf, B., Koch, M.H.J., Parfait, R., and Crichton, R.P. New low resolution model for 50s subunit of Escherichia-Coli ribosomes. *Proc. Natl. Acad. Sci. USA* **1976b**, *73*, 2379.

Svergun, D.I., Barberato, C., and Koch, M.H.J. CRYSOL - A program to evaluate x-ray solution scattering of biological macromolecules from atomic coordinates. *J. Appl. Cryst.* **1995**, *28*, 768.

Svergun, D.I., Richard, S., Koch, M.H.J., Sayers, Z., Kuprin, S., and Zaccai, G. Protein hydration in solution: Experimental observation by x-ray and neutron scattering. *Proc. Natl. Acad. Sci. USA* **1998**, *95*, 2267.

Tardieu, A., Mateu, L., Sardet, C., Weiss, B., Luzzati, V., Aggerbeck, L., and Scanu, A. M. Structure of human-serum lipoproteins in solution. 2. Samll-angle X-ray-scattering study of HDL3 and LDL. *J. Mol. Biol.* **1976**, *101*, 129.

van den Brandt, B., Glättli, H., Grillo, I., Hautle, P., Jouve, H., Kohlbrecher, J., Konter, J.A., Leymarie, E., Mango, S., May, R.P., Stuhrmann, H.B., and Zimmer, O. Neutron scattering from polarized proton domains. *Physica B* **2003**, *335*, 193.

van den Brandt, B., Glättli, H., Grillo, I., Hautle, P., Jouve, H., Kohlbrecher, J., Konter, J.A., Leymarie, E., Mango, S., May, R.P., Stuhrmann, H.B., and Zimmer, O. An experimental approach to the dynamics of nuclear polarization. *Nucl. Instr. Meth. A* **2004**, *526*, 81.

van den Brandt, B., Glättli, H., Grillo, I., Hautle, P., Jouve, H., Kohlbrecher, J., Konter, J.A., Leymarie, E., Mango, S., May, R.P., Stuhrmann, H.B., and Zimmer, O. Time-resolved nuclear spin-dependent small-angle neutron scattering from polarised proton domains in deuterated solutions. *Eur. Phys. J. B* **2006**, *49*, 157.

van den Brandt, B., Glättli, H., Hautle, P., Kohlbrecher, J., Konter, J.A., Michels, A., Stuhrmann, H.B., and Zimmer, O. Creating local contrast in small-angle neutron scattering by dynamic nuclear polarization. *J. Appl. Cryst.* **2007**, *40*, s106.

Willumeit, R., Forthmann, S., Beckmann, J., Diedrich, G., Ratering, R., Stuhrmann, H.B., and Nierhaus, K.H. Localization of the protein L2 in the 50 S subunit and the 70 S E-coli ribosome. *J. Mol. Biol.* **2001**, *305*, 167.

# III Data Treatment and Sample Environment

# III.5 Sample Environment: Soft Matter Sample Environment for Small-Angle Neutron Scattering and Neutron Reflectometry

*Peter Lindner,* Ralf Schweins, and Richard A. Campbell*

## III.5.1 SAMPLE ENVIRONMENT FOR SMALL-ANGLE NEUTRON SCATTERING

Modern small-angle neutron scattering (SANS) instruments provide with their variable wavelengths, and sample-to-detector and collimation distances a range of momentum transfer that is suitable for studying molecular structures on a length scale from about 1 nm up to some hundreds of nanometers. SANS, as well as other neutron techniques, thus can be used to contribute considerably to the understanding of the structure (and dynamics) of soft matter, for example, polymeric and colloidal systems, both in solution and in the bulk state.

Other scattering methods, using probes like laser light and X-rays, are—of course—employed alternatively and are in principle complementary for structure determination (Lindner and Zemb, 2002). However, the possibility of isotopic labeling of hydrogen-rich compounds in soft matter systems is the first and outstanding advantage of neutron scattering when applied to soft matter: the difference in neutron scattering lengths of hydrogen isotopes allows for an accentuation, that is, a "coloring" of molecules and specific molecular sites by replacing $^{1}H$ with $^{2}H$ , either in the solvent or in the solute (for the example of a solution). This so-called "contrast variation technique" does not change (significantly) the chemical nature of the soft matter system but has the advantage of enhancing the contrast in the

*Corresponding author.

*Neutrons in Soft Matter* Edited by Toyoko Imae, Toshiji Kanaya, Michihiro Furusaka, and Naoya Torikai

neutron scattering experiment, that is, providing a reasonably intense and calculable scattering signal as a result of the interaction of the neutron beam with the system under investigation.

A second important advantage of the neutron techniques—compared to alternative techniques using sources such as light and X-rays—is that for a neutron experiment there are fewer restrictions with respect to optical cleanness of the solution (dust problem in light scattering) and absorption (a well-known problem in small-angle X-ray scattering, SAXS).

Last, but not least, the transparency of the sample container material is crucial: soft matter systems are often in a liquid state and therefore have to be contained and sealed in sample holders. Very convenient window materials for sample holders in neutron experiments are optically transparent materials such as quartz glass (with the lowest possible Boron content; the isotope $^{10}B$ is a very strong neutron absorber) or sapphire, but metals such as vanadium, niobium, or even aluminum are also suitable in some cases.

Traditionally SANS experiments employ a variety of different equipments, which can be ranked in several classes. SANS sample environment for *equilibrium* conditions is described in Section III.5.1.1 (sample holders, furnaces, vacuum and pressure). *Nonequilibrium* sample environment is illustrated with the example of shear and flow experiments in Section III.5.1.2, followed by Section III.5.1.3 on *time-resolved and kinetic* studies (stopped-flow technique). Finally, Section III.5.1.4 describes recent, new developments by the combination of *complementary techniques* (*in situ* experiments with light scattering).

## III.5.1.1 Equilibrium Sample Environment

In what follows we are concerned with SANS experiments using neutrons with wavelengths typically in the range between 0.45 and 2.5 nm. Their high penetrating power into matter allows for typical sample thicknesses on the order of 1 mm, with a typical beam cross section on the order of 0.5–1.5 $cm^2$. These macroscopic dimensions (compared for instance to SAXS experiments) facilitate handling of the sample and its alignment with respect to the beam path. If the sample is a liquid, it is in most cases confined in an optically transparent quartz cell with defined path length (or another material transparent to neutrons, e.g., vanadium). In a standard equilibrium experiment the sample will usually be in thermodynamic equilibrium with its environment, at given chemical composition (concentration), pressure, and temperature. Without external constraints, the scattered intensity is a result of the space and time average of all molecular conformations and orientations and an (azimuthally) isotropic scattering pattern is observed on a two-dimensional multidetector.

### *III.5.1.1.1 Sample Changers*

**Standard Multiple-Position Sample Changers** The majority of SANS experiments studying soft matter are performed with motorized multiple-position

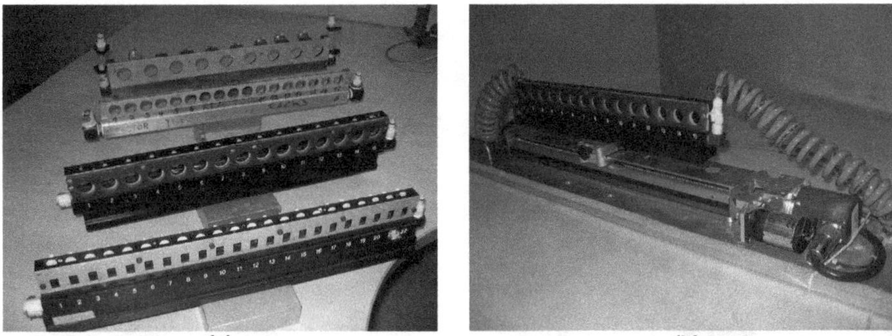

(a) (b)

**Figure III.5.1** (a) The four standard SANS sample racks at ILL; cell thicknesses up to 5 mm can be used. (b) The 15-position rack on its translation table.

sample changer racks. The racks are designed to hold a large number of standard quartz cuvettes in equidistant positions. They are fixed on a computer controlled horizontal translation table, which allows for the precise positioning of each sample in the neutron beam. Figure III.5.1a and b shows the equipment used at Institut Laue-Langevin, France (hereby abbreviated ILL). From front to back in Figure III.5.1a, the 22-position changer for HELLMA type 110 QS and 100 QS cells, the 15-position changer for HELLMA type 404 QS cells, the 17-position changer for HELLMA type 120 QS cells, and the 9-position multipurpose holder for bulk samples. Sample cell thicknesses up to 5 mm can be used.

The sample holders are made of aluminum, except the 9-position multipurpose holder that is made of brass. All racks have an inner closed flow circuit, which can be connected to a thermostat and allows for moderate temperature control in a range of ambient temperature to 90°C. Figure III.5.1b shows the 15-position rack on the motorized translation table of the D11 small-angle scattering instrument.

**Precise Temperature Control Sample Changers** Some experiments require more precise temperature control than is possible with the standard sample changer racks. Also, in some cases it is important to investigate samples at lower temperatures than ambient: the problem with the standard racks is that the thermal insulation is insufficient and at lower temperatures, typically below 8–10°C, depending on air humidity, the condensation of moisture on the sample container troubles the experiment. At ILL, the SANS instrument scientists have therefore developed a 10-position sample changer made from a massive copper block with excellent thermal contact to the sample (see Figure III.5.2a and b).

The device holds standard HELLMA 110 QS, 100 QS, and 404 QS cells (thicknesses 1–5 mm) and allows for precise temperature control ($\leq$0.1° for sample positions along the rack). The sample position can be doused in compressed air or dry nitrogen gas and experiments at temperatures down to –20°C are thus possible.

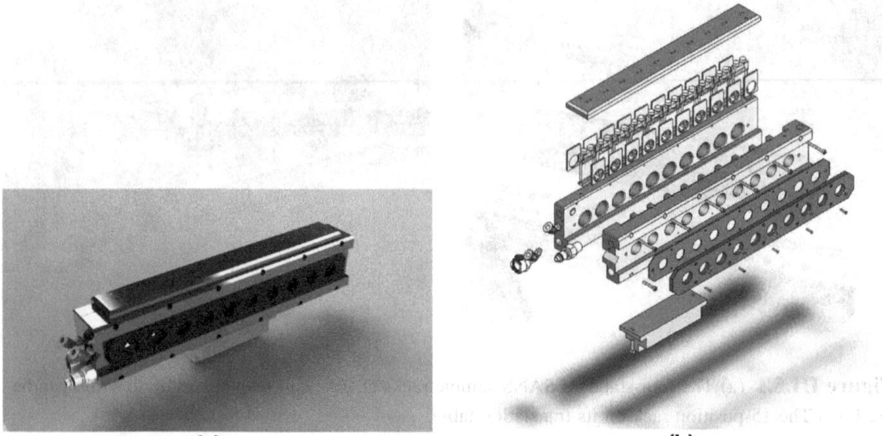


**Figure III.5.2** (a) Precise temperature SANS sample holder. (b) Detailed view of the high-temperature precision SANS sample holder.

**The "Tumbling Rack" Sample Changer** A frequently encountered problem in SANS studies is the instability of liquid suspensions. For solutions containing large particles, the particles may precipitate during the time of data acquisition in the scattering experiment. The resulting concentration changes inside the scattering volume can often make it difficult to interpret SANS data quantitatively.

In order to keep the sample dispersed in a solvent, a continuously rotating sample container is required. Such a device, without temperature control, exists at the ISIS spallation neutron source in the United Kingdom. A new version, including temperature control, has recently been developed at ILL. Figure III.5.3 shows a 6-position

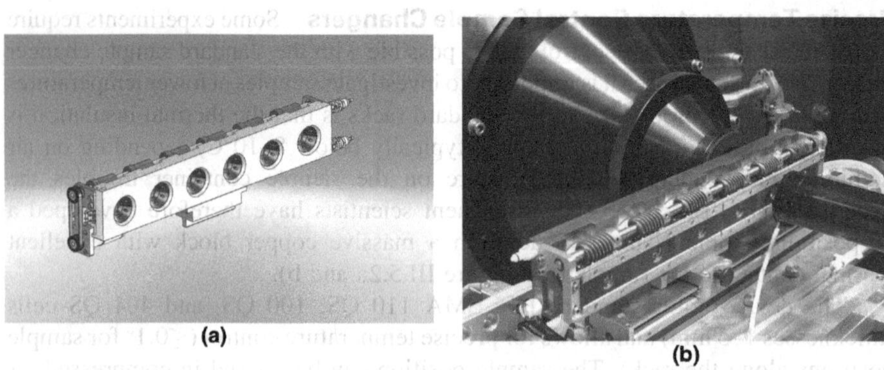


**Figure III.5.3** (a) Sketch of the tumbling rack (with the detector side facing the observer). (b) Photo of the tumbling rack, installed at D11 at ILL.

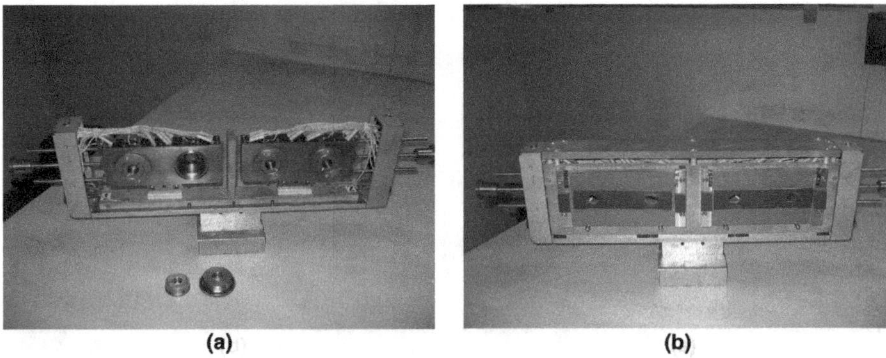


**Figure III.5.4** (a) Heatable sample changer (up to 250°C) with an example of the sample holder seen in front of the main unit. (b) Heatable sample changer (up to 250°C) with thermal insulation cover.

changer rack used at ILL for liquid samples in standard quartz cells, with the sample cells rotating at variable speed around an axis parallel to the neutron beam, in order to avoid precipitation. The prototype works with large 404 QS HELLMA cells but adapters for other cell types are foreseen. The copper block holding the cell adapters has an inner flow circuit, which can be connected to a thermostat and allows for temperature control.

**Heatable Sample Changers/Furnaces** In some cases it is interesting to investigate bulk polymer samples at elevated temperatures beyond the glass transition temperature. Figure III.5.4 shows a 4-position furnace with two separated stainless steel heating blocks.

Pellet-like bulk samples of diameter 13 mm and thicknesses in the range 0.8–1.2 mm can be mounted between quartz plates inside small stainless steel holders. These sample holders are screwed inside the heating block. Each block can be heated by electrical resistors, and the temperature is controlled by means of an Eurotherm temperature controller up to a maximum of 250°C. In order to protect the sample and to avoid thermal degradation of polymer samples at high temperatures it is possible to evacuate the sample position by connecting the heating blocks together with the sample holders to a vacuum pump.

### *III.5.1.1.2 Vacuum*

For cases where the coherent scattering signal is very weak, or when the signal-to-noise ratio between the sample and the background is unfavorable, it is convenient to perform the experiment under vacuum in order to suppress the air scattering and the window scattering. Figure III.5.5a and b shows a vacuum box that has been developed at ILL for the SANS instruments D11 and D22. At the bottom, inside the box, is a motorized horizontal translation table on which the standard multiposition sample changer racks (see above) can be mounted.

Figure III.5.5 (a) The D11/D22 SANS vacuum box open, shown with the motorized translation table for sample changers. (b) The D11/D22 SANS vacuum box closed.

#### *III.5.1.1.3 Pressure*

**Standard Pressure Cell for $p \leq 2.5$ kbar** Soft matter samples that are used for neutron scattering experiments are often nonequilibrium samples. In many cases, phase transitions shall be investigated by neutron scattering. Like the temperature $T$, the pressure $p$ is also a thermodynamic parameter that can be used to induce phase transitions. The investigated pressure range is quite diverse and depends on the sample and on the pressure cell available. As a common general feature, most pressure cells are equipped with thick sapphire windows. The sample volume often is variable by using spacer sapphire windows in order to reduce the path length.

Pressure cells are standard sample environment equipment and exist at almost every neutron research center. They are not only used for soft matter samples. The pressure cell available at the ILL is shown in Figure III.5.6. It has been constructed in the workshop of Prof. Dr. Lechner (University of Osnabrück, Germany) (Vennemann et al., 1987).

Even though the two sapphire windows have altogether a thickness of 24 mm, the transmission of neutrons at a wavelength of 0.6 nm is still as high as 85%. Tightness is achieved by using four gold sealings, two per window. The gold sealings in the form

Figure III.5.6 Photo of the ILL SANS pressure cell, manufactured by the Lechner workshop, University of Osnabrück. This stainless steel cell has two removable side plates where the sapphire windows and gold sealings are placed. Down right one of the two entries is visible.

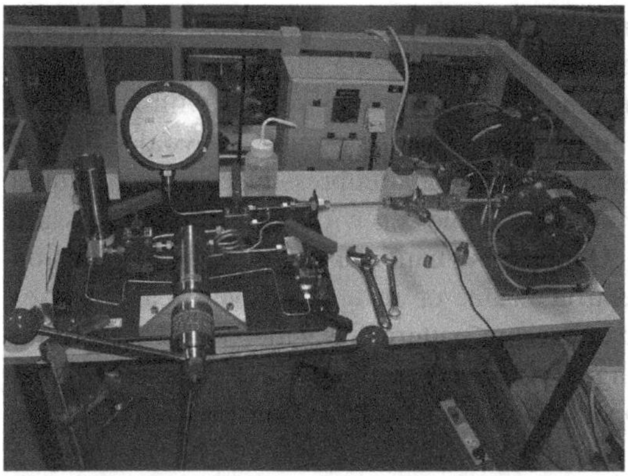

**Figure III.5.7** On the left side, the manual pressure pump is seen that is connected via a capillary with the two NovaSwiss electrovalves on the right side. The front valve serves to apply the pressure to the cell, and the back valve releases the pressure after the experiment. Right from the valve a thin flexible capillary is going down to the sample zone of D11 (ILL), allowing for precise adjustment of the pressure cell with respect to the neutron beam.

of rings are inserted in grooves; afterwards they are knocked with a soft brass hammer to make them stay in the groove. Eventually, the window is positioned and the steel plate (with the grooves and the window) is screwed on the cell corpus. Thereby, the screws must be tightened with a torque screwdriver subsequently, that is, not like when changing a car tire where the screws are tightened in the sequence of facing screws. The applied torque must be slowly increased by 1 $N/cm^2$; then after having done one round of all screws, the torque is increased stepwise until 45 $N/cm^2$. This procedure avoids breakage of the gold sealings and thus leaks of the cell. The same procedure must be applied to the other side of the pressure cell, too. This setup gives access to pressures up to 2500 bar. Temperature control is granted by an electrical heating jacket allowing a maximum temperature of approximately 230°C. The pressure cell has two connection entries; one is normally connected to the pressure pipe whereas the other one accommodates a pressure sensor. Figure III.5.7 shows a part of the pressure setup available at the ILL.

The pressure can be applied using manual pumps or automatic pumps. The setup described here and available at the ILL is equipped with two NovaSwiss electrovalves that open and close within 100 ms. Using this setup, pressure jumps can be performed between 1 and 2500 bar. Once the pressure has been built up with the pump, the valve is opened via computer control. The pressure is exposed to the sample cell, to which a pressure sensor is connected, which is read out electronically. When the pressure reaches the desired value, the pressure sensor readout generates a 5 V signal that is sent to the computer and serves to start the neutron scattering data acquisition. Depending on the scattering intensity, short time slices in the range of 100 ms can be chosen for data acquisition. In case of reversible processes, the pressure jump can be repeated several times. Then the same time slices of the different p-jump series (i.e., every first, every second, etc.) can be added together in order to improve the signal-to-noise ratio. If kinetic processes are followed where the sample evolves in a way that is irreversible, the mentioned procedure is not easily applicable. In this case, the time slices could also be adapted with respect to the scattering intensity.

An important step when commencing a pressure experiment with any given setup is the filling of the cell without air bubbles. Air bubbles must be avoided for two reasons: on the one hand the bubbles have a different compressibility than the liquid sample and therefore can reduce the pressure in the cell with respect to the pressure that has been built up with the pump. On the other hand, an air bubble, depending on the size, can be situated in the neutron beam, or in a less severe case, causes reflections that perturb the recorded scattering patterns. A widely used way to fill pressure cells consists of connecting a pump via a three way T-connector that evacuates the cell volume before opening the valve and letting the liquid fill the pressure cell completely.

The efficient use of pressure cells in terms of sample volume is more important for (rather expensive) biological samples than for soft matter samples. In the latter case, the sample can quite often be used to fill the whole pressure setup circuit, which, depending on the capillary volume and the pump cylinder volume, may add up to 15 mL. Precious biological samples however profit from a so-called pressure separator/multiplier. The concept is that of a translational piston, sealed in a pressure pipe. On the side towards the pressure pump, a standard pressure liquid like "Fluorinert" can be used, whereas the sample only fills the part of the circuit "behind" the separator. When the surface area of the piston on the side facing the pump ($A_p$) is smaller than on the side facing the pressure cell ($A_c$), the device acts as a multiplier, too. Obviously, the multiplication factor is defined by the ratio $A_p/A_c$.

Pressure cells are not restricted to use only for research on dissolved samples. A project carried out with Justin D. Holmes et al. (Cork University, Ireland) concerned the tailored engineering of pore sizes in mesoporous silicas during the silica hydrolysis process (Hanrahan et al., 2005). Mesoporous materials are used as column material for chromatography, where a well-defined pore size distribution is crucial for the efficiency of the separation quality. When the surfactant concentration was higher than 50%, a paste was applied onto the sapphire window of the pressure cell before screwing the cell together. Supercritical $CO_2$ was used as the pressure medium. SANS measurements were performed on D11 at ILL, following the controlled swelling of the added surfactants thus governing the pore sizes and pore distribution.

**"Cologne" Pressure Cell for $p \leq 0.3$ kbar and Precise T-Control** A different type of pressure cell has been developed by Thomas Sottmann et al. (Cologne University, Germany). This group is looking at phase transitions occurring in microemulsions, like for example, a temperature-induced change from a lamellar alpha ($L\alpha$) phase into a sponge phase. SANS is a highly appropriate tool to investigate these structural changes. Like temperature, pressure can be used as well in order to induce such phase transitions (Holderer et al., 2010). However, in such experiments it is crucial to control the temperature to within ±0.1°C, which is guaranteed by a surrounding cooling water circuit. The pressure cell has a sample volume of approximately 15 mL and is limited to pressures up to 300 bar (Figure III.5.8).

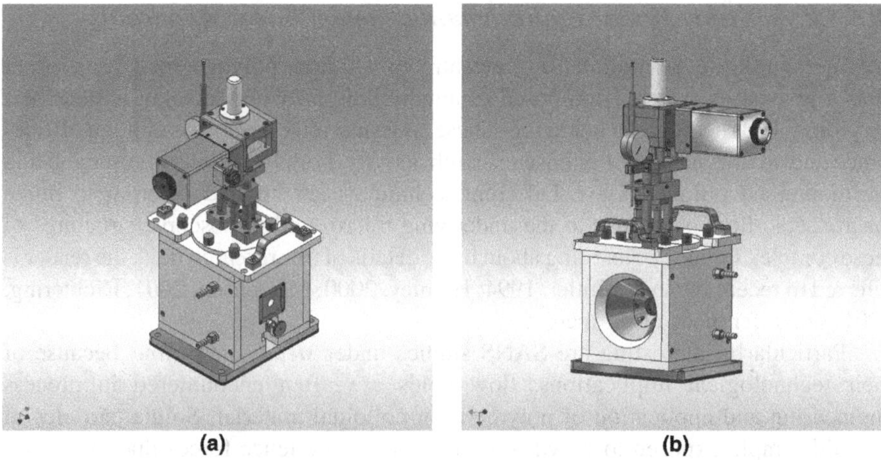


**Figure III.5.8** Schematic view of the Cologne pressure cell, developed jointly together with D11instrument team (ILL) showing the sides facing (a) the neutrons and (b) the detector. (*Courtesy*: T. Sottmann et al., Cologne University, Germany.)

This cell does not need a pressure circuit like the one presented in the "Standard Pressure Cell for $p \leq 2.5$ kbar" section. It is filled by turning the cell in a specially designed holder in order to access the tab situated underneath the cell. When filled completely, the tab is screwed on, and the pressure is applied with a handle (not seen on the shown sketch), which sits on the white rod on top of the cell. Handling this cell is very user-friendly, but it is limited in the accessible pressure range.

## III.5.1.2 Nonequilibrium Sample Environment

In contrast to the equilibrium conditions described in Section III.5.1.1, we now turn our attention to nonequilibrium experiments, where an external field is applied to the sample during the scattering experiment.

Structural changes can be induced, for instance, by submitting the sample to external constraints, such as magnetic, electric, or hydrodynamic fields. Furthermore, these fields can be imposed in different modes of operation: in a kinetic experiment, an activation process of the sample can be studied after a short perturbation pulse (or the inverse: the sample is allowed to relax back to equilibrium after an externally imposed perturbation of its equilibrium configuration). Second, in a cyclic experiment the sample is periodically distorted around its equilibrium state. A third type of nonequilibrium experiment is the steady-state experiment with keeping the external constraint constant during the measuring time.

### III.5.1.2.1 Study Under Hydrodynamic Fields: Shear Apparatus

Polymers in dilute, semidilute, or concentrated solution, polymer melts, surfactant phases as well as systems composed of amphiphilic polymers display in general a very complex rheological behavior. These systems have great technological relevance and are the subject of intense research activity, both from the experimental and the theoretical point of view. Different techniques are applied in order to relate macroscopic flow properties to the underlying micro- and mesoscopic structures of these complex fluids. For learning about more details of this research field, the reader is referred to recent reviews (Butler, 1994; Hamley, 2000; Mortensen, 2001; Richtering, 2001) and references given therein.

Particularly interesting are SANS studies under *steady-state flow*, because of their technological implications: flow fields are often encountered in process engineering and application of polymeric or colloidal material. Solute particles of a liquid sample exposed to a hydrodynamic field experience forces due to viscous drag in the streaming fluid, which tend to orient, to deform, or to order them. The intensity distribution on a two-dimensional multidetector might in this case become anisotropic with respect to the direction of flow. Although complex features, such as viscoelasticity, thixotropy, rheopexy, or drag reduction, are well known from (macroscopic) rheological measurements, the underlying mechanisms on the microscopic molecular scale are not completely understood. Experimentally important, for practical reasons as well as from theoretical aspects, is linear laminar shear flow between two parallel plates, one at rest and the other one moving with a constant velocity due to the action of an external force. The Couette-type shear apparatus constructed at ILL for SANS experiments with liquid systems in a constant shear gradient (Lindner and Oberthür, 1984) is a practical example for this kind of sample environment. The sample container consists of an inner fixed piston and a concentric outer rotating beaker. Both are made of quartz glass, which is highly transparent for thermal and cold neutrons and shows a very low small-angle scattering background. The sample is confined in the annular gap between rotor and stator (see Figure III.5.9). The outer cylinder is rotating at a constant speed. The inner static cylinder is under temperature control. The gap width is sufficiently small ($d \approx 0.5$ mm) compared to the cylinder diameter ($d \approx 48$ mm) and to a good approximation a plane Couette flow with a constant transverse (or shear) gradient of up to 12,500 $s^{-1}$ can be realized in the annulus (sample volume $\approx$4.5 mL).

With this equipment it has been possible, for instance, to study systematically the shear-induced structural changes of polymers in dilute solutions (Lindner and Oberthür, 1988) as well as in semidilute solutions (Saito et al., 2002). The Couette-type shear apparatus has become over the years routine equipment for SANS experiments at ILL with sheared complex fluids. Similar flow cells and modifications for special requirements (such as a variation of the shear gradient direction with respect to the direction of the primary neutron beam or an airtight flow cell for investigation of volatile liquid systems) have been constructed by other groups and are successfully used nowadays at almost every neutron research center all over the world.

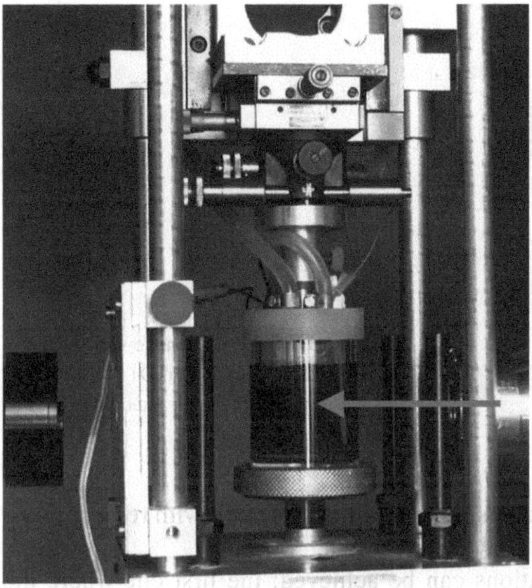

**Figure III.5.9** The Couette-type shear cell used at ILL. The arrow indicates the direction of the neutron beam.

#### *III.5.1.2.2 Combination with Rheology: rheo-SANS*

The combination of SANS with online rheology is another powerful technique that can be used to obtain an even more detailed molecular picture of the structural changes of complex fluids under shear. A typical example for rheo-SANS equipment is the ILL Searle-type shear cell with a variable gap width of 1 or 0.5 mm, depending on the diameter of the inner rotating cylinder. This shear cell had been developed to fit to a commercial Bohlin 120 CVO rheometer (see Figure III.5.10).

The outer, fixed quartz cylinder is mounted inside a stainless steel heating jacket, which can be connected to a thermostat for temperature control. Since high-quality quartz cylinders are used for the shear cell, excellent scattering data can be obtained even in the range of very low momentum transfer (see Figure III.5.11). The rheometer is mounted at the sample position on a horizontal translation table, which

**Figure III.5.10** Rheo-SANS equipment at D11 (ILL): the Bohlin 120 CVO rheometer.

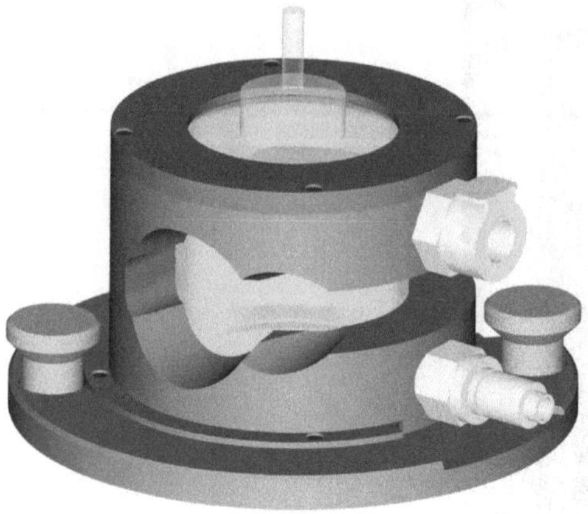

**Figure III.5.11** Schematic view of the Rheo-SANS equipment at D11 heating jacket in which the outer fixed quartz cylinder is mounted.

enables precise computer-controlled positioning of the shear cell with respect to the neutron beam.

Two different beam configurations can be achieved: the first one where the incident beam is aligned along the direction of the velocity gradient (called "radial" beam configuration) and the second where the beam is along the flow direction ("tangential" beam configuration). The latter is extremely important in studies, for instance, on samples with lamellar morphology (Nettesheim et al., 2003; Nettesheim et al., 2004). A further example for a typical rheo-SANS study is the investigation of the shear thinning and orientation of cylindrical surfactant and block copolymer micelles as a function of concentration and shear rate (Förster et al., 2005). In sheared solutions, elongated objects such as wormlike micelles align in the flow direction, with an orientational distribution that can be well described by an Onsager-type distribution function. The intensity distribution of the sheared solution, as recorded on the two-dimensional multidetector, becomes increasingly anisotropic. The order parameter $S$ as calculated from the anisotropic scattering pattern can be related to the shear viscosity and reveals important features of the molecular orientation process. The study provides direct relations between bulk properties like shear rate and shear viscosity, and molecular properties such as micellar thickness and orientation distribution of long wormlike micelles or fibrous structures.

### *III.5.1.2.3 Example of Other Geometries: Stretching Devices*

In contrast to the well-known shear thinning process of polymers, that is, an increased ease of processing at higher shear rates, the elongational viscosity can provide extensional strain hardening. This is an extra stabilization process in which a rubber-like behavior can be induced at high shear rates. The latter is very much dependent on the strain rate, temperature, and on the architecture of the polymer on a molecular scale.

The uniaxial nonlinear stretching of polymer melts or networks can be studied with a stretching device available at the Jülich Centre for Neutron Scattering (JCNS,

**Figure III.5.12** The JCNS strain rig, installed in the neutron beam at KWS2 (JCNS). The neutron beam passes the cell from right to left. A frozen sample in stretched state in the clamp is shown. The load cell is positioned in the lower section.

FRM2, Germany). With this stretching rheometer the sample is held inside a quartz cell, which consists of a vacuum double-walled cell (transmission about 85%) that itself is fitted in between two furnace parts (see Figure III.5.12). The lower part of the furnace is fed with liquid nitrogen and the gas is evaporated inside through a spiral heater to the required sample temperature. This gas is blown from four nozzles onto the sample and leaves the cell from the top. Both lower and upper parts of the furnace are connected to a vacuum pump to provide an isolation vacuum. The temperature of the gas close to the sample is measured as the average from four thermocouples, and this value is taken to be equal to the temperature of the sample. The quartz cell itself can be connected to a liquid nitrogen nozzle, which on request by opening a magnetic valve can be directed on the sample to freeze the present state and decouple microscopic relaxation times and macroscopic counting times. The lower clamp of the machine is fixed whereas the upper one is connected to a high-resolution servomotor. The maximum Hencky strain is about 6 and the strain rate resulting from an exponentially increasing speed is between 0.00001 and 2 $s^{-1}$ at room temperature. The temperature range itself is practically limited to $-100$ to $+200$°C. The machine allows a shift of the sample dynamics to both longer and shorter times, compared to room temperature, simply by changing the temperature and relying on the well-known time–temperature superposition (TTS) principle. The obtained strain rates can extend then to several thousands per second.

Therefore, fast processes typically on the order of 1 ms can be slowed down to the range of minutes so that the quenching times on the order of 1 s are negligible.

Stepwise relaxations can be observed by submitting the sample subsequently to temperature ramps and quenching the intermediate steps. The final corresponding relaxation time at a chosen reference temperature is then given by the sum of all relaxation times at every temperature during the temperature ramp, corrected by the TTS shift factors. In the off-line mode a small, thermally isolated force transducer, located below the furnace can be used to measure the mechanical response of relaxation, that is, the Young relaxation modulus $E(t)$ after the applied step strain as a function of time.

The inverse strain rate corresponds to the shortest time observable or likewise the highest frequency in the complex dynamic moduli $G^*$ or $E^*$. Using this technique it has become possible to track the first Doi relaxation process after step strain, that is, the chain retraction or contour length equilibration. This had been predicted and assumed all over the tube model but could not be verified until 2005 (Blanchard et al., 2005). Also, this stretching apparatus provided the first direct structural proof of all details of the molecular rheology in the description of hierarchical relaxations in branched polymers. H-type polymers in different labeling structure as well as recent blends of linear with hyperbranched polymers or dendrimers were investigated to study the time domain of the separate levels in the hierarchy: polymers relax from the outside inward and thereby several momentum transfer regions are affected in different ways (Heinrich et al., 2004). The strain rig can be modified to apply shear as well and is currently being upgraded.

#### *III.5.1.2.4 Studies Under Electric Fields: Capacitors*

In recent work, the group of Alexander Boeker (RWTH Aachen, Germany) has studied with SANS and SAXS the single-chain form factor of polystyrene–polyisoprene block copolymers subjected to an electric field (Boeker, A., et al., 2009, unpublished results). Figure III.5.13 shows the electric field–SANS capacitor as used at the instrument D11 at ILL. It is made of PTFE/PEEK and has dimensions of $120 \times 60 \times 20\,\text{mm}^3$ ($l \times w \times h$).

The path length inside the cell is 4.95 mm with a sample height of 9.85 mm. The sample volume (shown in the center) is sealed with neutron-transparent quartz windows. The electrodes are made of gold and the electrode spacing is between 3 and

**Figure III.5.13** An electric field capacitor for SANS.

6 mm. With this device a maximum electric field intensity of $\leq$6 kV/mm can be obtained. It had been recently demonstrated that by deuterium labeling the SANS signal can reveal the polymer chain scattering, and that a weak anisotropy of the two-dimensional SANS pattern can be observed at 1 kV/mm for deuterium-labeled blends of block copolymers diluted by deuterated toluene. A more thorough investigation at these sample conditions, with an improved $E$-field set up and higher field strength of 3 kV/mm led to the observation of polymer stretching perpendicular to the electric field lines.

## III.5.1.3 Time-Resolved and Kinetic Studies: Stopped-Flow Technique

The stopped-flow apparatus serves to fill a scattering cuvette placed into the neutron beam and to mix the sample at the same time. Such a commercial device, sold by BioLogic (France) has been adapted for neutron research at the ILL and is now used in many centers all over the world. The precise filling in various mixing ratios is the fundamental requirement for a time-resolved experiment. The compact device is shown in Figure III.5.14 as mounted at the sample position of D11 at ILL.

An example of recent interest is the family of amphiphilic block copolymers, which represent a tuneable model-system to study concepts of self-assembly. Very recent work by Reidar Lund et al. (a collaboration involving DIPC-San Sebastián, FZ Jülich and ILL) is focused on the shape transition kinetics between cylindrical and

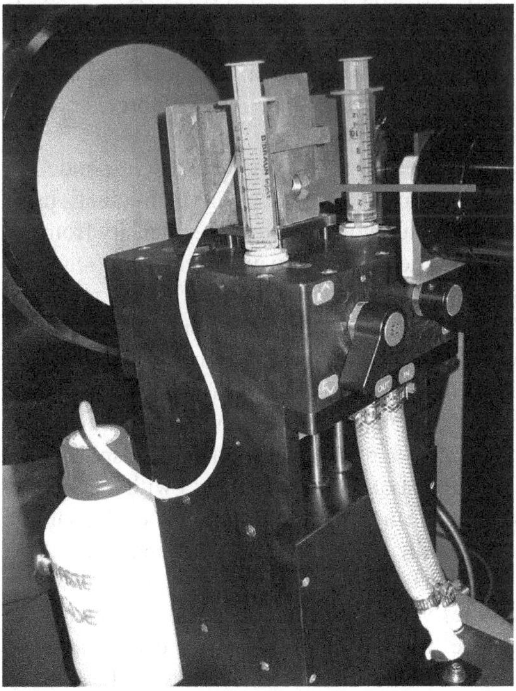

**Figure III.5.14** The BioLogic stopped-flow apparatus on the SANS instrument of D11 at ILL.

spherical micelles and vice versa (Lund, R., et al., 2009, unpublished results). These transitions occur in PEP–PEO micelles: cylindrical structures are favored for water-rich samples while spheres are formed in solvent mixtures containing predominantly dimethylformamide (DMF). By using the stopped-flow apparatus for rapid mixing, these cylinder-to-sphere transitions were monitored both by time-resolved (TR) SANS on D11 at ILL as well as TR SAXS at ID02 (ESRF, Grenoble, France) on a timescale between 2 and 100 ms. Results show that the cylinders are typically broken into spherical-like structures within a few hundred milliseconds and thereafter slowly reorganize to the final equilibrium spherical micelles. Interestingly, the opposite transition is not observed—at least over a period of days. Thus, investigating the early stages of such formation processes can also help to understand the equilibrium properties of these materials. For further information and an overview of other fields of science where the stopped-flow technique is used, the reader may refer to the literature (Grillo, 2009).

## III.5.1.4 *In Situ* Complementary Techniques

### *III.5.1.4.1* In Situ *Dynamic Light Scattering*

There is currently a great interest in the combination of various complementary techniques with neutron scattering. A project that has just started on D11 at ILL aims at providing simultaneous UV–VIS spectroscopy measurements during a variety of dynamic and static neutron scattering experiments. This project is pursued jointly between ILL and the group of Frank Schreiber (University Tübingen, Germany). *In situ* dynamic light scattering (DLS) is also available at Paul Scherrer Institut (PSI, Villigen, Switzerland). The setup was developed by Peter Schurtenberger et al. (University of Fribourg, Switzerland) together with Joachim Kohlbrecher (PSI, Villigen, Switzerland).

Like neutrons, light is scattered when interacting with particles suspended in solution. DLS allows characterization of a sample by measuring the fluctuations of the scattered light intensity over the time. An autocorrelation of the intensity from the different time slices is performed (a measurement is cut into many short time slices on the order of 100 ns, initial sampling times are of the order of 3.125 ns to 25 ns, depending on the used correlator). The initial slope of the autocorrelation function yields the diffusion coefficient, which can be transformed into a hydrodynamic effective radius $R_H$ via the Stokes–Einstein equation. SANS on the contrary measures the radius of gyration $R_G$. These two radii are related to each other, depending on the shape of particles. A value of 0.78 for $R_G/R_H$, for example, accounts for a sphere-like shape of the investigated particles. The interest of combining SANS with DLS is, for example, to track the stability of a sample during the SANS beam time. If a sample tends to aggregate, DLS will be able to show the growth of aggregates and detect them as well as the individual particles. SANS can only measure a mean $R_G$ in such a case, and the characterization of single particles via the form factor and determination of $R_G$ is no longer possible. In such a case the SANS measurement can be stopped immediately, allowing one to use the unperturbed data obtained so far and to

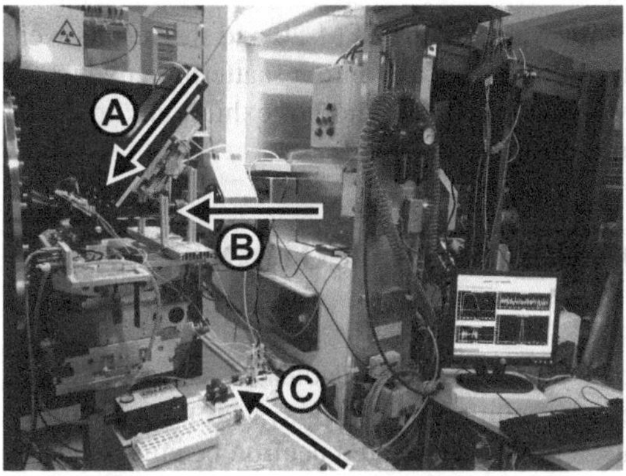


**Figure III.5.15** Photo of the *in situ* DLS–SANS setup on D11 at ILL (courtesy of Thomas Nawroth U. Mainz, Germany). Arrow A marks the incident laser light direction, arrow B shows the incident neutron beam direction, and arrow C highlights the stopped-flow mixing device.

exchange the sample. It thus saves precious beam time on highly overbooked SANS instruments. Furthermore, if a sample is evolving over a measurement time (like during an irreversible phase transition), DLS gives access to additional information like aggregation processes, and so on. If a sample evolves irreversibly, the same sample cannot be measured with an off-line DLS instrument after the SANS measurement.

In collaboration with Thomas Nawroth et al. (University Mainz, Germany), we did a successful experiment with combined DLS and SANS on D11. Figure III.5.15 shows a photo of the setup. The standard sample changer setup for HELLMA cells was used, and the cells were filled using a simple stopped-flow device. The laser is mounted at an angle of 45° with respect to the neutron beam. The scattered light intensity is recorded at 170° in backscattering. The setup is the property of Thomas Nawroth, which he put together using commercial components.

Time-resolved studies of processing of nanoparticles for biomedical applications in the cancer therapy could be tracked at different length scales with the dual-beam instrumentation. DLS looks at wide distributions up to 20 μm particle sizes at 20% precision, but cannot distinguish between particle components. This is where SANS comes into the game, by using contrast variation of solvents or single particle components. The structure and shape is elucidated precisely in the range 1–300 nm. SANS is also the faster method, where time slices of 1 s are feasible; in contrast DLS needs a measurement time of at least 10 s.

#### *III.5.1.4.2 Stopped-Flow and Light Scattering*

At ILL we succeeded recently in connecting a stopped-flow apparatus with a light scattering instrument. An SFM-20 and an ALV CGS-3 SLS/DLS have been used. The SFM-20 has been used to fill a light scattering cell that has been mounted in the ALV CGS-3 beforehand. The tab has a hole in which a capillary is introduced with a direct connection to the SFM-20. The crucial part is to avoid any dust particles in the scattering cuvette, as dust scatters in the angular range being detected by light scattering. Therefore, the SFM-20 was equipped with Millipore filters of 200 nm pore size. As a test, a cuvette was filled with water by the SFM-20 and the results were compared to a measurement of water filled classically into a scattering cuvette. As a result of the comparison, the SFM-20 filled cuvette did not exhibit any traces of impurities coming, for example, from dust.

## III.5.2 SAMPLE ENVIRONMENT FOR NEUTRON REFLECTOMETRY

Like with SANS, neutron reflectometry (NR) requires dedicated sample environments to control precisely the ambient conditions for the physical measurements. Typically for soft matter experiments, a highly collimated neutron beam is reflected off a fairly large planar surface. For air/liquid measurements there may be an illuminated area on the order of $100\text{–}150 \times 30\text{–}40\ \text{mm}^2$, yet for solid/liquid measurements the footprint may be reduced to a half or a third of that length due to the requirement in most cases to use planar single crystals. The collimating slits before the sample need to balance flux, the footprint size and the angular resolution. It is essential not to overilluminate the sample during reflectivity measurements as the reflectivity then becomes corrupted.

To summarize the salient points of the technique, measurements can be carried out in monochromatic mode or time-of-flight mode. The former option concerns where the momentum transfer is varied by changing routinely the incident angle of neutrons on the sample at fixed wavelength. The latter option concerns where the momentum transfer is varied by a wavelength distribution incident on the sample at fixed incident angle (or at least a limited distribution of incident angles). For time-of-flight reflectometry one may record data at a few different incident angles and the data are cojoined in the reduction process. With a 1D detector, the background may be measured by rotating the detector slightly away from the specular reflection angle. With a 2D detector, the background may be measured simultaneously to the specular reflectivity with the added advantage that off-specular scattering may also be recorded.

Typically for measurements at low incident angles ($<1°$), the final collimating slit before the sample may have an opening perpendicular to the sample of $<1$ mm to prevent overillumination of the sample and/or limit the angular resolution appropriately. As the reflectivity is highest at low incident angles, the restricted flux is normally not a problem, although in some kinetic studies maximum flux may be more important. At higher incident angles, the collimating slits may be opened up to several millimeters, often without a particularly detrimental effect on the footprint or angular resolution, to give the required high flux to make low reflectivity measurements on a reasonable timescale. In general, for standard reflectivity measurements, information is usually not resolved along the width of the sample, that is, neutrons that reflect from the left side are equivalent to those that reflect from the right side. Therefore in the direction parallel to the sample, the collimating slits may have an opening of several tens of millimeters; note that other applications such as grazing incidence small-angle neutron scattering (GI-SANS) resolves the lateral information and therefore has additional collimation requirements.

The following description of sample environments for NR measurements of soft matter systems falls into four sections. Sections III.5.2.1 and III.5.2.2 cover standard pieces of equipment for measurements at the *air/liquid interface* and the *solid/liquid interface*, respectively. Section III.5.2.3 considers some examples of *specialist sample environments and applications* that are not used routinely (liquid/liquid interface, overflowing cylinder, humidity chamber, shear cells, and equipment for lipid bilayer preparation). Lastly, Section III.5.2.4 touches on five noninvasive *complementary techniques* that either have been used *in situ* on neutron reflectometry beamlines or at least are rather suited to that purpose (ellipsometry, Brewster angle microscopy (BAM), X-ray reflectometry, electrochemistry and reflection absorption infrared spectroscopy (RAIRS), and total internal reflection (TIR) Raman spectroscopy).

## III.5.2.1 Air/Liquid Interface

### *III.5.2.1.1 Adsorption Troughs*

Adsorption troughs are containers in which to pour solutions so that the structure and composition of adsorption layers may be quantified by NR. Typically the containers may be 200–250 mm long × 40–50 mm wide. The most common material is poly(tetrafluoroethylene) (PTFE) as the material is chemically inert in most applications and is relatively easy to clean. Other troughs have been made out of stainless steel at ISIS to eliminate unwanted adsorption on PTFE and Delrin at ILL to use a more hydrophilic material so that a lower liquid volume may be used. Typical volumes for experiments may be on the order of 20–40 $cm^3$ depending on the surface tension of the solution. With an increasing number of biological applications, the demand for much lower volume troughs (at the expense of flux on the sample) is growing. The benefit of having many troughs on a sample changer is the option to automate measurements and also measure some samples while others equilibrate.

An adsorption trough assembly refined over many years at ISIS, in particular by John Webster and Jeff Penfold, has become routinely used (Lu et al., 2000; Taylor et al., 2007). Each trough sits in its own inner box, complete with sapphire windows before and after the sample. A strip heater is fixed to the length of the underside of each inner box to allow measurements above ambient temperature. Each inner box sits in a single large outer box also with sapphire windows before and after the sample. Large strip heaters traverse the whole length of the outer box so that just a few degrees can be given to the outer windows to avoid condensation problems. An optical sensor positioned above the assembly can be used to regulate the height of each liquid sample, even those weakly reflecting to neutrons.

A similar set of adsorption troughs were commissioned at ILL with the launch of the new horizontal reflectometer FIGARO (Fluid Interfaces Grazing Angles ReflectOmeter) in April 2009. Figure III.5.16 shows a photo of the assembly comprising six troughs. A special feature is the presence of a heated, tilted, rotatable quartz window incorporated in a narrow fixed part of the lid of the outer box above each trough. The windows allow the optical sensor to be used while the inner boxes are sealed.

**Figure III.5.16** Photo of an assembly of six PTFE adsorption troughs shown on the FIGARO reflectometer at ILL together with the optical sensor used to automate sample alignment through optical windows. (See the color version of this figure in Color Plates section.)

Furthermore, routines have been incorporated into the control software so that one can simply pour in a fresh sample, select the trough number and the alignment will be done automatically in just a few seconds. This approach to running adsorption trough experiments with no neutron alignment scans carried out by the users has proved to be very popular as the instrument control is simple and a lot of beam time is saved.

The "flow-trough method" of subphase exchange has recently been developed by John White (Australia National University, Australia) and coworkers to study structural effects of protein–protein and protein–inorganic interactions in thin films (Perriman et al., 2008). In the cited article the trough is used for X-ray reflectivity measurements, but already the trough has been used for NR measurements on the new INTER reflectometer at ISIS. The principle of the trough is that a mechanically robust surface film is formed at the air/liquid interface under which the subphase can be exchanged for another solution with minimal perturbation to the film itself. This methodology is useful for the recording of multiple reflectivity profiles in different isotopic solution contrasts, and it can be extended to the exposure of spontaneously formed surface films to chemical changes in the subphase.

#### *III.5.2.1.2 Langmuir Trough*

A Langmuir trough is a container of liquid (usually aqueous media) in which movable barriers regulate the surface pressure of spread films, as shown in Figure III.5.17. The

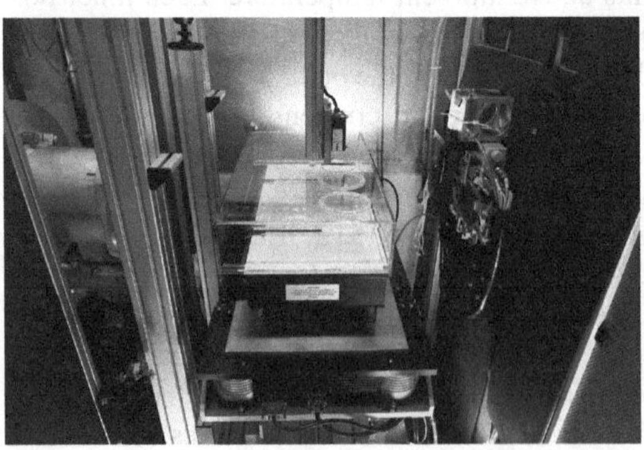

**Figure III.5.17** Photo of a Langmuir trough (Nima Ltd., UK) with dimensions $500 \times 250\ mm^2$ ($l \times w$) dedicated to high flux measurements shown on the FIGARO reflectometer at ILL; low-volume inserts are available for the conservation of materials.

device is used in the study of insoluble monolayers where the surface pressure, and in cases the resulting phase behavior of the monolayer, may be controlled precisely. One can drive directly to a surface pressure of interest or cyclic isotherms may be carried out where the surface pressure is increased then decreased repeatedly. The surface pressure is often limited to 40–50 mN/m before the monolayer collapses and wets the rims or the contact between the rim and a barrier, although there are designs that go to special efforts to reduce this effect, for example with the PTFE ribbon barrier as used in work on pulmonary surfactants by Schief et al. (2003). Further applications include studies of the interaction with insoluble monolayers of materials either dissolved or injected into the aqueous subphase (Miller et al., 2004) or present in the gas phase (King et al., 2009).

In a Langmuir trough, the liquid sits in a shallow well and the barriers make contact with the rim, so with a positive meniscus of liquid there is minimal transfer of insoluble molecules from one side of a barrier to the other. The result is that barrier movements are used to make controlled changes in the surface pressure, which is measured by a sensor. The sensor is typically a Wilhelmy plate dipped in the liquid and connected to a balance allowing the measurement of the pulling force of the liquid and the modification of its surface tension caused by the presence of and changes to a surface monolayer. The monolayer is typically formed by the spreading of insoluble molecules onto the surface by gentle dripping of a solution where the solvent is a volatile organic liquid such as chloroform.

The usual material for the trough is PTFE that is hydrophobic and produces a large contact angle with the liquid. The barriers are also usually made of PTFE but can be made of Delrin. The volume required to fill the trough varies according to the dimensions, and is typically on the order of 50–600 $cm^3$. The advantage of a long trough is that a higher surface compression ratio may be reached; the advantages of a wide trough are that large slits may be used to increase flux and reduce measurement times, and small incident angles may be used without overillumination of the sample. However, when carrying out measurements in $D_2O$ or inducing interactions with insoluble monolayers of precious materials dissolved or injected into the subphase, then there is a strong motivation to limit the trough volume and compromise on compression ratio or neutron flux. One option that can help, however, is the insertion of borated glass plates into the trough itself. It is also very important for the Langmuir trough to have a lid or be placed in a sealed box to reduce contamination of the surface from dust and to minimize disturbances from air currents.

### III.5.2.2 Solid/Liquid Interface

Measurements at the solid/liquid interface typically involve the transmission of a collimated beam of neutrons through a solid crystal (entering by a side face), followed by reflection off a liquid surface, then further transmission through the solid crystal (exiting by the opposite side face) (Fragneto et al., 1996; Fragneto-Cusani, 2001). The most common experimental approach is to investigate phenomena at the silica/liquid interface through using a single crystal of silicon with a

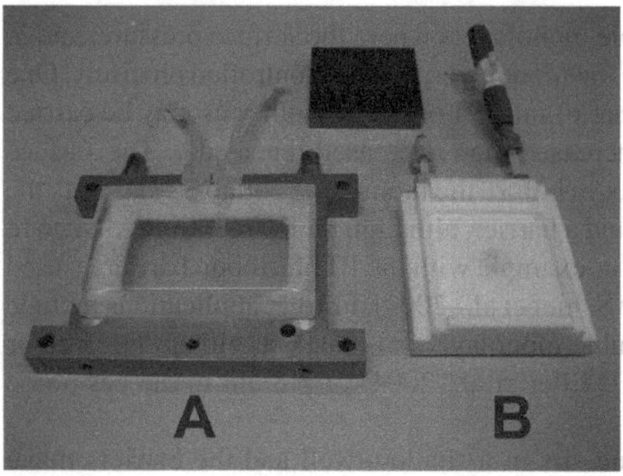


**Figure III.5.18** Two examples of solid/liquid sample cells: low-volume quartz cell and mechanical-mixing PTFE cell, with a silicon crystal of typical dimensions $50 \times 50 \times 10\,\text{mm}^3$ shown in the background.

native oxide layer. Silicon crystals are highly transparent to neutrons, and typical dimensions may be on the order of $50–100 \times 50 \times 10\,\text{mm}^3$ ($l \times w \times h$). A large face in contact with the liquid must be highly polished with a roughness of just a few Å as a higher roughness may be detrimental to the surface chemistry or the extraction of structural data from reflectivity profiles. Other solid substrates measured include quartz, sapphire, and mica.

There are a number of considerations when designing cells, such as minimizing the volume of liquid required during solution changes, reducing the background scattering, and limiting activation of the materials used. Accordingly, there have been different approaches employed in the design of sample cells. We describe two assemblies here, and Figure III.5.18 shows a photograph of both types of sample cell with a polished silicon crystal of characteristic dimensions in the background.

Sample cell A, developed by Giovanna Fragneto (ILL), involves a single crystal pressed mechanically against the base of a rigid material that has a shallow well to contain the liquid volume. The crystal is clamped against the rigid base using blocks of aluminum in which there are drilled circuits for water circulation to thermostat the samples. In the example shown, the base is made of quartz but other suitable materials include poly(etheretherketone) (PEEK) or stainless steel. Quartz was chosen here for several reasons: it is transparent to light so the accumulation or formation of air bubbles can be easily detected, it is hydrophilic and therefore useful when using protein solutions that adsorb preferentially to hydrophobic substrates, and it scatters neutrons less than the classically used PTFE, which results in a lower background. The main disadvantages are the fragility and high cost. The liquid injection inlet leads to a channel spanning most of the length of one end of the well while the outlet channel spans part of the length of the opposite end. The basis of the solution exchange is laminar flow. The principal design feature is the low volume of liquid required during solution exchanges—the well has a liquid depth of just 100 μm. An alternative to a shallow well is the use of an *o*-ring or gasket between the crystal and the flat rigid base. The volume of the cell can be minimized to $<1\,\text{cm}^3$ meaning that sample changes can be performed with just $3–4\,\text{cm}^3$ of solution. The cells were developed with biological applications in mind for which the availability of deuterated materials is limited (Al-Jawad et al., 2009).

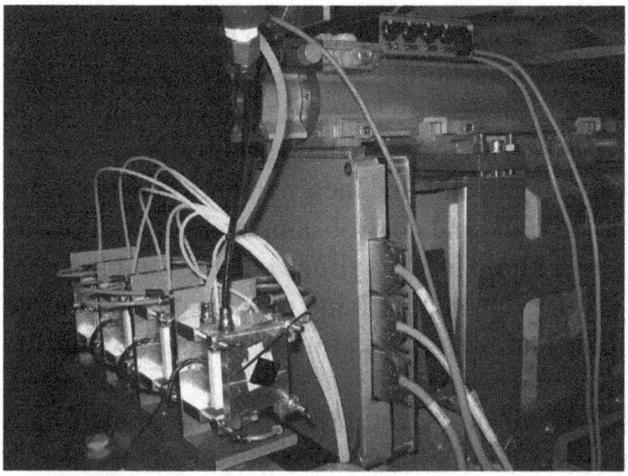

**Figure III.5.19** A sample changer with four efficient-mixing sample cells shown on the D17 reflectometer at ILL: the black cables are electrical connections for the magnetic stirrers, the gray tubing provides the thermostating water circuit, and the white tubing is for the sample injections.

Sample cell B, developed by Adrian Rennie (Uppsala University, Sweden) and Tommy Nylander (Lund University, Sweden), involves a single crystal pressed mechanically against a base of PTFE, which has a deeper machined well to contain the liquid volume (Vandoolaeghe et al., 2009a). The well has an additional indentation in which there sits a PTFE-coated magnetic stirrer bar. Again, there is a water circuit for thermostating purposes inside aluminum blocks that are used to clamp the crystal mechanically against the base. In the aluminum block adjacent to the PTFE base there is an indentation for a motor that drives the rotation of the magnetic stirrer bar and thus provides mixing of the liquid contained in the cell. In the example shown, the volume of the liquid well is $4\,cm^3$ and with short tubing the volume of solution required to exchange the contents of the cell is on the order of 10–12 $cm^3$. The principle of the exchange process in this design is one of continuous dilution involving the efficient mixing of the solution during the exchange process. Figure III.5.19 shows a sample changer with four of these cells positioned on the D17 reflectometer at ILL.

It can be important to take care not to trap air bubbles when mounting the crystal onto the sample cell. One approach can be to start with a positive meniscus of solution in the well, and with the tubing also prefilled with solution, then place the crystal—with the polished crystal face wetted—firmly down onto its support. The excess solution then spills as mechanical force is used to seal the crystal against the mount already with the well filled with solution. An alternative approach is to mount the crystal completely dry then fill the cells through a controlled injection of solution—or first with organic solvent then with the solution of choice. The issue of trapped air bubbles is particularly pertinent for hydrophobic surfaces where it is important to use degassed solutions and can be beneficial to limit the problem by using a reflectometer with a vertical sample geometry or (better still) with a horizontal sample geometry where the crystal is positioned below the liquid and the neutron beam reflects downward off the solid/liquid interface toward the detector.

For cells that employ either a liquid well or a gasket, typically there is tubing to provide an inlet and outlet for the solution exchanges, which can be performed manually using a syringe or automatically using a pump. It is also important to avoid the introduction of air bubbles into the tubing during this exchange process. One approach can be to use flexible tubing with stoppers where the tubing is squeezed to eliminate any trapped volume of air as the syringe makes the initial contact. Another approach can be to use rigid tubing with a connector that incorporates a tap. There are various pumps on the market that can simplify and control sample changes. Options range from simple motorized expulsion of liquid from a syringe into the sample cell at a predetermined constant flow rate, to the controlled mixing of selected amounts of different solutions to give precise concentrations and isotopic solution contrasts delivered at the right time.

Whether a sample cell exploits the well or gasket design to contain the liquid volume, the beam-defining collimation slits of the instrument must be chosen such that the illumination footprint on the liquid surface does not result in the scattering of neutrons from the sample cell itself. A typical illumination over the length of the sample may be on the order of 70–80% where the umbra and penumbra of the beam must be considered in the calculation of slit openings. Additionally, some degree of shielding, such as cadmium positioned toward the outer sides of the crystal, may be appropriate to limit further background scattering from edges of the sample cell. It is important to record a direct beam measurement through the crystal itself to correct the reflectivity for the crystal transmission, but also to consider any additional attenuation if the edges of the sample cell are shielded.

Experiments of a few days often take place with several blocks mounted adjacent to one another on a motorized sample changer. The alignment procedures can be done for each sample cell in turn, and with reliable motor/coder assemblies the appropriate positions of each sample cell can be recorded and incorporated into command files in the instrument control computer. This approach has the added advantage that one can carefully plan beam time so that chemical changes take place in some sample cells while other equilibrated samples are being measured. In addition to a sample changer, motorized goniometers to make precise changes to the incident angle and sideways tilt of the samples are now standard components of most reflectometers.

An important characteristic of neutron reflectivity experiments is the possibility to model simultaneously the same chemical interface under different isotopic conditions. With this capability comes a necessity to ensure that reflectivity from parallel measurements on different isotopic contrasts result from the same chemical interface. One issue to consider with exchanging the bulk solution to model simultaneously different solution contrasts (e.g., a peptide/bilayer interaction measured first in $D_2O$ and then in $H_2O$) is that the exchange process must not change the chemical integrity of the interface either through reequilibration or mechanical perturbations. For a series of measurements in different solution contrasts, the interfacial effects of solution exchange may be validated by a final measurement repeating the starting solution contrast. Such difficulties in the exploitation of isotopic contrast variation provided a principal motivation for the development of magnetic contrast variation in soft matter systems.

## III.5.2.3 Specialist Sample Environments

In the following paragraphs we mention some specialist sample environments that have been implemented to good effect on neutron reflectometry beamlines. The account gives the reader a feeling for some of the different types of NR experiments and applications possible and is not intended to be exhaustive. It is no coincidence that in all five cases successful execution of the ambitious experiments resulted from good collaborations between external users and the beamline scientists.

### *III.5.2.3.1 Liquid/Liquid Interface*

These measurements were pioneered by Ali Zarbakhsh (Durham University, UK) and coworkers at ISIS in the early 2000s (Zarbakhsh et al., 1999). A dedicated setup is required due to the low transmission to neutrons of $H_2O$ and $D_2O$. The apparatus is not very different to that of a standard solid/liquid sample cell. The methodology is to spin-coat a deuterated oil film onto a silicon crystal prior to its mounting in the sample cell, cool down the crystal until the oil freezes, put the oil in contact with the other liquid phase (usually an aqueous solution), seal the sample cell, and lastly warm up the system to melt the oil. The neutron beam then transmits through the crystal, as usual, and also transmits through the deuterated oil film (with some degree of attenuation that needs to be calibrated), before reflection at the oil/water interface. This approach has, for example, allowed the group to carry out studies of the structure and composition of mixed surfactant films at the oil/water interface.

### *III.5.2.3.2 Overflowing Cylinder*

A dynamic flow cell called the overflowing cylinder was developed by Colin Bain (University of Oxford, UK) in the 1990s based on previous versions of overflowing weirs. Liquid that is supplied by gravity from a reservoir flows up a vertical cylinder until it spills radially over the horizontal rim. The expanding liquid surface has been modeled mathematically and several reflectometry techniques (e.g., ellipsometry, laser Doppler scattering, surface light scattering, and infrared spectroscopy) have been applied to model the adsorption kinetics of surfactant solutions on short timescales ($<1$ s) and determine the concentration regimes of diffusion control or electrostatic barriers to adsorption. NR was first applied to the device in the mid-1990s for the study of surfactant solutions on the SURF beamline at ISIS where direct measures of the adsorbed amount were used in kinetic adsorption models (Manning-Benson et al., 1998). Recently the overflowing cylinder has been used on the new FIGARO reflectometer at ILL (see Figure III.5.20) where a project is underway to rationalize the dynamic behavior of a range of weakly to strongly interacting polyelectrolyte/surfactant mixtures.

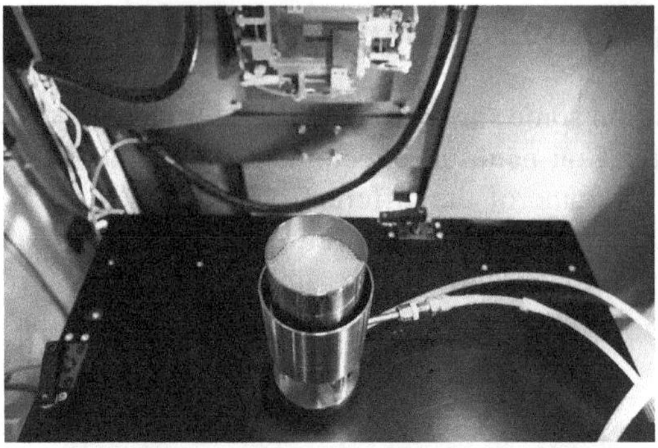

**Figure III.5.20** Custom-made dynamic flow cell called the overflowing cylinder shown on an active antivibration table ready for neutron reflectometry measurements on the FIGARO reflectometer at ILL.

**Figure III.5.21** Custom-made humidity chamber developed for the D16 diffractometer at ILL.

### *III.5.2.3.3 Humidity Chamber*

NR measurements at the solid/air interface have been carried out on hydrated samples in a temperature-controlled humidity chamber on the D16 diffractometer—used in reflectivity mode—at ILL (Perino-Gallice et al., 2002). This work focused on the stability of thin multilamellar assemblies of neutral lipid bilayers, which are unstable upon hydration, deposited on silicon crystals. Figure III.5.21 shows a photo of the experimental setup. The sample sits in one of two thermally insulated compartments that are connected by a small aperture, and in the other compartment is a water reservoir. The osmotic pressure of the aperture may be regulated by the temperature difference between the two compartments. The compartments are thermostated with water circulation channels, and warm water is used first to heat the compartment with the reservoir. As a result, with no osmostic stress, the vapor phase above the sample, which is the coolest point of the system, has a relative humidity of 100%.

### *III.5.2.3.4 Shear Cells*

Back in the early 1990s, a neutron reflectometry sample cell was made to investigate the effect of shear stress at the solid/liquid interface (Baker et al., 1994). Isomers of hexadecyltrimethylammonium dichlorobenzoate were studied in this work. In the cell, a single crystal of silicon or quartz is used as the substrate, which is held against a piece of PTFE milled with a shallow well. There is a series of holes at both short ends of the well so that during measurements laminar flow can be maintained over the entire surface illuminated by the neutron footprint. The range of shear rates possible can be controlled by changing the flow rate or the depth of the well. At the start of the following decade another cell was designed called the neutron confinement cell, where there is a thin film of liquid between a stationary upper quartz crystal and a moving lower quartz or sapphire crystal; the shear force is applied laterally (Kuhl et al., 2001). In this case, the upper quartz crystal is housed in a stainless steel holder while the lower crystal is mounted on a mechanical slider for shearing the thin liquid volume between the crystals.

#### *III.5.2.3.5 Equipment for Lipid Bilayer Preparation*

There are several different approaches for making supported lipid bilayers. For example, there is the vesicle fusion method where vesicles from solution rupture on a solid support to form a continuous lipid bilayer (Kalb et al., 1992). The procedure can be carried out directly in a sealed solid/liquid sample cell, but a tip sonicator is useful for the preparation of the vesicle solutions. Another method is to deposit a bilayer from mixed lipid/surfactant solutions also *in situ* (Tiberg et al., 2000). In this case, a sugar surfactant can be used to solubilize the lipid and the mixture adsorbs to the silica/liquid interface. Successive rinses and additions of mixtures at decreasing surfactant concentration allow elimination of the surfactant and buildup of a complete lipid bilayer. A third option is to deposit one or more bilayers on a silicon crystal using the Langmuir–Blodgett or Langmuir–Shaeffer dipping method (Tamm and McConnell, 1985). In this case, the silicon crystal is withdrawn through the surface of solution on which a Langmuir monolayer is maintained at a well-defined surface pressure using a compression barrier.

### III.5.2.4 *In Situ* Complementary Techniques

Particularly in soft matter research, there can often be several experimental techniques that give complementary information to help solve a particular scientific problem. Accordingly, there is a current drive to use techniques *in situ* with NR so that the maximum amount of information can be gained. Not only can it be efficient to record data simultaneously on samples that perhaps are not stable during transit, but the immediate extra information on a sample, in cases, can help the execution of NR measurements and thus increase the efficiency and success rate of experiments. In this section we touch on a selection of complementary techniques that each fall into one of the following categories: currently available for routine *in situ* measurements, currently under development after tests on a beamline, or simply well suited to the possibility of measurements *in situ* with NR in the future.

**Figure III.5.22** Photo of a custom-made thermostated sample cell used for ellipsometry measurements of adsorption layers at the air/liquid interface (Beaglehole, New Zealand).

### *III.5.2.4.1 Ellipsometry*

Ellipsometry is a noninvasive optical technique that is highly sensitive to thin films at interfaces. There are different ways of representing the amplitude and phase change of light upon reflection, the most common of which is the notation psi and delta, respectively. At solid/liquid interfaces where the solid absorbs light (e.g., a silicon crystal), both of these ellipsometric parameters are sensitive to thin homogenous adsorption layers, and they can be modeled simply to give the adsorbed amount and thickness. Whereas neutron reflectometry is useful to reveal the structure and composition of interfaces, the precision and kinetic resolution of ellipsometry—not forgetting less competitive access to time using such machines—can lead to a thorough mechanistic understanding of complex problems (Vandoolaeghe et al., 2009b). At the air/liquid interface, only the phase parameter delta is sensitive to the presence of isotropic thin films. As a result, the technique is limited in the depth of information it provides, but its precision and immediacy can be extremely useful, and its high spatial and temporal resolution can lead to the detection of lateral inhomogeneity in surface layers (Tonigold et al., 2009). Such information is useful to know before neutron beam time both for planning the experimental protocols and for the data fitting. Figure III.5.22 shows a thermostated cell made at ILL for adsorption measurements on a phase modulated ellipsometer (Beaglehole, New Zealand) to prepare for and complement data from NR experiments.

### *III.5.2.4.2 Brewster Angle Microscopy*

BAM is a technique in which *p*-polarized light is reflected at the Brewster angle of an interface. A clean surface or even a thin film reflects little and the image will be dark, but anisotropic domains or macroscopic aggregates reflect light to produce white regions in the image. In certain cases, the brightness can be interpreted as a quantitative measure of film thickness. The lateral structure from BAM can complement well the depth profiles determined from NR measurements, and hence many studies have been carried out on the structure Langmuir monolayers, such as the recent work on distearoylphosphatidylcholine (DSPC) (Hollinshead et al., 2009).

**Figure III.5.23** Brewster angle microscope (Nanofilm, Germany) shown with a Langmuir trough (Nima Ltd., UK) to be used *in situ* during neutron reflectometry experiments.

Figure III.5.23 shows a BAM machine (Nanofilm Ltd., Germany) set up with a Langmuir trough (Nima Ltd., UK). Applications also extend to adsorption layers at the air/liquid interface such as ordered polymer/surfactant films (Comas-Rojas et al., 2007). Measurements have been carried out *in situ* on the SURF reflectometer at ISIS and a BAM machine is currently under adaptation for routine measurement on the FIGARO reflectometer at ILL.

#### *III.5.2.4.3 X-Ray Reflectometry*

There are many similarities and several important differences between the principles of NR and X-ray reflectometry (Stamm et al., 1990). Both techniques can be used to gain quantitative structural information about interfacial phenomena. However, while neutrons interact with nuclei of a sample, X-rays interact with the electrons, and as a result the method of isotopic substitution—or contrast variation—is not applicable. Also, at synchrotron radiation facilities, the X-ray beam is so high in energy that heating the sample can be problematic; with a much smaller illuminated area on the sample, it can be possible to circumvent the problem by repeatedly moving the sample. Recent work by John White (ANO, Australia) and coworkers exploits neutron and X-ray reflectometry in investigations of the structure of protein/nanoparticle films at the air/liquid interface: they use the data from the two techniques to produce a self-consistent interfacial model (Ang et al., 2010). Combined NR and X-ray reflectometry has been implemented at the N-REX+ reflectometer at JCNS (FRM2, Germany).

#### *III.5.2.4.4 Electrochemistry and Reflection Absorption Infrared Spectroscopy*

Over the last few years, Andrew Glidle (University of Glagow, UK) and coworkers have developed a setup for *in situ* NR measurements of electrochemical reactions (Cooper et al., 2004). Work has been carried out on the CRISP reflectometer at ISIS and the D17 reflectometer at ILL. It has been possible to show that under dynamic redox switching conditions, polymer films at the solid/liquid interface show hysteresis in swelling, incomplete desolvation upon reduction, and transient salt retention under thermodynamically permselective conditions. The same group has carried out simultaneous data acquisition using *in situ* RAIRS (Glidle et al., 2005). In this case, infrared signatures allowed monitoring of electrochemical reactions of solution-based amine-tagged species such as amino-terminated poly(propylene glycol), ferrocene ethylamine, and lysine with film-based ester functionalities. The carbonyl bands are monitored in real time to show ester/amide interconversion and hydrolysis to acid.

#### *III.5.2.4.5 Total Internal Reflection-Raman Spectroscopy*

Recently, TIR-Raman spectroscopy has been used to supplement NR studies into lipid bilayer formation at the solid/liquid interface (Lee et al., 2009). TIR-Raman spectroscopy is a technique that enables the detection of the nonresonant Raman scattering signal from single lipid bilayers. Hydrogenated and deuterated materials can be distinguished due to their different resonant frequencies, and polarization variation can be exploited to determine changes in the individual packing densities and orientations of adsorbed molecules.

## III.5.3 BRIEF SUMMARY AND OUTLOOK

A large array of sample environments has been developed for SANS and NR measurements at large-scale facilities over the years. In the main, the facilities stock standard pieces of equipment while specialist items can be provided by the users. From the description of neutron sample environments for soft matter research in this chapter alone it is clear that the development of advanced sample environments leads to the highest quality of scientific research.

As time goes by there will be increasing demand to use smaller amounts of precious chemicals, automate highly controlled sample changes, perform faster experiments using straightforward instrument control software, and optimize the amount of information gained from a single experiment by carrying out *in situ* measurements using complementary techniques. Continued strong collaborations between beamline scientists and external users, and in particular healthy communication between scientists working at different facilities, will no doubt hasten the advancement of sophisticated sample environments to meet the challenges of tomorrow.

## ACKNOWLEDGMENTS

The authors gratefully acknowledge fruitful discussions with many colleagues and contributions from T. Sottmann (U. Köln, Germany), M. Ruppel (ORNL Oakridge, USA), R. Lund (DIPC-San Sebastián, Spain), W. Pyckhout-Hintzen (JCNS Jülich, Switzerland), J. Webster (ISIS, UK) Th. Nawroth (U. Mainz, Germany), R. Heenan

(ISIS, UK), and G. Fragneto, I. Grillo, L. Porcar, and H. Wacklin (Institut Laue-Langevin, France). 

## REFERENCES

Al-Jawad, M., Fragneto, G., Liu, J., Chang, S.R., and Clarkson, B. *Eur. Phys. J. E* **2009**, *30*, 175.

Ang, J.C., Lin, J.M., Yaron, P.N., and White, J.W. *Soft Matter* **2010**, *6*, 383.

Baker, S.M., Smith, G., Pynn, R., Butler, P., Hayter, J., Hamilton, W., and Magid, L. *Rev. Sci. Instrum.* **1994**, *65*, 412.

Blanchard, A., Graham, R.S., Heinrich, M., Pyckhout-Hintzen, W., Richter, D., Likhtman, A.E., McLeish, T.C.B., Read, D.J., Straube, E., and Kohlbrecher, J. *Phys. Rev. Lett.* **2005**, *95*, 166001.

Butler, P. *Curr. Opin. Colloid Interface Sci.* **1994**, *4*, 214.

Comas-Rojas, H., Aluicio-Sarduy, E., Rodríguez-Calvo, S., Pérez-Gramatges, A., Roser, S.J., and Edler, K.J. *Soft Matter* **2007**, *3*, 747.

Cooper, J.M., Cubitt, R., Dalgliesh, R.M., Gadegaard, N., Glidle, A., Hillman, A.R., Mortimer, R.J., Ryder, K.S., and Smith, E.L. *J. Am. Chem. Soc.* **2004**, *126*, 15362.

Förster, S., Konrad, M., Lindner, P. *Phys. Rev. Lett.* **2005**, *94*, 017803.

Fragneto, G., Lu, J.R., McDermott, D.C., Thomas, R.K., Rennie, A.R., Gallagher, P.D., and Satija, S.K. *Langmuir* **1996**, *12*, 477.

Fragneto-Cusani, G. *J. Phys.: Condens. Mat.* **2001**, *13*, 4973.

Glidle, A., Hadyoon, C.S., Gadegaard, N., Cooper, J.M., Hillman, A.R., Wilson, A.R., Ryder, K.S., Webster, J.R.P., and Cubitt, R. J. *Phys. Chem. B* **2005**, *109*, 14335.

Grillo, I. *Curr. Opin. Colloid Interface Sci.* **2009**, *14*, 402.

Hamley, I. *Curr. Opin. Colloid Interface Sci.* **2000**, *5*, 342.

Hanrahan, J.P., Copley, M.P., Ziegler, K.J., Spalding, T.R., Morris, M.A., Steytler, D.C., Heenan, R.K., Schweins, R., and Holmes, J.D. *Langmuir* **2005**, *21*, 4163.

Heinrich, M., Pyckhout-Hintzen, W., Allgaier, J., Richter, D., Straube, E., McLeish, T.C.B., Wiedenmann, A., Blackwell, R.J., and Read, D.J. *Macromolecules* **2004**, *37*, 5054.

Holderer, O., Klostermann, M., Monkenbusch, M., Schweins, R., Lindner, P., Strey, R., Richter, D., and Sottmann, T. *Soft Matter, Phys. Chem. Chem. Phys.* **2010**, DOI 10.1039/C0CP01199A.

Hollinshead, C.M., Harvey, R.D., Barlow, D.J., Webster, J.R.P., Hughes, A.V., Weston, A., and Lawrence, M.J. *Langmuir* **2009**, *25*, 4070.

Kalb, E., Frey, S., and Tamm, L.K. *Biochim. Biophys. Acta* **1992**, *1103*, 307.

King, M.D., Rennie, A.R., Thompson, K.C., Fisher, F.N., Dong, C.C., Thomas, R.K., Pfrang, C., and Hughes, A.V. *Phys. Chem. Chem. Phys.* **2009**, *11*, 7699.

Kuhl, T.L., Smith, G.S., Israelachvili, J.N., Majewski, J., and Hamilton, W. *Rev. Sci. Instrum.* **2001**, *72*, 1715.

Lee, C., Wacklin, H.P., and Bain, C.D. *Soft Matter* **2009**, *5*, 568.

Lindner, P.and Oberthür, R.C. *Rev. Phys. Appl.* **1984**, *19*, 759.

Lindner, P.and Oberthür, R.C. *Colloid Polym. Sci.* **1988**, *266*, 886.

Lindner, P.and Zemb, T. (editors). *Neutrons, X-Rays and Light: Scattering Methods Applied to Soft Condensed Matter*, Elsevier, Amsterdam–North Holland, 2002.

Lu, J.R., Thomas, R.K., and Penfold, J. *Adv. Colloid Interface Sci.* **2000**, *84*, 143.

Manning-Benson, S., Parker, S.R.W., and Bain, C.D. *Langmuir* **1998**, *14*, 990.

Miller, C.E., Majewski, J., Faller, R., Satija, S., and Kuhl, T.L. *Biophys. J.* **2004**, *86*, 3700.

Mortensen, K. *Curr. Opin. Colloid Interface Sci.* **2001**, *6*, 140.

Nettesheim, F., Olsson, U., Lindner, P., and Richtering, W. *J. Appl. Cryst.* **2004**, *37*, 438.

Nettesheim, F., Zipfel, J., Olsson, U., Renth, F., Lindner, P., and Richtering, W. *Langmuir* **2003**, *19*, 3603.

Perino-Gallice, L., Fragneto, G., Mennicke, U., Salditt, T., and Rieutord, F. *Eur. Phys. J. E* **2002**, *8*, 275.

Perriman, A.W., McGillivray, D.J., and White, J.W. *Soft Matter* **2008**, *4*, 2192.
Richtering, W. *Curr. Opin. Colloid Interface Sci.* **2001**, *6*, 446.
Saito, S., Hashimoto, T., Morfin, I., Lindner, P., and Boué, F. *Macromolecules* **2002**, *35*, 445.
Schief, W.R., Antia, M., Discher, B.M., Hall, S.B., and Vogel, V. *Biophys. J.* **2003**, *84*, 3792.
Stamm, M., Reiter, G., and Kunz, K. *Physica B* **1990**, *173*, 35.
Tamm, L.K.and McConnell, H.M. *Biophys. J.* **1985**, *47*, 105.
Taylor, D.J.F., Thomas, R.K., and Penfold, J. *Adv. Colloid Interface Sci.* **2007**, *132*, 69.
Tiberg, F., Harwigsson, I., and Malmsten, M. *Eur. Biophys. J. Biophys. Lett.* **2000**, *29*, 196.
Tonigold, K., Varga, I., Nylander, T., and Campbell, R.A. *Langmuir* **2009**, *25*, 4036.
Vandoolaeghe, P., Rennie, A.R., Campbell, R.A., and Nylander, T. *Langmuir* **2009a**, *25*, 4009–4020.
Vandoolaeghe, P., Campbell, R.A., Rennie, A.R., and Nylander, T. *J. Phys. Chem. C* **2009b**, *113*, 4483.
Vennemann, N., Lechner, M.D., and Oberthür, R.C. *Polymer* **1987**, *28*, 1738.
Zarbakhsh, A., Bowers, J., and Webster, J. *Meas. Sci. Technol.* **1999**, *10*, 738.

# IV Applications

# IV.1 Hierarchical Structure of Small Molecules

*Tsang-Lang Lin*

## IV.1.1 INTRODUCTION

Surfactants and many biomolecules can self-assembly into great varieties of structures with different functionalities and applications. There are wide applications for various industries, including detergent, pharmaceutical, cosmetic, paint, food, nanomaterial, and environmental industries. It is important to understand how the microstructure changes with the types of molecules and sample conditions (concentration, temperature, pH, etc.) in order to tailor the microstructure for specific application. Surfactants and amphiphilic biomolecules can form aggregates in aqueous solutions due to hydrophobic interaction. The shape and the size of such aggregates may vary with species, concentration, temperature, composition, pH, salt concentration, and so on. Surfactant aggregates called micelles often have spherical/globular shapes that have a compact hydrophobic core formed by the hydrophobic tails of the surfactants and a hydrophilic layer containing the hydrophilic head groups of the surfactants. Globular micelles may grow into short rod-like micelles or even worm-like micelles upon increasing surfactant concentration, increasing the chain length of the hydrophobic tails (Lin et al., 1987), or adding salts to charged micelles (Hassan et al., 2003). Other than the zero-dimensional (globular micelles) and the one-dimensional (rod-like micelles) structures, two-dimensional (disk shape) structures can also be formed by mixed micelles (Lin et al., 1991; Almgren, 2000). More complicated structures can also be formed, such as vesicles and liposomes (Dubois et al., 2004). Figure IV.1.1 shows the schematic diagrams of the globular micelle, the rod-like micelle, the disk-like micelle, and the vesicle. Bilayer vesicles formed by lipids that mimic biomembranes could be used as model systems to investigate biological interactions (Jeng et al., 2005). Mixing anionic and cationic surfactants at suitable ratios can also form vesicles (Bergström et al., 1999). Other than the richness in shape and structure, the other important property of such aggregates is that their

*Neutrons in Soft Matter* Edited by Toyoko Imae, Toshiji Kanaya, Michihiro Furusaka, and Naoya Torikai

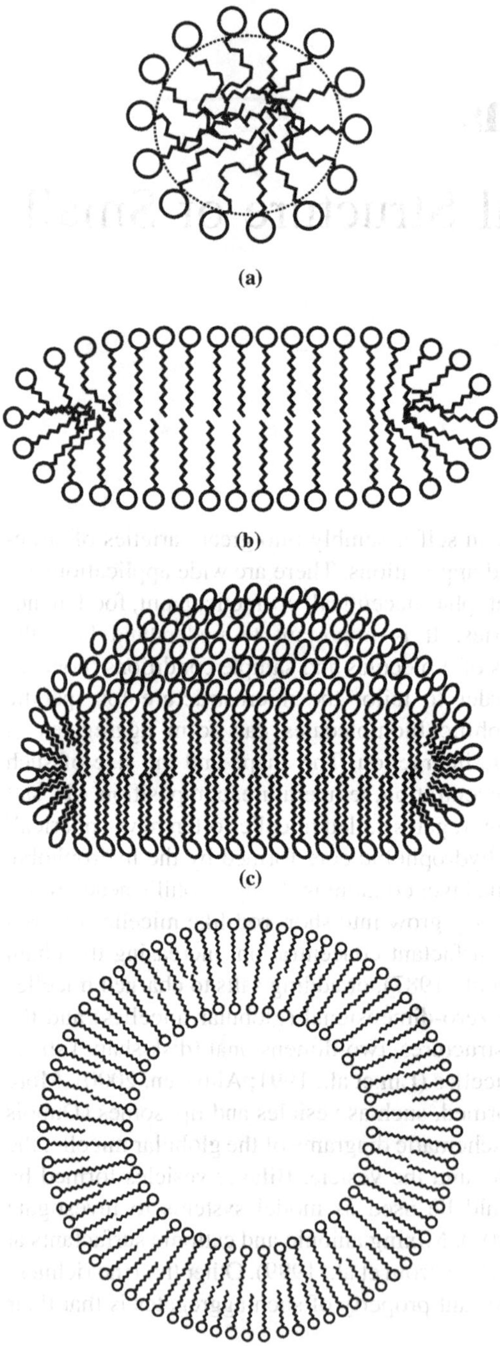


**Figure IV.1.1** The schematics of the (a) globular micelle, (b) rod-like (cylindrical) micelle, (c) disk-like micelle, and (d) vesicle.

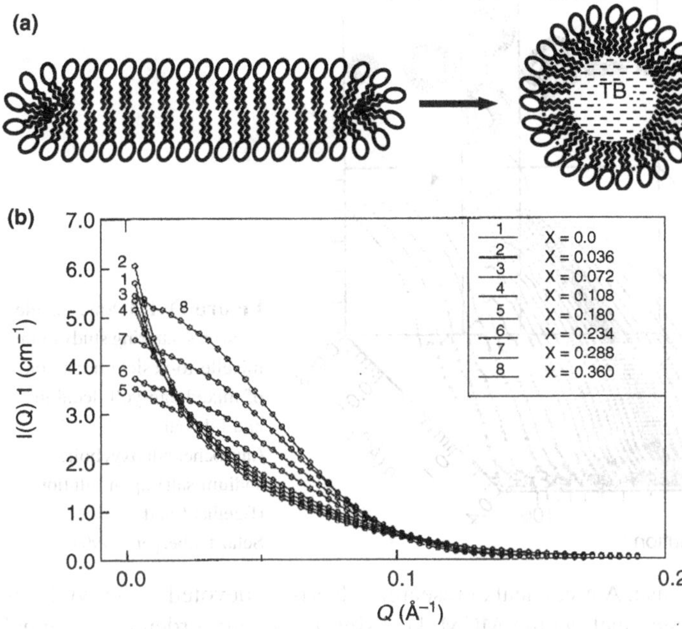


**Figure IV.1.2** (a) The schematic of the diheptanoyl-PC rod-like micelle to globular micelle transition upon solubilization of tributyrin; (b) small-angle neutron scattering profiles from the diheptanoyl-PC micellar solutions with increasing amounts of solubilized tributyrin. The concentration of diheptanoyl-PC is kept at 50 mM. The concentration ratio of tributyrin to diheptanoyl-PC is indicated as *X*. The rod-like diheptanoyl-PC micelles were found to transform into globular mixed micelles as *X* is raised beyond 0.18 (Lin et al., 1990).

size, shape, and structure may change upon interacting with other systems or change of the sample conditions. Such structural transitions are important in many applications; for example, the solubilization of oil molecules by micelles could turn the rod-like micelles into globular core–shell microemulsions (Lin et al., 1990). Figure IV.1.2 shows the schematic diagram of the diheptanoylphosphatidylcholine (diheptanoyl-PC) rod-like micelle to globular micelle transition upon solubilization of tributyrin and the corresponding small-angle neutron scattering (SANS) profiles from the diheptanoyl-PC micellar solutions with increasing amounts of solubilized tributyrin (Lin et al., 1990). The structural transitions can be well studied by using small-angle neutron scattering. The structural transition can be clearly viewed from the changes in the scattering profiles. A micelle to vesicle transition occurs when a mixture of lecithin and bile salt solution is diluted (Figure IV.1.3). The shape of the mixed lecithin and bile salt micelles could change from small spherical to elongated and flexible cylindrical micelles, then to the vesicle formation at high dilutions. The transition is caused by the decrease of the bile salt-to-lecithin ratio in the aggregates, which lowers the average spontaneous curvature. The decrease in the bile salt-to-lecithin ratio is due to the difference in the critical micellization concentrations (cmcs) of these two components (Egelhaaf and Schurtenberger, 1997; Egelhaaf and Schurtenberger, 1999). At high concentrations, the rod-like micelles may form

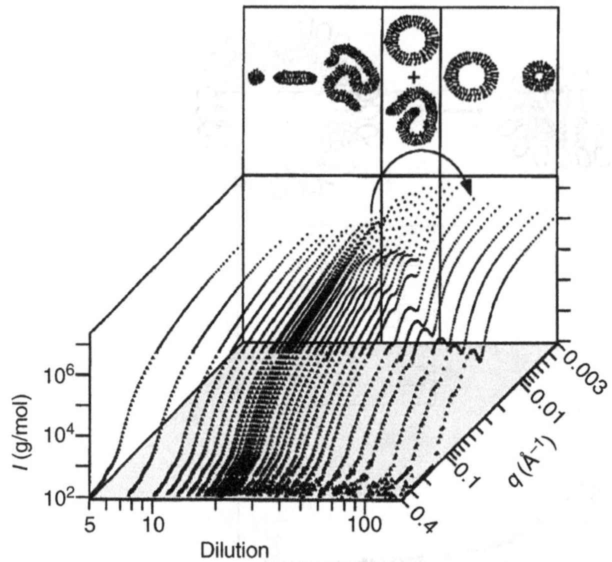


**Figure IV.1.3** Small-angle neutron scattering study of the micelle-to-vesicle transition of mixed egg yolk lecithin and bile salt (taurochenodeoxycholic acid sodium salt) upon dilution (Egelhaaf and Schurtenberger, 1999).

ordered hexagonal phase. A great deal of researches has been devoted to the synthesis of mesoporous silicate, such as the MCM 41, using the highly ordered hexagonal micellar structure as the template (Edler et al., 1998). These hierarchical structures have applications in catalysis, separation processes, molecular hosting processes, and so on. Very long and semiflexible worm-like micelles can also be formed that have special viscoelatic and rehological properties (Koehler et al., 2000; Jerke et al., 1998). Some types of aggregates could form superstructures through attractive interactions between aggregates, such as the fractal aggregates that have self-similarity with a constant fractal dimension over a certain length scales (Teixeira, 1988; Jeng et al., 1999).

Other than the aggregation of surfactants or small amphiphilic molecules, such as lipids and proteins, another important category is the microemulsions that are usually formed by mixing surfactants with oil and water (Chen et al., 1987). With the help of surfactant, either oil-in-water microemulsions or water-in-oil microemulsions can be formulated. They are also the subjects of intensive researches in recent decades for their scientific and technological importance. Besides the applications similar to the micelles, microemulsions can also be used as microreactors for synthesizing nanoparticles (Zhang and Chan, 2003).

Several tools can be used to characterize the nanostructure of such self-assembly aggregates. Dynamic light scattering (DLS) can conveniently measure the hydrodynamic radius of dispersed nanoparticles in solutions, but DLS could not provide the information about the exact shape and the exact structure. Scanning electronic microscope (SEM) and the transmission electronic microscope (TEM) could provide the image of nanoparticles and nanostructures in a dried or freeze state. They are useful complementary techniques as compared with the scattering methods. Scattering methods, such as the small-angle X-ray and small-angle neutron scattering could provide the structural information for samples in a solution state and could perform *in situ* and time-resolved measurements. X-ray scattering is more sensitive to heavy elements than the light elements of carbon and hydrogen that constitute most parts of the soft materials. On the other hand, for studying soft materials with neutrons, scattering contrast can be enhanced by using deuterated materials (Lin et al., 1986; Chen and Lin, 1987). This is due to that the neutron scattering from hydrogen and deuterium is very different. The contrast variation method is one of the key factors in choosing the neutron scattering as the investigation tool for studying the soft materials. Other important features of using neutrons include low radiation damage, good penetration capability, and comparable scattering cross sections for light elements as well as heavy elements. The contrast variation technique is especially useful in investigating complex structures or mixed systems (Endo, 2006). In aqueous solutions, one can easily change the contrast by varying the ratio of the $H_2O$ with $D_2O$ (Lin et al., 1986). For internal contrast variation, one can use deuterated materials or partially deuterated materials to enhance the contrast of a particular portion of the molecule (Lin et al., 1986). There were many successful neutron experiments that employed the neutron contrast variation methods to solve complex structures. Mixing deuterated lipid vesicles with proteo lipid vesicles enables the study of the mixing kinetics of lipid bilayer membranes by time-resolved neutron scattering (Nakano et al., 2007). Although neutron scattering is often the choice for studying the soft materials, X-ray scattering can still be a complementary tool. For example, the head group of the surfactant often contain heavier atoms and it is better be investigated by X-ray scattering if the interest is on the structure of the head group region of the surfactant aggregates. It is also advantageous to use X-ray scattering to investigate the ion distribution around the charged aggregates. In this chapter, details of methods of applying the small-angle neutron scattering to investigate the nanostructures formed by the aggregation of small molecules will be described.

## IV.1.2 NEUTRON SCATTERING ANALYSIS METHODS

The normalized neutron scattering intensity distribution (scattering cross section per unit sample volume) from samples containing monodisperse particles can be represented as

$$I(Q) = n_p P(Q) S(Q), \tag{IV.1.1}$$

where $n_p$ is the number of particles per unit sample volume (particle number density), $P(Q)$ is the particle form factor (intraparticle factor), and $S(Q)$ is the interparticle structure factor, or simply the structure factor. $Q$ is the scattering vector, and it is defined as $Q = (4\pi/\lambda)(\sin\theta/2)$, where $\lambda$ is the wavelength of the incident neutrons and $\theta$ is the scattering angle (the angle between the scattered neutron and the incident neutron). $P(Q)$ is given by

$$P(Q) = \left| \int \Delta\rho(\vec{r}) \cdot e^{-i\vec{Q}\cdot\vec{r}} d\vec{r} \right|^2, \tag{IV.1.2}$$

where $\Delta\rho(\vec{r})$ is the coherent neutron scattering length density (cnsld) difference between the cnsld of the particle, $\rho_p(\vec{r})$, and the cnsld of the solvent, $\rho_s$, and it is given by

$$\Delta\rho(\vec{r}) \equiv \rho_p(\vec{r}) - \rho_s. \tag{IV.1.3}$$

The integration in eq. (IV.1.2) is over a single particle. The shape and the structure of the particle determine the profile of $P(Q)$. The structure factor $S(Q)$ arises from the interference between the scattered waves and the interference depends on the relative positions of the particles. $S(Q)$ is given by

$$S(Q) = \frac{1}{N_p}\left\langle \left| \sum_{i=1}^{N_p}\sum_{j=1}^{N_p} e^{i\vec{Q}\cdot(\vec{R}_i-\vec{R}_j)} \right| \right\rangle, \tag{IV.1.4}$$

where $\vec{R}_i$ represents the position vector of $i$th particle. The average in eq. (IV.1.4) is the ensemble average of the particle configurations. If we define an amplitude term $I_0$ as

$$I_0 = n_p\, P(0) = n_p \left| \int \Delta\rho(\vec{r})\mathrm{d}\vec{r} \right|^2, \tag{IV.1.5}$$

the normalized neutron scattering intensity distribution eq. (IV.1.1) can be rewritten as

$$I(Q) = I_0 P'(Q)\, S(Q), \tag{IV.1.6}$$

where the normalized form factor $P'(Q)$ is defined as

$$P'(Q) = \frac{\left|\int \Delta\rho(\vec{r})\cdot e^{-i\vec{Q}\cdot\vec{r}}\mathrm{d}\vec{r}\right|^2}{\left|\int \Delta\rho(\vec{r})\mathrm{d}\vec{r}\right|^2}, \tag{IV.1.7}$$

where we have $P'(0) = 1$, but $S(0)$ is in general not equal to unity. At very dilute and very weak interparticle interaction conditions, we will have $S(Q) \approx 1$, and also $I(0) \cong I_0$. For uniform particles with particle volume $V_p$, the normalized form factor $P'(Q)$ would depend only on the shape of the particle and we have

$$I_0 = n_p(\Delta\rho)^2\, V_p^2. \tag{IV.1.8}$$

The contrast term $\Delta\rho$ is important since it affects the scattering intensity. The scattering from larger particles is much stronger than from smaller particles since the scattering amplitude is proportional to the square of the particle volume (or proportional to $R^6$ for uniform spherical particles with radius $R$). This means that small particles would be hard to detect if they are in coexistence with larger particles. For particles formed by the aggregation of small molecules, it is important to determine the aggregation number $N$ (the number of molecules that constitute one aggregate). For micellar systems, the micellar aggregates are in coexistence with free monomeric molecules, which is called the critical micelle concentration, $C_{\mathrm{cmc}}$. For a monodisperse system, the amplitude term $I_0$ of eq. (IV.1.5) for micellar systems can be rewritten as

$$I_0 = \frac{C - C_{\mathrm{cmc}}}{N}[N(b_{\mathrm{m}} - \rho_{\mathrm{s}} V_{\mathrm{m}})]^2 = (C - C_{\mathrm{cmc}})N(b_{\mathrm{m}} - \rho_{\mathrm{s}} V_{\mathrm{m}})^2, \qquad \text{(IV.1.9)}$$

where $C$ is the total surfactant molecular concentration in the solution, $b_{\mathrm{m}}$ is the total coherent neutron scattering length (cnsl) of one surfactant molecule, and $V_{\mathrm{m}}$ is the volume occupied by one surfactant molecule in the solution. The number density of aggregates $n_{\mathrm{p}}$ is equal to $(C - C_{\mathrm{cmc}})/N$. It is important to be able to measure the absolute scattering cross section in order to determine the aggregation number from eq. (IV.1.9). The scattering intensity is proportional to the molecular concentration in the aggregate $(C - C_{\mathrm{cmc}})$, the aggregation number $N$, and the square of the contrast between the molecule and the solvent. Contrast variation/enhancement can be achieved by varying either the $b_{\mathrm{m}}$ value (internal contrast variation) or the solvent $\rho_{\mathrm{s}}$ (external contrast variation). However, for uniform particles, varying the $\rho_{\mathrm{s}}$ value would not change the scattering profile $P'(Q)S(Q)$. If the scattering profile changes with varying the $\rho_{\mathrm{s}}$ value, the particle would not have a uniform structure (Lin et al., 1986) Although varying $\rho_{\mathrm{s}}$ would not change the scattering profile for uniform particles, it is still useful to vary $\rho_{\mathrm{s}}$ in order to determine the contrast match point (the $\rho_{\mathrm{s}}^*$ value that makes $I_0 = 0$, that is, $\rho_{\mathrm{s}}^* = b_{\mathrm{m}}/V_{\mathrm{m}}$), so as to determine $V_{\mathrm{m}}$ from the obtained average scattering length density of the particle, $\bar{\rho}_{\mathrm{p}}$, with $\bar{\rho}_{\mathrm{p}} \equiv b_{\mathrm{m}}/V_{\mathrm{m}}$. This is a useful scheme when the value of $V_{\mathrm{m}}$ is not available or hard to estimate. For cases with $S(Q) \approx 1$, plotting $\pm\sqrt{I(0)}$ versus $\rho_{\mathrm{s}}$ would give a straight line and it is easy to obtain the contrast match point from the zero crossing point. This works for uniform particles as well as nonuniform particles. As an example, Figure IV.1.4 shows the external contrast variation plot of the dihexanoylphosphatidylcholine (dihexanoyl-PC) at 50 mM. The micelle aggregation number $N$ and the $V_{\mathrm{m}}$ can be determined from such contrast variation analysis (Lin et al., 1986). The fact that the neutron scattering intensity is related to the scattering length density or $V_{\mathrm{m}}$ was successfully used in investigating the density minimum of supercooled water in nanosize channels (Liu et al., 2007). As for systems containing two kinds of particles with different average scattering length densities in coexistence, there is no contrast-matching point for the whole system, but it is still possible to determine the individual contrast-matching point (Lin, 1995).

The micellar aggregates can be conveniently modeled by a core–shell structural model for analyzing the scattering profile (Bendedouch and Chen, 1983; Lin et al., 1986). The shell consists of the head groups and the water molecules, while the core contains the hydrophobic tails. In $D_2O$ solution, there is a large scattering contrast between the hydrocarbon tails and the $D_2O$ solvent. The neutron scattering contribution from the head groups is also significant and it is as important as the scattering from the hydrophobic core. Hence, it is preferred to use proteo surfactants with $D_2O$ as the solvent for aqueous systems to gain scattering contrast and also to

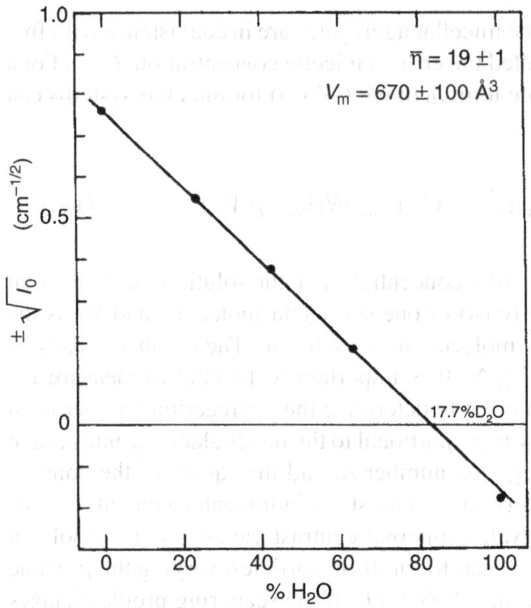


**Figure IV.1.4** Plot of $\pm\sqrt{I(0)}$ versus percentage of $H_2O$ in a mixture of $H_2O$ and $D_2O$ solvent for dihexanoylphosphatidylcholine (dihexanoyl-PC) at 50 mM. From the external contrast variation plot, the aggregation is found to be 19 ± 1 and $V_m = 670 \pm 100$ Å$^3$ (Lin et al., 1986).

reduce the incoherent scattering background ($D_2O$ has much lower incoherent scattering than $H_2O$). However, in $H_2O$ solution, typically the neutron scattering contribution from the hydrophobic core is much smaller than from the head groups since the cnsld of the hydrocarbons is very close to the cnsld of $H_2O$. Varying the solvent from pure $D_2O$ to $H_2O$ (external contrast variation) would be able to distinguish the core–shell structure (Lin et al., 1986). The scattering profile of a core–shell particle will vary with the changes of the solvent cnsld.

Other than the external contrast variation method, it is also advantageous to use the internal contrast variation. One can mix the proteo surfactants with per-deuterated surfactants of the same chemical structure. One may also use partially deuterated surfactants to determine the structure of a particular part of the aggregate, such as to study the distribution of the terminal methyl groups within the hydrophobic core of the micelle by only deuterating the terminal methyl groups and comparing the neutron scattering spectrum for this species with the proteo compound at the same concentration (Lin et al., 1986).

Contrast variation method is particularly useful in investigating mixed complex aggregates. For monodisperse mixed systems comprised of two different amphiphilic molecules, the scattering amplitude $I_0$ of the complex aggregates containing partial aggregation numbers $N_A$ for species $A$ and $N_B$ for species $B$ can be expressed as (Jeng et al., 2002a)

$$I_0 = n_p[N_A(b_A - \rho_s V_A) + N_B(b_B - \rho_s V_B)]^2, \qquad \text{(IV.1.10)}$$

where $b_A$, $b_B$ and $V_A$, $V_B$ are, respectively, the cnsl and the dry volume of the species $A$ and $B$ molecules. $n_p$ is the number density of the complex aggregate. One may replace one of the species by deuterated materials, such as the case of studying the association of fullerene-based ionomers $C_{60}[(CH_2)_4SO_3Na]_6$ (FC4S) with sodium dodecyl sulfate (SDS) by using deuteratede SDS (d-SDS). Denoting the cnsl of the deuterated species $A$ as $b'_A$, we have a different scattering amplitude $I'_0$ given by

$$I'_0 = n_p[N_A(b'_A - \rho_s V_A) + N_B(b_B - \rho_s V_B)]^2. \tag{IV.1.11}$$

Combining eqs. (IV.1.10) and (IV.1.11), the calculated values of $b_A$, $b'_A$, $b_B$, $V_A$, $V_B$, and $\rho_s$ and the measured values of $I_0$ and $I'_0$, we can determine the ratio of $N_A/N_B$ and subsequently the $N_A$ and $N_B$ values using the constraint on the number density of the mixed aggregates $n_p$. The number density of the mixed aggregates, $n_p$, is constrained by the molecular concentrations of species $A$ and $B$, and the partial aggregation numbers $N_A$ and $N_B$.

For some particular complex systems that can scatter both neutrons and X-rays very well, it is possible to combine the neutron scattering and the X-ray scattering to provide the scattering contrast (Lin et al., 2004; Jeng et al., 2002a).

The scattering profile is as important as the scattering amplitude since the information about the size and the shape is contained in the scattering profile. For uniform spherical particles with radius $R$, the particle form factor is given by

$$P(Q) = (\Delta\rho)^2 V_p^2 \left[\frac{3j_1(QR)}{QR}\right]^2, \tag{IV.1.12}$$

where $j_1(QR)$ is the spherical Bessel function and is given by

$$j_1(QR) = \frac{\sin(QR) - (QR)\cos(QR)}{(QR)^2}. \tag{IV.1.13}$$

For a spherical core–shell particle, the particle form factor is given by

$$P(Q) = \left\{(\rho_1 - \rho_2)\frac{4}{3}\pi R_1^3\left[\frac{3j_1(QR_1)}{QR_1}\right] + (\rho_2 - \rho_s)\frac{4}{3}\pi R_2^3\left[\frac{3j_1(QR_2)}{QR_2}\right]\right\}^2, \tag{IV.1.14}$$

where $\rho_1$ and $\rho_2$ are, respectively, the cnsld of the core and the shell regions of the particle. $R_1$ and $R_2$ are, respectively, the core radius and the outer radius of the shell. Equation (IV.1.14) can also be used to describe a shell particle by simply replacing the $\rho_1$ in eq. (IV.1.14) by $\rho_s$. The feature for shell particles is that their first interference bump in the particle form factor would be more significant than that for spherical particles.

For monodisperse elliptical particles, randomly oriented in the solution, the averaged particle form factor is given by

$$P(Q) = \int_0^1 d\mu |F(Q,\mu)|^2, \quad \text{(IV.1.15)}$$

$$F(Q,\mu) = (\Delta\rho)\left(\frac{4}{3}\pi ab^2\right)\frac{3j_1(u)}{u}, \quad \text{(IV.1.16)}$$

$$u = Q[a^2\mu^2 + b^2(1-\mu^2)]^{\frac{1}{2}}, \quad \text{(IV.1.17)}$$

Here, $a$ is the semimajor axis and $b$ is the semiminor axis of the elliptical particle. In order to compare the scattering profiles from elliptical particles of different $a/b$ ratios, here we introduce the radius of gyration $R_g$ of a particle by defining

$$R_g \equiv \left[\frac{\int r^2 \Delta\rho(\vec{r})d\vec{r}}{\int \Delta\rho(\vec{r})d\vec{r}}\right]^{\frac{1}{2}}. \quad \text{(IV.1.18)}$$

According to the Guinier approximation, the scattering profiles for particles with the same radius of gyration values but different shapes will coincide at very-small-$Q$ region. The Guinier approximation is given by

$$I(Q) \approx I_0 e^{-(1/3)Q^2 {R_g}^2}. \quad \text{(IV.1.19)}$$

Equation (IV.1.19) typically works for $Q < R_g$ and $S(Q) \approx 1$. According to the prediction of the Guinier approximation, the scattering data in the very-low-$Q$ region will fall on a straight in the plot of $\ln(I(Q))$ versus $Q^2$. The radius of gyration can be determined from such plots since the slope of the straight line is equal to $-R_g^2/3$. For elliptical particles, the radius of gyration is equal to $\sqrt{(a^2+2b^2)/5}$. Figure IV.1.5 shows the comparison of the scattering profiles from ellipsoidal particles with the same radius of gyration but different $a/b$ ratios. The normalized scattering profile $P'(Q)$ is plotted versus the dimensionless parameter $QR_g$. The scattering profile of spherical particles, the $a/b = 1$ case, is also shown in Figure IV.1.5. As shown in Figure IV.1.5, the scattering intensity decreases more slowly with increasing $QR_g$ for larger aspect ratio particles. It is not difficult to determine the aspect ratio of elliptical particles from the measured scattering profile.

Micelles could grow into rod-like aggregates with the same radius, but different lengths in coexistence (Lin et al., 1987). The particle form factor for monodisperse and uniform rod-like particles randomly oriented in the solution is given by

$$P(Q) = \int_0^1 d\mu \left|F_{\text{cyl}}(Q,\mu)\right|^2 \quad \text{(IV.1.20)}$$

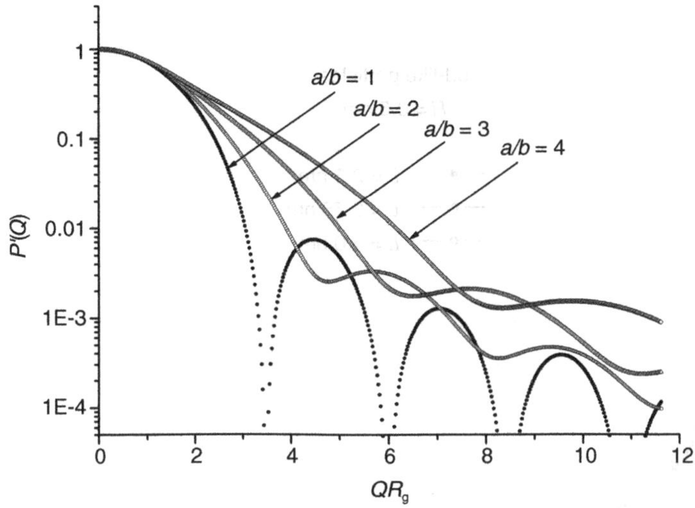


**Figure IV.1.5** Comparison of the scattering profiles from ellipsoidal particles with the same radius of gyration but different $a/b$ ratios, where $a$ and $b$ are, respectively, the semimajor axis and the semiminor axis. The normalized scattering profile $P'(Q)$ is plotted versus the dimensionless parameter $QR_g$. The case with $a/b = 1$ is the scattering profile of spherical particle.

$$F_{\text{cyl}}(Q,\mu) = (\Delta\rho)(\pi R^2 L)\left(\frac{2J_1(v)}{v}\frac{\sin w}{w}\right), \qquad \text{(IV.1.21)}$$

$$v = QR(1-\mu^2)^{\frac{1}{2}}, \qquad \text{(IV.1.22)}$$

$$w = \frac{1}{2}QL\mu, \qquad \text{(IV.1.23)}$$

where, $J_1(v)$ is the first-order Bessel function. The radius and the length of the rod-like particle are, respectively, denoted as $R$ and $L$. Equation (IV.1.20) can also used to describe disk-like particles with $R$ as the disk radius and $L$ as the disk thickness. Figure IV.1.6 shows the scattering profiles of rod-like particles of the same radius but different lengths. When the total sample volume $n_p V_p$ is kept the same, the scattering amplitude will be proportional to the length of the rod-like particle. The scattering profile of rod-like particles possesses some special features. First, there are two length scales of rod-like particles, the length $L$ and the radius $R$. In the low-$Q$ region, the scattering profile mainly depends on the larger length scale, the length $L$. In the middle-$Q$ region, the scattering profile mainly controlled by the smaller length scale, the radius $R$. These features can be noticed clearly when $L \gg R$. As shown in Figure IV.1.6, the scattering intensity drops steeply with increasing $Q$ in the low-$Q$ region due to the large length scale $L$ and the scattering intensity in the middle-$Q$ region decreases with increasing $Q$ more slowly, which reflects the small length

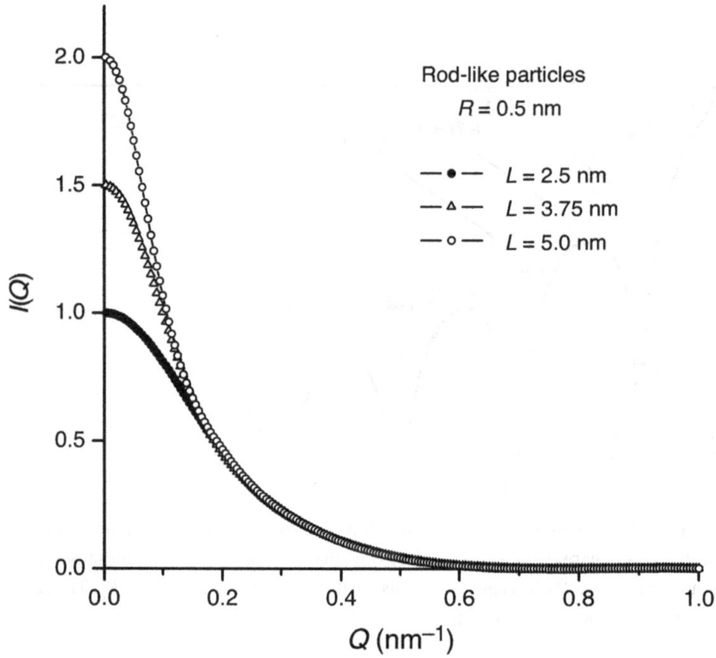


**Figure IV.1.6** Comparison of the scattering profiles for rod-like particles of the same radius but different lengths. Here, the scattering amplitude for the $L = 2.5$ nm case is normalized to 1.0 and it is assumed that $S(Q) \approx 1$. The total sample volume $n_p V_p$ is kept the same and the scattering amplitude is proportional to the length of the rod-like particle.

scale $R$. Assuming that $S(Q) \approx 1$ to simplify the discussions, the scattering intensity distribution in the middle-$Q$ region can be approximated by

$$I(Q) \approx n_p V_p (\Delta\rho)^2 (\pi R^2) \frac{\pi}{Q} e^{-(Q^2 R_c^2/2)}, \quad \frac{2\pi}{L} < Q < \frac{\pi}{R}, \qquad \text{(IV.1.24)}$$

where $R_c$ in is the radius of gyration across the cross section of the rod-like particle and it is equal to $R/\sqrt{2}$ for uniform rods. Equation (IV.1.24) shows that the scattering intensity in the middle-$Q$ region does not depend on the length of the rod-like particle, but depend on the molecular volume fraction in the solution, $n_p V_p$, and the rod radius $R$. The other feature is that the scattering intensity has $1/Q$ dependence. As shown in Figure IV.1.7 for five concentrations of diheptanoyl-PC rod-like micellar systems, the scattering data in the middle-$Q$ region, $2\pi/L < Q < \pi/R$, all fall on a straight line in $\ln(I(Q)Q)$ versus $Q^2$ plots (Lin et al., 1987). It is easy to identify rod-like particles by such plots. The slope of the fitted straight line is equal to $-R_c^2/2$ and it can be used to determine the rod radius (Lin et al., 1987). This method was also successfully used to determine the effective fibril radius of $\beta$-amyloids that form fibrils in aqueous solutions (Thiyagarajan et al., 2000; Jeng et al., 2006). The fibrils are formed by the aggregation of $\beta$-amyloids in $\beta$-sheet form and the formation of the fibrils is

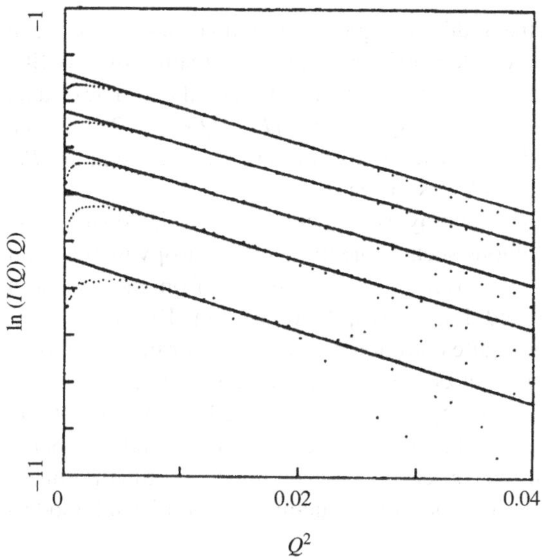


**Figure IV.1.7** Plots of $\ln(I(Q)Q)$ versus $Q^2$ for five concentrations, 2.2, 4.4, 8.7, 17.7, and 35.0 mM diheptanoyl-PC rod-like micellar systems. The solid lines are the fitted straight lines (Lin et al., 1987).

thought to be related to the Alzheimer's disease. For micellar aggregates, eq. (IV.1.24) can be expressed as (Lin et al., 1987)

$$I(Q) \approx (C - C_{\text{cmc}})\frac{\pi}{Q}\alpha(b_{\text{m}} - \rho_{\text{s}}V_{\text{m}})^2 e^{-(Q^2R_{\text{C}}^2/2)}, \quad \frac{2\pi}{L} < Q < \frac{\pi}{R}, \tag{IV.1.25}$$

where $\alpha$ is the average aggregation number per unit rod length of the rod-like micelles. This means that the middle-$Q$ region data can provide the information about the cross-sectional structure of the rod-like particles (the rod radius and the average aggregation number per unit rod length). It is also evident from eq. (IV.1.25) that the scattering intensity distribution would coincide in the middle-$Q$ region as that shown in Figure IV.1.6 for rod-like systems having the same total concentration and radial structure regardless of the difference in rod length. As for the semiflexible/worm-like micelles, other than the straight section for the middle-$Q$ region, there is an upturn in $\ln(I(Q)Q)$ versus $Q^2$ plots at lower values of $Q$, which is characteristic for worm-like micelles (Jerke et al., 1998). The persistence length of the worm-like micelles can be determined from the scattering profile.

Similar to the rod-like particles, the scattering intensity of the disk-like particles in the middle-$Q$ region can be approximated by

$$I(Q) \approx n_{\text{p}}(\Delta\rho)^2 V_{\text{p}} T \frac{2\pi}{Q^2} e^{-Q^2R_t^2}, \quad \frac{\pi}{R} < Q < \frac{2\pi}{T}. \tag{IV.1.26}$$

For simplicity, it is also assumed that $S(Q) \approx 1$. The thickness and the radius of the disk-like particle are, respectively, denoted as $T$ and $R$ with $R \gg T$. $R_{\text{t}}$ is the radius of gyration across the thickness of the disk, and it is equal to $T/\sqrt{12}$ for uniform disks. For disk-like particles, the scattering intensity in the middle-$Q$ region has $1/Q^2$

dependence. The scattering data in the middle-$Q$ region would fall on a straight line in $\ln(I(Q)Q^2)$ versus $Q^2$ plots. The $1/Q^2$ dependence is a special feature of disk-like particles (Lin et al., 1991) and sheet-like structures, such as unilamellar vesicles (Imae et al., 1996). The scattering data in the middle-$Q$ region, $\pi/R < Q < 2\pi/T$, fall on a straight line in $\ln(I(Q)Q^2)$ versus $Q^2$ plots. The slope of the straight line is equal to $R_t^{\,2}$ and it can be used to determine the thickness of the disk.

When there is a size distribution, one may use either preknown size distribution functions or model-independent methods such as the maximum entropy method and the indirect transformation method to recover the size distribution from the measured scattering data (Glatter, 1988; Lin and Tsao, 1996; Tsao and Lin, 1997).

Clustering of small particles is a quite common phenomenon for many complex fluid systems, such as protein–surfactant complex (Chen and Teixeira, 1986; Jeng et al., 2006), polymer–nanoparticle complex (Lin et al., 2007), water-soluble fullerene-derivative aggregates (Jeng et al., 1999; Jeng et al., 2002b), and nanoparticle aggregates (Freltoft et al., 1986). The ordering of the aggregates within the cluster can be described by the following interparticle structure factor (fractal model) (Teixeira, 1988):

$$S(Q) = 1 + \frac{1}{(QR)^D}\frac{D\Gamma(D-1)}{[1+(Q\xi)^{-2}]^{(D-1)/2}}\sin\left[(D-1)\tan^{-1}(Q\xi)\right], \qquad \text{(IV.1.27)}$$

where $\xi$ is the cutoff distance that roughly defines the characteristic distance (size) of the cluster that obeys the fractal law. $\Gamma(x)$ is the gamma function of argument $x$. The parameter $R$ in eq. (IV.1.27) is the radius of the aggregates that form the fractal cluster. Since eq. (IV.1.27) reduces to $S(Q) \sim Q^{-D}$ for $1/\xi < Q < 1/R$, the fractal dimension $D$ can be easily determined from the slope of the straight line in the plot of $\ln(I(Q))$ versus $\ln(Q)$. Here $S(Q)$ can be regarded as the form factor of the fractal cluster.

The analysis of the scattering data can be greatly simplified when $S(Q) \approx 1$, however, sometimes it is useful or necessary to include the interparticle structure factor in the analysis. $S(Q)$ can provide the information about the interparticle interactions, such as the charge repulsion between ionic micelles. The simplest model for $S(Q)$ is the Percus–Yevick hard sphere model. The particles are modeled as hard spheres that have no interactions when two particles are at a distance larger than their hard sphere diameter $\sigma$. There is an exact analytical solution to the Percus–Yevick hard sphere model (Wertheim, 1963):

$$S_{\mathrm{HS}}(Q) = \frac{1}{1-D(Q)}, \qquad \text{(IV.1.28)}$$

$$D(Q) = -\frac{24\phi}{K^6}\{\alpha_0 K^3(\sin K - K\cos K) + \beta_0 K^2\left[2K\sin K - (K^2-2)\cos K - 2\right]$$
$$+\gamma\left[(4K^3-24K)\sin K - (K^4-12K^2+24)\cos K + 24\right]\}, \qquad \text{(IV.1.29)}$$

$$\alpha_0 = (1+2\phi)^2/(1-\phi)^4, \tag{IV.1.30}$$

$$\beta_0 = -6\phi(1+0.5\phi)^2/(1-\phi)^4, \tag{IV.1.31}$$

$$\gamma = 0.5\phi\alpha_0, \tag{IV.1.32}$$

$$K = Q\sigma. \tag{IV.1.33}$$

The hard sphere structure factor $S_{HS}(Q)$ depends only on the volume fraction $\phi$, $\phi = n_p V_p$, and the hard sphere diameter $\sigma$. In fact, if it is plotted against the dimensionless parameter $Q\sigma$, the hard sphere interparticle structure factor would depend only on the volume fraction. Increasing the particle number density will increase the particle volume fraction and the particles will become more ordered. Interference peaks will appear due to the ordering of the particles at higher volume fractions. Figure IV.1.8 shows the interparticle structure factors $S_{HS}(Q)$, $P(Q)$, and $I(Q)$ for hard spheres of diameter 5 nm, and volume fraction 0.15. The scattering intensity distribution is significantly affected by the interparticle structure factor. The

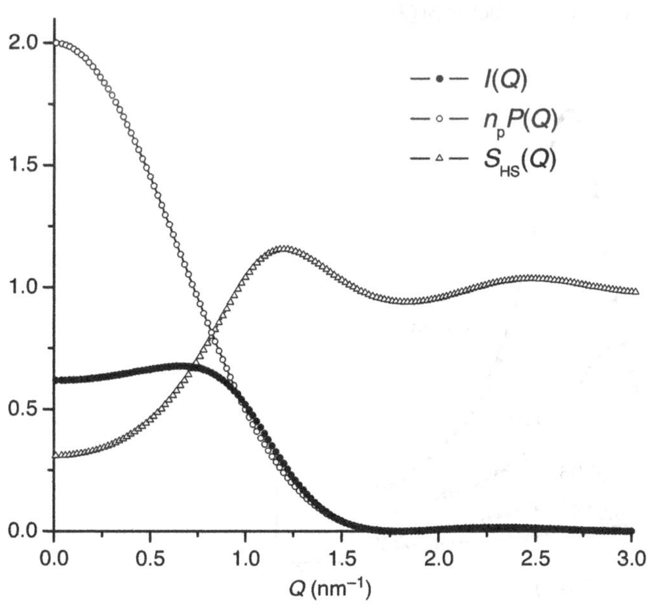


**Figure IV.1.8** The interparticle structure factor $S_{HS}(Q)$ based on the Percus–Yevick hard sphere model for hard spheres of diameter 5 nm, and volume fraction 0.15. The particle form factor $P(Q)$ and the scattering intensity distribution $I(Q) = n_p P(Q) S(Q)$ are also shown. The scattering amplitude is set equal to 2.0. Both $I(Q)$ and $n_p P(Q)$ are in relative units.

low-$Q$ intensity is suppressed due to that $S_{HS}(Q)$ is lower than unity in the low-$Q$ region. A diffraction bump appears in the scattering intensity distribution due to the interparticle interference peak. At high-$Q$ region, $S(Q) \sim 1$ and its effect on $I(Q)$ would not be so significant.

For ionic micellar systems and other charged systems, there is a Coulomb interaction between the aggregates. The interparticle structure factor can be calculated by the mean spherical approximation (MSA) (Hayter and Penfold, 1981a, 1981b) or the rescaled MSA (Sheu et al., 1985). Figure IV.1.9 shows the neutron scattering intensity distribution $I(Q)$ of the globular lithium dodecyl sulfate (LDS) micelles at LDS concentrations of 0.037, 0.074, 0.147, and 0.294 M in pure $D_2O$ at 37°C, which, respectively, correspond to the volume fraction of 0.015, 0.03, 0.07, and 0.14 as indicated in the plot (Bendedouch and Chen, 1983). For charged micellar systems without much salt in the solution to screen their Coulomb repulsive interactions, there will be a very intense interference peak in the scattering intensity distribution as shown in Figure IV.1.9. The scattering profile is dominated by the interference peak. The corresponding particle form factor and the $S(Q)$ are shown in Figure IV.1.10. Here, $S(Q)$ is calculated according to the MSA by assuming the globular micelle to be rigid charged spheres that interact through a screened Coulomb potential (Bendedouch and Chen, 1983). It is noticeable that there is marked depression at low $Q$, and the interference peak gets sharper and higher as the concentration increases. The depression at low $Q$ is much more significant as compared with the hard sphere cases. Attractive interactions can also be treated with the MSA or by other suitable models for predicting the interparticle structure factor $S(Q)$.

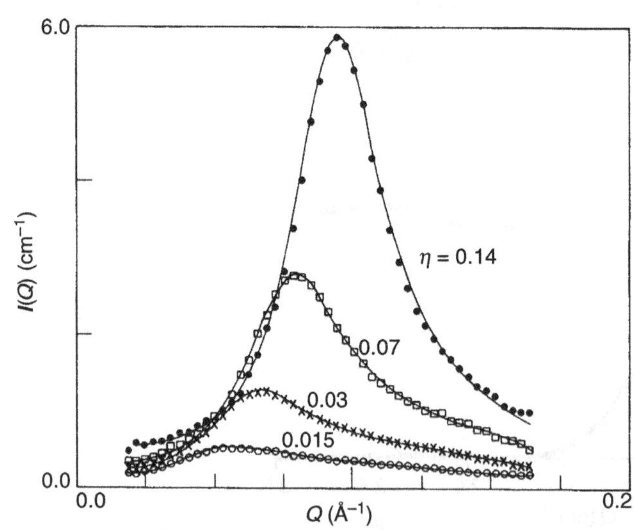


**Figure IV.1.9** The neutron scattering intensity distribution of the globular LDS micelles at LDS concentrations of 0.037, 0.074, 0.147, and 0.294 M in pure $D_2O$ at 37°C, which, respectively, correspond to the volume fraction of 0.015, 0.03, 0.07, and 0.14 as indicated in the plot (Bendedouch and Chen, 1983).

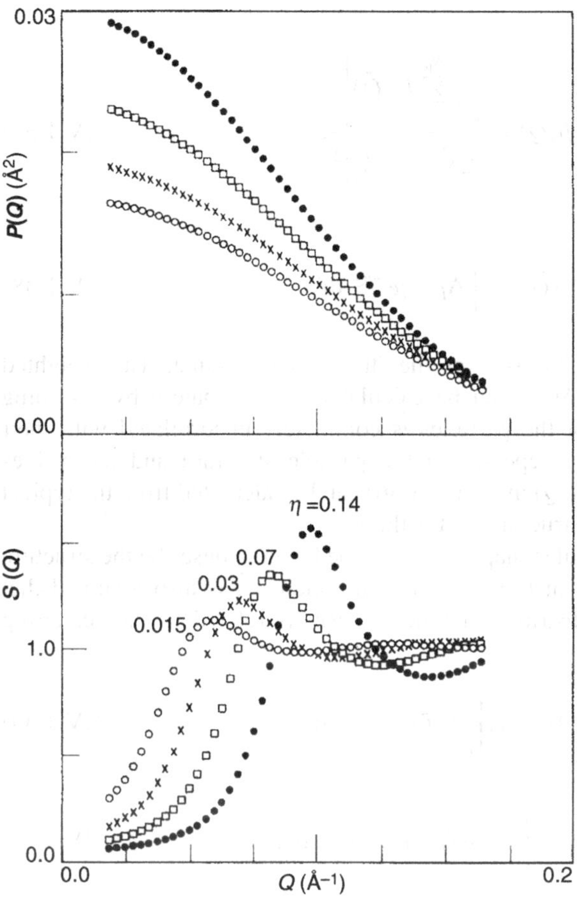


**Figure IV.1.10** The calculated $S(Q)$ (lower graph) and $P(Q)$ (upper graph) for the LDS concentrations of 0.037, 0.074, 0.147, and 0.294 M in $D_2O$ at 37°C, which correspond to the $I(Q)$ in Figure IV.1.9. $S(Q)$ is calculated according to the MSA by assuming the globular micelle to be rigid charged spheres that interact through a screened Coulomb potential (Bendedouch and Chen, 1983).

For polydisperse systems and nonspherical particles, the interparticle structure factor could not be separated completely from the particle form factor as given by eq. (IV.1.1). Instead, approximate equations may be used (Kotlarchyk and Chen, 1983):

$$I(Q) = n_p \tilde{P}(Q)\tilde{S}(Q), \qquad \text{(IV.1.34)}$$

$$\tilde{P}(Q) = \frac{1}{N_p}\sum_{i=1}^{N_p} |F_i(\vec{Q})|^2, \qquad \text{(IV.1.35)}$$

$$\tilde{S}(Q) = 1 + \beta(Q)(S(Q)-1), \qquad \text{(IV.1.36)}$$

$$\beta(Q) = \frac{\left|\frac{1}{N_p}\sum_{i=1}^{N_p} F_i(\vec{Q})\right|^2}{\frac{1}{N_p}\sum_{i=1}^{N_p} \left|F_i(\vec{Q})\right|^2}, \tag{IV.1.37}$$

$$F_i(\vec{Q}) = \int \Delta\rho(\vec{r}) e^{-i\vec{Q}\vec{r}} \mathrm{d}\vec{r}. \tag{IV.1.38}$$

The integration in eq. (IV.1.38) is over the $i$th particle volume. The weighted interparticle structure factor $\tilde{S}(Q)$ can be calculated approximately by assuming that the size or orientation of the particles is completely uncorrelated with their positions. The function $\beta(Q)$ depends on the particle structure and has values between unity and zero. The $S(Q)$ in eq. (IV.1.36) can be calculated from the typical one-component interparticle structure factor theory.

For particles having irregular shapes, it would be better to describe the structure by the distance distribution function $p(r)$. For simplicity, it is also assumed that $S(Q) \approx 1$ and the distance distribution function $p(r)$ is related to the scattering intensity distribution by

$$I(Q) = n_p \int_{V_p} \gamma(\vec{r}) \cdot \mathrm{e}^{-i\vec{Q}\cdot\vec{r}} \mathrm{d}\vec{r}, \tag{IV.1.39}$$

$$\gamma(\vec{r}) \equiv \int_{V_p} \Delta\rho(\vec{r}_i)\Delta\rho(\vec{r}_i + \vec{r})\mathrm{d}\vec{r}_i, \tag{IV.1.40}$$

$$p(r) = r^2\gamma(r), \tag{IV.1.41}$$

where $\gamma(\vec{r})$ is the radial distribution function, or alternatively the density–density correlation function. The distance distribution function $p(r)$ can be recovered from the measured scattering intensity distribution to reveal the characteristics of the particle structure.

It is also useful to obtain the information about the surface or interface by scattering method. At very-large-$Q$ region, the scattering intensity typically tends to vary with $Q^{-4}$. This is called the Porod's law:

$$I(Q)|_{Q\to\infty} \approx \frac{2\pi}{Q^4}(\Delta\rho)^2 S, \tag{IV.1.42}$$

where $S$ is the total surface area of the particles per unit sample (solution) volume. The Porod's law works only for systems with sharp interface. By plotting $I(Q)Q^4$ versus $Q$, at high-$Q$ region, $I(Q)Q^4$ will approach a constant value that can be used to estimate the total interface area of the system. This is especially useful to characterize

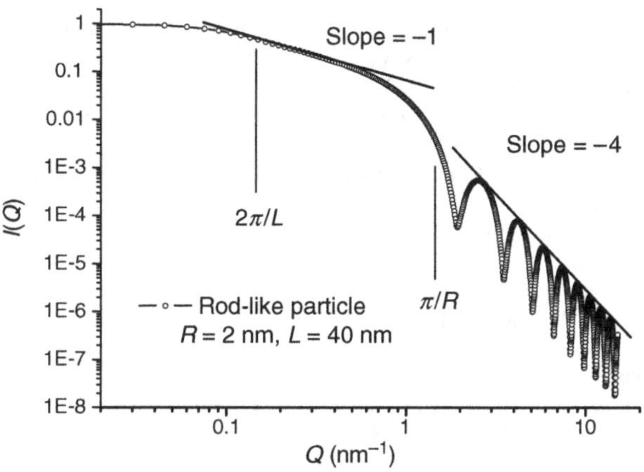


**Figure IV.1.11** The calculated intensity distribution of rod-like particles with radius 2 nm and length 40 nm to show the $Q^{-1}$ dependence in the middle-$Q$ region and the $Q^{-4}$ characteristics at the high-$Q$ region that obeys the Porod's law.

porous materials for catalytic and other applications. Figure IV.1.11 shows the calculated intensity distribution of rod-like particles with radius 2 nm and length 40 nm to show the $Q^{-1}$ dependence in the middle-$Q$ region and the $Q^{-4}$ characteristics at the high-$Q$ region that obeys the Porod's law. For rough interfaces, a modified form of eq. (IV.1.42) should be used.

## IV.1.3 SUMMARY

Neutron scattering has been successfully used in studying a wide range of molecular aggregates of various sizes and shapes. Analysis methods and models are developed in recent decades to recover the structural information as well as the interactions of molecular aggregates from neutron scattering data. Neutron scattering is particularly important in investigating soft materials since the neutron scattering contrast can be enhanced easily by using deuterated materials. Both the external neutron contrast variation and the internal neutron contrast variation are proven to be powerful methods in studying complex structures. In studying soft materials, it is also important that the neutrons have relatively low radiation damages as compared with the powerful synchrotron X-rays. With the installation of MW class spallation neutron sources (Fujii, 2007; Mason et al., 2005) and the development of new class of neutron scattering instruments that provide even more intense neutron flux and a wider accessible $Q$ range, we are equipped with better tools to solve the even more complex structures in many emerging new technologies.

## REFERENCES

Almgren, M. *Biochim. Biophys. Acta* **2000**, *1508*, 146.
Bendedouch, D. and Chen, S.H. *J. Phys. Chem.* **1983**, *87*, 2621.
Bergström, M., Pedersen, J.S., Schurtenberger, P., and Egelhaaf, S.U. *J. Phys. Chem. B* **1999**, *103*, 9888.
Chen, S.H. and Lin, T.L. Colloidal solutions. In: Sköld, K.and Price, D.L. (editors). *Methods of Experimental Physics: Neutron Scattering*, Vol. *23* B, Academic Press, New York, 1987, p. 489–543.
Chen, S.H. and Teixeira, J. *Phys. Rev. Lett.* **1986**, *57*, 2583.

CHEN, S.H., LIN, T.L., and HUANG, J.S. The structure and phase transitions of a three-component microemulsion system: AOT/water/alkane. In: SAFRAN, S.A.and CLARK, N. (editors). *Complex and Supermolecular Fluids*, Wiley, New York, 1987, p. 285–313.

DUBOIS, M., LIZUNOV, V., MEISTER, A., GULIK-KRZYWICKI, T., VERBAVATZ, J. M., PEREZ, E., ZIMMERBERG, J., and ZEMB, T. *Proc. Natl. Acad. Sci. USA* **2004**, *101*, 15082.

EDLER, K.J., REYNOLDS, P.A., BROWN, A.S. SLAWECKI, T.M., and WHITE, J.W. *J. Chem. Soc. Faraday Trans.* **1998**, *94*, 1287.

EGELHAAF, S.U. and SCHURTENBERGER, P. *Physica B* **1997**, *234–236*, 276–278.

EGELHAAF, S.U. and SCHURTENBERGER, P. *Phys. Rev. Lett.* **1999**, *82*, 2804.

ENDO, H. *Physica B Condens. Matter* **2006**, *385–386*, 682.

FRELTOFT, T., KJEMS, J.K., and SINHA, S.K. *Phys. Rev. B* **1986**, *33*, 269.

FUJII, Y. *Thin solid films* **2007**, *515*, 5696.

GLATTER, O. *J. Appl. Crystallogr.* **1988**, *21*, 886.

HASSAN, P.A., FRITZ, G., and KALER, E.W. *J. Colloid Interface Sci.* **2003**, *257*, 154.

HAYTER, J.B. and PENFOLD, J. *J. Chem. Soc. Faraday Trans. 1* **1981a**, *77*, 1851.

HAYTER, J.B. and PENFOLD, J. *Mol. Phys.* **1981b**, *42*, 109.

IMAE, T., KAKITANI, M., KATO M., and FURUSAKA, M. *J. Phys. Chem.* **1996**, *100*, 20051.

JENG, U.S., LIU, W.J., LIN, T.L., WANG, L.Y., and CHIANG, L.Y. *Fullerene Sci. Technol.* **1999**, *7*, 599.

JENG, U., LIN, T.L., HU, Y., CHANG, T.S., CANTEENWALA, T., CHIANG, L.Y., and FRIELINGHAUS, H. *J. Phys. Chem. A* **2002a**, *106*, 12209.

JENG, U.S., LIN, T.L., WANG, L.Y., CHIANG, L.Y., HO, D.L., and HAN, C.C. *Appl. Phys. A* **2002b**, *74*, S487.

JENG, U.S., HSU, C.H., LIN, T.L., WU, C.M., CHEN, H.L., TAI, L.A., and HWANG, K.C. *Physica B* **2005**, *357*, 193.

JENG, U.S., LIN, T.L., LIN, J.M., and HO, D.L. *Physica B* **2006**, *385–386*, 865.

JERKE, G., PEDERSEN, J.S., EGELHAAF, S.U., and SCHURTENBERGER, P. *Langmuir* **1998**, *14*, 6013.

KOEHLER, R.D., RAGHAVAN, S.R., and KALER, E.W. *J. Phys. Chem. B* **2000**, *104*, 11035.

KOTLARCHYK, M. and CHEN, S.H. *J. Chem. Phys.* **1983**, *79*, 2461.

LIN, T.L. *Physica B* **1995**, *213–214*, 594.

LIN, T.L. and TSAO, C.S. *J. Appl. Crystallogr.* **1996**, *29*, 170.

LIN, T.L., CHEN, S.H., GABRIEL, N.E., and ROBERTS, M.F. *J. Am. Chem. Soc.* **1986**, *108*, 3499.

LIN, T.L., CHEN, S.H., GABRIEL, N.E., and ROBERTS, M.F. *J. Phys. Chem.* **1987**, *91*, 406.

LIN, T.L., CHEN, S.H., GABRIEL, N.E., and ROBERTS, M.F. *J. Phys. Chem.* **1990**, *94*, 855.

LIN, T.L., LIU, C.C., ROBERTS, M.F., and CHEN, S.H. *J. Phys. Chem.* **1991**, *95*, 6020.

LIN, T.L., JENG, U., TSAO, C.S., LIU, W.J., CANTEENWALA, T., and CHIANG, L.Y. *J. Phys. Chem. B* **2004**, *108*, 14884.

LIN, J.M., LIN, T.L., JENG, U.S., ZHONG, Y.J., YEH, C.T., and CHEN, T.Y. *J. Appl. Crystallogr.* **2007**, *40*, s540.

LIU, D., ZHANG, Y., CHEN, C.C., MOU, C.Y., POOLE, P.H., and CHEN, S.H. *Proc. Natl. Acad. Sci. USA* **2007**, *104*, 9570.

MASON, T.E., ABERNATHY, D., ANKNER, J., EKKEBUS, A. GRANROTH, G., HAGEN, M., HERWIG, K., HOFFMANN, C., HORAK, C., KLOSE, F., MILLER, S., NEUEFEIND, J., TULK, C., and WANG, X. L. *AIP Conf. Proc.* **2005**, *773*, 21.

NAKANO, M., FUKUDA, M., KUDO, T., ENDO, H., and HANDA, T. *Phys. Rev. Lett.* **2007**, *98*, 238101.

SHEU, E.Y., WU, C.F., CHEN, S.H., and BLUM, L. *Phys. Rev. A* **1985**, *32*, 3807.

TEIXEIRA, J. *J. Appl. Crystallogr.* **1988**, *21*, 781.

THIYAGARAJAN, P. BURKOTH, T.S., URBAN, V., SEIFERT, S., BENZINGER, T.L.S., MORGAN, D. M., GORDON, D., MEREDITH, S.C., and LYNN, D.G. *J. Appl. Crystallogr.* **2000**, *33*, 535.

TSAO, C.S. and LIN, T.L. *J. Appl. Crystallogr.* **1997**, *30*, 353.

WERTHEIM, W.S. *Phys. Rev. Lett.* **1963**, *10*, 321.

ZHANG, X. and CHAN, K.Y. *Chem. Mater.* **2003**, *15*, 451.

# IV Applications

# IV.2 Structure of Dendritic Polymers and Their Films

*Koji Mitamura and Toyoko Imae**

## IV.2.1 INTRODUCTION

Dendritic polymers have well-designed branching architectures and dense functional groups in the inner core and the terminal shell. Owing to such characters, the dendritic polymers are gathering attention from a number of scientists as attractive functional materials. High interest is not only on the unique structure of dendritic polymers but also on their assembling. Since the investigation of dendritic polymers has been performed using many techniques such as spectroscopy, microscopy, NMR, rheology, thermometry, the structures of dendritic polymers and their assemblies are going to be elucidated. In this chapter, we introduce the analysis by means of small-angle neutron scattering (SANS) and reflection and discuss the structural analyses of dendritic polymers and their films.

## IV.2.2 OVERVIEW OF STRUCTURAL INVESTIGATION OF DENDRITIC POLYMERS

Structural concern is of the segment distribution in dendrimer and relevant dynamics of dendrimers. The pioneering theoretical prediction was addressed by de Gennes and Harvert (1983) and extended by some groups on the basis of different theories and computation. However, the general view has not been attainable. In time, the validity of theoretical simulation results are going to be proved by experiments of $^{13}C$ NMR relaxation, hydrodynamic radius, and viscosity. Although it is not easy to obtain the information of structure, morphology, and dynamics of dendrimers with a size from few nm to several dozen nm, SANS is recognized as one of the most valuable techniques for such purpose and utilized for research of dendrimers since early 1990. Here, the structural analyses of dendrimers using SANS are presented.

*Corresponding author.

*Neutrons in Soft Matter* Edited by Toyoko Imae, Toshiji Kanaya, Michihiro Furusaka, and Naoya Torikai

### IV.2.2.1 Static Structure of Dendritic Polymers

Total small-angle scattering intensity $I(q)$ radiated at scattering vector $q$ from monodisperse spherical molecule is described by

$$I(q) = n_{\mathrm{p}}P(q)S(q) \tag{IV.2.1}$$

and

$$q = \frac{4\pi}{\lambda}\sin\left(\frac{\theta}{2}\right), \tag{IV.2.2}$$

where $n_{\mathrm{p}}$, $P(q)$, and $S(q)$ are the number density of molecule, the intraparticle form factor, and the interparticle structure factor, respectively; $\lambda$ is the incident wavelength; and $\theta$ is the scattering angle. Structural investigation of dendrimer in solution by SANS has first been reported by Bauer and his coworkers (Bauer et al., 1992; Briber et al., 1992). They compare P($q$) curves of poly(amido amine) (PAMAM) dendrimer with those of various shapes and determine that PAMAM dendrimer takes intermediate structures between Gauss chain and sphere. Meanwhile, poly(propyreneimine) (PPI) dendrimer behaves like hard sphere (Scherrenberg et al., 1998; Ramzi et al., 1998). This result is reasonable because of shorter spacer distance of PPI.

The polyelectrostatic behavior of PAMAM dendrimers in solution is also confirmed from SANS analysis (Bauer et al., 1992; Briber et al., 1992). That is, acidification or concentration increase of dendrimers brings long-distance interaction, but the excess addition of acid or the addition of salt shield the electrostatic interaction. However, the size change by pH variation is not observed contrary to the expectation (Nisato et al., 2000). Meanwhile, dendrimer size varies with alkyl chain length of alcohol as a medium, but it does not change with mixing ratio of methanol and acetone and temperature (Topp et al., 1999a). Similar behavior is recently reported (Chen et al., 2007; Porcar et al., 2008). Tertiary amines of PPI dendrimer protonate with addition of HCl and simultaneously the correlation peak arises in the scattering curve, indicating the mutual arrangement of molecules in space by electrostatic repulsion (Ramzi et al., 2002). Then the variation of correlation peaks depending on the concentrations of acid and dendrimer is equivalent to the variation of $S(q)$ function (Topp et al., 1999b).

If the concentration of molecules is enough dilute (less than 1%), the approximation of $S(q) \approx 1$ is applicable. Then, $I(q)$ function is expanded at small scattering vector and $P(q)$ function can be obtained as a Guinier equation, in relation to radius of gyration $R_{\mathrm{G}}$, as below.

$$P(q) = V^2(\rho - \rho_{\mathrm{S}})^2 \exp\left(-\frac{R_{\mathrm{G}}^2}{3}\right) \tag{IV.2.3}$$

$$R_{\mathrm{SANS}}^2 = \left(\frac{5}{3}\right)R_{\mathrm{G}}^2, \quad \text{for small hard sphere} \tag{IV.2.4}$$

where $V$, $\rho$, and $\rho_s$ are the total volume of molecule, mean coherent neutron scattering length densities of molecule and solvent, respectively. Then molecular radius $R_{SANS}$ can be calculated from eq. (IV.2.4).

The analysis is applied to the evaluation of dendrimers with different structures. For poly(2-methyl-2-oxazoline)-block-poly(amido amine) dendrimer, a head-to-tail-type amphiphilic block copolymer, which consists of linear polymer and dendron, average diameters of 6.4 and 54 nm are evaluated from different $q$ ranges of Guinier plot (Aoi et al., 1997; Aoi et al., 1999a). The former is a molecular size in consistence with the estimated value from molecular structure and the latter is a size of aggregate that corresponds to assembly of about 1000 molecules. For star polymer of PPI dendrimer with polysarcosine terminal chains, a molecular size evaluated from Guinier plot of SANS is considerably small from a size of extended polysarcosine chain structure, but close to that of coiled polymer structure (Aoi et al., 1999b). This behavior is like linear polymer in poor solvent. The molecular structure of poly (benzylether) dendron with linear polystylene tail is quantitatively analyzed from SANS result using cylindrical molecular shape and excluded volume chain structure (Förster et al., 1999). That is, since bulky high-generation dendron is combined with linear polymer chain to be *trans*-zigzag, whole structure of dendritic polymer becomes cylinder-like. When the dendritic polymer is charged, the polymer chain extends by electrostatic repulsion. The results support the increase in stiffness.

The scattering curve $I(q)$ was Fourier-transformed into the three-dimensionally averaged pair distance distribution function $P(r)$ as a function of radial distance $r$ (see eq. (IV.2.5)) and shapes of electrostatic self-assembly consisting of PAMAM dendrimer and dye were determined to be sphere, ellipsoid, cylinder, or core–shell type (Willerich et al., 2009). Results reveal that the charge ratio of the components is a crucial factor in the aggregation.

$$I(q) = 4\pi \int_{r=0}^{\infty} P(r) \frac{\sin(qr)}{qr} dr \qquad \text{(IV.2.5)}$$

The calculation of molecular weight and scattering length density of dendrimer has been carried out by using external contrast variation result on SANS, where $H_2O$ and $D_2O$ are mixed with different ratios (Pötschke et al., 1999). The variation in contrast of solvent varies scattering profile depending on the internal structure of molecules. On the other hand, in the contrast matching method, labeled dendrimers are solved in mixture of nondeuterated and deuterated solvents. According to the analysis, the terminal groups of PAMAM dendrimer distributes in the periphery (Topp et al., 1999c), but the urea-modified terminal groups of PPI dendrimer are buried in the internal (Rosenfeldt et al., 2002).

The contrast variation examination of SANS has been carried out for three kinds of dendrimers with different chemical and core–shell structures and analyzed by computer simulation of SANS intensity curves (Imae et al., 1999; Funayama and Imae, 1999; Funayama et al., 2001; Funayama et al., 2003a). The scattering intensity profile remarkably changes with contrast ($D_2O$ content) of solvent and the variation

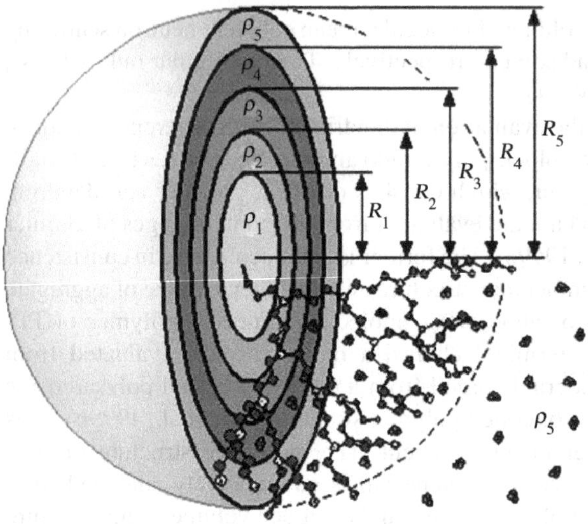


**Figure IV.2.1** Geometric definition of a spherical "five-layers model" consisting of a concentric structure applied for a G5 dendrimer in solution.

depends on dendrimers. Therefore, it is evident that the coherent neutron scattering length density in the dendrimer varies depending on the variation of internal structure of dendrimers. Meanwhile, dendrimer is assumed to encapsulate solvent (water) in its interior due to its structural character. Supposing that the segment density and the water penetration are different at each generation layer in dendrimer, as shown in Figure IV.2.1, the intraparticle form factor for fifth-generation dendrimer is written as below.

$$P(q) = \left| \sum_{i=1}^{4} \frac{4\pi R_i}{3} (\rho_i - \rho_{i+1}) \frac{3J_1(qR_i)}{qR_i} + \frac{4\pi R_{i=5}}{3} (\rho_{i=5} - \rho_S) \frac{3J_1(qR_{i=5})}{qR_{i=5}} \right|^2, \tag{IV.2.6}$$

where $J_1 = (\sin x - \cos x)/x^2$.

$$\rho_i = \rho_{\text{den},i} + A_i \rho_S \tag{IV.2.7}$$

where $\rho_i$ is a coherent neutron scattering length density of the $i$th layer and described as a summation of contributions from dendrimer segment $\rho_{\text{den},i}$ and solvent $A_i\rho_S$, where $A_i$ is a number density of solvent in the $i$th layer.

When the calculated curve of SANS intensity is compared with experimental curve, the curve based on five-layer model is better fit to the observed one than the calculation by a sphere model with homogeneous internal segment density and water penetration. Then, the set of optimum parameters (scattering length density and penetration amount of water) is obtained from optimum fitting to observed curve. Figure IV.2.2 compares the radial density and the water penetration profiles of fifth-generation hydroxyl-terminated PAMAM dendrimer, hydroxyl-terminated PPI/monoamide amine(AMAM) dendrimer, and glucopeptide-terminated PAMAM (Sugar Ball, SB) dendrimer. It can be elucidated that PAMAM dendrimer turns

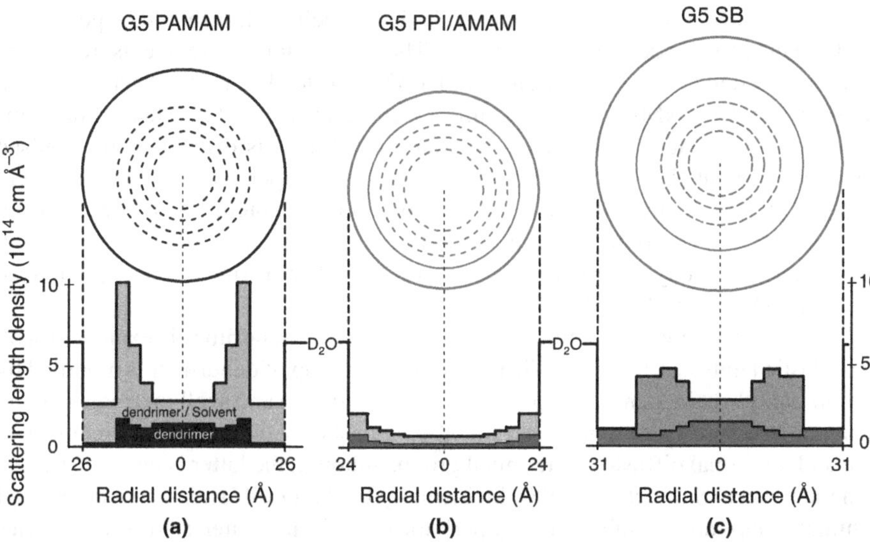


**Figure IV.2.2** Layer geometry (upper) and coherent neutron scattering length density profile (lower) of G5 dendrimers as a function of the radial distance from a concentric center. (a) OH-type PAMAM dendrimer; (b) OH-type PPI/AMAM dendrimer; (c) sugar ball. Thick and thin solid lines represent the scattering length densities of the segments in the dendrimers with and without penetrated solvents, respectively.

hydroxy terminals to its interior, but hydroxy terminals of PPI/AMAM dendrimer locates in periphery. While hydroxy groups are hydrogen-bonded with amide groups in internal of PAMAM dendrimer, hydroxy groups on PPI/AMAM dendrimer preferably exists in hydrophilic periphery of dendrimer. Bulky glucopeptide terminals of SB also lie in periphery due to steric hindrance. Water (solvent) can penetrate in proportion to number of hydrophilic segment such as AMAM, but less in highly dense glucopeptide outermost layer of SB. It should be noticed that density distribution and molecule penetration in dendrimers are dependent on their chemical structure, but not defined in common for dendrimer species.

Incidentally, dendrimer structure at high dendrimer concentration in medium has scarcely been investigated by SANS. Rietveld et al. (2000) have reported the inverse osmotic compressibility of PPI dendrimers in deuterated methanol. The dendrimers behave like hard-sphere in a very dilute region and the solvation layers overlap at a denser region. The maximum compressibility occurs at the concentration where the distance between centers of two dendrimers is twice their radius of gyration. This concentration is the crossover to a semidilute phase where dendrimers shrink, but do not interpenetrate each other. Finally, the dendrimers are collapsed at high concentration and the compressibility increases again.

## IV.2.2.2 Dynamic Structure of Dendritic Polymers

In the previous section, one should have known the result that water can penetrate even in the dendrimer with peripheral layer of dense segment density. Then one may suspect why solvent can pass through highly dense shell wall. It should be possible if whole segments in dendrimer fluctuate. The fluctuation of segments relates to dynamics of dendrimer. The dynamics of PAMAM dendrimer has been discussed by means of $^{13}$C NMR relaxation time (Meltzer et al., 1992). Meanwhile, self-diffusion coefficient of poly(allylcarbosilane) dendrimer is evaluated from pulsed magnetic gradient NMR (Sagidullin et al., 2002). The dependency of the coefficient on dendrimer concentration is similar to that of globular proteins. It is observed from solid NMR investigation that there are fast and slow modes associating with fluctuation of phenyl terminal and substituted phenyl ring of polyphenylene dendrimer (Wind et al., 2002).

Moreover, segment dynamics of poly(carbosilane) dendrimer is examined with quasielastic neutron scattering. When the terminal group of dendrimer is modified by perfluoroalkyl chain, dendrimer phase-separates into core and shell in dendrimer and dynamic structure factors are contributed by translational diffusion of segment in core and rotational diffusion of terminal group, although the latter is one order larger than the former (Stark et al., 1998). Incidentally the latter diffusion is not detected if terminal groups are not substituted by perfluoroalkyl chain. Later, same group carries out a study using neutron spin echo spectroscopy as well as SANS (Stark et al., 2003). The overall shape of the low-generation molecules is like core–shell particle and the fourth-generation molecule is regarded as a compact sphere. Neutron spin echo spectroscopy reveals a relaxation time that is attributed to the form fluctuation of this particle.

Neutron spin echo is one of the valuable tools to examine dynamics of dendrimer and normalized intermediated scattering function is obtained as a function of time $t$ from spin echo (Funayama et al., 2001; Funayama et al., 2003b). If the function obeys a typical single decaying exponential function, it can be described as

$$\frac{I(q,t)}{I(q,0)} = \exp(-\Gamma(q)t) \tag{IV.2.8}$$

where $I(q, t)$ is a intermediate correlation function at time $t$. Decay rate $\Gamma(q)$ is relating to diffusion coefficient $D_{\text{eff}}$ as below.

$$\Gamma(q) = D_{\text{eff}}q^2 \tag{IV.2.9}$$

If the intermediated scattering function consists of two modes, the function must be two exponential relaxation function of fast (F) and slow (S) modes.

$$\frac{I(q,t)}{I(q,0)} = f_S \exp(\Gamma_S t) + f_F \exp(\Gamma_F t)$$

$$f_S + f_F = 1$$

$$\Gamma_S = D_S q^2, \quad \Gamma_F = D_F q^2 \tag{IV.2.10}$$

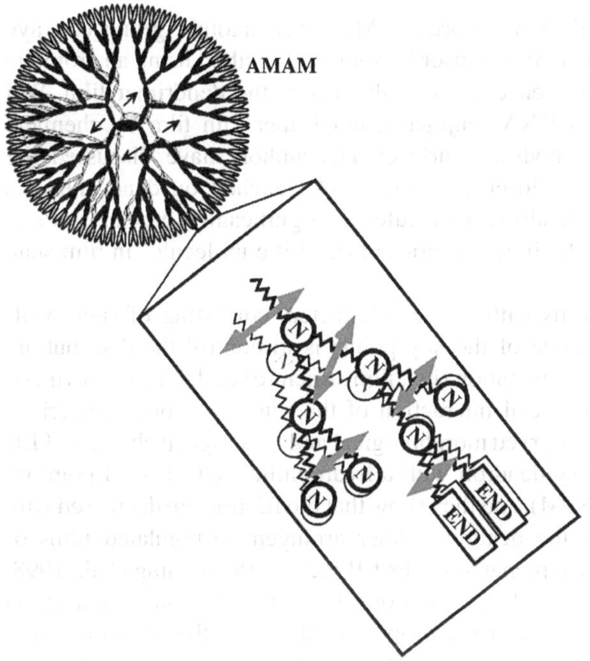


**Figure IV.2.3** Schematic representation of segmental motions in fifth-generation dendrimers with internal segments of amido-amine repeating units.

According to eq. (IV.2.10), the ratio $f$ and the diffusion coefficient $D$ of two relaxation modes can be evaluated.

The experimental results display single mode decay for a concentrated aqueous solution (10 wt%) of PAMAM dendrimer and two-mode decay for a dilute solution (1 wt%). The slow mode is observed at both concentrations and is assigned to translational diffusion of dendrimer. The fast mode, which is obtained only in a dilute solution, should be the deformation motion of dendrimer, cooperative to harmonic motion of segments (see Figure IV.2.3).

## IV.2.3 OVERVIEW OF STRUCTURAL INVESTIGATION OF DENDRITIC POLYMER FILMS

The scientists have designed novel dendritic polymers and fabricated them into thin film on various surfaces: Till today, the dendritic polymers have been directed to nanotechnology as building blocks for chemical sensors, surface wettability, surface protection, separation membranes, and bioapplication (biochip, drug delivery system, etc.). Zhao et al. (1999) have conducted hyperbranching polymerization that is initiated from OH groups of OH-terminated PAMAM dendrimer immobilized on a metal surface and they have obtained the effectively passivated surface. Tanaka et al. (2003) have fabricated a monolayer of amphiphilic PAMAM-type dendron via Langmuir–Blodgett (LB) method and discussed about its nanofrictional and glass transition behavior. Hirano et al. (2006) have designed various generations (G1.5–3.5) of PAMAM dendrons with a fullerene focal point and utilized them as protection groups on a silver nanoparticle. Moreover, another group (Cavaye et al., 2009) has fabricated a chemical sensor by spin–cast method using an aromatic dendrimer, which entraps nitrobenzene molecules inside the dendrimer film. Wei et al. (2009) have prepared a DNA-conjugated dendrimer thin film by chemical adsorption on a polypyrrole-modified surface: The authors have discussed the fabrication of electrochemical biosensor using DNA attached on dendrimer. As shown here, the design of dendritic molecules is significantly crucial since it remarkably has influence on the functionalities of dendritic molecules in film state as well as in bulk.

On the preparation of films with desired characters consisting of such well-designed dendritic polymers, one of the key points is to control the distribution, orientation, deformation, and aggregation (i.e., film structure) of dendritic polymers, since these largely affect interfacial interaction of the films with other materials. Indeed, Xu et al. (2003) have observed morphological and tribological changes of LB films of amphiphilic benzyloxy dendron with a hydrophilic COOH focal point by scanning probe microscopy (SPM), and they show that the friction are decreased with increasing the generation of the dendron. Other arrangement-regulated films of amphiphilic dendrimers are also prepared by the LB method (Schenning et al., 1998; Kampf et al., 1999). It has been elucidated from the surface pressure–area ($\pi$–$A$) isotherms and optical spectroscopy that hydrophobic moiety in the dendritic molecules is supposed to be directed to air phase and hydrophilic one is to water phase at air/water interface. Another technique for the film preparation of the dendritic polymer is a self-assembled monolayer (SAM) method, which allows strong attachment and reasonable organization of molecules onto substrates. For instance, Singh et al. (2009) have chemically immobilized a fourth-generation (G4) amine terminated-PAMAM dendrimer by the SAM method on a gold surface, for utilizing as a platform of the following protein adsorption, which is characterized by surface plasmon resonance (SPR) measurements. Ujihara and Imae (2006) have investigated kinetics of SAM formation of dendrimer with an azacrown core by attenuated total reflectance-infrared (ATR-IR) absorption spectroscopy and SPR measurements. Other than these studies, there are many investigations on characterization of film structures of dendritic polymers by various methods such as SPM and $\pi$–$A$ isotherms (Hierlemann et al., 1998; Sade et al., 1993; Peyser-Capadona et al., 2005). However, the detailed film structures such as molecular distribution, orientation, and aggregation in a film are hardly discussed by these ways.

As one of the innovative methods for detailed characterization of film structures, neutron reflection is a powerful tool especially for organic materials. As described in detail in Chapter 3, neutron reflection data are generable a profile of scattering length density (SLD) along depth direction (Parratt, 1954) that provides the information of layered structures in a film.

The advantages of neutron reflection are now summarized in the following: (1) Fair neutron reflection arises from interfaces between matrices of light elements represented by organic matters and water through adjustment of SLD by the isotropic substitution (e.g., deuteration). (2) Neutrons penetrate deeply into a film and are reflected at buried interfaces inside the film. (3) Neutron reflectometry can be applied in *in situ* condition to measure on various interfaces (gas/liquid, gas/solid, liquid/liquid, and so on) and at various circumstances (high humidity, high temperature, etc.). These features result from dependency of scattering lengths (SLs) on atomic species and low absorption coefficient for most atomic species to neutrons (i.e., less disturbance from outer circumstance). These are characteristic and unique in neutron reflection or scattering phenomenon, unlike X-ray reflection as a similar method.

Nowadays, using the neutron reflection, more accurate arguments on the film structure have been largely promoted. Especially, the detailed investigations on film structures of dendritic polymers by the aid of the neutron reflection are increasing and producing fruits. In the following sections, brief overview of neutron reflection analysis and investigations of dendritic polymer films using neutron reflectometry are presented.

## IV.2.3.1 Brief Overview of Neutron Reflection Analysis

Before the overview of the researches, let us make broadly a survey of the theoretical and the analytical background of neutron reflection. Assuming several layers with different SLD in a film, the normalized intensity (reflectivity: $R(q)$) of reflected neutron from the film at a scattering vector $q$ ($=4\pi\sin\theta/\lambda$) where $\theta$ is a scattering angle and a neutron wave length, $\lambda$, is basically given as

$$R(q) = R_{\mathrm{F}} \int_0^{\infty} \frac{\mathrm{d}\rho(z)}{\mathrm{d}z} \exp(iqz)\mathrm{d}z \qquad \text{(IV.2.11)}$$

where $R_{\mathrm{F}}(q)$ is a Fresnel's reflectivity ($\sim q^{-4}$), $z$ is a distance from a substrate and $\rho(z)$ is a SLD distribution along $z$-axis (van der Lee, 2000). Since such SLD distribution ($\rho(z)$) provides the structural information of a film, the aim of neutron reflection analysis is to elucidate the SLD distribution along depth direction ($z$-axis).

Although the $\rho(z)$ is mathematically obtained as Fourier transformation of $R(q)$, in actual manner, we calculate a theoretical reflectivity profile from a certain $\rho(z)$ profile and fit it to an experimental reflectivity. The $\rho(z)$ profile (i.e., depth–SLD profile) is usually displayed as illustrated in Figure IV.2.4. It is described by combination of three kinds of structural parameters: (1) SLD, (2) thickness ($d$), and (3) interfacial roughness ($\sigma$) in each layer. The SLD is given as a summation of production of SL with number density of each component in a layer, that is, SLD of a layer $i$ (SLD($i$)) is obtained as

$$SLD(i) = \sum\nolimits_j b_{ij} n_{ij} \qquad \text{(IV.2.12)}$$

where $b_{ij}$ and $n_{ij}$ are SLs and number density of component ($j$) in layer ($i$), respectively. Here, it is obvious that SLD value of each layer reflects atomic species of components included in its layer. On the other hand, thickness ($d$) is literally a layer

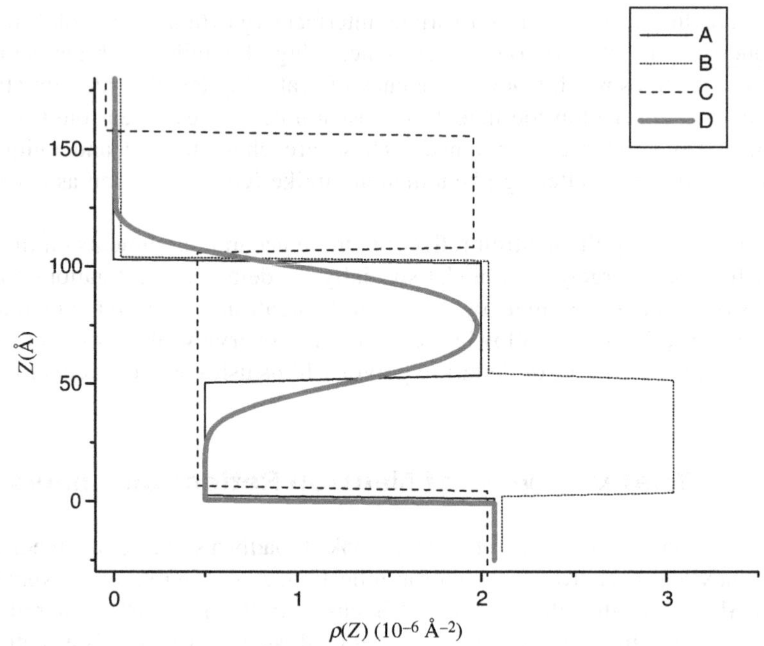


| Model | SLD(1) ($10^{-6}$Å$^{-2}$) | *d*1 (Å) | σ1 (Å) | SLD(2) ($10^{-6}$Å$^{-2}$) | *d*2 (Å) | σ1 (Å) |
|---|---|---|---|---|---|---|
| A | 0.5 | 40 | 0.0 | 2.0 | 40 | 0.0 |
| B | 3.0 | 40 | 0.0 | 2.0 | 40 | 0.0 |
| C | 0.5 | 100 | 0.0 | 2.0 | 40 | 0.0 |
| D | 0.5 | 40 | 10.0 | 2.0 | 40 | 10.0 |

SLD(0) = $2.0 \times 10^{-6}$Å$^{-2}$; *d*0 = ∞; σ0 = 0 as a substrate.

SLD(3) = $0.0 \times 10^{-6}$Å$^{-2}$; d3 = ∞ as air.

**Figure IV.2.4** Depth–SLD profiles ($z$ versus $\rho(z)$) and structural parameters for evaluation of the profiles.

thickness with a definite SLD value determined by eq. (IV.2.12). By combination of SLD and thickness, we can describe $\rho(z)$ as a layered model with discrete interface as shown in Figure IV.2.4, curves A–C, where SLD(1) and d1 are varied from A to B and C, respectively.

However, for real cases, SLDs at the interfaces are not ideally discrete but are blurred by mixing of components between different layers and surface roughness. In order to express such irregularity at the interfaces in the $\rho(z)$ profile, the parameter $\sigma$ is introduced. Since interfacial roughness (or mixing) causes continuous and gradual change of SLD values along $z$-axis at an interface as shown in Figure IV.2.4, curve D, the gradation of the SLD value is approached by an error function using the parameter $\sigma$, that is,

$$\rho(z) = A \cdot erf\left(\frac{z}{2\sigma}\right) + B \qquad \text{(IV.2.13)}$$

where A and B are constants assuring the continuity of $\rho(z)$. The parameter $\sigma$ declares degree of gradual change of SLD. In calculation of a neutron reflectivity, we interpret it as "interfacial roughness" including the meaning of interfacial mixing. Therefore, we interpret that the larger the parameter $\sigma$, the more is the interfacial roughness.

Thus, by setting three kinds of structural parameters (SLD, thickness, and interfacial roughness), we can establish the $\rho(z)$ profile matching experimental systems and calculate a theoretical reflectivity based on eq. (IV.2.11). Since the SLD distribution is changeable by combination of three parameters, in the actual analysis, we must choose an optimal set of these parameters in order to evaluate the most-fitting calculated curve in a good agreement with experimentally obtained reflectivity. The optimal model is usually confirmed with additional experiments such as SPM, optical spectra, and so on.

In order to realize clearly the reflectivity profile, the model calculation can be illustrated. Figure IV.2.5 displays reflectivity profiles calculated on the basis of the model shown in Figure IV.2.4, curves A–D. It is obvious from Figure IV.2.4 that when each structural parameter is changed, the reflectivity profile is also drastically varied. Then, (1) SLD generally affects the depth or shape of fringes (namely, the undulation of the reflectivity) (compare A and B), (2) the value of thickness ($d$) induces the change on the frequency of the fringes (compare A and C), and (3) the increase of interfacial roughness ($\sigma$) enhances the reflectivity decay with increment of $q$ value

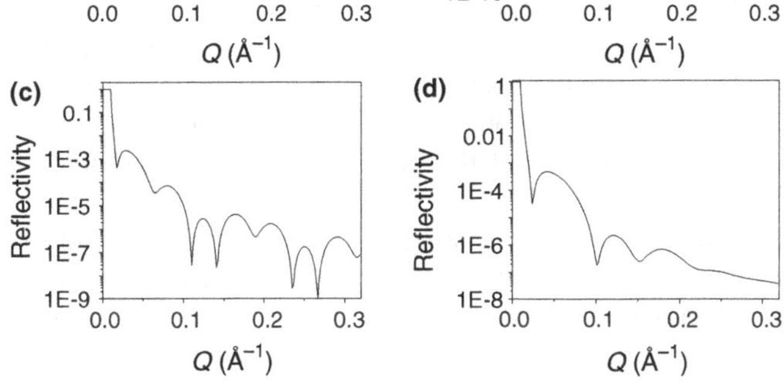


**Figure IV.2.5** (a)–(d) Demonstration of calculated reflectivity profiles based on the depth–SLD profiles of the curves A–D shown in Figure IV.2.4.

(compare A and D). The analysis of neutron reflection comes down to an operation of choosing the most optimum model with varying the parameters (SLD, $d$, and $\sigma$) of each layer. On the practical calculation of the reflectivity based on the assumed model, free or commercial softwares specialized for neutron reflection are available. Taking the analytical background in mind, we introduce the practical examples applying the neutron reflection in the next sections.

### IV.2.3.2 Molecular Structure in Dendritic Polymer Monolayers

As introduced above, some dendritic polymers have been fabricated on the interface by means of SAM, LB, and other techniques. Then neutron reflection enables the strict analysis of molecular arrangements and aggregation on such interfaces. In fact, several researches have elucidated film structures or molecular behaviors of monolayers fabricated by dendritic polymers.

Saville et al. (1995) have applied the neutron reflection for elucidating the molecular orientation of poly(aryl ether) dendrons with various generations (G2–G6) on water subphase. The neutron reflection profile reveals that the molecular configuration is changed to ellipsoidal shape with compression of the dendron Langmuir film. The dendrons exhibit the high resistance against compression and the scaling relationship between molecular weight and molecular occupied area. Directly, the authors have performed experiments on various SLD-adjusted subphases by mixing $D_2O$ with $H_2O$, and done the analysis with higher accuracy.

Kirton et al. (1998) have investigated, by neutron reflectometry, about molecular orientation of similar types of dendrons, that is, terminal-unsubstituted and -substituted ($CH_3O-$ and $NC-$) G4 poly(aryl ether) dendrons with an OH focal point, on aqueous subphase and also studied effects of the terminal groups on monolayer configurations. In this system, the neutron reflectivity profiles are varied with the compression of the Langmuir film, indicating the structural variation of their films with the compression. It is elucidated from their results that the unsubstituted dendron forms a monolayer with prolate conformation and a partial second layer. On the other hand, both of the terminal-substituted dendrons develop more compact and thinner monolayers than the unsubstituted molecule, due to the increased attractive interaction between terminals of dendrons and water in the subphase, because of the introduction of the polar groups ($CH_3O-$ and $NC-$) in the terminals. This work indicates the dependency of dendron chemical structures in monolayer on the film compression and the crucial role of the polarities of terminal groups in control of the film structures.

In addition, molecular arrangement consisting of another type of dendritic polymer with an alkyl chain-terminated G4 PAMAM dendron head and a poly (ethylene oxide) (PEO) tail has been evaluated on a water surface by neutron reflection (Johnson et al., 2002). In this study, the analysis of neutron reflection is carried out on the assumption of the finely discrete layer model in order to obtain more precise depth–SLD profile (nine layers in total including air and subphase are assumed). It is revealed that, upon the compression on water surface, monolayer of the dendron arranges in order with alkyl chain terminals pointing to air, while the PEO tail is expanded into water. However, in highly compressed condition, the alkyl terminal groups cannot arrange in order, but are mixed with hydrophilic dendron internal because of the surface curvature of G4 dendron. This chapter provides more accurate analysis about the variation of molecular arrangements along the compression.

These three investigations employ amphiphilic dendritic polymers possessing both hydrophilic and hydrophobic segments and study the molecular orientation on water subphase. In these systems, the molecular orientation can be somewhat easily expected by considering the characteristics of each segment, that is, the hydrophilic segments turn to water subphase and hydrophobic ones point into air. In the following paragraph, we take up the study using different types of dendritic polymers without water-compatible segment.

Mitamura and Imae, 2004 have designed novel block copolymers with poly(aryl ether) dendron (Den: lipophilic) and perfluoroalkyl (PFA: solvophobic) segments as illustrated in Figure IV.2.6 and investigated the molecular arrangement on air/water interface. Since the molecule has segments of solvophobic and hydrophobic(lipophilic), the molecular orientation is not easily expected because of absence of hydrophilic group. The authors have prepared two kinds of copolymers with varied polymerization degree of each block (namely, M2den2 and M2den3; see Figure IV.2.6) and investigated the effect of the polymerization degree on their molecular orientations. The neutron reflectivity profile of each copolymer and the depth–SLD profile are shown in Figure IV.2.7. In consideration of higher SLD of PFA than that of Den, it is obvious that molecular axis of M2den2 (with relatively low

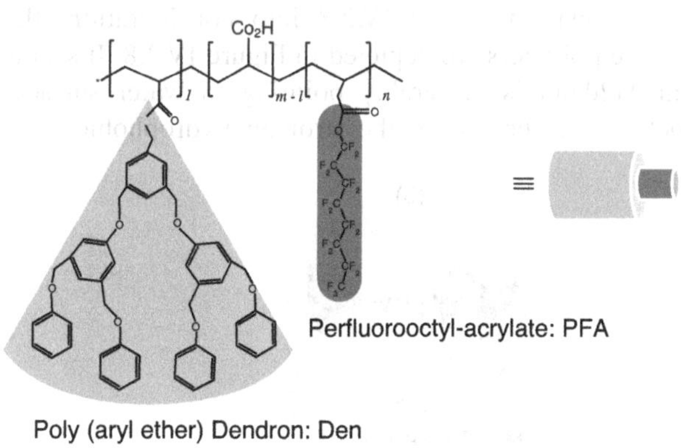


| | Composition | | | |
|---|---|---|---|---|
| Lot. No. | $l$ | $m-l$ | $n$ | $m+n$ |
| M2den2 | 24 | 12 | 11 | 47 |
| M2den3 | 28 | 2 | 8 | 38 |

**Figure IV.2.6** Chemical structure and composition of M2den2 and M2den3.

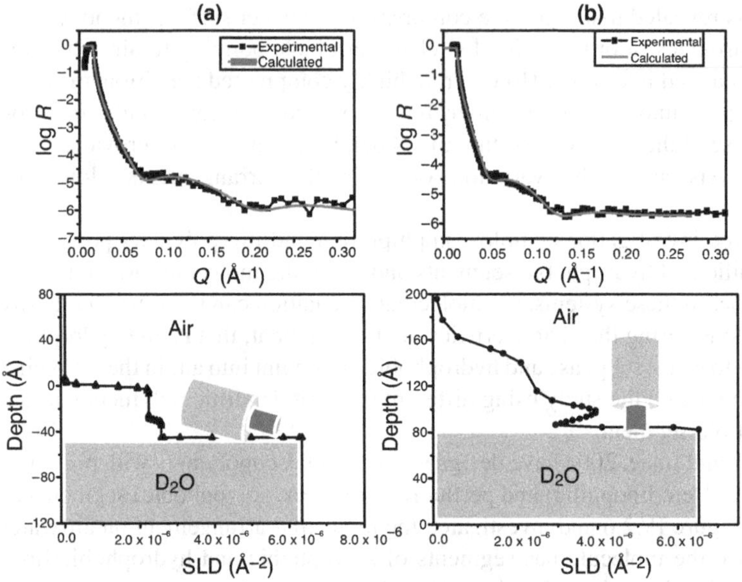


**Figure IV.2.7** Neutron reflectivity (upper) and depth–SLD (below) profiles of (a) M2den2 and (b) M2den3 with calculated reflectivity curves (solid line) and estimated molecular orientation.

Den ratio to PFA) is located flat on water surface. On the other hand, the axis of M2den3 (with relatively high Den ratio to PFA) are almost vertically orientated to water surface, where PFA blocks are pointing to water phase and Den blocks to air phase. Taking AFM images shown in Figure IV.2.8 into consideration, the molecular arrangements of these polymers are depicted in Figure IV.2.8. It seems unlikely that PFA block in M2den3 is preferably pointing to water surface, different from the Den block. It can be due to the stronger hydrophobicity of

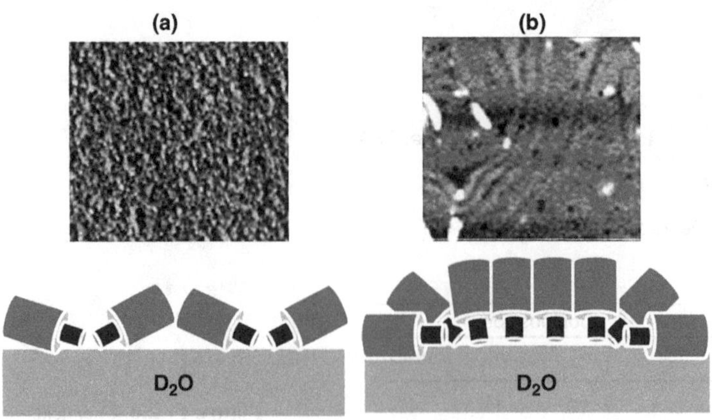


**Figure IV.2.8** AFM images (upper) and molecular orientation models (below) of (a) M2den3 and (b) M2den3.

Den block than that of PFA block. Indeed, M2den3 with the lower Den ratio is laid on water surface because hydrophobicity of Den block in M2den2 is comparable to that of PFA block. The neutron reflection reveals that these copolymers display the unique molecular orientations, which are varied dependently on the ratio of each block.

The M2den3 molecule has been mixed with perfluorooctadecanoic acid (PFOA) to examine behaviors of large M2den3 and small PFOA molecules on water surface (Mitamura et al., 2006). The areas occupied by a molecule, which are obtained from the surface pressure–area ($\pi$–$A$) isotherms are decreased with increasing mixing ratio of PFOA to M2den3 (Table IV.2.1). Although the AFM image also displays the phase separation texture, the experimental molecular areas shown in Table IV.2.1 do not completely match with the calculation assuming the additivity (model 1). In order to explain this discrepancy, the authors have assumed another model that is a complete incorporation model (model 2). While the molecular areas of PFOA and M2den3 are independently contributing to total molecular area in the model 1, PFOA is not contributing to molecular area in the model 2 because of the incorporation of PFOA under the occupied area of M2den3 molecules (therefore the molecular area should be controlled only by M2den3). To elucidate molecular arrangement in Langmuir film and contribution of each molecule to the molecular area, neutron reflectivity has been measured at different mixing ratios as shown in Figure IV.2.9. It is clear that the M2den3 molecules are almost vertically arranged to water surface in the same manner as the previous study shown above. Then the experimental SLD values of Den layer at the various mixing ratios are evaluated and listed in Table IV.2.1 with the values calculated on the basis of models 1 and 2. The experimental value is decreased with PFOA ratio similarly to the result from the ($\pi$–$A$) isotherms. The SLDs are almost the same, but slightly differ from model 1 (phase separation model). This indicates that some PFOA molecules are incorporated into the M2den3 monolayer domain, but the remains form PFOA monolayer domain as illustrated in Figure IV.2.9. This work proves that the neutron reflection reveals the molecular orientation and the miscibility of component molecules.

### IV.2.3.3 Structure of Dendritic Polymer Hybrids on Interfaces

In the previous section, several attempts using neutron reflection have been conducted for clarifying film structure and molecular arrangement of dendritic polymers on water surface. Here, the investigations confirming the interaction or hybridization of dendritic polymers with other materials on interfaces are digested. The dendritic polymers have an abundance of functional groups on terminals and internals. Owing to such characteristics, they are capable of strong interaction with other functional materials. The hybridizations of dendrimers on substrates have been achieved with lamellar liquid crystals (Li et al., 2002), linear polymers (Chun et al., 2003), colloidal particles (Mitamura and Imae, 2003), and so on. In such cases, X-ray reflection or scattering (diffraction) measurements are employed for fine structural evaluation of

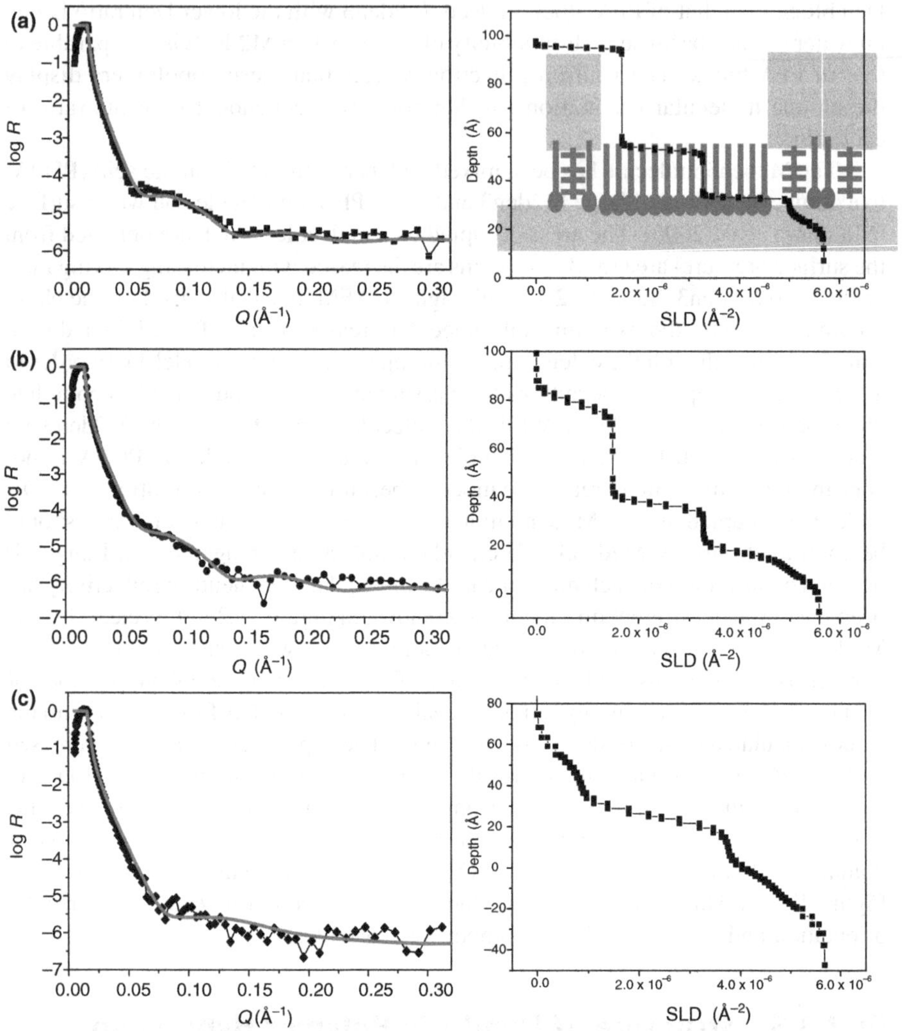


**Figure IV.2.9** Neutron reflectivity profiles and depth–SLD profiles of [PFOA]:[M2den3] = (a) 1:1; (b) 4:1; and (c) 15:1. Solid line: calculated reflectivity curve. Inset: estimated molecular arrangement in the hybrid film.

the hybrid matters. However, there are some issues in the use of X-ray: (1) Soft matters interacting with each other are hard to distinguish by X-ray technique because of the relatively low contrast in electron density, and (2) ultrastrong X-ray intensity, which is available for the fine structure evaluation, can give fatal damages on the samples. On the other hand, neutron technique does not accompany these problems, because (1) SLD adjustment is possible by isotope substitutions (e.g., $^1$H and $^2$H (=D)) and (2) neutron has high transmittance for most materials (energy transfer from neutron to a material is relatively small). Owing to such principal

**Table IV.2.1** Molecular Area Obtained by Experiment and Calculation Based on (a) Phase Separation Model (Model 1) and (b) Complete Incorporation Model (Model 2)

| Mol Ratio [PFOA]:[M2den3] | Experimental Molecular Area Å$^2$/molecule | Calculation Å$^2$/molecule | Calculation Å$^2$/molecule |
|---|---|---|---|
| 1:0 | 42 | – | – |
| 8:1 | 103 | 119 | 82 |
| 4:1 | 174 | 181 | 151 |
| 0:1 | 174 | – | – |

| Mol Ratio [PFOA]:[M2den3] | Experimental SLD of Den Å$^{-2}$ | Calculation Å$^{-2}$ | Calculation Å$^{-2}$ |
|---|---|---|---|
| 15:1 | 0.90 | 0.97 | 1.80 |
| 4:1 | 1.50 | 1.47 | 1.80 |
| 1:1 | 1.70 | 1.70 | 1.80 |
| 0:1 | 1.80 | – | – |

(a) Model 1: Phase separation model (b) Model 2: Complete incorporation model

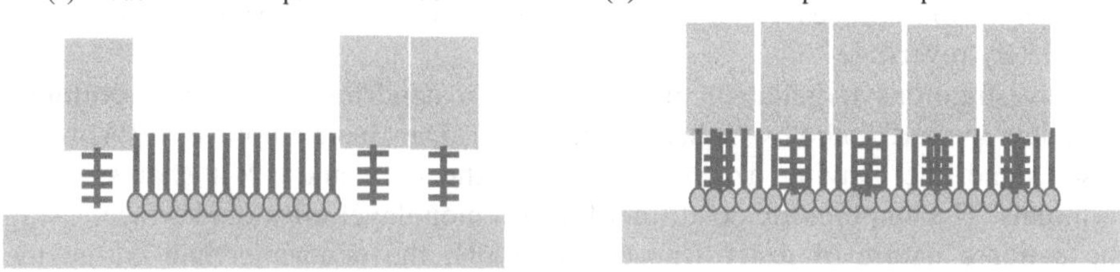

background of neutron, studies using the neutron reflection are recently increasing in order to argue finely the structurally complicated films. Some researches are targeted on neutron analysis of the hybrids of dendritic polymers.

Krasteva et al. (2003) have attempted the hybridization of PAMAM dendrimer with dodecylamine-stabilized gold nanoparticles (AuNPs) by layer-by-layer method on solid surface. In this work, the authors have provided the neutron reflectivity and the successive depth–SLD profiles before/after exposure to solvent vapor such as water and toluene, and demonstrated the possible capture of the vaporized low molecules by the hybridized dendrimer. They have confirmed the capturing of the deuterated molecules into the hybrid dendrimer film and clarified the analyte distribution across the film. The depth–SLD profiles indicate that the solvent molecules penetrate with exponential decaying into the film. It should be noted that the neutron reflection provides the more accurate information about depth profile of solvent or analyte.

Krasteva et al. (2009) have made another advanced study using the dendrimer's capturing capability of low molecules. The investigation has been attempted for poly (propylene imine) dendrimer as well as PAMAM dendrimer. In addition of depth–SLD profiles before/after vapor exposure, the time-resolved shift of the reflectivity is taken at a certain *q* value during the exposure of the vapor. The time-resolved profile is initially increased with time and reached to equilibrium (such as a Langmuir isotherm), which indicates the adsorption of the vapor in the film. The adsorption process takes place at the period of minute. Similarly, desorption process is observed by flowing dry nitrogen gas, where the reflectivity monotonically diminishes with duration. New probability of neutron reflection is enlarged by applying in time-resolved analysis.

Cavaye et al. (2009) have studied the adsorption process on thin films of an arylic dendrimer using *p*-nitro toluene as an analyte by neutron reflection. While the dendrimer displays photoluminescence (PL), the PL is quenched by incorporation of *p*-nitro toluene. Complementally with the PL measurements, Cavaye et al. have provided clear depth–SLD profiles and finely discussed the analyte distribution across the films, which enable the quantitative estimation with high accuracy. They have also compared two films with different thickness (23 and 75 nm) and found that *p*-nitro toluene doped in the thick film is greater in the SLD value and more dispersed all over the film than that in the thin one. These results indicate that the efficiency of the molecular entrapment depends on the film thickness. They have also showed the desorption process and proved that adsorption/desorption process is completely reversible.

Investigations targeted on interaction of a dendrimer with large colloidal materials are performed by Ujihara et al. (2006). They have employed a PAMAM-based dendrimer having azacrown ether in the core and alkyl chains in the terminals. The amphiphilic dendrimer forms a monolayer on water surface. Using the cationic nature of azacrown at neutral pH, the dendrimer can entrap the negatively charged materials. The authors have dispersed COOH-capped AuNPs in water phase and monitored the incorporation of the AuNPs on the dendrimer monolayer by neutron reflection and complementary X-ray reflection. Indeed, the dendrimer interacts with the AuNPs and make a complex on the interface, where upper half on the AuNPs is covered with the dendrimers. On the other hand, for the complementary X-ray reflection, the hybridized monolayer film is transferred on a solid substrate and, then the reflectivity and the corresponding depth–mass density profiles are consistent with results from the neutron reflection. The authors could make a precise discussion by considering X-ray reflection profile along neutron reflection.

## IV.2.4 CONCLUSIONS

Many scientists have been clarifying the structures of dendritic polymers and their affinities to thin films using SANS, quasielastic neutron scattering, neutron spin echo, and neutron reflectometry. The studies introduced in this chapter utilize adequately the advantages of these methods. However, for further fine analysis and discussion, the improvements are required in points of the neutron intensity, resolution, and measurement mode. The intensity of neutron is mainly dependent on a facility, but, fortunately, the neutron source with ultrahigh power has been establishing at some facilities in the world. In respect of resolution, high resolution ($q/q < 1\%$) is achieved by development of optical systems (neutron mirror, collimator, chopper, etc.). The measurement mode has also been fulfilled. For example, equipments for in-plane neutron scattering and neutron spin echo measurements accompanying neutron reflection have been developed. The more the developments of the reflectometers are proceeded, the discussion and understanding on interfacial structures of soft materials represented by dendritic polymers should be substantially deepened, which should drastically promote the studies on soft materials.

## REFERENCES


Aoi, K., Motoda, A., Okada, M., and Imae, T. *Macromol. Rapid. Commun.* **1997**, *18*, 945.

Aoi, K., Motoda, A., Ohno, M., Tsutsumiuchi, K., Okada, M., and Imae, T. *Polymer J.* **1999a**, *31*, 1071.

Aoi, K., Hatanaka, T., Tsutsumiuchi, K., Okada, M., and Imae, T. *Macromol. Rapid Commun.* **1999b**, *20*, 378.

Bauer, B.J., Briber, R.M., Hammouda, B., and Tomalia, D.A. *Polym. Mater. Sci. Eng.* **1992**, *66*, 428.

Briber, R.M., Bauer, B.J., Hammouda, B., and Tomalia, D.A. *Polym. Mater. Sci. Eng.* **1992**, *66*, 430.

Cavaye, H., Smith, A.R.G., James, M., Nelson, A., Burn, P.L., Gentle, I.R., Lo, S.-C., and Meredith, P. *Langmuir* **2009**, *25*, 12800.

Chen, W-R., Porcar, L., Liu, Y., Butler, P.D., and Magid, L. J. *Macromolecules* **2007**, *40*, 5887.

Chun, L., Mitamura, K., and Imae, T. *Macromolecules* **2003**, *36*, 9957.

de Gennes, P.-G. and Harvert, H. *J. Phys.* **1983**, *44*, L351.

Förster, S., Neubert, I., Schlüter, A.D., and Lindner, P. *Macromolecules* **1999**, *32*, 4043.

Funayama, K. and Imae, T. *J. Phys. Chem. Solids* **1999**, *60*, 1355.

Funayama, K., Imae, T., Aoi, K., Tsutsumiuchi, K., Okada, M., Seto, H., and Nagao, M. *J. Phys. Soc. Jpn.* **2001**, *70*, 326.

Funayama, K., Imae, T., Aoi, K., Tsutsumiuchi, K., Okada, M., Furusaka, M., and Nagao, M. *J. Phys. Chem. B* **2003a**, *107*, 1532.

Funayama, K., Imae, T., Seto, H., Aoi, K., Tsutsumiuchi, K., Okada, M., Nagao, M., and Furusaka, M. *J. Phys. Chem. B* **2003b**, *107*, 1353.

Hierlemann, A., Campbell, J.K., Baker, L.A., Crooks, R.M., and Ricco, A. J. *J. Am. Chem. Soc.* **1998**, *120*, 5323.

Hirano, C., Imae, T., Yanagimoto, Y., and Yamaguchi, Y. *Polymer J.* **2006**, *38*, 44.

Imae, T., Funayama, K., Aoi, K., Tsutsumiuchi, K., Okada, M., and Furusaka, M. *Langmuir* **1999**, *15*, 4076.

Johnson, M.A., Santini, C.M.B., Iyer, J., Satija, S., Ivkov, R., and Hammond, P.T. *Macromolecules* **2002**, *35*, 231.

Kampf, J.P., Frank Eva, C.W., Malmström, E., and Hawker, C.J. *Langmuir* **1999**, *15*, 227.

Kirton, G.F., Brown, A.S., Hawker, C.J., Reynolds, P.A., and White, J.W. *Physica B* **1998**, *248*, 184.

Krasteva, N. Krustev, R., Yasuda, A., and Vossmeyer, T. *Langmuir* **2003**, *19*, 7754.

Krasteva, N., Möhwald, H., and Krastev, R. *Comptes Rendus Chimie* **2009**, *12*, 129.

Li, X., Imae, T., Leisner, D., and Lopez-Quintera, M.A. *J. Phys. Chem. B* **2002**, *106*, 12170.

Meltzer, A.D., Tirrell, D.A., Jones, A.A., Inglefield, P.T., Hedstrand, D.M., and Tomalia, D.A. *Macromolecules* **1992**, *25*, 4541.

Mitamura, K. and Imae, T. *Trans. MRS-J* **2003**, *28*, 71.

Mitamura, K., Imae, T., Mouri, E., Torikai, N., Matsuoka, H., and Nakamura, T. *J. Nanosci. Nanotech.* **2006**, *6*, 36.

Mitamura, K., Takahashi, M., Hamaguchi, S. and Imae, T. *Trans. MRS-J*, **2004**, *9*, 255.

Nisato, G., Ivkov, R., and Amis, E.J. *Macromolecules* **2000**, *33*, 4172.

SINGH, P., ONODERA, T., MIZUTA, Y., MATSUMOTO, K., MIURA, N., and TOKO, K. *Sens Actuators B Chem* **2009**, *137*, 403.

PARRATT, L.G. *Phys. Rev.* **1954**, *95*, 359.

PEYSER-CAPADONA, L., ZHENG, J., GONZÁLEZ, J.I., LEE, T-H., PATEL, S.A., and DICKSON, R.M. *Phys. Rev. Lett.* **2005**, *94*, 058301.

PORCAR, L., LIU, Y., VERDUZCO, R., HONG, K., BUTLER, P.D., MAGID, L.J., SMITH, G.S., and CHEN, W.-R. *J. Phys. Chem. B* **2008**, *112*, 14772.

PÖTSCHKE, D., BALLAUFF, M., LINDNER, P., FISCHER, M., and VÖGTLE, F. *Macromolecles* **1999**, *32*, 4079.

RAMZI, A., SCHERRENBERG, R., BRACKMAN, J., JOOSTEN, J., and MORTENSEN, K. *Macromolecules* **1998**, *31*, 1621.

RAMZI, A., SCHERRENBERG, R., JOOSTEN, J., LEMSTRA, P., and MORTENSEN, K. *Macromolecules* **2002**, *35*, 827.

RIETVELD, I.B., BEDEAUX, D., and SMIT, J.A.M. *J. Colloid Interface Sci.* **2000**, *232*, 317.

ROSENFELDT, S., DINGENOUTS, N., BALLAUFF, M., WERNER, N., VÖGTLE, F., and LINDNER, P. *Macromolecles* **2002**, *35*, 8098.

SADE, P.M., WHITE'C, J.W., HAWKER, J., WOOLEY, L., and FRÉCHET, J.M.J. *J. Phys. Chem*, **1993**, *97*, 293.

SAGIDULLIN, A.I., MUZAFAROV, A.M., KRYKIN, M.A., OZERIN, A.N., SKIRDA, V.D. and IGNAT'EVA, G.M. *Macromolecules* **2002**, *35*, 9472.

SAVILLE, P.M., REYNOLDS, P.A., WHITE, J.W., HAWKER, C.J., FRÉCHET, J.M.J., WOOLEY, K.L., PENFOLD, J., and WEBSTER, J.R.P. *J. Phys. Chem.* **1995**, *95*, 8283.

SCHENNING, A.P.H.J., ELISSEN-ROMÁN, C., WEENER, J.-W., BAARS, M.W.P.L., van der GAAST, S.J., and MEIJER, E.W. *J. Am. Chem. Soc.* **1998**, *120*, 8199.

SCHERRENBERG, R., COUSSENS, B., van VLIET, P., EDOUARD, G., BRACKMAN, J., de BRABANDER, E., and MORTENSEN, K. *Macromolecles* **1998**, *31*, 456.

STARK, B., STÜHN, B., FREY, H., LACH, C., LORENZ, K., and FRICK, B. *Macromolecules* **1998**, *31*, 5415.

STARK, B., LACH, C., FARAGO, B., FREY, H., SCHLENK, C., and STÜHN, B. *Colloid Polym. Sci.* **2003**, *28*, 593.

TANAKA, K., DAI, S., KAJIYAMA, T., AOI, K., and OKADA, M. *Langmuir* **2003**, *19*, 1196.

TOPP, A., BAUER, B.J., TOMALIA, D.A., and AMIS, E.J. *Macromolecules* **1999a**, *32*, 7232.

TOPP, A., BAUER, B.J., PROSA, T.J., SCHERRENBERG, R., and AMIS, E.J. *Macromolecules* **1999b**, *32*, 8923.

TOPP, A., BAUER, B.J., KLIMASH, J.W., SPINDLER, R., TOMALIA, D.A., and AMIS, E.J. *Macromolecules* **1999c**, *32*, 7226.

UJIHARA, M. and IMAE, T. *J. Colloid Interfacial Sci.* **2006**, *293*, 333.

UJIHARA, M., MITAMURA, K., TORIKAI, N., and IMAE, T. *Langmuir* **2006**, *22*, 3656.

VAN DER LEE, A. *Solid State Sci.* **2000**, *2*, 257.

WEI, F., LIAO, W., XU, Z., YANG, Y., WONG, D.T., and HO, C.-M. *Small* **2009**, *5*, 1784–1790.

WILLERICH, I., RITTER, H. and GRÖHN, R. *J. Phys. Chem. B* **2009**, *113*, 3339.

WIND, M., SAALWÄCHTER, K., WIESLER, U.-M., MÜLLEN, K. and SPIESS, H.M. *Macromolecules* **2002**, *35*, 10071.

XU, S.-L., KANG, S.-Z., DENG, G.-J., WU, P., FAN, Q.-H., WANG, C., WAN, L.-J., and BAI, C.-L. *Chem. Lett.* **2003**, *32*, 290.

ZHAO, M., LIU, Y., CROOKS, R.M., and BERGBREITER, D. E. *J. Am. Chem. Soc.* **1999**, *121*, 923.

# IV Applications

## IV.3 Dynamics of Polymers

*Toshiji Kanaya* and Barbara J. Gabrys*

In this chapter, we review previous inelastic and quasi-elastic neutron scattering experiments on polymer systems. First, we outline the basics of inelastic and quasi-elastic scattering to give sound foundations to this chapter. Second, we bring out characteristic features of inelastic and quasi-elastic neutron scattering for the studies of soft matter and polymers. Finally, we review some typical polymer motions revealed by these techniques, ranging from local motion in picosecond (ps) to large-scale motion in the order of several hundred nanoseconds (ns). Inelastic and quasi-elastic neutron scattering is also a powerful tool to reveal the dynamics of emulsions, micelles, surfactants, liquid crystals, proteins, or supercooled liquids. More examples of such studies are reviewed in Chapters II.3.1–II.3.3 and IV.5. The readers are advised to refer these chapters for a more complete picture of inelastic and quasi-elastic neutron scattering studies on dynamics of soft matter.

### IV.3.1 BASICS OF INELASTIC AND QUASI-ELASTIC NEUTRON SCATTERING

In previous chapters, the neutron scattering theory and the instrumentation have been summarized for soft matter studies. To avoid an overlap, here we outline only the essentials of inelastic and quasi-elastic neutron scattering necessary for understanding of the contents of this chapter.

#### IV.3.1.1 Elastic and Inelastic Scattering

Figure IV.3.1 illustrates schematically the neutron scattering process in real and reciprocal space. In experiments, we observe a double-differential scattering cross section $\partial^2\sigma/\partial\Omega\partial E$, which is a probability that an incident neutron with wave vector $\mathbf{k}_i$ and energy $E_i$ is scattered by a scattering angle $2\Theta$ into a solid angle element $d\Omega$ and an energy interval between $E_f$ and $E_f + dE$. The energy and wave vector of the

*Corresponding author.

*Neutrons in Soft Matter* Edited by Toyoko Imae, Toshiji Kanaya, Michihiro Furusaka, and Naoya Torikai

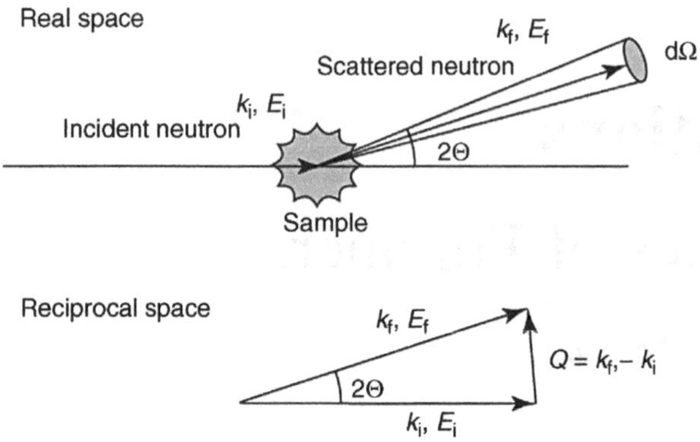


**Figure IV.3.1** Neutron scattering process in real and reciprocal space. An incident neutron with wave vector $\mathbf{k}_i$ and energy $E_i$ is scattered by a scattering angle 2Θ into a solid angle element dΩ. Wave vector and energy of the scattered neutron are $\mathbf{k}_f$ and $E_f$, respectively.

scattered neutron are $E_f$ and $\mathbf{k}_f$, respectively. The momentum transfer $\hbar\mathbf{Q}$ and energy transfer $\Delta E$ of the neutron are given by

$$\hbar\mathbf{Q} = \hbar(\mathbf{k}_f - \mathbf{k}_i), \qquad \text{(IV.3.1)}$$

$$\Delta E = \hbar\omega = E_f - E_i = \frac{\hbar}{2m}(k_f^2 - k_i^2). \qquad \text{(IV.3.2)}$$

Elastic scattering is a scattering process without energy transfer ($\Delta E = 0$) between the neutron and a sample—atoms or groups of atoms—while inelastic and quasi-elastic scattering includes energy transfer between the neutron and sample materials.

## IV.3.1.2 Double-Differential Scattering Cross section, Dynamic Scattering Law, and Intermediate Scattering Function

For a system consisting of one kind of atom, the double-differential scattering cross section $\partial^2\sigma/\partial\Omega\partial E$ is simply related to the scattering law

$$\frac{\partial^2\sigma}{\partial\Omega\partial E} = \frac{k_f}{4\pi\hbar k_i} N[\sigma_{inc}S_{inc}(\mathbf{Q},\omega) + \sigma_{coh}S_{coh}(\mathbf{Q},\omega)], \qquad \text{(IV.3.3)}$$

where $S_{inc}(\mathbf{Q},\omega)$ and $S_{coh}(\mathbf{Q},\omega)$ are, respectively, incoherent and coherent scattering laws, $\sigma_{inc}$ and $\sigma_{coh}$ are incoherent and coherent atomic scattering cross sections, respectively, and $N$ is the number of atoms. The incoherent and coherent dynamic scattering laws are related to incoherent and coherent intermediate scattering functions $I_s(Q,t)$ and $I(Q,t)$, respectively, through the Fourier transform with angular frequency $\omega$.

$$S_{inc}(\mathbf{Q},\omega) = \frac{1}{2\pi}\int_{-\infty}^{+\infty} I_s(\mathbf{Q},t)\exp(-i\omega t)\mathrm{d}t, \qquad \text{(IV.3.4)}$$

$$S_{\text{coh}}(\mathbf{Q}, \omega) = \frac{1}{2\pi} \int_{-\infty}^{+\infty} I(\mathbf{Q}, t) \exp(-i\omega t) \mathrm{d}t. \tag{IV.3.5}$$

Hence, the incoherent and coherent intermediate scattering functions $I_s(Q, t)$ and $I(Q, t)$ are, respectively, defined by

$$I_s(\mathbf{Q}, t) = \frac{1}{N} \sum_k \langle \exp\{-iQr_k(0)\} \exp\{iQr_k(t)\} \rangle, \tag{IV.3.6}$$

$$I(Q, t) = \frac{1}{N} \sum_k \sum_l \langle \exp\{-iQr_k(0)\} \exp\{iQr_l(t)\} \rangle. \tag{IV.3.7}$$

### IV.3.1.3 Time–Space Correlation Function

Following the van Hove approach (Hove, 1954), incoherent and coherent dynamic scattering laws $S_{\text{inc}}(\mathbf{Q}, \omega)$ and $S_{\text{coh}}(\mathbf{Q}, \omega)$ are given by the time–space Fourier transforms of the time–space self- and pair correlation functions $G_s(\mathbf{r}, t)$ and $G(\mathbf{r}, t)$, respectively,

$$S_{\text{inc}}(\mathbf{Q}, \omega) = \frac{1}{2\pi} \int_{-\infty}^{+\infty} \int_{0}^{+\infty} G_s(\mathbf{r}, t) \exp(i(\mathbf{Qr} - \omega t)) \mathrm{d}\mathbf{r}\, \mathrm{d}t, \tag{IV.3.8}$$

$$S_{\text{coh}}(\mathbf{Q}, \omega) = \frac{1}{2\pi} \int_{-\infty}^{+\infty} \int_{0}^{+\infty} G(\mathbf{r}, t) \exp(i(\mathbf{Qr} - \omega t)) \mathrm{d}\mathbf{r}\, \mathrm{d}t. \tag{IV.3.9}$$

It is useful to separate the pair correlation function $G(\mathbf{r}, t)$ into a self- and distinct part:

$$G(\mathbf{r}, t) = G_s(\mathbf{r}, t) + G_d(\mathbf{r}, t). \tag{IV.3.10}$$

Assuming that there is a particle at a position $\mathbf{r} = 0$ at time $t = 0$, the physical meaning of the self-correlation function $G_s(\mathbf{r}, t)$ is a probability of finding the same particle in an interval $\mathrm{d}\mathbf{r}$ at a position $\mathbf{r}$ at a later time $t$. In analogy, the distinct correlation function is a probability of finding another particle at $\mathbf{r}$ at a time $t$ (Figure IV.3.2).

### IV.3.1.4 Gaussian Approximation

If the distribution of particle positions $r_l(t)$ and $r_k(0)$ are described by a Gaussian distribution (the Gaussian approximation), then incoherent and coherent intermediate

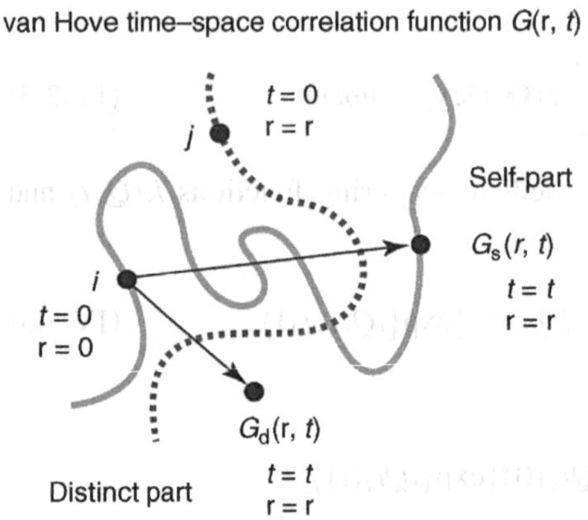


**Figure IV.3.2** Schematic sketch of time–space correlation functions. $G_s(r, t)$, self-correlation function; $G_d(\mathbf{r}, t)$, distinct correlation function.

scattering functions can be written, respectively, in terms of $Q^2$ only as

$$I_s(\mathbf{Q}, t) = \frac{1}{N}\sum_k \exp\left\{-\frac{Q^2}{6}\langle(r_k(t)-r_k(0))^2\rangle\right\}, \qquad \text{(IV.3.11)}$$

$$I(\mathbf{Q}, t) = \frac{1}{N}\sum_k\sum_l \exp\left\{-\frac{Q^2}{6}\langle(r_k(t)-r_l(0))^2\rangle\right\}. \qquad \text{(IV.3.12)}$$

Here, it is worth pointing out that the time-dependent mean square displacement (MSD) is observed from the incoherent scattering measurements. It is also noted that the frequency-dependent mean square displacement can be obtained from the measurements of incoherent dynamic scattering law within the Gaussian approximation.

### IV.3.1.5 Coherent and Incoherent Cross Sections

In most cases, samples are a mixture of isotopes $j$ with different scattering lengths $b_j$, and it is assumed that they are randomly distributed over the sample. Such random distribution of different isotopes results in an incoherent cross section. In analogy with isotope incoherence, spin incoherence is also observed. For nuclei with spin $I \neq 1$, the interaction depends on the orientation between the neutron and nuclear spins; the scattering lengths $b_+$ and $b_-$ for parallel and perpendicular spin, respectively, are different and the orientations of spin are randomly distributed in the nuclei even if they are the same kinds of nuclei. Then, incoherent and coherent atomic cross sections are given by

$$\sigma_{\text{inc}} = 4\pi(\langle b^2\rangle - \langle b\rangle^2), \qquad \text{(IV.3.13)}$$

$$\sigma_{\text{coh}} = 4\pi\langle b\rangle^2, \qquad \text{(IV.3.14)}$$

where $\langle b^2 \rangle$ and $\langle b \rangle^2$ are the mean square scattering length and the square of mean scattering length. For organic polymers that usually include many hydrogen atoms, contribution from spin incoherence is very large: for hydrogen atom, $\sigma_{\text{inc}}$ and $\sigma_{\text{coh}}$ are 80.3 and 1.8 barns, respectively (Sears, 1992). Therefore, incoherent scattering is dominant for many organic polymers at large momentum transfers.

### IV.3.1.6 Inelastic Scattering and Simple Models

Inelastic scattering of neutrons is caused by an oscillatory motion. An example is the inelastic scattering of neutrons by phonons. The coherent double-differential scattering cross section of neutrons for one phonon process is given by (Willis, 1973)

$$\begin{aligned}\frac{\partial^2\sigma^1_{\text{coh}}}{\partial\Omega\partial E} &= \frac{(2\pi)^3}{v}\sum_{q,j}\frac{k_1}{k_0}\delta(\hbar\omega-\hbar\omega_j(\boldsymbol{q}))\sum_i\delta(\mathbf{Q}\pm\mathbf{q}-t)\\ &\quad\times\frac{\hbar\left(n_s+\frac{1}{2}\pm\frac{1}{2}\right)}{2\omega_j(\mathbf{q})}\left|\sum_r\frac{\langle b\rangle_\rho}{M_\rho}e^{iQr}\mathbf{Q}\cdot\mathbf{U}^j_\rho(\mathbf{q})e^{-W_\rho}\right|. \end{aligned} \tag{IV.3.15}$$

Here, $\mathbf{q}$ is the wave vector of phonon, $\omega_j(\mathbf{q})$ is the characteristic frequency of the mode $j$ at $\mathbf{q}$ and $\tau$ is the reciprocal lattice vector. $\langle b\rangle_\rho$ and $M_r$ are the coherent scattering length and the mass of the atom at $\mathbf{r}$ in the unit cell, respectively, and $e^{-W_\rho}$ is its Debye–Waller factor. $\mathbf{U}^j_\rho(\mathbf{q})$ is the polarization vector, corresponding to the displacement vector of a simple harmonic oscillator. The upper and the lower values of double signs in the equation refer to neutron energy loss or phonon creation and to neutron energy gain or phonon annihilation, respectively. All phonon modes are harmonic oscillators, and hence they have Bose–Einstein population factors $n_s+1$ for energy loss and $n_s$ for energy gain.

$$n_s = [\exp(\hbar\omega_j(q)/k_BT)-1]^{-1}. \tag{IV.3.16}$$

Two $\delta$-functions in eq. (IV.3.15) mean that the scattering occurs under the conditions of energy and momentum conservation. One obtains a phonon peak in experiments, not phonon distributions.

The incoherent double-differential scattering cross section for a one phonon process is given by

$$\frac{\partial^2\sigma^1_{\text{inc}}}{\partial\Omega\partial E} = \sum_{q,j}\frac{k_1}{k_0}\delta(\hbar\omega-\hbar\omega_j(\mathbf{q}))\frac{\hbar(n_s+\frac{1}{2}\pm\frac{1}{2})}{2\omega_j(\mathbf{q})}\sum_\rho\frac{(\langle b^2\rangle_\rho-\langle b\rangle^2_\rho)}{M_\rho}|\mathbf{Q}\cdot\mathbf{U}^j_\rho(\mathbf{q})|^2e^{-W_\rho}. \tag{IV.3.17}$$

$4\pi(\langle b^2\rangle_\rho-\langle b\rangle^2_\rho)$ corresponds to incoherent atomic scattering cross section. Eq. (IV.3.17) includes one $\delta$-function for energy conservation, meaning that the

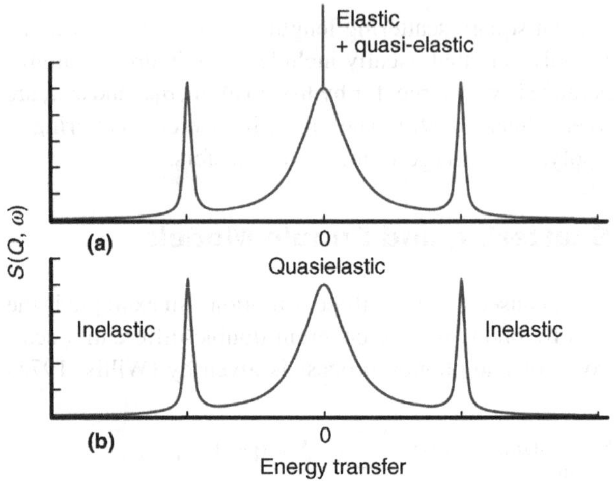


**Figure IV.3.3** Schematic sketch of inelastic and quasi-elastic scattering spectra $S(Q, \omega)$. (a) Quasi-elastic scattering spectrum from moving particles in confined space and (b) quasi-elastic scattering spectrum from moving particles in infinite space.

scattering consists of a broad distribution in energy. For a special case of a crystal with cubic symmetry, containing one atom in the unit cell, the relation $\langle(\mathbf{Q}\cdot\mathbf{U})^2\rangle = (1/3)Q^2\langle u^2\rangle$ is sustained, where $\langle u^2\rangle$ is the mean square amplitude of vibration of the atom. Assuming continuous distribution of the frequency, eq. (IV.3.17) is reduced to

$$\frac{\partial^2\sigma^1_{\text{inc}}}{\partial\Omega\partial E} = \frac{k_1}{k_0}\left(\langle b^2\rangle - \langle b\rangle^2\right)\frac{Q^2\langle u^2\rangle}{2M}\left(n_s + \frac{1}{2} \pm \frac{1}{2}\right)e^{-2W}N\frac{G(\omega)}{\omega}, \tag{IV.3.18}$$

where $G(\omega)$ is the normalized density of phonon states and $3N$ is the total number of phonon states. Such inelastic scattering is observed as a peak in double-differential cross section or the dynamic scattering law $S(Q, \omega)$. This is schematically illustrated in Figure IV.3.3a and b.

### IV.3.1.7 Quasi-Elastic Scattering and Some Models

Quasi-elastic scattering is caused by random motions or energy dissipation processes. One of the typical examples of such motions is a simple diffusion of independent particles described by Fick's law. The incoherent intermediate scattering function is given by

$$I_{\text{inc}}(\mathbf{Q}, t) = \int_0^{\infty} G_s(\mathbf{r}, t)\exp(-\mathbf{Q}\cdot\mathbf{r})d\mathbf{r} = \exp(-DQ^2t), \tag{IV.3.19}$$

where $D$ is a diffusion coefficient. The corresponding incoherent scattering law is given by a Lorentzian function whose half-width at half-maximum (HWHM) increases with the momentum transfer according to a $DQ^2$ law:

$$S_{\text{inc}}(Q, \omega) = \frac{1}{\pi}\frac{DQ^2}{\omega^2 + (DQ^2)^2}. \tag{IV.3.20}$$

Such quasi-elastic scattering is observed as broadening of an elastic scattering peak in the double-differential scattering cross section or the dynamic scattering law $S(Q, \omega)$ as shown in Figure IV.3.3b.

In cases of real systems, random motions are often limited in a finite space. Diffusion motions on a sphere and jump motions between two sites (Bee, 1988) are such cases. For the jump motion between two equivalent sites separated by a distance $d$ with the mean residence time $\tau_r$ in each site, the scattering law after a powder average is given by

$$S(Q,\omega) = F_0(Q)\delta(\omega) + F_1(Q)\frac{1}{\pi}\frac{2\tau_r}{4+\omega^2\tau_r^2}, \qquad \text{(IV.3.21)}$$

$$\begin{aligned} F_0(Q) &= [1 + j_0\{Qd\}]/2, \\ F_0(Q) &= [1 - j_0\{Qd\}]/2, \end{aligned} \qquad \text{(IV.3.22)}$$

where $j_0$ is a spherical Bessel function of zeroth order. Due to the spatial restriction, the dynamic scattering law $S(Q, \omega)$ includes a $\delta$-function in addition to the quasi-elastic component, giving a geometrical information of the motion. This is also schematically illustrated in Figure IV.3.3a.

## IV.3.2 CHARACTERISTIC FEATURES OF INELASTIC AND QUASI-ELASTIC NEUTRON SCATTERING FOR POLYMER STUDIES

### IV.3.2.1 Neutron and Electromagnetic Wave

The neutron is described as a particle wave that obeys the de Broglie relation. The relation between the energy $E$ and the wavelength $\lambda$ is given by

$$E = \frac{h^2}{2m\lambda^2}. \qquad \text{(IV.3.23)}$$

On the other hand, the relation between the energy $E$ and the wavelength $\lambda$ for an electromagnetic wave such as X-ray is written in the following form:

$$E = h\frac{c}{\lambda}. \qquad \text{(IV.3.24)}$$

These relations are shown in Figure IV.3.4. When we want to study the microscopic structure of materials in Å or nm scale, we need neutrons or X-rays with wavelength of Å or nm scale. Such X-rays, for example, with $\lambda = 1.5$ Å have energy of 8 keV, which is higher than the thermal energy (~30 meV) by about five orders of magnitude. Therefore, it is hard to investigate the thermal motions of materials using X-rays that are suitable for structural studies. On the other hand, a neutron with wavelength of 1.5–5 Å has energy of 36.4~3.2 meV, which is comparable to thermal energy of materials. This means that neutrons with wavelengths suitable to study microscopic structure can be used to study dynamics of materials,

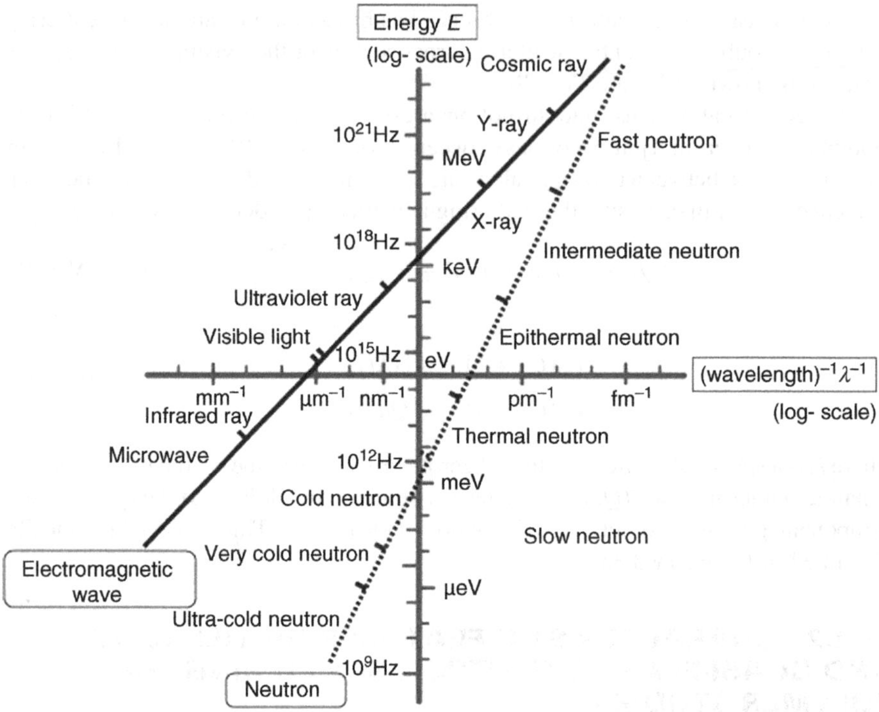


**Figure IV.3.4** Some typical motions in amorphous polymers in time–length space. The time–length ranges covered by inelastic and quasi-elastic neutron scattering, dynamic light scattering, and X-ray photon correlation spectroscopy are approximately shown. The time ranges accessed by mechanical relaxation (MR), dielectric relaxation (DR), and NMR are indicated by lines.

but X-rays with wavelengths suitable for microscopic structure are not applicable for studies of dynamics. Hence, by using neutrons we can study the microscopic structure and dynamics of materials simultaneously. This is one of the biggest advantages of neutrons in materials research. Note that recent advances in X-ray experimental techniques including synchrotron radiation (SR) make it possible to measure inelastic scattering to resolve the very small energy transfer of scattered X-rays ($\Delta E/E = 10^{-6}$ to $10^{-7}$) (Sette and Krisch, 2006).

## IV.3.2.2 Which Polymer Motions are Observed in Which Time–Length Space?

There are many kinds of modes of motions present in polymer systems from picosecond to several days. In Figure IV.3.5, some typical motions of amorphous polymers are illustrated on the time–length space. In the same figure, we also show the time–length space covered by various experimental methods, including inelastic and quasi-elastic neutron scattering technique. As seen in the figure and explained in Chapters II.3.1–II.3.3, inelastic and quasi-elastic neutron scattering can cover a time

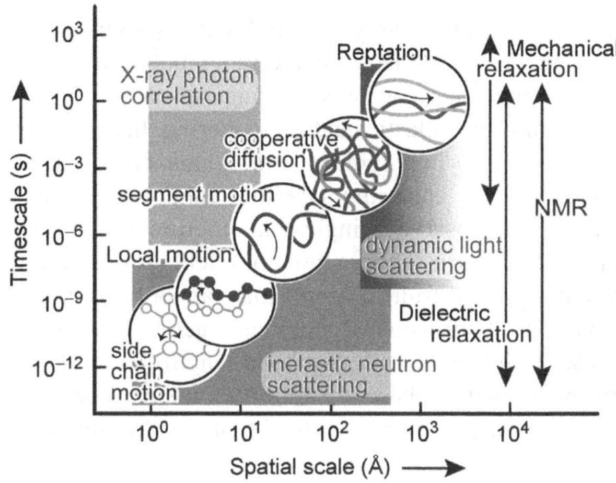


**Figure IV.3.5** Relations between energy and wavelength for neutrons and electromagnetic waves.

range from ~0.01 ps to several hundred nanoseconds (ns) and a $Q$ range from 0.001 to 30 $\text{Å}^{-1}$. This depends on a type of neutron spectrometers used, such as neutron spin echo (NSE) (Chapter II.3.1), backscattering (BS) (Chapter II.3.2), and time of flight (TOF) (Chapter II.3.3). This time–length space is very wide. The time range accessible to neutrons can cover molecular vibration motions in ~0.01–1 ps range, which are usually studied by infrared (IR) absorption spectroscopy and Raman spectroscopy, and stochastic motions (or energy dissipative motions) in a slower than picosecond time range. With respect to the stochastic motions in amorphous polymer in melts and solutions shown in Figure IV.3.5, we see some side chain motions in polymers such as methyl group rotation in picosecond region and local conformational transition of a polymer chain in subnanosecond region. Segmental motions (described by the Rouse and Zimm models) are detected by quasi-elastic neutron scattering in a region from a nanosecond to several tens of nanoseconds. Cooperative diffusion motion in polymer networks is also detectable in a range from a nanosecond to tens of nanoseconds region. Very slow reptation motion is also detectable by quasi-elastic neutron scattering in several tens to hundreds of nanoseconds in a $Q$ range of 0.01–0.3 $\text{Å}^{-1}$, depending on the type of polymer and the temperature of measurement. It is noted that such cooperative diffusion motion and reptation motions are very slow on a large spatial scale in the $\mu$m region. They can be observed by dynamic light scattering, while in neutron scattering, we observe such motions in the higher $Q$ range than available in light scattering. Moreover, the characteristic timescale is much faster in neutron scattering than that accessed by light scattering. In the next section, we show how inelastic and quasi-elastic scattering detects such motions and how the neutron data are analyzed.

## IV.3.2.3 Deuterium Labeling

As shown in Chapter I, atomic scattering length of hydrogen (H) atom is $-3.74 \times 10^{-15}$ m while that of deuterium (D) is $6.671 \times 10^{-15}$ m. They are very different although chemical natures of H and D atoms are very similar. Most of polymers and soft matters include many hydrogen atoms, and hence deuterium labeling is one of the largest advantages in structure studies by neutron scattering. Of course, it also provides unique feature to inelastic and quasi-elastic neutron scattering. In density of phonon (or vibration) states in measurements of polymers, an assignment of vibration motion could be facilitated by selective deuterium labeling. In polymer melts where polymer chains are entangled, we observe a single chain motion of hydrogenated polymer in a deuterated polymer matrix. Thus, a judicious deuterium labeling in polymer research can produce unique and fruitful results that cannot be obtain in other experiments. Note, however, that hydrogenated and deuterated polymers are not usually cocrystallized because of the small, different bond lengths and interactions. Furthermore, even in amorphous states, phase separation sometimes occurs in blends of hydrogenated and deuterated polymers due to the small difference in interactions.

## IV.3.3 STUDIES OF DYNAMICS OF POLYMERS

Polymer systems have a very complicated and hierarchical structure on a very wide spatial scale from Å to several tens of micrometers. Corresponding to such hierarchical structure, the dynamics of polymer systems also lies in a wide timescale ranging from ps to several days. As mentioned in the previous section, inelastic and quasi-elastic neutron scattering techniques cover the timescale faster than $\sim 10^{-6}$ s and the spatial scale from Å to 1 μm. Even these large time and spatial scales are not necessarily sufficient for the study of dynamics of polymer systems. Nevertheless, extensive studies on polymer dynamics by inelastic and quasi-elastic scattering techniques are worthwhile because of unique features offered compared to other methods. In this section, we will discuss exemplary results on polymer dynamics from ps to several hundred ns that demonstrate characteristic features of inelastic and quasi-elastic neutron scattering techniques.

### IV.3.3.1 Phonon Dispersion Curve in Polytetrafluoroethylene

One of the advantages of inelastic neutron scattering is the determination of phonon dispersion curves in crystals. For this purpose, we have to measure the coherent inelastic neutron scattering from single crystals. However, polymers and soft materials usually include many hydrogen atoms so that strong spin incoherent scattering is observed at large $Q$ values, which makes it difficult to measure the coherent scattering. Furthermore, it is very hard to prepare a polymer single crystal. Hence, there are not many experiments to study phonon dispersion curves to date. Twisleton and White (1972) determined the $[10\bar{1}0]$ phonon dispersion curve for longitudinal lattice vibrations perpendicular to the chain axes in hexagonal polytetrafluoroethylene (PTFE) powder and fiber samples using a time-of-flight spectrometer. These samples have no hydrogen atoms and have very small incoherent

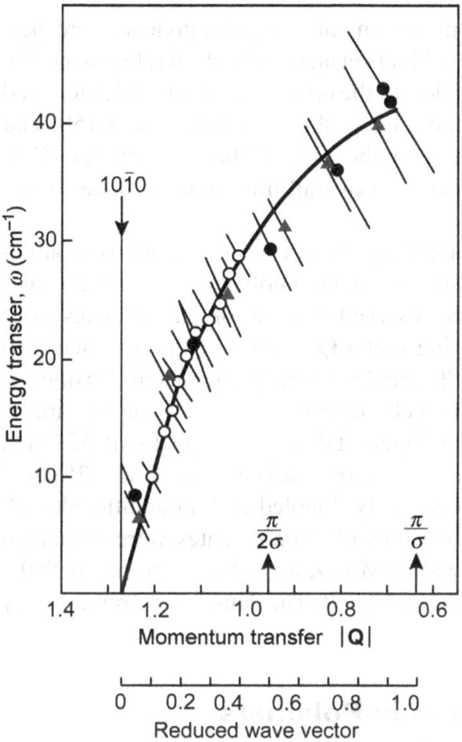


**Figure IV.3.6** Dispersion curve for the longitudinal acoustic mode along $[10\bar{1}0]$ in hexagonal polytetrafluoroethylene (Twisleton and White, 1972).

scattering contributions. The authors characterized the coherent features in the time-of-flight spectra of PTFE as follows: the anisotropy and high symmetry of the unit cell lead to a diffraction picture with only one strong reflection at low-momentum transfers, which corresponds to the interchain spacing; acoustic phonons are only observable when excited in the region of a strong Bragg reflection and there is but one here; the longitudinal phonons are alone characterized by a singularity, on account of the scalar product term in eq. (IV.3.15). The obtained dispersion curve is shown in Figure IV.3.6. From the slope at the low reduced wave vector of phonon $q$, they found that the sound velocity was $2.77 \pm 10.10 \times 10^5$ cm/s. Using the crystal density (=2.342 g/cm$^3$), they also evaluated the crystalline elastic constant $C_{11}$ of $18.2 \pm 10.2 \times 10^{10}$ dyn/cm$^2$.

### IV.3.3.2 Density of Phonon States of Polyethylene Crystal by Incoherent Inelastic Neutron Scattering

As shown in Section IV.3.1.6, the incoherent double-differential scattering cross section is related to the density of phonon states or the vibration frequency distribution $G(\omega)$ (see eq. (IV.3.18)). In the early stage of inelastic neutron scattering studies on polymers, extensive studies have been performed on polyethylene crystals to evaluate the density of phonon states and compared with the results of normal mode calculations. Inelastic scattering measurements on polyethylene were performed by Danner et al. (1964) followed by Myers et al. (1965). With reference to the frequency distribution (and vibrational modes) of the orthorhombic crystal calculated by Kitagawa and Miyazawa (1965), scattering peaks at ~560, 190, ~150, and 90 cm$^{-1}$ (1 meV ~ 8 cm$^{-1}$) were assigned to the C–C–C bending modes, C–C internal rotation mode, overall rotary mode, and overall translational mode (in the ab plane), respectively.

More detailed information was obtained by measuring anisotropy of double-differential scattering cross section of a stretch-oriented polyethylene with a triple-axis spectrometer (Myers et al., 1966). The observed densities of phonon states $G(\omega)$ are shown in Figure IV.3.7a for the scattering vector **Q**, parallel and perpendicular to $c$-axis (stretch direction). In Figure IV.3.7b, the theoretically calculated densities of phonon states of orthorhombic crystal of polyethylene are also shown for the parallel and perpendicular geometry. As seen in Figure IV.3.7a, two peaks at 525 and 190 cm$^{-1}$ were observed in the parallel geometry, while the peak at 525 cm$^{-1}$ disappeared but the peak at 190 cm$^{-1}$ was nearly doubled and an additional peak appeared at 150 cm$^{-1}$. These observed densities of phonon states were compared with the theoretical calculation by Kitagawa and Miyazawa (1967) to discuss both the multiphonon effects and the Debye–Waller coefficient. The details were reviewed by Kitagawa and Miyazawa (1973).

## IV.3.3.3 Methyl Group Motion in Polymers

In polymers and other soft matter, there are many dynamic processes occurring in a wide timescale from ps to several days. One of the simplest motions is the methyl

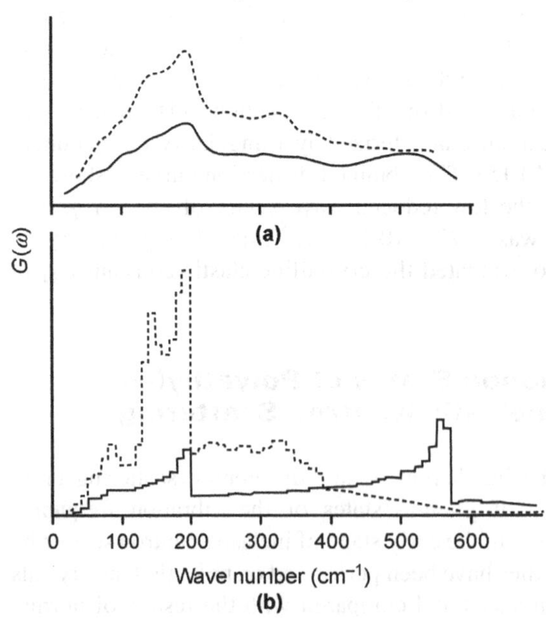


**Figure IV.3.7** Directional density of phonon states $G(\omega)$ of orthorhombic polyethylene at 100 K. Solid lines and dashed lines correspond to **c**//**Q** and **c**⊥**Q**, respectively. (a) Experimental curves from Myers et al. (1966) and (b) theoretical curves from Kitagawa and Miyazawa (1973).

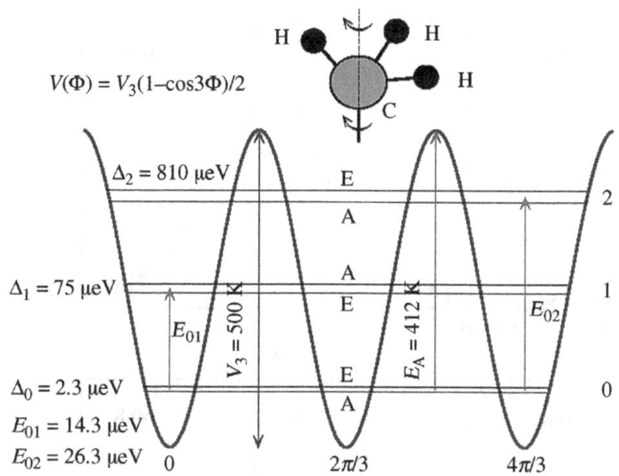


**Figure IV.3.8** Picture of methyl group rotation about its $C_3$-axis, and level scheme for a purely threefold potential with barrier height $V_3 = 500\text{K}$ (Colmenero et al., 2005).

group rotation in natural and synthetic macromolecules. It is assumed that in the crystalline and glassy states the main chains are frozen in: only methyl groups are still mobile and the interactions between methyl groups are negligible. In such conditions, the methyl group motion can be described by a single particle rotation in a threefold potential on an angular coordinate $\phi$, which is measured in the plane perpendicular to $C_3$ symmetry axis of methyl group (see Figure IV.3.8).

In the simplest case, the potential $V(\phi)$ governing the motion is given by

$$V(\phi) = V_3(1 - \cos 3\phi)/2, \tag{IV.3.25}$$

where $V_3$ is a potential barrier height. The potential function is illustrated in Figure IV.3.8 for $V_3 = 500\text{K}$ (Colmenero et al., 2005) (here, $V_3$ is given as $V_3/k_B$, $k_B$ being the Boltzmann constant). By inserting the threefold potential $V(\phi)$ into the stationary Schrödinger equation, we obtain the well-known Mathieu equation. By solving this equation, the energy levels corresponding to the torsional levels $E_{0i}$ and splitting levels $\Delta_i$ due to the coupling between the single-well wave functions are obtained. These energy levels $E_{0i}$ and $\Delta_i$ are shown in Figure IV.3.8 (Colmenero et al., 2005). We can expect three types of methyl group motions. The first is the quantum tunneling motion, corresponding to the energy splitting $\Delta_i$ ($\Delta_0 = 2.3\ \mu\text{eV}$ for $V_3 = 500\text{K}$) due to the single-well wave function coupling. The second is the torsional vibration (or librational motion), corresponding to the energy level $E_{0i}$ ($E_{01} = 14.3\ \text{meV}$ for $V_3 = 500\text{K}$). The third is the classical hopping motion over activation energy $E_A$. According to the scheme in Figure IV.3.8, the methyl group dynamics at very low temperature (<20K) is dominated by the quantum tunneling motion, whereas classical hopping motions over rotational barrier control the nonvibrational dynamics at high temperatures (>80K). Thus, methyl rotational motion is the simplest example to see the quantum tunneling to the classical hopping.

In Figure IV.3.9, inelastic neutron scattering spectra of crystalline sodium acetate trihydrate are shown at four temperatures from 2 to 50K (Moreno et al., 2002). Very sharp quantum tunneling excitation peaks are observed

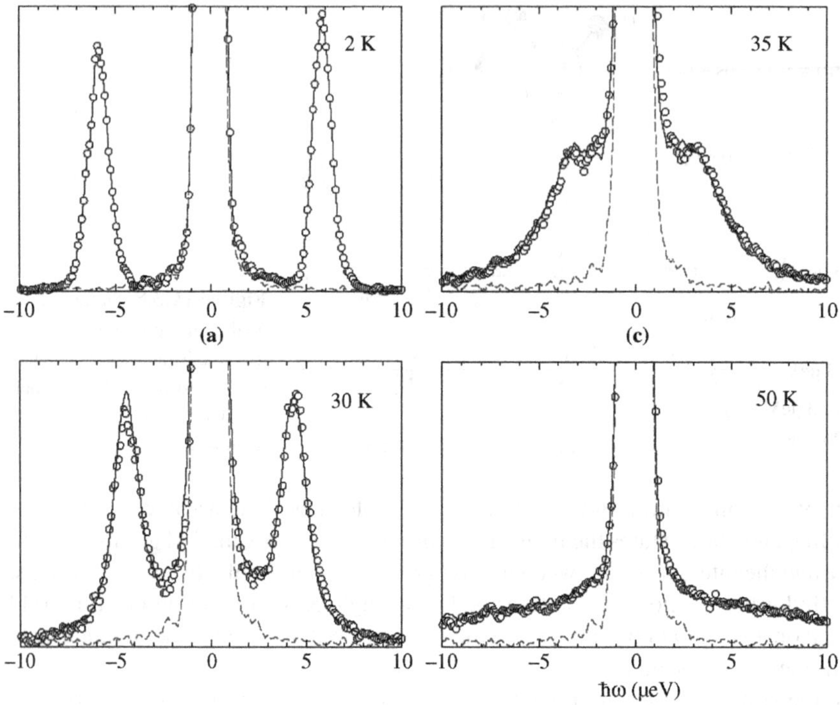


**Figure IV.3.9** Experimental spectra (circles) for methyl group dynamics in crystalline sodium acetate trihydrate. Solid lines are the fits to the model for crystalline systems. Dashed lines correspond to the experimental resolution, $Q = 1.8\,\text{Å}^{-1}$. Scales (referred to the maximum):(a) 8%, (b)–(d) 5% (Colmenero et al., 2005).

at $\Delta = \pm 6\,\mu$eV at 2K. With increasing temperature, the peaks are gradually damped and no peaks are observed at 50K. On the other hand, a very broad quasi-elastic spectrum is observed over the experimentally recorded resolution function. This broadening is due to the classical hopping motion. This experiment clearly showed a transition from the quantum tunneling to the classical hopping with increasing temperature; in polymeric sample, such a sharp transition would not be observed.

In the high-energy region above $\sim$10 meV, the rotational excitation $E_{01}$ is observed using inelastic scattering spectrometers; sometimes it overlaps with the other intramolecular vibrational modes and the lattice phonons. This fact often complicates the identification of methyl group rotational transition, but anyway the measurements are useful for the accurate determination of the potential barrier height. An example is shown in Figure IV.3.10 for poly(methyl methacrylate) (PMMA) although the rotational excitation peak is not so sharp.

In the last decades, inelastic neutron scattering techniques have been applied frequently to investigate the methyl rotation in polymers. It is noted that polymers are usually very disordered systems, suggesting a wide distribution of potential barriers. The broad rotational excitation peak is due to the distribution of barrier heights. The

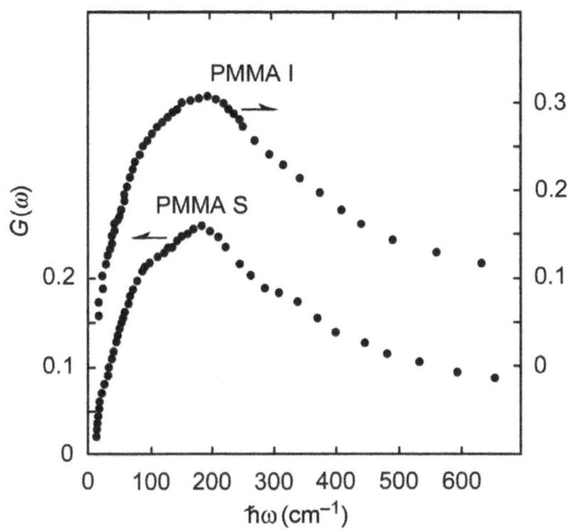


**Figure IV.3.10** Density of phonon states $G(\omega)$ summed over all scattering angles for isotactic and syndiotactic poly(methyl methacrylate) (PMMA I and PMMA S) (Gabrys et al., 1984).

quantum tunneling motion of methyl groups has been predicted for a long time, but observations were not reported. This is also due to a wide distribution of potential barriers. The rotational rate distribution model (RRDM) was introduced to explain the distribution of potential barrier height, and the same barrier distribution explained the quantum tunneling and the classical hopping. In this model, it is assumed that the only effect of the structure disorder on methyl group dynamics is to introduce a distribution of rotational barriers $g(V_3)$. The incoherent dynamic scattering law observed for methyl group dynamics is given by

$$S_{\text{inc}}^{MG}(Q,\omega) = \int_0^{\infty} g(V_3) S_{\text{inc}}^{MG}(Q,\omega,V_3)\mathrm{d}V_3. \qquad \text{(IV.3.26)}$$

Here, $S_{\text{inc}}^{MG}(Q,\omega,V_3)$ is a single-crystal spectrum. Relating the barrier height $V_3$ to other parameters $\Delta_i$, $E_{01}$, and $E_A$, describing the $S_{\text{inc}}^{MG}(Q,\omega,V_3)$ through the Schrödinger equation, and assuming a Gaussian distribution of $V_3$, the quantum tunneling spectra, the rotational vibration spectra, and the classical hopping spectra were described using the same distribution of potential barriers (Colmenero et al., 1998; Moreno et al., 2001; Mukhopadhyay et al., 1998) (see Figure IV.3.11).

### IV.3.3.4 Dynamics of Glass-Forming Polymers

In the past two decades, dynamics of glass-forming materials, including polymers, organic and inorganic compounds, and metallic glasses, were extensively studied using many kinds of experimental methods (Affouard et al., 2006). Results coming from inelastic and quasi-elastic neutron scattering techniques elucidated the characteristics of glass-forming polymers. These features are common to all glass-forming materials. In this section, we review some important experiments on

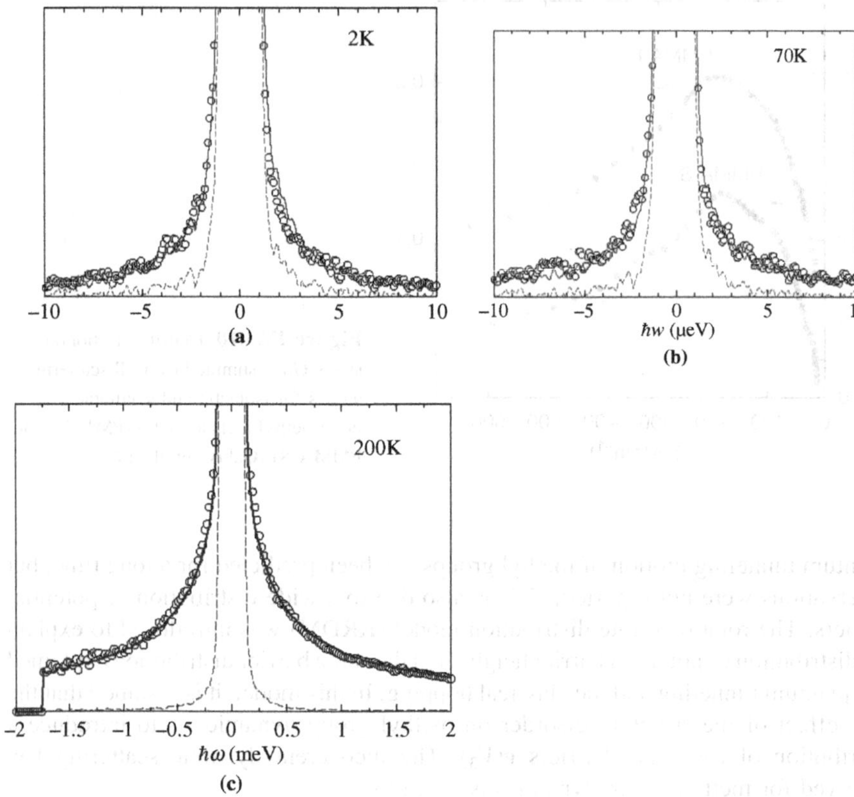


**Figure IV.3.11** Inelastic and quasi-elastic spectra for poly(vinyl acetate) (PVAc) at 2, 70, and 230K. Circles are experimental data. Solid lines correspond to the RRDM description. Dashed lines are the resolution function. (a, b) Observed by a backscattering spectrometer (IN16 at ILL) at $Q = 1.8\,\text{Å}^{-1}$; (c) observed by a time-of-flight spectrometer (IN6 at ILL) at $Q = 1.5\,\text{Å}^{-1}$ and the scale is a 3% of the maximum. Note that the energy scale in (a) and (b) is much lower than that in (c) (Colmenero et al., 2005).

dynamics of glass-forming polymers performed with inelastic and quasi-elastic neutron scattering techniques.

#### *IV.3.3.4.1 Dynamics in Glassy States: Boson Peak*

It is well know that glassy materials show anomalous thermal properties at low temperatures. A typical manifestation of the anomalies is heat capacity. As pointed out by Zeller and Pohl (1971), heat capacities of amorphous materials show an excess value compared to those of crystalline materials in two low temperature regions: one is below 1K where heat capacity is proportional to temperature $T$ and the other is an excess heat capacity at around 10–20K. The former has been interpreted in terms of a tunneling motion in an asymmetric double-well potential proposed independently by Phillips (1972) and Anderson et al. (1972). However, an excess heat capacity at around 10–20K is less understood. In the past two decades, extensive studies of this phenomenon have been performed by inelastic neutron scattering and Raman scattering techniques (Kanaya and Kaji, 2001), revealing that an excess excitation peak exists at 2–3 meV in spectra of all amorphous materials well below the glass transition temperature $T_g$. This corresponds to the excess heat capacity at around 10–20K and this peak is called the "boson peak." In the following, we present neutron scattering investigations of the boson peak, with a focus on amorphous polymers.

Since thermal anomalies at low temperatures at 10–20K are observed in all reported amorphous materials, this suggests that the boson peak should be a generic feature of amorphous materials. To confirm this hypothesis, inelastic neutron scattering investigations were carried out on various amorphous polymers as well as inorganic glasses (Inoue et al., 1991). Observed dynamic scattering laws $S(Q, \omega)$ are shown for 10 organic polymers and 3 inorganic glasses in Figure IV.3.12.

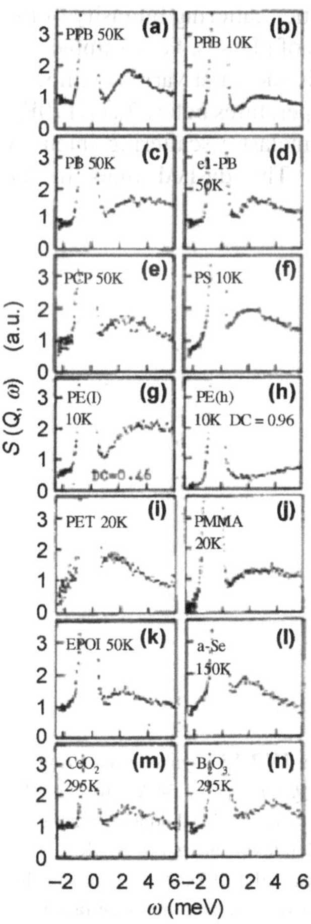


**Figure IV.3.12** Dynamic scattering laws $S(Q, \omega)$ for various amorphous polymers. (a) Polyisobutylene (PIB) at 50K; (b) PIB at 10K; (c) *cis*-1,4-polybutadiene at 50K; (d) cross-linked PB at 50K; (e) *trans*-1,4-polychloroprene (PCP) at 50K; (f) atactic polystyrene (PS) at 10K; (g) semicrystalline polyethylene (s-PE) with degree of crystallinity of 0.46 at 10K; (h) highly crystalline polyethylene with degree of crystallinity of 0.96 at 10K; (i) poly(ethylene terephthalate) (PET) at 18K; (j) atactic poly (methyl methacrylate) at 18K; (k) epoxy resin (EXPO) at 50K; (l) amorphous selenium (a-Se) at 150K; (m) germanium ($GeO_2$) glass at 295K; (n) boric oxide ($B_2O_3$) glass at 295K. Spectra of (a)–(k) are obtained by summing up the seven spectra at scattering angles of 8°, 24°, 40°, 56°, 72°, 88°, and 104°. Spectra of (l)–(n) are at $Q = 2.07\ \text{Å}^{-1}$ (Inoue et al., 1991).

All the samples except highly crystalline polyethylene (h-PE) show a broad excitation peak in the energy range of 1.5–4.0 meV. These data present compelling evidence that the boson peak is a universal property of amorphous materials and is absent in the crystalline phase in polyethylene. The density of phonon states $G(\omega)$ for the crystalline phase of polyethylene was proportional to $\omega^2$ below ~10 meV, meaning that the excitations in the crystalline phase can be described by the Debye theory. For the amorphous phase, $G(\omega)/\omega^2$ shows a peak at about 2.5 meV and the value of $G(\omega)$ is 7–8 times larger than that of the crystalline phase. The heat capacity $C(T)$ was calculated from the density of states $G(\omega)$ for the amorphous and crystalline phases in a temperature range of 2–15K, which agreed with the observed law $C(T)/T^3$, confirming that the boson peak at around 2–3 meV is the origin of the excess heat capacity of the amorphous phase.

#### IV.3.3.4.2 Dynamics of Glass-Forming Polymers near Glass Transition Temperature $T_g$

**Fast Process** Temperature dependence of the inelastic scattering intensity of the boson peak has been thoroughly studied for many kinds of glasses. For example, the dynamic scattering law of *cis*-1,4-polybutadiene (PB) is shown at various temperatures in Figure IV.3.13 (Kanaya et al., 1993). At low temperatures below $T_g$ (=170K), the broad boson peak was observed at 2–3 meV; the inelastic scattering intensity increased according to the Bose population factor. The dashed lines in the

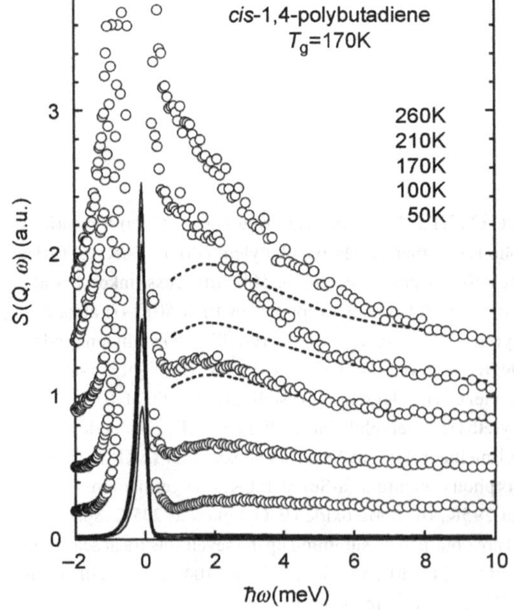


**Figure IV.3.13** Dynamic scattering law $S(Q, \omega)$ at $Q = 1.54\,\text{Å}^{-1}$ for *cis*-1,4-polybutadiene below and above the glass transition temperature $T_g = 170\text{K}$ (Kanaya et al., 1993). Dashed lines are the values expected from the Bose factor.

Figure IV.3.13 represent the expected values calculated from the Bose factor. On the other hand, the excess scattering intensity was observed at around $T_g$ over the expected value. This excess scattering is attributed to the so-called fast process. The characteristic time of the fast process is $\sim 10^{-12}$ s and is independent of temperature, suggesting that the fast process is a kind of case motion. Onset temperature of the fast process is close to the Vogel–Fulcher temperature, which is ~50K below $T_g$, implying that the fast process is a precursor of the glass transition.

**α-Process** In the low-frequency region below the fast process, the so-called α-process is observed, which governs the viscosity and is directly related to the glass transition. The relaxation time drastically increases more than several orders of magnitude with decreasing temperature toward $T_g$. Hence, it is necessary to employ many kinds of experimental techniques to cover the very wide frequency region. Neutron scattering techniques can cover a frequency range faster than $\sim 10^6$ Hz, meaning that we observe rather fast motions in glass-forming polymers.

Richter and coworkers (Frick et al., 1990) have measured normalized intermediate scattering functions $I(Q, t)/I(Q, 0)$ from the α-process of PB using a NSE technique in a temperature range between 200 and 280K above the glass transition temperature ($T_g = 181$K). These measurements were performed at the first maximum position $Q_m$ ($=1.48\,\text{Å}^{-1}$) of the structure factor $S(Q)$. They proposed that the intermediate scattering function $I(Q, t)$ is not described by a single exponential function but by a stretched exponential function or KWW function (Figure IV.3.14),

$$I(Q,t) = \exp[-(t/\tau_\alpha)^\beta] \quad (0 < \beta < 1). \tag{IV.3.27}$$

This anomalous relaxation behavior is often observed in complex systems including glass-forming materials and interpreted in terms of a wide distribution of the relaxation times. Using the distribution function $g(\ln \tau_\alpha)$, the observed stretched exponential function can be described by

$$I(Q,t) = \int_{-\infty}^{\infty} g(\ln \tau_\alpha)\exp(-t/\tau_\alpha)\mathrm{d}(\ln \tau_\alpha). \tag{IV.3.28}$$

It is believed that the wide distribution of the relaxation time is caused by the dynamic heterogeneity due to the cooperative motion (Ediger, 2000). It was also found (Frick et al., 1990) that the intermediate scattering functions $I(Q, t)$ can be scaled to a master curve using shift factors $a_T$ determined from the temperature dependence of the viscosity as shown in Figure IV.3.14. This result suggests that the neutron scattering can see the same α-process as that observed in macroscopic viscosity measurements and the relaxation mechanism does not change in the temperature range examined.

**Johari–Goldstein Process** Richter et al. also carried out NSE measurements at the minimum position $Q_{min}$ of $S(Q)$ on the same polymer (PB) as that described previously (Richter et al., 1992b). The intermediate scattering functions were

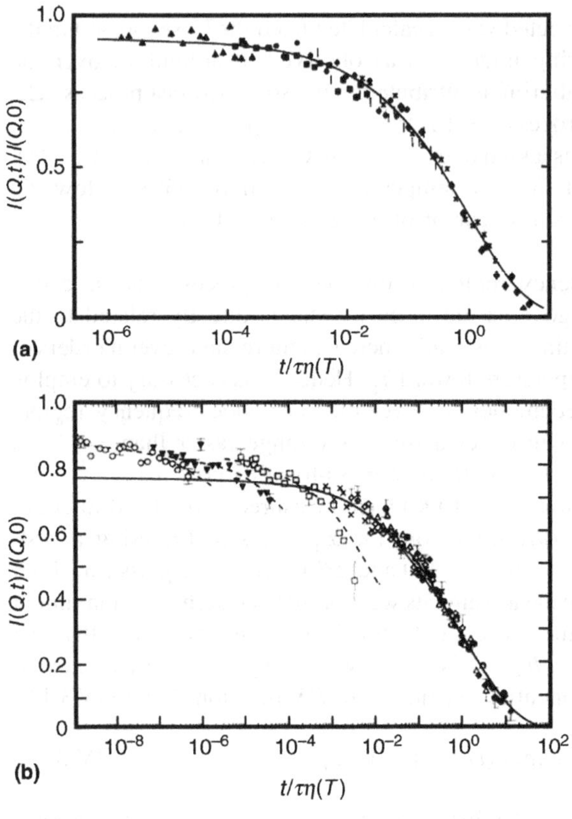


**Figure IV.3.14** Scaling representation of normalized intermediate scattering functions $I(Q, t)/I(Q, 0)$ of polybutadiene at various temperatures above $T_g$. (a) Near the maximum of $S(Q)$ at $Q = 1.48\ \text{Å}^{-1}$ (Richter et al., 1988) and (b) near the minimum of $S(Q)$ at $Q = 1.88\ \text{Å}^{-1}$ (Richter, 1992b).

described by a stretched exponential function as well, but could not be scaled to a master curve using a shift factor $a_T$. The relaxation times extracted from the observed stretched exponential functions deviates from the relaxation time of the α-process and the temperature dependence of $\tau_{JG}$ is well described by the Arrhenius formula. It was confirmed that the process observed in PB at the minimum position $Q_{min}$ in $S(Q)$ is the JG process. The fact that the JG process is observed at $Q_{min}$ suggests that the process is not a cooperative motion but an isolated one.

It is considered that the fast process, the Johari–Goldstein (JG) process, and the α-process are commonly observed in glass-forming materials including polymers and small molecules. In fact, the relaxation time map of ortho-terphenyl (OTP) (Roessler et al., 1994), which is a typical glass-forming material, is very similar to that of polybutadiene (Kanaya, 1999a). Therefore, these three processes should not be considered as special features of polymers but as common features of glass-forming materials. On the other hand, we know from traditional polymer science that the glass transition is related to some typical characteristics of polymers such as chain flexibility. In what follows, we focus on dynamical properties related to characteristic features of polymers.

## IV.3.3.5 Dynamics of Polymers in Melts

As described above, dynamics of glass-forming polymers near $T_g$ are dominated by cooperative motions of molecules or segments and their nature is in short wavelength, typically 0.5–3 nm. On the other hand, the length scale of polymer motions in melts or solutions encompasses a wide range, that is, from microscopic (~0.5 nm) to mesoscopic (several tens of nanometers or more), and the linear connectivity in a polymer chain plays an important role in the dynamics at various length scales.

It is well known that long linear polymer chains show a plateau in the time dependence of the dynamic modulus (Ferry, 1980). In the plateau region, stress is proportional to strain (Hook's law), meaning that polymer melts behave elastically although a melt is a liquid. This behavior can be understood in terms of an idea that polymer chains are entangled in the melt and the entanglement points act as temporal cross-linking points, resulting in rubber elasticity. On the basis of this assumption, it is possible to estimate the mean distance between the temporal entanglement points. It is typically several tens to hundreds Å and its value is between two characteristic length scales of polymers: the segment length (1–3 nm) and the end-to-end distance $R_e$ (~100 nm) of a polymer coil. On the scale larger than the entanglement distance, it is impossible to neglect the restriction due to the entanglements. The tube model (Doi and Edwards, 1986) was successful in describing this motion. On the other hand, on the length scale shorter than the entanglement distance and larger than the segment length, it is expected that the entropy-driven dynamics is dominant. This can be described by a coarse-grained model, that is, the Rouse model or the bead-spring model (Rouse, 1953). On the length scale shorter than the segment length, however, coarse graining of polymer chain does not work, and hence realistic models are necessary to describe local motions that are related to conformational transition between different rotational isomeric states. These three motions are typical for polymer melts. In what follows, we describe some quasi-elastic neutron scattering experiments on three types of polymer motions in melts: the local conformational transition, the Rouse motion, and the reptation motion.

### *IV.3.3.5.1 Local Conformational Transitions in Melt*

In a field of polymer science, extensive theoretical and experimental studies were carried out on local dynamics of polymers in solution and bulk to elucidate the mechanism of conformational transitions (Ediger, 1991). Initially, it was believed that the most reasonable mechanism for the conformational transitions was a crankshaft-like motion such as the Schatzki crankshaft (Schatzki, 1962) or three-bond motions (Monnerie and Geny, 1969) in which two bonds in a main chain rotate simultaneously. However, computer simulations (Ediger, 1991) revealed that single bond rotations were possible in a polymer chain for both conformational transitions and cooperative rotations. The latter are mainly manifested in the counterrotations of two second neighboring bonds separated by a *trans* bond. The key to understand the conformational transitions is that distortions in a polymer chain due to the strain brought about by the single bond rotation can be relaxed through deformations of degrees of freedom in the neighborhood of conformation-transforming bond (localization of conformational transition).

Focusing attention on conformational transitions of polymer chains, quasi-elastic neutron scattering studies (Kanaya et al., 1991; Kanaya et al., 1999a) were done on some polymers in an energy range from 0.016 to 0.5 meV above $T_g$. The observed spectrum of polybutadiene at 260K is shown in Figure IV.3.15a. The spectrum consists of at least two quasi-elastic components, broad and narrow; the broad component is the fast process having an energy of 1–2 meV, and the narrow component is a new slow process having an energy of ~100 μeV. The latter is termed E-process. This motion is observed only in polymers, suggesting that the E-process is characteristic to polymers and related to local conformational transitions.

In order to analyze the E-process in terms of local conformational transitions, a jump diffusion model with damped vibrations was employed (Kanaya et al., 1991, 1999a). In the model, conformational transitions are represented by a jump motion from one conformation (a rotational isomeric state) to another. In each rotational isomeric state, the polymer segment performs the damped vibrations that assist distortions of degrees of freedom in the neighborhood of the conformation-transforming bond and keep the transition coordinate localized. This model gives a very good correspondence between physical picture and that predicted by the computer simulations.

The $Q$ dependence of the HWHM of the spectra calculated from the model agrees well with the observed ones (Figure IV.3.15b), showing the localization of conformational transitions in a polymer chain. The activation energy of the relaxation time is 3–4 kcal/mol. This also agrees with the prediction of the computer simulations. It follows that a single bond rotation is possible in local conformational transitions as predicted in the MD simulations.

### *IV.3.3.5.2 Rouse Dynamics in Polymer Melt*

Some slow mode motions in polymers occur on a large length scale, larger than the segment length. Computer simulations using atomistic models are not feasible for such length scale motions. Hence, coarse graining of polymer chains is necessary to describe such slow motions. On the length scale between the segment length and the entanglement distance, polymer motions are driven by entropic forces. This motion was modeled by Rouse (1953) more than 50 years ago. In this model, a polymer chain is described by $N_R$ beads with mass $m$ connected by Hook's springs with the length $l$. A number of studies on macroscopic scale demonstrated that the Rouse model describes well many relaxation phenomena in polymer systems. NSE investigations of the relaxation phenomena on microscopic level confirmed the validity of the Rouse model for description of the microscopic data.

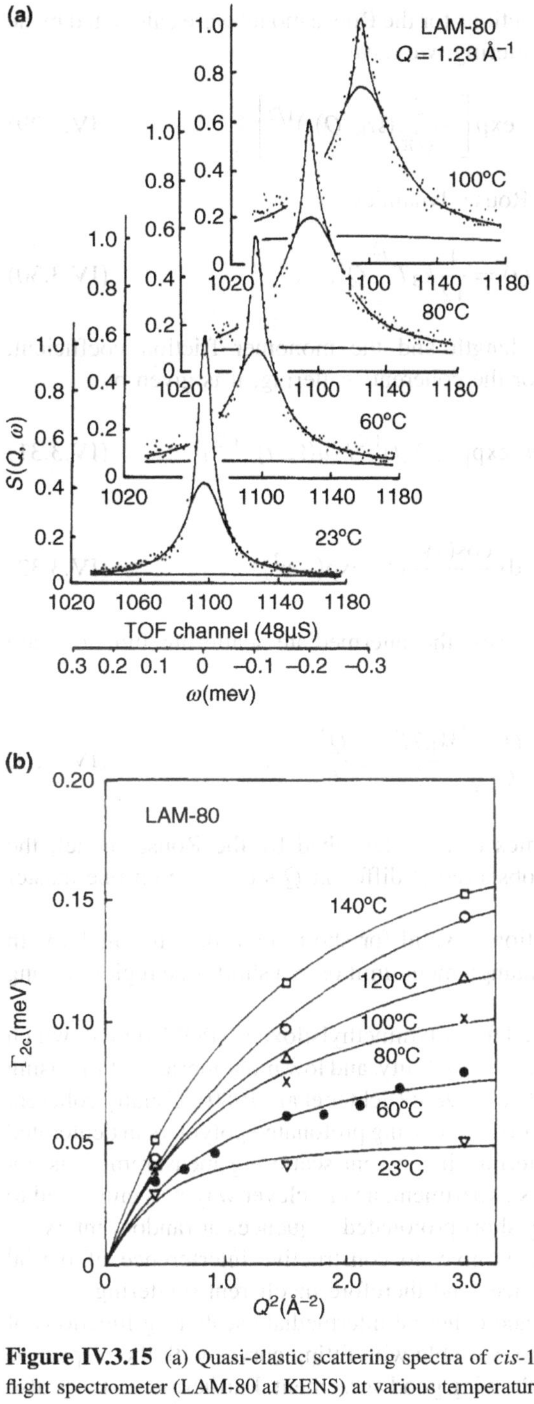


**Figure IV.3.15** (a) Quasi-elastic scattering spectra of *cis*-1,4-polybutadiene observed by a time-of-flight spectrometer (LAM-80 at KENS) at various temperatures far above $T_g$. (b) $Q^2$ dependence of the HWHM $\Gamma$ of the Lorentzian function fitted to the quasi-elastic spectra (Kanaya et al., 1991).

The intermediate scattering functions for the Rouse model were calculated by de Gennes (1967). For incoherent scattering, we have

$$\frac{I_{\text{inc}}(Q,t)}{I_{\text{inc}}(Q,0)} = \exp\left[-\frac{2}{\sqrt{\pi}}(\Omega_{\text{R}}(Q)t)^{1/2}\right] \tag{IV.3.29}$$

with the characteristic rate of the Rouse dynamics

$$\Omega_{\text{R}}(Q) = \frac{1}{12}k_{\text{B}}T\frac{l^2}{\zeta}Q^4, \tag{IV.3.30}$$

where $l$ and $\zeta$ are the segment length and the monomer friction coefficient, respectively. On the other hand, for the coherent scattering, it is given by

$$\frac{I_{\text{coh}}(Q,t)}{I_{\text{coh}}(Q,0)} = \int_0^\infty du \exp(-u)\exp[-(\Omega_{\text{R}}t)^{1/2}h(u(\Omega_{\text{R}}t)^{-1/2})], \tag{IV.3.31}$$

$$h(y) = \frac{2}{\pi}\int_0^\infty dx\frac{\cos(xy)}{x^2}(1-\exp(-x^2)). \tag{IV.3.32}$$

In both coherent and incoherent cases, the intermediate scattering functions are scaled by the Rouse variable

$$(\Omega_{\text{R}}t)^{1/2} = \frac{Q^2}{6}\sqrt{\frac{3k_{\text{B}}Tl^2t}{\zeta}} \equiv \frac{Q^2}{6}\sqrt{Wt}. \tag{IV.3.33}$$

Therefore, if polymer dynamics can be described by the Rouse model, the intermediate scattering functions observed at different $Q$'s collapse on one master curve.

The Rouse dynamics description is valid for short chain polymer melts with molecular weight lower than the entanglement limit or in a short time region of long chain polymer melts.

The Rouse motion was observed for polydimethylsiloxane (PDMS) melt, which has little entanglement constrains, high flexibility, and low monomeric friction, using a NSE spectrometer (Richter and Ewen, 1989; Richter et al., 1989). Usually coherent scattering in NSE measurements is observed using protonated polymers in deuterated solvents because of spin-flip scattering; incoherent scattering measurement is not usual in NSE measurements. In this experiment, a very clever way was employed to use a deuterated PDMS containing short protonated sequences at random intervals. The randomly distributed sequences give no constructive interference of partial waves produced at different sequences and therefore incoherent scattering.

In Figure IV.3.16, incoherent and coherent intermediate scattering functions of PDMS are plotted against the Rouse variable with fitting curves with the respective intermediate scattering functions, showing good agreement. Furthermore, it was also shown that the characteristic frequency $\Omega_{\text{R}}$ was proportional to $Q^4$ (Richter et al.,

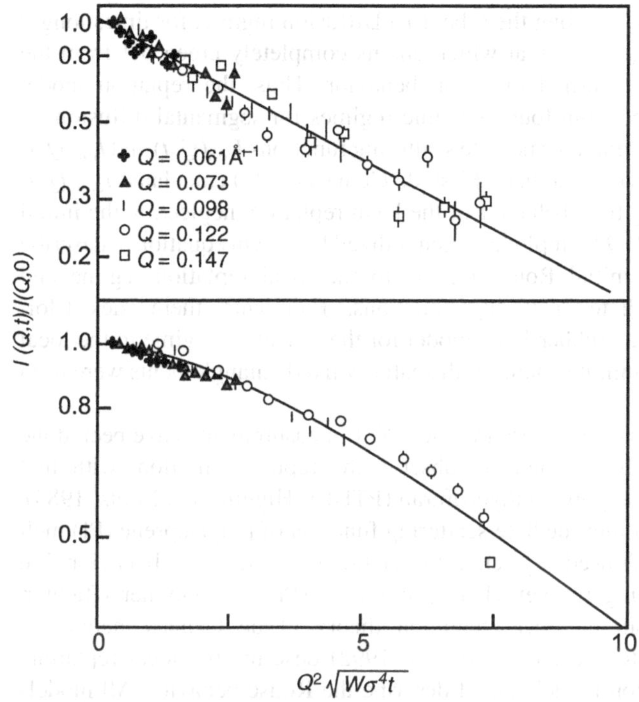


**Figure IV.3.16** Incoherent (above) and coherent (below) intermediate scattering functions measured by a NSE spectrometer for PDMS melts at 100°C. The data are scaled to the Rouse variable. Solid curves are fit curves with the respective dynamic structure factors (Richter et al., 1989).

1989) and the time dependence of the incoherent intermediate function $I_{\rm inc}(Q, t)$ was described by a stretched exponential function with exponent of 1/2, confirming that the Rouse model can well describe the dynamics of short chain polymer melts.

### *IV.3.3.5.3 Reptation in Melt*

For long chain polymer melts, the plateau modulus is described by rubber elasticity of a temporary network consisting of entanglements of mutually interpenetrating chains. In such situation, polymer chains cannot move across each other, and hence must move like a snake along a tube produced by surrounding chains. This is schematically illustrated in Figure IV.3.17a. This motion is called the reptation motion—the theoretical description was given by de Gennes (1971) and Doi and Edwards (1978).

The reptation motion is classified into three regimes according to characteristic times: (i) Rouse regime: a short time region where chains do not feel the restriction due to a tube with a diameter $d$. This regime is limited by a time $\tau_e$ at which the mean square displacement of the segment reaches $d^2$ ($t < \tau_e$). (ii) Local reptation regime: it is divided into two subregimes by the Rouse time $\tau_R$. In the time interval of $\tau_e < t < \tau_R$, the Rouse modes of polymer chains equilibrate within the tube while for $t > \tau_R$, the chain diffuses along the tube. (iii) Diffusion regime: for times longer than the terminal time $\tau_d$ ($t > \tau_d$), at which chains completely creep out from the original tube, we expect normal diffusion behavior. Thus, the reptation model predicts three (strictly speaking four) dynamic regimes for segmental diffusion.

Coherent and incoherent intermediate scattering functions $I_{coh}(Q, t)$ and $I_{inc}(Q, t)$ were calculated by several researchers. First, de Gennes (1981) calculated $I_{coh}(Q, t)$ and $I_{inc}(Q, t)$ for the long time behavior of the local reptation, neglecting the initial Rouse behavior. Ronca (1992) employed a generalized Langevin equation to describe the crossover from the initial Rouse regime to the local reptation regime and calculated the intermediate scattering functions. Following them, des Cloizeaux (1993) developed a "rubber-like" model for the transition regime to the local reptation. In this theory, infinite chains with spatially fixed entanglements were used at intermediate time.

Following predictions from these theories, NSE measurements have been done on some long chain polymer melts to observe the reptation motion, with first experiments carried out on polytetrahydrofuran (PTHF) (Higgins and Roots, 1985). The normalized coherent intermediate scattering function of polyisoprene (PI) melt (Richter, 1992a) clearly showed a systematic deviation from the Rouse behavior. The NSE results on alternating (polyethylene–propylene) (PEP) copolymer (Richter, 1992a) showed that both the local reptation theory of de Gennes theory (de Gennes, 1981) and the Ronca theory (Ronca, 1992) describe the local reptation, whereas the local reptation model cannot describe the Rouse behavior. All models presented can describe the local reptation mode in the time range shorter than 25 ns, meaning that conventional NSE measurements cannot conclude which model is the best one for the description of the reptation motion. A novel NSE spectrometer (IN15 at ILL) can cover a time range up to ~400 ns. Using this spectrometer, NSE measurements were done on polyethylene (Schleger et al., 1998) in a time range from 0.3 to 175 ns. In Figure IV.3.17b, the observed intermediate scattering functions are shown at $Q = 0.05$ and $0.077$ Å$^{-1}$. In the same figure, the theoretical curves are also plotted for comparison, concluding that the reptation model by de Gennes (1981) is the best one to describe the chain motion in the PE melt.

## IV.3.3.6 Dynamics of Polymers in Solution

### *IV.3.3.6.1 Linear Polymers in Dilute Solution*

In the Rouse model, the entropic force dominates the dynamics of polymer melts in short time region. On the other hand, polymer chains in solutions must feel hydrodynamic interactions from solvent. A segment moving relative to the surrounding solvent generates a flow field that influences the movements of all other segments. These interactions are incorporated into the Rouse model in terms of the Oseen tensor. This is the so-called Zimm model (Zimm, 1956).

The intermediate scattering functions (de Gennes, 1967) were theoretically calculated in a way similar to the Rouse model. For the incoherent scattering,

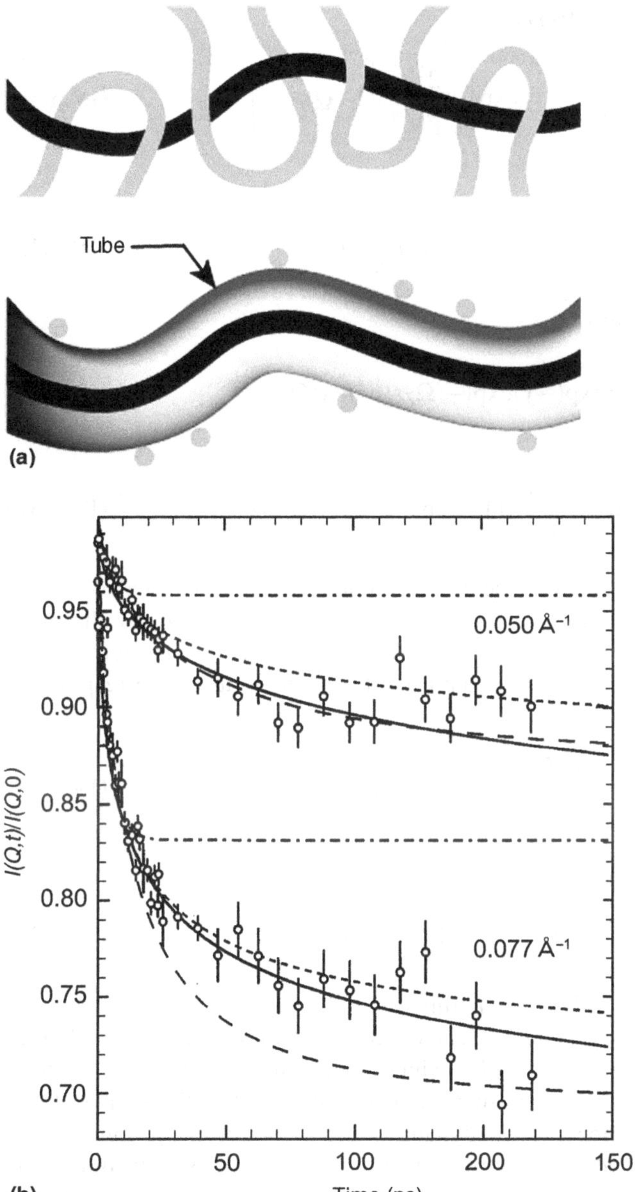


**Figure IV.3.17** (a) Schematic representation of reptation motion. (b) Normalized intermediate scattering functions $I(Q, t)/I(Q, 0)$ of polyethylene with four theoretical predictions: the predictions of reptation (solid lines), local reptation (dotted lines), the model of des Cloizeaux (1993) (dashed lines), and the Ronca model (Ronca, 1992) (dot-dashed lines).

we have (Ewen and Richter, 1997)

$$\frac{I_{\mathrm{inc}}(Q,t)}{I_{\mathrm{inc}}(Q,0)} = \exp\left[-\frac{2^{11/2}}{9\sqrt{\pi}}[\Omega_Z(Q)t]^{3/2}\right] \qquad \text{(IV.3.34)}$$

with

$$\Omega_Z(Q) = \frac{1}{6\pi} k_B T \frac{l^2}{\eta_s} Q^3 \qquad \text{(IV.3.35)}$$

and for the coherent scattering,

$$\frac{I_{\mathrm{coh}}(Q,t)}{I_{\mathrm{coh}}(Q,0)} = \int_0^\infty \mathrm{d}y \exp(-y)\exp[-(\Omega_Z t)^{2/3} h(y(\Omega_Z t)^{-2/3})], \qquad \text{(IV.3.36)}$$

$$h(u) = \frac{2}{\pi}\int_0^\infty \mathrm{d}x \frac{\cos(xu)}{x^2}(1-\exp(-2^{-1/2}-x^{3/2})). \qquad \text{(IV.3.37)}$$

The incoherent and coherent scattering functions are both scaled by the variable $(\Omega_Z t)^{2/3}$, as in the case of the Rouse model. It is noteworthy that long time behavior $((\Omega t) > 1)$ of the coherent intermediate scattering functions for the Rouse and Zimm models shows the same time dependences as the corresponding incoherent scattering:

$$\frac{I_{\mathrm{coh}}(Q,t)}{I_{\mathrm{coh}}(Q,0)} = \exp\left[-\frac{2}{\sqrt{2}}[\Omega_R(Q)t]^{1/2}\right] \qquad \text{(IV.3.38)}$$

for the Rouse model and

$$\frac{I_{\mathrm{coh}}(Q,t)}{I_{\mathrm{coh}}(Q,0)} = \exp\left[-\frac{2^{2/3}}{\pi}\Gamma(1/3)[\Omega_Z(Q)t]^{3/2}\right] \qquad \text{(IV.3.39)}$$

for the Zimm model, where $\Gamma$ is the $\Gamma$-function.

To confirm the Zimm dynamics, NSE measurements were performed on PDMS solution in *d*-bromobenzene at 357K, which meets the $\Theta$-condition (Richter et al., 1984). The observed coherent intermediate scattering functions suggest that the Zimm model can describe well the dynamics of polymer chains in solutions. In Figure IV.3.18, the intermediate scattering functions are plotted as a function of the scaling variable $[\Omega_Z(Q)t]^{2/3}$ (Ewen and Richter, 1997). The data follow the master curve. It was also found that the characteristic frequency $\Omega_Z(Q)$ obeys the theoretical $Q^3$ law when the scattering function of the Zimm model is fitted to the spectra at different $Q$ values. In contrast to the $\Theta$-condition, theoretical intermediate scattering function is not available for the good solvent condition. However, the observed intermediate scattering functions are well described by that derived for the $\Theta$-condition (Ewen and Richter, 1997; Richter et al., 1980; Nicholson et al., 1981; Ewen, 1984).

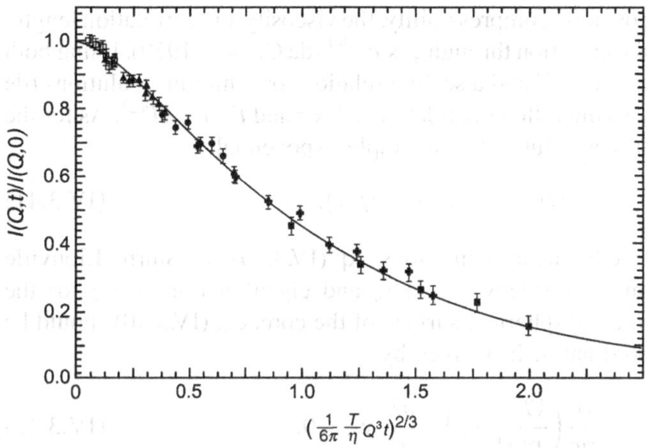


**Figure IV.3.18** Intermediate scattering function of a dilute *d*-bromobenzene solution of PDMS under the Θ-condition. The data are scaled to the Zimm variable (Richter et al., 1984).

#### *IV.3.3.6.2 Collective Dynamics of Polymer Chains in Corona of Polymer Micelle*

As shown in Sections IV.3.3.5.2 and IV.3.3.6.1, the Rouse and Zimm models described well the chain motions in melts and solutions, respectively. It should be noted that the motion described by these models are an isolated motion or a single chain motion. However, in polymer systems there are many collective motions due to various interactions. In this section, we will discuss collective chain motion in a corona of a polymer micelle as an example of collective dynamics.

When diblock copolymers are added in a selective solvent, aggregation occurs and a polymeric micelle is formed in analogy to micellization of surfactant molecules. Such polymeric micelles are isolated in a dilute region while they form a lattice in a concentrated region. In the dilute solution, the polymer chains in the corona part of the micelle belong to a family of tethered chains that are attached to microstructures by their ends (Halperin et al., 1992).

Until now, theoretical and experimental studies were mainly performed on the structure and forces of tethered chains. As for dynamics, de Gennes (1986) first formulated motion of tethered polymer chains on a flat surface, and the experimental work followed: Richter and coworkers (Farago et al. 1993, Monkenbusch et al. 1995) and Kanaya et al. (1999b) using a NSE technique. The theory and the experiments have shown that in the corona of polymer micelle, the motion is driven by the osmotic compressibility, which is the so-called breathing mode.

For the tethered chains on a flat surface, the breathing mode has been formulated by de Gennes (1986). According to him, the equation of local displacement **u** of tethered chains along the normal direction $z$ to the surface, that is, for the breathing mode, is given by

$$\frac{\partial}{\partial z}\left(E(z)\frac{\partial u}{\partial z}\right) = \frac{\eta}{\xi^2}\frac{\partial u}{\partial t}, \tag{IV.3.40}$$

where $E$, $\eta$, and $\xi$ are the osmotic compressibility, the viscosity, the correlation length, which is related to the concentration through $\xi \propto c^{-3/4}$ (de Gennes, 1979). Using both the concentration profile $c \propto z^{-4/3}$ and a scaling relation for semidilute solutions (de Gennes, 1979), one obtains the following relations: $\xi \propto z$ and $E(z) \propto kT/z^3$. As for the time decay, eq. (IV.3.40) has solutions of a simple exponential.

$$u(t) = u_n(z)\exp(-t/\tau_n). \qquad \text{(IV.3.41)}$$

In addition, due to the boundary conditions, eq. (IV.3.40) is a Sturm–Liouville boundary value problem with eigenvalues $1/\tau_n$ and eigenfunctions $u_n(z)$ for the displacement. Considering the spherical surface of the core, eq. (IV.3.40) should be changed in spherical coordinates. It is given by

$$\frac{\partial}{\partial r}\left(\frac{kT}{r^5}\frac{\partial}{\partial r}y_n\right) - \frac{\eta\lambda_n}{r^4}y_n = 0, \qquad \text{(IV.3.42)}$$

where $y_n$ is the radial part of the displacement $u_n$ ($y_n = u_{r,n}r^2$). Furthermore, according to Farago et al. (1993), the influence of the plane parallel modulation was taken into account only for the eigenvalue. Finally, the intermediate scattering function is obtained as

$$I(Q,t)/I(Q,0) \propto \sum_{n,l}\left[\langle\alpha_{n,l}^2\rangle 4\pi(2l+1)\left|\int_0^{\infty} j_l(Qr)[2u_{r,n}cr + (u_{r,n}c)'r^2]\mathrm{d}r\right|^2 \exp(-t/\lambda_n)\right] \qquad \text{(IV.3.43)}$$

with $\langle\alpha_{n,l}^2\rangle = \gamma kT/\eta\lambda_{n,l}$, where $g$ is a numerical constant and $j_l(Qr)$ is a spherical Bessel function of order $l$.

NSE measurements were performed on a dilute solution of a deuterated $n$-decane solution of block copolymers consisting of deuterated polystyrene (PSD) and protonated polyisoprene (PI) (Farago et al., 1993; Kanaya et al., 1999b). The solvent (deuterated $n$-decane) is poor for PSD and good for PI, and hence the core and the corona of the polymer micelle consist of PSD and PI, respectively. The scattering length density of deuterated $n$-decane is almost identical to that of the core (PSD) so that neutrons see only PI chains in the corona. The observed coherent intermediate scattering function of the 2% solution are shown in Figure IV.3.19a at various $Q$ values (Kanaya, 1999b). The solid curves in the figure are the intermediate scattering functions predicted by the theory. The agreements are rather good except for the data at $Q = 0.05$ Å$^{-1}$. The intermediate functions above about 0.07 Å$^{-1}$ show a multi-component decay—the fast decay in a shorter time region than about 3 ns and the slow decay in longer time region than about 5 ns. This behavior is very different from that of the Zimm mode, confirming that, at least, at high $Q$ values the polymer motion in the corona is dominated by the osmotic compressibility.

NSE measurements on a similar polymer micelle consist of deuterated polystyrene–protonated butadiene diblock copolymer in deuterated $n$-decane (Kanaya et al., 2005a). Although the polybutadiene block had similar molecular weight with that of protonated isoprene (PI), the breathing mode driven by the osmotic

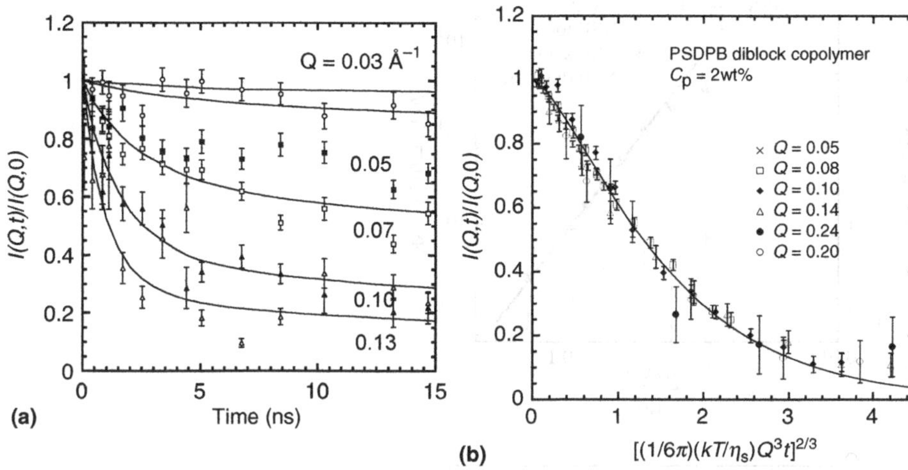


**Figure IV.3.19** Intermediate scattering function of two polymer micelles in deuterated *n*-decane. (a) deuterated polystyrene–protonated isoprene diblock copolymer micelle (Kanaya, 1999a, 1999b). (b) deuterated polystyrene–protonated butadiene diblock copolymer micelle. The data are scaled to the Zimm variable in (b) (Kanaya et al., 2005a).

compressibility completely disappeared in this experiment, but the Zimm mode, which is a single chain mode, was observed (Figure IV.3.19b). This suggests that the interactions among PB chains in *n*-decane are weaker than those among PI chains, resulting in the suppression of the collective dynamics of PB chains. In other words, it predicts that the second virial coefficient $A_2$ of PB is smaller than that of PI.

#### *IV.3.3.6.3 Structure and Dynamics of Poly(vinyl alcohol) Gels*

A polymer gel is a three-dimensional network swollen by a solvent and shows many interesting properties. Poly(vinyl alcohol) (PVA) is one of the most interesting gel-forming polymers due to its water solubility and biocompatibility. This polymer forms physically cross-linked gels in many solvents and also chemically cross-linked gels through a reaction with many kinds of cross-linkers. The resultant gels are very different in both appearance and properties, showing a wide variety of dynamics. We will review dynamics studies of three types of PVA gels using a NSE technique.

**PVA Gel Formed in a Mixture of Dimethyl Sulfoxide and Water with Volume Ratio 60/40** Small-angle and wide-angle neutron scattering studies on the PVA gels in a mixture of dimethyl sulfoxide (DMSO) and water (60/40) (Kanaya et al., 1994, 1995) showed that the cross-linking points of the network are crystallites. The distance between the nearest neighboring crystallites is approximately 200 Å and the size is approximately 70 Å in radius. In the $Q$ range between 0.02 and 0.1 $Å^{-1}$, the so-called Porod's law ($I(Q) \sim Q^{-4}$) was found to fit the results of the SANS measurements (see Figure IV.3.20a); this was assigned to smooth surface of the

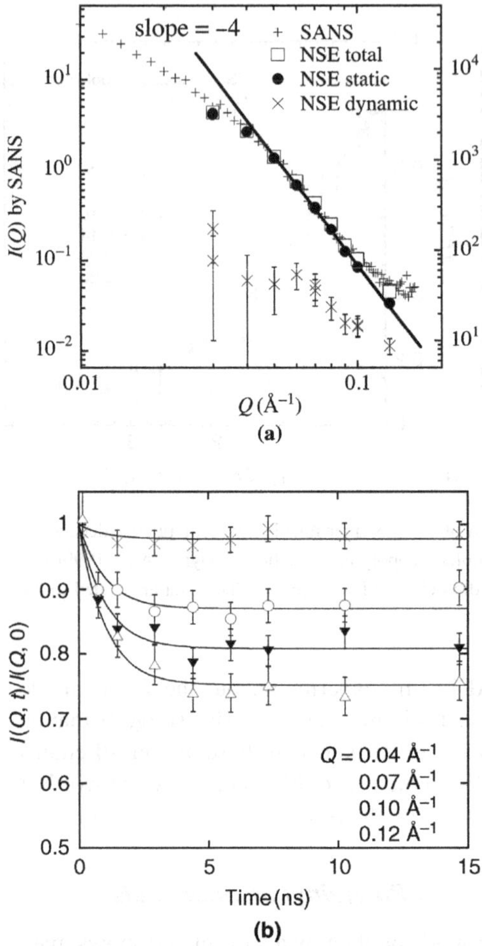


**Figure IV.3.20** (a) Small-angle neutron scattering intensity of PVA gel in a mixture of DMSO and water (60/40). SANS intensity is separated into static and dynamic components based on the NSE data. (b) Normalized intermediate scattering function $I(Q, t)/I(Q, 0)$ of PVA gel in a mixture of DMSO and water (60/40). Solid curves are the results of fit (see text) (Kanaya et al., 2005b).

crystallites. In NSE measurements in this $Q$ range, the dynamics of the cross-linking points or the crystallites in the gel were mainly observed.

In the SANS measurements, we observe the concentration fluctuations in the PVA gel. If the fluctuations are time dependent, $I(Q, t)$ decays with time. On the other hand, if they are frozen, it is independent of time, remaining constant. The former and the latter are called static (or frozen) and dynamic fluctuations, respectively. In the SANS measurements, we cannot distinguish the dynamic and static fluctuations.

To see the time-dependent and/or time-independent fluctuations, the normalized intermediate scattering function $I(Q, t)/I(Q, 0)$ was measured using the NSE. The observed $I(Q, t)/I(Q, 0)$ is shown in Figure IV.3.20b at $Q = 0.04$, 0.07, 0.10, and

0.12 Å$^{-1}$. In a short time region below about 3 ns, $I(Q, t)/I(Q, 0)$ decays rapidly, while it does not decay and remains constant in the longer time range. The observed intermediate scattering function can be phenomenologically described by $I(Q, t)/I(Q, 0) = f_n(Q) + [1 - f_n(Q)]F(Q, t)$, where $f_n(Q)$ is a fraction of the nondecaying component and $F(Q, t)$ is a generalized decay function. Assuming that the decay function could be expressed by a single exponential function, the equation was fitted to the observed intermediate functions to evaluate the relaxation time $\tau_f$ and the fraction of the nondecaying component $f_n(Q)$. The fractions of the nondecaying component $f_n(Q)$ and the rest $(1 - f_n(Q))$ correspond to those of the static and dynamic fluctuations, respectively. The observed total SANS intensity $I(Q)$ is then divided into the static $f_n(Q)I(Q)$ and dynamic fluctuations $[1 - f_n(Q)]I(Q)$. It is evident that the total SANS intensity of the PVA gel is mainly governed by the static fluctuations, indicating that the molecular motions in the PVA gel are almost frozen, at least in the present $Q$ range.

**PVA Gels in an Aqueous Borax Solution** Dynamics of PVA with molecular weight $M_w = 26{,}400$ in aqueous borax solutions were also studied by NSE. The viscosity of the aqueous solution with the PVA concentration $C_p = 2.4$ wt% is about $2 \times 10^{-3}$ Pa s, and this abruptly begins to increase with the polymer concentration at around $C_p = 3.0$ wt%. The viscosity is about $4 \times 10^2$ Pa s at $C_p = 4.8$ wt%, being more than five orders of magnitude larger than that of $C_p = 2.4$ wt% (Takada et al., 1998). The aqueous borax solutions of PVA with the concentration below and above 3.0 wt% are termed as PVA–borax sol and gel, respectively. In the aqueous borax solutions of PVA, the gelation process occurs owing to cross-linking of PVA chains through hydrogen bonds between hydroxyl groups on PVA chains and a borate anion.

Normalized intermediate scattering functions $I(Q, t)/I(Q, 0)$ were measured by a NSE spectrometer for the PVA–borax sol with $C_p = 2.4$ wt% and the PVA–borax gel with $C_p = 4.8$ wt%. In contrast to $I(Q, t)/I(Q, 0)$ of the PVA gel formed in a mixture of DMSO and water, they appear to decay to zero at infinite time, even in the gel. The PVA–borax sol is a solution and the Zimm model was applied to analyze the data. The observed intermediate scattering functions $I(Q, t)/I(Q, t = 0)$ were scaled by the Zimm time $(Q^3t)^{2/3}$ in Figure IV.3.21a for the sol and the gel. The data points fall on the theoretically calculated master curve, suggesting that the Zimm scaling works well for the sol, and even for the gel. The ratio of the Zimm decay rate $\Gamma_z$ of the gel to the sol is only 2.2 although the viscosity of the gel is more than five orders of magnitude larger than that of the sol. Surprisingly, the PVA–borax gel appears as the sol. Why can the Zimm model describe the dynamics of the gel whose viscosity is more than five orders of magnitude larger than that of the sol? To consider this problem, the first cumulant or the initial decay rate of $\ln[I(Q, t)/I(Q, 0)]$ $(= -d\{\ln[I(Q, t)/I(Q, 0)]\}/dt)$ is plotted in Figure IV.3.21b. In the $Q$ range above about a certain critical value $Q_c$, the initial decay rate $\Gamma_i$ is proportional to $Q^3$ for both the sol and the gel, confirming that the Zimm model is appropriate to describe the dynamics of both the sol and the gel, at least above $Q_c$. On the other hand, the initial decay rate $\Gamma_i$ is proportional to $Q^2$ below the critical value $Q_c$, showing the so-called gel mode (Tanaka et al., 1973). The gel mode is a propagating mode on elastic gel network,

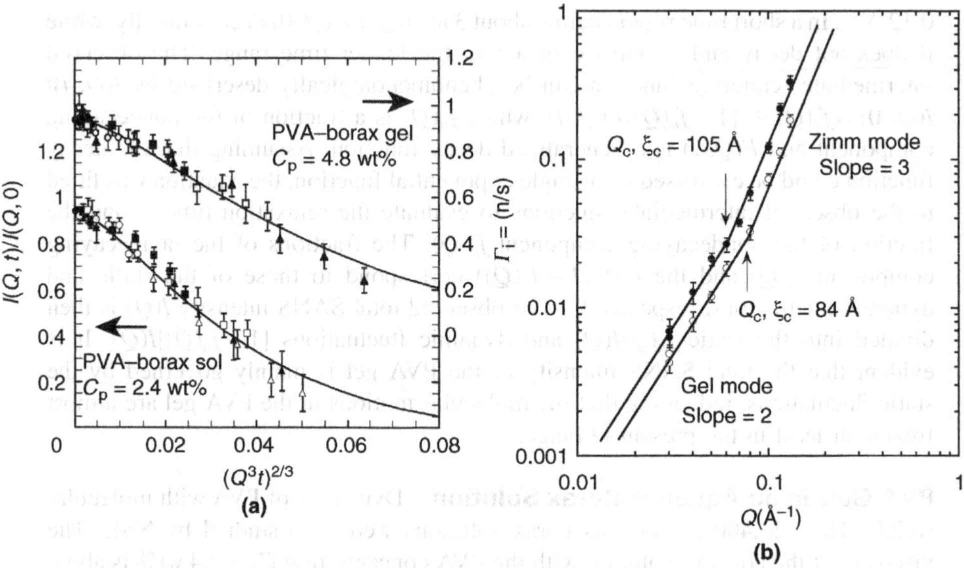


**Figure IV.3.21** (a) Zimm scaling of normalized intermediate scattering functions $I(Q, t)/I(Q, 0)$ of PVA–borax sol with $C_p = 2.4$ wt% and gel with $C_p = 4.8$ wt%. Solid curves are the Zimm master curve. (b) $Q$ dependence of first cumulant $\Gamma_i$ (initial decay of $\ln[I(Q, t)/I(Q, 0)]$) for PVA–borax sol with $C_p = 2.4$ wt% ($\lambda$) and gel with $C_p = 4.8$ wt% ($\mu$) (Kanaya et al., 2005b).

permitting a range larger than the mesh size of the network ($\xi_c$), which can be approximately evaluated from the critical $Q_c$ at the crossover from the Zimm to the gel mode ($\xi_c \sim 2\pi/Q_c$). Analyzing the critical $Q_c$, it was concluded that the gel mode originates from a temporal network owing to chain entanglements. Thus, for the gel mode to occur cross-linking through borax is not necessary, but overlap of polymer chains or entanglements is essential.

**Chemically Cross-Linking PVA Gels** PVA gels chemically cross-linked by glutaraldehyde (GA) were studied by small-angle neutron scattering (SANS) and NSE (Kanaya et al., 2006). For the PVA gels with PVA concentration of $C_p = 8$ g/dL, the SANS intensity is independent of the GA concentration $C_g$ (see Figure IV.3.22a). This result suggests that the cross-linking between the polymer chains does not change the structure in the semidilute solution of $C_p = 8$ g/dL because chains are overlapped before cross-linking. This means that the structure of the semidilute polymer solution is not different from that of the gel as found in the static SANS measurements. However, it is obvious that the dynamic properties of the gel must be different from those of the solution due to the cross-linking.

Normalized intermediate scattering functions $I(Q, t)/I(Q, 0)$ of the gels were measured using a NSE spectrometer as a function of the concentrations of PVA $C_p$ and glutaraldehyde $C_g$ (Kanaya et al., 2006). It was found that the Zimm scaling works well for the observed $I(Q, t)/I(Q, 0)$ in a short time region below about $(Q^3t)^{2/3} = 0.04$ ns$^{2/3}$/Å$^2$, while it deviates from the Zimm master curve and slows down in the long time region. In the short time region, polymer chains do not feel the

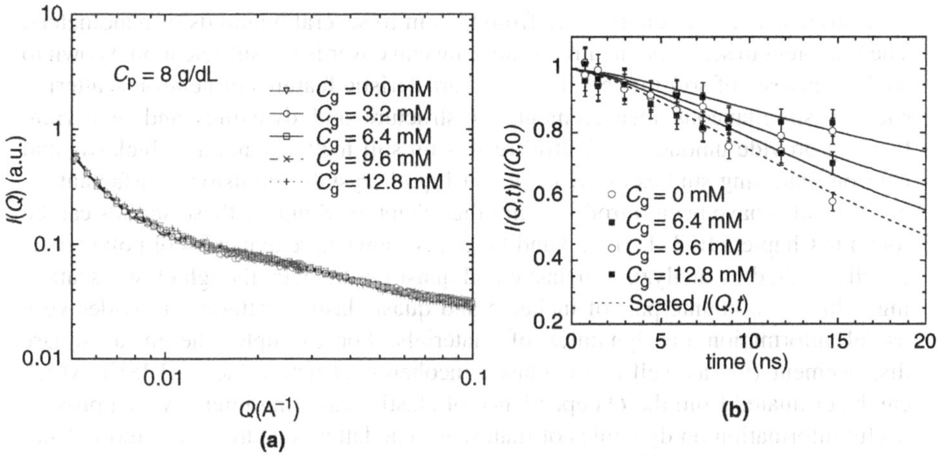


**Figure IV.3.22** (a) Small-angle scattering (SANS) intensity of chemically cross-linked PVA gels for the PVA concentration of $C_p = 8$ g/dL. (b) Normalized intermediate scattering function of chemically cross-linked PVA gel as a function of cross-linking density. Dashed line is Zimm master curve (Kanaya et al., 2006).

restriction of the cross-linking, and hence they behave as Zimm chains. On the other hand, they feel the restriction of the cross-linking in the long time region and the motions becomes slower in the gel than in the solution. To confirm that the slowing down is due to the cross-linking, the NSE measurements were performed on the chemically cross-linked gels ($C_p = 8$ g/dL) as a function of the cross-linking density (Figure IV.3.22b). The intermediate scattering function $I(Q, t)/I(Q, 0)$ decays according to the Zimm master curve (dashed line in Figure IV.3.22b) in the solution ($C_g = 0$ g/dL), while it does decay slower with increasing the cross-linking density or the concentration of glutaraldehyde $C_g$. It is evident that the polymer motions in the gels are restricted by the cross-linking although the static structure elucidated by the SANS measurement is independent of the cross-linking density.

As shown above, the NSE studies on the three types PVA gels demonstrated quite different dynamics and revealed the crucial role of cross-linking points in the gel dynamics.

## IV.3.4 CONCLUSION

In this chapter, we described some experiments of inelastic and quasi-elastic neutron scattering studies on polymer systems: phonon dispersion curve of polytetrafluoroethylene crystal, density of phonon states of polyethylene crystal, methyl group quantum tunneling, libration and classical hopping motions, glassy dynamics including the boson peak, fast process, α-process, and Johari–Goldstein process, melt dynamics of chain conformation transition, and Rouse and reptation motions. For polymer dynamics in solutions, we also described the Zimm motion, the breathing mode in corona of polymer micelles, and gel dynamics. The state-of-art neutron scattering technique can cover a time range from 0.01 ps to several hundreds of nanoseconds and a length scale from 0.1 nm to several hundreds of nanometers. The time–length scale that neutron scattering can cover is not sufficient on its own to study dynamics of soft matters, but the characteristic features of neutron scattering such as simultaneous measurements of structure and dynamics and deuterium labeling provide unique and fruitful results on soft matter dynamics. Inelastic and neutron scattering studies on dynamics of liquid crystals, emulsions, surfactant, or biomolecules have been introduced in other chapters. Some of these studies can be found in Chapters II. 3. 1–II. 3. 3 and IV. 5. As shown here dynamics of polymers is usually studied by analyzing inelastic and quasi-elastic data, though elastic scattering, which is a counterpart of inelastic and quasi-elastic scattering, provides very useful information on dynamics of materials. For example, the mean square displacement $\langle u^2 \rangle$ as well as the elastic incoherent structure factor (EISF), which can be evaluated from the $Q$ dependence of elastic scattering intensity, can provide useful information on dynamics of materials. The latter is addressed in more detail elsewhere (Gabrys and Kanaya, in preparation).

## REFERENCES


Affouard, F., Descamps, M., and Ngai, K. L. (editors). *J. Non-Cryst. Solids*, **2006**, 352.

Anderson, P.W., Halperin, B.I., and Varma, C.M. *Philos. Mag.* **1972**, *25*, 1.

Bee, M. *Quasielastic Neutron Scattering: Principles and Applications in Solid State Chemistry, Biology and Materials Science*, Adam Hilger, Bristol, 1988.

Colmenero, J., Mukhopadhyay, R., Alegria, A., and Frick, B. *Phys. Rev. Lett.* **1998**, *80*, 2350.

Colmenero, J., Morenoa, A.J., and Alegría, A. *Prog. Polym. Sci.* **2005**, *30*, 1147.

Danner, H.R., Safford, G.J., Boutin, H., and Berger, M. *J. Chem. Phys.* **1964**, *40*, 1417.

de Gennes, P.G. *Physics* **1967**, *3*, 37.

de Gennes, P.G. *J. Chem. Phys.* **1971**, *55*, 572.

de Gennes, P.G. Scaling Concepts in Polymer Physics, Cornell University Press, Ithaca, 1979.

de Gennes, P.G. *J. Phys. (Paris)* **1981**, *42*, 735.

de Gennes, P.G. *C. R. Acad. Sci. Paris, Ser. II* **1986**, *302*, 765.

de Gennes, P.G. and Dubois-Violette, E. *Physics* **1967**, *3*, 181.

des Cloizeaux, J. *J. Phys. I (France)* **1993**, *3*, 1523.

Doi, M. and Edwards, S.F. *J. Chem. Soc., Faraday Trans.* **1978**, *74*, 1787.

Doi, M. and Edwards, S.F. *The Theory of Polymer Dynamics*, Clarendon Press, Oxford, 1986.

Ediger, M.D. *Annu. Rev. Phys. Chem.* **1991**, *42*, 225.

Ediger, M.D. *Annu. Rev. Phys. Chem.* **2000**, *51*, 99.

Ewen, B. *Pure Appl. Chem.* **1984**, *56*, 1407.

Ewen, B. and Richter, D. *Adv. Polym. Sci.* **1997**, *134*, 1.

Farago, B., Monkenbusch, M., Richter, D., Huang, J.S., Fetters, L.J., and Gast, A.P. *Phys. Rev. Lett.* **1993**, *71*, 1015.

Ferry, J. Viscoelastic Properties of Polymers, Wiley, New York, 1980.

Frick, B., Farago, B., and Richter, D. *Phys. Rev. Lett.* **1990**, *64*, 2921.

Gabrys, B., Higgins, J.S., Ma, K.T., and Roots, J.E. *Macromolecules* **1984**, *17*, 560.

Halperin, A., Tirrell, M., and Lodge, T.P. *Adv. Polym. Sci.* **1992**, *100*, 31.

Higgins, J.S. and Roots, J.E. *J. Chem. Soc., Faraday Trans. II* **1985**, *81*, 757.

Inoue, K., Kanaya, T., Ikeda, S., Kaji, K., Shibata, K., Misawa, M., and Kiyanagi, Y. *J. Chem. Phys.* **1991**, *95*, 5332.

Kanaya, T. and Kaji, K. *Adv. Polym. Sci.* **2001**, *154*, 87.

Kanaya, T., Kaji, K., and Inoue, K. *Macromolecules* **1991**, *24*, 1826.

KANAYA, T., KAWAGUCHI, T., and KAJI, K. *J. Chem. Phys.* **1993**, *98*, 8262.

KANAYA, T., OHKURA, M., KAJI, K., FURUSAKA, M., and MISAWA, M. *Macromolecules* **1994**, *27*, 5609.

KANAYA, T., OHKURA, M., TAKESHITA, H., KAJI, K., FURUSAKA, M., YAMAOKA, H., and WIGNALL, G.D. *Macromolecules* **1995**, *28*, 3168.

KANAYA, T., KAWAGUCHI, T., and KAJI, K. *Macromolecules* **1999a**, *32*, 1672.

KANAYA, T., WATANABE, H., MATSUSHITA, Y., TAKEDA, T., SETO, H., NAGAO, M., FUJII, Y., and KAJI, K. *J. Phys. Chem. Solids* **1999b**, *60*, 1367.

KANAYA, T., MONKENBUSCH, M., WATANABE, H., NAGAO, M., and RICHTER, D. *J. Chem. Phys.* **2005a**, *122*, 14905.

KANAYA, T., TAKAHASHI, N., NISHIDA, K., SETO, H., NAGAO, M., and TAKEDA, T. *Phys. Rev. E* **2005b**, *71*, 011801.

KANAYA, T., TAKAHASHI, N., NISHIDA, K., SETO, H., NAGAO, M., and TAKEDA, Y. *Physica B* **2006**, *385/386*, 676.

KITAGAWA, T. and MIYAZAWA, T. *Rep. Prog. Polym. Phys. Jpn.* **1965**, *8*, 53.

KITAGAWA, T. and MIYAZAWA, T. *J. Chem. Phys.* **1967**, *47*, 337.

KITAGAWA, T. and MIYAZAWA, T. *Adv. Polym. Sci.* **1973**, *9*, 335.

MONKENBUSCH, M., SCHNEIDERS, D., RICHTER, D., FARAGO, B., FETTERS, L., and HUANG, J. *Physica B* **1995**, *213–214*, 707.

MONNERIE, L. and GENY, J. *J. Chim. Phys. Phys.-Chim. Biol.* **1969**, *66*, 1691.

MORENO, A.J., ALEGRIA, A., and COLMENERO, J. *Phys. Rev. B* **2001**, *63*, R60201.

MORENO, A.J., ALEGRIA, A., COLMENERO, J., and FRICK, B. *Phys. Rev. B* **2002**, *65*, 134202.

MUKHOPADHYAY, R., ALEGRIA, A., COMENERO, J., and FRICK, B. *Macromolecules* **1998**, *31*, 3985.

MYERS, W., DONOVAN, J.L., and KING, J.S. *J. Chem. Phys.* **1965**, *42*, 4299.

MYERS, W., SUMMERFIELD, G.C., and KING, J.S. *J. Chem. Phys.* **1966**, *44*, 184.

NICHOLSON, L.K., HIGGINS, J.S., and HYTER, J.B. *Macromolecules* **1981**, *14*, 836.

PHILLIPS, W.A. *J. Low Temp. Phys.* **1972**, *7*, 351.

RICHTER, D. and EWEN, B. *Prog. Colloid Polym. Sci.* **1989**, *80*, 53.

RICHTER, D., EWEN, B., and HYTER, J.B. *Phys. Rev. Lett.* **1980**, *45*, 2121.

RICHTER, D., BINDER, K., EWEN, B., and STUHN, B. *J. Phys. Chem.* **1984**, *88*, 6618.

RICHTER, D., FRICK, B., and FARAGO, B. *Phys. Rev. Lett.* **1988**, *61*, 2465.

RICHTER, D., EWEN, B., FARAGO, B., and WAGNER, T. *Phys. Rev. Lett.* **1989**, *62*, 2140.

RICHTER, D., BUTERA, R., FETTERS, L.J., HUANG, J.S., FARAGO, B., and EWEN, B. *Macromolecules* **1992a**, *25*, 6156.

RICHTER, D., ZORN, R., FARAGO, B., FRICK, B., and FETTERS, L.J. *Phys. Rev. Lett.* **1992b**, *68*, 71.

ROESSLER, E., WARSCHEWSKE, U., EIERMANN, P., SOKOLOV, A.P., and QUITMANN, D. *J. Non-Cryst. Solids* **1994**, *172–174*, 113.

RONCA, G.J. *J. Chem. Phys.* **1983**, *79*, 79.

ROUSE, P.E. *J. Chem. Phys.* **1953**, *12*, 1272.

SCHATZKI, T.F. *J. Polym. Sci.* **1962**, *57*, 337.

SCHLEGER, P., FARAGO, B., LARTIGUE, C., KOLLMAR, A., and RICHTER, D. *Phys. Rev. Lett.* **1998**, *81*, 124.

SEARS, V.F. *Neutron News* **1992**, *3*, 26.

SETTE, F., and KRISCH, M.,In: HIPPERT, F., GEISSLER, E., HODEAU, J.L., LELIEVRE-BERNA, E., and REGNARD J.-R. (editors). Neutron and X-ray Spectroscopy, Springer, Netherlands, 2006, p. 169.

TAKADA, A., NISHIMURA, M., KOIKE, A., and NEMOTO, N. *Macromolecules* **1998**, *31*, 436.

TANAKA, T., HOCKER, L.O., and BENEDEK, G.B. *J. Chem. Phys.* **1973**, *59*, 5151.

TWISLETON, J.F. and WHITE, J.W. *Polymer* **1972**, *13*, 40.

van HOVE, L. *Phy. Rev.* **1954**, *95*, 249.

WILLIS, B.T.M. Chemical Applications of Thermal Scattering, Oxford University Press, London, 1973.

ZELLER, R.C. and POHL, R.O. *Phys. Rev. B* **1971**, *6*, 2029.

ZIMM, B.H. *J. Chem. Phys.* **1956**, *24*, 269.

# IV Applications

## IV.4 Inhomogeneous Structure and Dynamics of Condensed Soft Matter

*Mitsuhiro Shibayama*

### IV.4.1 INTRODUCTION

Unlike ordinary liquids consisting of small molecules, some soft matters have inhomogeneous structures. This is partially because of its complexity in chemical structures, delicate balance of molecular interactions, and large internal degrees of freedom (Kleman and Lavrentovich, 2003). Even crystals and glasses often have inhomogeneous structures, such as grain boundary structures and heterogeneities formed during glass formation, respectively. However, crystals and glasses are usually not classified to soft matter. Typical examples of soft matters that have inhomogeneous structures are polymer gels (Cohen Addad, 1996; Shibayama, 1998). Figure IV.4.1 shows a Venn diagram of the publication on neutron scattering searched by ISI Web of Knowledge$^{SM}$ in May 2009. According to this diagram, the number of publication related to "neutron scattering" is over 38,000. Among them, papers on "polymer" are about 3400, which include the pioneering work by Jannink in 1968 on polystyrene (PS) solutions in carbon disulfide (Jannink, 1969). On the other hand, 326 hits were obtained by the key words of "neutron scattering" *and* "inhomogeneities *or* heterogeneities." These numbers, of course, include papers not only of soft matter but also of hard matter. Search with "neutron scattering" *and* "polymer blends," "block copolymers," and "polymer gels" gave 614, 823, and 411, respectively. Interestingly, by adding a keyword of "inhomogeneities *or* heterogeneities," the product set of the Venn diagram is dominantly occupied by "polymer gels" with 19%. This statistics means that polymer blends and block copolymers are not regarded as systems having inhomogeneities. Polymer blends may have inhomogeneous structures when they are in a phase-separated state or under deformation

*Neutrons in Soft Matter* Edited by Toyoko Imae, Toshiji Kanaya, Michihiro Furusaka, and Naoya Torikai

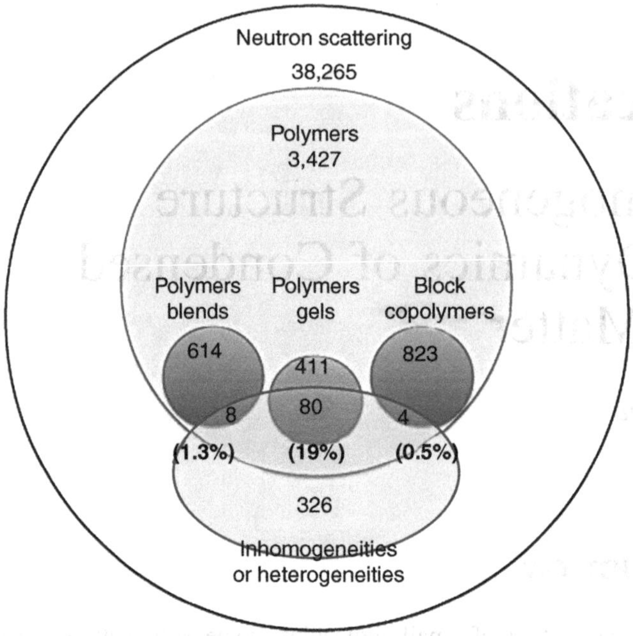


**Figure IV.4.1** Venn diagram showing the neutron scattering, inhomogeneities, and soft matter.

(Koizumi, 2004). However, inhomogeneities disappear by annealing. In the case of polymer gels, on the other hand, the presence of inhomogeneities is rather inherent. This is why "the inhomogeneous structures" in soft matter have been exclusively used in polymer gels (Pines and Prins, 1972; Bastide and Candau, 1996; Shibayama, 1998). In this section, we discuss the nature and the origins of inhomogeneities, the methodologies to investigate inhomogeneities, and recent topics of polymer gels from the viewpoints of inhomogeneities.

## IV.4.2 CLASSIFICATION OF GELS AND INHOMOGENEITIES

### IV.4.2.1 Chemical Gels Versus Physical Gels

Polymer gels are classified to chemical gels and physical gels. Chemical gels are, in most cases, formed by copolymerizing monomer and cross-linker, or introducing cross-links to a polymer solution by irradiation or by some means (de Rossi et al., 1991). During chemical reaction or cross-linking, the structure of the gel becomes frozen at least topologically or sometimes even mechanically. Hence, highly inhomogeneous structures are formed. Physical gels, on the other hand, have a tendency to undergo spontaneous gelation by crystallization, by helix-formation, or by microphase separation, and so on (Guenet, 1992). The structures and physical properties are dependent on the nature of gels.

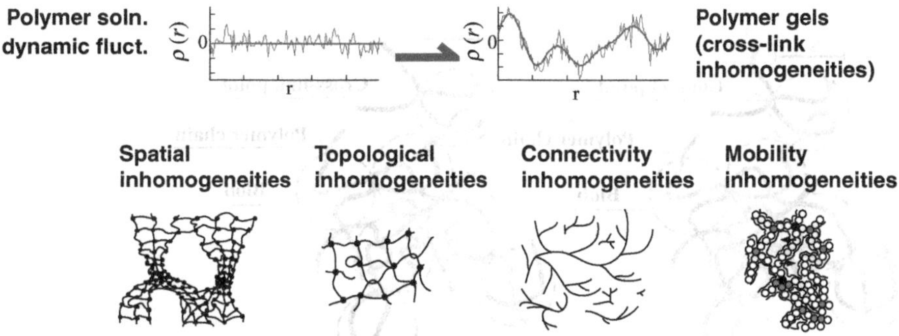


**Figure IV.4.2** Inhomogeneities in polymer gels. The upper figures illustrate the difference in concentration fluctuations between polymer solutions (left) and polymer gels (right). For polymer gels, in addition to thermal fluctuations, frozen inhomogeneities are superimposed. (Shibayama and Norisuye, 2002, Copyright, Chemical Society of Japan.)

## IV.4.2.2 Classification of Inhomogeneities

It should be noted here that there exist several types of *inhomogeneities* in gels, which play significant roles to characterize gels and gelation thresholds. The top part of Figure IV.4.2 indicates concentration fluctuations in polymer solutions (left) and in gels (right) (Shibayama and Norisuye, 2002). In polymer solutions, only thermal concentration fluctuations exist, of which the average is zero. On the other hand, gels contain both frozen concentration fluctuations (the low frequency component in this figure) introduced by cross-linking and thermal concentration fluctuations (high-frequency component). The introduction of cross-links brings about various types of inhomogeneities, as shown in the lower cartoons. The *spatial inhomogeneities* are nonrandom spatial variations of cross-link density in a gel, which result in anomalous scattering (Mendes et al., 1991). The *topological inhomogeneities* represent defects of network, such as dangling chains, loops, chain entrapment, and so on. These inhomogeneities affect the dynamics and swelling behavior of gels (Shibayama et al., 1997). Third, the *connectivity inhomogeneities* are dependent on cluster size, distribution, and architecture of polymer chains. It is not an exaggeration to claim that these connectivity inhomogeneities govern the dynamics of the system and become significant at the sol–gel transition threshold as *critical dynamics* (Norisuye et al., 1998). The *mobility inhomogeneities* correspond to variations of local degree of mobility by introduction of cross-links. The mobility inhomogeneities are the reason why scattering speckle appears exclusively in gel state (Pusey and van Megen, 1989). It is well known that static scattering is useful to study the spatial inhomogeneities and *spatial correlation*.

## IV.4.2.3 Cross-Linking Inhomogeneities

If cross-links are introduced instantaneously to a polymer solution by physical cross-linking (physical gels) or by photoirradiation or gamma-ray irradiation without disturbing the conformation of the polymer chains (chemical gels), the spectrum of

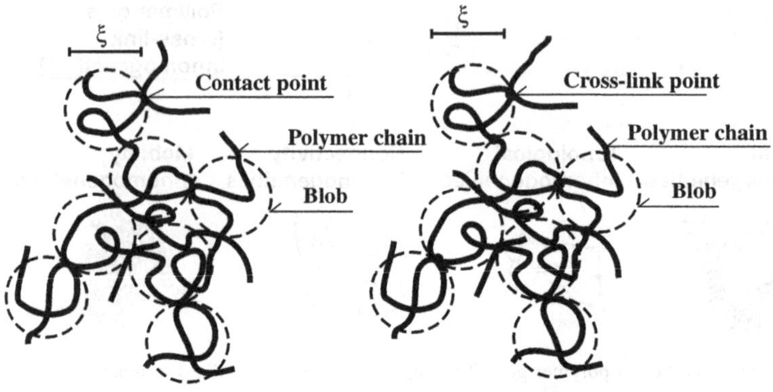


**Figure IV.4.3** Comparison of the microscopic structures of (a) polymer solution in semidilute regime and (b) polymer gel. Both are characterized by blobs with size $\xi$.

the concentration fluctuations may not change from those before cross-linking. Such a situation may be depicted as Figure IV.4.3. Here, the polymer solutions are characterized by an ensemble of blobs with size $\xi$ (de Gennes, 1979. In the case of gelatin gels (physical gels), such kind of immobilization are indeed occurs. Figure IV.4.4 shows the small-angle neutron scattering (SANS) intensity function of gelatin gel by simply cooling gelation aqueous solutions prepared at 50°C or higher,

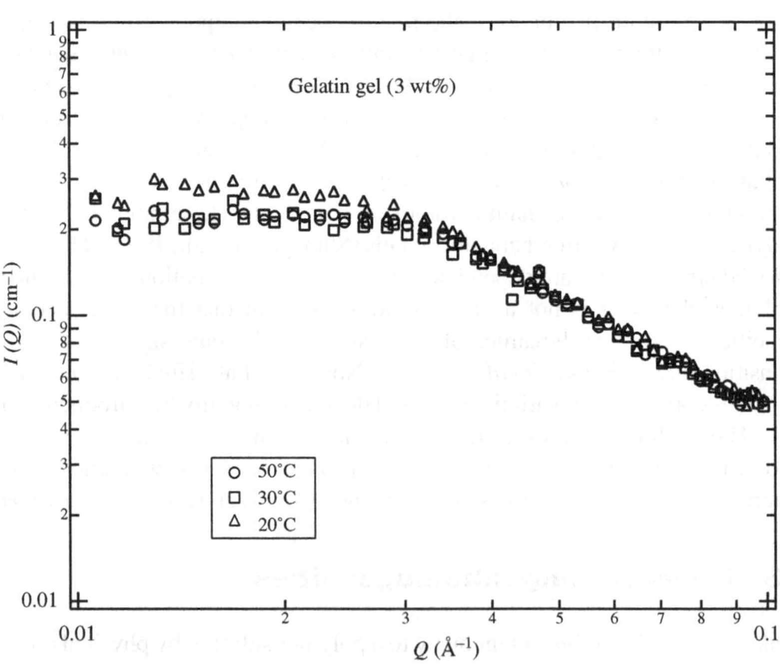


**Figure IV.4.4** SANS intensity functions of gelatin gels of 3 wt% measured at 20, 30, and 50°C. The sol–gel transition temperature was about 25°C.

where the variable $Q$ is the magnitude of the scattering vector. The gelatin was alkali-treated gelatin (type B; Lot No. P-3201, Nitta Gelatin Co., Osaka) with the molecular weight of $M_w = 1.45 \times 10^5$ Da and the isoelectric point of pH 4.97. According to rheological and dynamic light scattering (DLS) measurement, the gelation temperature, $T_{gel}$, was dependent on the gelatin concentration, and $T_{gel}$ was ca. 25°C for a 3 wt % gelatin gel (Matsunaga and Shibayama, 2007). Hence, the gelatin was in gel state at 20°C, but in sol state at 30 and 50°C. This figure clearly suggests that structure inhomogeneities do not appear by sol-to-gel transition in the case of physical gels. Note that a drastic change in dynamics, originating from connectivity and mobility inhomogeneities, is observed in the same gelatin gel by dynamic light scattering and by rheological measurement (Matsunaga and Shibayama, 2007).

The situation of inhomogeneities is quite different in chemically cross-linked gels. Typical polymer gels, such as acrylamide gels, are prepared by polymerizing monomer in the presence of cross-linker. In this case, cross-linking reaction and polymerization occurs concurrently with strong disturbance of the polymer segment distribution and immobilization of the system topologically as well as spatially. As a result, the structure of the gel is progressively disturbed and immobilized as reaction goes. Mallam et al. (1989) reported a significant change in small-angle X-ray and light scattering (LS) intensity functions with increasing cross-linker concentration. Figure IV.4.5

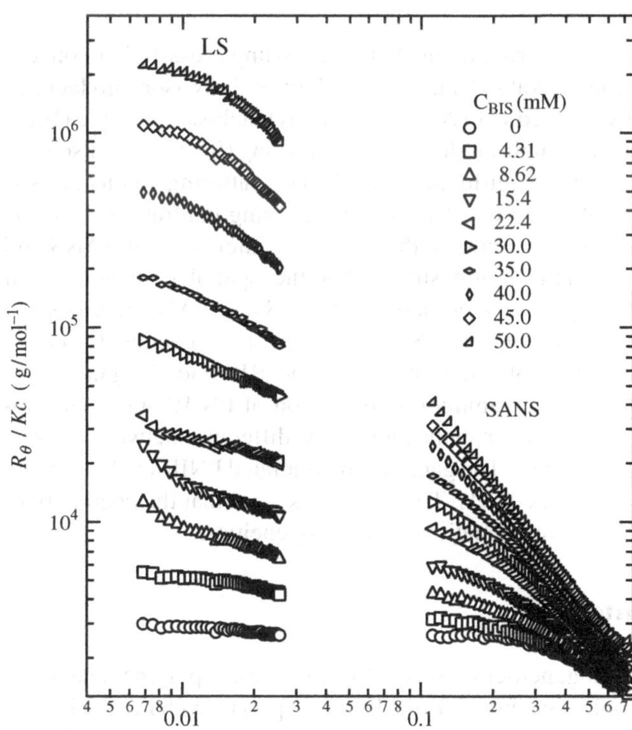


**Figure IV.4.5** SANS intensity functions of PNIPA gels prepared with various cross-linker concentrations, $C_{BIS}$. The NIPA monomer concentration was 700 mM.

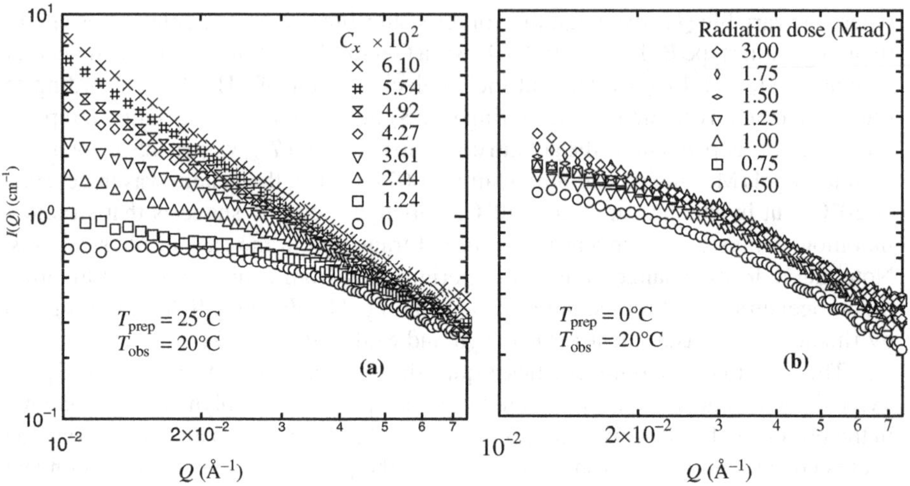


**Figure IV.4.6** SANS intensity functions of PNIPA gels prepared by (a) copolymerization of NIPA and BIS (chemically cross-linked gels) and (b) gamma-ray cross-linking of PNIPA polymer solutions. The cross-linker concentration, $C_x$, and the irradiation doze are given in the figures. (Norisuye et al., 2002, Copyright, Elsevier.)

demonstrates an increase of scattering intensity by increasing cross-linker concentration, $C_{BIS}$, at preparation (Takata et al., 2002). Here, poly(*N*-isopropylacrylamide) (PNIPA) was copolymerized with *N,N,N′,N′*-methylenebisacrylamide (BIS). The value of $Q$ covers not only small-angle neutron scattering (SANS) but also light scattering (LS) regime. Here, the ordinate, that is, the scattering intensity, was depicted by the Rayleigh ratio, $R_\theta$, divided by the scattering contrast, $K$, and the polymer concentration, $c$, so as to compare the scattering intensities of SANS and LS without arbitrary shift. This figure shows that the spatial distributions of inhomogeneities are smoothly connected between the LS and SANS regimes.

Figure IV.4.6 shows a comparison of SANS curves of (a) chemically cross-linked gels by copolymerization of NIPA monomer and BIS and (b) gamma-ray irradiated PNIPA gels obtained by gamma-ray irradiation of PNIPA aqueous solutions (Norisuye et al., 2002). As shown in the figures, the difference between the gels prepared by the two methods is clear. The gamma-ray irradiated PNIPA gels are more homogeneous than BIS cross-linked gels. Therefore, it is clear that the chemistry of gelation plays a dominant role in formation of inhomogeneities.

## IV.4.2.4 Clustering

Another possibility of inhomogeneities is ascribed to clustering of polymer chains via slight difference in the molecular interactions between polymer chains and polymer–solvent. Presence of long-range inhomogeneities was observed even in homopolymer solutions, for example, in concentrated polystyrene solutions (Benoit and Picot, 1966; Bastide and Candau, 1996). These are now called Picot–Benoit effect,

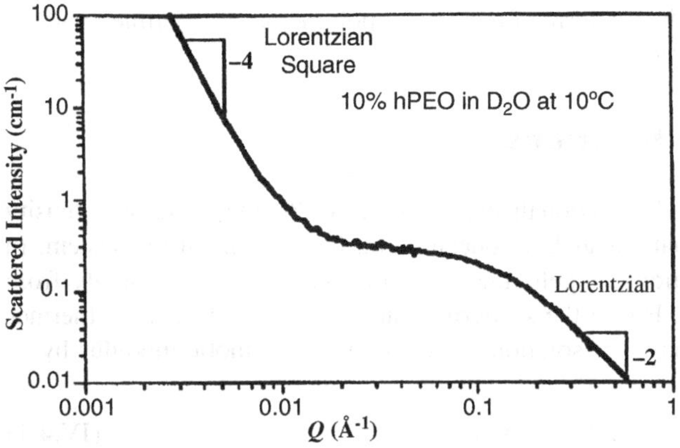


**Figure IV.4.7** SANS curves of hydrogeneous poly(ethylene oxide) (PEO) in $D_2O$. The scattering intensity shows an upturn at low $Q$. (Hammouda et al., 2002, Copyright, ACS.)

and are well represented by Debye–Bueche type function (Debye and Bueche, 1949). Such clustering can also be observed in aqueous solutions of hydrophilic polymers, for example, poly(ethylene glycol) (PEG) or equivalently poly(ethylene oxide) (PEO). Hammouda et al. (2002, 2004) reported that hydrogenous PEO (hPEO) in $D_2O$ exhibited a large-scale of clusters of the order of 500 nm by SANS as shown in Figure IV.4.7 (Hammouda et al., 2002, 2004).

### IV.4.2.5 Frozen Structures by Vitrification

Glassy polymers also exhibit strong upturn in low $Q$ region. This is ascribed to frozen-concentration fluctuations during polymerization process. Koike et al. (1989) investigated the origins of long-range heterogeneities with the dimension of ca. 1000 Å in poly(methyl methacrylate) (PMMA) by light scattering and found that the large excess scattering is caused mainly by the isotropic strain inhomogeneities caused during polymerization and not by a small amount of remaining monomers or additives, the molecular weight of polymers, the stereoregularity due to the specific tacticity of PMMA, or cross-linking as a result of the gel effect.

These examples clearly show that polymeric systems often contain inhomogeneous structures due to thermal freezing, topological freezing, specific molecular interactions, such as hydrogen bonding and hydrophobic interactions. In the next section are reviewed the scattering intensity functions characterizing these inhomogeneities.

## IV.4.3 THEORETICAL BACKGROUND OF SCATTERING FUNCTIONS

As already shown in the previous section, scattering methods are quite useful to investigate the inhomogeneities in polymer gels. In this section, a theoretical background dealing with the scattering functions of polymer gels is outlined with an emphasis of inhomogeneities.

### IV.4.3.1 Polymer Solutions

According to the Einstein's fluctuation theory (Einstein, 1926), the scattering intensity is related to density fluctuations and/or concentration fluctuations of the system. In the case of polymer solutions or gels, the scattering intensity arises mainly from concentration fluctuations. Hence the scattering intensity at $Q=0$ (i.e., the thermodynamic limit), $I(0)$, for polymer solutions is related to the osmotic modulus by

$$I(0) = (\Delta\rho)^2 \frac{k_B T \phi^2}{K_{os}}, \tag{IV.4.1}$$

where $k_B$ is the Boltzmann constant, $T$ is the absolute temperature, $\phi$ is the volume fraction of the polymer, and $K_{os}$ is the osmotic compressibility. $(\Delta\rho)^2$ is the scattering length density difference square between the polymer (2) and solvent (1). In the case of neutron scattering, $(\Delta\rho)^2$ is given by

$$(\Delta\rho)^2 = (\rho_2 - \rho_1)^2 = \left[\left(\frac{b_2}{\tilde{V}_2}\right) - \left(\frac{b_1}{\tilde{V}_1}\right)\right]^2. \tag{IV.4.2}$$

Here, $b_i$ and $\tilde{V}_i$ are the scattering length and the monomeric volume of component $i$ ($i = 1$ or 2), respectively. In the context of Flory–Huggins theory for polymer solutions (Flory, 1953), the osmotic pressure of the polymer solution, $\Pi$, is given by

$$\Pi = -\frac{RT}{V_1}\left[\ln(1-\phi) + \left(1 - \frac{1}{z}\right)\phi + \chi\phi^2\right], \tag{IV.4.3}$$

where $V_1$ is the molar volume of the solvent, $R$ is the gas constant, and $\chi$ is the Flory–Huggins interaction parameter. $z$ is the reduced degree of polymerization of the solute polymer normalized to the molar volume of the solvent, $V_1$,

$$z = \left(\frac{V_2}{V_1}\right) Z. \tag{IV.4.4}$$

Here, $Z$ is the degree of polymerization of the polymer based on the monomer unit of the polymer and $V_2$ is the molar volume of the monomeric unit of the solute (polymer). The osmotic modulus is given by (Onuki, 1993),

$$K_{os} \equiv \phi\frac{\partial\Pi}{\partial\phi} = \frac{RT\phi}{V_2 Z}\left[1 + \left(\frac{1}{1-\phi} - 2\chi\right)\left(\frac{V_2}{V_1}\right)\phi Z\right]. \tag{IV.4.5}$$

Note that the factor $(V_2/V_1)$ is necessary for the case where the molar volume of the monomeric unit of the solute is different from that of the solvent.

### IV.4.3.2 Polymer Gels

In the case of polymer gels, there are two contributions to the osmotic pressure. One is from the free energy of mixing, $\Pi_{mix}$, and the other is from the elasticity of network chains, $\Pi_{el}$. Since the degree of polymerization is infinite ($Z \to \infty$ or $z \to \infty$) for gels, $\Pi_{mix}$ is given by taking $z \to \infty$ in eq. (IV.4.3), that is,

$$\Pi_{mix} = -\frac{RT}{V_1}\left[\ln(1-\phi) + \phi + \chi\phi^2\right] \quad \text{(polymer gel)}. \tag{IV.4.6}$$

On the other hand, $\Pi_{el}$ is obtained by (Flory and Rehner, 1943)

$$\Pi_{el} = \nu_e RT\left[\frac{1}{2}\left(\frac{\phi}{\phi_0}\right) - \left(\frac{\phi}{\phi_0}\right)^{1/3}\right], \tag{IV.4.7}$$

where $\nu_e$ is the number density of the effective elastic chains in the network and $\phi_0$ is the volume fraction of the polymer at preparation. At swelling equilibrium, the following equation holds (Shibayama and Tanaka, 1993).

$$\Pi = \Pi_{mix} + \Pi_{el} = 0. \tag{IV.4.8}$$

Hence, $\nu_e$ is obtained for affine-model networks

$$\nu_{e,aff} = \frac{\left[\ln(1-\phi) + \phi + \chi\phi^2\right]}{V_1\left[\frac{1}{2}\left(\frac{\phi}{\phi_0}\right) - \left(\frac{\phi}{\phi_0}\right)^{1/3}\right]} \quad \text{(affine)}. \tag{IV.4.9}$$

This is the so-called Flory–Rehner equation and is used to determine the cross-link density (or the number density of effective polymer chains in a polymer network) (Flory and Rehner, 1943). In the case of phantom chains, the number density of effective polymer chains is modified to (James and Guth, 1947)

$$\nu_{e,ph} = \nu_{e,aff}\left(1 - \frac{2}{f_x}\right), \tag{IV.4.10}$$

where $f_x$ is the functionality of the cross-link. Since the phantom network does not have the translational entropy term of cross-links, the equation for swelling equilibrium is given by (Mark and Erman, 1988)

$$\nu_{e,ph} = -\frac{\left[\ln(1-\phi) + \phi + \chi\phi^2\right]}{V_1\left(\frac{\phi}{\phi_0}\right)^{1/3}} \text{(phantom)}. \tag{IV.4.11}$$

The osmotic modulus of polymer gels is given by

$$K_{os} = K_{os,mix} + K_{os,el} = \frac{RT\phi^2}{V_1}\left(\frac{1}{1-\phi} - 2\chi\right) + \nu_e RT\left[\frac{1}{2}\left(\frac{\phi}{\phi_0}\right) - \frac{1}{3}\left(\frac{\phi}{\phi_0}\right)^{1/3}\right]. \tag{IV.4.12}$$

The longitudinal modulus is obtained by (Tanaka et al., 1973)

$$M_{\mathrm{os}} = K_{\mathrm{os}} + \frac{4}{3}\mu. \tag{IV.4.13}$$

Here, $\mu$ is the shear modulus given by (Onuki, 1993)

$$\mu = \nu_{\mathrm{e}} RT \left(\frac{\phi}{\phi_0}\right)^{1/3}. \tag{IV.4.14}$$

Hence,

$$M_{\mathrm{os}} = \frac{RT\phi^2}{V_1}\left(\frac{1}{1-\phi} - 2\chi\right) + \nu_{\mathrm{e}} RT\left[\frac{1}{2}\left(\frac{\phi}{\phi_0}\right) + \left(\frac{\phi}{\phi_0}\right)^{1/3}\right]. \tag{IV.4.15}$$

The scattering function for polymer gels is given as a sum of Ornstein–Zernike (OZ) function (Ornstein and Zernike, 1914) and an excess scattering function, $A_{\mathrm{ex}}(Q)$, squared-Lorentzian (SL) function as written by (Shibayama, 1998; Shibayama et al., 2004)

$$I(Q) = \frac{(\Delta\rho)^2 RT\phi^2}{N_{\mathrm{A}} M_{\mathrm{os}}}\left[\frac{1}{1+\xi^2 Q^2} + A_{\mathrm{ex}}(Q)\right]. \tag{IV.4.16}$$

Here, $\xi$ is the correlation length of the network and $N_{\mathrm{A}}$ is the Avogadro number. $A_{\mathrm{ex}}(Q)$ represents the contribution of frozen inhomogeneities.

## IV.4.3.3 Excess Scattering Functions of Polymer Gels

An intuitive understanding of the excess scattering can be given by the cartoons in Figure IV.4.8. The left cartoons show that an invisible cross-link distribution (marked by dots) becomes visible by swelling due to inhomogeneous swelling (Bastide and Leibler, 1988). Here, highly cross-linked regions less swell than other regions. As a result, concentration fluctuations originating from the polymer segment density fluctuations (mesh) become large. Such concentration fluctuations, $\rho(\mathbf{r})$, are depicted in the right figure as a superposition of two types of concentration fluctuations, that is, thermal fluctuations by Brownian motion, $\rho^{\mathrm{th}}(\mathbf{r})$, and frozen inhomogeneities introduced by cross-linking, $\rho^{\mathrm{eq}}(\mathbf{r})$ (Panyukov and Rabin, 1996). The concentration fluctuations are thus obtained by,

$$\rho(\mathbf{r}) = \rho^{\mathrm{th}}(\mathbf{r}) + \rho^{\mathrm{eq}}(\mathbf{r}). \tag{IV.4.17}$$

The structure factor of the gels, $S(Q)$, is given by the Fourier conjugates of $\rho(\mathbf{r})$ as follows,

$$\begin{aligned} S(Q) &\equiv \overline{\langle \rho(Q)\rho(-Q)\rangle} = G(Q) + C(Q), \\ &= G(Q)\langle \rho^{\mathrm{th}}(Q)\rho^{\mathrm{th}}(-Q)\rangle, \quad C(Q) = \overline{\rho^{\mathrm{th}}(Q)\rho^{\mathrm{th}}(-Q)}. \end{aligned} \tag{IV.4.18}$$

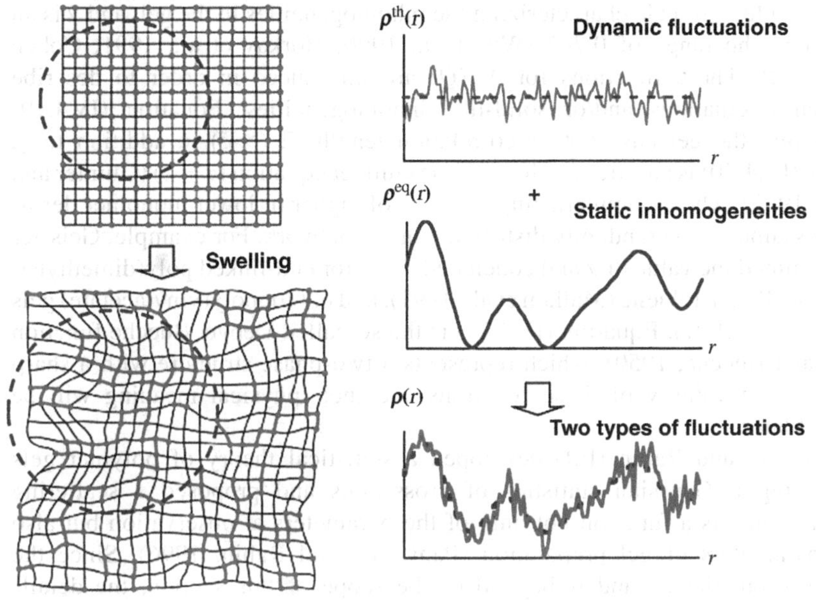


**Figure IV.4.8** (Left) Schematic representation of polymer networks where cross-links (dots) are nonrandomly introduced into the network. After swelling, the network swells inhomogeneously due to the nonrandom distribution of cross-links. (Right) Concentration fluctuations representing of thermal fluctuations (top) and frozen inhomogeneities (middle). The bottom figure represents the superposition of the two.

Here, $G(Q)$ is the thermal correlator representing the thermal fluctuations in gels ($\approx 1/(1+\xi^2Q^2)$) and $C(Q)$ is the static correlator (the excess scattering, $A_{ex}(Q)$). Note that $\langle X \rangle$ and $\overline{X}$ are the thermal average and ensemble average of $X$, respectively.

The excess scattering appearing at low $Q$ region is a steeply decreasing function with $Q$. Phenomenologically, several functional forms of $A_{ex}(Q)$ have been proposed in the literature, that is,

**(i)** a Lorentz function (Candau et al., 1982)

$$A_{ex}(Q) = \frac{A_{ex}(0)}{1+\Xi^2Q^2}(\Xi >> \xi); \qquad \text{(IV.4.19)}$$

**(ii)** a stretched exponential function (Mallam et al., 1989)

$$A_{ex}(Q) = A_{ex}(0)\exp[-(Q\Xi)^{\alpha}]; \qquad \text{(IV.4.20)}$$

**(iii)** a Debye–Bueche function (Wu et al., 1990)

$$A_{ex}(Q) = \frac{A_{ex}(0)}{(1+\Xi^2Q^2)^2}, \qquad \text{(IV.4.21)}$$

where $\Xi$ is a length scale characterizing the inhomogeneities in the gel, and $\alpha$ is an exponent in the range of 0.7–2 (Wu et al., 1990; Horkay et al., 1991; Cohen et al., 1992). These functions for $A_{ex}(Q)$ are introduced in order to describe additional fluctuations and/or solid-like inhomogeneities. Equation (IV.4.19) assumes that the gel has another correlation length, $\Xi$ ($\gg\xi$) in addition to $\xi$. Equation (IV.4.20) is an extended form of the Guinier equation ($\alpha = 2$) (Guinier and Fournet, 1955), where noninteracting domains of higher or lower monomer densities are assumed to be randomly distributed in the network. For example, Geissler et al. examined the value of $\alpha$ and concluded $\alpha = 2$ for end-linked poly(dimethylsiloxane) swollen in toluene (Mallam et al., 1991), and 0.7 for poly(vinylacetate) gels (Horkay et al., 1991). Equation (IV.4.21) is the so-called Debye–Bueche function (Debye and Bueche, 1950), which represents a two-phase structure with a sharp boundary. The validity of these functions and their physical meaning will be discussed later.

Panyukov and Rabin (PR) developed a statistical theory of polymer gels by assuming a Gaussian statistics of cross-links and proposed a scattering function, which is a function not only of the parameters at observation but also of the parameters of gel preparation (Panyukov and Rabin, 1996). Since the theory is sophisticated and is beyond of the scope of this section, the details and the experimental studies employing the PR theory are referred to elsewhere (Shibayama et al., 1998a, 1998b; Takata et al., 2002; Ikkai and Shibayama, 1999).

## IV.4.4 METHODOLOGIES OF INHOMOGENEITY CHARACTERIZATION

### IV.4.4.1 Small-Angle Neutron Scattering

As was already discussed in the preceding sections, small-angle neutron scattering is one of the best methods to investigate the inhomogeneities in soft matter. Here, an overview of one of the active SANS instruments, SANS-U, is given for reference. SANS-U is a multipurpose small-angle neutron scattering instrument owned by the University of Tokyo, located at JRR-3 Research Rector, Japan Atomic Energy Agency, Tokai, Ibaraki, Japan. Monochromated cold neutron beam with the average neutron wavelength of 7.00 Å and 10% wavelength distribution was irradiated to the samples. The scattered neutrons were counted with a two-dimensional position detector (Ordela 2660N, Oak Ridge, USA). The sample-to-detector distance can be varied from 1 to 16 m. After necessary corrections for open beam scattering, transmission and detector inhomogeneities, corrected scattering intensity functions were normalized to the absolute intensity scale with a polyethylene secondary standard. The details of the instrument are reported elsewhere (Okabe et al., 2005, 2007). Incoherent scattering subtraction was made with the method reported by Shibayama et al. 2005, 2009.

### IV.4.4.2 Neutron Spin Echo Spectroscopy

In order to investigate the structure and dynamics of soft matter, it is important to employ a probe covering the spatiotemporal range of the soft matter. Neutron spin echo (NSE) spectroscopy is a suitable means for such purpose. The dynamics of soft matter can be studied in either the frequency domain ($\omega$-domain) or the time domain ($t$-domain). These domains are connected with Wiener–Khinchin theorem. In the case of neutron spin echo spectroscopy, the intermediate scattering function, $S(Q, \omega)$ is directly obtained, from which the $t_F$-dependent scattering intensity, $I(Q, t_F)$, is evaluated as follows:

$$I_{NSE}(Q, t_F) = \frac{1}{2\pi} \int_{-\infty}^{\infty} S(Q, \omega) \frac{1 + \cos(\omega t_F)}{2} d\omega \propto S(Q, 0) f(t_F), \qquad (IV.4.22)$$

where $t_F$ is the Fourier time and $f(t_F)$ is the relaxation function. If the dynamics is described by unimodal diffusion, $f(t_F)$ is given by

$$f(t_F) = \exp[-DQ^2 t_F], \qquad (IV.4.23)$$

where $D$ is the collective diffusion coefficient. NSE measurements on polymer gels covering a wide range of $Q$ were carried out by Koizumi et al. (2004) at FRJ2-NSE at the Forschungszentrum Jülich (Germany) and at IN15 at the Institute of Laue-Langevin (France).

### IV.4.4.3 Static and Dynamic Light Scattering

Inhomogeneities in polymer gels can be extensively investigated by light scattering methods, namely, static light scattering (SLS) and dynamic light scattering. These are complementary methods of SANS and NSE (Shibayama et al., 2007). This is because that the spatial range of inhomogeneities covers not only nanometer orders but also micrometers and above as shown in Figure IV.4.5. In addition, inhomogeneities in polymer gels are closely coupled with nonergodicity (Pusey and van Megen, 1989). Hence, SLS/DLS is sometimes more effective means to study inhomogeneities of gels. However, discussions on light scattering are beyond the scope of this book, the reader may consult with relevant references (Joosten et al., 1991; Shibayama, 1998, 2006; Shibayama and Norisuye, 2002).

## IV.4.5 STUDIES ON INHOMOGENEITIES BY SANS

### IV.4.5.1 Conventional Chemical Gels

As shown in Figures IV.4.5 and IV.4.6, inhomogeneities in chemically cross-linked gels are characterized by a strong upturn in SANS (or SAXS) intensity at low $Q$

region. The presence of strong forward scattering in gels was reported by many groups. Mallam and coworkers investigated cross-linker concentration dependence of small-angle X-ray scattering (SAXS) and light scattering intensity for poly (acrylamide) gels prepared with fixed acrylamide concentrations but varied bisacrylamide (BIS) concentrations (Mallam et al., 1989) and for poly(dimethylsiloxane) in toluene (Mallam et al., 1991). This is ascribed to frozen inhomogeneities introduced by the cross-linker. They analyzed the SAXS intensity functions with a combination of Ornstein–Zernike and Gauss function (i.e., eq. (IV.4.20)). Shibayama et al. also employed eq. (IV.4.20) with $\alpha = 2$ for analyzing the critical phenomena of volume phase transition of poly(*N*-isopropylacrylamide) hydrogels (Shibayama et al., 1992). As already shown in Figure IV.4.6a, the effect of cross-linker on inhomogeneities is remarkable.

Koizumi et al. studied the structure of PNIPA gels by SANS as well as NSE. They observed a significant retardation of decay in PNIPA gels as compared to the corresponding PNIPA aqueous solutions as shown in Figure IV.4.9 (Koizumi et al., 2004). Here, the ratio of the intermediate scattering functions at $t_F$ and $t_F = 0$ is plotted as a function of the Fourier time $t_F$. The difference between gels and solutions became negligible as $Q$ increased to $Q = 0.16\ \text{Å}^{-1}$. This result indicates that the static inhomogeneities become less important at large $Q$ and thermal

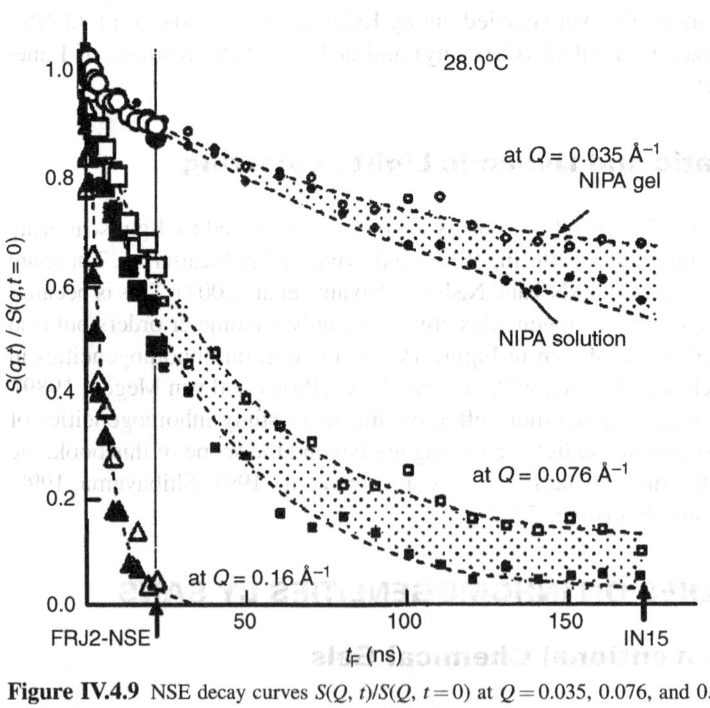


**Figure IV.4.9** NSE decay curves $S(Q, t)/S(Q, t = 0)$ at $Q = 0.035$, 0.076, and $0.16\ \text{Å}^{-1}$, obtained for swollen NIPA gel (open symbols) and NIPA solution (filled symbols) at 28°C. The larger and smaller symbols indicate those obtained by FRJ2-NSE with $t_F = 22$ ns and by IN15 with $t_F = 180$ ns. (Koizumi et al., 2004, Copyright, American Institute of Physics.)

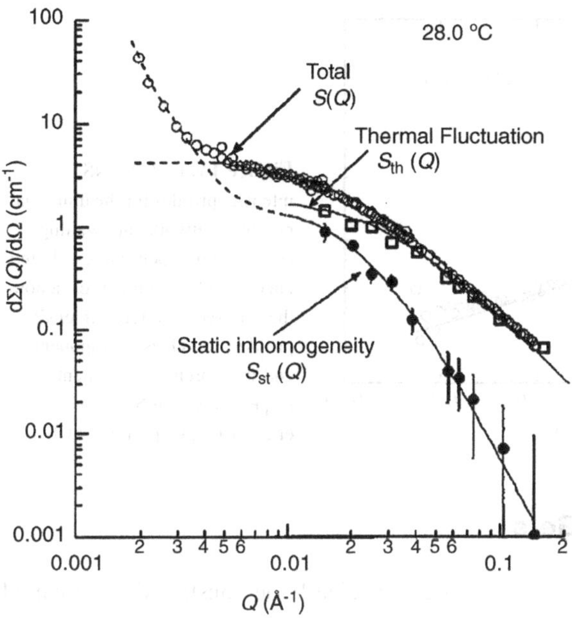


**Figure IV.4.10** SANS intensity function, $S(Q)$ (open circles) by SANS, $S_{st}(Q)$ (filled circles), and $S_{th}(Q)$ (open squares) determined by NSE for swollen PNIPA gel at 28°C. (Koizumi et al., 2004, Copyright, AIP.)

concentration fluctuations are dominant. Figure IV.4.10 shows the scattering function, $d\Sigma(Q)/d\Omega$, of PNIPA gels at 28°C and those decomposed to static and dynamic parts (Koizumi, 2004).

## IV.4.5.2 End-Linked Model Network

In order to reduce inhomogeneities and/or to prepare an "ideal" polymer networks, model networks were prepared by end-coupling of telechelic polymers with four-functional cross-linkers (Beltzung and Herz, 1983). We prepared a series of telechelic poly(tetrahydrofuran) (PTHF) prepolymers having allyl groups at both ends by living cationic polymerization. The telechelic PTHFs were end-cross-linked by a four-functional cross-linker. The details of sample preparation are described elsewhere (Shibayama et al., 1994). Figure IV.4.11 shows SANS curves of PTHF network in tetrahydrofuran. The SANS curve consists of solution-like (dashed line) and solid-like concentration fluctuations (dotted line). The former is simply given by a Lorentz (Ornstein–Zernike) type function (de Gennes, 1979). On the other hand, there is no deterministic function to represent the solid-like fluctuations since these are frozen-in fluctuations, dependent on the manner of network preparation, dispersity of the network, and so on. Therefore, we simply employ a combination of Gauss and Lorentz functions, originally proposed by Geissler and coworkers (Hecht et al., 1985; Mallam et al., 1989, 1991; Horkay et al., 1991). $\Xi$ represents the characteristic mean size of the static (frozen) inhomogeneity (assemblies of monomer-rich "domains"), respectively.

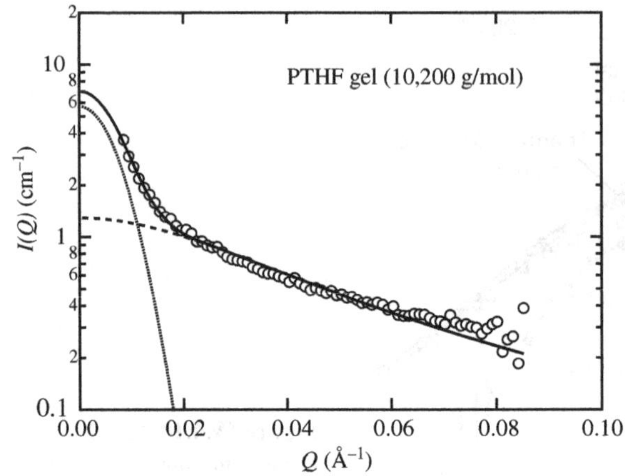


**Figure IV.4.11** SANS intensity profiles for the unimodal polymer networks at swelling equilibrium. Open circles, dotted curves, and dashed curves denote the observed scattered intensity profiles, the Gauss component and the Lorentz component, respectively. (Shibayama et al., 1995, Copyright, ACS.)

### IV.4.5.3 Deformed Gels

The presence of static inhomogeneities in gels is clearly demonstrated as so-called butterfly pattern. By uniaxially stretching a gel, an increase in the scattered intensity appears in the stretching direction. Since this elliptic contour pattern is opposite to that theoretically predicted on the basis of thermal fluctuations, this pattern is sometimes called "abnormal" butterfly pattern (Mendes et al., 1991). Rabin and Bruinsma (1992) explained the origin of the abnormal butterfly pattern with thermal fluctuations of concentration. However, it is now commonly accepted that quasistatic fluctuations (also called as frozen or quenched inhomogeneities) present inherently in gels are the origin of the "abnormal" butterfly pattern (Rouf et al., 1994). The "abnormal" butterfly pattern is characterized by an increase in the scattering intensity along the stretched direction.

Figure IV.4.12 shows examples of observed contour maps of a polystyrene gel uniaxially stretched with different elongation ratios, $\lambda$ (Mendes et al., 1996). The PS samples were randomly cross-linked and were immersed in deuterated toluene. As shown here, the contour pattern becomes elliptic; prolate and oblate with respect to the stretching direction at low and large $Q$s, respectively. The low $Q$ behavior, that is, a prolate shape for $\lambda \geq 1.33$, is opposite to that expected by theories (thermal fluctuations). In order to account for this type of abnormal scattering found in stretched gels, Onuki proposed a structure factor for deformed gels, which is given by

$$I(Q;\beta) = (\Delta\rho)^2\phi^2 k_B T\left\{\frac{1}{M^*_{os}(\beta)[1+\xi^2(\beta)Q^2]} + p\frac{[J(\beta)-(\phi_0/\phi)^{-2/3}]^2(\phi_0/\phi)^{2/3}\mu}{M^{*2}_{os}(\beta)[1+\xi^2(\beta)Q^2]^2}\right\}, \quad \text{(IV.4.24)}$$

where $\beta$ is the angle between $\mathbf{Q}$ and the stretching direction, $\phi_0$ is the volume fraction of the polymer in the reference state, and $p$ $(0 < p \ll 1)$ is the degree of irregularity of the network. $J(\beta)$ represents the modification of the shear modulus by deformation

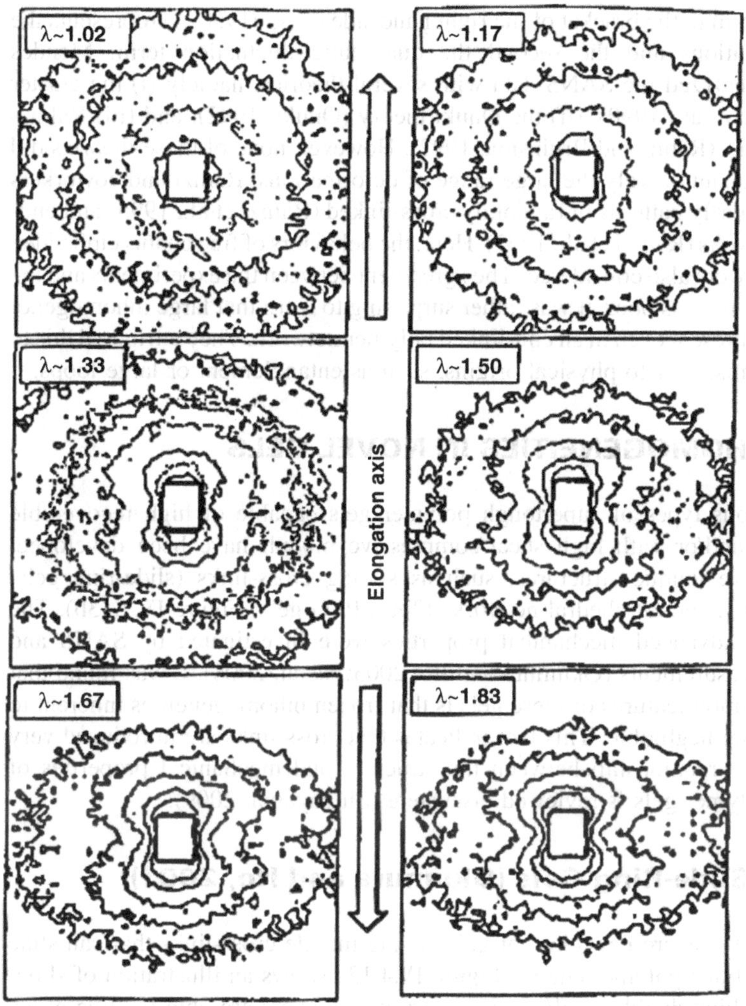


**Figure IV.4.12** Isointensity curves as a function of the elongation ratio, $\lambda$. (Mendes et al., 1996, Copyright, ACS.)

and $J(\beta)$ is a function of the stretching ratio, $\lambda$, given by

$$J(\beta) = \left(\lambda^2 - \frac{1}{\lambda}\right)\cos^2\beta + \frac{1}{\lambda}. \qquad \text{(IV.4.25)}$$

Thus, $J(\beta) = \lambda^2$ and $\lambda^{-1}$ for the stretching ($\beta = 0°$) and the perpendicular directions ($\beta = 90°$), respectively. $M^*_{\text{os}}(\beta)$ is the longitudinal modulus dependent on the stretching direction given by

$$M^*_{\text{os}}(\beta) = K_{\text{os}} + \left[\frac{1}{3} + J(\beta)\right]\mu. \qquad \text{(IV.4.26)}$$

The first term in the bracket of the right hand side of eq. (IV.4.24) represents the thermal fluctuations and the second the quasistatic fluctuation term. Mendes et al. (1996) analyzed the SANS data with several theories, namely (i) the cluster model (Bastide et al., 1990), (ii) the Onuki theory (Onuki, 1992), and (iii) Rabin–Bruinsma theory (Rabin and Bruinsma, 1992). However, none of these theories did not described quantitatively the large range of deformations. Ramzi and coworkers studied the butterfly patterns of randomly cross-linked (Ramzi et al., 1995) and end-linked PS networks (Ramzi et al., 1997). Here, the behaviors of free chains embedded in the network were also considered. The agreement between the experiments and the theories, however, was poor. It was rather surprising to learn that large inhomogeneities were also present even in an end-linked polymer network. They attributed this to imperfect chemistry or to physical origins, such as entanglement or large loops.

## IV.4.6 INHOMOGENEITIES IN NOVEL GELS

Recently, various types of supertough polymer gels capable of high recoverable deformability and/or with high shear/compressive moduli have been developed. These gels have unique structures, such as sliding cross-links (slide-ring gels; Figure IV.4.13a), and tetrahedral networks (Tetra-PEG gels; Figure IV.4.13b). The origins of the advanced mechanical properties were investigated by SANS and mechanical measurements (Okumura and Ito, 2003; Sakai, 2008). It was found that one of the common features of these gels is that frozen inhomogeneities inherent to polymer gels are negligible. This fact indicates that cross-links are introduced very effectively. The relationship between the structure and mechanical properties of supertough polymer gels is reviewed elsewhere (Shibayama, 2009).

### IV.4.6.1 Slide-Ring Gels (Okumura and Ito, 2001)

Slide-ring (SR) gels are new class of gels having mobile cross-links that can slide along the backbone polymer chains. Figure IV.4.13a shows an illustration of slide-ring gels, in which double- and/or triple-ring cross-linkers are able to slide along

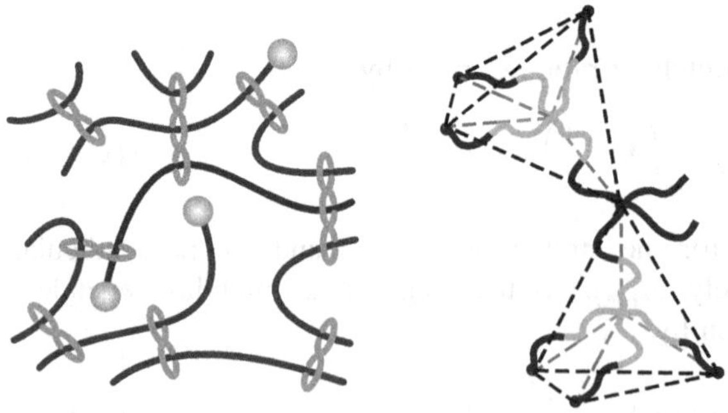

(a) Slide-ring (SR) gel

(b) Tetra-PEG gel

**Figure IV.4.13** Illustration of (a) slide-ring gels and (b) Tetra-PEG gels.

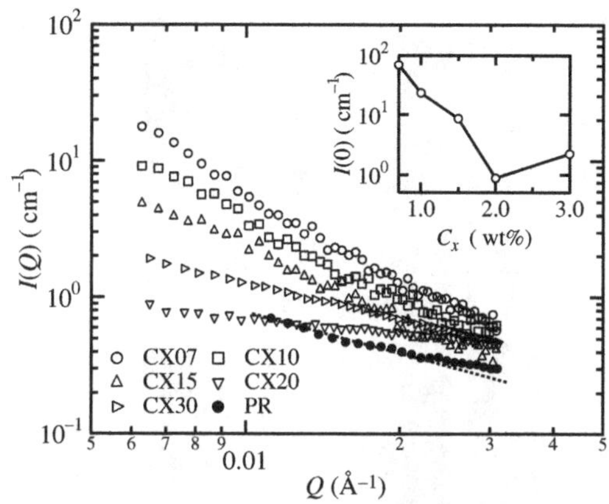


**Figure IV.4.14** Scattering intensity, $I(Q)$, of the SR gels in d-DMSO with various cross-link concentrations, $C_x$s. $I(Q)$ decreases with increasing $C_x$, which is opposite to the case of conventional covalent-bonded chemical gels. The filled circles with the line denote $I(Q)$ for the precursor of the SR gel, that is, polyrotaxane solution in d-DMSO. The inset shows the $C_x$ dependence of $I(0)$. (Karino et al., 2005, Copyright, ACS.)

polymer chains. These mobile cross-links allows (1) high extensibility, (2) large degree of swelling, and (3) large reversible deformability. Figure IV.4.14 shows SANS curves of SR gels with different cross-link densities together with polyrotaxane solution (PR). Contrary to the case of conventional chemical gels (Figure IV.4.5), $I(Q)$ decreased with increasing cross-linker concentration from $C_x = 0.7$ wt% (CX07) to $C_x = 2.0$ wt% (CX20) and then increased at $C_x = 3.0$ wt% (CX30). This behavior is explained by a competition of (1) suppression of concentration fluctuations by introducing cross-links to a polymer solution and of (2) an increase of frozen inhomogeneities (Karino et al., 2005).

Figure IV.4.15 shows the two-dimensional SANS isointensity patterns of deformed SR gels (CX10). The stretching direction was horizontal. As clearly shown, the scattering patterns at $\lambda = 1.0$ are isotropic, but become elliptic by increasing $\lambda$ with the long axis perpendicular to the stretching direction ("normal" butterfly pattern). This was the first observation of normal butterfly patterns in polymer gels (Karino et al., 2005). As was discussed in Section IV.4.5.3, conventional chemical gels show "abnormal" butterfly patterns in which the scattering intensity increased in the parallel direction (see Figure IV.4.12) (Mendes et al., 1991, 1995; Rouf et al., 1994; Shibayama et al., 1998a). The origin of the appearance of abnormal butterfly patterns is explained by inhomogeneities. The SR gels, on the other hand, exhibit normal butterfly patterns, indicating that sliding cross-links effectively lower the inhomogeneities in the gel. Figure IV.4.16 shows the isointensity scattering patterns of CX10 ($C_x = 1.0$ wt%) (upper) and CX20 ($C_x = 2.0$ wt%) (lower). As clearly shown in the figure, the scattering intensity for CX20 is much lower than that of CX10. More interestingly, CX20 shows "abnormal" butterfly patterns with increasing $\lambda$. This suggests that cross-link inhomogeneities increase even in SR gels by increasing $C_x$.

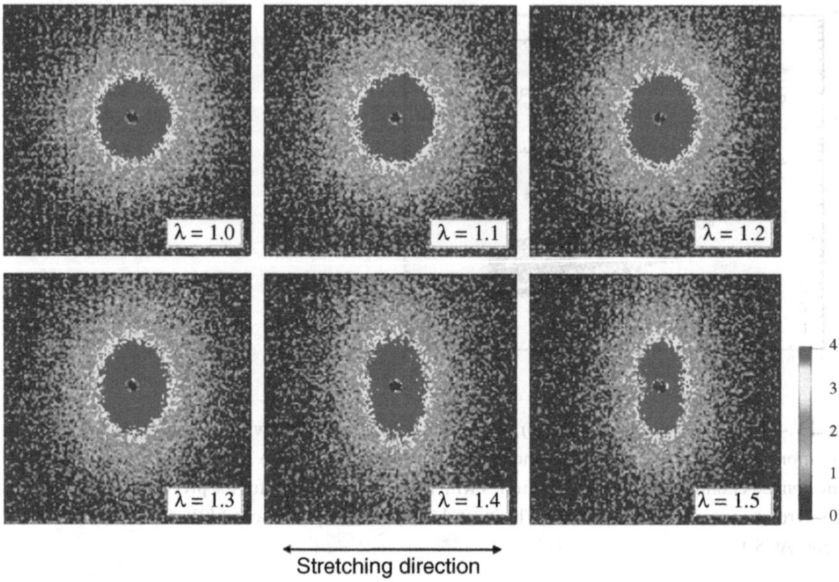


**Figure IV.4.15** SANS isointensity patterns of the SR gel (CX10) with various stretching ratios, $\lambda$s. The stretching direction is horizontal. (Karino et al., 2005, Copyright, ACS.)

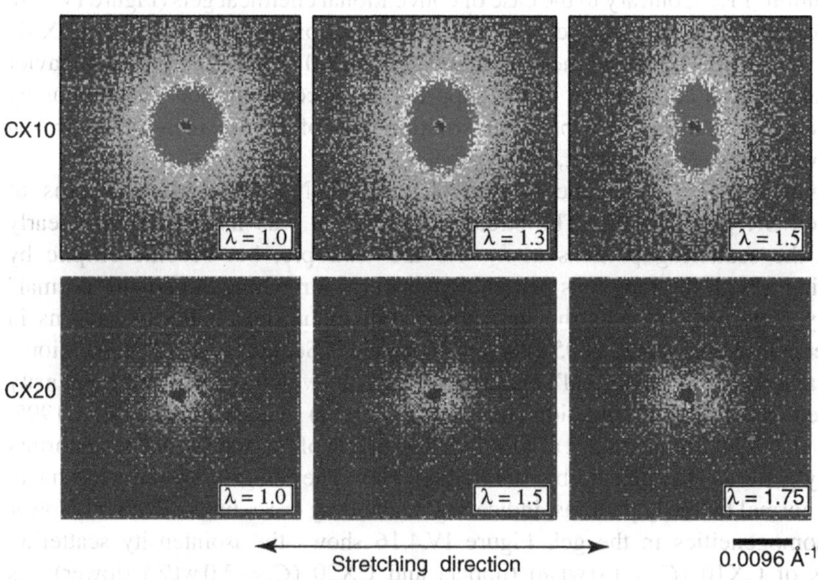


**Figure IV.4.16** SANS isointensity patterns of the SR gel with CX10 ($C_x = 1.0\%$) and CX20 ($C_x = 2.0\%$). Upon stretching, CX10 exhibits normal butterfly patterns, while CX20 does abnormal butterfly patterns. (Karino et al., 2005, Copyright, ACS.) (See the color version of this figure in Color Plates section.)

## IV.4.6.2 Tetra-PEG Gels: Polymer Gels with Negligibly Inhomogeneous Gels

Sakai et al. (2008) succeeded in preparation of extremely homogeneous polymer gels called Tetra-PEG gels. As shown in Figure IV.4.13b, Tetra-PEG gels consist of cross-end-coupling of tetraamine-terminated PEG (TAPEG) and tetra-NHS-glutarate-terminated PEG (TNPEG) were prepared from tetrahydroxyl-terminated PEG (THPEG) having equal arm lengths. Here, NHS represents for *N*-hydroxysuccinimide. The molecular weights ($M_w$) of TAPEG and TNPEG were matched to each other, and four sets of samples having different $M_w$s were prepared, that is, $M_w = 5000$, 10,000, 20,000, and 40,000 g/mol. By simply mixing TAPEG and TNPEG aqueous solutions, Tetra-PEG gels can be obtained instantly. Tetra-PEG gels are (1) quick and easy made, (2) nontoxic, and (3) mechanically tough. Surprisingly, Tetra-PEG gels exhibit high mechanical moduli even superior to those of native articular cartilage. The details of TAPEG and TNPEG preparation and physical properties are reported elsewhere (Sakai et al., 2008).

Figure IV.4.17 shows SANS intensity curves of (a) Tetra-PEG-10k (b) Tetra-PEG-20k at various polymer concentrations, $\phi_0$ (Matsunaga et al., 2009). The solid lines represent fitting curves with OZ functions (eq. (IV.4.16)) (de Gennes, 1979) with $A_{\text{inhom}} = 0$ and the dashed lines are obtained with the star polymer function (eq. (IV.4.16)). Surprisingly, all curves for gels are nicely fitted with the theoretical function (eq. (IV.4.16)) without the inhomogeneity term (i.e., $A_{\text{inhom}} = 0$). The correlation length, $\xi$, was a decreasing function of $\phi_0$. Note that the PTHF model network prepared by end-coupling of telechelic polymer chains exhibited significant inhomogeneities (see Figure IV.4.11). Comparison of the PTHF gels and Tetra-PEG gels indicates that Tetra-PEG gels, prepared by cross-end-coupling two kinds of four-arm PEG macromers having different functional groups at the ends, are extremely homogeneous, and an "ideal" network free from defects is formed.

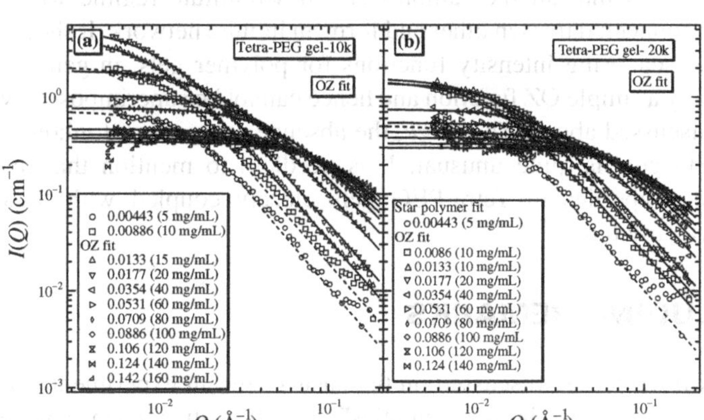


**Figure IV.4.17** SANS intensity functions of Tetra-PEG gels. (a) Tetra-PEG-10k and (b) Tetra-PEG-20k. The solid lines are obtained by fitting with eq. (IV.4.16) with $A_{\text{inho}} = 0$. The dashed lines are fitted functions for a star polymer chains. (Matsunaga et al., 2009, Copyright, ACS.)

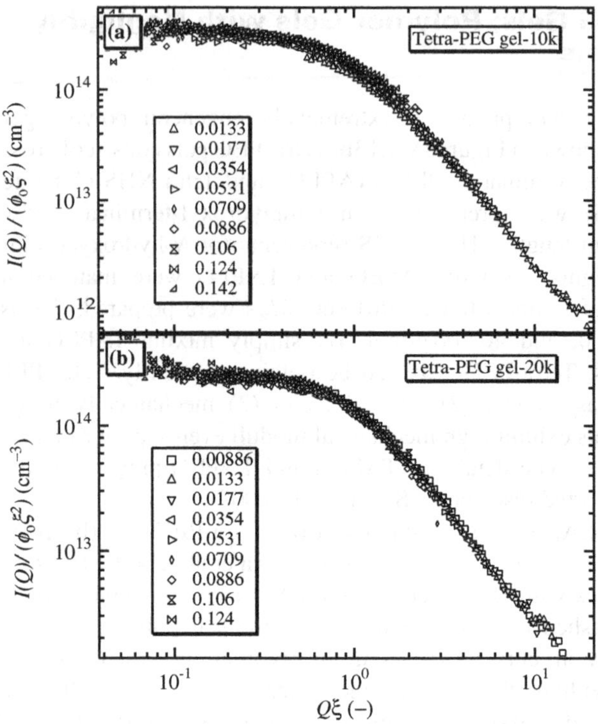


**Figure IV.4.18** Master curve of Figure IV.4.17 for (a) Tetra-PEG-10k and (b) Tetra-PEG-20k. (Matsunaga et al., 2009, Copyright, ACS.)

Figure IV.4.18 shows plots of $I(Q)/\phi_0\xi^2$ versus $\xi Q$. All the curves for gels are superimposed to a single master curve for (a) Tetra-PEG-10k and (b) Tetra-PEG-20k, respectively. This indicates that all the samples are in semidilute regime where macromers chains are connected to each other and form an infinite network. It should be noted here that the scattering intensity functions for polymer gels, in general, cannot be represented by a simple OZ function and hence cannot be superimposed by a single function as discussed above. Therefore, the absence in significant inhomogeneities in Tetra-PEG gels is quite unusual. It is needless to mention that the advanced mechanical properties in Tetra-PEG gels closely coupled with their homogeneous network structure.

## IV.4.7 CONCLUDING REMARKS

Inhomogeneities in soft matter are discussed on the basis of recent experimental as well as theoretical works. First, "inhomogeneities" in soft matter science were discussion based on a literature survey. The survey clearly indicates that "inhomogeneities" are more or less exclusively used to discuss polymer gels. This is because polymer gels inherently possess inhomogeneities by introduction of cross-links either chemically or physically. Other origins of inhomogeneities, such as clustering and vitrification are pointed out. Then, various scattering theories to deal with inhomogeneities are reviewed.

Characteristic features of the inhomogeneities in conventional chemical and physical gels are frozen inhomogeneities, which are observed as an upturn in the low $Q$ region in scattering functions. Different from these classic gels, supertough gels developed recently with novel concepts show extremely small inhomogeneities. Deformation studies of slide-ring gels exhibit normal butterfly patterns, indicating that SR gels behave like a polymer solution under shear. Furthermore, normal-to-abnormal butterfly pattern transition suggests that even SR gels loose their capability of sliding cross-link motion by increasing the cross-link density. Tetra-PEG gels consist of tetra-arm chains constructing a diamond-like three-dimensional structure. This is the reason why Tetra-PEG gels exhibit high mechanical moduli superior to those of native articular cartilage. It is concluded that the mechanical properties of polymer gels are closely coupled with the structural inhomogeneities.

## REFERENCES


Bastide, J. and Candau, S.J. Structure of gels as investigated by means of static scattering techniques. In: Cohen Addad, JP (editor). *The Physical Properties of Polymer Gels*, Wiley, New York, 1996, Chapter 9, p. 143.

Bastide, J. and Leibler, L. *Macromolecules* **1988**, *21*, 2647.

Bastide, J. Leibler, L., and Prost, J. *Macromolecules* **1990**, *23*, 1821.

Beltzung, M., Herz, J., and Picot, C. *Macromolecules* **1983**, *16*, 580.

Benoit, H. and Picot, C. *Pure Appl. Chem.* **1966**, *12*, 545.

Candau, S., Bastide, J., and Delsanti, M. *Adv. Polym. Sci.* **1982**, *44*, 27.

Cohen Addad, J.P. (editor). *Physical Properties of Polymer Gels*, Wiley, New York, 1996.

Cohen, Y., Ramon, O., Kopelman, I.J., and Mizrahi, S. *J. Polym. Sci. Polym. Phys. Ed.* **1992**, *30*, 1055.

Debye, P. and Bueche, A.M. *J. Appl. Phys.* **1949**, *20*, 518.

Debye, P. and Bueche, A.M. *J. Chem. Phys.* **1950**, *18*, 1423.

de Gennes, P.G. *Scaling Concepts in Polymer Physics*, Cornell University, Ithaca, 1979.

de Rossi, D., Kajiwara, K., Osada, Y.,and Yamauchi, A. (editors). *Polymer Gels*, Plenum Press, New York, 1991.

Einstein, A. Theory of the opalescence of homogeneous liquids and mixtures of liquids in the vicinity of the critical state. In: Alexander, J. (editor). *Colliod Chemistry*, Vol. *1*, Reinhold, New York, 1926, p. 323.

Flory, P.J. *Principles of Polymer Chemistry*, Cornell University, Ithaca, 1953.

Flory, P.J. and Rehner, J., Jr. *J. Chem. Phys.* **1943**, *11*, 521.

Guenet, J.M. *Thermoreversible Gelation of Polymers and Biopolymers*, Academic Press, New York, 1992.

Guinier, A. and Fournet, G. *Small-Angle Scattering of X-Rays*, John Wiley & Sons, Inc., New York, 1955.

Hammouda, B., Ho, D.L., and Kline, S. *Macrmolecules* **2002**, *35*, 8578.

Hammouda, B., Ho, D.L., and Kline, S. *Macrmolecules* **2004**, *37*, 6932.

Hecht, A.M., Duplessix, R., and Geissler, E. *Macromolecules* **1985**, *18*, 2167.

Horkay, F., Hecht, A.M., Mallam, S., Geissler, E., and Rennie, A.R. *Macromolecules* **1991** *24*, 2896.

Ikkai, F. and Shibayama, M.In: Stokke, B.T. (editor). *Anomalous Crosslink Density Dependence of Spatial Inhomogeneities in Polymer Gels*, Vol. *30*, Wiley Polymer Networks Group Review Series, John Wiley & Sons, Ltd., Elgsaeter, 1999, pp. 383–392.

James, H.M. and Guth, E. *J. Chem. Phys.* **1947**, *15*, 669.

Jannink, G. *J. Polym. Sci. Part B, Polym. Lett.* **1969**, *7*, 865.

Joosten, J.G.H., McCarthy, J.L., and Pusey, P.N. *Macromolecules* **1991**, *24*, 6690.

KARINO, T., OKUMURA, Y., ZHAO, C., KATAOKA, T., ITO, K., and SHIBAYAMA, M. *Macromolecules* **2005**, *38*, 6161.
KLEMAN, M. and LAVRENTOVICH, O.D. *Soft Matter Physics*, Springer-Verlag, New York, 2003.
KOIKE, Y., TANIO, N., and OHTSUKA, Y. *Macrmolecules* **1989**, *22*, 1367.
KOIZUMI, S. *J. Polym. Sci. Part B, Polym. Phys.* **2004**, *42*, 3148–3164.
KOIZUMI, S., MONKENBUSCH, M., RICHITER, D., SCHWAHN, D., and FARAGO, B. *J. Chem. Phys.* **2004**, *121*, 12721.
MALLAM, S., HORKAY, F., HECHT, A.M., and GEISSLER, E. *Macromolecules* **1989**, *22*, 3356.
MALLAM, S., HORKAY, F., HECHT, A.M., RENNIE, A.R., and GEISSLER, E. *Macromolecules* **1991**, *24*, 543.
MARK, J.E. and ERMAN, B. *Rubberlike Elasticity: A Molecular Primer*, Wiley, New York, 1988.
MATSUNAGA, T. and SHIBAYAMA, M. *Phys. Rev. E, Rapid. Commun.* **2007**, *76*, 030401.
MATSUNAGA, T., SAKAI, T., AKAGI, Y., CHUNG, U., and SHIBAYAMA, M. *Macromolecules* **2009**, *42*, 1344.
MENDES, E.J., LINDNER, P., BUZIER, M., BOUE, F., and BASTIDE, J. *Phys. Rev. Lett.* **1991**, *66*, 1595.
MENDES, E., SCHOSSELER, F., ISEL, F., BOUE, F., BASTIDE, J., and CANDAU, S.J. *Europhys. Lett.* **1995**, *32*, 273.
MENDES, E., OESER, R., HAYES, C., BOUE, F., and BASTIDE, J. *Macromolecules* **1996**, *29*, 5574.
NORISUYE, T., TAKEDA, M., and SHIBAYAMA, M. *Macromolecules* **1998**, *31*, 5316.
NORISUYE, T., MASUI, N., KIDA, Y., SHIBAYAMA, M., IKUTA, D., KOKUFUTA, E., ITO, S., and PANYUKOV, S. *Polymer* **2002**, *43*, 5289.
OKABE, S., NAGAO, M., KARINO, T., WATANABE, S., ADACHI, T., SHIMIZU, H., and SHIBAYAMA, M. *J. Appl. Cryst.* **2005**, *38*, 1035.
OKABE, S., KARINO, T., NAGAO, M., WATANABE, S., and SHIBAYAMA, M. *Nucl. Inst. Meth. Phys. Res. A* **2007**, *572*, 853–858.
OKUMURA, Y. and ITO, K. *Adv. Mater.* **2001**, *13*, 485.
OKUMURA, Y. and ITO, K. *Nippon Gomu Kyokaishi* (*Jpn*), **2003**, *76*, 31.
ONUKI, A. *J. Phys. II France* **1992**, *2*, 45.
ONUKI, A. *Adv. Polym. Sci.* **1993**, *109*, 63.
ORNSTEIN, L.S. and ZERNIKE, F. *Proc. Acad. Sci., Amsterdam* **1914**, *17*, 793.
PANYUKOV, S. and RABIN, Y. *Phys. Rep.* **1996**, *269*, 1.
PINES, E. and PRINS, W. *J. Polym. Sci., Polym. Phys. Ed.* **1972**, *B10*, 719.
PUSEY, P.N. and van MEGEN, W. *Physica A* **1989**, *157*, 705.
RABIN, Y., and BRUINSMA, R. *Europhys. Lett.* **1992**, *20*, 79.
RAMZI, A., ZIELINSKI, F., BASTIDE, J., and BOUE, F. *Macromolecules* **1995**, *28*, 3570.
RAMZI, A., HAKIKI, A., BASTIDE, J., and BOUE, F. *Macromolecules* **1997** *30*, 2963.
ROUF, C., BASTIDE, J., PUJOL, J.M., SCHOSSELER, F., and MUNCH, J.P. *Phys. Rev. Lett.* **1994**, *73*, 830.
T. SAKAI, T. MATSUNAGA, Y. YAMAMOTO, C. ITO, R. YOSHIDA, S. SUZUKI, N. SASAKI, M. SHIBAYAMA and U. CHUNG *Macromoleules* **2008**, *41*, 5379.
SHIBAYAMA, M. *Macromol. Chem. Phys.* **1998**, *199*, 1.
SHIBAYAMA, M. *Bull. Chem. Soc. Jpn.* **2006**, *79*, 1799.
SHIBAYAMA, M. *J. Phys. Soc. Jpn.* 2009, *78*, 041008.
SHIBAYAMA, M. and NORISUYE, T. *Bull. Chem. Soc. Jpn.* **2002**, *75*, 641.
SHIBAYAMA, M. and TANAKA, T. *Adv. Polym. Sci.* **1993**, *109*, 1.
SHIBAYAMA, M., ISONO, K., OKABE, S., KARINO, T., and NAGAO, M. *Macromolecules* **2004**, *37*, 2909.
SHIBAYAMA, M., KARINO, T., DOMON, Y., and ITO, K. *J. Appl. Cryst.* **2007**, *40*, s43.
SHIBAYAMA, M., KAWAKUBO, K., IKKAI, F., and IMAI, M. *Macromolecules* **1998a**, *31*, 2586.
SHIBAYAMA, M., KAWAKUBO, K., and NORISUYE, T. *Macromolecules* **1998b**, *31*, 1608.
SHIBAYAMA, M., MATSUNAGA, T., and NAGAO, M. *J. Appl. Cryst.* **2009**, *42*, 628.
SHIBAYAMA, M., NAGAO, M., OKABE, S., and KARINO, T. *J. Phy. Soc. Jpn.* **2005**, *74*, 2728.
SHIBAYAMA, M., SHIROTANI, Y., HIROSE, H., and NOMURA, S. *Macromolecules* **1997**, *30*, 7307.
SHIBAYAMA, M., TAKAHASHI, H., and NOMURA, S. *Macromolecules* **1995**, *28*, 6860–6864.
SHIBAYAMA, M., TAKAHASHI, H., YAMAGUCHI, H., SAKURAI, S., and NOMURA, S. *Polymer* **1994** *35*, 2944.
SHIBAYAMA, M., TANAKA, T. and HAN, C.C. *J. Chem. Phys.* **1992**, *97*, 6829.
TAKATA, S., NORISUYE, T., and SHIBAYAMA, M. *Macromolecules* **2002**, *35*, 4779.
TANAKA, T., HOCKER, L.O., and BENEDEK, G.B. *J. Chem. Phys.* **1973**, *59*, 5151.
WU, W., SHIBAYAMA, M., ROY, S., KUROKAWA, H., COYEN, L.D., NOMURA, S., and STEIN, R.S. *Macromolecules* **1990**, *23*, 2245.

# IV Applications

## IV.5 Protein Dynamics Studied by Neutron Incoherent Scattering

*Mikio Kataoka* and Hiroshi Nakagawa*

### IV.5.1 INTRODUCTION

Proteins are functional elements in all living organisms. Almost all vital phenomena are mediated by specific proteins. A protein is a heteropolymer comprising 20 types of amino acids. The sequence of amino acids of a protein is strictly determined by genetic code. When the polypeptide chain with the amino acid sequence encoded in the gene is biosynthesized, the polypeptide chain folds spontaneously into the unique tertiary structure (Figure IV.5.1a). The tertiary structure of a protein can be determined by X-ray crystallography. Protein works in an aqueous environment at ambient temperature, indicating that protein cannot escape from thermal fluctuations. In fact, proteins are fluctuating thermally and can take some conformational substates (Ansari et al., 1985; Frauenfelder and McMahon, 1998; Frauenfelder et al., 2001). The magnitude of physiologically relevant input is at the same level as that of the thermal fluctuations. The understanding of protein dynamics is important to clarify how protein can discriminate physiologically relevant motions from random fluctuations. It is generally recognized that the internal motion of protein is essential for a protein function (Ikeguchi et al., 2005; Karplus, 2000). Protein internal dynamics is characterized in the timescale from picosecond to nanosecond and in the space scale on the order of approximately angstrom (Joti et al., 2008a; Kataoka et al., 2003; Kitao and Go, 1999; Smith, 1991). Inelastic neutron scattering (INS) measurement is a powerful and unique technique for studying the protein dynamics (Gabel et al., 2002; Kataoka et al., 1999; Smith, 1991; Zaccai, 2000). A neutron from the nuclear reactor and spallation source has wavelengths of a few angstroms and energies of a few millielectron volts, which correspond to the interatomic distances and the energy of excitations in the protein, respectively. Thus, the scattered neutron from the molecule has both timescale and space scale information of the protein conformational dynamics (Smith, 1991).

*Corresponding author.

*Neutrons in Soft Matter* Edited by Toyoko Imae, Toshiji Kanaya, Michihiro Furusaka, and Naoya Torikai

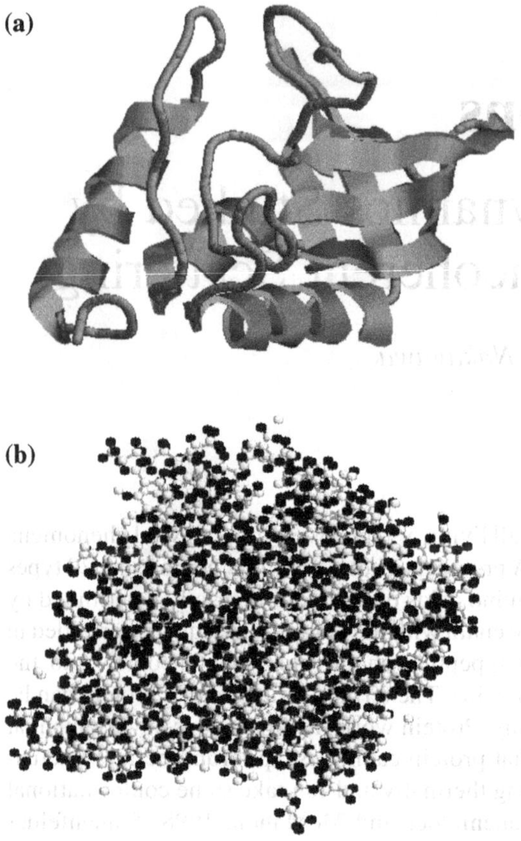

**Figure IV.5.1** Crystal structure of SNase PDB (Protein Data Bank) code: 1STN (Loll and Lattman, 1989). (a) Ribbon model of the structure. (b) Every atom is described by ball and the hydrogen atom is shown in black.

Protein dynamics study by INS first appeared in 1982 (Jacrot et al., 1982). Doster et al. (1989) published the pioneering and important paper on a protein dynamical transition in 1989, which indicates the effectiveness of INS. INS studies of the protein dynamics were reviewed in 1991 by Smith (1991), who compared the experimental data with the molecular dynamics (MD) simulation. At that time, the applications of INS to biology were very limited, regardless of its effectiveness for the protein dynamics study. But in the past two decades, the number of related works has gradually increased with the advanced neutron technology (Gabel et al., 2002). In the postgenome study, knowledge of not only the protein structure but also its fluctuation becomes more important in relation to protein function.

In this chapter, our protein dynamics works by INS will be introduced, including the practical aspects of the experiment: the biological sample and instrument. These dynamical properties of proteins are closely related to the understanding of molecular mechanism of biological functions. The INS studies on protein dynamics, thus, contribute to protein science and biophysics. Most observable phenomena such as Boson peak and dynamical transition are also observed in other soft matter materials. Therefore, our studies are expected to also contribute to the soft matter physics.

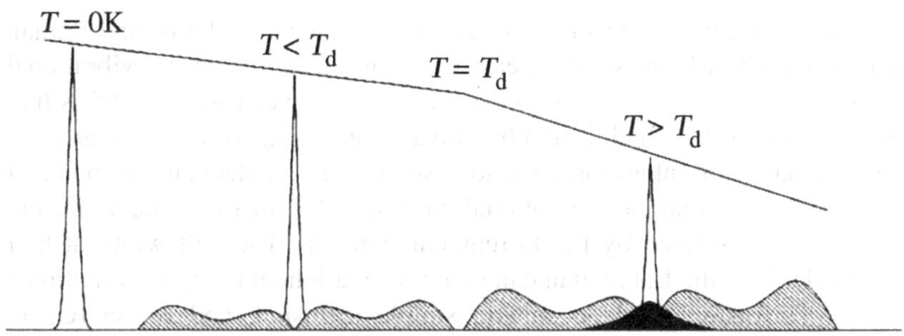


**Figure IV.5.2** Schematic diagram of three observed types of incoherent neutron scattering: elastic, quasi-elastic, and inelastic scattering at three typical temperatures (Cusack, 1989; Smith, 1991).

## IV.5.2 NEUTRON SCATTERING IN PROTEIN DYNAMICS: GENERAL ASPECTS OF THE EXPERIMENT

The incoherent neutron scattering cross section for hydrogen atom is larger than any other atoms. Since half of atoms in a protein are hydrogen atoms, the signal from the hydrogen atoms in the protein dominates the total neutron scattering intensity. Because hydrogen atoms are evenly distributed in a protein (Figure IV.5.1b), the scattering gives a global view of protein motions (Smith, 1991). The typical incoherent scattering at three different temperatures is schematically shown in Figure IV.5.2, where scattering can be classified into three regions along the energy axis: elastic, quasi-elastic, and inelastic scattering (Cusack, 1989; Smith 1991). At $T = 0\text{K}$, the protein motions are essentially frozen and only strong elastic scattering can be observed. As the temperature is raised, the elastic intensity decreases, accompanying the appearance of inelastic scattering and, at the higher temperature, quasi-elastic scattering. The elastic scattering provides information on correlation between the initial atomic position and the position at infinite time. The correlation time $\tau$ is determined by the instrumental energy resolution $\Delta\omega$, with the uncertainty principle. When the atoms fluctuate around the equilibrium position, the elastic scattering gives information on geometries of the motions involved, as is the case with the Debye–Waller factor (B-factor) in X-ray crystallography (Chong et al., 2001; Joti et al., 2002). In the Gaussian approximation, the elastic intensity can be described with the atomic mean square displacement (MSD) $\langle u^2 \rangle$ as follows:

$$S(Q, \omega = 0 \pm \Delta\omega) \propto \exp\left( -\frac{1}{3} \langle u^2 \rangle Q^2 \right). \qquad \text{(IV.5.1)}$$

Inelastic scattering gives the excitation of the vibrational spectrum extended up to several hundreds $\text{cm}^{-1}$ (Goupil-Lamy et al., 1997; Kataoka et al., 2003). INS is complementary to the IR or Raman spectroscopy. One advantage of INS over these spectroscopic methods is that the observed spectrum is quantitatively compared with the theoretically calculated spectrum. The estimation of transition dipole moment and changes in electronic polarizability is necessary to calculate the IR and Raman spectrum, respectively, which is not an easy calculation. Moreover, all vibrational modes are not necessarily observed by optical vibrational spectroscopy. INS is free from such a selection rule, which is another advantage. The quasi-elastic scattering comes from stochastic or anharmonic motions such as intramolecular motions that cross energy barrier, overdamped vibrational modes, and diffusions. The scattering intensity is usually described by the Lorentzian function. The half width at half maximum (HWHM) of the Lorentzian can give information on the type of motions: jump diffusion, diffusion inside a sphere, simple diffusion, and so on (Gabel et al., 2002). Anharmonic protein motion is relatively slow, which is characterized by nanosecond. Therefore, INS experiment with high-energy resolution of approximately microelectron volt is necessary to analyze the quasi-elastic scattering of the proteins.

Protein dynamics are spread over a wide range of time and space. A single-neutron instrument cannot cover all the dynamics. There are also some other types of instruments for the protein dynamics study: backscattering, time of flight (TOF), and neutron spin echo (NSE) instruments. The backscattering instrument is usually installed at the reactor. The quasi-elastic scattering measurement with high-energy resolution is available by the cold source. With the thermal source, scattering in the wide $Q$ range is accessible. TOF instruments are often installed at the spallation source. Wide spectrum range can be measured using white neutron. NSE can observe intermediate scattering function directly and analyze the very slow dynamics with nanosecond to microsecond timescale. But the application to the protein dynamics remains quite rare cases because of the limitation of accessible $Q$ and $\omega$ range (Bellissent-Funel et al., 1998; Bu et al., 2005).

The disadvantages of INS come from the fact that we need a reactor or an accelerator for the neutron source. The places where we can carry out INS experiment are quite limited. Although we can have access to the neutron facilities, the allocated machine time is quite limited. Furthermore, the neutron experiments usually require a large amount of biological samples, approximately a few 100 mg, and long measurement time, about a week. This is because the scattering signals from the samples are very weak due to the low incident neutron flux (this situation would be improved by construction of new neutron sources such as J-PARC in Japan and SNS in the United States). For the incoherent neutron scattering, the low incident flux is also a disadvantage in obtaining good energy resolution. The $\Delta E/E$ is generally more than 10%. With the optical vibrational spectroscopy, we can reach energy resolution of 1 $cm^{-1}$. Therefore, INS may not be a universal method, but a quite unique method for quantitative analysis of the protein dynamics. Recent development of the neutron source and the neutron spectrometer as well as the method to purify a large amount of protein in molecular biology allows us to obtain the good quality INS data.

We have been applying INS to Staphylococcal nuclease (SNase) to understand the dynamical properties of proteins (Goupil-Lamy et al., 1997; Kataoka et al., 1999, 2003; Nakagawa et al., 2004). SNase is a widely used model protein for the study of protein folding (Anfinsen, 1972). The advantages of SNase for INS study can be summarized as follows: (1) The high-resolution crystal structure is available

(Figure IV.5.1a) (Loll and Lattman, 1989). We can carry out MD simulations or normal mode analysis for the analysis of INS spectrum (Goupil-Lamy et al., 1997; Tokuhisa et al., 2007). (2) A molecular genetic system is available. We can obtain a fairly large amount of protein easily. The proteins were overexpressed in *Escherichia coli* and purified (Shortle and Meeker, 1989; Nakagawa et al., 2008; Onitsuka et al., 2008). In general, we could obtain 100–120 mg proteins with 1 L culture. We also have interesting sets of mutants (Flanagan et al., 1993; Onitsuka et al., 2008). Especially, the large fragment that lacks 13 amino acids from its C-terminus cannot take a unique tertiary structure; however, it takes nonnative conformations under physiological solvent conditions (Flanagan et al., 1992; Goupil-Lamy and Smith, 1996; Hirano et al., 2002, 2005). The fragment retains enzymatic activity and can fold into native conformation in the presence of substrate or inhibitor (Flanagan et al., 1992; Hirano et al., 2002; Onitsuka et al., 2008). The fragment is considered to be a good model for natively unfolding proteins (Onitsuka et al., 2008). This is a great advantage for the INS experiment to examine whether there are specific dynamical properties of the folded native protein. Unfolded state of protein is usually realized as thermal denaturation, acid denaturation, or denaturant-induced denaturation. These procedures are not necessarily adequate for INS measurement. Since INS measures thermal motions, thermal denaturation is inadequate to examine the effect of denaturation. Addition of acid or denaturant may affect the hydrogen composition of the solvent and INS spectrum, since INS mainly comes from hydrogen. We can examine the effect of the conformational states on dynamics under the same condition using the mutant.

The environment of the protein is another factor for determining the internal motions of proteins (Bellissent-Funel, 2000; Fitter et al., 1999b; Nakagawa et al., 2008; Settles and Doster, 1996; Tournier and Smith, 2003a). It has been reported that proteins cannot function at cryogenic temperatures or at low hydration levels (Doster and Settles, 2005; Ferrand et al., 1993), which strongly implies that both thermal fluctuation and hydration affect protein function. To understand the molecular mechanisms that underlie cellular biology, it is essential to examine protein dynamics in hydration environments. To study the hydration effect on protein dynamics, lyophilized powder of protein was equilibrated with 90% relative humidity of $D_2O$ to ensure hydration. The hydration level was examined by a thermal balance (Nakagawa et al., 2008).

## IV.5.3 COMPUTATIONAL CALCULATION AND NEUTRON SCATTERING

MD simulation is another powerful method for the analysis of the protein internal dynamics (Joti et al., 2008b; Kitao and Go, 1999; Karplus and McCammon, 2002). Time and space range in MD simulation is almost consistent with that accessible by INS. The INS spectrum can be calculated quantitatively (Goupil-Lamy et al., 1997; Joti et al., 2008a; Tarek and Tobias, 2000a; Tarek et al., 2000b). The experimentally observable quantity is the dynamical structural factor $S(Q, \omega)$. $S(Q, \omega)$ is the time

Fourier transform of the intermediate scattering function $I(Q, t)$:

$$S(Q,\omega) = \frac{1}{2\pi}\int_{-\infty}^{\infty} I(Q,t)\exp(-i\omega t)\mathrm{d}t. \qquad \text{(IV.5.2)}$$

$I(Q, t)$ is the space Fourier transform of self-correlation function $g(r, t)$, which is described as

$$I(Q,t) = \int_{-\infty}^{\infty} g(r,t)\exp(-iQr)\mathrm{d}t = \sum_{l=1}^{N}\langle \mathrm{e}^{iQr_l(t)}\mathrm{e}^{-iQr_l(0)}\rangle. \qquad \text{(IV.5.3)}$$

Thus, $S(Q, \omega)$ can be calculated directly from the atomic trajectories $r_i(t)$ in MD simulation. The combination of MD simulation with the INS allows the characterization of the protein dynamics in detail. For the purpose, we can obtain simulation packages for biomolecules such as Amber (Case et al., 2006) and CHARMM (Brooks et al., 1983). Kneller developed the neutron data calculation software, nMoldyn (Kneller, 1995). This software can calculate the neutron spectrum and structural factor using the atomic trajectory from MD simulation.

## IV.5.4 MOLECULAR VIBRATION IN HIGH-ENERGY SPECTRUM

The molecular vibrational INS in the wide energy range between 100 and 4000 cm$^{-1}$ can be obtained by the time-focusing crystal analyzer spectrometer such as TOSCA at the ISIS spallation source (Goupil-Lamy et al., 1997; Kataoka et al., 2003). This paves the way for detailed comparisons of the results of dynamical calculations and optical spectroscopic experiments. Figure IV.5.3 is the INS spectrum of SNase at 25 K. This spectrum is the best quality INS data of a globular protein reported so far.

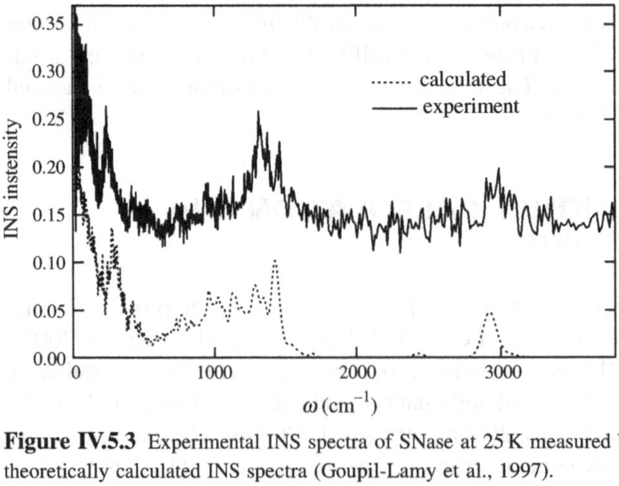


**Figure IV.5.3** Experimental INS spectra of SNase at 25 K measured by TFXA at ISIS and the theoretically calculated INS spectra (Goupil-Lamy et al., 1997).

The theoretical spectrum was calculated with CHARMM. The calculated spectra were convoluted with the instrumental energy resolution, and thus the obtained normal mode calculation was compared with the experimental spectrum quantitatively, including the spectral intensity. Using the results of normal modes calculation, we can assign the peaks of INS spectrum. The typical assignment for the spectrum is given in Table IV.5.1. For example, the CH stretch band is assigned at around 2900 $cm^{-1}$, and the CN stretch band is also visible at around 1600 $cm^{-1}$. The vibrations between 700 and 1000 $cm^{-1}$ are mostly assigned as $CH_3$ rotation, $CH_2$ rotation, and CC stretch. The sharp peak at 235 $cm^{-1}$ is identified as methyl torsions. In the range between 350 and 500 $cm^{-1}$, the vibrations are delocalized over the protein, which are coupled skeletal angle and dihedral displacements. Although the peaks can be assigned qualitatively, the quantitative coincidence between the experiment and the theory is not necessarily sufficient. The observed peak positions are shifted from the expected positions, and intensity distribution between the experiment and the theory is completely different. This fact strongly indicates that the improvement of theoretical model including the force field is required for the entire understanding of experimental spectrum of a protein. Consequently, INS can contribute to the improvement of the theoretical dynamical model of proteins.

**Table IV.5.1** Vibrational Peaks for INS Spectrum of Staphylococcal nuclease at 25 K and the Peak Assignment of the Experiment with the Theoretical Normal Mode Analysis (Goupil-Lamy et al.,1997)

| Experiment ($cm^{-1}$) | Theory ($cm^{-1}$) | Assignment[a] |
|---|---|---|
| 235 | 269 | $CH_3$-t |
| 400–450 | 395–435 | CCC-def, CCN-def, skeletal |
| 477 | 471 | CCC-def, CCN-def, skeletal |
| 470–590 | – | water O–H. . .O-b |
| 720–775 | 720–795 | $CH_2$-r, CH-b |
| 837 | 835 | $CH_2$-r |
| 936 | 958 | $CH_2$-r, $CH_3$-r |
| 1136 | 1136 | CH-b, $CH_2$-tw, $CH_3$-r |
| 1284 | 1291 | $CH_2$-tw, $CH_2$-w, CH-b, $CH_3$-sb |
| 1326 | | |
| | 1352 | $CH_2$-w, $CH_2$-tw, CH-b, $CH_3$-sb |
| 1386 | | |
| 1455 | 1431 | $CH_2$-b, $CH_3$-ab, CH-ip |
| 1555 | 1530 | CN-s |
| | 1686 | CN-s |
| | 2430 | ND-s |
| 2952 | 2920 | CH-s |

[a] The abbreviation used are as follows: *t*, torsion; def, deformation; b, bend; r, rock, tw, twist; w, wag; sb, symmetric bend; ab, antisymmetric bend; ip, in-plane bend; s, stretch.

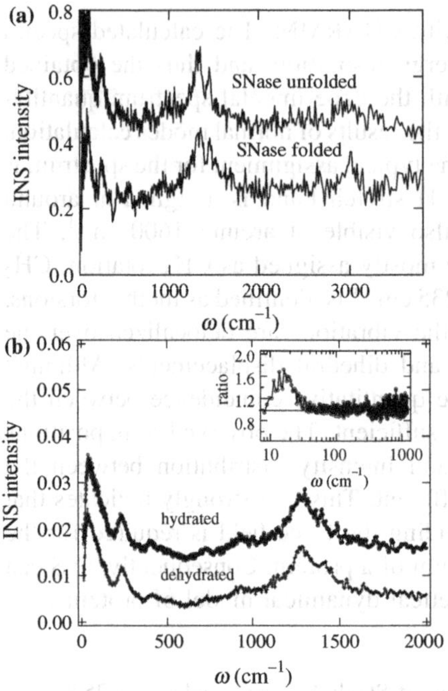


**Figure IV.5.4** (a) Comparison of INS spectra between the wild-type SNase (folded form) and the truncated mutant (unfolded form). For the sake of clarity, the spectrum of mutant is shifted 0.2 unit along the ordinate (Kataoka et al., 2003). (b) INS spectra of dehydrated and hydrated wild-type SNase. The spectrum of the hydrated sample is shifted 0.01 unit along the ordinate. Inset is the ratio of the dehydrated spectrum to the hydrated one (Nakagawa et al., 2008).

We are interested in the dynamical properties specific to the folded protein. In order to approach this problem, the INS spectra of the folded and the unfolded SNase were measured (Kataoka et al., 2003). The INS spectra of the wild-type SNase and the truncated mutant are shown in Figure IV.5.4a. The CD spectrum of the truncated mutant of the hydrated powder ensures that the mutant takes nonnative conformation. The folded and the unfolded SNase gave quite similar spectra, especially in higher energy region. There are some minor differences. For example, the peak at about 3000 $cm^{-1}$ is sharper for the folded state than for the unfolded state. The peak at about 150 $cm^{-1}$ is distinct for the folded state, while the peak at about 770 $cm^{-1}$ is distinct for the unfolded state. The latter would be due to the secondary structures. However, as a whole, within the present energy resolution, we concluded that these two spectra are essentially identical. This is rather reasonable, because these vibrations are originated from covalent bond natures. The covalent bonds are unchanged upon unfolding. Since the local environment of each amino acid residue formed by noncovalent bonds should be different for the folded and unfolded states, peak shifts and/or peak broadening can be expected. In order to reveal such differences, we need more precise measurement with higher energy resolution. We also examined whether the protein individuality appears in the high-energy INS spectrum by comparing the spectrum of photoactive yellow protein (PYP) with that of SNase. Both proteins are categorized into α/β proteins (Loll and Lattman, 1989; Borgstahl et al., 1995). PYP contains *p*-coumaric acid as a chromophore (Borgstahl et al., 1995). Although the two spectra are quite similar in their appearances, especially in high-energy region, their intensity distribution is different. The difference would reflect the differences in the composition of amino acids, and thus the composition of covalent bonds. Especially, the differences in the range 100–600 $cm^{-1}$ may be due to the differences in the composition of secondary and tertiary structures (Kataoka et al., 2003).

We also examined the effect of hydration on the INS spectrum (Figure IV.5.4b). These spectra were observed at 20K with the LAM-D spectrometer at KENS in Tsukuba, Japan (Nakagawa et al., 2008). The hydration levels were 0.12 and 0.44 g $D_2O$/g protein for the dehydrated and the hydrated specimens, respectively. The ratio of the dehydrated spectrum to the hydrated spectrum is given in the inset of Figure IV.5.4b. The ratio confirms that the high-energy vibrational modes are almost identical between them, indicating that the covalent bond vibrations were not markedly affected by hydration. These modes are rather localized and not damped significantly by hydration friction. The small differences between 320 and 530 $cm^{-1}$ are likely due to the bending motion of hydrated heavy water on the protein. The inset also shows the significant effect of the hydration on the modes below 40 $cm^{-1}$, indicating that the harmonic modes of the protein in the low-energy range were highly affected by hydration. According to the theoretical studies, the vibrational modes with higher energy than 30 $cm^{-1}$ are evenly contributed to the atomic displacements of each amino acid residue, while the contribution from the modes with the energy lower than 30 $cm^{-1}$ shows the residue number dependence (Nishikawa and Go, 1987). Therefore, the present experimental observation confirms that the low-energy modes that are delocalized and collective motions over the protein structure are highly coupled with hydration.

## IV.5.5 BOSON PEAK

In the low-energy region of INS spectrum of proteins at cryogenic temperature, deviation of the excess density of states from the Debye frequency law is observed (Kanaya and Kaji, 1988). This is called a Boson peak, which is also commonly observed in synthetic polymers and glassy materials (Kanaya and Kaji, 1988; Yamamuro et al., 2002). This vibrational anomaly has been extensively studied, but its origin is still debated. As described above, the low-energy vibrational spectrum is affected by hydration, suggesting that the low-energy dynamics is closely related with protein function. The question arises whether the protein individuality appears in the INS spectrum in the low-energy region.

Figure IV.5.5a shows INS spectra up to 120 $cm^{-1}$ of native SNase, its truncated mutant, and myoglobin at 100K (Kataoka et al., 1999). For each specimen, spectrum of 100 K shows a Boson peak around 32 $cm^{-1}$. Myoglobin is a typical $\alpha$-protein, while SNase contains three $\alpha$-helices and a core composed of five $\beta$-sheets. The truncated mutant of SNase is in a compact denatured state, where $\alpha$-helical components are almost lost (Flanagan et al., 1992; Goupil-Lamy and Smith, 1996; Hirano et al., 2002). We expect that the collective modes are different from each other in these specimens. However, no significant differences are observed for the three spectra, indicating no essential differences in the density of states despite the

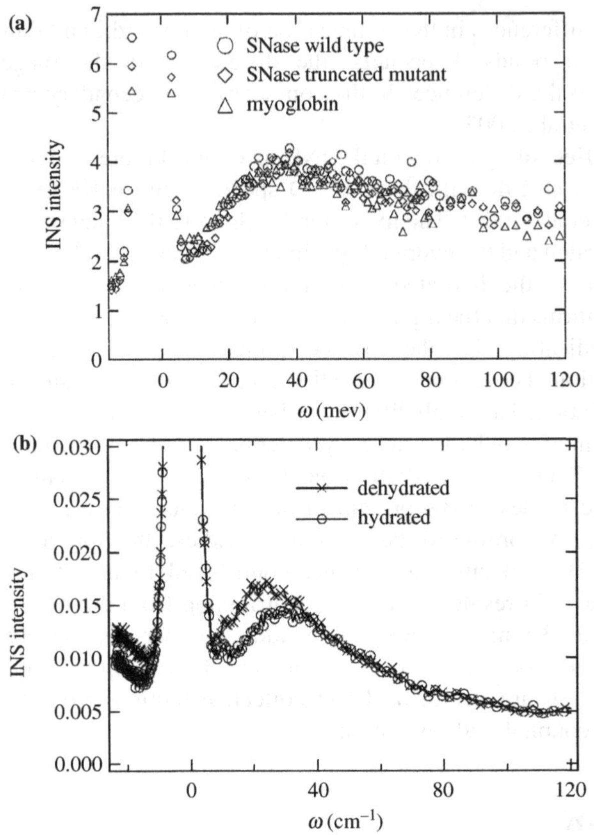


**Figure IV.5.5** Protein Boson peak observed at cryogenic temperature. (a) Comparison among wild-type SNase, its truncated mutant, and myoglobin observed with LAM-40 (Kataoka et al., 1999). (b) Boson peak of both the hydrated and the dehydrated SNase observed with LAM-40 (Nakagawa et al., 2008).

secondary structure components. The similarity in the Boson peak among the three specimens is suggestive of the origin of the Boson peak. Brown et al. (1972) suggested that the Boson peak was sensitive to protein secondary or tertiary structure. Chou (1984) suggested that low-frequency modes correspond to the internal deformation of secondary structure, such as accordion-like modes of α-helices. Cusack and Doster (1990) suggested that the denatured myoglobin also showed the low-frequency peak. They concluded that the peak is linked to modes involving cross-chain interactions such as relative motions of α-helices. The results shown in Figure IV.5.5a deny these possibilities. The question arises whether the Boson peak comes from modes based on secondary structure elements or from modes extended over the whole molecule. If the modes based on secondary structure elements are dominant, the position of the Boson peak remains at almost the same energy, while the modes extended over molecular size would show molecular weight dependence for the Boson peak position (Kataoka et al., 1999). We showed the molecular weight dependence of Boson peak position of globular protein, indicating that the low-frequency modes are extended over the whole molecule (Kataoka et al., 1999). At low temperatures, harmonic properties of protein dynamics are dominated in the low-energy region. In other words, a protein molecule can be regarded as a continuous elastic body like rubber in terms of low-energy dynamics (Kataoka et al., 1999). Tarek and Tobias (2001) showed that the protein backbone, nonpolar side chains in the interior of the protein, and polar side chains in the protein surface equivalently contribute to the Boson peak. This also indicates that Boson peak arises from motions distributed throughout the protein. For the origin of the Boson peak of molecular glasses, Yamamuro and Matsuo (1996) interpreted that the overall motions of the local cooperative domain composed of several molecules are softened owing to the frozen density fluctuations and local strain fields. The molecular weight dependence of the Boson peak position is an opposite way as proteins. We consider that the low-energy excitation of protein comes from a single molecule. Interestingly, the Boson peak positions of globular proteins are located in almost the same range as those of molecular glasses (15–30 $cm^{-1}$). As the tertiary structure of a protein can be rigorously determined, more detailed studies on low-energy dynamics of proteins may give deep insight into the low-energy dynamics of amorphous materials. We cannot find dynamical properties specific to the folded functional protein nor individuality of protein species at 100 K. We demonstrated that the differences in dynamical properties between the folded and the unfolded states are observed with INS spectrum below 150 $cm^{-1}$ at room temperature for hydrated specimen (Kataoka et al., 1999). This implies that protein character in dynamics should appear in the physiological condition. In fact, we observed that quasi-elastic scattering is more for the unfolded state than for the folded state at room temperature, and that inelastic intensity above 40 $cm^{-1}$ is higher for the folded state (Kataoka et al., 1999). We consider that the unfolded state is more flexible than the folded state, but collective motions are remarkable for the folded state (Kataoka et al., 1999).

Theoretical studies have showed that the collective motions of proteins are affected by hydration (Kitao et al., 1991). To examine the hydration effect on the Boson peak, the INS spectra in the lower energy range were measured with hydrated and dehydrated proteins at 100 K, as shown in Figure IV.5.5b (Nakagawa et al., 2006a, 2008). For the hydrated sample, the scattering intensity lower than 40 $cm^{-1}$ is depressed, and then the Boson peak position shifts from 25 to 30 $cm^{-1}$ by hydration. A hydration-induced shift of Boson peak was observed for the other proteins (Diehl et al., 1997; Fitter, 1999a; Kurkal et al., 2005). At the cryogenic temperature, protein motions are harmonic (Joti et al., 2005; Kataoka et al., 1999). The frequency upshift of the Boson peak by hydration suggests the increase of the average force constant of harmonic motions due to an increase in the number and the strength of protein–water hydrogen bonds. A computational study showed that solvent water affects the shape of the potential energy surface of the collective modes (Kitao et al., 1991, 1993). The hydrogen bond interaction should deform the harmonic potential at the local minima, causing the characteristic frequency to shift higher. The protein–water interaction restricts the protein fluctuation, indicating the hardening of potential surface (Nakagawa et al., 2008). MD simulation demonstrated that hydration makes the energy landscape rugged and protein fluctuation is restricted in the small space of a local minimum by hydration, and that the Boson peak originates in the rugged energy landscape (Joti et al., 2005). Glycerol and trehalose also cause protein hardening at a low temperature (Caliskan et al., 2003). This suggests that the hydrogen bond interaction of glassy solvent with the embedded protein affects the flexibility of the protein structure. In this sense, the origin of the protein Boson peak is the hydrogen bond-coupled low-energy mode. A computational study showed that MD simulation for a single protein molecule *in vacuo* did not result in a Boson peak (Joti et al., 2005). MD simulation in crystal, however, reproduces the Boson peak quantitatively (Joti et al., 2008a). This indicates that the Boson peak is also sensitive to the protein– protein interaction. Thus, protein environments such as neighbor proteins as well as hydration affect the low-frequency collective modes. Such interactions deform the energy landscape to become more rugged and to generate the Boson peak.

## IV.5.6 DYNAMICAL TRANSITION

Upon cooling glass-forming materials, the heat capacity and the viscosity change at a glass transition temperature $T_g$ (Sokolov, 1996). Under $T_g$, the molecules are in glassy state and the molecular fluctuations are frozen. At $T_g$, the molecular conformation starts to relax. As the temperature increases, kinetics of the relaxation becomes faster. In the INS spectrum, when the relaxation time crosses experimental timescales, the glass transition is observed. In the case of protein, similar phenomenon is observed (Ringe and Petsko, 2003). It used to be called protein glass transition. However, recently it is rather called glass-like transition or dynamical transition.

The abrupt increase of MSDs at a temperature of about 200K was first observed by Mössbauer spectroscopy of myoglobin (Parak et al., 1982). Such anomaly in the temperature dependence of MSD, the dynamical transition, was observed for several proteins (Doster et al., 1989; Ferrand et al., 1993; Gabel et al., 2002; Zaccai, 2000, 2003). It is well known that the functions of proteins are supressed with the loss of anharmonic dynamics by dehydration or by cooling down below the transition temperture (Ferrand et al., 1993; Rasmussen et al., 1992). The anharmonic motion activated above the dynamical transition temperature has been discussed in relation to the protein function (Ferrand et al., 1993; Roh et al., 2005).

Figure IV.5.6 is the temperature dependence of MSD of hydrated and dehydrated SNase estimated by incoherent neutron scattering (Nakagawa et al., 2008). Two transitions are observed around 150 and 240 K. It should be noted that the transition around 150 K is hydration independent, while the transition around 240 K is observed only for the hydrated sample. These dynamical transitions indicate the onset of certain types of anharmonic motions. The 150 K transition was attributed to the β-process originated from local motions of atoms (Roh et al., 2005). The local dynamics in the polypeptide chain at high-energy region, such as methyl rotation and $CH_2$-twisting and CH-bending motions, should contribute to the observed transition (Nakagawa et al., 2008; Roh et al., 2005). The 240 K transition indicates that anharmonic motions are activated by hydration.

A few models have been proposed for the origin of the 240 K transition. The asymmetric double-well model was introduced by Doster et al. (1989). Above the dynamical transition temperature, thermal motions can pass over the energy barrier

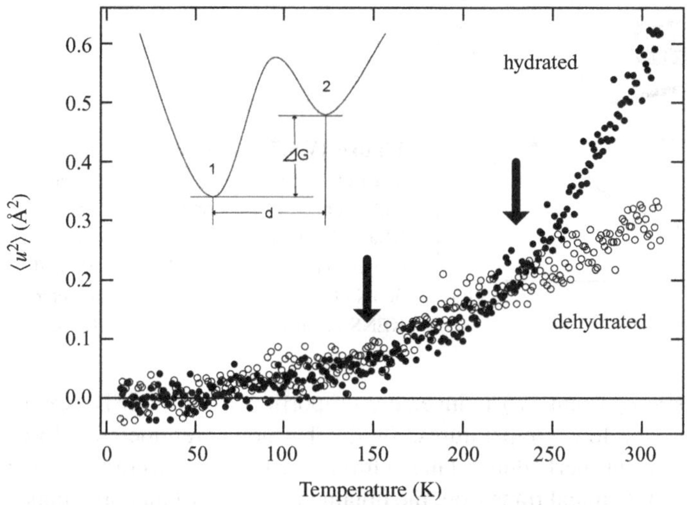


**Figure IV.5.6** Temperature dependence of MSDs of dehydrated and hydrated SNase. The dynamical transitions are observed at around 150 and 240 K. The MSDs were obtained from the $Q$-dependence of elastic scattering observed with IN10. Inset shows the asymmetric double-well potential model (Doster et al., 1989). The free energy of the ground state (1) is lower than the excited state (2) by $\Delta G$. The states 1 and 2 are separated by a distance $d$.

between two sites with a distance $d$. Given the free energy difference between two wells, $\Delta G$ (inset in Figure IV.5.6), MSD is given by

$$\langle u^2 \rangle = \langle u^2 \rangle_{\mathrm{vib}} + \frac{p_1 p_2 d^2}{3}, \tag{IV.5.4}$$

where $p_1$ and $p_2$ denote the probability of finding a hydrogen in the ground and excited states, respectively. In this model, the long time dynamics (equilibrium position distribution) changes with temperature. Another model is a frequency window model (Becker et al., 2004). When the motions cross over the energy barrier, the kinetics would obey the Arrhenius equation:

$$\tau = A \exp\left(-\frac{E}{RT}\right), \tag{IV.5.5}$$

where $E$ is the activation energy. As the temperature increases, the relaxation frequencies move into the instrumental energy resolution range, resulting in the decrease of elastic intensity. Since MSD is estimated by the $Q$-dependence of elastic intensity, the dynamical transition is observed. This model successfully interpreted the timescale-dependent dynamical transition (Becker et al., 2004; Daniel et al., 1999).

In order to quantify the temperature dependence of the MSDs obtained by neutron scattering, an effective force constant associated with protein resilience has been introduced (Zaccai, 2000):

$$\langle k' \rangle = 2k_{\mathrm{B}} \left(\frac{\mathrm{d}\langle u^2 \rangle}{\mathrm{d}T}\right)^{-1} \mathrm{N/m}. \tag{IV.5.6}$$

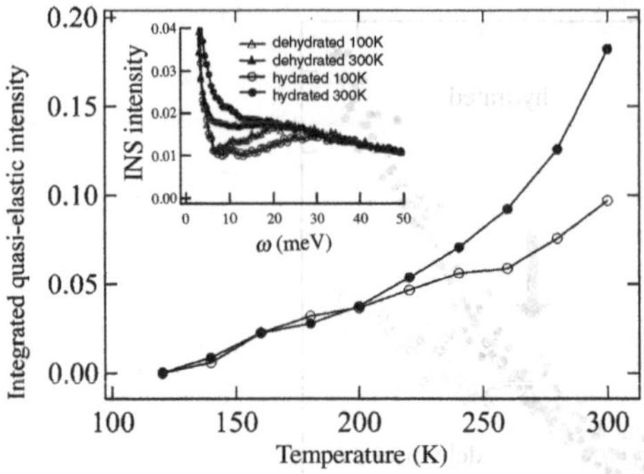


**Figure IV.5.7** Temperature dependence of quasi-elastic intensity integrated between 8 and 32 $cm^{-1}$ of dehydrated and hydrated SNase. Inset shows INS spectra of the dehydrated and hydrated SNase at 100 and 300 K observed with LAM-40 spectrometer at KENS in Japan (Nakagawa et al., 2008).

The effective force constant $\langle k' \rangle$ is inversely proportional to the MSDs versus temperature gradient. At low temperatures, in the harmonic regime, the force constant corresponds to a properly defined mean force constant for a set of harmonic oscillators. Above the dynamical transition, the obtained force constant corresponds to a force constant in a quasi-harmonic approximation. The force constant is the indicator for the protein softness or resilience.

The activation of anharmonicity should be reflected in the appearance of quasi-elastic scattering. At 100 K, protein dynamics is harmonic; therefore, the excess intensity over the spectrum at 100 K represents the contribution from quasi-elastic scattering (Diehl et al., 1997; Nakagawa et al., 2008). Figure IV.5.7 shows the contribution of the quasi-elastic scattering at various temperatures. Inset in Figure IV.5.7 is the INS in the low-energy region with the hydrated and dehydrated samples at 100 and 300 K. The integrated quasi-elastic scattering intensities were estimated as follows:

$$S_{\mathrm{qel}}(Q,E)|_T = S(Q,E)|_T - S(Q,E)|_{T_0} \frac{\mathrm{e}^{-Q^2 U_{\mathrm{vib}}(T)}\left(\mathrm{e}^{E/k_{\mathrm{B}}T_0}-1\right)}{\mathrm{e}^{-Q^2 U_{\mathrm{vib}}(T_0)}\left(\mathrm{e}^{E/k_{\mathrm{B}}T}-1\right)}. \qquad \text{(IV.5.7)}$$

The quasi-elastic intensities for both states increased when the temperature was increased, suggesting the appearance of some anharmonic motions. The quasi-elastic intensities up to approximately 200 K were not markedly affected by hydration. The anharmonicity is responsible for the transition around 140 K commonly observed for the hydrated and dehydrated proteins. Above 200 K, the increase in the quasi-elastic intensity of the dehydrated sample was roughly proportional to the temperature. On the other hand, the hydrated sample gained additional strong quasi-elastic scattering above 200 K. This increase in the quasi-elastic scattering indicates the appearance of hydration-dependent relaxation processes. This corresponds to the onset of anharmonicity responsible for the transition around 240 K. This dynamical transition originates from the activation of anharmonic motions driven by translational hydration water diffusion (Tournier et al., 2003b). To characterize these anharmonic motions quantitatively using the proper dynamical model, the high-energy resolution instrument, submicroelectron volt, covered up to high $Q$ is required. Although such types of spectrometers are already constructed, the neutron flux is not necessarily sufficient for the systematical study. The new high-flux neutron source will pave the way for the intensive protein dynamics study by INS.

## IV.5.7 NON-GAUSSIANITY IN ELASTIC SCATTERING AND DYNAMICAL HETEROGENEITY

The $Q$-dependence of elastic scattering gives the MSD. Under the Gaussian approximation, incoherent elastic scattering $S_{\text{inc}}(Q, 0)$ at small $Q$ region can be described as

$$S_{\text{inc}}(Q,0) = b^2 \mathrm{e}^{-(Q^2/3)\langle u^2 \rangle}. \tag{IV.5.8}$$

This is strictly valid for atoms in ideal gases, and is also valid for short times or small values of $Q$ (Rahman et al., 1962; Rahman, 1963). Figure IV.5.8 is the $Q$-dependence of incoherent elastic neutron scattering profiles from SNase in the wide $Q$ range. The low $Q$ profile is given in the bottom figure, where the Gaussian approximation holds. In the higher $Q$ region, deviation from the Gaussian approximation is clearly realized. Information on protein dynamics should be obtained through the analysis of the non-Gaussianity. In the case of proteins, hydrogen atoms are equally distributed throughout the molecule. Thus, the total scattering intensity is the summation of scattering of hydrogen. The incoherent dynamical structure factor is a summation of Gaussian as follows:

$$S_{\text{inc}}(Q,0) = \frac{1}{N_{\text{atom}}} \sum_{\alpha} b_\alpha^2 \mathrm{e}^{-(Q^2/3)\langle u_\alpha^2 \rangle}. \tag{IV.5.9}$$

Therefore, even if the Gaussian approximation were valid for individual hydrogen atoms, $S(Q, 0)$ will not have a Gaussian form because a summation of Gaussian is not necessarily Gaussian. However, Figure IV.5.8 gives exactly the Gaussian form at the limit of $Q \to 0$. We can consider various possibilities of the origin of the non-Gaussianity, such as anharmonicities (Doster et al., 1989), dynamical heterogeneity (Lehnert et al., 1998; Nakagawa et al., 2004; Smith, 1991), and dynamical anisotropy. In the case of polymers, it is shown that the dynamical heterogeneity due to the variety in local environments is a plausible origin, and the heterogeneity is well correlated with the fragility (Kanaya et al., 1997, 1998). However, for proteins, the non-Gaussianity has not been analyzed intensively, except for a few cases. Doster et al. (1989) analyzed the non-Gaussianity in terms of anharmonicity with an asymmetric double-well model. The anharmonicity is unambiguously one of the origins for the non-Gaussianity above the transition temperature, because anharmonic motions are activated. It is reasonable to consider that all hydrogen atoms in proteins are not dynamically equivalent, for example, the hydrogen atoms in the protein surface exposed to solvent are expected to be more mobile than the atoms highly packed in the protein interior. This suggests that non-Gaussianity is also derived from the dynamical heterogeneity.

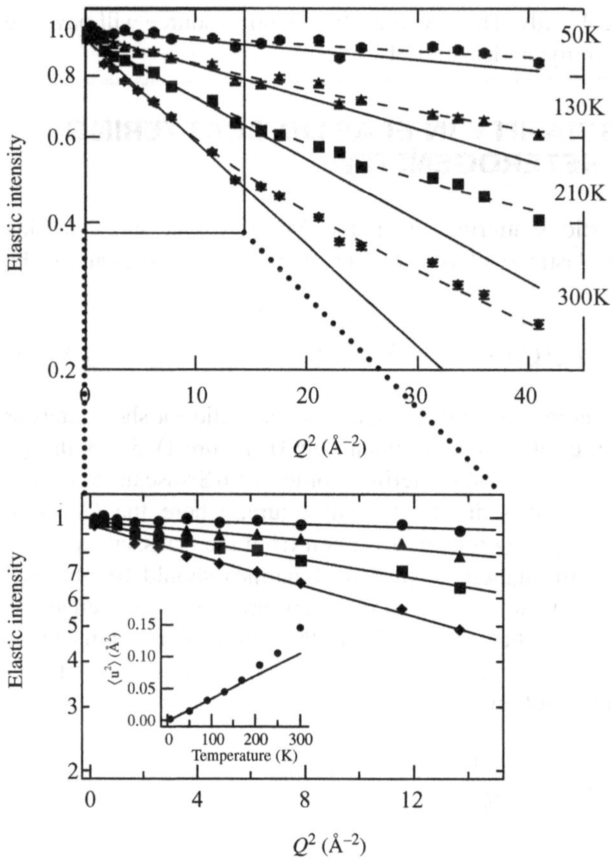


**Figure IV.5.8** Elastic neutron scattering intensity with a wide $Q$ range from SNase at various temperatures measured by GPTAS at JRR-3 M in Japan. The solid and dashed lines show a Gaussian approximation in the low $Q$ region and the fitting with the bimodal distribution, respectively. Inset shows the temperature-dependent MSDs by Gaussian approximation.

Experimentally, it cannot be determined to what extent both possible origins contribute to the non-Gaussianity. We examined the origin of non-Gaussianity by MD simulation. If anharmonic contribution is not negligible, the non-Gaussian behavior depends on the instrumental energy resolution. Figure IV.5.9 shows the non-Gaussian profiles in two energy resolutions, 1 meV and 10 μeV, calculated by MD simulation (Nakagawa et al., 2006b). We can estimate the contributions of anharmonicity and heterogeneity separately by the comparison of two types of simulations: the full calculation without any approximation that corresponds to the elastic intensity observable experimentally and the calculation with the dynamical heterogeneity only. At 100 K, both simulations give almost identical results independent of instrumental energy resolution, indicating that the heterogeneity contributes only to the non-Gaussianity. At 300 K, although the dynamical heterogeneity is still

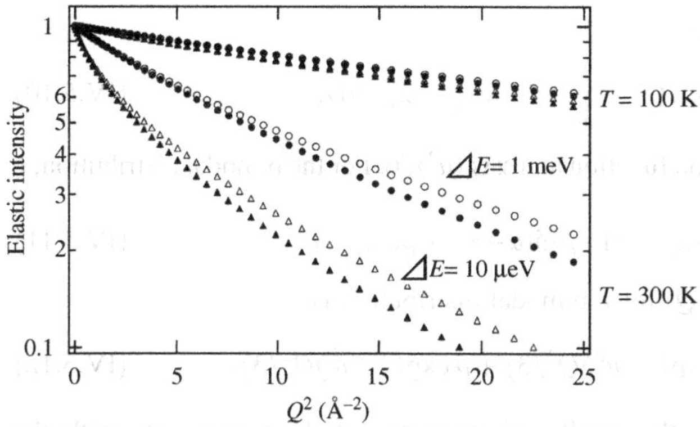


**Figure IV.5.9** Calculated incoherent neutron elastic scattering convoluted with the energy resolution of 1 meV and 10 μeV by MD simulation at 100 and 300 K. The open circles (○) show the full calculation without an approximation that corresponds to the expected experimental data. The triangles show the calculation only with dynamical heterogeneity (Tokuhisa et al., 2007).

dominant for both energy resolutions, contribution of anharmonicity becomes remarkable. For the further quantitative analysis, we separate the elastic intensity into three components: heterogeneity, anisotropy, and anharmonics contributions (Figure IV.5.10) (Tokuhisa et al., 2007). We elucidated that the latter two contributions to total scattering are negligibly smaller than the heterogeneity. This result was also valid for other proteins: BPTI, SNase, hen egg white lysozyme, and T4 lysozyme. We conclude that the analysis of the non-Gaussian behavior of experimental data in terms of dynamical heterogeneity is reliable and reasonable, and as the initial approximation, the dynamical heterogeneity model is effective for the practical analysis of the non-Gaussianity.

We analyzed the protein dynamical heterogeneity by the non-Gaussianity with the experimental data. In order to incorporate the dynamical heterogeneity, the distribution function for MSD should be explicitly included for the calculation of

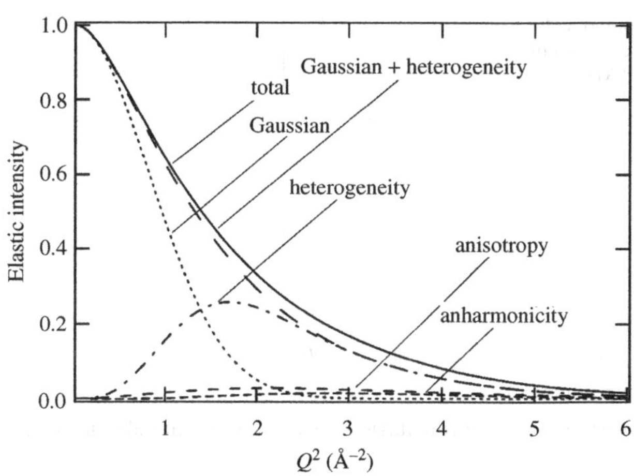


**Figure IV.5.10** Elastic intensity as a function of $Q$ calculated by MD simulation at 300 K. Total intensity and its components, Gaussian, heterogeneity, anisotropy, and anharmonicity, are shown (Tokuhisa et al., 2007).

elastic scattering $S_{el}(Q)$ as

$$S_{el}(Q) = \int_0^{\infty} g(\alpha) \exp(-\alpha Q^2) d\alpha, \quad \text{(IV.5.10)}$$

where $g(\alpha)$ is the distribution function and $\alpha$ is $\langle u^2 \rangle / 6$. For the bimodal distribution, $g(\alpha)$ is given by

$$g_B(\alpha) = (1-p)\delta(\alpha-\alpha_s) + p\delta(\alpha-\alpha_l). \quad \text{(IV.5.11)}$$

The final form of $S_{el}(Q)$ for a bimodal distribution is

$$S_{el}(Q) = (1-p)\exp(-\langle u_s^2 \rangle Q^2/3) + p\exp(-\langle u_l^2 \rangle Q^2/3), \quad \text{(IV.5.12)}$$

where $\langle u_s^2 \rangle$ and $\langle u_l^2 \rangle$ are the small and large amplitude components with the corresponding population fractions, $1-p$ and $p$, respectively.

The bimodal distribution model was applied to the experimental data with GPTAS. The results of fits by the bimodal distribution model are shown as broken curves in Figure IV.5.8. Three parameters can be obtained with the bimodal distribution model: two MSDs and their ratio. A very rough picture of protein is that a protein is composed of flexible regions and tightly packed cores. The flexible parts have a larger MSD and the cores have a smaller value. Thus, the bimodal distribution describes the essence of the protein dynamical heterogeneity despite its simplicity. Figure IV.5.11 shows the temperature-dependent changes of two MSDs and the averaged MSDs. The fraction of a motion with a larger MSD is obtained as $p = 0.35$ for each temperature. The value is roughly equal to the ratio of possible flexible regions, judged from the crystal structure (Nakagawa et al., 2004). Motions with both the larger and the smaller MSDs show the dynamical transition. This suggests that the origin of the dynamical transition would be the anharmonic motions distributed throughout the whole protein. Theoretical analysis insisted that the collective motions dominantly contribute to the protein anharmonic motions (Kitao

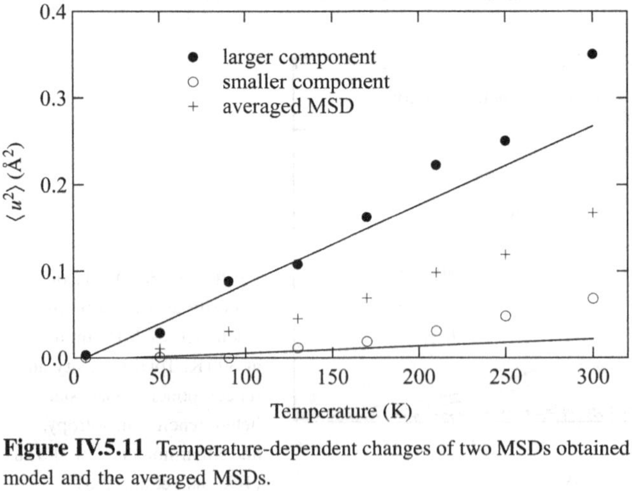


**Figure IV.5.11** Temperature-dependent changes of two MSDs obtained with the bimodal distribution model and the averaged MSDs.

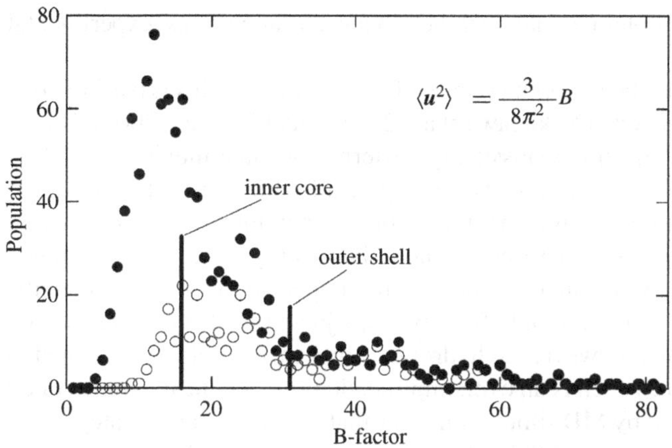


**Figure IV.5.12** Distribution of crystallographic B-factor is derived from the PDB code 1STN. The filled circles (●) and the open circles (○) indicate the distribution of all atoms and the atoms in the outer shell of the protein (35 % of total atoms in the protein), respectively. The vertical bars indicate the average B-factor values of the inner core and the outer shell. The length of the vertical bars is proportional to the ratio (Nakagawa, et al., 2004).

and Go, 1999). The bimodal distribution analysis supports this dynamical picture because both components contribute to anharmonicity.

Figure IV.5.12 shows the distribution of B-factor based on the SNase crystal structure. The B-factor is closely related with the atomic MSD, as shown in the inset of Figure IV.5.12. Since the crystallographic B-factor includes statistical disorder as well as intrinsic dynamics, a direct comparison with the distribution function obtained by the analysis of non-Gaussianity is rather impossible. However, the comparison still gives useful insights into the dynamical heterogeneity. The B-factor distribution of all atoms is asymmetric with a major peak in the lower value and the long tail in the higher value. The two vertical bars in Figure IV.5.12 indicate the average B-factors of the atoms inside and outside the protein, respectively, which represents a bimodal distribution function. The bimodal distribution catches essentially the characteristics of the distribution of the B-factor. Other than the bimodal distribution, we considered two distribution functions: the Gaussian distribution function and the exponential distribution function (Nakagawa et al., 2004). The Gaussian function explained the major peak of the B-factor distribution but not the long tail (Nakagawa et al., 2004). On the other hand, the exponential function explained the long tail (Nakagawa et al., 2004). The exponential distribution function had the maximum at 0, which was completely different from the B-factor distribution. The B-factor distribution of the atoms in the outer shell of the protein (35% of total atoms in the protein) shown in Figure IV.5.12 indicates that the long tail in B-factor distribution of all atoms is due to the residues located outside the protein. The merit of the B-factor analysis is that the molecular fluctuation can be analyzed with the atomic resolution. But the drawback of the analysis is that it is difficult or even impossible to examine the B-factor in various environments, such as temperature, hydration, and so on. Therefore, the analysis of non-Gaussianity with INS is effective for characterizing the protein dynamical heterogeneity at various experimental conditions.

We can conclude that the non-Gaussianity of $S(Q, 0)$ for proteins is partially due to the dynamical heterogeneity (Nakagawa et al., 2004; Tokuhisa et al., 2007). Doster et al. (1989) explained the non-Gaussianity in terms of anharmonicity. In their analysis, each hydrogen atom can take two energetically different states: a ground state and an excited state. Above the transition temperature, anharmonicity is undoubtedly one of the major reasons for non-Gaussianity. The two-site model (Doster et al., 1989) is an adequate model for the explanation of anharmonicity, although anharmonicity comes not only from two-site jump but also from relaxation and diffusion. In the model, however, all hydrogen atoms in a protein are assumed to undergo the same motion, which is an oversimplification. In fact, the dynamics of all hydrogen atoms examined by MD simulation show that dynamics can be categorized into four groups (Tokuhisa et al., 2007). However, the two-site model represents the typical motion of anharmonic hydrogen atoms. The combination of anharmonicity with dynamical heterogeneity would be more realistic for the explanation of non-Gaussianity. The combination of a two-state model with a bimodal distribution is more appropriate for the analysis of non-Gaussianity.

## IV.5.8 CONCLUSION

Inelastic scattering in higher energy region gives vibrational spectrum of a protein, which is directly calculated by normal mode analysis and MD simulation. The inelastic scattering spectrum of SNase at 25K was directly compared with the result of the normal mode analysis to assign the obtained peaks. The comparison of the results of computation and the experiment further shows that the improvement of dynamical model or potential energy for calculation is required. Inelastic scattering in lower energy region gives information on collective modes. Boson peak observed at cryogenic temperature is a common property for soft matters. The origin of protein Boson peak is examined by the molecular weight dependence and hydration dependence. The modes responsible for the Boson peak are spread over a whole protein molecule. Hydration brings hardening of the harmonic property of proteins at cryogenic temperature. Partial hydration is sufficient for the Boson peak shift. Thus, the Boson peak is closely related with the local hydrogen bond between hydration water and protein surface. The theoretical study demonstrated that hydration makes protein potential surface rugged. The observed Boson peak shift coincides with the theoretical prediction. Dynamical transition is another important property of protein dynamics. Two transitions around 150 and 240 K are generally observed for proteins. The 150 K transition is not affected by hydration, while the 240 K transition is specific to the hydrated protein. There is a threshold hydration level to generate the 240 K transition, suggesting the cooperative property of hydrogen bond between the hydration waters. Finally, the non-Gaussian behavior of $Q$-dependence of elastic scattering gives information on dynamical heterogeneity. Neutron scattering provides us useful and unique information on protein dynamics. It becomes more powerful when it is combined with the computational biology. The operation of the next generation of neutron source such as J-PARC and SNS will open up a new era in protein dynamical study.

## REFERENCES


Anfinsen, C.B. *Biochem. J.* **1972**, *128*, 737.

Ansari, A., Berendzen, J., Bowne, S.F., Frauenfelder, H., Iben, I.E.T., Sauke, T.B., Shyamsunder, E., and Young, R.D. *Proc. Natl. Acad. Sci. USA* **1985**, *82*, 5000.

Becker, T., Hayward, J.A., Finney, J.L., Daniel, R.M., and Smith, J.C. *Biophys. J.* **2004**, *87*, 1436.

Bellissent-Funel, M.C. *J. Mol. Liquids* **2000**, *84*, 39.

Bellissent-Funel, M.C., Daniel, R., Durand, D., Ferrand, M., Finney, J.L., Pouget, S., Reat, V., and Smith, J.C. *J. Am. Chem. Soc.* **1998**, *120*, 7347.

Borgstahl, G.E.O., Williams, D.R., and Getzoff, E.D. *Biochemistry* **1995**, *34*, 6278.

Brooks, B., Bruccoleri, R., Olafson, B., States, D., Swaminathan, S., and Karplus, M. *J. Comput. Chem.* **1983**, *4*, 187.

Brown, K.G., Erfurth, S.C., Small, E.W., and Peticolas, W.L. *Proc. Natl. Acad. Sci. USA* **1972**, *69*, 1467.

Bu, Z., Biehl, R., Monkenbusch, M., Richter, D., and Callaway, D.J.E. *Proc. Natl. Acad. Sci. USA* **2005**, *102*, 17646.

Caliskan, G., Kisliuk, A., Tsai, A.M., Soles, C.L., and Sokolov, A.P. *J. Chem. Phys.* **2003**, *118*, 4230.

Case, D.A., Darden, T.A., Cheatham, T.E., Simmerling, C.L., Wang, J., Duke, R.E., Luo, R., Merz, K.M., Pearlman, D.A., Crowley, M., Walker, R.C., Zhang, W., Wang, B., Hayik, S., Roitberg, A., Seabra, G., Wong, K.F., Paesani, F., Wu, X., Brozell, S., Tsui, V., Gohlke, H., Yang, L., Tan, C., Mongan, J., Hornak, V., Cui, G., Beroza, P., Mathews, D.H., Schafmeister, C., Ross, W.S., and Kollman, P.A. AMBER9, University of California, San Francisco, 2006.

Chong, S.H., Joti, Y., Kidera, A., Go, N., Ostermann, A., Gassmann, A., and Parak, F. *Eur. Biophys. J.* **2001**, *30*, 319.

Chou, K.C. *Biophys. J.* **1984**, *45*, 881.

Cusack, S.In: Cooper, A., Houben, J.L., and Chien, L.C. (editors). *The Enzyme Catalysis Process*, Plenum, 1989, p. 103.

Cusack, S. and Doster, W. *Biophys. J.* **1990**, *58*, 243.

Daniel, R.M., Finney, J.L., Reat, V., Dunn, R., Ferrand, M., and Smith, J.C. *Biophys. J.* **1999**, *77*, 2184.

Diehl, M., Doster, W., Petry, W., and Schober, H. *Biophys. J.* **1997**, *73*, 2726.

Doster, W. and Settles, M. *Biochim. Biophys. Acta* **2005**, *1749*, 173.

Doster, W., Cusak, S., and Petry, W. *Nature* **1989**, *337*, 754.

Ferrand, M., Dianoux, A.J., Petry, W., and Zaccai, G. *Proc. Natl. Acad. Sci. USA* **1993**, *90*, 9668.

Fitter, J. *Biophys. J.* **1999a**, *76*, 1034.

Fitter, J., Lechner, R.E., and Dencher, N.A. *J. Phys. Chem. B* **1999b**, *103*, 8036.

Flanagan, J.M., Kataoka, M., Shortle, D., and Engelman, D.M. *Proc. Natl. Acad. Sci. USA* **1992**, *89*, 748.

Flanagan, J.M., Kataoka, M., Fujisawa, T., and Engelman, D.M. *Biochemistry* **1993**, *32*, 10359.

Frauenfelder, H. and McMahon, B.H. *Proc. Natl. Acad. Sci. USA* **1998**, *95*, 4795.

Frauenfelder, H., McMahon, B.H., Austin, R.H., Chu, K., and Groves, J.T. *Proc. Natl. Acad. Sci. USA* **2001**, *98*, 2370.

Gabel, F., Bicout, D., Lehnert, U., Tehei, M., Weik, M., and Zaccai, G. *Q. Rev. Biophys.* **2002**, *35*, 327.

Goupil-Lamy, A.V. and Smith, J.C. *J. Am. Chem. Soc.* **1996**, *118*, 7326.

Goupil-Lamy, A.V., Smith, J.C., Yunoki, J., Parker, S.F., and Kataoka, M. *J. Am. Chem. Soc.* **1997**, *119*, 9268.

Hirano, S., Mihara, K., Kamikubo, H., Yamazaki, Y., Imamoto, Y., and Kataoka, M. *Proteins Struct. Funct. Genet.* **2002**, *49*, 255.

Hirano, S., Kamikubo, H., Yamazaki, Y., and Kataoka, M. *Proteins Struct. Funct. Bioinform.* **2005**, *58*, 271.

Ikeguchi, M., Ueno, J., Sato, M., and Kidera, A. *Phys. Rev. Lett.* **2005**, *94*, 078102.

JACROT, B., CUSAK, S. DIANOUX, A.J., and ENGELMAN, D.M. *Nature* **1982**, *300*, 84.
JOTI, Y., NAKASAKO, M., KIDERA, A., and GO, N. *Acta. Crystallogr. D* **2002**, *58*, 1421.
JOTI, Y., KITAO, A., and GO, N. *J. Am. Chem. Soc.* **2005**, *127*, 8705.
JOTI, Y., NAKAGAWA, H., KATAOKA, M., and KITAO, A. *Biophys. J.* **2008a**, *94*, 4435.
JOTI, Y., NAKAGAWA, H., KATAOKA, M., and KITAO, A. *J. Phys. Chem. B* **2008b**, *112*, 3522.
KANAYA, T. and KAJI, K *Chem. Phys. Lett.* **1988**, *150*, 334.
KANAYA, T., TSUKUSHI, I., and KAJI, K. *Prog. Theor. Phys. Suppl.* **1997**, *126*, 93.
KANAYA, T., TSUKUSHI, I., KAJI, K., GABRYS, B., and BENNINGTON, S.M. *J. Non-Cryst. Solids.* **1998**, *235–237*, 212.
KARPLUS, M. *J. Phys. Chem. B* **2000**, *104*, 11.
KARPLUS, M. and MCCAMMON, A. *Nat. Struct. Biol.* **2002**, *9*, 646.
KATAOKA, M., KAMIKUBO, H., YUNOKI, J., TOKUNAGA, F., KANAYA, T., IZUMI, Y., and SHIBATA, K. *J. Phys. Chem. Solids* **1999**, *60*, 1285.
KATAOKA, M., KAMIKUBO, H., NAKAGAWA, H., PARKER, S.F., and SMITH, J.C. *Spectroscopy* **2003**, *17*, 529.
KITAO, A. and GO, N. *Curr. Opin. Struct. Biol.* **1999**, *9*, 164.
KITAO, A., HIRATA, F., and GO, N. *Chem. Phys.* **1991**, *158*, 447.
KITAO, A., HIRATA, F., and GO, N. *J. Phys. Chem.* **1993**, *97*, 10231.
KNELLER, G.R. *Comput. Phys. Commun.* **1995**, *91*, 191.
KURKAL, V., DANIEL, R.M., FINNEY, J.L., TEHEI, M., DUNN, R.V., and SMITH, J.C. *Biophys. J.* **2005**, *89*, 1282.
LEHNERT, U., REAT, V., WEIK, M., ZACCAI, G., and PFISTER, C. *Biophys. J.* **1998**, *75*, 1945.
LOLL, P.J. and LATTMAN, E.E. *Proteins Struct. Funct. Genet.* **1989**, *5*, 183.
NAKAGAWA, H., KAMIKUBO, H., TSUKUSHI, I., KANAYA, T., and KATAOKA, M. *J. Phys. Soc. Jpn.* **2004**, *73*, 491.
NAKAGAWA, H., KATAOKA, M., JOTI, Y., KITAO, A., SHIBATA, K., TOKUHISA, A., TSUKUSHI, I., and GO, N. *Physica B* **2006a**, *385–386*, 871.
NAKAGAWA, H., TOKUHISA, A., KAMIKUBO, H., JOTI, Y., KITAO, A., and KATAOKA, M. *Mater. Sci. Eng. A* **2006b**, *442*, 356.
NAKAGAWA, H., JOTI, Y., KITAO, A., and KATAOKA, M. *Biophys. J.* **2008**, *95*, 2916.
NISHIKAWA, T. and GO, N. *Proteins Struct. Funct. Genet.* **1987**, *2*, 308.
ONITSUKA, M., KAMIKUBO, H., YAMAZAKI, Y., and KATAOKA, M. *Proteins Struct. Funct. Bioinform.* **2008**, *72*, 837.
PARAK, F., KNAPP, E.W., and KUCHEIDA, D. *J. Mol. Biol.* **1982**, *161*, 177.
RAHMAN, A. *Phys. Rev.* **1963**, *130*, 1334.
RAHMAN, A., SINGWI, K.S., and SJOLANDER, A. *Phys. Rev.* **1962**, *126*, 986.
RASMUSSEN, B.F., STOCK, A.M., RINGE, D., and PETSKO, G.A. *Nature* **1992**, *357*, 423.
RINGE, D. and PETSKO, G.A. *Biophys. Chem.* **2003**, *105*, 667.
ROH, J.H., NOVIKOV, V.N., GREGORY, R.B., CURTIS, J.E., CHOWDHURI, Z., and SOKOLOV, A.P. *Phys. Rev. Lett.* **2005**, *95*, 038101.
SETTLES, M. and DOSTER, W. *Faraday Discuss.* **1996**, *103*, 269.
SHORTLE, D. and MEEKER, A.K. *Biochemistry* **1989**, *28*, 936.
SMITH, J.C. *Q. Rev. Biophys.* **1991**, *24*, 227.
SOKOLOV, A.P. *Science* **1996**, *273*, 1675.
TAREK, M. and TOBIAS, D.J. *Biopys. J.* **2000a**, *79*, 3244.
TAREK, M. and TOBIAS, D.J. *J. Chem. Phys.* **2001**, *115*, 1607.
TAREK, M., MARTYNA, G.J., and TOBIAS, D.J. *J. Am. Chem. Soc.* **2000b**, *122*, 10450.
TOKUHISA, A., JOTI, Y., NAKAGAWA, H., KITAO, A., and KATAOKA, M. *Phys. Rev. E* **2007**, *75*, 041912.
TOURNIER, A.L. and SMITH, J.C. *Phys. Rev. Lett.* **2003a**, *91*, 208106.
TOURNIER, A.L., XU, J., and SMITH, J.C. *Biophys. J.* **2003b**, *85*, 1871.
YAMAMURO, O. and MATSUO, T. *J. Chem. Phys.* **1996**, *105*, 732.
YAMAMURO, O., TAKEDA, K., TSUKUSHI, I., and MATUSO, T. *Physica B* **2002**, *311*, 84.
ZACCAI, G. *Science* **2000**, *288*, 1604.
ZACCAI, G. *J. Phys. Condens. Matter* **2003**, *15*, 1673.

# IV Applications

# IV.6 Polymer Interfaces and Thin Films

*David G. Bucknall*

## IV.6.1 INTRODUCTION

The use of neutron reflection (NR) is in many ways an ideal technique for studying polymer interfaces and surfaces. Coupled with careful deuteration and the ability to use a very wide range of sample environments, the types of interfaces that can be studied are limitless and effectively only restricted by access to the reflectometers to make the measurements. Neutrons are highly penetrating and consequently NR measurements have been applied in many different sample environments. In general, all of the various sample environment equipment are essentially subsets of two geometries. The most commonly used geometry is where the polymer(s) is supported on a substrate, which could either be a solid such as polished silicon or a fluid surface such as water. In this geometry, the incident neutron beam is projected on to the sample from the "free" surface. The bulk media above the "free" surface can of course be vacuum, a gas, or, in a limited number of cases, a fluid. In the second geometry, the incident beam is transmitted through the support substrate on to the "confined" surface of the sample. The choice of which geometry to use depends on the type of system under investigation, but consideration for sample configuration may also be affected by the neutron optics of the particular instrument being used. While these are practical considerations they do not change the capability of the technique. NR is a deceptively simple technique, although this belies the immense wealth of information that is contained in the reflectivity data. Although the data are often easily obtained, it is not always as easy to extract the information from the data. However, carefully designed experiments coupled with some very sophisticated fitting routines to extract the data, have made NR a technique of choice in many instances. Within this chapter, some of the areas in which neutrons have been used to study polymer interfaces are highlighted from among the many thousands of research articles that have appeared in the literature. Since it is impossible to mention all the

*Neutrons in Soft Matter* Edited by Toyoko Imae, Toshiji Kanaya, Michihiro Furusaka, and Naoya Torikai

work NR has been applied in polymer research, the following can only summarize the possibilities NR and associated techniques offer in studying polymer interfaces. NR is most often taken to mean specular reflection, where the angle of incidence and reflection are the same, and other reflection angles are not measured, so that the specular NR measurements provides composition data normal to the film. Data relating to the lateral structure of the film, that is in the plane of the film, can of course be obtained by looking at the "off-specular" scattering angles. Techniques to capture the lateral structure, such as off-specular reflectivity and grazing incidence small-angle neutron scattering are discussed later in the chapter.

## IV.6.2 POLYMER–POLYMER INTERFACES

The technological use of blends has driven extensive studies not only of the phase behavior but also of the interfaces between the polymers. In blends containing immiscible or partially miscible polymers the interface is a critical factor in determining the bulk mechanical properties of the polymer. The reason for these weak interfaces is the narrow interface formed between the two polymers resulting from the balance in enthalpic and entropic contributions that govern the degree of intermixing across the interface. Using self-consistent mean-field (SCMF) theory, Helfand and Tagami derived the stunningly simple equation for determining the theoretical interfacial width for symmetric infinite molecular weight homopolymers in the strong segregation limit, $w_\mathrm{I} = (2b)/\sqrt{6\chi}$, where $b$ is the statistical segment length and $\chi$ is the Flory–Huggins interaction parameter (Helfand and Tagami, 1996). Inspection of this equation shows that for a system such as polystyrene (PS) and poly (methyl methacrylate) (PMMA), the maximum size of the interface width can be determined by substituting $b = 0.71\,\mathrm{nm}$ (Kirste, 1967) and $\chi = 0.037$ (Russell et al., 1990), giving $w_\mathrm{I} = 3.0\,\mathrm{nm}$. These narrow interfacial widths are characteristic of such blends, which make them very difficult to measure experimentally, and consequently have led to a wide scatter of interfacial properties being reported in the older literature. However, NR is ideally suited to investigation of such narrow interfaces and the technique has therefore been widely exploited in the study of these types of interfaces.

The interfacial width between bilayers of PS ($M_\mathrm{w} = 220{,}000$ g/mol) and dPMMA ($M_\mathrm{w} = 19.2{,}000$ g/mol) annealed at 120°C, have been determined by fitting the NR data with different functional forms for the interfacial profile (Fernandez et al., 1988). For both Gaussian and exponential density gradient profiles, the interfacial roughness, $\Delta$, was found to be $2 \pm 0.5\,\mathrm{nm}$, which is equivalent to an interfacial width $w = \sqrt{2\pi} \cdot \Delta = 5.0 \pm 1.2\,\mathrm{nm}$. Subsequently, the interface width between dPS ($M_\mathrm{w} = 110{,}000$ g/mol) and PMMA ($M_\mathrm{w} = 107{,}000$ g/mol) bilayers annealed at 170°C was also measured (Anastasiadis et al., 1990). Fitting the data using a hyperbolic tangent function, interfacial profile they obtained $w = 5.0 \pm 0.5\,\mathrm{nm}$. Both these sets of results are consistent even though the annealing temperatures and molecular weights used in both sets of measurements were remarkably different. However, both NR measurements are much larger than the value predicted by SCMF theory as described above. This discrepancy can be accounted for by considering capillary wave fluctuations at the interface (Shull et al., 1993). Although thermally excited capillary waves were considered a plausible explanation of the differences between measured and intrinsic interfacial widths, it was not until the NR studies of Sferrazza et al. that full details of the capillary wave amplitudes were derived (Sferrazza et al., 1997). For strongly immiscible polymer pairs, the interfacial width measured by NR is a Gaussian quadrature sum of the SCMF intrinsic interfacial width, $w_I$, and the contribution associated with the thermally excited capillary waves, and can be written as (Carelli et al., 2005b):

$$w = \left(4w_I^2 + 2\pi\sigma_\zeta^2\right)^{\frac{1}{2}}, \qquad \text{(IV.6.1)}$$

where $\sigma_\zeta^2 = (k_B T/4\pi\gamma)\ln\left[(2\pi/\pi w_I)^2/(2\pi/\lambda_{coh})^2 + (2\pi/\lambda)^2\right]$ is the capillary wave mean dispersive term, $k_B$ is the Boltzmann constant, $\gamma$ is the interfacial tension, $\lambda_{coh}$ is the neutron coherence length, $A$ is the Hamaker constant, $d$ is the film thickness, and $\lambda$ is a wave vector cutoff given by the disperse capillary length $a_{disp} = \sqrt{\pi\gamma d^4/A}$. The short wave vector cutoff of the logarithmic divergence was shown not to be given by the coherence length of the neutrons but by the effect of dispersion forces acting across the film. In this case, as shown by eq. (IV.6.1) there is expected to be a logarithmic dependence of the interfacial width on film thickness. By measuring the interfacial widths between relatively thick (>400 nm) PMMA layers and dPS films of thickness in the ranging from 5 to 2000 nm, Sferrazza et al. demonstrated the logarithmic dependence of interfacial width on the dPS film thickness as predicted by theory (see Figure IV.6.1) (Sferrazza et al., 1997).

An interesting question arises as to how polymer–polymer interfaces approach their equilibrium state. Initially, single-chain mobility is predicted to control the interface broadening, while at long times, hydrodynamic flow would be expected. Exploiting slow polymer chain dynamics NR measurements of polyolefin blends of various degrees of incompatibility were used to follow the broadening of interfaces to

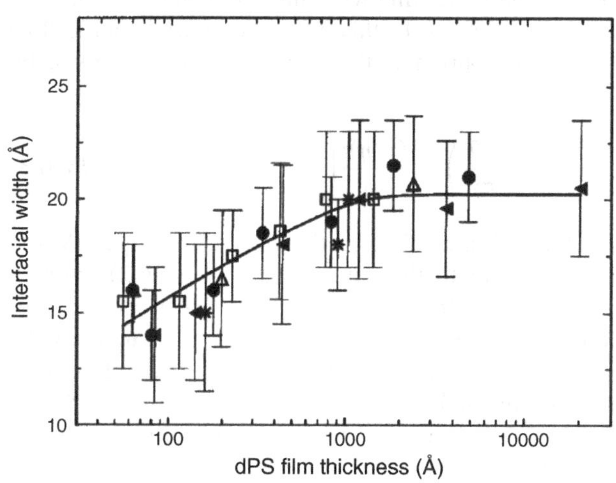


**Figure IV.6.1** Measured interfacial width (w) between a thick PMMA film and various film thickness of dPS. The solid line is a fit to eq. (IV.6.1) assuming an intrinsic interfacial width 23.3 ± 3.5 Å, an interfacial tension of $\gamma = 2.7 \pm 0.3$ mJ/m$^2$, and a neutron coherence length of 20 μm. Reproduced from the study by Sferrazza et al. (1997).

their equilibrium state (Beziel et al., 2008). Using partially deuterated random copolymers of controlled ratios of ethylene and ethyl-ethylene, the Flory–Huggins interaction parameter was controlled by the ratio of the components, to give systems with $\chi N$ of 3.7, 6.8, and 18, for molecular weights in all cases of ~150,000 g/mol. To avoid confinement effects that decrease the apparent interfacial width (Sferrazza et al., 1997), relatively thick layers of between 200 and 300 nm for both bilayers were utilized in these NR measurements. To observe the earliest stages of interfacial broadening, the top polymer film was spun cast onto a glass slide and then floated onto the Si previously coated with the copolymer layer, and the system rapidly quenched to −80°C, which is 19°C below the glass-transition temperature, $T_g$, for the system. To investigate how the interface reaches equilibrium, the samples were subsequently annealed for various lengths of time up to 5 days at 85°C, before NR measurements were taken. In the strong segregation limit (SSL) and for these finite molecular weight polymers, the intrinsic interfacial width is given by (Tang and Freed, 1991; Broseta et al., 1990)

$$w_I = \frac{b}{\sqrt{6\chi}}\left[\frac{3}{4}\left(1-\frac{2}{\chi N}\right)+\frac{1}{4}\left(1-\frac{2}{\chi N}\right)^2\right]^{\frac{1}{2}}. \qquad \text{(IV.6.2)}$$

For $\chi N = 18$, the measured interfacial roughness, $\Delta$, for these polyolefin blends was found to be ~4 nm, compared to a predicted interfacial roughness ($\Delta_0 = 2b\sqrt{2\pi}/\sqrt{6\chi}$) of 2.2 nm. The difference in these values was shown to be due to capillary wave broadening with experimental estimates for the capillary wave width equal to ~3.3 nm in excellent agreement with a theoretical estimate of 3.0 nm. When $\chi N = 6.8$, the equilibrium interface roughness of ~6.5 nm was shown to lie somewhere between the SSL and the weak segregation limit (WSL), in which $w_I = (b\sqrt{N}/3)(\chi/\chi_{crit}-1)^{-1/2}$, where $\chi_{crit}$ is the interaction parameter at the critical point. For the system where $\chi N = 3.7$, very different behavior was observed, with the interfacial roughness increasing with time from an initial value of 9.1 nm to an equilibrium value of ~18 nm. In this case, the hydrodynamic flow means that the capillary wave contribution is time dependent and is given by (Beziel et al., 2008)

$$\sigma_\zeta^2 = \left(\frac{k_B T}{\pi\gamma}\right)\ln\left[\frac{(\gamma/\eta)t}{w_I}\right] \qquad \text{(IV.6.3)}$$

where $\eta$ is the viscosity of the polymer. Fits to the data using eqs. (IV.6.1) and (IV.6.3) gave excellent agreement to the data (see Figure IV.6.2), with fit parameters of $\Delta_0 = 8.6 \pm 0.3$ nm, $\gamma = (2.0 \pm 0.5) \times 10^{-5}$ J/m$^2$, and $\eta = (3.0 \pm 0.5) \times 10^5$ Pa·s. The same random polyolefin copolymers were also used to study the interfacial width as a function of the distance from the critical point (Carelli et al., 2005a). At high degrees of immiscibility for $\chi N > 8$, the data follow SCMF theory, but as $\chi N$ decreases a crossover occurs to a region of miscibility in which a WSL square gradient theory applies.

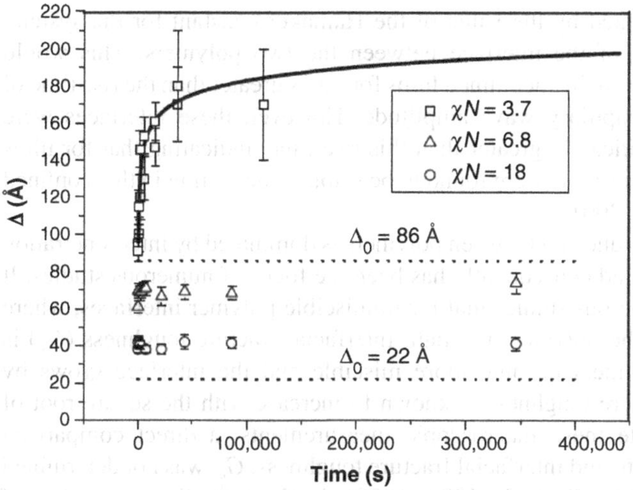


**Figure IV.6.2** Interfacial roughness as a function of time measured by NR between partially deuterated polyolefin blends where control of the comonomer ratio gives $\chi N$ of 3.7, 6.8, and 18. The dotted lines are for the SSL and the WSL limits and the solid line is a fit assuming hydrodynamic flow. Reproduced from the study by Beziel et al. (2008).

Many of the most important industrial polymers, such as polyethylene terephthalate (PET), polyethylene (PE), and polypropylene (PP), are either crystalline or semicrystalline. However, at temperatures well below the crystalline melt temperature, $T_m$, the crystallinity causes the film surface of semicrystalline polymers to be molecularly rough. NR studies of such interfaces at room temperature (i.e., $T \ll T_m$) therefore are extremely problematic since there is a drastic loss of specular reflectivity associated with this surface roughness (Bucknall et al., 1999a). At best, the measurements are extremely difficult, but in most cases, the measurements are more likely to lead to reflectivity profiles that cannot be interpreted. Heating a thin crystalline polymer film supported on a substrate into the melt regime ($T > T_m$) is typically insufficient to produce meaningful reflectivity profiles, since remnant long range waviness remains even after the local molecular roughness is removed (Bucknall et al., 1999a). To overcome these problems, a sample cell based on the design of solid–liquid cells has been utilized (Hermes et al., 1997). The sample cell was heated and held within an inert gas atmosphere to prevent any polymer degradation during the NR measurements. The system used an inverted geometry where the incident neutron beam passed through the silicon block to the polymer layers beneath. This approach has been successfully utilized to study a number of crystalline polymer interfaces including PS-PE (Hermes et al., 1997), PS-PP (Hermes et al., 1998a; Bucknall et al., 1999a), and PS-PE interfaces modified by PS-*b*-PB copolymers (Hermes et al., 1998b). In these systems, once again thermally excited capillary waves had to be accounted for. Whereas the capillary wave mechanism accounts for films that have a free surface, in this melt cell the interface is mechanically confined between two bulk media (Sferrazza et al., 1997). In such cases, the interaction between the polymers (for instance, PE and dPS) and the silicon substrate, which is dictated by the value of the Hamaker constant for the system, predicts destabilization of the interface between the two polymers. This would ultimately lead to dewetting in unconfined films for times greater than the rise time of the fastest interfacial capillary wave amplitude. However, these interfaces were measured at times significantly greater than this rise time, indicating that for films that are predicted to be unstable were found to be stable to dewetting in this confined geometry (Butler et al., 2001).

The mechanism of adhesion between polymers is dominated by interpenetration between of the chains, and consequently has been the focus of numerous studies. It has been shown in numerous studies that for immiscible polymer interfaces, where the interface between the polymers is small, interfacial fracture toughness ($G_c$) is negligible. As the polymers become more miscible and the interface grows by interdiffusion, the fracture toughness is known to increase with the square root of annealing time. Despite these macroscopic measurements, a direct comparison between interfacial width and interfacial fracture toughness, $G_c$, was not determined until fairly recently (Schnell et al., 1998, 1999). In these studies, a number of styrenic-based polymers were evaluated, so that not only could the interfacial widths be carefully controlled but also there were minimal differences in the mechanical characteristics of the polymers so that direct comparison in fracture toughness could be made. The interfacial fracture toughness ($G_c$) was determined by asymmetric double cantilever beam (ADCB) measurements, while the interfacial width was determined by NR. Initial studies were made using dPS and poly(*p*-methyl styrene) (P*p*MS) as well as dPS and PS systems, and direct correlation between interfacial width and fracture toughness were made (Schnell et al., 1998). Subsequently, PS, P*p*MS, and statistically brominated PS ($PBr_xS$) systems were compared allowing a wider range of interfacial widths to be studied (Schnell et al., 1999). The interfacial size dependence on $\chi$ for PS-$PBr_xS$ interfaces had previously been studied using NR, which agreed with the dependence predicted by SCMF theory in the strong segregation limit (Guckenbiehl et al., 1994). Using these styrenic polymers, systems with interfacial widths from ~3 to 20 nm were evaluated. Fracture toughness dependence on the interfacial width was seen to occur in three distinct regimes. For interfacial widths less than approximately 6 nm, the fracture toughness was essentially independent of width and negligible as predicted for highly immiscible systems, where interfacial mechanical failure is caused by chain pullout. For interfacial widths between ~6 and ~12 nm, a steep increase in fracture toughness was observed, as the polymer chains on either side of the interface begin to form effective mechanical entanglements, and produce a plastic zone ahead of the crack tip during crack propagation. For interfacial widths greater than ~12 nm, the fracture toughness asymptotes to values close to those for the bulk toughness as interdiffusion between the chains are able to form entanglements that are as mechanically effective as in the bulk phase.

In these and other studies, the interfaces investigated were between glassy polymers where the interface was formed at temperatures above $T_g$, while the mechanical testing were undertaken at room temperature (i.e., below $T_g$ of the polymers). Adhesion between immiscible polymers in the melt, where both the interface formation and the mechanical testing occur above the glass-transition temperature, has been studied (Schach et al., 2007). The systems investigated were between deuterated polybutadiene (dPB; $M_n = 120{,}000$ g/mol) and hydrogenous polymer melts of styrene-butadiene random (SBR) copolymer, polyisobutylene (PIB), ethylene-propylene-diene copolymer (EPDM) and poly(dimethylsiloxane) (PDMS). For these viscoelastic polymers, ADCB fracture measurements were not possible. Therefore, probe tests were used, in which the two polymers were prepared on separate surfaces and after they were pushed together, the stress–strain curve for their detachment was used to determine adhesion energy from the total work of detachment ($W_{adh}$). NR measurements were made of the same polymers as the mechanical tests to determine interfacial widths. The adhesion energy was found to increase with interfacial width (see Figure IV.6.3). For large interfacial widths, of the order of the radius of gyration of the polymer, an asymptotic value of $W_{adh}$ was reached, that is equivalent to the cohesive fracture energy for polybutadiene. These measurements also showed that the degree of interpenetration necessary to achieve bulk strength of the interface between the polymer melts was larger than that in polymer glasses. In fact, bulk strength was only achieved when the interface was of the order of several entanglements rather than one entanglement found for glassy polymers. This behavior is perhaps not surprising since it reflects the ease with which polymer chains above their $T_g$ can be removed from a melt.

Several methods are utilized for strengthening the interface between immiscible polymers, including addition of block copolymers, or reactive processing where a fraction of the two polymers are functionalized and combine at the interface (Brown, 2000). At immiscible polymer interfaces the copolymers segregate so that blocks penetrate into the respective homopolymer phases, and due to the covalent bond between the copolymer blocks the effective increase of entanglement density across the interface improves the adhesion and toughness of the interface. The technological implications of this have led to numerous studies to determine the distribution and therefore basis for the compatibilizing effect of copolymers. NR has

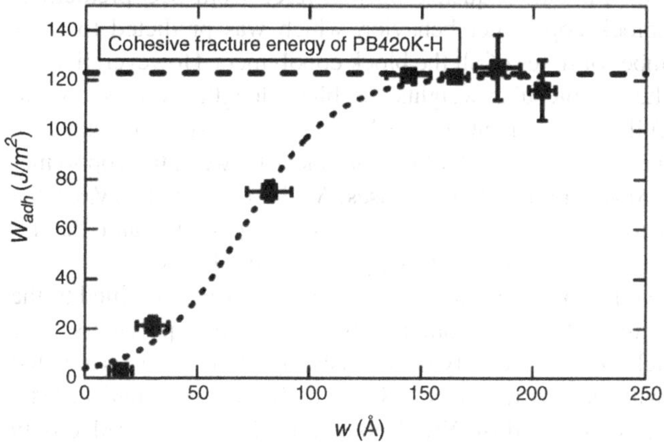


**Figure IV.6.3** Adhesion energy of polymer melt interfaces as a function of the interfacial width measured by NR between dPB and various polymers. Reproduced from the study by Schach et al. (2007).

shown that with its excellent depth resolution it is an ideal technique for revealing in detail the segment distribution at the interface. Some of the earliest theoretical and experimental studies were on diblock copolymer segregation behavior at highly immiscible polymer interfaces. SCFT theories predict the block copolymer location at the polymer–polymer interface and also the extension of the blocks into the respective bulk phases (Noolandi and Hong, 1982; Leibler, 1988; Semenov, 1992). The detailed distribution of a P(S-*b*-MMA) diblock copolymer at an interface between PS and PMMA homopolymers has been studied using NR measurements (Russell et al., 1991). Similarly, Dai et al. investigated a PS and poly(vinyl pyridine) (PVP) interface modified with a P(dS-*b*-VP) diblock copolymer (Dai et al., 1994). In both these systems, it was found that at equilibrium the copolymer segregated to the homopolymer–homopolymer interface, with composition profiles for the homopolymers and copolymer blocks that agree well with predictions from SCFT theory (Shull and Kramer, 1990). More recently, Eastwood et al. have studied both multiblock and random copolymers at immiscible polymer interfaces computationally with Monte Carlo (MC) calculations (Dadmun, 1996, 2000), and experimentally with NR and ADCB measurements (Eastwood et al., 2005). The MC calculations predicted that the relative strengthening ability of the copolymer at the interface would follow the sequence: heptablock > pentablock > triblock > diblock > random. This behavior was expected since the multiblock copolymers are able to create better entanglement with the homopolymers than diblock copolymers due to the formation of loops. Random copolymers are able to strengthen the interface if they are "blocky" and able to form extended loops. If the random copolymers are more "alternating" in nature, sufficiently long loops cannot extend either side of the interface and consequently do not form entanglements. Therefore, the blocky random copolymers are expected to be significantly better copolymers than alternating ones for interfacial compatibilization. However, using copolymers of PS-PMMA ($M_w \sim 150{,}000$–$200{,}000$ g/mol) at PS/PMMA interfaces of similar molecular weight, the order of interfacial toughening, determined by ADCB measurements, was found to be pentablock > triblock > diblock > heptablock > random. This sequence only differed from MC predictions with regard to the heptablock copolymer behavior, which was predicted to cross the interface the most number of times of all the block copolymers. However, like the random copolymer for these molecular weights the block lengths were shown to be insufficient to allow full entanglement. Using NR, it was possible to relate the interfacial fracture toughness with the width of the interfaces between the copolymer segments and the PS and PMMA homopolymer phases. As shown in Figure IV.6.4, the width of the interface between the PS segments of the copolymer and the PS homopolymer do follow the $G_c \propto w^2$ dependence predicted by theory.

Reactions between two immiscible functionalized polymers are confined to the interfacial region due to the intrinsic incompatibility of the two polymers. The interface between two such immiscible polymers consisting of an amineterminated PS (dPS-$NH_2$) blended with unfunctionalized PS and carboxy terminated PMMA (PMMA-$CO_2H$) has been investigated by NR (Coote et al., 2003). The reflectivity data showed that dPS-$NH_2$ formed an interfacial excess layer at the PS/PMMA-$CO_2H$ interface with a hyperbolic tangent composition profile. From the values of

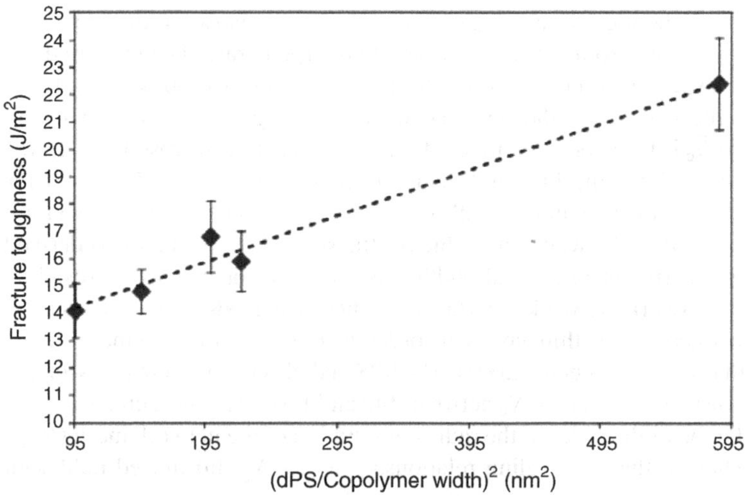


**Figure IV.6.4** Interfacial toughness of a PS-co-PMMA copolymer compatibilized PS-PMMA interface as a function of the square of the interfacial width between the PS copolymer segment and the PS homopolymer. The dashed best fit agrees well with scaling behavior predicted by theory. Reproduced from the study by Eastwood et al. (2005).

interfacial excess ($z^*$) of dPS-$NH_2$ derived from the composition profiles, the extent of grafting, $\Sigma$, and the excess layer thickness ($L$) were shown to increase with time, reaching saturated values of $\Sigma$ ($= \Sigma_{sat}$) and $L$ at long reaction times. The values of $\Sigma$, $\Sigma_{sat}$, and $L$ were also shown to be dependent on molecular weight and initial concentration of dPS-$NH_2$ blended with the hPS. The reactive interface between polyamide (PA) and polysulfone (PSU) blended with a phthalic anhydride reactive terminated polysulfone (PSU-R) was studied by NR and SANS (Hayashi et al., 2000). When the system was reacted at 210°C, the interfacial width grew to a maximum within 2 min, with the creation of the copolymers at the interface preventing coarsening of the phase-separated domains that were observed by SANS if no PSU-R was included. Since PA and PSU are immiscible, SANS data were analyzed using Porod's laws, which assumes that the interfaces between the immiscible domains are sharp, with widths, $\sigma$, determined by the $q = 0$ intercept of a Porod plot, that is, $\ln[S(q)q^4]$ versus $q^2$. For the PA-PSU system, the interfacial width $\sigma$ derived from the Porod analysis was found to be directly related to the interfacial width parameter, $w = \sigma\sqrt{2\pi}$, of a hyperbolic tangent function used to fit the NR data. In order to compare the values of $\sigma$ from SANS with that from the NR data, it was necessary to derive the intrinsic interfacial width by deconvoluting the capillary wave broadening from the NR experimentally determined value.

The interfaces of chemically cross-linked PS networks have been studied using NR (Geoghegan et al., 1998). The interfacial width between two networks was found to increase with annealing time, but very rapidly formed a relatively narrow equilibrium interfacial width, $w_{eq}$. This behavior is not surprising since the cross-links prevent macroscopic interdiffusion, but local chain rearrangements at the interface, driven by the local free energy, causes the chains between the cross-link points and dangling ends from the two networks to interpenetrate. The magnitude of $w_{eq}$ is therefore expected to be controlled by the degree of cross-linking and will depend on the mesh size for the network with a scaling relationship given by $w_{eq} \propto \bar{N}_c^{\alpha}$, where $\bar{N}_c$ is the average number of repeat units between cross-links. It was found that for $N = 32$ for the hPS and 35 for dPS networks, $w_{eq} \sim 5$ nm and for networks with $N = 584$ (hPS) and 145 (dPS) $w_{eq}$ increased to between 7 and 11 nm (Geoghegan et al., 1998). However, no value for the scaling exponent $\alpha$ was derived although clearly a narrower interfacial width was observed for increased cross-link densities. These network–network interfacial studies using NR were revisited by Briber and coworkers using thin layers in order to optimize measurement of the interfacial width (Perez-Salas et al., 2002). The hPS and dPS layers were cross-linked using $\gamma$-irradiation to give gels with $\bar{N}_c$ between 140 and 1700. The measured values of $w_{eq}$ were found to be of the order of the radius of gyration of the network mesh ($R_{g,c}$). The data also showed that the scaling relationship $w_{eq} \propto \bar{N}_c^{\alpha}$ did indeed hold with $\alpha = 0.47 \pm 0.1$.

The discovery of conducting polymers has ultimately led to the growth of polymer-based electronics, where thin polymer films are widely utilized. The performance of such polymeric-based electronic devices is often limited by the interfacial properties of these film interfaces, requiring a detailed fundamental understanding of their behavior. Organic field-effect transistors consist of a semiconductor layer deposited onto a dielectric-coated gate electrode. Pinto et al. have used NR and GISAXS to study the structure of the pentacene and poly(3-hexylthiophene) (P3HT) semiconducting layers both deposited onto a PMMA dielectric film (Pinto et al., 2008). Use of NR showed that the interdiffusion between these organic layers could be controlled in two ways. First, by grafting the PMMA onto the gate electrode using atom transfer radical polymerization (ATRP) very thin (50 nm) stable dielectric layers could be formed, from which better crystal growth of the semiconductor layer was observed. Second, by cross-linking the PMMA brush, interdiffusion with the semiconductor was found to be reduced. Interfaces between conducting polymer films are also important in the performance of polymer-based photovoltaic cells or light emitting diodes. Higgins et al. have used NR and nuclear reaction analysis (NRA) to demonstrate the ability to control the interfacial width from $\sim$1 and 30 nm between thin films of poly(9,9-dioctylfluorene) and poly(9,9-dioctylfluorene-alt-benzothiadiazole) (Higgins et al., 2006).

## IV.6.3 POLYMER–POLYMER INTERDIFFUSION

NR has enjoyed significant success in many areas of polymer research, a notable example of which is the study of polymer diffusion not only to test polymer–polymer diffusion theories but also to address technologically important questions such as adhesion, interfacial welding, and dissolution. NR is an extremely effective technique for studying the early stages of interdiffusion, allowing details of not only the interfacial width but also its compositional profile between the diffusing polymers to be determined with subnanometer resolution. The basic measurement is to determine the growth of the interface between the interdiffusing polymers as a function of time. However, it is often difficult to measure late stages of diffusion using NR, since interfacial diffusion information manifests itself most obviously in the dampening of the interference fringes associated with the polymer film thickness. As the interface grows the dampening effectively destroys the fringes making large interfaces difficult to determine with any accuracy.

The diffusion between bilayers of linear PS and dPS with molecular weights in the range of between $\sim 2 \times 10^5$ and $1 \times 10^6$ g/mol have been investigated with NR to derive the mutual diffusion coefficient ($D_{\mathrm{M}}$) (Karim et al., 1994). Although PS is in many ways a model system for studying diffusion, at high molecular weights, there is a small repulsive interaction between the deuterated and hydrogenous chains, giving rise to an interaction parameter, $\chi = 0.2/T - 2.9 \times 10^{-4}$, and consequently an upper critical solution temperature (UCST) phase behavior for PS/dPS blends. As a result, $D_{\mathrm{M}}$ is not constant but varies with concentration ($\phi$) and depends on the degree of polymerization ($N$) and $\chi$ (Green and Doyle, 1986, 1987). Making allowances for these thermodynamic effects, it was shown that the interfacial profile for PS/dPS samples measured were represented by a weighted sum of two Gaussian functions. The mean square interfacial widths increased with a $t^{1/2}$ dependency, which may naively simply be assumed to indicate Fickian diffusion. However, due to the sensitivity of the NR measurements it was possible to demonstrate the difference between the fits assuming interfacial profiles with two Gaussian functions and a conventional error function, which would be anticipated if the diffusion were occurring by Fickian dynamics. These Gaussian functions with two different mean square interfacial thickness contributions were identified with the reptation and Rouse interfacial broadening. However, although the initial interfacial broadening was consistent with Rouse broadening, the sparsity of experimental points did not allow detailed comparison to theory.

The initial stages of interdiffusion in PMMA-dPMMA systems have also been studied by NR (Kunz and Stamm, 1996). The thermodynamic slowing down observed in PS-dPS blends is negligible for PMMA polymers and was consequently not necessary to be accounted for. For the very short and longer annealing times, the interface between PMMA-dPMMA bilayers could be described by a simple single error function. At intermediate annealing times, a more complicated functional form for the scattering length density profile at the interface was required to fit the data consisting of convolution with a Gaussian and an error function. In these measurements, the experimentally determined interfacial widths, $\sigma_{\exp}$, were simply corrected by the initial roughness of the as-made sample, $\sigma_0$, so that the interfacial broadening $\Delta\sigma = \sqrt{\sigma_{\exp}^2 + \sigma_0^2}$. The magnitude of $\Delta\sigma$ as a function of time effectively showed three regimes, although the sparsity of data in the first two regions was difficult to define with any reliability. However, beginning at times approximately equal to the calculated Rouse time $\Delta\sigma$ increased according to $\Delta\sigma \propto t^{0.3}$, which agrees reasonably well with the $t^{1/4}$ prediction from reptation theory. At longer annealing times, greater than the reptation time, $\tau_d$, the predicted Fickian diffusion with a $\Delta\sigma \propto t^{1/2}$ dependency was not observed. This was thought to be due to the limitations in the accuracy of NR in determining large interfacial widths (Kunz and Stamm, 1996). In these cases, other techniques such as NRA or dynamic secondary ion mass spectroscopy (DSIMS) are better able to measure these larger interfacial widths.

Measurement of the interfacial widths at the very earliest stages of polymer interdiffusion is experimentally a demanding exercise. The broadening of the interface occurs over length scales that are less than the overall chain dimensions. In these cases, since the interfacial broadening is not achieved by whole chain motion, a length scale-dependent diffusion coefficient, $D(k)$, is expected to be observed. To test this theory (Sivaniah et al., 2003), multilayer samples of eight alternating repeat layers of PS and its deuterated analog, dPS, were analyzed by NR (Sivaniah et al., 1999). The mean layer thickness in these multilayers were in the range ~7–19 nm, which produce a series of Bragg peaks in the NR data associated with the characteristic repeat layers. However, since the Born approximation becomes progressively less accurate as it approaches the critical condition for total reflection it is not possible to measure the interdiffusion directly from the decay in Bragg peak intensities in the NR data. For this reason, maximum entropy fitting methods were used to initially determine the compositional profiles, which were subsequently fast Fourier transformed back into (Fourier) scattering space. The resulting idealized power spectra consequently gave intensities from an ideal scattering experiment. For multilayer systems that undergo Fickian diffusion between the layers, the evolution of an arbitrary concentration profile is given by

$$\phi(z,t) = \sum_{n=1} [A_n \cos(nkz) + B_n \sin(nkz)] \exp\left(-\frac{t}{\tau_k}\right) \qquad \text{(IV.6.4)}$$

where $A_n$ and $B_n$ are determined from the Fourier components of a square wave profile (Sivaniah et al., 1999), the characteristic time $\tau_k^{-1} = D(k)n^2k^2$, $k = 2\pi/\lambda$, and $D(k)$ is the wave vector-dependent diffusion coefficient. Therefore, from the NR profiles used to determine the decrease in the "idealized" Bragg peak intensities with annealing time as the PS-dPS multilayers interdiffuse, it was possible to extract $D(k)$ from the gradient of the logarithmic decay of these peak intensities. The PS multilayer samples were annealed for times less than the calculated reptation time for the polymers in order to study the very early stages of annealing. By deriving the $D(k)$ behavior for these multilayer samples, these results were used to reconstruct the interfacial behavior that would exist in semi-infinite bilayer samples (Sivaniah et al., 2003). These interfaces showed a sharp interfacial core (at the center of the interface) with tails diffusing into the bulk of the sample. With further annealing, the interface was shown to become more diffuse, as would be expected as the polymers interpenetrate, but the sharp core persisted and simply reduced in size and effectively only disappeared as time approached the reptation time ($\tau_d$). Several diffusive stages were observed over the annealing time of the experiments. The biggest changes in Bragg peak intensities occurred at longer annealing times, that were still significantly shorter than $\tau_\mathrm{d}$, indicating the existence of collective diffusional transport at times well below the reptation time. These studies thereby raise questions as to the validity of the previously accepted understanding of the early stages of interfacial diffusion.

Arguably one of the most elegant studies of polymer diffusion is the so-called "ripple experiment" (Russell et al., 1993; Welp et al., 1998). In these experiments, the interdiffusion of oppositely labeled triblock polystyrene chains (HDH and DHD) were measured using DSIMS and NR. The HDH chains had a central deuterated block that amounted to ~50% of the chain with the remainder consisting of symmetrically sized chain ends of hydrogenous PS. The DHD was oppositely labeled but retained the same block length sequence. Measurement of the interfacial development between these two polymers is therefore extremely sensitive to any topographic constraints and/or anisotropic motion of the chains. The concept in these experiments is that since motion of chain ends is preferred over the centers, once interdiffusion starts this difference in chain motion would establish a deuterium enrichment on the HDH side and a depletion of the DHD side of the interface. The deuterated concentration profile therefore displays a characteristic "ripple" (see Figure IV.6.5), hence the name these experiments have acquired. Using the change in peak height, position and areas of these ripples it is possible to uniquely test features predicted for different dynamic models. The development of these parameters, as determined by the NR measurements, allowed all possible models to be tested. The results were found to be entirely consistent with predictions from reptation theory. A later study by the same group, developed this experiment further and used NR measurements to make comparisons between the central and the end section chain dynamics (Welp et al., 1999). In these experiments, interdiffusion between bilayers of a HDH triblock against both dPS and hPS films were investigated. Using bilayers of HDH and dPS it was possible to observe the development of the initially sharp interface by the exchange of hydrogenous chain end segments across the interface. For the HDH and hPS bilayer samples, the development of the interface is characterized only by exchange of the deuterated chain center segments. Using these two sample geometries, it was possible to isolate the dynamics of the chain ends

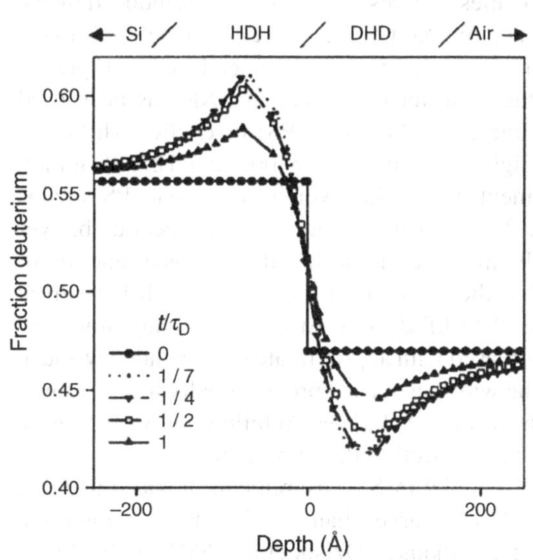


**Figure IV.6.5** Interfacial deuterium concentration profile determined from fits to the NR data for HDH/DHD bilayer interdiffusion, showing the characteristic "ripple." Reproduced from the study by Welp et al. (1998).

and central segments. In both Rouse and reptation dynamics, both the mechanisms predict a gradual increase in the chain end segment interpenetration with increasing annealing time. By comparison, the behavior for the central segment of the chain is predicted to be quite different between these models. In the Rouse model, both the central segment of the chain and the chain ends are predicted to continuously increase their interpenetration depth with time. The reptation model, on the other hand, predicts that due to the tube constraint implicit in this diffusion mechanism the central segment does not initially move across the interface. This causes a delay before any increase in the central segments that are seen to diffuse across the interface. By careful determination of the interpenetration distance across the interface in the HDH/hPS samples, a delay in central segment diffusion was observed, which confirms that reptation was the underlying diffusion mechanism (Welp et al., 1999).

The effects that polymer architecture play on polymer diffusion make diffusion of nonlinear chain architectures an interesting area of study. Cyclic polymers are particularly interesting systems to study since the diffusion theories of chains are heavily dominated by the existence of chain ends, as described under the framework of the reptation theory (Doi and Edwards, 1986). Taking this theory at face value, cyclic molecules without chain ends would therefore not be expected to diffuse. Using forward recoil elastic spectrometry (FRES) Kramer and coworkers studied diffusion processes between cyclic and linear PS, revealing how cyclic polymers move in entangled media (Mills et al., 1987; Tead et al., 1992). In these cases, reptation and constraint release processes observed as may be expected for linear polymer diffusion. In addition, two further diffusion modes for the cyclic PS were observed, which restricted the cyclic diffusion either by entanglements of the linear chains or by threading of the cyclic onto linear chains. It was believed that the slower diffusion of cyclic molecules observed by Kramer and others was simply associated with the higher densities of these molecules with respect to linear ones induced by the reduction of free volume in the absence of chain ends. This effect would of course become negligible at high molecular weight where chain ends cannot play a dominant role. Following on from these earlier measurements, NR has been used to compare interdiffusion between linear (l-hPS and l-dPS) or cyclic (c-hPS and c-dPS) polymers with molecular weights of between 109,000 and 127,000 g/mol, well above the nominal entanglement molecular weight of linear PS chains ($\sim$30,000 g/mol) (Kawaguchi et al., 2006). Using deuterated-hydrogenous bilayer PS films, NR was used to measure the increase in interfacial width as a function of annealing at 120°C. It was found that the increase in interfacial width for c-hPS/c-dPS bilayers was faster than that for the l-hPS/l-dPS. At short annealing times, the interfacial widths for both systems increased with approximate $t^{1/2}$ dependency and at later annealing times ($t > \sim 6000$ s) the width increased proportionally to $t^{1/2}$, as was predicted by theory (see Figure IV.6.6) (Karim et al., 1990; Whitlow and Wool, 1991). The crossover time was found to also agree well with the reptation time for l-PS of 4000 s at this annealing temperature. Using DSIMS measurements, it was found that the diffusion coefficient for c-PS is 2.2 times larger than that for l-PS, contradicting results obtained for low molecular weight alkanes (Ozisik et al., 2002), indicating a

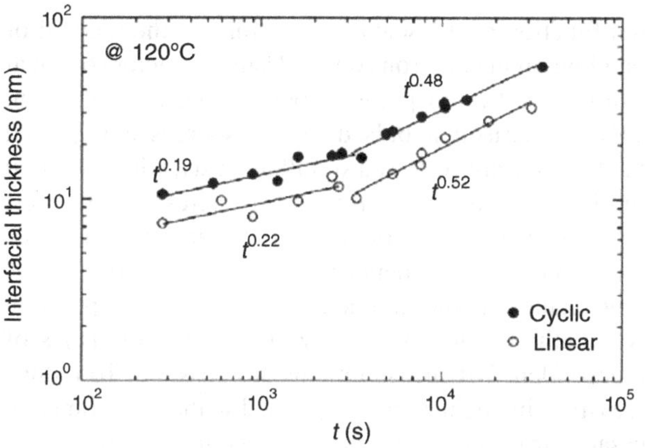


**Figure IV.6.6** Log–log plot of the change interfacial thickness with time for l-PS/l-dPS and c-PS/c-dPS bilayers. Reproduced from the study by Kawaguchi et al. (2006).

clear molecular weight dependence on cyclic molecule dynamics. However, further work remains to be undertaken to resolve this molecular weight dependency.

Diffusion and surface enrichment of nitroplastizicers consisting of bis(dinitropropyl)acetal or formal in estane thermoplastic polyurethane has been investigated by NR (Smith et al., 2004). The determination of film composition showed that plasticizer diffusion caused enrichment at both the polymer/substrate and the polymer/air interfaces. The enrichment at the silicon surface once established remained constant with time, but at the air surface, the weak attractive driving force for enrichment was balanced by evaporation of the plasticizer, the rate of which was shown to follow Fick's law. Low-pressure $CO_2$ has also been shown through NR interfacial measurements to enhance chain mobility, increasing PS-dPS and PMMA-dPMMA interdiffusion rates as well as increasing the surface layer mobility (Yang et al., 2007).

## IV.6.4 REAL-TIME MEASUREMENTS OF FILM KINETICS

NR measurements of polymer interfaces even when studying interdiffusion have typically been undertaken at room temperature (RT) where most polymer systems measured are below their glass-transition temperatures, and any kinetic processes are essentially quenched. An "anneal-quench" procedure is repeated for each successive annealing time. Although this measurement procedure has proved very successful for amorphous polymer systems where the $T_g$ values of both polymers are well above RT, it is not applicable for the study of systems where the $T_g$ for one, or both, of the polymers is close to or less than RT, or if small-molecule ingress into a polymer layer is to be studied. Since the interface remains mobile at RT an extremely rapid quench to below RT must be made where the diffusing species are immobilized for the length scale of the NR measurement. Difficulties may arise either if the quench is not rapid enough or if the interface is disturbed by the quench process—especially, when heating the sample for the next annealing.

Time invariant NR measurements can be performed very efficiently at both pulsed and reactor-based neutron sources. Although the data obtained from either source provide reflectivity as a function of the scattering vector ($q$), the method of collecting this data varies depending on the neutron source. Using a reactor source at fixed wavelength, the NR profile is collected point-by-point using a $\theta$–$2\theta$ scan, where $\theta$ is the incident angle. By comparison, at pulsed neutron sources, at any fixed incident angle data within a fixed $q$ range are collected simultaneously, with often two or three incident angles required to cover the entire $q$ range of interest. For the best statistics and therefore ability to analyze the data, count times are important and since even at the current highest flux sources measurement times are counted more in tens of minutes, consequently kinetic changes in the sample in most cases are impossible to follow using traditional NR methods. However, utilizing the characteristics of pulsed neutron sources, time-dependent NR measurements are possible. To reduce collection times, a fixed $q$ range (fixed incident angle) is selected so that the $q$ range of the reflectivity profile is necessarily truncated from that achieved using two or three angles. However, this means that the data collection times can be reduced by an order of magnitude or better. By choosing the correct incident angle so that the significant features of the reflectivity are within the obtainable $q$ range, changes observed in the reflectivity profile as a function of time can be observed. If and when the significant features in the reflectivity move outside of the observable $q$ window, the angle can be changed to shift the observable range.

Utilizing this approach, real-time NR methods were first demonstrated using polystyrene–deuterated polystyrene (dPS) interdiffusion at 115°C, that is, above the $T_g$ of either polymer (Bucknall et al., 1999b). The system and the temperature were chosen so that the bilayer could be quenched to room temperature and a full reflectivity profile measured over an extended $q$ range to check the validity of the technique. The data measured in time bins of 6 min over 225 min were able to be analyzed despite the truncation of the lowest and the highest $q$ data. A diffusion coefficient for the system using this approach was determined from the time-dependent increase in interfacial widths to be $(1.7 \pm 0.2) \times 10^{-17}\,\text{cm}^{-2}/\text{s}$, which is entirely consisted with values obtained from other researchers. The technique was subsequently extended to measure diffusion of low molecular weight (oligomeric) methyl methacrylate (OMMA) and ethylene glycol (PEG) into thin layers of dPMMA (Bucknall et al., 2001), as well as oligo-styrene into dPS films (Bucknall et al., 1999a). Figure IV.6.7 shows an example of real-time NR data for OMMA diffusion into dPMMA plotted as a contour plot with time plotted against $q$ (Bucknall et al., 2001). Data collection times per reflectivity profile using the SURF reflectometers at ISIS were reduced down to 20 s, which was a practical limit to the technique using the flux available when the data were collected. The data clearly demonstrate the restricted $q$ window ranges that the data were collected over at any fixed incident angle. As the interference fringes dampen due to interdiffusion, the incident angle was reduced to shift the $q$ window to lower ranges in order to capture the most important features of the data at increasing annealing times. From these data, both interfacial width and deuterated layer thickness values were obtained from which the kinetics of the interdiffusion for the various systems were determined. Inspection of the NR data (Figure IV.6.7) also shows that with increasing time the dPMMA layer decreases in thickness associated with OMMA diffusing into the layer,

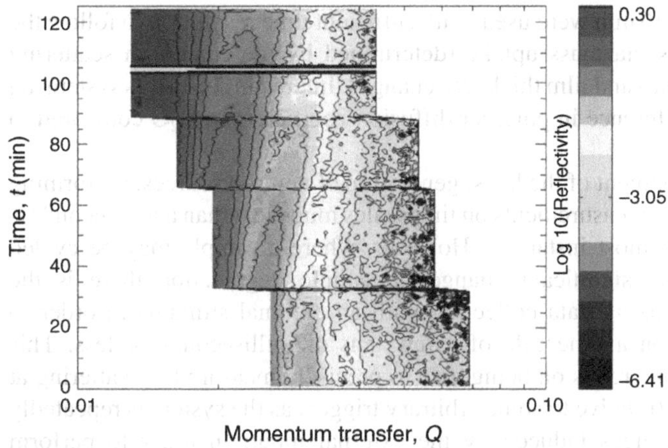


**Figure IV.6.7** 2D contour plot of time-dependent change in reflectivity for an OMMA (510 g/mol) dissolving away a film of dPMMA (115,000 g/mol). Reproduced from the study by Bucknall et al. (2001).

while simultaneously the interfacial width between the OMMA and dPMMA also increases. This is clearly observed qualitatively in these NR contour plots, by the fringe spacing increasing with time, that is, the contours fan out. Utilizing data collection times of ~20 s, real-time NR was also used to determine the kinetics of three different molecular weight phthalate plasticisers (diisononyl phthalate (DINP), dioctyl phthalate (DOP), and dibutyl phthalate (DBP)) diffusing into dPMMA (Bucknall et al., 2004). Analysis of these data showed that DINP and DOP, the two highest molecular weight phthalates, swell the dPMMA, whilst the DBP that has the lowest molecular weight dissolves the dPMMA.

*In situ* real-time NR measurements with 30 s time resolution has also been used to study the swelling kinetics of a dPS film in toluene vapor (Muller-Buschbaum et al., 2006a). Both specular and off-specular reflectivity data were collected. Initially, a decrease in the off-specular Yoneda peak intensity normalized to that of the dry film was found indicating that the film thickness increased. This was associated with swelling of the film by the toluene vapor. As the film neared saturation point by the toluene vapor, the film thickness did not change but increases in surface roughness were, as observed by an increase in the off-specular intensity. Real-time NR measurements (3–5 min per reflectivity profile) have also been used to study PS diffusion into a dPS swollen by supercritical $CO_2$ ($scCO_2$) by determination of the growth in interfacial width with annealing time (Gupta et al., 2003). These measurements were performed in a custom built temperature controlled high-pressure cell within which the polymer films on a silicon substrate are held. The effect on diffusion rate as a function of molecular weight, $CO_2$ concentration, and temperature were studied. These results showed that the $scCO_2$ is an effective plasticizing fluid and that the self-diffusivity ($D_s$) scales as $\sim M^{-2.38}$, which agrees with the behavior observed in good solvents. The swelling kinetics of hydrophilic polyacrylamide (PAM) films in water vapor have also been investigated by real-time reflectivity (neutron and X-ray) measurements (Mukherjee et al., 2007). Data acquisition times of 5–30 min were used, with sufficient time resolution to follow the PAM swelling kinetics via mass uptake (determined by the change in scattering length density of the film) and film thickness changes. Interestingly, in this system, an order of magnitude difference in polymer diffusion coefficient in $D_2O$ compared to $H_2O$ was observed.

Despite the development of the latest generation of neutron sources, performing time-resolved scattering measurements on timescales much less than a few seconds is virtually impossible in most instances. However, where a sample may be cycled repeatedly without any significant change or sample degradation there is the possibility of synchronizing data collection with an external stimulus in order to obtain measurements on a timescale of a few tens of milliseconds or less. This "stroboscopic" technique relies on being able to precisely measure the scattering at exactly the same point (relative to some arbitrary trigger) as the system is repeatedly cycled through the changes induced by the external input. In order to perform synchronized data collection on such short timescales, the standard data acquisition electronics (DAE) have to be modified to enable precise synchronization of time-resolved data collection with changes in sample environment equipment, so that the DAE also triggers the sample environment equipment. An initial proof of principle of this stroboscopic NR technique was made studying the reorientation of a nematic liquid crystal (LC), 4,4'-octyl cyanobiphenyl (8CB), in an electric field (Dalgliesh et al., 2004). LCs were chosen because of their stability under the repeated application of an electric field. Additionally, the director reorientation of LCs has been extensively studied using optical microscopy, X-ray scattering and nuclear magnetic resonance in numerous sample geometries, meaning the NR results could be readily compared to theoretical models.

A simple sample cell was constructed to allow a field to be applied across a 65 $\mu$m gap containing 8CB. The expected diffuse nematic ordering peak $q = 0.19$ A$^{-1}$ was observed on the application of an AC field at a temperature of 35°C. Further dynamic measurements were then carried out by applying an AC field of 1 kHz frequency lasting 320 ms to the 8CB. Decay in the measured intensity resulting from the relaxation of the LC ordering was clearly observed, and the fitted relaxation time agreed well with existing theory. These preliminary LC experiments showed the feasibility of stroboscopic NR measurements in more complex dynamic systems. Of particular interest were electrochemical systems, which have been studied in depth by the groups of Hillman and Glidle (Hillman et al., 1998, 2002; Glidle et al., 2001, 2003). Generally, measurements on such systems have been carried out either in the fully oxidized or in the fully reduced states. The reason for this being that these states may be more readily held under the steady state conditions required to perform a neutron reflection measurement taking up to 2 h. Although these measurements were extremely valuable, they failed to allow the full characterization of the often rich voltammetric behavior observed using standard electrochemical techniques. The effect of solvation in a film of poly(vinylferrocence) (PVF) was therefore studied using stroboscopic NR (Cooper et al., 2004). A spin-coated PVF film was coated onto a 20 nm thick Au working electrode deposited onto a polished quartz substrate and exposed to aqueous $NaClO_4$ in an *in situ* three-electrode cell. The sample was voltammetically cycled at varying sweep rates of up to 10 mV/s over a range from 0.6 to 0.16 V and neutron reflectivity data collected in steps of 20 mV at various points around the redox cycle. By collecting data in evenly distributed potential bins the length of time counted in each DAE period varied leading to a significant variation in the counting statistics across the data set. This form of cycling required an alternative approach to data collection to that used for the LC studies. In this case, it was necessary for the sample environment equipment to control the data acquisition system. The resulting NR data from the redox cycling of a film of PVF in 0.1 M $NaClO_4$ in $D_2O$ showed clear changes in fringe periodicity indicating swelling of the film on oxidation with subsequent shrinkage on reduction. It was also observed that the swelling of the film occurs over a much narrower potential range than during shrinkage. On closer examination, the data revealed that the film showed hysteresis in deswelling, incomplete desolvation upon reduction, and transient salt retention under thermodynamically permselective conditions. The information obtained from stroboscopic NR could not be obtained through any other method as other probes such as infrared techniques spatially integrate through the depth of the film.

## IV.6.5 POLYMER BRUSHES AND THIN FILMS

In thin films, confinement of the polymer chains occurs due to the presence of the surfaces. This affects, for example, the apparent $T_g$ of the polymer in thin films. This was initially studied by Keddie et al. (1994) using ellipsometry, but Kanaya and coworkers have also investigated the $T_g$ dependence on dPS film thickness using NR (Kanaya et al., 2003). In these NR measurements, films of different initial thickness were preannealed either below (at 80°C) or above (135°C) the bulk $T_g$ for dPS. The thickness of the films was then measured as a function of temperature and the $T_g$ evaluated for each film from the change in gradient of a thermal expansivity plot. The data obtained agree well with the earlier predictions of Keddie et al. where the thickness-dependent glass-transition temperature, $T_g(d) = T_g^{\text{bulk}}\left[1 + (\beta/d)^{\delta}\right]$, where $T_g^{\text{bulk}}$ is the bulk $T_g$ of the polymer, $d$ is the initial film thickness, and $\beta$ and $\delta$ are constants. In addition to dynamic changes in thin films, the interfaces of thin films also provide attractive surfaces for segregation phenomena. In melt blends of polymers on solid substrates, this difference in surface energy drives segregation of one of the polymers to the surface to reduce the system energy (Jones and Richards, 1998). Segregation of melt blend films of end-functionalized PS and PS has been studied using NR by a number of groups (Hopkinson et al., 1997; Tanaka et al., 2003; Kawaguchi et al., 2007b). For $\alpha,\omega$-hPS$(R_f)_2$/dPS blend films, the $\alpha,\omega$-hPS$(R_f)_2$ chains were seen to be preferentially enriched at the air/polymer surface region via the strong surface localization of the fluoroalkyl ($R_f$) chain ends. By comparison, for $\alpha,\omega$-hPS$(COOH)_2$/dPS blend films, depletion of the COOH chain ends at the surface created dPS enrichment at the surface owing to intermolecular association via hydrogen bonding of the $\alpha,\omega$-hPS$(COOH)_2$ chains (Kawaguchi et al., 2007b). Surface segregation has also been observed using NR in blends of PB and polyisoprene (PI), where the PI enriched the air surface in both the 1- and the 2-phase regions, with lateral separation or surface enrichment depending on layer thickness (Grull et al., 2004). Surface segregation behavior in miscible polymer blend thin films of poly(4-trimethyl silylstyrene)(PT)/PI as a function of temperatures showed PT segregated at the surface of the blend film at all temperatures studied (Kawaguchi et al., 2007a). Using NR, the decay length and the surface excess amount, $z^*$, were shown to increase with increasing temperature.

Neutron reflectivity has been used extensively to study thin films of block copolymers (Russell, 1996). While studies have largely focused on diblock copolymers, recently multiblock copolymers have also been studied. An example of the studies of copolymer thin films has been a study of the effect of composition distribution in phase separated thin films of PS-*b*-PMMA diblock copolymers (Mayes et al., 1994), and PS-*b*-P2VP diblock and P2VP-*b*-PS-*b*-P2VP triblock copolymers have been investigated using NR (Noro et al., 2006). The thin film microphase structure of blends of copolymers in all these systems formed alternating lamellar microdomains oriented parallel to the substrate surface. In the studies of PS-*b*-P2VP diblock and P2VP-*b*-PS-*b*-P2VP triblock copolymers, mixtures were prepared with copolymers of different PS weight fractions, but approximately similar molecular weights (Noro et al., 2006). The lamellar structure causes characteristic Bragg peaks in the NR profiles (see Figure IV.6.8). Although the lamellar structure is retained for all compositions, the domain repeat thickness and interfacial width for both diblock and triblock copolymer systems increased with polydispersity. This increase was associated with constraint effects of junction points at the lamellar interfaces. In addition, using careful deuteration schemes, it was shown that longer block chains were localized more at the center of the domains, while shorter block

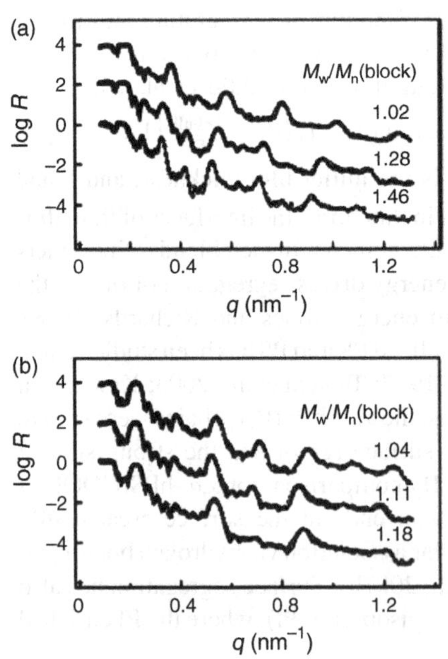


**Figure IV.6.8** NR profiles from blends ofdiblock and triblock copolymers with different effective polydispersity values. Reproduced from the study by Noro et al. (2006).

chain segments are localized near the domain interface. Orientation of the copolymer is directly related to the substrate interfacial energy. Using random copolymer films on a silicon substrate creates a neutral surface, allowing perpendicular orientation of a block copolymer film on this surface. However, it was shown using NR and scanning force microscopy that the thickness of the underlying random copolymer must be above a minimum value (~5.5 nm for a PS-PMMA system on silicon), otherwise the selective interactions with the substrate are not screened (Ryu et al., 2007).

Polymer brushes have been extensively studied at both solid–liquid and air–liquid interfaces (Fragneto-Cusani, 2001; Lu and Thomas, 1998; Field et al., 1992; Lee et al., 1999). Polymerization from a silica substrate has been shown to create highly controllable PS brushes, with controlled grafting density, $\Sigma$, and brush thickness, $h$, by the chain molecular weight (Devaux et al., 2005). The very high degree of $\Sigma$ of these brushes in good solvents were measured using NR and gave values of $h$, which did not follow classical scaling laws, which are only applicable for low graft density systems. The structure and the swelling of pH-responsive random copolymer polyampholyte brushes containing a weak acid (methacrylic acid) and a weak base ((2-dimethylamino) ethyl methacrylate) in aqueous solutions at different pH values have been evaluated using neutron reflectivity (Sanjuan et al., 2007; Sanjuan and Tran, 2008). The profiles of the polyampholyte brushes on solid surfaces were shown to depend strongly on the net charge of the chains. When an excess of charge exists the brush swelling behaved like a polyelectrolyte brush, while at the isoelectric point (zero net charge), the polyampholyte effect dominates, resulting in the collapse of the chains. Recently, absorption of an ABA triblock copolymer of poly (2-vinylpyridine)-*b*-deuterated polystyrene-*b*-poly(2-vinylpyridine) (PVP-*b*-dPS-*b*-PVP) onto silicon has been investigated (Huang et al., 2008). The copolymer forms brushes that are tethered to the silicon by the PVP end blocks forming dPS loops. Using NR, the swelling behavior and the segment density profiles of these brushes in toluene and cyclohexane was evaluated. Different solvation creates brush profiles that differ not only on solvent quality but also on tethering density.

The structure of a low molecular weight water soluble biocompatible block copolymer of poly(2-(methacryloyloxy ethyl phosphorylcholine)-*b*-2-(dimethylamino ethyl methacrylate) (MPC30-DMA60) at the air/water interface has been studied using NR as a function of solution pH and copolymer concentration (Mu et al., 2006). With increasing copolymer concentration and pH, an increase in surface excess was observed. The surface structure of the adsorbed diblock copolymer film was shown to consist of a dense layer of 1–1.5 nm on the air side, accompanied by a less dense layer of 2–3 nm extending into the aqueous phase. Salt concentration can also affect the nanostructure of a polymer monolayer at the air–water interface, as observed for an ionic amphiphilic diblock copolymer of (diethyl silacyclobutane)-*b*-(methacrylic acid) ($PEt_2SB$-*b*-PMAA), at the air–water interface (Mouri et al., 2004). As shown by Figure IV.6.9, a dense $PEt_2SR$ was found at the air surface that tethers the PMAA brush extending in to the water subphase, the extent of which was shown to be directly related to the salt concentration associated with Coulomb interactions between the salt and the charged polymer chain. In uncharged systems, temperature

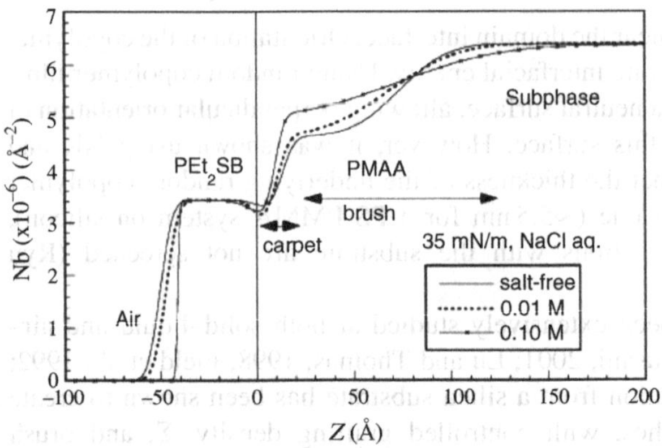


**Figure IV.6.9** Scattering length density profiles obtained from fitting NR measurements of $PEt_2SB$-PMMA at the air–water interface as a function of salt concentration. Reproduced from the study by Mouri et al. (2004).

and concentration have been shown to affect the structure of the polymer at the air–water interface. Adsorbed layers of two ABA triblock copolymers of poly (ethylene oxide)-*b*-poly(propylene oxide)-*b*-poly(ethylene oxide) (PEO-*b*-PPO-*b*-PEO) at the air/water interface have been studied as a function of concentration and temperature (Vieira et al., 2002). In all cases, an upper water free layer consisting of PPO is formed with PEO extending into the water subphase. Using selective deuteration and contrast matching of the water phase, the copolymer structure was shown to be more complex than this simplistic model suggested, with the copolymer defined in total by four individual layers.

Polyelectrolyte multilayers (PEMs) have received a great deal of attention recently, since they have potential in areas such as sensors, separation membranes, and drug delivery vehicles. NR has been shown to be a powerful method for elucidating the polymer density and the water content in such films (Schmitt et al., 1993; Losche et al., 1998). Other studies have shown water vapor induced density gradients within the PEM layers (Steitz et al., 2000), as well as allowing Flory–Huggins parameters to be determined (Kugler et al., 2002). However, until recently, little was known about the water distribution in the PEMs. NR was therefore used to determine the water concentration profile in PEMs consisting of poly(acrylic acid) (PAA) and poly(allylamine hydrochloride) (PAH) (Tanchak et al., 2006). The water was found to be asymmetrically distributed through the multilayer film with a larger concentration at the air–film surface and an exclusion zone where no water is found at the silicon substrate–film interface. Swelling behavior has also been studied in polybase brushes of poly(2-(dimethylamino)ethyl methacrylate) (PDMAEMA) and its quaternized analog of poly(2-(trimethylamino)ethyl methacrylate) (PTMAEMA) (Sanjuan et al., 2007). The degree of swelling in methanol and water as a function of grafting density of the polymer brushes covalently bonded to a silicon substrate were both shown to follow predicted mean-field scaling laws. It was also possible to calculate the effective charge ratio of the PDMAEMA brushes as a function of pH from the swollen layer thickness. The so-called weak polyelectrolyte multilayers (WPEMs) containing weak acidic or basic polyelectrolytes are known to be responsive to pH, although there is a lack of understanding of the fundamental mechanisms surrounding the response of WPEMs to their environment. One study to investigate the pH response of WPEMs consisting of alternating layers of poly (methacrylic acid) (PMAA) and partially quarternized poly(4-vinyl pyridine) (P4VP) was undertaken using NR (Kharlampieva et al., 2008). The work showed that there was a pH-induced release of polyanions from the WPEMs, and that after PMAA release from the films the well-defined multilayer structure of the initial film completely disappears. Reabsorption of the PMAA was seen to swell the films back to the film thickness in the as-made sample, but the PMAA was shown to be uniformly distributed through the film thickness.

Supercritical $CO_2$ has been widely studied largely due to its potential as an environmentally benign "green" solvent for chemical reactions, recycling or aiding processing since pressure and temperature can be controlled to manipulate the solvating properties of the $scCO_2$. However, only a limited number of polymers are "$CO_2$-philic" in easily accessible temperature/pressure regimes. Despite this, it has been shown by NR measurements that up to 60% $CO_2$ could be absorbed in thin styrene-butadiene rubber (SBR) polymer films even when the polymer and $scCO_2$ are essentially immiscible in the bulk (Koga et al., 2002). In SBR, and subsequent measurements on PS brushes, the degree of swelling quantified through the linear dilation ($S_f = L{-}L_0/L_0$, where $L_0$ and $L$ are, respectively, the dry and the $scCO_2$ solvated brush thicknesses) was shown to produce a maximum at the density fluctuation ridge of the $scCO_2$ associated with the critical condition for solvent–solute molecules (Koga et al., 2004). The NR data from the solvated films could all be described using a step concentration profile indicating that the solvent quality of $scCO_2$ was poor at all pressures and temperatures. The anomalous swelling was found to be independent of whether the PS brush was covalently or physically adsorbed onto the substrate. Studies of polymer films of SBR, PB, and PS, and silicon rubber showed that the magnitude of the swelling was a function of the elasticity of the films rather than the bulk solubility of $CO_2$ (Koga et al., 2003). The enhanced miscibility of the rubber/$CO_2$ systems that are immiscible in the bulk was found to be almost identical to that of the highly miscible silicon rubber/$CO_2$ mixture.

Perhaps the most experimentally difficult interfaces to study with neutron reflectivity are liquid–liquid interfaces. The most significant problem is associated with minimizing the very significant loss of neutron intensity upon transmission through one of the liquid mediums in order to probe the "buried" liquid–liquid interface. To overcome this problem, it is necessary to create and maintain a thin film of the upper liquid phase through which the incident and the reflected neutron beams are transmitted. A number of approaches including spreading, and condensation of a volatile oil on to an aqueous subphase have been demonstrated in NR experiments (Cosgrove et al., 1992; Lee et al., 1994). A more versatile method has been developed, in which the oil layer is trapped between a surface-modified silicon substrate and the bulk aqueous subphase (Bowers et al., 2001; Zarbakhsh et al., 2005). This method is not restricted to studies of oil–water interfaces but, in principle, can be adapted to any liquid–liquid interface, as long as the silicon surface is wet by the thin film fluid. This experimental approach has been used to study the detailed interfacial structure of PB-*b*-PEO diblock copolymers (Bowers et al., 2001), and also a PEO-*b*-PPO-*b*-PEO triblock copolymer (Pluronic L64) at hexadecane–water interfaces (Zarbakhsh et al., 2005).

## IV.6.6 LATERAL STRUCTURE IN THIN FILMS

In many thin films, the lateral structure of the film is not homogeneous. Measurement of these laterally inhomogeneous films using specular NR measurements produces data that simply averages out the structure in the normal direction. In these cases, off-specular and grazing incidence small-angle neutron scattering (GISANS) measurements can be utilized to determine the lateral structure. In essence, the results of both off-specular reflectivity and GISANS are the same, with differences resulting from the incident beam geometry with GISANS providing both $q_x$ and $q_y$ data, with only one of these vectors obtained in off-specular reflectivity. The similarity (or differences) between off-specular reflection and GISANS has been discussed (Wolff et al., 2005a). Off-specular NR has been used to study the capillary wave roughening at PS-dPMMA interfaces (Sferrazza et al., 1998). The off-specular intensity was measured as a function of scattering angle, that is the in-plane momentum transfer, $q_x$. The off-specular intensity for the unannealed (as-made) PS-PMMA sample showed a rapid monotonic drop in intensity from the specular ridge. After annealing, a prominent maximum that grows in intensity with annealing time was observed (see Figure IV.6.10). The peak corresponded to a scattering dimension of ~1 μm, which was interpreted as the fastest growing capillary wave at the PS-PMMA interface. The rise time ($\tau_m$) of the capillary wave for these samples is predicted to be dominated by dissipation forces of the substrate, and is given by $\tau_m = h^6\eta_s\gamma/h_s|A|^2$, where $h$ and $h_s$

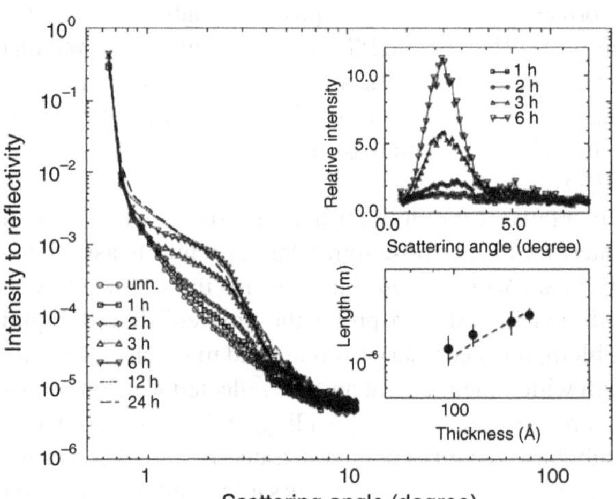


**Figure IV.6.10** Off-specular neutron intensities versus scattering angle away from specular direction for a thin (9.7 nm) dPMMA film on a PS "substrate" film as a function of annealing time at 155°C. The top inset figure is the excess scattering intensity relative to the unannealed sample for the first 6 h of annealing. The lower inset figure shows the characteristic wavelength of the fastest growing unstable capillary wave as a function of dPMMA thickness, with the dashed line a theoretical prediction given by $\lambda_m = 4(\pi^3\gamma/A)^{1/2}h^2$. Reproduced from the study by Sferrazza et al. (1998).

are the thickness of the thin and the "substrate" polymer films, respectively, $\eta_s$ is the viscosity of the "substrate" film, $\gamma$ is the interfacial tension, and $A$ is the Hamaker constant. Capillary wave contributions to the interfaces between the dPS and the poly (2-vinyl pyridine) (P2VP) blocks in a lamellar forming P2VP-*b*-dPS-*b*-P2VP triblock copolymer also give rise to off-specular scattering (Torikai et al., 2007). Due to the lamellar formation parallel to the silicon substrate, the specular reflectivity consists not only of thickness fringes but also of Bragg peaks associated with the copolymer lamella repeat length. Off-specular scattering associated with interfacial capillary wave roughening was observed in the $q_x$ directions orthogonal to the specular ridge. The off-specular was most obvious at the Bragg positions at $q_z = 0.65$ and $0.9\,\text{nm}^{-1}$. The interfacial roughening associated with inclusion of nanoparticles in to lamella forming diblock copolymers of dPS and poly(butyl methacrylate), dPS-*b*-PBMA have been determined by off-specular NR measurements (Lauter-Pasyuk et al., 2003). The copolymer, both with and without nanoparticles added, formed perfect multilayer stacks with the segment layers parallel to the substrate. The roughness and the lateral correlation lengths for the buried interfaces and also the free copolymer surface were determined using off-specular reflectivity measurements, and the latter compared to AFM measurements. For a 150 nm thick film, the lateral correlation lengths for the pure copolymer film were $\xi_i = 300 \pm 5$ nm for the buried interfaces and $\xi_s = 400 \pm 50$ nm for the free surface. When 14% of 5 nm PS-coated maghemite ($Fe_2O_3$) nanoparticles were added to the copolymer, the PS segment layers of the multilayer stack increased in thickness by 12% to accommodate the PS-coated nanoparticles which segregated to the PS blocks. For the copolymer plus nanoparticles films, $\xi_i$ decreased to $240 \pm 5$ nm, while the surface correlation length ($\xi_s$) increased significantly to $1000 \pm 50$ nm. The values of $\xi_s$ were found to be consistent with AFM results. The lateral roughness measured by specular and off-specular NR has also been used to determine the reaction–diffusion front in a model extreme UV-lithography photoresist (Lavery et al., 2008). In these measurements, a polymer–polymer bilayer geometry was used to mimic an ideal exposure line edge in a model extreme UV-lithography (EUVL) photoresist, consisting of poly(hydroxystyrene-co-$d_9$-*tert*-butyl acrylate) (PHOST-co-$d_9$TBA) and poly(hydroxyl adamantly methacrylate) (PHAdMA). The latter contained 6% by mass of a photoacid generator triphenyl sulfonium perfluorobutane sulfonate. The films were measured by specular and off-specular NR before and after exposure to 248 nm light and a postexposure bake (90°C for 180 s). The diffuse scattering was analyzed using a modified distorted-wave Born approximation (Sinha et al., 1988; Wormington et al., 1996). This model describes the interfacial roughness with an amplitude, $\sigma$, and lateral length $\xi$. The amplitude $\sigma$ of order 8 nm was determined independently by specular reflectivity and confirmed by off-specular NR. From the off-specular analysis, it was also possible to separate the contributions to the interfacial width from a gradient in material density and physical roughness. The lateral length $\xi$ was determined from the off-specular data to be 1.5 μm before reaction and 0.8 μm after exposure and postbake.

Use of GISANS has not been as widely applied as its X-ray analog of GISAXS. This has arisen primarily due to the lower fluxes on neutron beam lines compared to those of X-rays so that poor signal-to-noise and long collection times often make GISANS measurements prohibitive. However, at, for example, the ILL (Grenoble, France) where the time integrated flux of the beam lines are significant, GISANS has been used to study a number of systems, including, the lateral structure within thin films of diblock copolymers of dPS-*b*-P*p*MS (Muller-Buschbaum et al., 2000, 2003, 2004), and ABA triblock copolymers of P*p*MS-*b*-dPS-*b*-P*p*MS (Muller-Buschbaum et al., 2008, 2006b), pluronic micelle organization at the solid–aqueous solution interfaces (Wolff et al., 2005a, 2005b), and solution flow behavior over PS films (Van Der Grinten et al., 2008). A number of techniques including NR and GISANS have been used to study the thermal stability of thin polymer bilayers against dewetting (Wunnicke et al., 2003). An amorphous PA thin film on a silicon substrate was shown to be stable to dewetting. However, a dPS layer on the PA film was shown to be thermally unstable and dewets to form characteristic macroscopic droplets. Addition of a small amount (5%) of poly(styrene-co-maleic anhydride) (SMA2) containing 2% maleic anhydride groups, prevented dewetting of the dPS film. The mechanism for this dewetting suppression was revealed by the neutron measurements, which showed that the SMA2 segregated to the PA-dPS interface thereby reducing the effective interfacial tension between the films. Surface interactions with a substrate also modify the structure of copolymers, as shown by the study of dependence of surface energy on thin films of triblock copolymers of P*p*MS-*b*-dPS-*b*-P*p*MS with a lamellar microphase separated structure in the bulk (Muller-Buschbaum et al., 2008). For thick films on neutral, nonpreferential wetting surfaces, the lamellar formed perpendicular to the substrate since there are unfavorable interactions with both blocks. From the GISANS measurements, it was possible to show that the degree of order in these films is greater than in the bulk, and that the order extends up to 51 nm away from the surface. When the surface was tuned to be wet by PS, extended stretching of the mid-block of the copolymer was shown to increase the lamellar spacing by 8% with respect to the bulk value. Similarly, when the surface energy favors the P*p*MS, a slight increase of the lamellar spacing of between 2–4% was observed. Unlike diblock copolymers, where selective segregation of one block of a lamellar copolymer will lead to parallel alignment, these ABA triblock copolymers formed lateral structures on selective surfaces.

An alternative method to determine the nanostructure within the film thickness is the recently developed method of rotational small-angle neutron scattering (RSANS) (Zhang et al., 2008). In this method, the scattering intensity is accumulated as a function of sample rotational angle, so that the scattering pattern becomes sensitive to order along different planes through the scattering volume. Figure IV.6.11 shows the measured as well as the reconstructed reciprocal-space scattering intensity pattern for a block copolymer film where the $z$-axis points along the film normal, and the sample lies in the $x$–$y$ plane. RSANS allows determination of the block copolymer (BCP) orientation within the film. If all the cylinders are oriented randomly with respect to the substrate the scattering would exhibit a uniform scattering ring at $q=\sqrt{(q_x^2+q_z^2)}$ (Zhang et al., 2008). The RSANS pattern for a dPS-*b*-PMMA diblock copolymer (see Figure IV.6.11) displays a weak peak at $q_z=0$ and $q_x=0.19\,\text{nm}^{-1}$, and two intense vertical peaks at $q_z=\pm 0.1\,\text{nm}^{-1}$, $q_x=0.16\,\text{nm}^{-1}$, derive

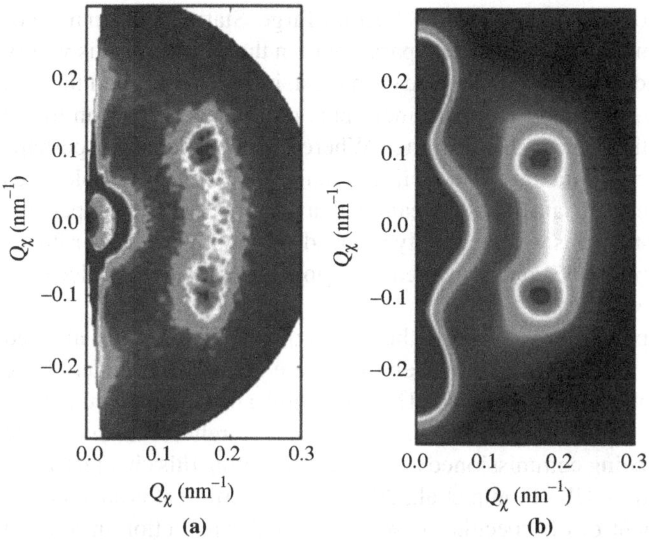


**Figure IV.6.11** Rotational SANS patternfrom a 141 nm thick dPS-*b*-PMMA film annealed for 15 h at 147°C. The plot is calculated assuming a film consisting of hybrid cylinder phase morphologies aligned perpendicular and parallel to the substrate. Reproduced from the study by Zhang et al. (2008). (See the color version of this figure in Color Plates section.)

from hexagonally packed cylindrical morphology, which is consistent with both parallel and perpendicular alignment of the copolymer domains. By mapping the in-plane structure of the films, it is possible to derive a "phase diagram" as a function of film thickness and annealing temperature, where the occurrence of parallel, perpendicular, or mixed mode aligned copolymer domains can be observed.

## IV.6.7 FUTURE PROSPECTS

In the panoply of techniques for studying polymer surfaces, interfaces and thin films, neutron reflection arguably must rank as one of the most powerful techniques. However, the technique is not without fault. Being a scattering technique, neutron reflectivity is an indirect method of structural determination, and this makes data analysis a continual challenge to interpret. The development of sophisticated software and more powerful computers has, however, largely removed these obstacles. More difficult to overcome is the current problem of access to the neutron reflectometers. Here again, things are improving with new instruments being installed at neutron facilities around the globe, making it easier than ever to obtain measuring time. Much more significant is some of the developments in a number of neutron sites where some of the world's most powerful neutron sources are being (or have recently been) upgraded or in some showpiece engineering projects being newly developed. Included in this latter category are the Japan Proton Accelerator Research Complex (J-PARC) at Tokai, Japan; the Spallation Neutron Source (SNS) at Oak

Ridge National Laboratory, USA; and the ISIS Second Target Station, UK. In most of the new developments, neutron flux has been paramount in the design. This is vitally important if time-dependent measurements are of interest, in which case total flux is a hugely important factor. Clearly if the instrument puts more neutron flux on to the sample then data collection times will be shorter. Where time dependency is perhaps not paramount, the increased flux at these facilities can also mean that sample sizes can be reduced without loss of data quality, meaning that problems with the expensive or difficult isotopic substitutions will be greatly reduced. Alternatively, higher fluxes may simply mean, in principle, one can measure more samples during the time allotted on the instrument.

While increased flux is clearly a benefit, the design of new facilities has also led to rethinking about the reflectometers themselves. Where new reflectometers have been built from the ground up, this has allowed a total rethink about the reflectometers design and capabilities. An example of this new breed of reflectometers is the OffSpec reflectometer (being commissioned at the time of writing this chapter) at the ISIS Second Target Station, UK (Plomp et al., 2007). This instrument is specifically designed for measurement of off-specular as well as specular reflection in a novel approach utilizing neutron spin echo methods that were first developed for inelastic scattering measurements. OffSpec will be able to measure in several reflectivity modes, as well as allowing spin echo small-angle neutron scattering (SESANS) as well as spin echo inelastic neutron scattering (SEINS) measurements. With these and other new developments at instruments around the globe, neutron reflectometers with more flux, and greater capabilities will undoubtedly enter new areas of interfacial studies that have not previously been possible. These new instrument are a truly exciting prospect for opening up new areas of research.

## REFERENCES

Anastasiadis, S.H., Russell, T.P., Satija, S.K., and Majkrzak, C.F. *J. Chem. Phys.* **1990**, *92*, 5677.

Beziel, W., Fragneto, G., Cousin, F., and Sferrazza, M. *Phys. Rev. E* **2008**, *78*, 022801.

Bowers, J., Zarbakhsh, A., Webster, J.R.P., Hutchings, L.R., and Richards, R.W. *Langmuir* **2001**, *17*, 140.

Broseta, D., Fredrickson, G.H., Helfand, E., and Leibler, L. *Macromolecules* **1990**, *23*, 132.

Brown, H.R. In: Paul, D.R. and Bucknall, C.B. (editors). *Polymer Blends: Formulation and Performance*, John Wiley & Sons, Inc., New York, 2000.

Bucknall, D.G., Butler, S.A., and Higgins, J.S. *J. Phys. Chem. Solids* **1999a**, *60*, 1273.

Bucknall, D.G., Butler, S.A., and Higgins, J.S. *Macromolecules* **1999b**, *32*, 5453.

Bucknall, D.G., Higgins, J.S., and Butler, S.A. *Chem. Eng. Sci.* **2001**, *56*, 5473.

Bucknall, D.G., Higgins, J.S., and Butler, S.A. *J. Polym. Sci. B Polym. Phys.* **2004**, *42*, 3267.

Butler, S.A., Higgins, J.S., Bucknall, D.G., and Sferrazza, M. *Macromol. Chem. Phys.* **2001**, *202*, 2275.

Carelli, C., Jones, R.A.L., Young, R.N., Cubitt, R., Dalgliesh, R., Schmid, F., Sferrazza, M. *Physical Review E* **2005a**, *72*, 031807.

Carelli, C., Jones, R.A.L., Young, R.N., Cubitt, R., Krastev, R., Gutberlet, T., Dalgliesh, R., Schmid, F., and Sferrazza, M. *Europhys. Lett.* **2005b**, *71*, 763.

Cooper, J.M., Cubitt, R., Dalgliesh, R.M., Gadegaard, N., Glidle, A., Hillman, A.R., Mortimer, R.J., Ryder, K.S., and Smith, E.L. *J. Am. Chem. Soc.* **2004**, *126*, 15362.

Coote, M.L., Gordon, D.H., Hutchings, L.R., Richards, R.W., and Dalgliesh, R.M. *Polymer* **2003**, *44*, 7689.

COSGROVE, T., PHIPPS, J.S., and RICHARDSON, R.M. *Colloids Surf.* **1992**, *62*, 199.
DADMUN, M. *Macromolecules* **1996**, *29*, 3868.
DADMUN, M.D. *Macromolecules* **2000**, *33*, 9122.
DAI, K.H., NORTON, L.J., and KRAMER, E.J. *Macromolecules* **1994**, *27*, 1949.
DALGLIESH, R.M., LAU, Y.G.J., RICHARDSON, R.M., and RILEY, D.J. *Rev. Sci. Instrum.* **2004**, *75*, 2955.
DEVAUX, C., COUSIN, F., BEYOU, E., and CHAPEL, J.P. *Macromolecules* **2005**, *38*, 4296.
DOI, M., and EDWARDS, S.F. *The Theory of Polymer Dynamics*, Oxford University Press, Oxford, 1986.
EASTWOOD, E., VISWANATHAN, S., O'BRIEN, C.P., KUMAR, D., DADMUN, and M.D. *Polymer* **2005**, *46*, 3957.
FERNANDEZ, M.L., HIGGINS, J.S., PENFOLD, J., WARD, R.C., SHACKLETON, C., and WALSH, D.J. *Polymer* **1988**, *29*, 1923.
FIELD, J.B., TOPRAKCIOGLU, C., BALL, R.C., STANLEY, H.B., DAI, L., BARFORD, W., PENFOLD, J., SMITH, G., and HAMILTON, W. *Macromolecules* **1992**, *25*, 434.
FRAGNETO-CUSANI, G. *J. Phys. Condens. Matter* **2001**, *13*, 4973.
GEOGHEGAN, M., BOUE, F., BACRI, G., MENELLE, A., and BUCKNALL, D.G. *Eur. Phys. J. B* **1998**, *3*, 83.
GLIDLE, A., BAILEY, L., HADYOON, C.S., HILLMAN, A.R., JACKSON, A., RYDER, K.S., SAVILLE, P.M., SWANN, M. J., WEBSTER, J.R.P., WILSON, and R.W., COOPER, J.M. *Anal. Chem.* **2001**, *73*, 5596.
GLIDLE, A., COOPER, J., HILLMAN, A.R., BAILEY, L., JACKSON, A., WEBSTER, J.R.P. *Langmuir* **2003**, *19*, 7746.
GREEN, P.F. and DOYLE, B.L. *Phys. Rev. Lett.* **1986**, *57*, 2407.
GREEN, P.F. and DOYLE, B.L. *Macromolecules* **1987**, *20*, 2471.
GRULL, H., SUNG, L., KARIM, A., DOUGLAS, J.F., SATIJA, S.K., HAYASHI, M., JINNAI, H., HASHIMOTO, T., and HAN, C.C. *Europhys. Lett.* **2004**, *65*, 671.
GUCKENBIEHL, B., STAMM, M., and SPRINGER, T. Physica B **1994**, *198*, 127.
GUPTA, R.R., LAVERY, K.A., FRANCIS, T.J., WEBSTER, J.R.P., SMITH, G.S., RUSSELL, T.P., and WATKINS, J.J. *Macromolecules* **2003**, *36*, 346.
HAYASHI, M., HASHIMOTO, T., HASEGAWA, H., TAKENAKA, M., GRULL, H., ESKER, A.R., WEBER, M., SATIJA, S.K., HAN, C.C., and NAGAO, M. *Macromolecules* **2000**, *33*, 8375.
HELFAND E. and TAGAMI, Y. *J. Polym. Sci. B Polym. Phys.* **1996**, *34*, 1947.
HERMES, H.E., HIGGINS, J.S., and BUCKNALL, D.G. *Polymer* **1997**, *38*, 985.
HERMES, H.E., BUCKNALL, D.G., BUTLER, S.A., and HIGGINS, J.S. *Macromol. Symp.* **1998a**, *126*, 331.
HERMES, H.E., BUCKNALL, D.G., HIGGINS, J.S., and SCHERRENBERG, R.L. *Polymer* **1998b**, *39*, 3099.
HIGGINS, A.M., MARTIN, S.J., GEOGHEGAN, M., HERIOT, S.Y., THOMPSON, R.L., CUBITT, R., DALGLIESH, R.M., GRIZZI, I., and JONES, R.A.L. *Macromolecules* **2006**, *39*, 6699.
HILLMAN, A.R., SAVILLE, P.M., GLIDLE, A., RICHARDSON, R.M., ROSER, S.J., SWANN, M.J., and WEBSTER, J.R.P. *J. Am. Chem. Soc.* **1998**, *120*, 12882.
HILLMAN, A.R., BAILEY, L., GLIDLE, A., COOPER, J.M., GADEGAARD, N., and WEBSTER, J.R.P. *J. Electroanal. Chem.* **2002**, *532*, 269.
HOPKINSON, I., KIFF, F.T., RICHARDS, R.W., BUCKNALL, D.G., and CLOUGH, A.S. *Polymer* **1997**, *38*, 87.
HUANG, Z., ALONZO, J., LIU, M., JI, H., YIN, F., SMITH, G.D., MAYS, J.W., KILBEY, S.M., and DADMUN, M.D. *Macromolecules* **2008**, *41*, 1745.
JONES, R.A.L. and RICHARDS, R.W. *Polymers at Surfaces and Interfaces*, Cambridge University Press, Cambridge, 1998.
KANAYA, T., MIYAZAKI, T., WATANABE, H., NISHIDA, K., YAMANA, H., TASAKI, S., and BUCKNALL, D.G. *Polymer* **2003**, *44*, 3769.
KARIM, A., MANSOUR, A., FELCHER, G.P., and RUSSELL, T.P. *Phys. Rev. B Condens. Matter* **1990**, *42*, 6846.
KARIM, A., FELCHER, G.P., and RUSSELL, T.P. *Macromolecules* **1994**, *27*, 6973.
KAWAGUCHI, D., MASUOKA, K., TAKANO, A., TANAKA, K., NAGAMURA, T., TORIKAI, N., DALGLIESH, R.M., LANGRIDGE, S., and MATSUSHITA, Y. *Macromolecules* **2006**, *39*, 5180.
KAWAGUCHI, D., OHYA, M., TORIKAI, N., TAKANO, A., and MATSUSHITA, Y. *Polym. J.* **2007a**, *39*, 1274.
KAWAGUCHI, D., TANAKA, K., TORIKAI, N., TAKAHARA, A., and KAJIYAMA, T. *Langmuir* **2007b**, *23*, 7269.
KEDDIE, J.L., JONES, R.A.L., and CORY, R.A. *Europhys. Lett.* **1994**, *27*, 59.
KHARLAMPIEVA, E., ANKNER, J.F., RUBINSTEIN, M., and SUKHISHVILI, S.A. *Phys. Rev. Lett.* **2008**, *100*, 128303.
KIRSTE, R.G. *Makromol. Chem.* **1967**, *101*, 91.
KOGA, T., SEO, Y.S., ZHANG, Y.M., SHIN, K., KUSANO, K., NISHIKAWA, K., RAFAILOVICH, M.H., SOKOLOV, J.C., CHU, B., PEIFFER, D., OCCHIOGROSSO, R., and SATIJA, S.K. *Phys. Rev. Lett.* **2002**, *89*, 125506.

KOGA, T., SEO, Y.S., SHIN, K., ZHANG, Y., RAFAILOVICH, M.H., SOKOLOV, J.C., CHU, B., and SATIJA, S.K. *Macromolecules* **2003**, *36*, 5236.

KOGA, T., JI, Y., SEO, Y.S., GORDON, C., QU, F., RAFAILOVICH, M.H., SOKOLOV, J.C., and SATIJA, S.K. *J. Polym. Sci. B Polym. Phys.* **2004**, *42*, 3282.

KUGLER, R. SCHMITT, J., and KNOLL, W. *Macromol. Chem. Phys.* **2002**, *203*, 413.

KUNZ, K. and STAMM, M. *Macromolecules* **1996**, *29*, 2548.

LAUTER-PASYUK, V., LAUTER, H.J., GORDEEV, G.P., MULLER-BUSCHBAUM, P., TOPERVERG, B.P., JERNENKOV, M., and PETRY, W. *Langmuir* **2003**, *19*, 7783.

LAVERY, K.A., PRABHU, V.M., LIN, E.K., WU, W.L., SATIJA, S.K., CHOI, K.W., and WORMINGTON, M. *Appl. Phys. Lett.* **2008**, *92*, 064106.

LEE, L.T., JEAN, B., and MENELLE, A. *Langmuir* **1999**, *15*, 3267.

LEE, L.T., LANGEVIN, D., MANN, E.K., and FARNOUX, B. *Physica B* **1994**, *198*, 83.

LEIBLER, L. *Makromol. Chem. Macromol. Symp.* **1988**, *16*, 1.

LOSCHE, M., SCHMITT, J., DECHER, G., BOUWMAN, W.G., and KJAER, K. *Macromolecules* **1998**, *31*, 8893.

LU, J.R. and THOMAS, R.K. *J. Chem. Soc. Faraday Trans.* **1998**, *94*, 995.

MAYES, A.M., RUSSELL, T.P., DELINE, V.R., SATIJA, S.K., and MAJKRZAK, C.F. *Macromolecules* **1994**, *27*, 7447.

MILLS, P.J., MAYER, J.W., KRAMER, E.J., HADZIIOANNOU, G., LUTZ, P., STRAZIELLE, C., REMPP, P., and KOVACS, A.J. *Macromolecules* **1987**, *20*, 513.

MOURI, E., KAEWSAIHA, P., MATSUMOTO, K., MATSUOKA, H., TORIKAI, N. *Langmuir* **2004**, *20*, 10604.

MU, Q.S., LU, J.R., MA, Y.H., De BANEZ, M.V.P., ROBINSON, K.L., ARMES, S.P., LEWIS, A.L., and THOMAS, R. K. *Langmuir* **2006**, *22*, 6153.

MULLER-BUSCHBAUM, P., GUTMANN, J.S., STAMM, M., CUBITT, R., CUNIS, S., Von KROSIGK, G., GEHRKE, R., and PETRY, W. *Physica B* **2000**, *283*, 53.

MULLER-BUSCHBAUM, P., CUBITT, R., and PETRY, W. *Langmuir* **2003**, *19*, 7778.

MULLER-BUSCHBAUM, P., HERMSDORF, N., ROTH, S.V., WIEDERSICH, J., CUNIS, S., and GEHRKE, R. *Spectrochim. Acta B* **2004**, *59*, 1789.

MULLER-BUSCHBAUM, P., BAUER, E., MAURER, E., and CUBITT, R. *Physica B* **2006a**, *385*, 703.

MULLER-BUSCHBAUM, P., MAURER, E., BAUER, E., and CUBITT, R. *Langmuir* **2006b**, *22*, 9295.

MUKHERJEE, M., SINGH, A., DAILLANT, J., MENELLE, A., and COUSIN, F. *Macromolecules* **2007**, *40*, 1073.

MULLER-BUSCHBAUM, P., SCHULZ, L., METWALLI, E., MOULIN, J.F., and CUBITT, R. *Langmuir* **2008**, *24*, 7639.

NOOLANDI, J. and HONG, K.M. *Macromolecules* **1982**, *15*, 482.

NORO, A., OKUDA, M., ODAMAKI, F., KAWAGUCHI, D., TORIKAI, N., TAKANO, A., and MATSUSHITA, Y. *Macromolecules* **2006**, *39*, 7654.

OZISIK, R., Von MEERWALL, E.D., MATTICE, W.L. *Polymer* **2002**, *43*, 629.

PEREZ-SALAS, U., BRIBER, R.M., HAMILTON, W.A., RAFAILOVICH, M.H., SOKOLOV, J., and NASSER, L. *Macromolecules* **2002**, *35*, 6638.

PINTO, J.C., WHITING, G.L., KHODABAKHSH, S., TORRE, L., RODRIGUEZ, A.B., DALGLIESH, R.M., HIGGINS, A.M., ANDREASEN, J.W., NIELSEN, M.M., GEOGHEGAN, M., HUCK, W.T.S., SIRRINGHAUS, H. *Adv. Funct. Mater.* **2008**, *18*, 36.

PLOMP, J., De HAAN, V.O., DALGLIESH, R.M., LANGRIDGE, S., and Van WELL, A.A. *Thin Solid Films* **2007**, *515*, 5732.

RUSSELL, T.P. *Curr. Opin. Colloid Interface Sci.* **1996**, *1*, 107.

RUSSELL, T.P., HJELM, R.P., and SEEGER, P.A. *Macromolecules* **1990**, *23*, 890.

RUSSELL, T.P., ANASTASIADIS, S.H., MENELLE, A., FELCHER, G.P., and SATIJA, S.K. *Macromolecules* **1991**, *24*, 1575.

RUSSELL, T.P., DELINE, V.R., DOZIER, W.D., FELCHER, G.P., AGRAWAL, G., WOOL, R.P., and MAYS, J.W. *Nature* **1993**, *365*, 235.

RYU, D.Y., WANG, J.Y., LAVERY, K.A., DROCKENMULLER, E., SATIJA, S.K., HAWKER, C.J., RUSSELL, T.P. *Macromolecules* **2007**, *40*, 4296.

SANJUAN, S. and TRAN, Y. *Macromolecules* **2008**, *41*, 8721.

SANJUAN, S., PERRIN, P., PANTOUSTIER, N., and TRAN, Y. *Langmuir* **2007**, *23*, 5769.

SCHACH, R., TRAN, Y., MENELLE, A., and CRETON, C. *Macromolecules* **2007**, *40*, 6325.

SCHMITT, J., GRUNEWALD, T., DECHER, G., PERSHAN, P.S., KJAER, K., and LOSCHE, M. *Macromolecules* **1993**, *26*, 7058.

SCHNELL, R., STAMM, M., and CRETON, C. *Macromolecules* **1998**, *31*, 2284.

SCHNELL, R., STAMM, M., and CRETON, C. *Macromolecules* **1999**, *32*, 3420.

SEMENOV, A.N. *Macromolecules* **1992**, *25*, 4967.

SFERRAZZA, M., XIAO, C., JONES, R.A.L., BUCKNALL, D.G., WEBSTER, J.R.P., and PENFOLD, J. *Phys. Rev. Lett.* **1997**, *78*, 3693.

SFERRAZZA, M., HEPPENSTALLBUTLER, M., CUBITT, R., BUCKNALL, D.G., WEBSTER, J.R.P., and JONES, R.A.L. *Phys. Rev. Lett.* **1998**, *81*, 5173.

SHULL K.R. and KRAMER, E.J. *Macromolecules* **1990**, *23*, 4769.

SHULL, K.R., MAYES, A.M., and RUSSELL, T.P. *Macromolecules* **1993**, *26*, 3929.

SINHA, S.K., SIROTA, E.B., GAROFF, S., and STANLEY, H.B. *Phys. Rev. B* **1988**, *38*, 2297.

SIVANIAH, E., SFERRAZZA, M., JONES, R.A.L., and BUCKNALL, D.G. *Phys. Rev. E* **1999**, *59*, 885.

SIVANIAH, E., JONES, R.A.L., and SFERRAZZA, M. *Phys. Rev. E* **2003**, *67*, 052801.

SMITH, G.S., SKIDMORE, C.B., HOWE, P.M., and MAJEWSKI, J. *J. Polym. Sci. B Polym. Phys.* **2004**, *42*, 3258.

STEITZ, R., LEINER, V., SIEBRECHT, R., and Von KLITZING, R. *Colloids Surf. A Physicochem. Eng. Asp.* **2000**, *163*, 63.

TANAKA, K., KAWAGUCHI, D., YOKOE, Y., KAJIYAMA, T., TAKAHARA, A., and TASAKI, S. *Polymer* **2003**, *44*, 4171.

TANCHAK, O.M., YAGER, K.G., FRITZSCHE, H., HARROUN, T., KATSARAS, J., and BARRETT, C.J. *Langmuir* **2006**, *22*, 5137.

TANG, H. and FREED, K.F. *J. Chem. Phys.* **1991**, *94*, 6307.

TEAD, S.F., KRAMER, E.J., HADZIIOANNOU, G., ANTONIETTI, M., SILLESCU, H., LUTZ, P., and STRAZIELLE, C. *Macromolecules* **1992**, *25*, 3942.

TORIKAI, N., YAMADA, N.L., NORO, A., HARADA, M., KAWAGUCHI, D., TAKANO, A., and MATSUSHITAZ, Y. *Polym. J.* **2007**, *39*, 1238.

Van Der GRINTEN, D., WOLFF, M., ZABEL, H., and MAGERL, A. *Meas. Sci. Technol.* **2008**, *19*, 034016.

VIEIRA, J.B., LI, Z.X., THOMAS, R.K., and PENFOLD, J. *J. Phys. Chem. B* **2002**, *106*, 10641.

WELP, K.A., WOOL, R.P., SATIJA, S.K., PISPAS, S., and MAYS, J. *Macromolecules* **1998**, *31*, 4915.

WELP, K.A., WOOL, R.P., AGRAWAL, G., SATIJA, S.K., PISPAS, S., and MAYS, J. *Macromolecules* **1999**, *32*, 5127.

WHITLOW S.J. and WOOL, R.P. *Macromolecules* **1991**, *24*, 5926.

WOLFF, M., MAGERL, A., and ZABEL, H. *Eur. Phys. J. E* **2005a**, *16*, 141.

WOLFF, M., MAGERL, A., and ZABEL, H. *Physica B* **2005b**, *357*, 84.

WORMINGTON, M., PAPE, I., HASE, T.P.A., TANNER, B.K., BOWEN, D.K. *Philos. Mag. Lett.* **1996**, *74*, 211.

WUNNICKE, O., MULLER-BUSCHBAUM, P., WOLKENHAUER, M., LORENZ-HAAS, C., CUBITT, R., LEINER, V., STAMM, M. *Langmuir* **2003**, *19*, 8511.

YANG, Y., CHENG, M.M.C., HU, X., LIU, D.H., GOYETTE, R.J., LEE, L.J., and FERRARI, M. *Macromolecules* **2007**, *40*, 1108.

ZARBAKHSH, A., QUEROL, A., BOWERS, J., WEBSTER, J.R.P. *Faraday Disc.* **2005**, *129*, 155.

ZHANG, X.H., BERRY, B.C., YAGER, K.G., KIM, S., JONES, R.L., SATIJA, S., PICKEL, D.L., DOUGLAS, J.F., and KARIM, A. *ACS Nano* **2008**, *2*, 2331.

# IV Applications

## IV.7 Neutron Diffraction from Polymers and Other Soft Matter

*Geoffrey R. Mitchell*

### IV.7.1 INTRODUCTION

This chapter focuses on the use of neutron diffraction techniques in the so-called wide-angle scattering regime to explore the molecular organization in polymers and other soft matter. For reasons that will become apparent when detailing studies performed at instruments based at pulsed neutron sources, we prefer the term broad $Q$ neutron diffraction. The chapter seeks to highlight the areas of soft matter science where broad $Q$ neutron diffraction brings a strong, if not unique, contribution to unraveling the complexities of the structure of soft matter. As with many areas of science, often such studies are not stand-alone but form part of a multiprobe research program involving both experimental activity and computational modeling. Increasingly structural studies extend over many length and timescales and the simple division by experimental technique becomes increasing arbitrary. We will highlight where there are synergies to be gained from multiprobe and multitechnique studies and we will conclude with a look to the future when the available data requires a multiscale model to fully understand the structure.

### IV.7.2 BASICS

Neutron diffraction experiments yield the differential scattering cross section, essentially the ratio of neutrons scattered into a solid angle element $\Omega$ to $\Omega + d\Omega$ relative to the number of incident neutrons (Lovesey, 1984; Squires, 1996; Roe, 1999). The differential cross section contains both coherent and incoherent

*Neutrons in Soft Matter* Edited by Toyoko Imae, Toshiji Kanaya, Michihiro Furusaka, and Naoya Torikai

**Table IV.7.1** Scattering Lengths and Cross Sections for Isotopes Common to Soft Matter (Higgins and Benoit, 1994)

| Isotope | $b$ (fm) | $\sigma_{coh}$ (barn/atom) | $\sigma_{inc}$ (barn/atom) |
|---|---|---|---|
| $^{1}H$ | −3.74 | 1.76 | 80.27 |
| $^{2}D$ | +6.67 | 5.59 | 2.05 |
| $^{12}C$ | +6.65 | 5.56 | 0 |
| $^{14}N$ | +9.40 | 11.01 | 0.5 |
| $^{16}O$ | +5.80 | 4.23 | 0 |
| $^{19}F$ | +5.70 | 4.02 | 0 |
| $^{32}S$ | +2.80 | 1.02 | 0 |
| $^{35}Cl$ | +9.60 | 11.53 | 5.3 |

scattering and most experiments will record the total scattering cross section (see Section IV.7.3). The coherent scattering component can be written as:

$$\left(\frac{\partial\sigma}{\partial\Omega}\right)_{coh} = \frac{1}{N}\left\langle N\sum_{i,j=1}^{N} b_i b_j\, e^{i\mathbf{Q}\mathbf{r}_{ij}}\right\rangle, \tag{IV.7.1}$$

where $b_j$ is the coherent scattering length for atom $j$, $N$ is the number of atoms and $\mathbf{r}_{ij}$ is the instantaneous vector between the positions of atoms $i$ and $j$. The angular brackets indicate the thermal average. $\mathbf{Q}$ is the momentum transfer or scattering vector with a modulus given by

$$|\mathbf{Q}| = 4\pi \sin(\theta)/\lambda \tag{IV.7.2}$$

and $2\theta$ is the scattering angle and $\lambda$ is the wavelength of the incident neutrons. We will refer to the modulus of $\mathbf{Q}$ in the remainder of this chapter by the symbol $Q$ as is the common practice.

Equation (IV.7.1) shows the relationship between the structure and the scattering and that it depends on both the positions of the atoms and the coherent scattering lengths. Values for the scattering lengths and cross sections for some elements typically found in soft matter are listed in Table IV.7.1.

As the values in Table IV.7.1 show, some isotopes exhibit a significant incoherent cross section, which contains no static structural information and is $Q$ independent. The incoherent component can be written as

$$\left(\frac{\partial\sigma}{\partial\Omega}\right)_{inc} = \frac{1}{4\pi N}\sum_{i=1}^{N}\sigma_{inc}^{j}. \tag{IV.7.3}$$

It is clear that for samples containing hydrogen there will be a significant incoherent component and minimizing any problems that ensue are discussed in Section IV.7.3.

Extracting structural information from neutron diffraction data can proceed using the data presented in either reciprocal space or in real space. For the latter, the scattering data may be Fourier transformed to yield a radial distribution function. If we consider a system with $N$ identical nuclei we can write

$$\left(\frac{\partial\sigma}{\partial\Omega}\right)_{\text{coh}} = Nb^2 S(Q) = 1 + \rho_0 \int_0^\infty [g(r) - 1] e^{i\mathbf{Qr}}\, d\mathbf{r}, \qquad \text{(IV.7.4)}$$

where $\rho_0$ is the average density of the system and $g(r)$ is the pair distribution function expressing the probability that there exists an atom at a distance $r$ from the origin atom. $S(Q)$ is often referred to as the structure factor. For a system with different chemical species it is straightforward to split $g(r)$ and $S(Q)$ into partial terms, for example, for deuterated polystyrene containing carbon and deuterium, the distributions functions would be $g_{CC}(r)$, $g_{CD}(r)$, and $g_{DD}(r)$. We will also find it useful to separate the distribution functions and hence the structure factors in terms of specific types of atoms, so for example, in polystyrene there will be carbon and deuterium atoms in the skeletal chain and in the phenyl side group (Mitchell et al., 1994).

Figure IV.7.1 shows the experiment structure factors for selectively deuterated samples of atactic polystyrene. These curves are typical for disordered materials including polymers in showing a series of broad diffuse peaks. In spite of the broad features the data show the considerable variation in the structure factor that takes place with different selective deuteration. Typically in a disordered molecular material such as a molten or glassy polymer, the scattering at $Q$ values ~1–1.5 Å$^{-1}$ is associated with correlations between molecular segments on neighboring chains or molecules (Lovell et al., 1979; Mitchell, 1989). In contrast, the scattering at higher $Q$

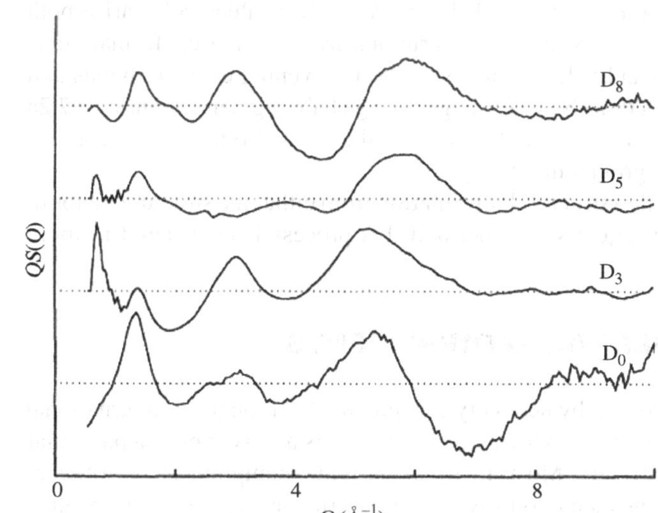


**Figure IV.7.1** The structure factors for selectively deuterated glassy atactic polystyrene obtained using neutron diffraction techniques. The label $D_x$ indicated the number of hydrogen positions in a repeat unit, which was deuterated as described in the text. This graph is reproduced from a reference (Mitchell et al., 1994).

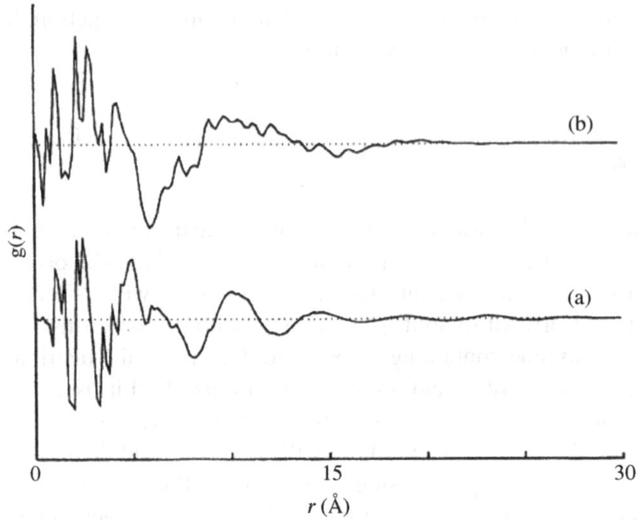


**Figure IV.7.2** Real space correlation functions of glassy polystyrene: (a) total correlation function for $D_8$ and (b) partial correlation function $g_{bb}$ describing the correlations involving the backbone atoms only. This graph is reproduced from a reference (Mitchell et al., 1994).

arises from correlations within those molecular segments or chains. Figure IV.7.2a shows the Fourier transform of the data for the perdeuterated polystyrene ($D_8$) and now the order of features is reversed. The peaks in the $g(r)$ function for $r < 4$ Å relate to distances determined by the covalent bonding, for example, the C–D bond with $r \sim 1.1$ Å and the C–C bond with $r \sim 1.3$–$1.5$ Å. Peaks at $r$ values >4 Å arise both from the intrachain correlations that are determined by the chain conformation as well as the bond lengths and valence angles of the intervening covalent bonds and from the interchain correlations between neighboring chain segments. Figure IV.7.2a shows a clear series of peaks at ~5, 10, 15, 20 and 25 Å that relate to correlations between the phenyl side groups in polystyrene.

To extract more detailed structural information from either the structure factor or the distribution function requires a model and this process is considered in more detail in Section IV.7.5.

## IV.7.3 EXPERIMENTAL REQUIREMENTS

Neutron diffraction studies are by necessity carried out at National and International Facilities and hence many of the experimental parameters are fixed once a particular instrument has been selected. Most facilities provide comprehensive software packages for processing the data into fully corrected structure factors on an absolute scale, for example (Hannon et al., 1990; McLain et al., 2009) and these procedures will not be considered further in this chapter. As discussed in Section IV.7.2, most broad $Q$ diffraction experiments are total scattering experiments, in that all the scattered neutrons that reach the detectors are considered. Soft matter, by its very nature, usually contains a significant fraction of hydrogen atoms and the large incoherent scattering length for hydrogen (Table IV.7.1) may cause major problems as it will produce a substantial "background" signal. This may lead to a poor signal-to-noise ratio for the extracted coherent contribution and generate problems in the data analysis stage. There are three approaches to overcoming this central problem for broad $Q$ neutron diffraction that are considered in Sections IV.7.3.1–IV.7.3.3.

### IV.7.3.1 Deuterated Samples

The most direct solution to the "hydrogen" problem is to "replace" the hydrogen atoms with deuterium. This process may range from very straightforward to the nearly impossible. In many cases, for example, biological systems or technologically important polymers, the requirement will be to exchange the hydrogens for deuterium atoms in existing materials. However, this can only be achieved in specific conditions with hydrogens attached to particular groups, such as acidic hydrogen in the carboxylic acid group, or the basic amino group within a protein. Similarly aromatic hydrogens can be exchanged under acidic conditions in those aromatic groups that are highly activated toward electrophilic substitution, for example, the hydrogens in pyrrole can be exchanged with a solution of DCl in $D_2O$. This option is not possible with saturated hydrocarbons such as polyethylene or polypropylene. For some polymers such as polystyrene, monomers with differing deuterium labeling are readily available, for others, such poly(ε-caprolactone), only a partially deuterated monomer is available from commercial suppliers and for most polymers preparation of the deuterated form will require extensive chemical work for both the monomer and the polymer. The situation is particular challenging where there is a requirement to prepare materials with very specific molecular weight distributions or stereoregularity to mimic industrially produced polymers and other soft matter.

### IV.7.3.2 Scattering at Small Angles

Diffractometers with a monochromatic incident neutron beam achieve high $Q$ values through the use of a short wavelength and scanning to high values of the scattering angle $2\theta$. At a pulsed neutron source, a relatively high proportion of short wavelength epithermal neutrons, which are not present in neutron beams derived from a reactor, is available. The short wavelength neutrons enable a wide range of $Q$ values to be explored at relatively small angles. The advantage of this approach is that the correction for the inelasticity effects is more straightforward, essentially it reduces with the scattering angle (Rosi-Schwartz et al., 1992). SANDALS at the ISIS pulsed neutron source is an example of such an instrument and it is configured to scan all scattering angles with a continuous coverage of detectors. The disadvantage of using detectors at small angles is that the resolution $\Delta Q/Q$ is reduced; on SANDALS the resolution is $\sim$2.5% that is adequate for studies of most disordered materials.

Figure IV.7.3 shows the $Q$ weight structure factors recorded for perdeuterated and perhydrogenated atactic polystyrene after normalization for the incident beam

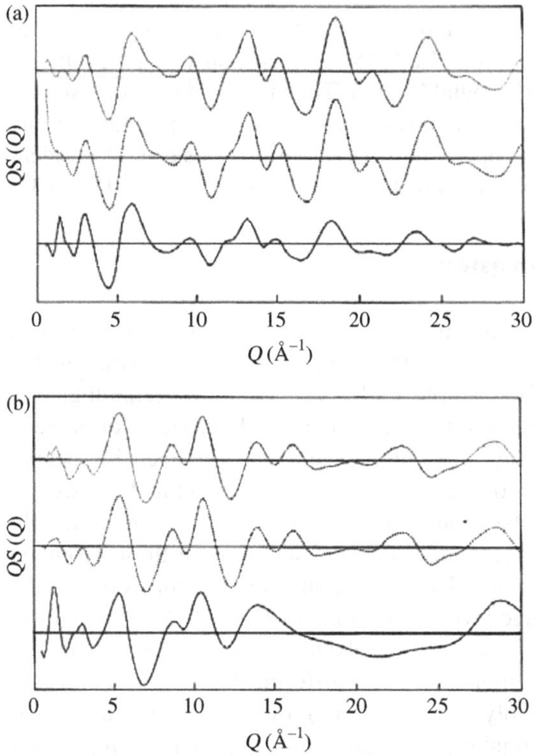


**Figure IV.7.3** The experimental and calculated $Q$ weighted structure factors for (a) perdeuterated atactic polystyrene and (b) perhydrogenated polystyrene obtained using the SANDALS diffractometer at ISIS. The full line is the experimental functions, the dotted curves are calculated structure factors based on a full atomistic model and the dashed curves are calculated structure factors for a full atomistic model using intrachain correlations only. This figure is reproduced from a reference (Rosi-Schwartz et al., 1992).

flux, converted to an absolute scale using a standard vanadium sample and corrected for absorption, multiple scattering and inelastic effects (Rosi-Schwartz et al., 1992). The curves shown in Figure IV.7.3 demonstrate very clearly that high quality data can be readily obtained for hydrogenated polymers (and hence other soft matter) using instruments such as SANDALS. This opens up the possibility of using neutron diffraction from materials with their natural composition with clear advantages for soft matter of technological or commercial importance as well as biological systems. The data also underlines the remarkable variation in scattering features with deuteration that offers considerable potential in structural studies that will be considered later in this chapter.

### IV.7.3.3 Polarization Analysis

The third approach to reducing the "hydrogen" problem is to utilize polarization analysis to attempt to eliminate the incoherent component experimentally. As with the approach of scattering at low angles, the use of polarization analysis is limited to very specific instruments. For example, Genix et al. (2006) have used the diffuse scattering spectrometer DNS at the FZ Julich to obtain structure factors for selectively deuterated syndiotactic polymethylmethacrylate. As highlighted above, a standard diffractometer provides the total differential cross section that includes both the coherent and the incoherent contributions. The DNS instrument exploits a spin polarized neutron beam and polarization analysis of the scattering neutrons. If the incoherent scattering arises from spin disorder only, the neutron spin is flipped with a probability of 2/3 while the coherent scattering process leaves the spin unchanged. The number of neutrons, $I_{\mathrm{NSF}}$ that have not been spin flipped is given by (Gabrys, 2000)

$$I_{\mathrm{NSF}} = I_0 \left[ \left( \frac{\partial \sigma}{\partial \Omega} \right)_{\mathrm{coh}} + \left( \frac{\partial \sigma}{\partial \Omega} \right)_{\mathrm{inc}} \right]. \tag{IV.7.5}$$

And the number of spin flipped neutrons $I_{\mathrm{SF}}$ is given by

$$I_{\mathrm{SF}} = I_0 \left[ \frac{2}{3} \left( \frac{\partial \sigma}{\partial \Omega} \right)_{\mathrm{inc}} \right]. \tag{IV.7.6}$$

The procedure involves evaluating $I_0$ from the $I_{\mathrm{SF}}$ measurement and a calculation of incoherent scattering cross section $\sigma_{\mathrm{inc}}$. Thereafter, the coherent partial structure factor can be simply obtained from the measurement of $I_{\mathrm{NSF}}$. The technique has been utilized by a number of successful studies of polymers, for example (Gabrys, 2000; Gabrys et al., 1993; Zajac et al., 2002; Lamers et al., 1992; Narros et al., 2005; Perez-Aparicio et al., 2009). One possible limitation is that the instruments equipped with polarization analysis offer a limited $Q$ range, in the case of DNS the value of $Q_{\mathrm{max}}$ is $\sim$3.3 Å$^{-1}$. Early experiments using this approach often exhibited rather low signal-to-noise ratios, but the more recent work appears to yield high-quality structure factors.

## IV.7.4 EXPLOITING SCATTERING LENGTHS

Broad $Q$ neutron diffraction offers three clear benefits over X-ray scattering. The first is that due to the $Q$ independence of the scattering lengths it is straightforward to obtain high quality data up to $Q$ values of 50 or even 100 Å$^{-1}$. The second benefit arises from the weak interaction of neutrons with materials that allows complex environmental stages for heating, cooling, pressure, humidity, flow, and so on, to be used in the instrument without any great problems. For polymers and other soft matter, perhaps the greatest benefit comes from the more or less equally weighted scattering lengths, other than particular isotopes such as hydrogen and deuterium (Table IV.7.1). In this section we consider the possibilities that arise from the characteristics of the scattering lengths.

### IV.7.4.1 Deuterated and Hydrogenated Polymers

We have already seen in Figures IV.7.1 and IV.7.3 how the substitution of deuterium in place of hydrogen leads to considerable variation in the position and strengths of features in the experimental structure factors. Clearly we need to be confident that the two different samples exhibit a similar molecular organization. Factors such as molecular weight distribution, stereoregularity and purity as well as prior thermal and mechanical treatment will be important here. For a glassy polymer this requires some serious attention but for the liquid state there is less opportunity for labeled samples to exhibit significantly different structures. The use of models can be very helpful here. Figure IV.7.3 shows the structure factors calculated for polystyrene from the same molecular model in which the scattering has been weighted for either deuterium or hydrogen labeling. Even though this model is not quantitatively correct, the changes observed in moving from the model structure factor for the hydrogenated polystyrene to that for the deuterated polystyrene are indeed replicated in the experimental factors. If we are confident that these samples have the same structure then both structure factors should match those calculated from a single model with the appropriate weightings for the scattering; this is a particular stringent test of any model.

The availability of neutron diffraction data from selectively labeled molecules enables partial structure factors and hence partial radial distribution functions to be derived and this is considered in Section IV.7.4.3.

### IV.7.4.2 Revealing Structure

The scattering factors relevant to X-ray scattering are proportional to the atomic number and thus in some materials, one elemental type dominates. PVC is an important industrial polymer that in its usual form is largely amorphous with little or no crystallinity. Despite its technological importance, experimental structural studies have failed to establish the details of the local molecular organization of this polymer; information vital for an understanding of physical properties.

The X-ray structure factors obtained from samples of chlorinated polyethylene with various chlorine content are shown in Figure IV.7.4a (Chiou, 2004). The features in the scattering pattern from samples with high chlorine content are difficult to understand. The scattering patterns appear to suggest a highly disordered structure. Some insight can be gained by considering the scattering from a chlorinated polyethylene over an extended $Q$ range as shown in Figure IV.7.4b. The full line corresponds to the scattering expected for a dilute gas of chlorine atoms. The relatively small deviations away from this line by the scaled experimental data indicate the low level of spatial correlations between chlorine atoms. The "hole" is the experimental scattering at $Q < 2\,\text{Å}^{-1}$ arises as a consequence of a minimum separation of chlorine atoms. Figure IV.7.5 compares the X-ray and neutron scattering for a sample of perdeuterated poly(vinyl chloride) (Mitchell, G.R., Chiou, Y.S., Starnes, W.H., and Zaikov, V., submitted). The neutron diffraction pattern is radically different to the X-ray scattering curve. For the neutron scattering, the contributions from the C, D, and Cl scattering centers in the d-PVC are more equally weighted, while in the X-ray scattering case the pattern is entirely dominated by the Cl atoms.

The rather continuous and broad peaks in the X-ray scattering data are replaced with distinctive peaks rather similar to that displayed by molten polyethylene

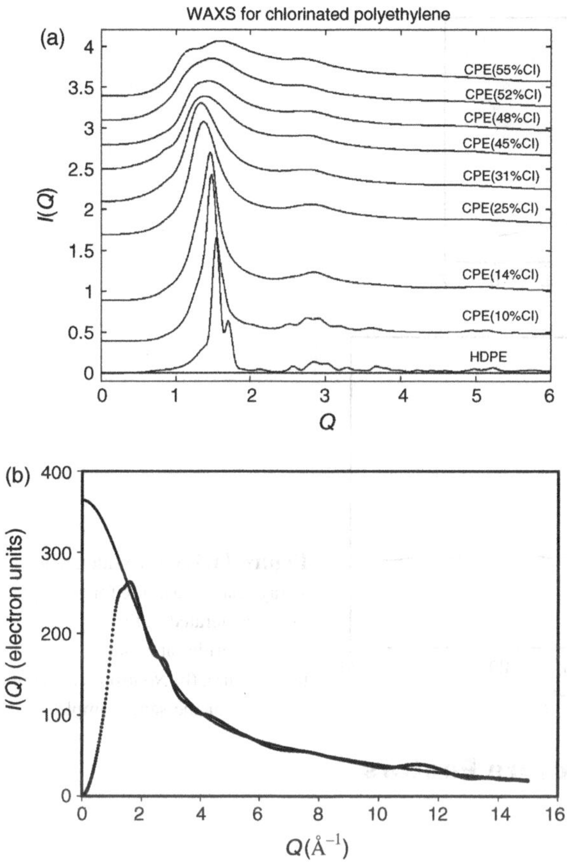


**Figure IV.7.4** (a) Wide-angle X-ray scattering data for a series of polymers with differing chlorine contents as indicated on the curves. (b) The fully corrected and scaled experimental intensity function (broken line) for a sample of chlorinated polyethylene (58% w/w chlorine) compared to the self-scattering.

(Rosi-Schwartz and Mitchell, 1994). In common with other highly disordered polymers, there is a relatively intense peak at $Q \sim 1.15\,\text{Å}^{-1}$ with a width at half the maximum height of $\Delta Q \sim 0.23\,\text{Å}^{-1}$. This contrast with the data for molten perdeuterated polyethylene in which there is a relatively intense peak at $Q \sim 1.26\,\text{Å}^{-1}$ with a width $\Delta Q \sim 0.3\,\text{Å}^{-1}$ (Rosi-Schwartz and Mitchell, 1994). This peak may be attributed to the correlations between chain segments. The shift in the peak position is related to the increased effective diameter of the PVC chain. The peak width ($\Delta Q$) is inversely related to the extent of spatial correlations between chain segments, for d-PVC, this length, $L = 2\pi/\Delta Q \sim 27$ Å, while for d-PE melt it is ~20 Å. In other words the d-PVC is rather more ordered than molten polyethylene, in contrast to the impression given by the X-ray scattering data! The availability of the neutron scattering for the perdeuterated PVC has revealed considerable detail in the structure.

Rosi-Schwartz and Mitchell (1992, 1994) used neutron diffraction to develop a detailed molecular model for polytetrafluoroethylene and again the study was possible because of the more evenly weighted contributions from the fluorine and carbon atoms in contrast to the fluorine dominated X-ray scattering data.

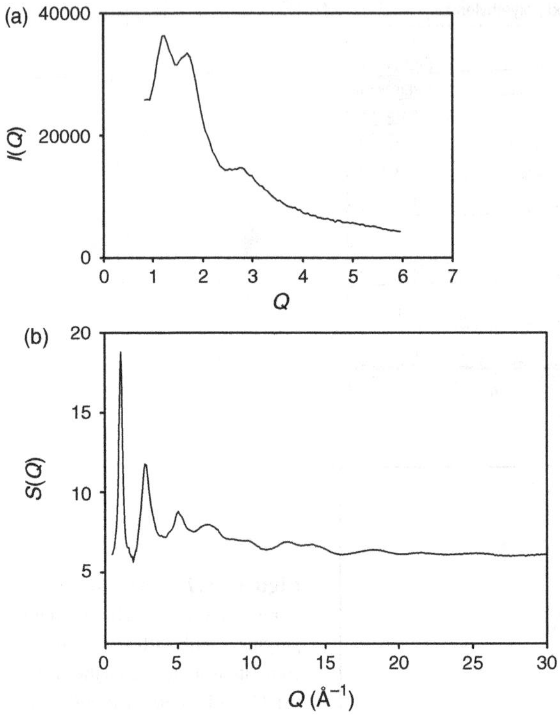


**Figure IV.7.5** (a) Wide-angle X-ray scattering curve for a sample of perdeuterated poly (vinylchloride) at room temperature. (b) Neutron scattering function for the same sample.

## IV.7.4.3 Partial Structure Factors

### *IV.7.4.3.1 Basics*

Enderby et al. (1966) pioneered the use of neutron diffraction combined with isotope substitution to extract detailed structural information in multicomponent disordered materials. The essence of the technique can be written:

$$D(Q) = \sum_{\alpha,\beta\geq\alpha} (2-\delta_{\alpha\beta})c_\alpha c_\beta b_\alpha b_\beta S_{\alpha\beta}(Q), \qquad \text{(IV.7.7)}$$

where $D(Q)$ is the measured structure factor and $S_{\alpha\beta}$ are the partial structure factors with weights determined by the products of the atomic fraction $c_\alpha$ and the scattering length $b_\alpha$. The Kronecker $\delta_{\alpha\beta}$ is introduced to avoid double counting of the same type. The partial structure factors are related to the corresponding partial distribution function $g_{\alpha\beta}(r)$ by

$$S_{\alpha\beta}(Q) = \rho \int_0^\infty [g_{\alpha\beta}(r-1)]e^{i\mathbf{Qr}}\, \mathbf{d}\mathbf{r}. \qquad \text{(IV.7.8)}$$

If a suitable set of isostructural materials are available with a sufficient number of isotope variants, the set of total diffraction functions $D(Q)$ can be inverted to produce a complete set of partial structure factors and accompanying radial distribution functions. For a complex molecular systems such as polymers, the elemental pair distribution functions are not so useful as for relatively chemically simple materials such as liquid $ZnCl_2$ extensively studied by Enderby et al. (1966). Of more interest is the possibility of site-specific pair distribution functions. For example, consider polystyrene and the possibility of functions describing the correlations between atoms on neighboring skeletal chains or between atoms in adjacent phenyl side groups. This information would greatly facilitate the analysis. We describe three approaches to using or deriving this information.

#### *IV.7.4.3.2 Using Partial Structure Factors Derived from Simulations*

Genix et al. (2006) have used partial structure factors calculated from simulations to study the local structure of syndiotactic polymethylmethacrylate using the neutron diffraction from five samples with different selective deuteration. They carried out a fully atomistic molecular dynamics simulation using the amorphous cell approach with four chains each containing 25 monomer units, a total of 1500 atoms in total. The simulation cell had a cell size of 25.16 Å. They used the results of the simulation to calculate partial structure factors and the corresponding distribution functions in terms of correlations involving atoms in the skeletal chains only, the side group and the α-methyl side group. The functions obtained from the simulations showed that the five experimental structure factors were each sensitive to particular functional components of the molecular organization. They used this information to conclude that there was a strong anticorrelation between the skeletal chain and the ester side groups suggesting a precursor effect of the nanophase separation observed in methacrylates with larger side groups as well as identifying that the ester side groups on neighboring chains appear to interdigitated. A similar coupled approach has been employed with other polymers and these are considered in Section IV.7.5.1

#### *IV.7.4.3.3 Experimentally Derived Partial Functions*

Mitchell et al. (1994) utilized neutron diffraction data in a more direct manner to obtain a partial distribution function for the skeletal chains. For atactic polystyrene, they partitioned the structure into the polymer backbone and the side group. As a consequence the structural correlations could be represented by several partial correlation functions each related to interactions between the specific sites, that is, $g_{bb}(r)$, $g_{bs}(r)$, $g_{ss}(r)$ where b denotes the backbone site and s is the side group site. They accessed these partial correlations in the manner described in eqs. (IV.7.7) and (IV.7.8). They used selective deuteration to prepare four glassy polystyrene samples with differing levels of deuteration labeled as $D_0$, $D_3$, $D_5$, and $D_8$. The number denotes the level and type of deuteration in terms of a monomer unit. $D_3$ involves deuteration of the skeletal chain hydrogens, while $D_5$ refers to the deuteration of the side group hydrogens. The resultant structure factors are shown in Figure IV.7.1. These quantitative measurements over a broad $Q$ range for polymers with significant hydrogen content are only possible with the use of small-angle scattering as discussed in Section IV.7.3.2. Manipulation of these functions yields a partial correlation function in which only the correlations involving atoms in the backbone unit are included, which was labeled $S_{bb}(Q)$ and the corresponding real space function as $g_{bb}(r)$. Figure IV.7.2b shows $g_{bb}(r)$ obtained by Fourier transformation compared to the total $g(r)$ obtained from the structure factor for $D_8$; the latter includes all correlations.

The partial correlation function shows a broad feature at $r \sim 10$–12 Å, which is clearly not discernible in the total correlation function obtained from the structure factor for $D_8$ and illustrates the particular power of this approach to structural studies. The authors attribute this feature to the correlations between the polymer backbones. It is noticeable that it is a broad and weak peak indicating very limited backbone correlations. This feature is much weaker than the corresponding features in simple polymers such as molten poly(ethylene) or poly(tetrafluoroethylene) (Rosi-Schwartz and Mitchell, 1992, 1994). They concluded that no substantial chain segment correlations are present in glassy polystyrene. Moreover, it suggests that the correlations between the side groups are the overriding features as previously proposed (Mitchell and Windle, 1984) and thus the polymer chains are forced into highly disordered conformations with limited spatial correlations. It is clear that this type of neutron scattering has powerful advantages when unraveling the complex structural correlations present in polymeric materials with both backbone and side group units.

#### *IV.7.4.3.4 Local Mixing in Miscible Blends*

Gkourmpis et al. performed broad $Q$ neutron and X-ray scattering experiments in order to evaluate the local structure of mixtures of chemically different polymers in the miscible state (Gkourmpis, 2003; Gkourmpis, T.E. and Mitchell, G.R., submitted). The objective of these experiments was to obtain information on the extent of segmental mixing at the local level. Simple theory assumes that mixing is random at the segmental level, whereas experimental evidence suggests otherwise. They examined blends with strong specific interactions that are chemically similar but topologically different in light of the proposal from Lohse and coworkers that the local chemical topology has a strong role in the local mixing (Krishnamoorti et al., 1998). Gkourmpis (2003) considered 50/50 (w/w) blends of 1,4-polybutadiene and 1,2-polybutadiene namely (a) d-14PBD/d-12PBD, (b) d-14PBD/h-12PBD, (c) h-14PBD/d-12PBD, and (d) h-14PBD/h-12PBD.

For a binary mixture they wrote the interchain component in terms of the partial structure factors $P_{AA}(Q)$, $P_{AB}(Q)$ and $P_{BB}(Q)$, where A and B are segments in the two chemically different polymers as shown in the following equation:

$$\begin{aligned} S_i(Q) = {} & c_A c_A F^2_{A_{H,D}} P_{AA}(Q) \\ & + 2 c_A c_B F_{A_{H,D}} F_{B_{H,D}} P_{AB}(Q) \\ & + c_B c_B F^2_{B_{H,D}} P_{BB}(Q) \end{aligned} \qquad \text{(IV.7.9)}$$

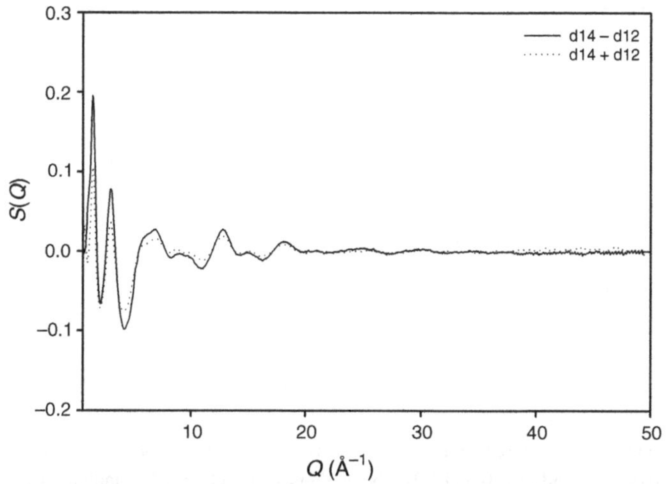


**Figure IV.7.6** Broad $Q$ neutron scattering data for a 50/50 (w/w) blend of perdeuterated 1,4-*trans*-polybutadiene and perdeuterated 1,2-polybutadiene (solid line) compared with a synthesized structure factor for 50/50 (w/w) mixture calculated by the appropriate weighting of the structure factors for the parent polymers (dotted line).

where $c_A$ and $c_B$ are the concentrations of A and B segments and $F_A(Q)$, and $F_B(Q)$ are the molecular form or scattering factors for segments A and B. This relationship is only valid for an isotropic sample in which the local mixing is also isotropic. To extract $P_{XX}(Q)$ from such a structure factor, a series of isostructural blends is required in which either the composition is systematically varied or in which the molecular form factors are altered through selective deuteration. Broad $Q$ X-ray scattering that is insensitive to the isotope content confirms the isostructural nature of the blends. Figure IV.7.6 shows the broad $Q$ scattering for a blend of the perdeuterated polybutadienes compared with a structure factor synthesized by adding the structure factors of the parent polymers in the correct proportions. In this work only the intense interchain peak at $Q \sim 1.4\,\text{Å}^{-1}$ is of interest. The approach is first to decompose the observed structure factor into intrachain and interchain scattering components by developing an atomistic level model, whose structure factor matches the observed scattering at high $Q$ that is then subtracted from the observed data to yield a function that arises from intersegmental scattering alone. The partial structure factors are obtained by inverting eq. (IV.7.9) and the results are shown in Figure IV.7.7. It is clear from the characteristics of the partial structure factors that there is little local mixing of A and B. This can be seen from the cross-term function. Using the width of the interchain peak, the extent of spatial correlations between A,A and B,B segments was estimated to be limited to ~2 chain segments as might be expected for a mixture of components.

Broad $Q$ neutron scattering on isotopically labeled blends provides a route to extracting partial structure factors arising from the correlations between like and unlike chain segments, which enables the extent of local mixing in a miscible blend to be quantified. This enables a molecular insight into many factors that are mapped on

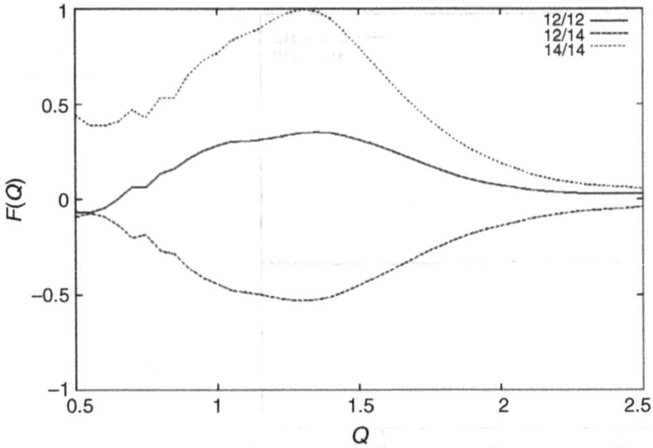


**Figure IV.7.7** The partial interchain structure factors for A–A, A–B and B–B correlations derived from experimental data for a 50/50 (w/w) blend of 1,4- and 1,2-polybutadienes. Solid line: 1,2-butadiene–1,2-butadiene; dotted line: 1,4-butadiene–1,4-butadiene; dashed line: 1,2-butadiene–1,4-butadiene.

to a single interaction parameter in standard blend theory. In the case of the polybutadiene blends considered in this work, the very local state is largely unmixed in contrast to larger lengths scales at which the system is homogenous, for example, the blends are optically clear.

## IV.7.5 COUPLED DIFFRACTION AND MODELING

It is clear from the discussion in the previous section that broad $Q$ neutron diffraction data contains considerable structural information and that progress is likely to be effective if the analysis is tightly coupled to molecular modeling. Of course we can see the diffraction data as simply some data with which to validate the results of a molecular dynamic simulation. However, when we compare the observed and experimental structure factors what do the differences indicate about the model? Is the model completely wrong, could we provide a better match by introducing/eliminating segmental orientation correlations? Without such information small deviations between experimental and predicted may be very important or may be limitations arising from the size of the model. It is vital to understand the model/scattering derivative, without this any validation of a model must be suspect. The essence of the problem is demonstrated in Figure IV.7.8 that shows two structure factors for samples of amorphous polyethylene prepared in two different ways (Gkourmpis, 1999; Gkourmpis, T.E. and Mitchell, G.R., unpublished). The first route involved irradiating a molten sample at 160°C with high-energy electrons (1 MeV, 50 MGy) that induced both cross-linking and chain scission (Vaughan and Stevens, 1995). The structure factor is very similar to that displayed by the molten sample (Rosi-Schwartz and Mitchell, 1994). The second sample was a semicrystalline polyethylene sample that was then extensively irradiated at 20°C in the absence of oxygen with high-energy electrons (1 MeV, 50 MGy). The irradiation of crystalline

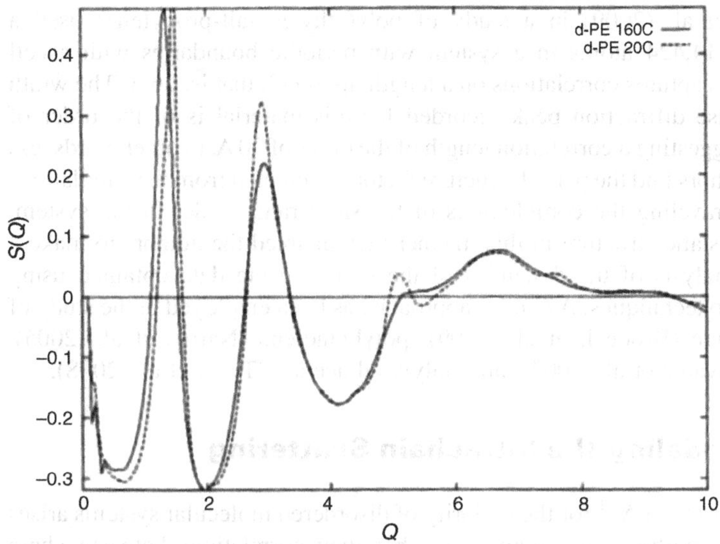


**Figure IV.7.8** The experimental structure factors for samples of perdeuterated polyethylene at room temperature after electron irradiation at 160°C (full line) and 20°C (broken line) as described in the text.

polyethylene with a dose of greater than 30 MGy leads to the destruction of the crystals (Keller, 1982). The structure factors show some significant differences that indicate that these two amorphous structures are locally different. The feature at ~5 Å$^{-1}$ for the 20°C sample is indicative of a high fraction of trans conformations (Mitchell, 1989; Mitchell et al., 1982). The interchain peak maximum occurs at $Q = 1.5$ Å$^{-1}$ compared with $Q = 1.35$ Å$^{-1}$ for the 160°C sample suggesting the chains are more closely packed. We interpret these factors as indicating that the 20°C sample contains small clusters of parallel chains segments in the trans conformation while the 60°C sample is representative of the random chain melt structure. In other words, the structure factors shown in Figure IV.7.8 represent two very different models with substantially different levels of segmental orientational correlation despite the broad similarities between the structure factors. In short, it underlines the fact that any quantitative comparison needs a clear understanding of the sensitivity to the structural parameters.

### IV.7.5.1 Comparison with Molecular Dynamic Simulations

A number of authors have utilized molecular dynamics to provide information to help interpret the complexities of structure factors obtained from synthetic polymer systems as well as using the neutron diffraction data to validate the output of the molecular dynamics simulation. To provide a simulation that can encompass the range of spatial correlations observed in the structure factors of disordered polymers a large simulation in terms of the number of atoms is required. For example

Perez-Aparicio et al. (2009) in a study of poly(ethylene-alt-propylene) used a simulation with 14,424 atoms in a system with periodic boundaries with a cell of $L \sim 52$ Å. This captures correlations on a length up to $L/2$, that is, 26 Å. The width of the first intense diffraction peak recorded for this material is of the order of $\Delta Q \sim 0.3$ Å$^{-1}$ suggesting a correlation length of the order of 20 Å, in other words less than $L/2$. The authors find the partial structure factors calculated from the simulations invaluable in unraveling the complexities of the short-range order in the system. Establishing the static structure in this manner then enabled the authors to make a more detailed analysis of the dynamics of the system using data obtained using neutron spin-echo techniques. A similar approach has been employed in the study of polyethylene oxide (Brodeck et al., 2009), polybutadiene (Narros et al., 2005), polyisoprene (Alvarez et al., 2003), and polyvinyl acetate (Tyagi et al., 2008).

### IV.7.5.2 Modeling the Intrachain Scattering

The scattering for $Q > 3$ Å$^{-1}$ for the majority of disordered molecular systems arises from correlations within chain segments rather than correlations between chain segments. If we restrict comparison between the model scattering and the observed scattering to $Q$ values $>3$ Å$^{-1}$ then the generation of models is considerably simplified. Gkourmpis et al. obtained broad $Q$ neutron scattering data from deuterated 1,4-polybutadiene over a substantial temperature range (20–300 K) using the SANDALS diffractometer at the ISIS pulsed neutron source (Gkourmpis, 2003; Gkourmpis, T.E. and Mitchell, G.R., submitted). Figure IV.7.9 shows the structure factor for deuterated polybutadiene at temperatures corresponding to the glassy state and the liquid state. In line with expectations, the data at high $Q$ is largely invariant with temperature since it is dominated by correlations between covalently bonded

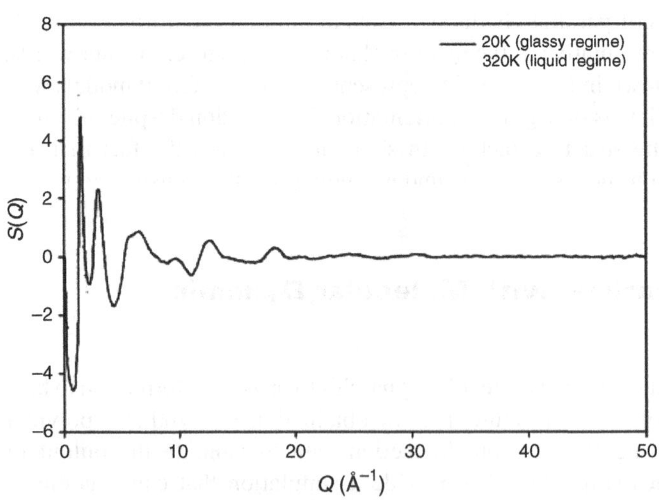


**Figure IV.7.9** Structure factors for perdeuterated 1,4-polybutadiene in the glassy (full line) and liquid state (broken line) obtained by neutron diffraction techniques.

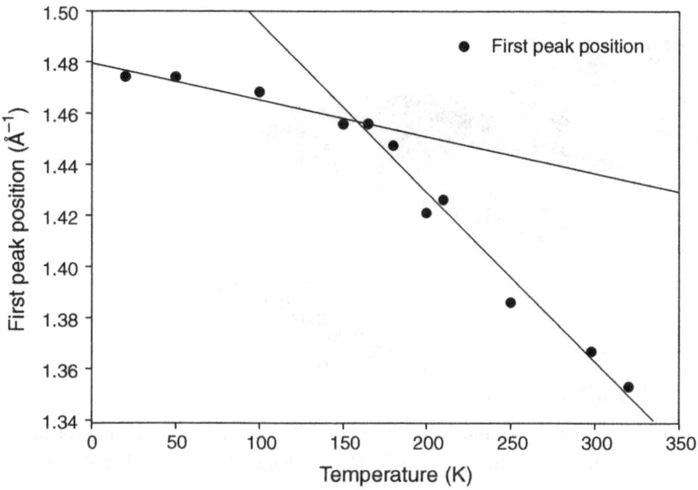


**Figure IV.7.10** A plot of the position of the first intense peak in the structure factor for perdeuterated 1,4-polybutadiene as a function of temperature. The solid lines represent best fits to the low- and high-temperature regions of the data.

atoms. In contrast, the scattering at $Q \sim 1.4\,\text{Å}^{-1}$ arises from correlations between chain segments and thus is strongly affected by changing temperature. The effect of temperature on the peak position is shown in Figure IV.7.10.

Gkourmpis et al. used internal coordinates to define a polybutadiene chain in terms of bond lengths, bond angles, and torsional angles (Gkourmpis, T.E. and Mitchell, G.R., submitted). Values were assigned to each internal coordinate within the chain using a stochastic Monte Carlo method, in which the probabilities were drawn from distributions representing the possible range of values. For example, the C–D bond length probability distribution was represented by a Gaussian distribution defined by two parameters, namely a mean value and a width. Using this approach random chain configurations could rapidly built and the intrachain structure factor calculated utilizing a small set of 25 parameters. At each stage the structure factor was averaged over a number of models each containing 15,000 atoms and a $\chi^2$ value derived where $\chi^2$ is given by

$$\chi^2 = \sum_{i=1}^{m} \frac{[S^{C}(Q_i) - S^{E}(Q_i)]^2}{\sigma^2(Q_i)}, \qquad \text{(IV.7.10)}$$

where E and C indicated experimental and calculated structure factors with $m$ points and standard deviation $\sigma$. The calculation of $\chi^2$ were made for values of $Q > 3\,\text{Å}^{-1}$. The number of the models used to calculate the average structure factor to be used in eq. (IV.7.10) for the final phase was 100. The parameters representing the probability distribution functions were systematically varied using a grid search technique to find the values that gave the lowest $\chi^2$ value. Figure IV.7.11 shows an example section through this multidimensional space with a plot of $\chi^2$ against the value of the width of

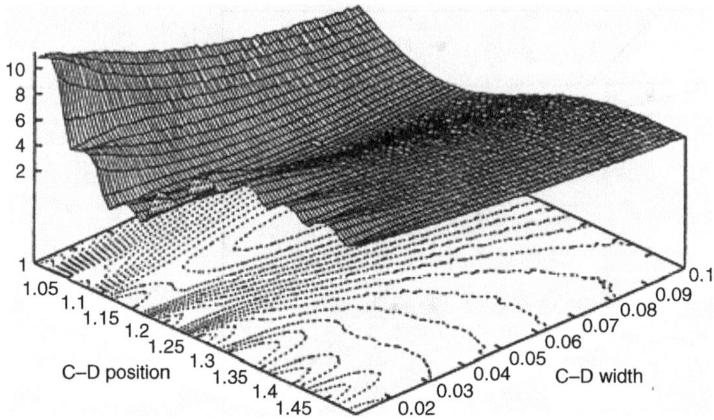


**Figure IV.7.11** A plot of the $\chi^2$ surface obtained using the procedure described in the text for the parameters describing the probability distribution for the C–D bond length for 1,4-polybutadiene.

the C–D length distribution. As an example of the structural detail available, Figure IV.7.12 shows the fraction of the sequence =C–C–C–C=, which was in the trans state as a function of data. The match between the final calculated structure factor and the observed data is shown in Figure IV.7.13 and has been annotated to indicate the regions of the structure factor which are most sensitive to which parameter.

### IV.7.5.3 Reverse Monte Carlo Procedures

A more direct approach is to use the data to "refine" an initial model and here we outline a process for achieving this based on reverse Monte Carlo procedures (Rosi-Schwartz and Mitchell, 1994). In such an approach, we carry out an off-lattice

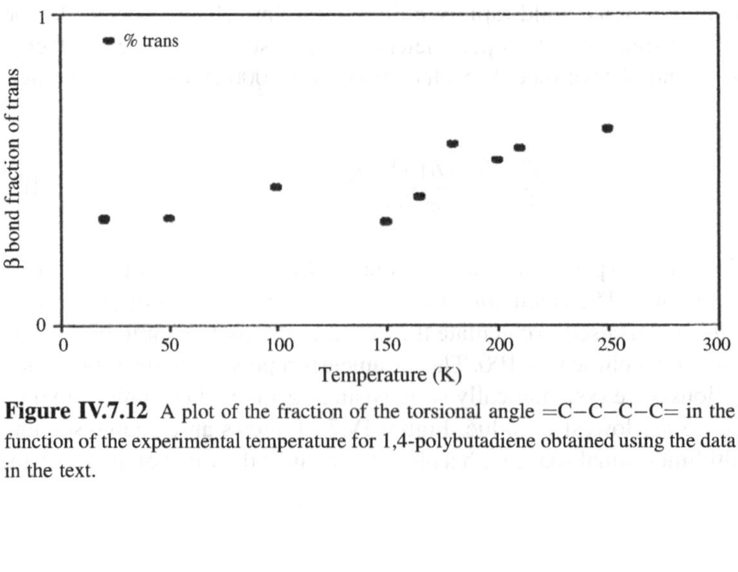


**Figure IV.7.12** A plot of the fraction of the torsional angle =C–C–C–C= in the trans state as a function of the experimental temperature for 1,4-polybutadiene obtained using the data analysis outlined in the text.

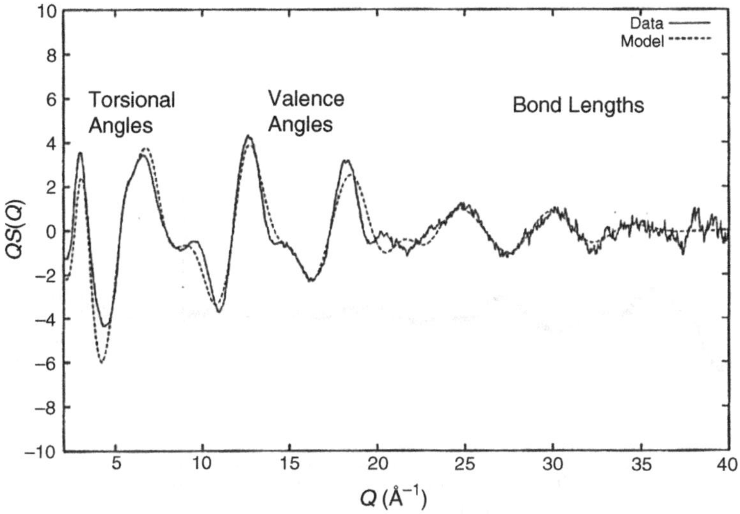


**Figure IV.7.13** A plot of the experimental structure factor for perdeuterated 1,4-polybutadiene at 210 K (full line) compared with the calculated structure factor based on the best-fit model for the intrachain scattering using the methodology as outlined in the text.

Monte Carlo atomistic simulation of the structure in which the fit between the calculated and experimental structure factors takes the role of the energy differences and the standard deviation in the data takes the role of the temperature. In particular, it is straightforward to show that the quantity $\chi^2$ defined in eq. (IV.7.10) plays the role of $U/k_BT$ in the Metropolis methodology (Rosi-Schwartz and Mitchell, 1994). The procedure is straightforward. The initial configuration is modified by a random move of one atom. The move is subjected to constraints imposed on the coordination number and the bond length of each atom type as well as the requirements for a fixed density and an excluded volume. The model structure factor is recalculated and a new $\chi^2$ is obtained. If this is less than the previous $\chi^2$ the move is accepted and if not the move is accepted with a probability given by

$$p = \exp[-(\chi_n^2 - \chi_0^2)/2]. \qquad \text{(IV.7.11)}$$

The procedure is repeated until $\chi^2$ reaches a value around which it oscillates. Key ingredients of the procedure are the maximum size of each random move $d$ and the chosen standard deviation $\sigma$. The parameter $d$ effectively determines both the ratio of accepted to rejected moves as well as amount of change in the structure with each move. We have found that it is preferable to choose small moves and preserve the chain connectivity.

Rosi-Schwartz and Mitchell (1994) used the reverse Monte Carlo procedures to develop a fully atomistic model of molten polyethylene. Figure IV.7.14 shows the final reverse Monte Carlo fit compared to the experimental structure factor. Of course a fit is expected and it is the values of the internal coordinates such as bond lengths and angles, which indicate if this is a reasonable model. Using this approach they were able to derive a series of structural parameters including the level of chain

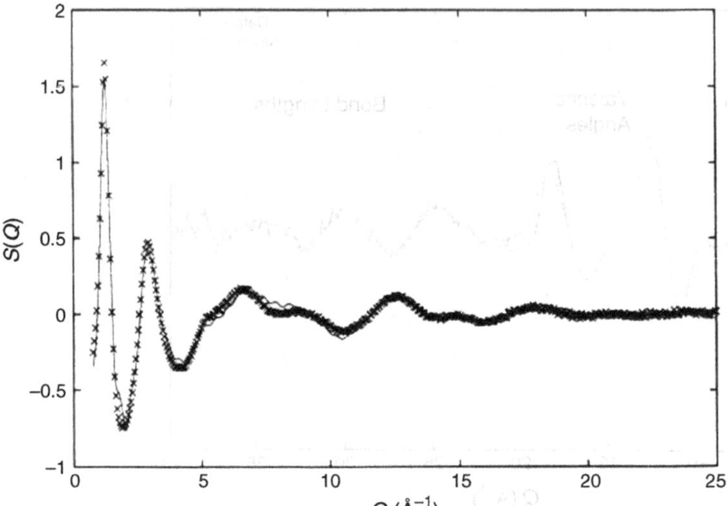


**Figure IV.7.14** A plot of the structure factor for a deuterated polyethylene at 160°C (symbols) compared with the calculated structure factor obtained using the reverse Monte Carlo methodology. This graph is reproduced from a reference (Rosi-Schwartz and Mitchell, 1994)

segmental orientational correlation as a function of separation between chain segments as shown in Figure IV.7.15. The figure compares the results from molten PE (Rosi-Schwartz and Mitchell, 1994) and PTFE (Mitchell et al., 1994). The broad peak at $r \sim 5$ Å for the functions obtained for polyethylene indicates a small level of

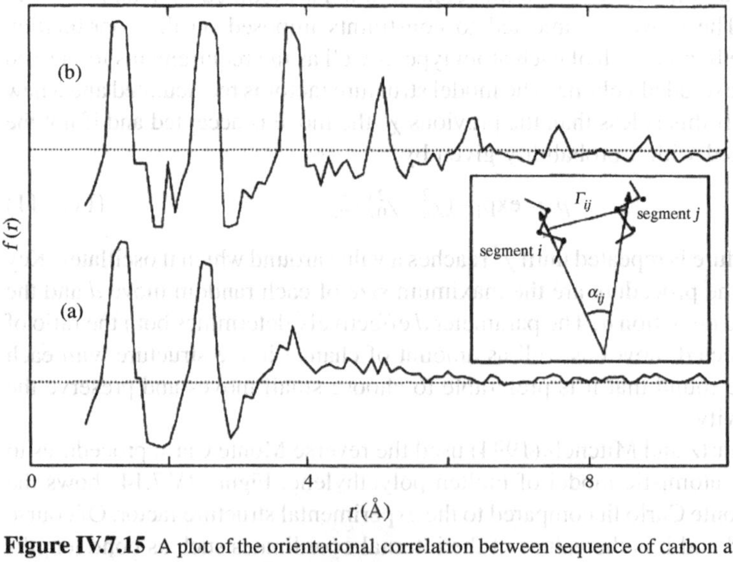


**Figure IV.7.15** A plot of the orientational correlation between sequence of carbon atoms as defined in the inset as a function of their separation distance for (a) molten polyethylene and (b) molten polytetrafluoroethylene. $f(r) = \langle (3\cos^2(\alpha_{ij}) - 1)/2 \rangle$ with average taken over all pairs in the model. This graph is reproduced from a reference (Mitchell et al., 1994).

near neighbor correlation between adjacent chain segments while the function for the PTFE only shows correlations that are of intrachain origin.

The reverse Monte Carlo approach provides a realistic route forward that automatically provides an understanding of the differences between calculated and experimental functions. Of course the results need careful consideration and refining models with different starting configurations both provides confidence and additional structural information. It provides a quantitative answer to the question what could those differences mean. It is also straightforward to "refine" against both X-ray and neutron scattering data including functions obtained with isotopic substitution at the same time, thereby both greatly increasing the information content and changing the relative emphasis of structural features within the refinement. The atomistic model gives access to structural parameters such as the segmental orientational correlations that cannot be deduced directly from the data.

## IV.7.6 FIBER DIFFRACTION

Fibrous materials are common place in both synthetic polymers and biological systems. A particular characteristic of recent studies involving neutron fiber diffraction techniques is that the neutron scattering data has been complemented by X-ray diffraction data, electron diffraction, and molecular modeling. Central to most of these studies has been the use of broad $Q$ neutron scattering to elucidate the positions of the hydrogen atoms in the crystal structure. The work of Tashiro et al. (2004) in this respect is an exemplar study in which they used a neutron imaging plate to record fiber patterns from samples of polyethylene-$d_4$ and polyethylene-$h_4$ prepared by mechanical alignment. The incident neutron beam had a wavelength of 1.51 Å and so the geometry was very equivalent to the classic X-ray cylindrical camera. There were 29 observed crystal reflections or diffraction maxima for the polyethylene-$d_4$ sample and these intensities were subjected to standard crystallographic analysis to produce "neutron density" maps that revealed the positions of the both carbon and deuterium atoms in the unit cell. Similar maps were also obtained for the polyethylene-$h_4$ sample, although in this case the density at the hydrogen position was negative due to the negative coherent neutron scattering length. Structural refinement on all the atom positions using standard least-squares methods gave a full crystal structure including the hydrogen atoms for polyethylene for the first time with a reliability of $R$ factor of 0.135. Earlier work by Takahashi and Kumano (2004) using neutron diffraction with the same instrument on polyethylene-$d_4$ but using a rigid body refinement achieved an $R$ Factor of 0.184. Tashiro et al. contrasted the advantages of neutron, X-ray and electron diffraction and concluded that the neutron diffraction was the most appropriate for yielding the hydrogen positions. As with any crystallographic analysis, the number of observed crystal reflections is key to achieving a high-resolution study and the authors propose using a shorter incident wavelength in future work to achieve this. Central to the analysis was the use of a fully deuterated material to reduce the incoherent background. Tashiro et al. (2009) have extended their work to the crystal structure of polyoxymethylene using crystals prepared by topological polymerization of tetraoxane crystals using $\gamma$-rays. For this system some 72 reflections were observed mainly due to the somewhat larger *c*-axis unit cell and a reliability factor of 0.22 was achieved.

Gardner et al. (2004) have used neutron fiber diffraction to unravel a long-standing issue with the structure of poly(*p*-phenylene-terephthalamide) known commercially as Kevlar and Twaron. This polymer is used to prepare light and strong synthetic fibers in which the high strength comes principally from the many intermolecular hydrogen bonds formed between the carbonyl groups and protons on neighboring polymer chains to form sheets. The basic chain conformation and sheet structure are well established but a number of different crystal structures have been proposed with varying arrangements for the displacements of the chains and the intersheet interactions. The similarity of the X-ray scattering from the terephthaloyl and diamine groups in the poly(*p*-phenylene-terephthalamide) chain has been thought to be the difficulty in differentiating between the various models.

Gardner et al. (2004) used the D19 Diffractometer at the ILL to obtain neutron diffraction fiber patterns for both deuterated and nondeuterated fibers of poly(*p*-phenylene-terephthalamide). The deuterated material gave a strong pattern that was only consistent with the model proposed by Liu et al. (1996). The unequivocal nature of the result arises from the much stronger scattering from the deuterium atoms with neutrons than an equivalent X-ray diffraction study. The authors emphasize the value of neutron fiber diffraction with deuterated samples for resolving the subtle aspects of a structure especially where these have a significant impact on the function of the material, for example, the mechanical properties.

The work of Sapede et al. (2005) extended the use of neutron-based fiber diffraction to the study of the structure of spider dragline silk. They utilized both small-angle and wide-angle neutron scattering techniques as well as quasielastic neutron scattering experiments. Neutron fiber diffraction patterns were obtained using D19 at the ILL using samples with ~150 mg of silk. The samples were held in a stream of humidified gas that was used to perform hydrogen/deuterium exchange. Prominent among their discoveries using neutron fiber diffraction was the presence of a diffuse peak at the position of the 001 reflection that is not allowed for the space groups ($C222_1$ or $P2_12_12_1$) derived from X-ray diffraction data. This reflection is not observed with the nondeuterated system. By comparing the azimuthual width of this reflection, they tentatively propose a smectic β-sheet structure for the glycine-rich chains as well as crystalline and amorphous phases as previously proposed.

It is almost a century ago that it was discovered that cellulose in the form of plant fibers was crystalline. Subsequent studies have revealed the complexities of cellulose structures both in native form (Cellulose-I) and in recrystallized or reconstituted form (Cellulose-II and Cellulose-III) but there have remained considerable uncertainties. Nishiyama et al. have recorded in a series of papers (Nishiyama et al., 1999, 2002, 2003, 2008; Langan et al., 1999; Wada et al., 2004) on how they have applied neutron fiber diffraction techniques to clarify the structure of a range of cellulose systems including details of the hydrogen boding. A particular problem with Cellulose-I is the nature of the disordered hydrogen bonding. Nishiyama et al. (2008) employed fiber diffraction data obtained at different temperatures using D19 at the ILL coupled with the results of molecular dynamics and quantum mechanical calculations to identify the precise form of the hydrogen bonding. These studies further underline the value of selective deuteration and the use of neutron fiber diffraction techniques.

Such studies focus attention on the possibility of deuteration or even selective deuteration. The study of Parrot et al. (2006) illustrates the future possibilities. They prepared a sample of DNA in which the alternating sequence of adenine and thymine residues was modified by the selective deuteration of the thymine residues. They were then able to obtain high-resolution fiber diffraction patterns of samples held in either $H_2O$ or $D_2O$ environments. The objective of this study is to understand the transitions that accompany hydration. The work emphasizes the coupled challenging requirements of sample preparation in the form of selective deuteration and high-quality instrumentation to obtain the required information.

## IV.7.7 *IN SITU* DIFFRACTION STUDIES

In contrast to synchrotron-based X-ray studies, neutron diffraction techniques are not normally seen as relevant to time-resolving studies. Thus, in general, neutron diffraction procedures have been used to evaluate the steady-state situation. This limitation partly arises from the lower neutron fluxes and partly from the weak interaction between the neutron and materials. Of course, this weak interaction has particular advantages in that the radiation does not damage the sample under study and it does not lead to a significant heat load. Moreover, neutrons have little difficulty in passing through the walls of sample environment equipment, making the use of cryostats and furnaces routine.

### IV.7.7.1 Deformation

Polymers offer a particular advantage in the study of samples subjected to mechanical and thermal treatments in that the particular structure induced by the treatment can be locked in by rapidly quenching to the glassy state. Casas et al. (2006) used this approach in a combined small-angle and broad $Q$ neutron scattering study of the sub-Tg deformation of perdeuterated polymethylmethacrylate. They were able to obtain static structure factors for $\mathbf{Q}$ parallel and perpendicular to the deformation axis. They found that the positions of the peaks in terms of $|\mathbf{Q}|$ for the deformed structure were unchanged from the undeformed sample but that there were significant changes in the intensities. In particular, the first peak was more intense with $\mathbf{Q}$ parallel to the extension axis, while the second and third peaks were more intense with $\mathbf{Q}$ perpendicular to the extension axis; these differences were attributed to local modifications of the partial structure factors involving the side groups. In contrast, the parallel small-angle scattering data on a mixture of isotopically labeled polymers displayed substantial differences that were shown to be due to the affine deformation of the polymer chains. Clearly the extent of anisotropy depended on the $Q$ scale examined.

Brown et al. (2007) utilized the particular advantage of pulsed neutron sources to simultaneously measure the scattering functions for **Q** parallel and perpendicular to the deformation axis for a polytetrafluoroethylene system using the basic geometry first exploited by Mitchell and Cywinski (1992) in the study of anisotropic films of electrically conducting polymers. A time-of-flight diffractometer usually contains a number of detectors each of which provides scattering data as a function of $|\mathbf{Q}|$, the particular range of $Q$ depends on the wavelength distribution of the neutrons and the scattering angle. By arranging the symmetry axis of the sample to be at 45° to the incident beam, detectors arranged at a scattering angle either side of the sample, data for **Q** parallel and perpendicular to the symmetry axis can be obtained in straightforward manner. Brown et al. (2007) used this technique coupled with extension and compression to show that polytetrafluoroethylene responded to uniaxial deformation by undergoing a crystalline phase transition, previously thought to only occur at very high hydrostatic pressures. These experiments were facilitated by the use of a polymer that contains no hydrogen and thus allowed thick samples of ~10 mm to be used.

### IV.7.7.2 Neutron Diffraction Cryoporometry

Beau et al. (2008) have used neutron diffraction data obtained as a function of temperature to develop information on the state of material confined in a mesoporus system. The technique is named to reflect the similarities with thermoporosimetry and NMR cryoporometry in which the melting point of a material confined in a mesoporus system is used to evaluate the size of the pores. The Gibbs–Thomson relationship inversely equates the melting point depression of a small crystal of a liquid to the size of that crystal. Hence by determining the melting point distribution of, for example, ice in a templated silica, a map of the pore sizes can in principle be determined. A particular advantage of neutron diffraction is that information about the phases of the confined material can be identified, for example the different phases of ice.

### IV.7.7.3 Absorption in Microporus Carbon

Steriotis et al. (2004) have used *in situ* neutron diffraction techniques to study the absorption of $CO_2$ in a mesoporus carbon as a function of temperature. The experiment used a pressure cell mounted on a cryofurnace. They calculated the difference between the $CO_2$ loaded and unloaded carbons to obtain the $CO_2$ diffraction pattern. This approach assumes that the interfaces make little contribution to the observed scattering. They observed an intense peak at $\sim 2\,\text{Å}^{-1}$ that shifts slightly up in $Q$ with increasing pressure. The authors argue that the presence of the peak that corresponds to intermolecular correlations shows that the $CO_2$ is present in the condensed state similar to that for the supercritical fluid. The densification is attributed to the strong interaction between the carbon walls and the $CO_2$ molecules.

## IV.7.7.4 Time-Resolved Crystallization Studies

The routes to increasing the time-resolution available in a neutron diffraction experiment center on optimizing the sample size to the neutron beam as the beam size is often substantial with a width and height of several centimeters and detecting as many neutrons as possible. The use of a position sensitive detector clearly provides one route, but a pulsed neutron facility using time-of-flight techniques offers the possibility of substantially increasing the solid angle over which neutrons are captured. An example is GEM at the ISIS pulsed neutron source (Hannon, 2005), which contains 7270 detectors arranged in time-focused banks that provide a coverage of ~30% of all scattering directions from the sample. Each detector provides a structure factor with a $Q$ range dependent on the scattering angle. Mitchell et al. have used this diffractometer to explore whether time-resolved studies of phase transitions such as crystallization in polymers is possible (Mitchell, G.R., Davis, F.J., and Siripitayananon, J., unpublished). They used a small furnace attached to a CCR system and used an exchange gas to quickly cool the sample.

Figure IV.7.16 shows scattering data recorded for a partially deuterated poly (ε-caprolactone) following a quench to below the crystallization experiment. The lowest curve corresponds to the melt at a temperature (80°C). The system was then quenched rapidly to a lower temperature selected so that the rate of crystallization was modest, in this case it was 50°C. The inset shows the peak intensity of the (110) crystal peak as a function of time after the quench. In the first few frames the intensity is zero as there is no crystallization. The time slices are 600 s in length while the half-time for crystallization is of the order of 16,000 s. The objective of these experiments

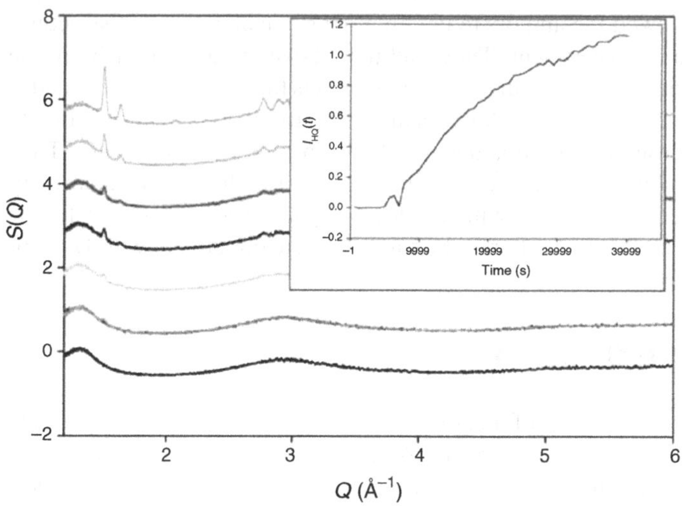


**Figure IV.7.16** Structure factors recorded using the GEM diffractometer for a sample of partially deuterated poly(ε-caprolactone) that had been held in the melt at 80°C (lowest curve) and then quenched to 50°C and held. The time slice is 600 s and successive time slices are offset vertically. The inset is a plot of the peak intensity of the (110) crystal reflection as a function of time. $t = 0$ is the point of the quench.

is to exploit the broad $Q$ data to follow the different length scales within the crystallization process.

## IV.7.8 POLYELECTROLYTES

Polymer electrolytes are currently attracting considerable attention due to applications in nontoxic and cheap batteries with high-energy density. They are complex material systems based on complexes of different polymers and salts. The ion conduction mechanism is not fully understood and neutron diffraction techniques have been employed to gain a greater insight into the molecular organization of these novel materials.

Carlsson et al. (1998, 2000, 2004) have employed both neutron and X-ray diffraction data with reverse Monte Carlo analysis procedures for electrolyte based on the amorphous polymer polypropylene oxide complexed with $LiClO_4$. They used structural configurations based on a cubic cell of length ~47 Å and first obtained a reverse Monte Carlo structure for the polypropylene oxide alone. This was then used as the matrix for the analysis of the polymer complexes. In a detailed analysis, the authors identified that the solvation of the salt induces increased ordering of the polymer matrix (Carlsson et al., 2004). They attribute this to well-defined structural arrangements within the salt rich regions. The reverse Monte Carlo approach shows that there are interanionic correlations at distances of ~5 Å involving the ether oxygens of the polymer chains.

Zajac et al. (2002) have utilized neutron diffraction with polarization analysis to study the polyethylene oxide and $LiSO_3CF_3$ system. In contrast to the polypropylene oxide system this complex exhibits a semicrystalline structure. The use of spin polarized neutrons allowed the authors to obtain more static and dynamic structural information in the same experiment. They find that above the glass transition, the matrix exhibits considerable short-range order that affects the dynamics of the matrix and hence the electrical conductivity. Annis et al. (2004) used isotopic substitution to study the Li ion environment in an aqueous solution of polyethylene oxide. They found that the local structure near the Li ion indicated a hydration structure very similar to that observed for Li in $D_2O$ with no added polymer. The results show for the solvent ratios employed that the polymer does not compete effectively with $D_2O$ for solvation sites near the Li ion.

## IV.7.9 GLIMPSE OF THE FUTURE

There is continual development of neutron diffraction facilities that enable and support new science and new experiments. At the ISIS pulsed neutron facility, a second target station has recently (2009) come on line with a variety of new instruments and in particular NIMROD that is a diffractometer tuned for the study of the structure of liquids and disordered matter (Bennington, 2009). It has a $Q$ range of between 0.2 and 100 $Å^{-1}$, enabling it to probe length scales ranging from the interatomic to the mesoscopic (Bowron, D., 2010). Figure IV.7.17 shows an

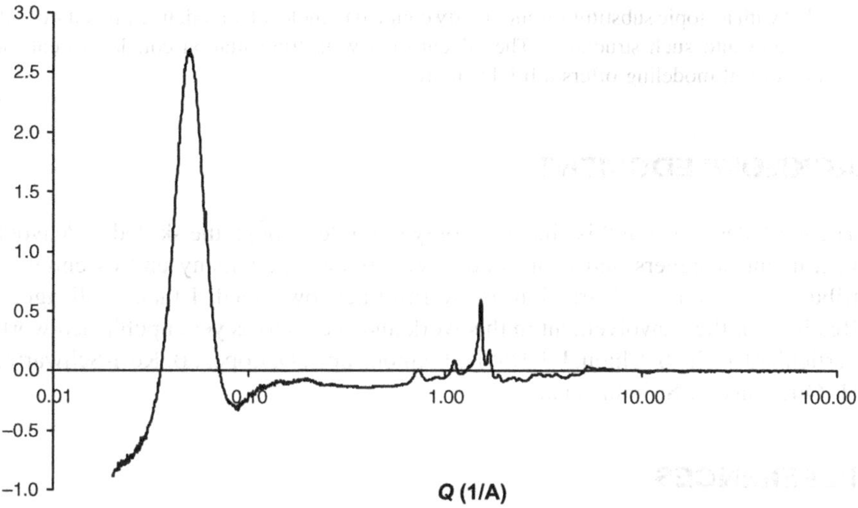


**Figure IV.7.17** A plot of the structure factor for a partially deuterated poly(ε-caprolactone) at 20°C measured using the new diffractometer NIMROD at ISIS. Note the logarithmic scale for the $Q$ axis. Data were obtained for $Q$ values from 0.02 to 80 $\text{Å}^{-1}$.

example structure factor obtained at room temperature using NIMROD for a sample of deuterated poly(ε-caprolactone) (Mitchell, G.R., Lopez, D., and Davis, F.J., unpublished). The large peak at low $Q$ arises from the lamellar crystals (~150 Å) in the polymer, while the "wide-angle" crystalline scattering peaks (~5 Å) occur midpoint on this logarithmic scale and at high $Q$ we reach a length scale of 0.1 Å. The ability to obtain a single structure factor with this range of structural information opens immense possibilities in the study of structural reorganization accompany phase transitions such as crystallization. In the current configuration, it proved possible to obtain data in ~450 s time slices that were of adequate quality to follow the crystallization process. Optimization of the geometry should enable a time slice of ~50–60 s.

The data shown in Figure IV.7.17 offers exciting prospects but poses considerable challenges in the development of suitable computational models to allow the evaluation of the structure on the range of length scales involved. Typical length scales in full atomistic simulations or in reverse Monte Carlo procedures are on the scale of 30–50 Å. The lamellar crystal peak corresponds to a length scale of the order of 150 Å.

Clearly multiple scale models with suitable coarse graining will be required and the challenge will be to provide an approach in which structural parameters from the differing length scales couple together.

## IV.7.10 SUMMARY

This chapter has explored the potential of broad $Q$ neutron diffraction to probe the complex structural organization of polymers and other soft matter. It is clear that neutron diffraction coupled with isotopic substitution and tightly coupled to molecular modeling procedures offers real insight into such structures. The advent of new instrumentation coupled to enhanced computational modeling offers a bright future.

## ACKNOWLEDGMENTS

The work described in this chapter is only possible through the skilled endeavors of instrument designers and beam-line scientists and their many and essential contributions to neutron diffraction are warmly acknowledged. I thank colleagues at Reading for their involvement in this work and access to as yet unpublished work in particular Drs. Y.-S. Chiou, F.J. Davis, T. Gkourmpis, D. Lopez, B. Rosi-Schwartz, R. H. Olley, and J. Siripitayananon.

## REFERENCES

Alvarez, F., Colmenero, J., Zorn, R., Willner, L., and Richter, D. *Macromolecules* **2003**, *36*, 238.

Annis, B.K., Badyal, Y.S., and Simonson, J.M. *J. Phys. Chem.* **2004**, *108*, 2554.

Beau, J., Webber, W., and Dore, J.C. *Nucl. Instrum. Meth. Phys. Res. A* **2008**, *586*, 356.

Bennington, S.M. *Nucl. Instrum. Meth. Phys Res. A* **2009**, *600*, 32.

Bowron, D.T., Soper, A.K., Jones, K., Ansell, S., Birch, S., Norris, J., Perrott, L., Riedel, D., Rhodes, N.J., Botti, A., Ricci, MA., Grazzi, F., Zoppi, M. *Rev. Sci. Insstr.* **2010**, *81*, 033905.

Brodeck, M., Alvarez, F., Arbe, A., Juranyi, F., Unruh, T., Holderer, O., Colmenero, J., and Richter, D. *J. Chem. Phys.* **2009**, *130*, 094908.

Brown, E.N., Dattlebaum, D.M., Brown, D.W., Rae, P.J., and Clausen, B. *Polymer* **2007**, *48*, 2531.

Carlsson, P., Mattsson, B., Swensson, J., Borjesson, L., Torell, L.M., McGreevy, R.L., and Howells, W.S. *Electrochim. Acta* **1998**, *43*, 1545.

Carlsson, P., Swensson, J., Borjesson, L., McGreevy, R.L., Jacobsson, P., Torell, L.M., and Howells, W.S. *Electrochim. Acta* **2000**, *44*, 1449.

Carlsson, P., Anderson, D., Swenson, J., McGreevy, R.L., Howells, W.S., and Borjesson, L. *J. Chem. Phys.* **2004**, *121*, 12026.

Casas, F., Alba-Simionesco, C., Lequeux, F., and Montes, H. *J. Non-Cryst. Solids* **2006**, *352*, 5076.

Chiou, Y.S. Ph. D. Thesis, University of Reading, 2004.

Enderby, J.E., North, D., and Egelstaff, P.A. *Philos. Mag.* **1966**, *14*, 961.

Gabrys, B.J. *Applications of Neutron Scattering to Soft Condensed Matter*, Gordon and Breach, Amsterdam, 2000.

Gabrys, B., Scharpf, O., and Peiffer, D.G. *J. Polym. Sci. Oart B Polym. Phys.* **1993**, *31*, 1891.

Gardner, K.H., English, A.D., and Forsyth, V.T. *Macromolecules* **2004**, *37*, 9654.

Genix, A.C., Arbe, A., Alvarez, F., Colmenero, J., Scheika, W., and Richter, D. *Macromolecules* **2006**, *39*, 3947.

Gkourmpis, T.E. M. Sc. Thesis, University of Reading, 1999.

Gkourmpis, T.E. Ph. D. Thesis, University of Reading, 2003.

Hannon, A.C. *Nucl. Instrum. Meth. Phys. Res. A* **2005**, *551*, 88.

Hannon, A.C., Howells, W.S., and Soper, A.K. In: Johnson, M.W. (editor). *Proceedings of the Conference on Neutron Scattering Data Analysis 1990*. IOP Conference Series, Vol. 107, 1990, p. 193.

Higgins, J.S. and Benoit, H.C. *Polymers and Neutron Scattering*, Clarendon Press, Oxford, 1994.

Keller, A. Developments in Crystalline Polymers–1, Applied Science, London, 1982.

Krishnamoorti, R., Graessley, W.W., Fetters, L.J., Garner, R.T., and Lohse, D.J. *Macromolecules* **1998**, *31*, 2312.

Lamers, C., Scharpf, O., Schweika, W., Batoulis, J., Sommer, K., and Richter, D. *Physica B* **1992**, *180/181*, 515.

Langan, P., Nishiyama, Y., and Chanzy, H. *J. Am. Chem. Soc.* **1999**, *121*, 9940.

Liu, J., Cheng, S.Z.D., and Geil, P.H. *Polymer* **1996**, *37*, 1413.

Lovell, R., Mitchell, G.R., and Windle, A.H. *Faraday Discuss. Chem. Soc.* **1979**, *68*, 47.

Lovesey, S.W. *Theory of Neutron Scattering from Condensed Matter*, Clarendon Press, Oxford 1984.

McLain, S.E., Bowron, D.T., Hannon, A.C., and Soper, A.K. Gudrun ISIS Report, 2009.

Mitchell, G.R. In: Allen, G., Bevington, J.C., Booth, C., and Price, C. (editors). *Comprehensive Polymer Science*, Vol. 1, Pergamon Press, Oxford, 1989, Chapter 31, p. 687.

Mitchell, G.R. and Cywinski, R. *J. Mater Sci.* **1992**, *27*, 2117.

Mitchell, G.R. and Windle, A.H. *Polymer* **1984**, *25*, 906.

Mitchell, G.R., Lovell, R., and Windle, A.H. *Polymer* **1982**, *23*, 1273.

Mitchell, G.R., B, Rosi-Schwartz., and Ward, D.J. *Phil. Trans. R. Soc. London A* **1994**, *348*, 97.

Narros, A., Arbe, A., Alvarez, F., Colmenero, J., Zorn, R., Schweika, W., and Richter, D. *Macromolecules* **2005**, *38*, 9847.

Nishiyama, Y., Okano, T., Langan, P., and Chanzy, H. *Int. J. Biol. Macromol.* **1999**, *26*, 279.

Nishiyama, Y., Langan, P., and Chanzy, H. *J. Am. Chem. Soc.* **2002**, *124*, 9074.

Nishiyama, Y., Sugiyama, J., Chanzy, H., and Langan, P. *J. Am. Chem. Soc.* **2003**, *125*, 14300.

Nishiyama, Y., Johnson, G.P., French, A.D., Forsyth, V.T., and Langan, P. *Biomacromolecules* **2008**, *9*, 3133–3140.

Parrot, I.M., Laux, V., Urban, V., Haertlein, M., and Forsyth, V.T. *Physica B* **2006**, *385/386*, 848.

Perez-Aparicio, R., Arbe, A., Alvarez, F., Colmenero, J., and Willner, L. *Macromolecules* **2009**, *42*, 8271.

Roe, R.-J. *Methods of X-Ray and Neutron Scattering in Polymer Science*, Oxford University Press, Oxford, 1999.

Rosi-Schwartz, B. and Mitchell, G.R. *Polymer* **1994**, *35*, 3139.

Rosi-Schwartz, B. and Mitchell, G.R. *Physica B* **1992**, *180/181*, 558.

Rosi-Schwartz, B. and Mitchell, G.R. *Polymer* **1994**, *35*, 5398.

Rosi-Schwartz, B., Mitchell, G.R., and Soper, A.K. *Polymer* **1992**, *33*, 3744.

Sapede, D., Seydel, T., Forsyth, V.T., Koza, M.M., Schweins, R., Vollrath, F., and Riekel, C. *Macromolecules* **2005**, *38*, 8447.

Squires, G.L. *Introduction to the Theory of Thermal Neutron Scattering*, Dover Publication Inc., New York, 1996.

Steriotis, Th.A., Stefanopoulos, K.L., Kanellopoulos, N.K., Mitropoulos, A.Ch. and Hoser, A. *Colloids Surf. A* **2004**, *241*, 239.

Takahashi, Y. and Kumano, T. *J. Polym. Sci. Part B Polym. Phys.* **2004**, *42*, 3836.

Tashiro, K., Tanaka, I., Oohara, T., Niimura, N., Fujiwara, S., and Kamae, T. *Macromolecules* **2004**, *37*, 4109.

Tashiro, K., Hanesaka, M., Yoshizawa, Y., Oohara, T., Kuroki, R., Kitano, T., Nishu, T., Sasaki, S., and Takata, M. *J. Phys. Conf. Ser.* **2009**, *184*, 012012.

Tyagi, M., Arbe, A., Alvarez, F., Comenero, J., and Gonzalez, M.A. *J. Chem. Phys.* **2008**, *129*, 224903.

Vaughan, A.S. and Stevens, G.C. *Polymer* **1995**, *36*, 1531.

Wada, M., Chanzy, H., Nishiyama, Y., and Langan, P. *Macromolecules* **2004**, *27*, 8548.

Zajac, W., Gabrys, B.J., Scharpf, O., Andersen, K.H., and Pasonage, E.E. *Solid State Ionics* **2002**, *147*, 213.

# V Current Facilities

## V.1 Pulsed Neutron Sources and Facilities

*Masatoshi Arai*

### V.1.1 INTRODUCTION

Accelerator-based neutron sources were developed after the 1960s, starting with electron accelerators using the bremsstrahlung photoneutron reaction. Tohoku University Linac and Harwell Linac were a pioneering development in this field. In the 1970s, feasibility and effectiveness of the spallation reaction, as a neutron source, induced by high-energy protons were intensively studied at the Argonne National Laboratory by J.M. Carpenter and M. Kimura. Then, a dawn of spallation neutron sources occurred in the 1980s. After a test facility of ZING at the Argonne National Laboratory, the KENS facility in Japan in 1980, the IPNS facility at the Argonne National Laboratory in 1981, the LANSCE facility at the Los Alamos National Laboratory in 1983, and the ISIS facility in the United Kingdom in 1985 started operations as user facilities. In those days, accelerators for nuclear physics had already been developed worldwide. Most of them were about finishing their mission and retiring, hence spallation sources were built either as satellite facilities or as refurbished facilities to the main nuclear physics facilities.

During the 20 years of operation of these facilities, the usefulness and effectiveness of the spallation sources have been established. Now, building spallation neutron sources has became a world trend because of engineering, cost efficiency, and environmental reasons; some examples are Spallation Neutron Source (SNS) in Oak Ridge, USA and the J-PARC project in Japan (Mason et al., 2003). The site of the European Spallation Source (ESS) project has been recently decided at Lund, Sweden, and it will be completed in 2019. China has also started planning it own facility called Chinese Spallation Neutron Source (CSNS).

Since the neutron yield is almost proportional to proton power, improvement of accelerator performance consequently increases neutron flux of spallation sources. The pulsed nature of spallation sources can give a huge advantage in various kinds of

*Neutrons in Soft Matter* Edited by Toyoko Imae, Toshiji Kanaya, Michihiro Furusaka, and Naoya Torikai

experiments using the time-of-flight (TOF) method. Moreover, because of the less heat deposition in target systems, it can ease very high instantaneous flux although there still remains a thermal shock problem in the target system to be overcome in the future.

At present, there are two kinds of spallation sources. One is the short pulsed source, typically delivering 1 μs proton pulse widths, on which we will concentrate in this chapter. Another is the long pulsed source that will be built in SNS as the second target station and in ESS, typically having 1 ms proton pulse widths (Mezei, 1993).

## V.1.2 NEUTRON CREATION IN SPALLATION SOURCES

High-energy particles such as deuteron, tritium, and other heavier particles can smash nuclei. Protons, however, are most commonly used in spallation sources because of engineering reasons: easier in acceleration with high current and effectiveness of spallation processes (Vassil'kov and Yurevich, 1990).

Spallation reactions occur above 100 MeV proton energy. High-energy neutrons, pions, and spalled nuclei cause internuclear cascades followed by low-energy neutron evaporation from excited nuclei having the maximum intensity at about 2 MeV (Watanabe, 2003) (Figure V.1.1). Protons can create a large number of neutrons from heavy nuclei: 1 GeV proton produces about 25 neutrons from a lead target or similar heavy metal target, with heat deposition in the target as low as only about half the proton beam power and another half is dissipated in the surrounding shields—one order of magnitude lower than fission reactions for the same number of time-averaged neutron fluxes.

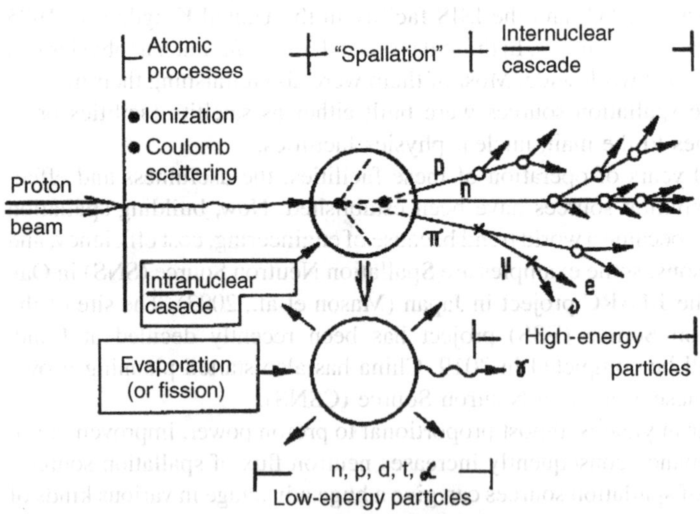


**Figure V.1.1** Nuclear spallation process by proton bombardment. Protons release energy through the ionization process at the front part of target and create intranuclear cascade followed by internuclear cascade. Segmented nuclei and excited nuclei trigger the evaporation process creating neutrons and protons associated with collapse of nuclei (Watanabe, 2003).

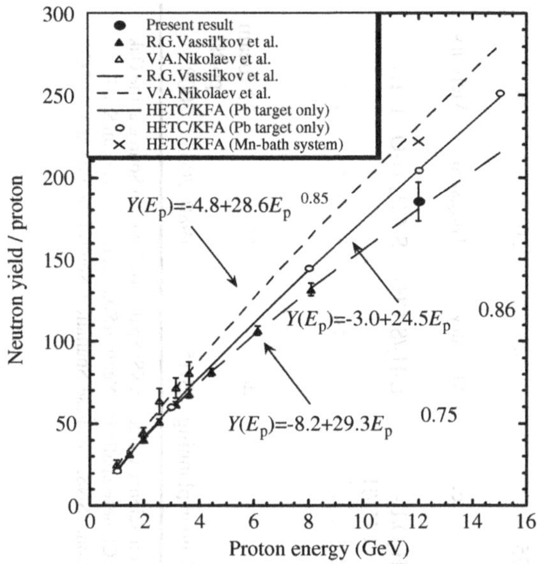


**Figure V.1.2** Neutron yield from a lead target as a function of proton energy. Neutron yield is almost proportional to proton energy up to 12 GeV (Arai et al., 1999).

In the early days of design work on spallation sources in the 1980s, proton energies considered to be suitable were in the region less than 1.0 GeV because of lack of experimental experience at higher energies. However, more sophisticated codes and experiments in the 1990s showed that neutron production rate is almost proportional to the accelerator power and even higher above 3 GeV (Arai et al., 1999) (Figure V.1.2). This resulted in flexibility in optimizing accelerator and neutron target design. Proton current and energy are equally optimized for beam experiments although it is necessary to take the proton stopping length $R$ into account for further thermalization process by moderators ($R$ is about 50 cm for 1 GeV and it is not drastically enlarged in energy). The performance of spallation sources is listed in Table V.1.1. A typical spallation neutron source is shown in Figure V.1.3.

## V.1.3 ACCELERATOR PERFORMANCE

As described earlier, accelerated proton power is the most important measure of the performance of a spallation source. Repetition rate of acceleration is another important parameter for a spallation source to be considered in combination with moderator performance. The repetition rate is usually set equivalent to power lines. Actually, the 50 Hz of ISIS and the 60 Hz of SNS are natural choices for an intensive source, and these repetition rates are sufficient for most instruments with shorter flight paths or instruments for thermal and epithermal neutrons. However, the usefulness of cold neutrons with low repetition rates in spallation sources has been established, which has led to the construction of the second target station of ISIS and the second target station of SNS with 10 Hz each.

A lower repetition rate at the same power makes engineering for accelerators and target systems more difficult. However, this should be determined and optimized to

**Table V.1.1** Performance of Spallation Sources in the World

| Country/region | USA | USA | USA | UK | Switzerland | China | Europe | Japan | Japan |
|---|---|---|---|---|---|---|---|---|---|
| Neutron source | IPNS | LANSCE | SNS | ISIS | SINQ | CSNS | ESS | KENS | JSNS |
| Proton energy (MeV)/current (μA) | 450/15 | 800/70 | 1000/1400 | 800/200 | 590/1500 | 1600 | 1333/7500 | 500/9 | 3000/333 |
| Proton beam power | 7 kW | 56 kW | 1.4 MW | 160 kW | 1 MW | 0.1 MW | 5 MW/5 MW | 4.5 kW | 1 MW |
| Repetition rate (Hz) | 30 | 20 | 60 | 50 | DC | 25 | 10/50 | 20 | 25 |
| Target material | Depleted uranium | Tungsten | Mercury | Tantalum | Zircaloy | Tungsten | Mercury | Tungsten | Mercury |
| Moderator | S-$CH_4$/L-$CH_4$ | S-$CH_4$/$H_2O$ | L-$H_2$/$H_2O$ | L-$H_2$/L-$CH_4$/$H_2O$ | L-$D_2$/$D_2O$ | $H_2O$/L-$CH_4$/$H_2$ | L-$H_2$/S-$CH_4$ | S-$CH_4$/$H_2O$ | L-$H_2$ |
| Time-averaged neutron flux | $5 \times 10^{15}$ | $6.7 \times 10^{15}$ | $1.8 \times 10^{17}$ | $1.8 \times 10^{16}$ | $1.25 \times 10^{17}$ | $1.5 \times 10^{16}$ | $6.3 \times 10^{17}$ | $5 \times 10^{14}$ | $1.25 \times 10^{17}$ |
| Number of instruments | 12 | 7 | 24 (beam ports) | 17 | 15 | | 40 | 15 | 23 (beam ports) |
| Working since | 1981–2006 | 1983 | 2006 | 1985/2007 | 1996 | Under planning | 2015 | 1980–2005 | 2008 |

IPNS: Intense Pulsed Neutron Source at the Argonne National Laboratory; LANSCE: Los Alamos Neutron Science Center; SNS: Spallation Neutron Source at the Oak Ridge National Laboratory; SINQ: Swiss Spallation Neutron Source at the Paul Scherrer Institute; CSNS: Chinese Spallation Neutron Source; ESS: European Spallation Source; KENS: Koh-Energy-ken Neutron Source at the High Energy Accelerator Research Organization; JSNS: Japanese Spallation Neutron Source at J-PARC.

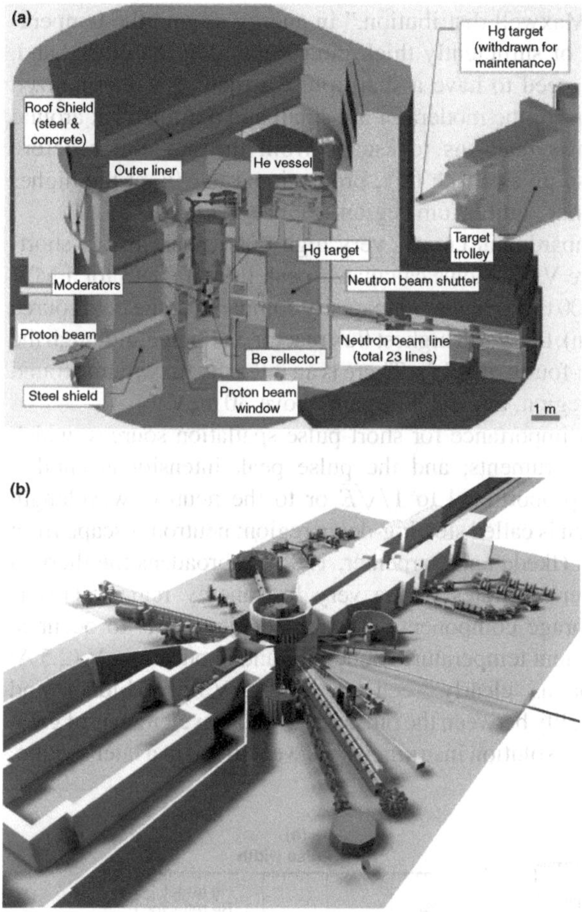


**Figure V.1.3** A typical spallation neutron source, Japan Spallation Neutron Source facility (JSNS, J-PARC), Japan Atomic Energy Agency. (a) Target station; (b) experimental hall.

purposes of the source. For a long flight path with slow repetition rate of source, a coupled moderator can realize compatibility in high resolution and high flux.

## V.1.4 USEFUL NEUTRONS FOR CONDENSED MATTER SCIENCE

For neutron scattering experiments for condensed matter research, it is necessary to reduce the energy of neutrons, in moderators, to less than 100 meV or 1 Å, comparable to atomic spacing. Moderators contain hydrogenous materials such as water/heavy water, methane, or hydrogen, which have large cross sections and light masses, so that neutrons are strongly scattered and effectively lose their energies. Neutrons in moderators reach thermal equilibrium after a large number of scatterings and have a thermal equilibrium, called "Maxwell distribution," in energy around the temperature of the moderator in case of sufficiently thick moderators. On the other hand, short-pulse spallation sources need to have a sharp pulse structure of thermalized neutrons; hence, the dimensions of the moderator are small and optimized at around $10 \times 10 \times 10\,cm^3$, which allows neutrons to escape from the moderator before equilibrium. This situation also gives a rich flux, proportional to $1/E$, at the higher energy region above the thermal equilibrium region (Figure V.1.4).

Instantaneous proton bombardment creates very high pulse peak flux in short-pulse spallation sources. Figure V.1.4 shows calculated peak flux of JSNS for 1 MW proton power. It can be about 100 times higher in intensity compared to the cold source in ILL (Institute Laue-Langevin), the world highest flux reactor, in France, although the time-averaged flux is about one-fourth of ILL's. There is an intensity maximum around 12 meV, thermal equilibrium region, and a $1/E$ region above 30 meV.

The pulse width is of key importance for short-pulse spallation sources, which determine performances of instruments, and the pulse peak intensity is another measure. The pulse width is proportional to $1/\sqrt{E}$ or to the neutron wavelength $\lambda$ in the $1/E$ region of flux, which is called slowing-down region: neutrons escape from moderators before equilibrium (Ikeda and Carpenter, 1985). It broadens the thermal equilibrium region, which then saturates in the very low-energy region. This is composed of the so-called storage components. The broadening starts to occur at about 300 meV (0.5 Å) for ambient temperature moderator and about 15 meV (2.5 Å) for 20K methane moderator as clearly seen in Figure V.1.5 (Mildner and Sinclair, 1979). The proportionality between the pulse width and wavelength is of great importance especially for high-resolution instruments, as we will see in a later section.

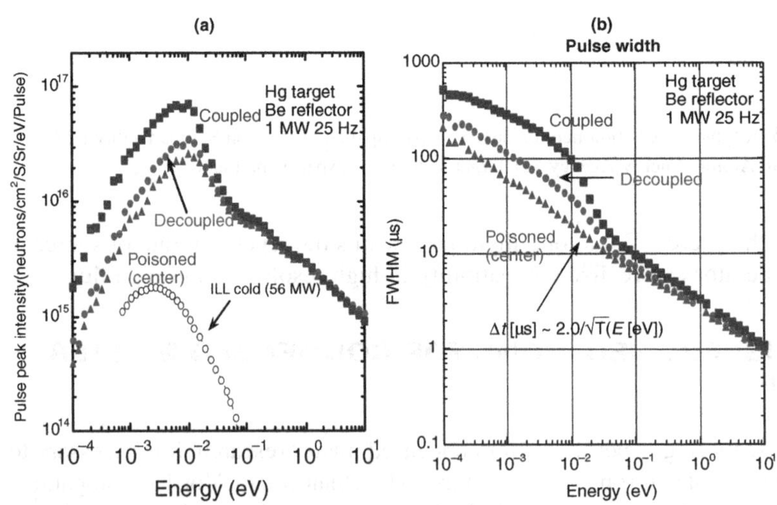


**Figure V.1.4** Calculated moderator performances of JSNS. (a) Pulse peak intensity of CM, DM, and PM in comparison with the cold source of ILL (Grenoble). (b) Pulse width of each moderator. The flux consists of a thermal equilibrium at low energies around 12 meV and a $1/E$ region at the higher energies (http://j-parc.jp/MatLife/en/instrumentation/data/Pulse_paper.pdf).

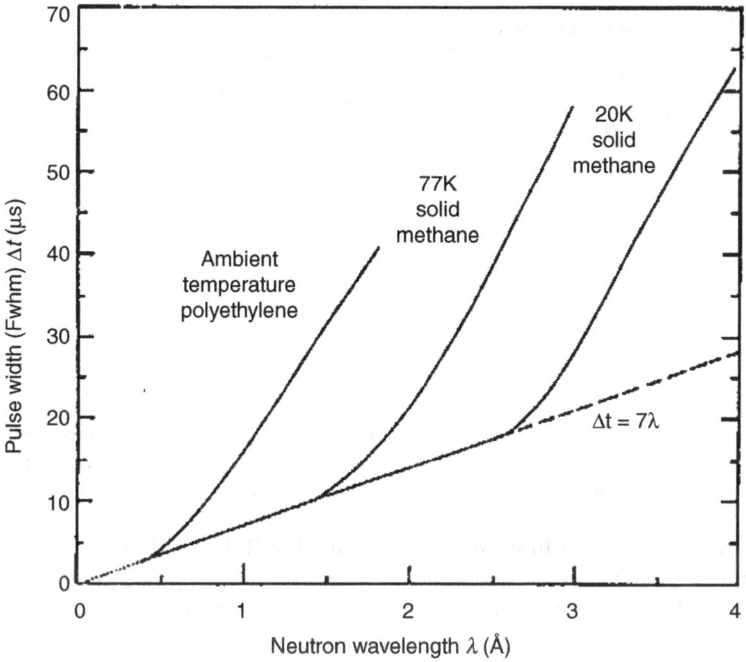


**Figure V.1.5** Pulse widths for typical moderators. In the $1/E$ region for each moderator, the pulse width $\Delta t$ (μs) $\sim 2/\sqrt{E\,(\mathrm{eV})} \sim 7\lambda$ (Å) (Mildner and Sinclair, 1979).

## V.1.5 CHOICE OF PERFORMANCES IN SPALLATION SOURCES

Design of instruments depends on the scientific and user demands. This is also true in designing source characters. Especially, characteristics of moderators dominate the performance of instruments, resolution, intensity, dynamic range, and so on. Hence, making a conceptual design of moderators is an important process because the moderator is a key component in spallation sources.

Temperature, neutronics structure, and materials are the key parameters to characterize moderators. Low temperature can naturally enhance cold neutron flux in spite of sacrificing thermal neutron flux. But more importantly, this concept can extend the slowing-down region to the lower energy and make sharp pulse structures in the extended energy range.

In the recent advanced neutronics technology, there are three kinds of moderator concepts: (1) coupled moderator (CM), for high flux while sacrificing peak width; (2) decoupled moderator (DM), offering good resolution; and (3) poisoned decoupled moderator (PM), which gives very high resolution with very sharp peak structure but sacrifices intensity.

The coupled moderator is a hybridized moderator, in which a cold moderator is surrounded by an ambient moderator layer. This concept is a great invention of Watanabe and Kiyanagi, which largely enhances neutron flux from a cold moderator

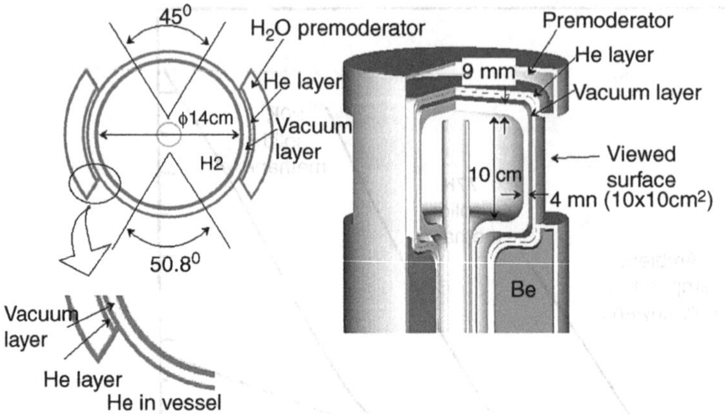


**Figure V.1.6** Design of a coupled moderator for JSNS, J-PARC. The hydrogen moderator is surrounded by an ambient water premoderator, which reduces neutron energy in the first stage and removes heat load before neutrons enter the hydrogen moderator (Teshigawara et al., 2005).

without serious heat deposition in it (Watanabe et al., 1988). Recently constructed spallation sources have benefited from this moderator and produce high flux (Figure V.1.6). A decoupled moderator has a neutron decoupling layer around moderator neutronically separating it from reflector and other components of target reflector assembly. A decoupler (absorber), decoupling energy typically 1 eV, cuts off neutrons from reflector after full thermalization and eventually sharpens the time structure of neutron pulses. To enhance the sharpness of the peak structure, an absorbing plate is inserted in the middle of the moderator, the so-called poisoned moderator. This type of moderator is used for extremely high-resolution instruments, such as a high-resolution powder diffractometer. By decoupling or poisoning, intensities can be considerably sacrificed as shown in Figure V.1.7.

## V.1.6 HIGH-ENERGY BACKGROUND

After the spallation reaction, nuclear evaporation is the major process to produce neutrons, which gives a maximum flux at around 2 MeV. After thermalization in moderators, a part of neutrons lose energy; however, still a majority of neutrons from moderators have energies above 1 eV as shown in Figure V.1.8. The neutrons above keV range can easily pass through steel shields and make backgrounds after

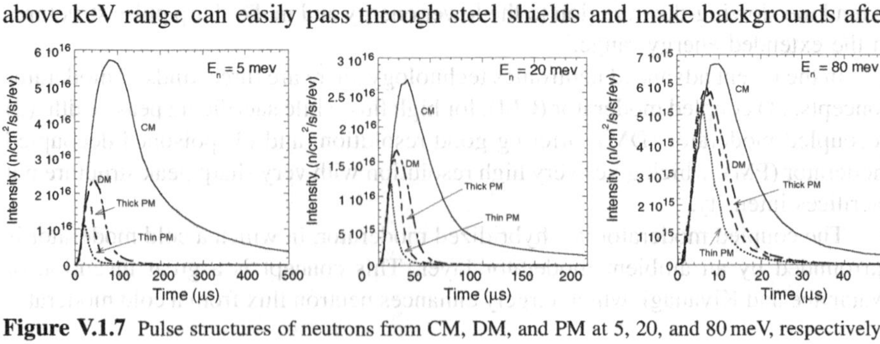


**Figure V.1.7** Pulse structures of neutrons from CM, DM, and PM at 5, 20, and 80 meV, respectively.

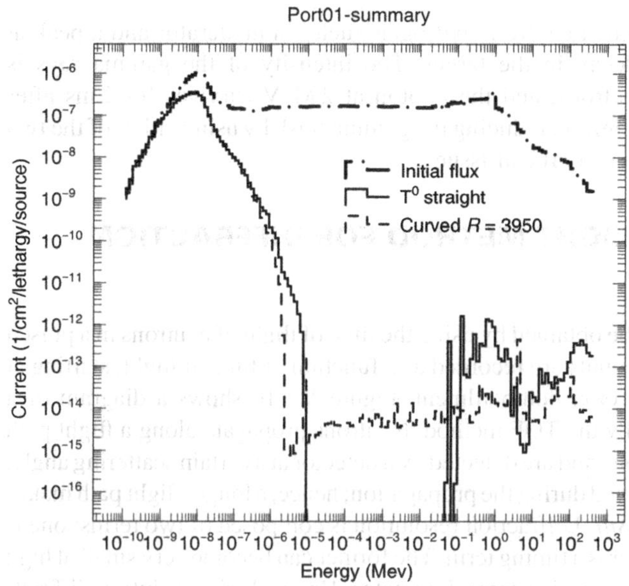


**Figure V.1.8** Neutron intensity at a beam port of JSNS. The number of neutrons as a function of energy and background suppression by either a curved guide or a $T_0$ chopper is clearly shown (Niita et al., 2005).

thermalization in instruments. There are two methods to dramatically reduce high-energy backgrounds. If an instrument has a long flight path and energy bandwidth could be limited, then curved guides are introduced. If this is not the case, a background suppression chopper (the so-called $T_0$ chopper) can be an alternative. Figure V.1.8 shows that these two methods can effectively suppress backgrounds (Niita et al., 2005).

Recent neutronics calculation can also give detailed information on photons, gamma rays, and so on (http://j-parc.jp/MatLife/en/instrumentation/data/Photon_paper.pdf). Figure V.1.9a shows gamma spectrum. A peak at around 2 MeV is caused

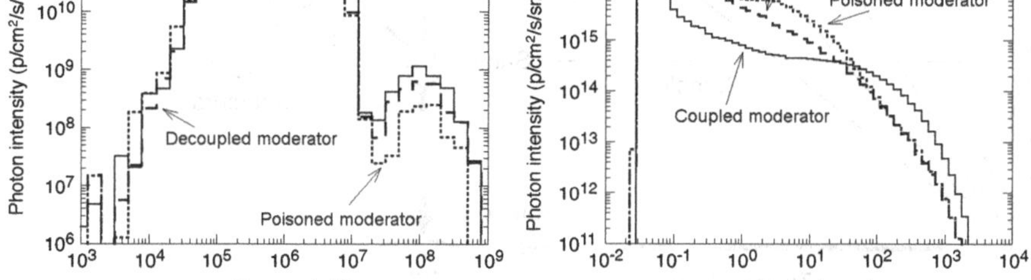


**Figure V.1.9** Gamma spectrum from JSNS moderators (a) and the time structure (b) (http://j-parc.jp/MatLife/en/instrumentation/data/Photon_paper.pdf).

by neutron-captured gamma rays from hydrogen nuclei in moderator and a peak at 100 MeV is due to $\pi_0$ decay in the target. The intensity of the gamma rays is comparable to that of neutrons, and the photon at 2 MeV survives for 2 ms after proton bombardment. Therefore, reducing the gamma flush by using either of the two methods described above is a crucial issue.

## V.1.7 TIME-OF-FLIGHT METHOD FOR DIFFRACTION MEASUREMENTS

The neutron energies can be obtained by using the time of flight of neutrons in a pulsed neutron source. Neutron counts are recorded as a function of time of flight, starting at neutron emission upon proton bombardment. Figure V.1.10 shows a diagram of a diffraction measurement by the TOF method. Neutrons propagate along a flight path $L_1$, are scattered by a sample, and are detected by a detector at a certain scattering angle. The pulse width is maintained during the propagation; hence, a longer flight path makes a higher time resolution, $\Delta t/t$. Diffraction resolution is composed of two terms: one is an angular term and another is a timing term. The former can become very small at high scattering angles, where the timing term dominates. Hence, high-resolution diffractometers eventually prefer a long flight path.

$$\left(\frac{\Delta d}{d}\right)^2 = \left(\frac{\Delta t}{t}\right)^2 + (\Delta\theta \cot(\theta))^2.$$

If one wants to have a 0.1% resolution diffractometer at 1 Å at backscattering position, where angular term can be neglected, $L_1$ should be about 30 m. By taking 7 μs in the time width and travel time of 7000 μs for 1 Å neutron (Figure V.1.5), we can compute as follows: 0.1% = 7 μs/7000 μs. Now supermirror neutron guides lead neutrons to long distances without considerable losses. Therefore, it is a practical way to elongate the flight path to realize high resolution.

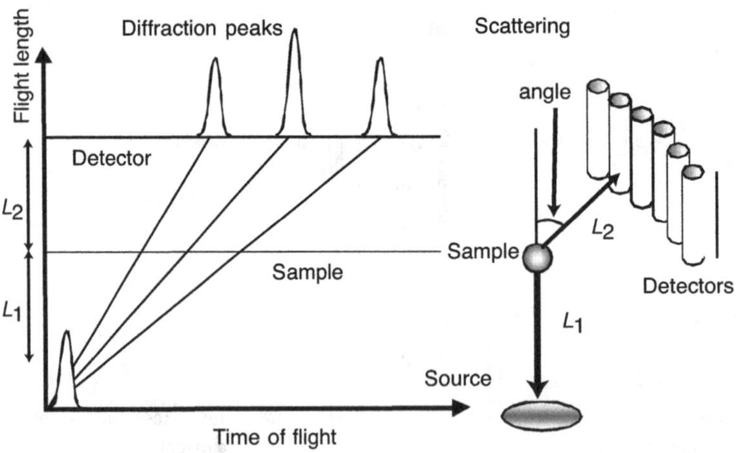


**Figure V.1.10** Flight length/time-of-flight diagram for a diffraction measurement from a crystalline material. Bragg reflections occur at a certain angle when wavelengths match with the Bragg condition.

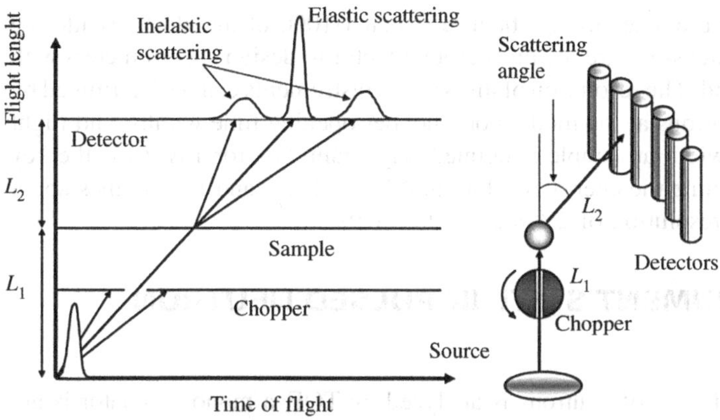


**Figure V.1.11** Flight length/time-of-flight diagram for a typical chopper spectrometer.

## V.1.8 TIME-OF-FLIGHT METHOD FOR INELASTIC SCATTERING MEASUREMENTS

Because of the relation between wavelength and energy of neutrons, thermal neutron, say 1 Å, has a suitable energy, 80 meV, to study dynamic properties of materials, such as phonon, spin wave, and so on, by inelastic scattering. One of the typical inelastic neutron scattering instruments is a chopper instrument. A fast rotating chopper is phased with proton bombardment, $T_0$, and monochromates neutrons. Scattered neutrons at the sample are detected by detectors, where the energy is analyzed by time of flight. Hence, together with angular information on momentum transfers at detector, two-dimensional data in a momentum–energy space can be easily obtained (Figure V.1.11). Figure V.1.12 shows a typical example of magnetic dynamic structure factor from one-dimensional antiferromagnet $CuGeO_3$ (Arai et al., 1996). The results established the effectiveness of pulsed neutron scattering.

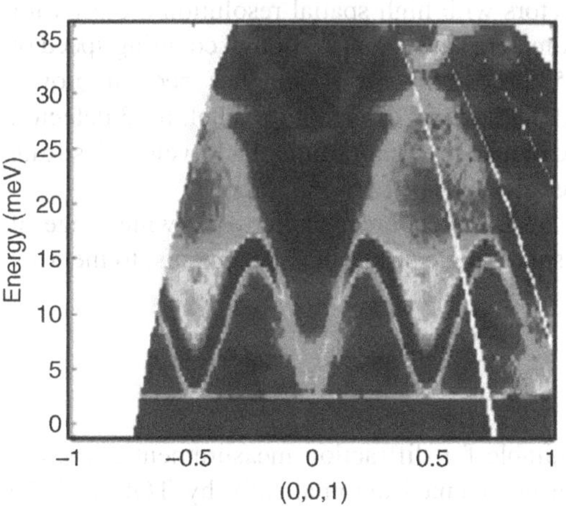


**Figure V.1.12** A typical example of magnetic dynamic structure factor for one-dimensional antiferromagnet $CuGeO_3$.

Inelastic scattering cross section is about less than 1/1000 of the elastic scattering intensity. Hence, intensities are the most crucial factor to design spectrometers with very low background. The resolution of this kind of instruments can be determined by pulse widths of neutrons at the moderator, chopper opening time widths, and flight paths. If the time width at sample is defined, say again 7 μs for 1 Å, $L_2$ is the key parameter to determine the energy resolution, $\Delta E/E = 2\Delta t/t$, and $L_2$ becomes about 3 m for an energy resolution of $2\% = 2 \times 7\ \mu s/700\ \mu s$.

## V.1.9 INSTRUMENT SUITE IN PULSED NEUTRON SOURCES

Since energy/wavelength of neutrons is analyzed by TOF, a monochromator is not needed for diffraction measurements, and detectors can be fixed at a certain angle to scan momentum transfer (Figure V.1.10). There is no movement of detectors, so scattering angles surrounding the sample are zero; therefore, they can be covered by large-area detectors to increase detecting efficiency as much as one can afford to have. This situation also provides valuable opportunities for experiments in extreme conditions, such as diffractions with a high-pressure cell, stress analysis measurements from a small gauge volume, and so on, where outgoing paths are very limited for scattered neutrons to pass through. It is also very useful for studying reflection from free surface of liquids and maintaining fixed values of scattering angle and scattering volume, momentum transfers ($Q$) can be automatically scanned in TOF. In addition, off-specular spectrum can be obtained by installing position-sensitive detectors.

These situations are also true for inelastic scattering instruments as seen in Figure V.1.11, typically chopper instruments. Although incoming neutrons should be monochromated by a chopper, scattering neutron energies are analyzed by TOF. Large detector coverage gives a wide range of information in the reciprocal space.

Nowadays, pixelized detectors are mainly used. An array of one-dimensional position-sensitive helium-3 detectors is mainly used to cover very wide solid angles with medium spatial resolution in chopper and powder instruments. One-dimensional or two-dimensional scintillation detectors with high spatial resolution are used for single-crystal diffractometers, reflectometers, and so on. Although counting speed of detectors is still about less than 25 kHz, spatial resolution has been improved significantly: one-third of the diameter in the axial direction for a helium-3 detector, for instance, 4 mm for a half inch detector, and less than 1 mm for a wavelength sifting fiber scintillation detector (Katagiri et al., 2004).

By choosing proper moderators, spallation sources can cover a wide range of instruments from diffractometers, to small-angle scattering instruments, to inelastic scattering instruments.

### V.1.9.1 Diffractometers

Pulsed spallation sources are very suitable for diffraction measurements, where a wide range of momentum space can be scanned automatically by TOF, and the performances can be easily controlled by the pulse time structure and flight path length of instruments. As described in Section V.1.7, a 100 m flight path gives a resolution of 0.03% (Kamiyama and Oikawa, 2003) and a 30 m flight path gives a resolution of 0.1%. However, a coupled moderator can provide much higher intensity at required resolution with a long flight path (Tanaka et al., 2005).

### V.1.9.2 Idea of Wide *Q* in SANS and Total Scattering Instruments

It is a natural consequence to have extended *Q*-range for liquid and amorphous diffractometers, where Fourier transformation is normally used for data analysis and needs extended *Q*-range for detailed atomic structural information (Figure V.1.13). Spallation neutron sources give a very wide range of neutron energies even above 1 eV, which can realize observations in a high-*Q* region up to 100 $Å^{-1}$. A cold neutron source extends the *Q*-range to a smaller region and widens the available wavelength band. By having detector array not only at high angles but also at small angles, the covering *Q*-ranges can be easily extended from 0.01 to 50 $Å^{-1}$ for all scattering instruments. An innovative design of this kind of instruments was first introduced in GEM, ISIS (Day et al., 2004), which has very high intensity and an extended *Q*-range. SANS instruments can also use this approach with enhancement in the small-angle regions, which can give structural information of macromolecules including atomic scale structures (Shinohara et al., 2009).

### V.1.9.3 Inelastic Scattering Instruments

Chopper instruments are very suitable for spallation sources and are typical inelastic scattering instruments in spallation sources. Development of this sort of instruments started at the Argonne National Laboratory in the 1970s (Price and Carpenter, 1987); the usefulness and outstanding efficiency of this kind of instruments were established in the ISIS with stimulative demonstrations by MARI and MAPS instruments (Arai

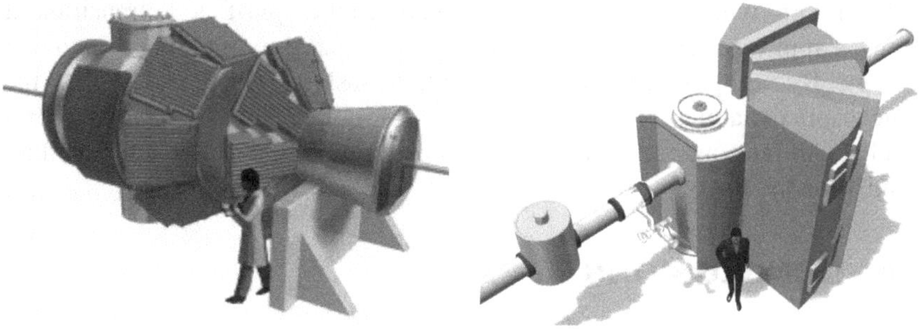

**Figure V.1.13** Images of a liquid–amorphous diffractometer and a chopper instrument. Scattering angles are covered by detectors from small angles to high angles (Otomo and Suzuya, 2008; Kajimoto et al., 2007).

et al., 1992; Perring and Frost, 2004). Large-area pixelized detectors showed broad quantum magnetic excitations, which had never been clearly observed previously (Figure V.1.12) (Arai et al., 1996).

Backscattering instruments giving a μeV energy resolution, the so-called inverted geometry instruments, are also getting popular, whose technologies have been fostered in reactor sources so far. Instead of having a long flight path, introducing a pulse shaping chopper can realize a high resolution while keeping high intensity at a shorter flight path (Takahashi et al., 2007).

In Table V.1.2, instruments in JSNS, SNS, and ISIS are summarized. In spite of different proton power, frequency, and moderator performances of the three facilities, constructed or planned instrument suites are very similar to each other. This is because instrumentation plans are led by scientific trends and user demands, which are a global common feature.

### V.1.9.4 Innovative Methods in the Third-Generation Spallation Sources

The third-generation spallation sources such as ISIS, SNS, and JSNS have made a great improvement in the accelerator performances. The power of the accelerators is nearly 10–300 times higher than that of the facilities built in the 1980s. Target moderator reflector assemblies and instrumentation have also been improved to a great extent. Therefore, effectiveness of scattering experiments has improved more than 1000 times compared to that in the past. One of such improvements has been recently demonstrated by applying a multi-$E_i$ method (repetition rate multiplication, RRM) by a chopper instrument (Russina and Mezei, 2009; Nakamura et al., 2009), which was first proposed by F. Mezei in 1997 (Mezei, 1997).

In the conventional inelastic neutron scattering, only single incident energy has been used as described above. However, instead of the single-opening chopper, if one uses a multi-opening chopper as shown in Figure V.1.14, multi-incident energies can be utilized in one measurement.

RRM was applied to measure a scattering function of a quasi-one-dimensional antiferromagnet $CuGeO_3$ (Nakamura et al., 2009). This material contains "hierarchical spin dynamics": a spin gap formation by a spin-Peierls transition at low energy, a spin wave excitation at medium energy, and a continuum excitation at high energy. By using RRM, one can simultaneously obtain each part of the hierarchical spin dynamics as shown in Figure V.1.15. The observation was successful not only by the cutting-edge hardware components but also by the event recording data acquisition system.

## V.1.10 CONCLUSIONS

Due to environmental, economical, technical, and efficiency reasons, spallation neutron sources are becoming major neutron sources. They can have stepwise improvements of several orders of magnitude depending on the development in

**Table V.1.2** Instruments in JSNS, SNS, and ISIS

| | | JSNS (1 MW, 25 Hz) | | SNS (1.4 MW, 60 Hz) | | ISIS TS1 (160 kW, 50 Hz) | | ISIS TS2 (48 kW, 10 Hz) | |
|---|---|---|---|---|---|---|---|---|---|
| Elastic scattering instruments | Single-crystal diffraction | **iBIX** | **Protein** | **MaNDi** | **Protein** | | | **Lmx** | **Protein** |
| | | | High-resolution protein | | | | | | |
| | | | Versatile crystal | **TOPAZ** | **Versatile crystal** | **SXD** | **Single crystal** | | |
| | | | | **CORRELI** | **Diffuse scattering** | | | | |
| | Powder diffraction | **S-HRPD** | **High-resolution powder** | **POWGEN** | **High-resolution powder** | **HRPD** | **High resolution** | **WISH** | **Special environmental** |
| | | **TAKUMI** | **Stress analysis** | **VULCAN** | **Stress analysis** | **ENGIN-X** | **Stress analysis** | | |
| | | **iMATERIA** | **Versatile** | | | **POLARIS** | **Medium resolution** | | |
| | | | Special environment | | | **INES** | **Versatile** | | |
| | | **PLANET** | **High pressure** | **SNAP** | **High pressure** | **PEARL** | **High pressure** | **Exeed** | **High pressure** |
| | | | | | | **ROTAX** | **Wide band** | | |
| | Total scattering | **NOVA** | **High intensity** | **NOMAD** | $S(Q)$ | **GEM** | **High intensity $S(Q)$** | **Nimrod** | $S(Q)$ **(wide$Q$)** |
| | | | | | | **SANDALS** | **Liquid $S(Q)$** | | |

*(continued)*

**Table V.1.2** (*Continued*)

| | | JSNS (1 MW, 25 Hz) | | SNS (1.4 MW, 60 Hz) | | ISIS TS1 (160 kW, 50 Hz) | | ISIS TS2 (48 kW, 10 Hz) | |
|---|---|---|---|---|---|---|---|---|---|
| | SANS | **TAIKAN** | **SANS** | **EQ-SANS** | **SANS** | **LOQ** | **SANS** | **Sans-2d** | **SANS** |
| | | | Microfocusing SANS | **TOF-USANS** | **USANS** | | | Bounce | USANS |
| | | | | | | | | Spiral | SESANS |
| | | | | | | | | **Zoom** | **Focusing SANS** |
| | Reflecto-meters | **ARISA-II** | **Horizontal** | **LR** | **Horizontal** | **SURF** | **Horizontal** | **Inter** | **Horizontal** |
| | | | Vertical | **MR** | **Vertical** | **CRISP** | **Horizontal polarization** | **Polref Polarization** | |
| | | | | | | | | **Offspec** | **Spin echo** |
| Inelastic scattering instruments | Direct geometry | **4SEASONS** | **High intensity** | **ARCS** | **High intensity** | **Merlin** | **High intensity** | | |
| | | **HRC** | **High resolution** | **SEQUOIA** | **High resolution** | **MAPS** | **High resolution** | | |
| | | **AMATERAS** | **Cold neutron disk** | **CNCS** | **Cold neutron disk** | **HET** | **Chopper** | **Let** | **Cold neutron disk** |
| | | | | **HYSPEC** | **Crystal–Chopper** | **MARI** | **Chopper (wide angular)** | Exess | Extreme condition |
| | Inverted | **DNA** | **Backscattering** | **BASIS** | **Backscattering** | **IRIS** | **Backscattering** | Fires | Backscattering |
| | Geometry | | | | | **OSIRIS** | **Backscattering (polarization)** | | |
| | | | | **VISION** | **Molecular spectrometer** | **TOSCA** | **Molecular spectrometer** | | |
| | | | | | | **PRISMA** | **Versatile crystal analyzer** | | |
| | | | | | | **VESUVIO** | **eV spectroscopy** | | |
| | Spin echo | VINE-ROSE | Spin echo (MIEZE) | **NSE** | **Spin echo (MEZEI)** | | | NeSsiE | Spin Echo (MEZEI) |

| | | | | | | | | | |
|---|---|---|---|---|---|---|---|---|---|
| Fundamental physics, nuclear engineering, test port | | **NNRI** | **Neutron cross section** | **FNPB** | **Fundamental physics** | **ALF** | **Crystal alignment** | **ChipIR** | **Electronics irradiation** |
| | | **NOP** | **Fundamental physics** | | | | | **Larmor** | **Polarized neutrons** |
| | | **NOBORU** | **Test port** | | | | | | |
| Imaging instruments | | | Energy selective | | Energy selective | | | **Imat** | **Imaging** |

Bold denotes those in operation/constructed. The simple tests are in planning.

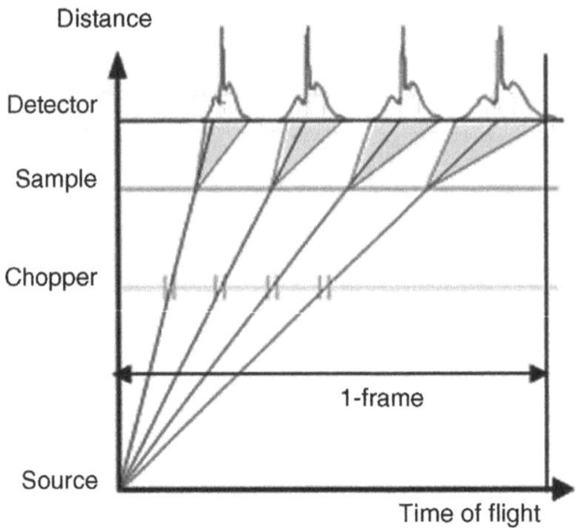


**Figure V.1.14** TOF–distance diagram. Multi-opening chopper lets neutrons pass through it if their speed (energy) matches with the opening times of the chopper.

accelerator, target, and instrumentation technologies. In the early 1980s, pulsed neutron scattering was not considered suitable for studies on single crystals. Reactor sources were considered much adequate to focus on spots in the momentum–energy space for detailed measurements. However, megawatt class spallation sources may change such a preoccupation of the past. However, time-averaged flux of a spallation source may not be always superior to that of a reactor even in the future, and complementarity of two kinds of sources should be mutually enhanced by sharing their roles. At present, there are three regional spallation neutron centers in the world: SNS in the American continents, ISIS and ESS in the European region, and JSNS, J-PARC in the Asia–Oceania region. They will serve as a neutron platform in those regions, contribute to relevant scientific developments, promote human exchange, and encourage worldwide mutual collaborations. The costs of facilities can easily exceed affordability of one country in the future, if more than 1 MW power is needed. Therefore, a worldwide participation in such a project, like ESS, will be indispensable to have sustainable developments in the neutron science fields.

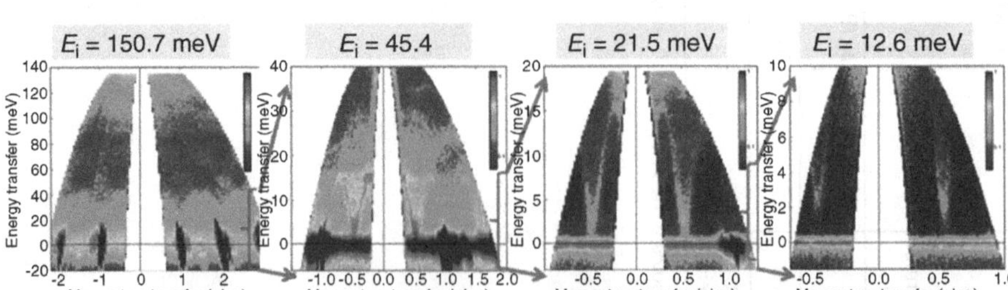


**Figure V.1.15** Plots showing the data taken from four different incident energies of neutron beam between 12.6 and 150.7 meV in one measurement.

## ACKNOWLEDGEMENTS

I acknowledge the staff of the Materials Life Science Facility (MLF) of the J-PARC Center for their indispensable efforts and support to construct JSNS, J-PARC by spending more than 10 years. I also acknowledge members of the International Collaboration on Advanced Neutron Sources (ICANS) for their sustained advice and collaboration to develop spallation sources worldwide.


## REFERENCES


Arai, M. Taylor, A.D., Bennington, S.M., and Bowden, Z.A.In: Howells, W.S.and Soper, A.K. (Eds.), *Recent Developments in the Physics of Fluids*, Adam Hilger, 1992, pp. 321–328.

Arai, M., Fujita, M., Motokawa, M., Akimitsu, J., and Bennington, S.M. *Phys. Rev. Lett.* **1996**, *77*, 3649.

Arai, M., Kiyanagi, Y., Watanabe, N., Takagi, R., Shibazaki, H., Numajiri, M., Itoh, S., Otomo, T., Furusaka, M., Inamura, Y., Ogawa, Y., Suda, Y., and Satoh, S. *J. Neutron Res.* **1999**, *8*, 71.

Carpenter, J.M., Gabriel, T.A., Iverson, E.B., and Jerng, D.W. *Physica B* **1999**, *270*, 272.

Day, P., Enderby, J.E., Williams, W.G., Chapon, L.C., Hannon, A.C., Radaelli, P.G., and Soper, A. K. *Neutron News* **2004**, *15*, 19.

Ikeda, S. and Carpenter, J. *Nucl. Instrum. Methods Phys. Res. A* **1985**, *239*, 536.

Kajimoto, R., Yokoo, T., Nakajima, K., Nakamura, M., Soyama, K., Ino, T., Shamoto, S., Fujita, M., Ohoyama, K., Hiraka, H., Yamada, K., and Arai, M. *J. Neutron Res.* **2007**, *15*, 5.

Kamiyama, T. and Oikawa, K. *Proceedings of the 16th Meeting of the International Collaboration on Advanced Neutron Sources (ICANS-XVII)*, Dusseldorf-Neuss, Germany, 2003, pp. 309–314.

Katagiri, M., Matsubayashi, M., Sakasai, K., Nakamura, T., Ebine, M., Birumachi, A., and Rhodes. N. *Nucl. Instrum. Methods Phys. Res. A* **2004**, *529*, 313.

Mason, T.E., Arai, M., and Clausen, K.N. *MRS Bull.* 2003, 923–928.

Mezei, F. *Proceedings of the 12th Meeting of International Collaboration on Advanced Neutron Sources (ICANS-XII)*, Abingdon, UK, 1993, pp. 377–384.

Mezei, F. *J. Neutron Res.* **1997**, *6*, 3.

Mildner, D.F.R. and Sinclair, R.N. *J. Nucl. Energy* **1979**, *6*, 225.

Nakamura, M., Kajimoto, R., Inamura, Y., Mizuno, F., Fujita, M., Yokoo, T., and Arai, M. *J. Phys. Soc. Jpn.* **2009**, *78*, 093002.

Niita, K., Suzuya, K., Nakajima, K., Kajimoto, R., Nakamura, M., Shibata, K., Soyama, K., Torii, S., Harjo, S., Aizawa, K., Kawai, M., Kamiyama, T., Itoh, S., Torikai, N., Maekawa, F., Oikawa, K., Tamura, M., Harada, M., and Arai, M. *Proceedings of the 17th Meeting of the International Collaboration on Advanced Neutron Sources (ICANS-XVII)*, Santa Fe, NM, 2005, pp. 640–647.

Otomo, T. and Suzuya, K. *J. Crystallogr Soc. Jpn.* **2008**, *50*, 13 (in Japanese).

Perring, T.G. and Frost, C.D. *Neutron News* **2004**, *15*, 30.

Price, D.L. and Carpenter, J.M. *J. Non-Cryst. Solids* **1987**, *92*, 153.

Russina, M. and Mezei, F. *Nucl. Instrum. Methods Phys. Res. A* **2009**, *604*, 624.

Shinohara, T., Takata, S., Suzuki, J., Oku, T., Suzuya, K., Aizawa, K., Arai, M., Otomo, T., and Sugiyama, M. *Nucl. Instrum. Methods Phys. Res. A* **2009**, *600*, 111.

Takahashi, N., Shibata, K., Sato, T.J., Tamua, I., Kajimoto, R., Harjo, S., Oikawa, K., Arai, M., and Mezei, F. *J. Phys. Chem. Solids* **2007**, *68*, 2199.

Tanaka, I., Niimura, N., Ozeki, T., Ohara, T., Kurihara, K., Kusaka, K., Morii, Y., Aizawa, K., Arai, M., Kasao, T., Ebata, K., and Takano, Y. *Proceedings of the 17th Meeting of the International Collaboration on Advanced Neutron Sources (ICANS-XVII)*, Santa Fe, NM, 2005, pp. 937–945.

Teshigawara, M., Oikawa, K., Harada, M., Maekawa, F., Kato, T., Watanabe, N., Ikeda, Y., Kikuchi, K., Oda, Y., and Hirota, T. *Proceedings of the 17th Meeting of the International Collaboration on Advanced Neutron Sources (ICANS-XVII)*, Santa Fe, NM, 2005, pp. 365–373.

Vassil'kov, R.G. and Yurevich, V.I. *Proceedings of the 11th Meeting of International Collaboration on Advanced Neutron Sources (ICANS-XI)*, Tsukuba, Japan, 1990, pp. 340–353.

Watanabe, N. *Rep. Prog. Phys.* **2003**, *66*, 339.

Watanabe, N., Kiyanagi, Y., Inoue, K., Furusaka, M., Ikeda, S., Arai, M., and Iwasa, H. *Proceedings of the 10th Meeting of International Collaboration on Advanced Neutron Sources (ICANS-X)*, Los Alamos, NM, 1988, pp. 787–797.

# V Current Facilities

## V.2 Reactor Overview

*Colin J. Carlile*

### V.2.1 INTRODUCTION

Much of the matter in the world around us consists of neutrons, bound into the nuclei of atoms. The free neutron, however, is an unstable particle with a half-life of about 10 min, which undergoes a three-body decay to a proton and an electron with an accompanying electron antineutrino. Neutrons can be produced in a variety of nuclear reactions such as the (α, n) and (γ, n) reactions, fission, photofission, spallation, and fusion. For scattering experiments aimed at understanding the structure and dynamics of the rich variety of materials, the most significant of these are fission and spallation.

Fission in a nuclear research reactor provides the most common source of slow neutrons for scattering experiments in soft condensed matter research, and there are many research reactors around the world.

Spallation of heavy atoms takes place in accelerator-based sources of neutrons. The construction of spallation sources began almost four decades after instruments for neutron scattering were first installed on research reactors. Reactors are, however, now approaching the maximum neutron fluxes that are technologically achievable, whereas pulsed spallation sources have not yet attained this limit and offer the most promising method of realizing even higher fluxes than are available at the brightest research reactors such as the Institut Laue-Langevin in Grenoble or the High Flux Isotope Reactor (HFIR) at Oak Ridge. The improvement of reactor facilities to provide quantitatively and qualitatively better measuring opportunities is now focused upon the improvement of beam delivery and the instrumentation where a lot of potential still exists.

Nevertheless, it seems likely that neutron sources will always remain relatively weak. At best, a monochromatic beam with a flux of around $10^8$ neutrons/cm$^2$/s strikes the sample in a neutron experiment: this compares with a photon flux at least three orders of magnitude higher produced by the characteristic spectral lines of a

*Neutrons in Soft Matter* Edited by Toyoko Imae, Toshiji Kanaya, Michihiro Furusaka, and Naoya Torikai

sealed-off X-ray tube. More than 10 orders of magnitude higher fluxes are provided by third-generation X-ray synchrotron sources. Nevertheless, the unique sensitivity of the neutron to low-mass atoms, in particular hydrogen, its ability to probe dynamics, its spin, allowing it to sense magnetic effects, and its neutrality, allowing it to penetrate into dense samples under extreme conditions, continue to counteract the raw intensity of synchrotron radiation sources.

Before the advent of fission reactors, the earliest scattering experiments were conducted using neutrons from radioactive sources. Such sources are still in use today as portable devices for calibrating and setting up detectors. The first ideas of using fusion as a source of slow neutrons are emerging but this must be regarded as a very long-term bet.

## V.2.2 THE DISCOVERY OF THE NEUTRON

James Chadwick modestly announced "the possible discovery of the neutron" in 1932 (Chadwick, 1932a), and he felt confident enough to confirm its existence later the same year (Chadwick, 1932b). Chadwick undertook a series of experiments in the 1920s in search of the neutron, whose existence had been predicted by Rutherford in 1920. It was known from the work of Bothe and Becker (1930) that an unusually penetrating radiation, assumed to be gamma rays, was produced by the $\alpha$-particle bombardment of beryllium, but Chadwick was not able to realize his final goal until Irène and Frédéric Joliot-Curie (Curie and Joliot, 1932) had reported that this radiation was capable of ejecting high-speed protons from hydrogenous materials such as paraffin or water. Joliot and Curie asserted that the radiation consisted of very high-frequency $\gamma$-rays, and that the protons were ejected by a process analogous to the Compton effect.

Chadwick recognized, however, that there were two serious difficulties with this $\gamma$-ray hypothesis. First, the number of protons observed by Joliot and Curie was two orders of magnitude higher than that calculated from the theoretical Klein–Nishina formula for the scattering of $\gamma$-rays by protons, and second, the energy calculated from the same formula was much less than the observed energy. He realized that both these difficulties could be overcome by assuming that the unknown radiation consisted of neutral mass one particles—neutrons.

Chadwick confirmed the neutron hypothesis after 10 days of intense activity in which he extended the measurements of Joliot and Curie using the apparatus illustrated in Figure V.2.1. A photograph of Chadwick's original chamber, in which neutrons were produced by bombarding beryllium with alpha particles, is shown in Figure V.2.2. Chadwick, after repeating the earlier experiments with paraffin wax, replaced the paraffin wax successively with a number of light elements (helium, lithium, beryllium, boron, carbon, nitrogen, oxygen, and argon) either in solid form or as a gas introduced into the ionization chamber. He found that the unknown radiation ejected particles from all these light elements, and he determined the maximum energy of these recoil particles from their range in air. By combining the measurements for hydrogen and nitrogen and assuming that the unknown radiation

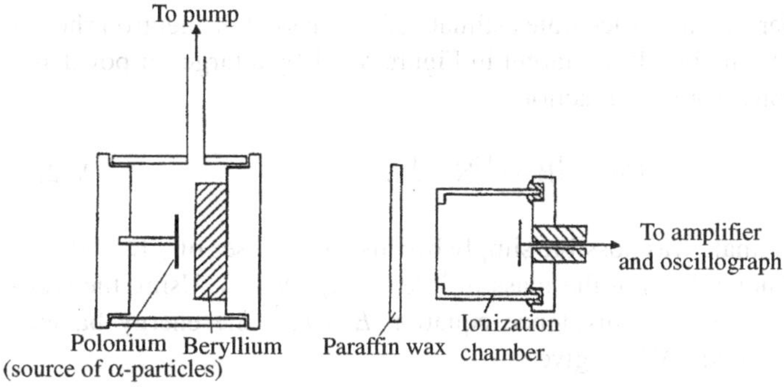


**Figure V.2.1** The apparatus used by James Chadwick in his discovery of the neutron. α-particles from a $Po^{210}$ source bombard a block of beryllium to produce neutrons by the (α, n) reaction. These neutrons then knock out protons from a sheet of paraffin wax that are counted in the ionization chamber.

consisted of neutral particles, Chadwick obtained an estimate for the mass of the neutral particle of

$$m_{\mathrm{n}} = 1.15\ \mathrm{amu},$$

with an accuracy of about 10%.

Independent but consistent values of $m_{\mathrm{n}}$ were obtained by using the energy measurements for other pairs of light elements. Chadwick concluded that the radiation consisted of neutrons ${}^{1}_{0}\mathrm{n}$ of mass one and zero charge that were ejected from the beryllium nucleus by α-particles in the following reaction.

$${}^{4}_{2}\mathrm{He} + {}^{9}_{4}\mathrm{Be} \rightarrow {}^{12}_{6}\mathrm{C} + {}^{1}_{0}\mathrm{n}. \qquad (\mathrm{V.2.1})$$

**Figure V.2.2** Chadwick's evacuated chamber in which α-particles from a $Po^{210}$ source at one end bombarded a beryllium target at the opposite end. The vertical pipe was attached to a vacuum pump. (Photograph from the Archives of the Cavendish Laboratory, Cambridge.)

He was able to derive a more accurate estimate of the mass of the neutron shortly afterward by replacing the beryllium target in Figure V.2.1 by a target of powdered boron to yield neutrons from the reaction

$$ {}^{4}_{2}\mathrm{He} + {}^{11}_{5}\mathrm{B} \rightarrow {}^{14}_{7}\mathrm{N} + {}^{1}_{0}\mathrm{n}. \tag{V.2.2}$$

The improved estimate was possible simply because the masses of ${}^{11}_{5}\mathrm{B}$ and ${}^{14}_{7}\mathrm{N}$ in eq. (V.2.2) were known but not the mass of ${}^{9}_{4}\mathrm{Be}$ in eq. (V.2.1). Using the well-known Einstein mass–energy equivalence equation $E = mc^2$, the energy balance requirements applied to eq. (V.2.2) give

$$(m_{\mathrm{He}} + m_{\mathrm{B}})c^2 + \mathrm{KE}_{\mathrm{He}} \rightarrow (m_{\mathrm{N}} + m_{\mathrm{n}})c^2 + \mathrm{KE}_{\mathrm{N}} + \mathrm{KE}_{\mathrm{n}}, \tag{V.2.3}$$

where KE denotes the kinetic energies of the different particles. The nuclear masses $m_{\mathrm{He}}$, $m_{\mathrm{B}}$, and $m_{\mathrm{N}}$ were known from mass spectrograph measurements, $\mathrm{KE}_{\mathrm{He}}$ was obtained from the range of the α-particles, $\mathrm{KE}_{\mathrm{N}}$ was calculated from the momentum conservation law, and $\mathrm{KE}_{\mathrm{n}}$ was obtained from the maximum recoil energy of the proton. The remaining unknown in eq. (V.2.1) is $m_{\mathrm{n}}$, which could therefore be evaluated as

$$m_{\mathrm{n}} = 1.0067 \pm 0.0012\ \mathrm{amu}. \tag{V.2.4}$$

This result compares remarkably well with the currently accepted value given by Eidelman (2004) of

$$m_{\mathrm{n}} = 1.008664915 \pm 0.000000001\ \mathrm{amu}. \tag{V.2.5}$$

Chadwick assumed that the neutron consists of a proton and an electron in close combination. We now know that the neutron consists of three smaller particles—quarks—that orbit the center of mass of the neutron and are held together by exchange forces.

It is interesting to note the speed of events in 1932. In January 1932, Chadwick received the paper of Joliot and Curie that had been published earlier that month in *Comptes Rendus*. From 2nd February to 12th February, Chadwick carried out his experimental measurements, and on 17th February he submitted his letter to *Nature* entitled "Possible existence of a neutron," which was published on 27th February. The full paper "The existence of a neutron" was published in the June 1932 issue of *The Proceedings of the Royal Society.* The events surrounding this period are described in the very readable book "The Neutron and the Bomb" by Andrew Brown (1997).

Early scattering experiments were carried out with a radium–beryllium neutron source, which produces a white beam of neutrons with an approximately Maxwellian distribution of velocities. For example, the wave nature of neutrons was first demonstrated by Mitchell and Powers in 1936 using a moderated radium–beryllium source in which neutrons were diffracted from single crystals of aluminum.

For quantitative studies, much more intense neutron sources were required, which were capable of providing collimated and monochromatic neutron beams. The construction of the first nuclear fission reactor in Chicago in 1942, only 10 years after the discovery of the neutron, opened up the way for the eventual development of such sources.

The reactor, referred to as a pile in those days, was built in strict secrecy inside a squash court on the campus of the University of Chicago under the leadership of Enrico Fermi. Channels were machined into 4500 graphite bricks, which were then assembled into a stack, and 5.5 tons of natural uranium metal and 36 tons of uranium oxide were inserted into the channels. The purpose of the graphite was to slow down, or moderate, the neutrons emitted by the uranium, whose fission cross section is enhanced at low neutron velocities.

Figure V.2.3 shows the neutron intensity in CP-1, as indicated by the chart recorder on that fateful afternoon of December 2, 1942, when a man-made self-sustaining fission reaction was achieved for the first time. The neutron intensity rises and then levels off after each successive neutron absorbing control rod is withdrawn, and eventually the pile achieves criticality with the neutron population doubling approximately every 2 min. The reaction came to a stop at 3.53 p.m. when the safety rods were inserted. Figure V.2.4 is a copy of an artist's illustration of the event, the original having been drawn with ink made from the graphite used in CP-1.

When the first "atomic pile" (Chicago Pile 1 or CP-1) was assembled by Enrico Fermi's team at the University of Chicago in 1942, there was every reason to believe that it was the first such nuclear fission reactor on Earth. It was subsequently discovered that Fermi's experiment had been anticipated in nature 2 billion years earlier! The extraordinary sequence of historical events leading to the natural occurrence of a nuclear reactor in a uranium deposit at Oklo in Gabon, Africa is

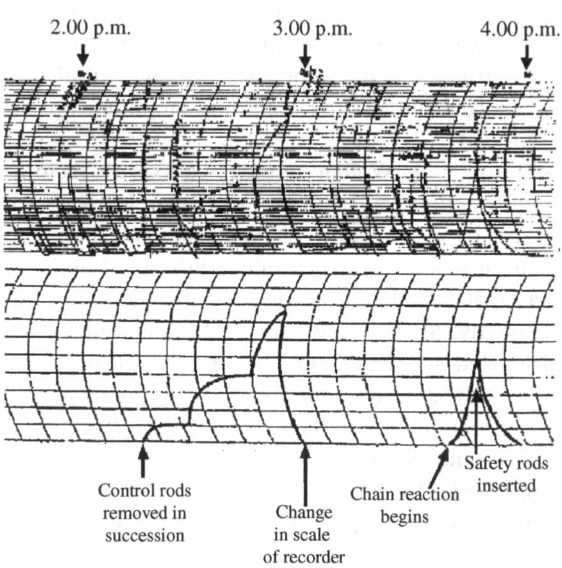


**Figure V.2.3** *Top*: Neutron intensity in Chicago Pile 1 (CP-1) as registered on a chart recorder trace taken during the afternoon of December 2, 1942. *Bottom*: A simplified representation of the recorder trace. (After Rhodes (1988).)

**Figure V.2.4** On December 2, 1942, man first achieved a self-sustaining nuclear chain reaction. This is a copy of the lithograph of the event made by Leo Vartanien. The overlaid portraits are, from left to right, Leo Szilard, Arthur Compton, Enrico Fermi, and Eugene Wigner. (Courtesy of Argonne National Laboratory.)

vividly described by Cowan (1976). Nevertheless, Fermi's experiment laid the foundation of one of the most important scientific developments of the twentieth century.

In 1943, additional reactors were constructed: a heavy-water moderated reactor CP-2 in the Argonne Forest close to Chicago (that was later to become the site of the Argonne National Laboratory), and a graphite-moderated reactor at the Oak Ridge National Laboratory (ORNL) in Tennessee, USA. The principal purpose of these reactors was to produce plutonium for the Manhattan project, but after the Second World War the first experiments were carried out with neutron beams. The neutron flux in the cores of these reactors was about $10^{12}$ neutrons/cm$^2$/s, so it was possible to carry out experiments with thermal neutrons, collimated into well-defined beams and encompassing an energy range that was reasonably narrow. These monoenergetic neutron beams were used primarily to collect neutron cross-section data for the nuclear weapons program, but some neutron beams were also available for scattering experiments. A neutron diffractometer was built by W.H. Zinn (1947) at the Argonne National Laboratory, and a similar instrument at the same laboratory was employed by Fermi and Marshall (1947) to measure the neutron scattering amplitudes of 22 elements ranging from hydrogen to lead.

At the Oak Ridge Graphite Reactor, Ernie Wollan set up a two-axis diffractometer and recorded the first neutron diffraction pattern of NaCl in 1946. ORNL soon became the pioneering research center in the world for neutron scattering. The powder diffraction technique was used to study magnetic materials, and Shull and Smart (1949) confirmed the prediction of Néel that the spins of the manganese ions in MnO are arranged antiferromagnetically. Shull and Wollan (1951) determined the magnetic structure of $Mn_3O_4$ and showed that the magnetic intensity is proportional to $\sin^2 \alpha$, where $\alpha$ is the angle between the normal to the Bragg scattering plane and the direction of magnetization. This observation was in agreement with the theories of Schwinger, Halpern, and Johnson and contradicted the theory of Bloch that predicted a variation with $\cos^2 \alpha$. Cliff Shull was awarded the Nobel Prize in 1994 together with Bertram Brockhouse who developed the techniques for neutron spectrometry.

## V.2.3 NEUTRONS FROM NUCLEAR REACTOR

Nuclear reactors consist of four basic and essential components: the fuel, the moderator, the coolant, and the shielding. The fuel comprises a fissile material, most frequently uranium, but occasionally thorium, or plutonium, either in their elemental form alloyed or sintered into a matrix with aluminum, for example, or as a compound, for example, uranium dioxide. The fissile material can be in its naturally occurring form (natural uranium or thorium), or else it can be isotopically enriched in the fissionable isotope ($^{235}$U or $^{239}$Pu ). The moderator is a material of low atomic mass, usually light water, heavy water, or graphite. The coolant is normally water but is sometimes a gas such as carbon dioxide or even helium in power reactors. Fast reactors (in which fast neutrons alone maintain the chain reaction) have no moderator and employ liquid metals, such as sodium or a sodium–potassium alloy, as coolant in order to avoid moderating the neutrons.

In a nuclear reactor fueled with uranium, thermal fission of $^{235}$U takes place with the production, on average, of 2.7 fast neutrons and the simultaneous disintegration of the uranium nucleus into two fission fragments of unequal mass (see Figure V.2.5). These fragments are predominantly rich in neutrons and are highly radioactive: they decay via a cascade of β-emissions (i.e., expulsion of electrons) to what is called the line of β-stability.

The decay chain of a single fission fragment tin-138 is illustrated in Figure V.2.6. Its sister fission fragment is molybdenum-103 whose decay chain is not shown. Five sequential decays lead, after only a few days, to the isotope xenon-131, which is stable. Another isotope of xenon—xenon-135—also results from fission fragment decay, which has a high neutron absorption cross section and is a serious reactor poison.

The line of β-stability is obtained by plotting the number of neutrons versus the atomic number for the naturally occurring stable isotopes, as shown in Figure V.2.7.

The fission reaction becomes self-supporting provided that, from this yield of 2.7 neutrons per fission, one neutron causes further fission in another $^{235}$U nucleus, producing a self-sustaining or critical chain reaction. The remaining 1.7 neutrons either are absorbed in components of the reactor or escape from the surface of the

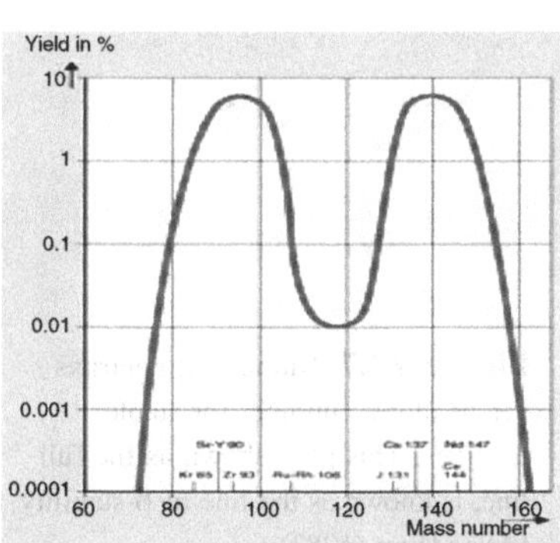


**Figure V.2.5** The idealized fission fragment distribution from the fission of uranium-235. The double-peaked distribution is characteristic of all fission reactions. Fission fragments are highly neutron rich, being created far from the line of β-stability where the stable isotopes sit. Fission fragments therefore decay by consecutive β-decays to that line, and are the source of the significant radioactivity of spent nuclear fuel. (Courtesy of the European Nuclear Society.)

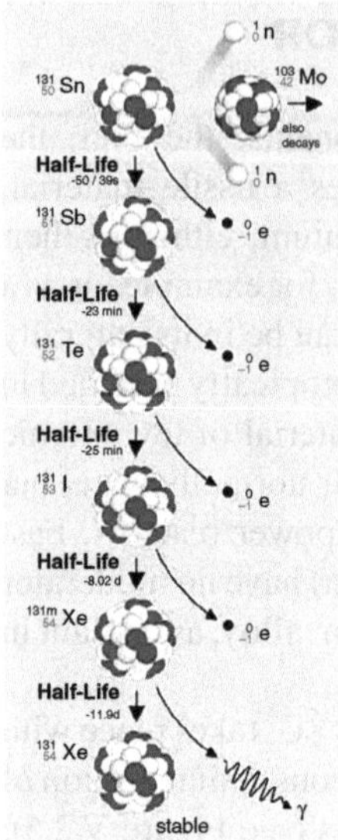


**Figure V.2.6** The decay chain of the fission fragment tin-131 that decays within days to the stable isotope xenon-131. (Courtesy of the European Nuclear Society.)

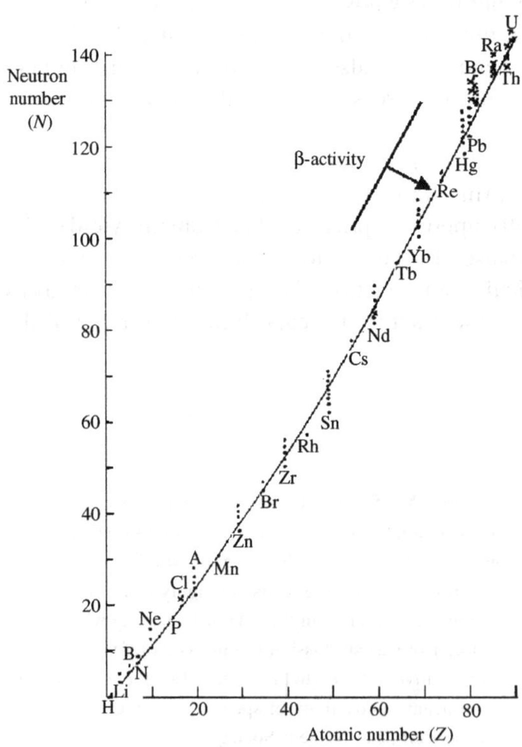


**Figure V.2.7** Number of neutrons versus atomic number for stable nuclides. This plot, shown as the full line, is known as the line of β-stability. (After Hunt (1987).)

reactor. It is these latter neutrons that form our source of slow neutrons for scattering experiments.

The purpose of the moderator in the reactor is to slow down the source neutrons from their initial MeV ($10^6$ eV) energies to meV ($10^{-3}$ eV) energies, where the fission cross section is significantly higher. The critical chain reaction is thereby more efficiently propagated. However, the neutron population is then approaching thermal equilibrium with the moderator. Note that $^{235}U$ is a *fissile* material, which will undergo fission at all neutron energies down to zero energy, but the neutron yield for thermal fission is higher than that for intermediate or fast fission. In contrast, $^{238}U$ is denominated as a *fertile* material that does not fission with slow neutrons but instead, following slow neutron absorption and subsequent α-decay, produces a new isotope $^{239}Pu$ that is fissionable at thermal energies and is therefore fissile. A so-called thermal reactor is thus a fission reactor employing a moderator as one of its essential components. Because of the presence of the moderator, the core of a power reactor can be several meters across, whereas a fast reactor core could measure less than 20 cm across.

In nuclear fission, the incident neutron is absorbed into the nucleus of the fissile atom to produce a *compound nucleus,* and within $10^{-14}$ s the compound nucleus disintegrates into two fission fragments, releasing fast neutrons with energies around 1 MeV. These are referred to as *prompt neutrons*. A small but significant fraction (≈0.7%) of fission neutrons are released at significantly longer times, even up to a few minutes after fission: these are called *delayed neutrons*, and are of the greatest importance in the control of reactors. The multiplication factor of neutron population $k$ from one generation of neutrons to the next originates from both the prompt neutron fraction and the delayed neutron fraction: a sustainable critical reaction occurs when $k = 1$. If $k > 1$, a *supercritical* reaction occurs, and for $k < 1$ the reaction is *subcritical.* By ensuring that $k$ for prompt neutrons is just less than unity—between 0.993 and 1—the rate of increase of neutron density in the reactor is determined essentially by the time constant of the delayed neutrons. Thus, the power output can be increased relatively slowly and controlled by the physical movement of neutron absorbers—the control rods—in and out of the reactor core.

The intense β-emission of fission fragments has two important practical consequences. The first is that the fuel elements in nuclear reactors rapidly become highly radioactive and a thick shield around the core is required, not only for the safety of personnel but also to reduce the background radiation recorded by the suite of instruments immediately around the reactor. The second consequence is that many of the fission fragments and their decay products are highly absorbing to neutrons and can stop the chain reaction. In particular, this is true of the isotope $^{135}Xe$ that has an absorption cross section for thermal neutrons of $3.0 \times 10^6$ barn, one of the highest known absorption cross sections for any nuclear reaction, and $^{135}Xe$ is the so-called *reactor poison* since it can reduce the reactivity of the reactor core below levels that can sustain a critical reaction. $^{135}Xe$ is produced both directly as a fission product of $^{235}U$, with a relatively low yield of 0.2% per fission, and indirectly as the granddaughter product of $^{135}Te$, itself produced as a fission product with a much higher yield of 6.1%. During steady operation of a reactor, the concentration of $^{135}Xe$

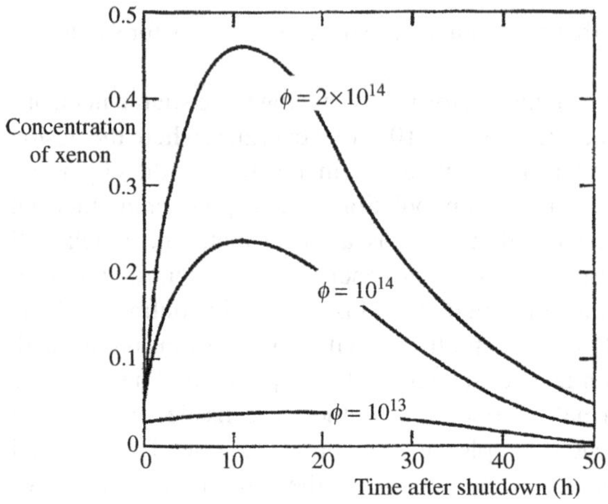


**Figure V.2.8** Rise in concentration of xenon after the shutdown. $\phi$ is the neutron flux in neutrons/cm$^2$/s.

reaches a constant value determined by the balance between its rate of production, its rate of radioactive decay, and its rate of removal by neutron absorption. When the reactor is shut down, the removal process by neutron absorption stops and the $^{135}$Xe concentration increases, since the half-life for its formation (6.7 h) is less than its half-life for decay (9.2 h). The xenon concentration reaches a maximum value about 10 h after the reactor has shut down, at which point the concentration begins to fall but does not reach its pre-shutdown value until much later.

Figure V.2.8 illustrates the variation of xenon concentration as a function of time after shutdown, following a period of operation at steady power. The higher the neutron flux in the reactor, the higher the concentration of xenon and the faster the rise in the concentration in the period immediately following shutdown. For low-flux reactors (flux $\sim 10^{13}$ neutrons/cm$^2$/s), the effect of xenon poisoning is not particularly important except perhaps toward the end of a reactor cycle when the fuel has been burned up and the excess reactivity in the core is small. However, for a high-flux reactor like that at the Institut Laue-Langevin in Grenoble, if a shutdown occurs at any time during the operating cycle, apart from the first few hours of operation, then within not many minutes the concentration of xenon rises to a level that exceeds the excess reactivity in the core. The reactor is then said to have "poisoned out" and cannot be restarted until the xenon concentration has fallen sufficiently, requiring typically 2 days of shutdown. A nuclear reactor can be designed with sufficient excess reactivity such that, at any time in the cycle, it can be brought back to criticality after a shutdown; this is an expensive contingency that is usually reserved for nuclear-powered submarines where poisoning out would have more severe consequences. This problem would have been even more serious for the Advanced Neutron Source (ANS) at Oak Ridge, USA, which was first proposed in 1985 but later discontinued. The ANS was designed to have a reactor power of 350 MW and a flux three or four times higher than the ILL flux. Fast reactors do not suffer from xenon poisoning.

High-flux beam reactors for neutron scattering purposes are specifically designed to provide the highest neutron density just outside the edge of the reactor core where the beam tubes terminate. Such reactors are said to be undermoderated. Earlier research reactors, built in the 1950s and 1960s, were designed to undertake a wide range of activities including isotope production, gamma-ray activation analysis, basic reactor physics experiments, and neutron cross-section measurements and were not optimized for neutron scattering, nor indeed any one application. Thus, the development of neutron beam reactors can be thought of in three distinct phases: the early experimental or prototype assemblies; the later multipurpose research reactors; and the modern dedicated high-flux neutron beam reactors, such as the high-flux reactor (HFR) at the Institut Laue-Langevin or the FRM-II reactor at Munich, Germany.

## V.2.4 THE ILL REACTOR

Since 1973, this reactor has been the world's most powerful source of neutrons for scattering experiments. The thermal flux at the beam tubes is $1.5 \times 10^{15}\,\text{cm}^{-2}\,\text{s}^{-1}$, supplying neutron beams to instruments in the main reactor hall and in the two *neutron guide halls*. The thermal neutron flux in equilibrium with the $D_2O$ moderator at 300K has a peak in the Maxwellian distribution at a wavelength of 1.2 Å. For some of the beams, this distribution is modified by a hot source of graphite at 2400K that enhances the flux at wavelengths below 0.8 Å and two cold sources of liquid hydrogen at 25K that enhance the flux at wavelengths exceeding 3 Å. A vertical guide supplies neutrons to an upper laboratory for storage as *ultracold neutrons* in neutron bottles; their wavelengths are around 1000 Å and they are used for studying the fundamental properties of the neutron itself.

The reactor core of HFR is a single fuel element of uranium, 93% enriched in $^{235}U$ . This fuel element has a central cylindrical cavity containing the neutron absorbing control rod made from a silver–indium–cadmium alloy. The coolant is water, which is pumped through the internal fins of the core at a velocity of 15.5 m/s to remove the 58 MW of heat generated by the fission process. The coolant acts as the primary moderator but encircling the core is a heavy-water moderator, which also serves to reflect neutrons back into the core for further moderation and continuation of the chain reaction. The presence of the heavy water acts as a neutron reflector and causes a peaking of the thermal neutron flux just outside the reactor core where the front ends of the beam tubes are located.

Thus, the neutron beam tubes penetrate the biological shield of the reactor to the position of maximum flux surrounding the core. A hollow beam tube causes a local depression in the flux distribution around the core and the resultant neutron density gradient is the source of the high neutron current passing down the beam tube to the instruments at the other end of the tube. Most of the tubes do not point radially toward the reactor core but are tangential to it. This ensures that the background effects of the high fluxes of $\gamma$-rays, generated by both the fission reaction and the decaying fission fragments, are minimized, and also that the ratio of thermalized neutrons to fast neutrons (sometimes called the cadmium ratio) passing along the beam tube is as high as possible.

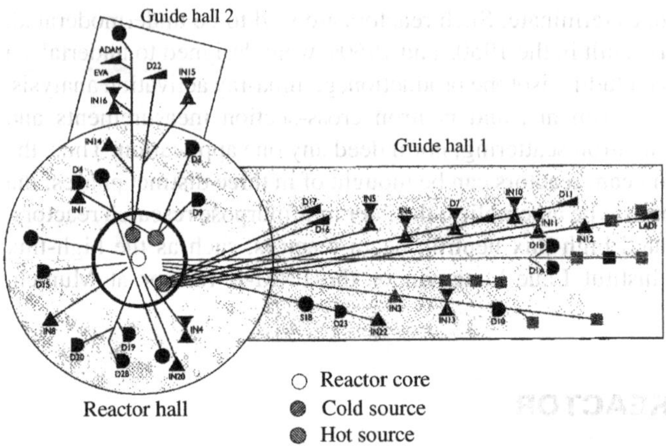


**Figure V.2.9** Layout of instruments in the main reactor hall and in the two guide halls at the ILL. Diffractometers are labeled D and inelastic scattering instruments are labeled IN.

There are several arrays of neutron guide tubes, which transport neutrons to the two low-background experimental halls without the loss in neutron intensity experienced with nonreflecting beam tubes. These guides are up to 120 m long, allowing the installation of many additional instruments, which can make measurements with a high thermal or cold neutron flux and a low background. The guides feed more than twice as many instruments as the beam tubes in the reactor hall. Figure V.2.9 illustrates the layout of the instruments in the reactor and the two guide tube halls.

## V.2.5 PULSED REACTORS

Time-of-flight methods using pulsed reactor beams offer an alternative procedure for carrying out neutron scattering experiments. Pulsed reactors are more compact than accelerator-based pulsed sources. Since 1960, they have been developed particularly at the Joint Institute for Nuclear Research at Dubna, Russia, with the construction of the IBR series of nuclear reactors.

A fission reactor can be pulsed in a number of ways:

(i) The reactivity for prompt neutrons of a subcritical assembly can be momentarily increased to become supercritical ($k > 1$), causing it to emit an intense burst of neutrons. This can be achieved mechanically by bringing either a piece of fissile material or a piece of reflector material close to the core of the subcritical assembly in a periodic manner. The neutron pulse length is determined by the speed with which the moving component can pass by the core. Because of the practical limitation on this speed, pulse lengths tend to be rather long (100–500 μs).

(ii) The reactivity of the assembly is pulsed to a supercritical level by the instantaneous generation of a burst of neutrons in the core, which is created by an electron pulse hitting a target in the center of the reactor and generating neutrons by the photofission process. These neutrons then multiply in the subcritical assembly. This is called a *static booster*. It has the disadvantage that *delayed neutrons* from the fission reaction undergo the same multiplication as the pulsed neutrons but are uncorrelated in time to the original pulse, so that scattering measurements are made on a relatively high time-independent background.

**(iii)** The combination of a pulsed particle beam and a rotating wheel overcomes, to some extent, the deleterious effects of multiplication of the delayed neutrons between pulses, since the reactivity of the subcritical assembly can be designed well below criticality. This combination is called a *dynamic booster* and can give pulse widths as short as 5 μs.

A useful distinction can be made between sources of long and short pulses. If the time required to generate the pulse (the proton pulse length) is significantly shorter than the moderation time of the neutrons, the source can be classified as a short-pulse source; when this pulse length is comparable to or greater than the moderation time, the source is a long-pulse source. For neutrons with energies less than 1 eV, accelerator-based spallation sources are normally short, whereas reactor-based pulsed sources are long. However, long-pulse spallation sources (LPSS) offer advantages over short-pulse spallation sources (SPSS) and particularly for cold neutron experiments with relaxed resolution, for example, in experiments using small-angle scattering, reflectometry, or neutron spin echo. There is by no means a consensus on this point, but the design for the 5 MW European Spallation Source (ESS) features a single long-pulse target fed by a nominally 2 ms proton pulse. The ESS is to be built in Lund in southern Sweden.

A diagram of the pulsed reactor IBR-2 at Dubna, about 100 km north of Moscow, is shown in Figure V.2.10. The average thermal power of the reactor is 2 MW but the peak power in the pulse is 1500 MW. The core is a compact fast reactor using highly enriched plutonium dioxide as fuel and liquid sodium (which produces negligible moderation) as coolant. Moderation for the neutron beams is provided by light water

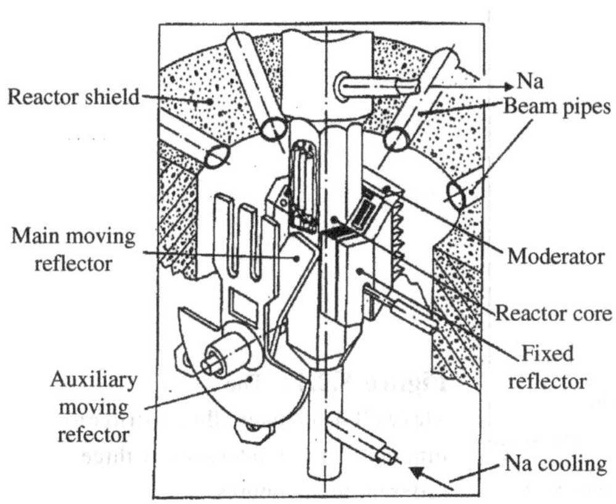


**Figure V.2.10** The pulsed reactor IBR-2 at Dubna, Russia. The reactivity is modulated by the two counterrotating neutron reflectors. At the center of the core is an electron target for possible use in the booster mode.

viewed by the beam tubes. The peak thermal neutron flux at the surface of the moderator is $1.5 \times 10^{16}$ neutrons/cm$^2$/s. The reactivity is pulsed by two rotating reflectors with an effective frequency of 5 Hz, providing a pulse length of 215 μs for thermal neutrons. Particular attention is paid to control circuitry and safety aspects, since fluctuations in reflector frequency and transverse vibrations of one component with respect to another can cause undesirable reactivity excursions. The gap between different components is only 10 μm and fluctuations in power level, pulse-to-pulse, can be of the order of 20–30%. The use of two reflectors rotating at different frequencies keeps the pulses relatively short and the period between pulses long, since only when both reflectors simultaneously pass close to the core does significant multiplication occur. A consequence of this is the presence of a series of satellite pulses that occur between the main pulses when only one reflector passes through the core. These satellite pulses are $10^4$ times weaker than the main pulse.

There is a possibility that the IBR-2 reactor can be converted to a dynamic booster mode by firing a short burst of electrons onto a target in the reactor core to initiate the neutron generation. In this way, the peak flux would be maintained while the sharpness of the pulse would be reduced to 5 μs, bringing the pulsed reactor into the short pulsed source regime for both thermal and cold neutrons. Nevertheless, it seems unlikely that pulsed reactors will ever be mainstream sources of neutrons for condensed matter research.

## V.2.6 NEUTRON MODERATION: HOT AND COLD SOURCES

The energy spectrum of neutrons emitted from a reactor or from an accelerator source depends on the temperature of the moderators surrounding the source, since the neutron "gas" is in thermal equilibrium with the materials of the moderators. Figure V.2.11 shows the Maxwellian flux for moderator temperatures of 25, 300, and 2000K. Most reactors are designed for a moderator temperature lying in the

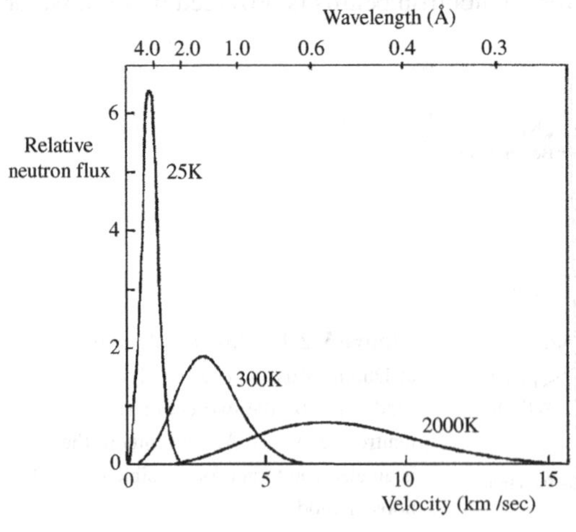


**Figure V.2.11** The Maxwell–Boltzmann flux distribution for infinitely large moderators at three different temperatures.

ambient range 300–350K, corresponding to the middle curve. However, for some experiments, such as high-resolution inelastic scattering, reflectometry, or small-angle scattering, there is a considerable advantage in using long-wavelength neutrons that require a cold source giving a spectrum similar to that shown in the left-hand curve. On the other hand, to extend the spectrum to shorter wavelengths or higher energies, a hot source can be employed: the hot source of the high-flux reactor at the Institut Laue-Langevin consists of a graphite block heated to around 2400K in the radiation environment of the reactor. The epithermal tail of the undermoderated spectrum from a pulsed accelerator source already provides intense beams of hot neutrons without the need to employ a heated moderator. Moderators on short pulsed sources, to achieve good coupling to the fast neutrons in the target and to maintain a narrow pulse and high neutron brightness, are considerably smaller than those on reactors where there is no time structure to preserve. The effect of this limited size is that the slow neutron spectra are appreciably undermoderated, and the characteristic temperature of the neutron spectrum is higher than the actual temperature of the moderator. Thus, the neutrons are of higher energy and there is an epithermal component to the spectrum, unlike a reactor spectrum. A second consequence of the small size of moderators on short pulsed sources is the greater difficulty of coupling to neutron guides for achieving full illumination, particularly at longer neutron wavelengths. These constraints are considerably reduced on long pulsed sources.

The energy and wavelength ranges covered by moderators at temperatures of 25, 300, and 2000K are given in Table V.2.1.

The material used for a low-temperature moderator must satisfy several conditions. It must have a low absorption cross section and a high scattering cross section. Also, the material must be of low atomic number in order to ensure that the number of collisions required for reaching thermal energies is relatively low, thus minimizing the chance of absorption: in this respect, hydrogen-containing moderators are the most effective. In addition, the moderator material must possess a suitable inelastic scattering mechanism for lowering the neutron energy.

Liquid hydrogen appears to be the most practical material for a cold source, in spite of its relatively high absorption cross section and its low density. Hydrogen has the advantage that it does not suffer from radiation damage. Methane, which has a proton density in the liquid phase twice that of liquid hydrogen and exhibits inelastic processes below 100 μeV, is neutronically very favorable but is less practical than hydrogen since it suffers severely from *radiolysis*. Radiolysis (i.e., the production of free radicals by the disruption of chemical bonds) results in chemical recombination reactions, and these produce tars leading to eventual blockage of the moderator.

**Table V.2.1** Energy and Wavelength Ranges of Neutron Moderators

| Moderator | Energy (meV) | Wavelength (Å) |
|---|---|---|
| Cold (20K) | 0.1–10 | 3–30 |
| Thermal (300K) | 5–80 | 1–4 |
| Hot (2000K) | 80–500 | 0.4–1 |

## V.2.7 THE RELATIVE MERITS OF CONTINUOUS AND PULSED SOURCES

An acceptable comparison of pulsed sources and continuous sources is fraught with difficulties, and while there have been several attempts to do this analytically, none have succeeded in demonstrating that there is a simple answer. Some general guidelines only can be drawn and these are changing constantly, since pulsed source technology is relatively new whereas reactor technology is more mature. In the end, it comes down to where the best science can be done—on which instruments, at which sources, and with which people. The body of knowledge accumulated by scientific endeavor is often subjective, so we shall confine ourselves to some general comments.

**(a)** *Politics*. Enriched uranium, that is, uranium highly concentrated in the fissile isotope $^{235}U$ , is the raw material of atomic weapons as well as the fuel for research reactors. Its production, availability, and use are the business of governments, which seek to control not only the supply of $^{235}U$ but also the information surrounding it. The U.S. government is the most powerful player in the game and is currently committed to converting the use of highly enriched uranium (HEU with $>20\%$ content of $^{235}U$ ) in research reactors to low enriched uranium (LEU with $<20\%$ content of $^{235}U$ ). Fuel supplies, even for existing reactors, are therefore uncertain and the prospect of lower fluxes, following core reconfigurations to adapt to the use of LEU, must be faced. Those reactors with the most compact cores—the high-flux reactors—will suffer most.

The reprocessing of used fuel elements—a source of plutonium, which is also a raw material for weapons—is another complex political and environmental issue. Even the transport of used fuel—let alone its reprocessing—is the focus of public dissent. Pulsed sources do not use fissile materials as targets and do not present the perceived risks that reactors sources do, and although the targets become highly radioactive, they contain little or no fissile material as a result of the spallation process. The target of a pulsed source attains, generally speaking, the same level of radioactivity during its total lifetime of several decades as does the core of a nuclear reactor after one operating cycle of perhaps 50 days, so there is a significant difference here.

On the other hand, governments committed to a dependence on electricity generated by nuclear power plants—France and India, for example—or in maintaining independent nuclear weapons need to maintain a trained workforce in reactor technology in all its forms. Because there is the possibility of converting radioactive waste to more benign materials using high-power accelerators, there is significant support for accelerator development at government level.

Environmentally, both power reactors and research reactors are perceived by the general public as being malign. Risks posed by supercriticality excursions, as in the Chernobyl and Three Mile Island incidents, are in fact low when set against other man-made risk factors in our daily lives. For example, deaths on the roads or from smoking tobacco are high risks but are well understood; they are apparently acceptable to the general public whereas the low risks from reactors are mysterious and unacceptable. It is unlikely that these perceptions will change significantly in the foreseeable future in favor of reactors. On the positive side, the generation of electricity by nuclear power stations does not produce greenhouse gases, and the nuclear fuel itself, unlike oil fuel, is not concentrated in politically unstable regions of the world. Also, the nuclear fuel represents a very small fraction of the total cost of energy production by nuclear power. There may yet be a revival of the fortunes of nuclear power, as governments struggle to meet the targets on carbon emissions set by the 1997 Kyoto Protocol on global warming and being debated in Copenhagen at the time of writing (December 2009).

**(b)** *Neutron Production*. Reactors generate neutrons continuously. The moderator is an intrinsic component of the reactor, helping to maintain criticality by thermalizing the fast neutrons. This results in a well-moderated spectrum for scattering measurements with a rich flux of low-energy neutrons (further enhanced by cold sources) and a relatively low flux of neutrons at energies exceeding 100 meV. In contrast, the moderators on a pulsed source serve no purpose in the neutron generation process and can therefore be tailored to suit the neutron instruments. Because there is a need to maintain good coupling to the target, in order to maximize the intensity and to retain the narrow pulse structure of the incident proton beam on a short pulsed source, moderators are small and thus the resultant spectrum is undermoderated. This results in a significant flux of epithermal neutrons and a relatively low flux of cold neutrons. This apparent complementarity—using reactors for low-energy neutrons and pulsed sources for high-energy neutrons—is oversimplistic and, in any case, is violated by the use of hot moderators on reactors and cold moderators on pulsed sources. On long pulsed sources, there is more freedom to have moderators more matched to the needs of the instrument suite.

Reactors are simpler to operate than pulsed sources and are intrinsically more reliable. In a reactor, the control of a single set of control rods to maintain reactor power requires the monitoring of a relatively small number of parameters, such as the temperature and the flow rate and radioactivity levels of the coolant. This contrasts with the large number of components to be controlled and monitored on a pulsed source, with perhaps up to 500 identifiable components on the accelerator itself. The reliability of pulsed sources is, therefore, noticeably poorer than that for reactor sources. This difference is further accentuated by the demand for the utmost security on reactors. Components and systems are built to the highest nuclear standards, and are often installed with triple redundancy, thereby enhancing overall reliability of the source for the users.

Because of its continuous operation, the power density levels generated in the core of a research reactor can be very significant—up to 10 MW/l, which must be removed by forced cooling to maintain steady temperatures. In contrast, on a pulsed source, power density levels in the target are much lower, around 5 kW 1 in present-day sources. This is partly a result of the pulsed nature of the neutron generation process and partly because the heat generated for each useful neutron (~30 neutrons generated per incident proton in the spallation process) is significantly lower than that in the fission process (~0.5 useful neutrons per fission reaction). Thus, in fission the heat released is ~200 MeV/neutron, whereas in spallation it is ~30 MeV/neutron. However, given that there is an "instantaneous" production of heat in a small volume when the proton beam strikes the target of a pulsed source, effective power densities are significantly higher than the time-averaged heat loads quoted here. Reactors are now at the limits of materials technology, so higher power densities, and hence higher neutron fluxes, are not foreseen. Pulsed sources, on the other hand, are operating well within cooling limits and sources are under construction that are 30 times more intense than those currently operating (e.g., the European Spallation Source). There are other problems to be considered with pulsed sources, such as the mechanical integrity of the target containment vessel as it is repeatedly struck by intense pulses of high-energy protons.

The possibility of designing a pulsed source that generates perhaps two orders of magnitude fewer fast neutrons than a reactor requires the utilization of the peak neutron flux for a large part of the time between the pulses. On a reactor, because of the need to remove at least 99% of all neutrons generated by using a monochromator, measurements are usually made continuously using a monochromatic beam. On a pulsed source, the well-defined time origin of the neutron burst allows the dispersion of neutrons of different energies before the beam strikes the sample, and if the instrument is properly designed these neutrons fill the whole measuring time frame with polychromatic (white) neutrons at peak intensity. The peak flux on a pulsed source can be compared to the average flux on a reactor to give one measure of comparability of source strength. This is illustrated in

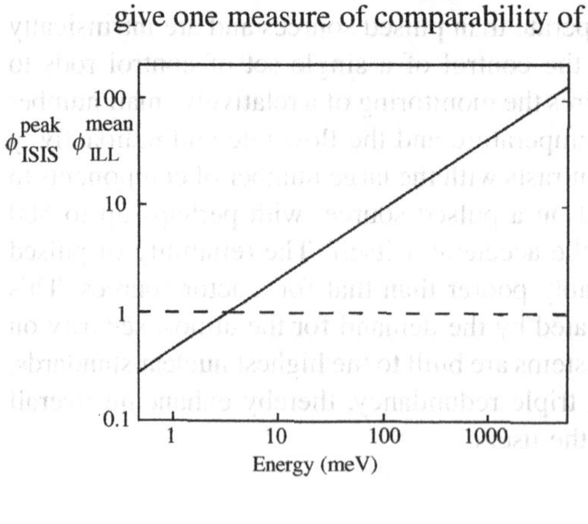


**Figure V.2.12** Variation with energy of the ratio of the peak flux at the ISIS source and the mean flux at the ILL reactor source (idealized).

Figure V.2.12 where the published values of the fluxes at ILL and ISIS are compared. It can be seen that pulsed sources compare particularly well at higher neutron energies. However, this simple comparison overlooks the fact that guides can be better coupled to larger moderators on reactor sources, that large-area focusing monochromators can be used to enhance the flux at the sample, and that on a white beam pulsed source whole frames are not filled optimally and the intensity within a frame can vary appreciably.

**(c)** *Neutron Utilization*. Whatever kind of source is used to record a diffraction pattern or an inelastic spectrum, the measurements are carried out in quite different ways. With a continuous source we can select a particular wavelength with a crystal monochromator and carry out measurements as a function of scattering angle, whereas with a pulsed source we use a wide band of wavelengths and analyze the neutrons by time-of-flight techniques. Because both the pulse width and the time of flight over a fixed distance are proportional to wavelength, short pulsed source instruments have resolutions that are relatively constant over a wide dynamic range. Coupled to the rich flux of short wavelength neutrons from a pulsed source, a short-pulse source is of particular advantage in powder diffraction when high resolution over a wide range of $Q$ ($=4\pi \sin \theta/\lambda$) is needed.

In general, pulsed source instruments are static during their operation whereas reactor instruments are mobile. This feature of pulsed sources can have advantages when using complex sample environment equipment, where, for example, complete scattering patterns from a sample under pressure in a fixed window cell can be obtained at a single scattering angle. On the other hand, the resolution function of a reactor instrument is invariably symmetric in time or in reciprocal space, whereas that from a pulsed source instrument is almost always asymmetric as a result of the sharp rise time and slow decay of the neutron pulse in the moderator. This can give rise to problems in data correction that are peculiar to pulsed source instruments. This disadvantage does not apply to long pulsed sources. Correction factors are more straightforward for reactor data when the measurements are made at fixed wavelength. When using beams from neutron guides, the distance of a pulsed source instrument from the moderator determines its resolution, whereas on a reactor it does not. There is therefore less flexibility in the siting of instruments on pulsed sources than there is on reactors.

In principle, because a pulsed source such as ISIS produces two orders of magnitude fewer neutrons than a reactor, instrumental backgrounds should be lower in the former case. In practice, this is only true for chopper spectrometers where very low signals can be sensed. The very fast neutrons generated by a pulsed source and the need to illuminate the sample with an intense white beam often militate against this apparent advantage. One further manifestation of this is the very large amount of steel and concrete shielding surrounding all pulsed source instruments, as well as the huge beamstops that are required. Combined with the need for long incident flight paths on white beam instruments, the instruments on short pulsed sources are appreciably more expensive than their reactor counterparts. On the ESS, the plans for guide bunches, as pioneered at ILL four decades ago, will reduce these problems.

A single-crystal diffraction study with a continuous source allows more flexibility than that with a pulsed source. The crystal and detector can move independent of one another, so there is no restriction on the choice of scan in reciprocal space. An analogous situation arises in comparing the triple-axis spectrometer on a steady source with a time-of-flight spectrometer on a pulsed source. With a triple-axis spectrometer, we can choose any type of scan in $(\mathbf{Q}, \omega)$ space, whereas this is not so for the pulsed source analog. On the other hand, the completeness of the pulsed source data set, whether from elastic or inelastic instruments, can reveal subtle effects that, if not expected, can remain unobserved on a reactor instrument.

In the field of liquid diffraction, the benefits of high fluxes of high-energy neutrons are that inelasticity corrections (so-called Placzek corrections) are minimized on a pulsed source instrument where measurements to very high momentum transfers can be made. However, the higher stability afforded by reactor-based instruments often counteracts these advantages.

With reflectivity measurements, it is a similar story. The fixed geometry of a pulsed source instrument yields great benefits in the study of surfaces (particularly so for liquids), whereas reactor instruments, in general, must scan in a $\theta/2\theta$ mode. However, the scanning process, which can be disadvantageous from a stability point of view, allows the experimentalist on a reactor instrument to increase the dwell time continuously point-by-point to match the fall in signal between the maximum and minimum intensities displayed in the reflectogram.

The above comments are not meant to be a definitive survey of pros and cons of the various instruments available on continuous and pulsed sources, but rather give an insight into the differences in technique. An advantage in one experiment can prove to be a disadvantage in another, on the same instrument. There is also no doubt that the area of overlap between what can be done on pulsed source instruments and what can be done on reactor-based instruments has increased considerably with time. This process will continue as more opportunities are exploited on pulsed sources, but the original, tidy concept of complementarity of the two kinds of sources has given way to a more realistic complementarity of instruments.

## V.2.8 FUTURE PROSPECTS

At the end of the first decade of the new millennium, we find that neutron scattering research is at cross-roads. Apart from FRM-II at Munich, ARR-II at Lucas Heights near Sydney, and a new facility in Indonesia, no new reactor source has been brought on line since ORPHÉE at Saclay, near Paris, and Dhruva at Trombay, India, in the 1980s. A few new reactor projects are funded, including one in China near Beijing. A reasonable estimate of the maximum lifetime of a reactor is 40–50 years, so there could be perhaps only 5 or 10 research reactors dedicated to neutron scattering still in operation by 2025. Despite this, the demand for neutron beam time continues to rise and the scientific disciplines that benefit from research using neutrons widen year by year. There is a long period, of perhaps 15 or more years, from the initial go-ahead for a new source to its final commissioning at full specification. A further 10 years would easily have been spent in making the scientific and political case before that and assembling the funding. The cancellation in 1995 of the Advanced Neutron Source at Oak Ridge, the technological successor to the ILL, which was to have been a 350 MW reactor, was a symbolic event for the future of reactor sources of slow neutrons. The supply of highly enriched uranium fuel, and its subsequent reprocessing or storage, is becoming increasingly difficult to secure for political reasons, and the environmental climate remains cautious toward the building of more nuclear power reactors, despite the dependence on them for electricity supplies both nowadays in the developed countries and in the future by developing nations. However, there are signs that, paradoxically for environmental reasons associated with the production of the greenhouse gas $CO_2$ in fossil-fueled power stations, this sentiment is slowly, and reluctantly, beginning to change.

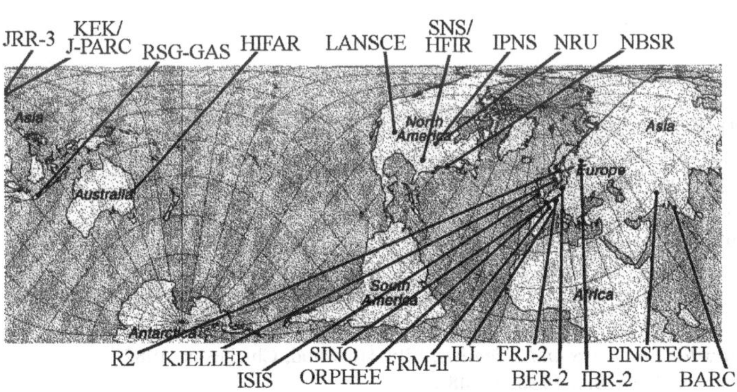


**Figure V.2.13** The geographical distribution of the principal centers of research using neutron scattering. Dhruva reactor, BARC, Mumbai, India; BER-II 10 MW reactor, Berlin, Germany; FRJ-2 23 MW reactor, Jülich, Germany (now closed down); FRM-II 20 MW reactor, Münich, Germany; ARR-II 10 MW reactor, Lucas Heights, Australia; HFIR 100 MW reactor, Oak Ridge, USA; IBR-2 pulsed reactor, Dubna, Russia; ILL 58 MW reactor, Grenoble, France; IPNS short pulsed source, ANL, USA (closed down); ISIS short-pulse spallation source, Chilton, UK; JRR-3 20 MW reactor JAERI, Japan; J-PARC 0.6 MW short-pulse spallation source, JAERI; KENS short-pulse spallation source, Tsukuba, Japan (closed down); LANSCE short-pulse spallation source, Los Alamos, USA; NIST 20 MW reactor, Washington, USA; NRU 130 MW reactor, Chalk River, Canada; ORPHÉE 14 MW reactor, Saclay, France; PINSTECH reactor, Islamabad, Pakistan; RSG-GAS 30 MW reactor, Serpong, Indonesia; R2 50 MW reactor, Studsvik, Sweden (closed down); SINQ 1 MW continuous spallation source, Villigen, Switzerland; SNS 1.4 MW short-pulse spallation source, Oak Ridge, USA.

Pulsed spallation sources, or continuous spallation sources such as SINQ at the Paul Scherrer Institute in Switzerland, offer a way forward out of these difficulties, because it is now recognized that pulsed source instrumentation in many areas of neutron scattering can not only equal but also exceed the best reactor instruments. Interestingly, the radioactive inventory in the targets of the new high-power spallation sources will exceed that in the core of the ILL reactor, but only after decades of operation. Figure V.2.13 shows the location of the principal pulsed and reactor sources throughout the world. The 1.4 MW Spallation Neutron Source is currently in the early years of commissioning at Oak Ridge, USA, having started operations in 2006. It has operated at 1 MW. In addition, the 0.6 MW spallation source at the Japanese Hadron facility J-PARC at Tokai, to the northeast of Tokyo, produced its first neutrons in 2007. The extremely powerful 5 MW long-pulse European Spallation Source has just passed the site decision phase, a painful and drawn-out procedure. It is to be built in Lund in southern Sweden and is scheduled for first operations in 2019.

The present state of neutron scattering research is apparently healthy. Claims made in the early 1990s that synchrotron radiation sources would render neutron scattering techniques irrelevant have matured and become more realistic. The two sources of radiation are complementary. The neutron user community is growing, thanks in part to organizations such as the European Neutron Scattering Association, and equivalent organizations in America and Oceania, which are drawing in new users and attracting researchers from new scientific fields. However, committed decisions for the construction of new sources are still the bottleneck.

## REFERENCES

Bothe, W. and Becker, H. *Z. Phys.* **1930**, *66*, 289.
Brown, A. *The Neutron and the Bomb: A Biography of Sir James Chadwick*, Oxford University Press, 1997.
Chadwick, J. *Nature* **1932a**, *129*, 312.
Chadwick, J. *Proc. R. Soc. Lond. A* **1932b**, *136*, 692.
Cowan, G.A. *Sci. Am.* **1976**, *235*, 36.
Curie, I. and Joliot, F. *C. R. Acad. Sci.* **1932**, *194*, 273.
Eidelman, S. *Phys. Lett. B* **2004**, *592*, 861.
Fermi, E. and Marshall, L. *Phys. Rev.* **1947**, *71*, 666.
Hunt, S.E. *Nuclear Physics for Engineers and Scientists*, Ellis Horwood, Chichester, 1987.
Mitchell, D.P. and Powers, N. *Phys. Rev.* **1936**, *50*, 486.
Rhodes, R. *The Making of the Atomic Bomb*, Penguin Books, 1988.
Shull, C.G. and Smart, J.S. *Phys. Rev.* **1949**, *76*, 1256.
Shull, C.G. and Wollan, E.O. *Phys. Rev.* **1951**, *81*, 527.
Zinn, W.H. *Phys. Rev.* **1947**, *71*, 752.

# Index



*Neutrons in Soft Matter* Edited by Toyoko Imae, Toshiji Kanaya, Michihiro Furusaka, and Naoya Torikai

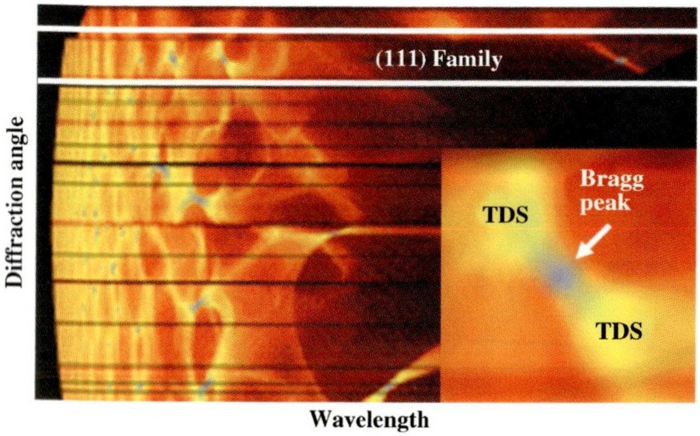


**Figure II.1.3.1.10** The 2D representation of Bragg diffraction and TDS from Si(111) crystal at $\theta_B = 45°$.

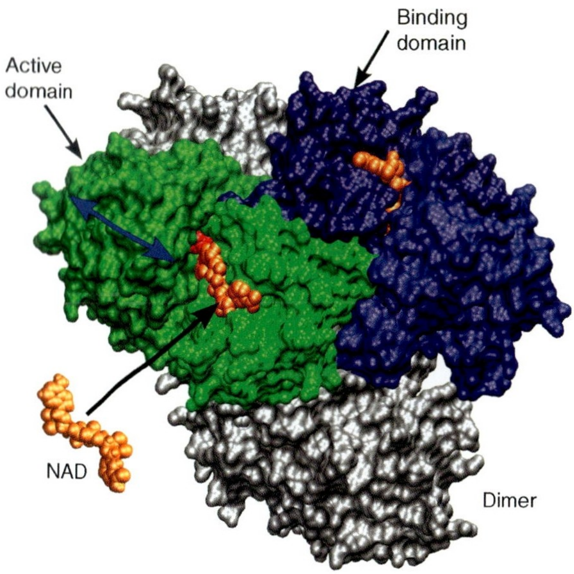


**Figure II.3.1.10** Space filing model of the protein ADH in its form as dimer. The binding of the cofactor NAD is indicated. The NSE experiment was performed on a tetrameric form of ADH, which corresponds to the association of two of these dimers.

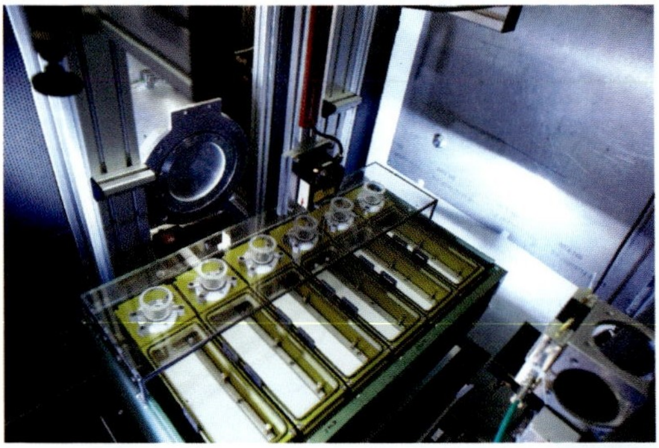

**Figure III.5.16** Photo of an assembly of six PTFE adsorption troughs shown on the FIGARO reflectometer at ILL together with the optical sensor used to automate sample alignment through optical windows.

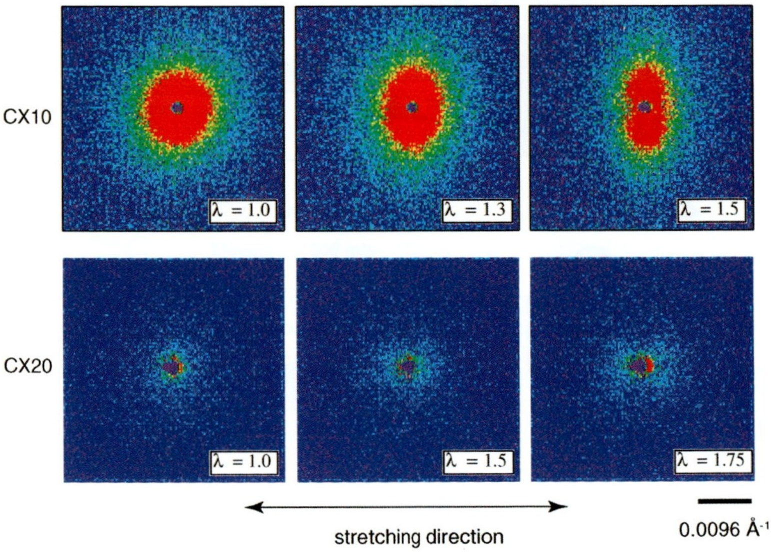


**Figure IV.4.16** SANS isointensity patterns of the SR gel with CX10 ($C_x = 1.0\%$) and CX20 ($C_x = 2.0\%$).

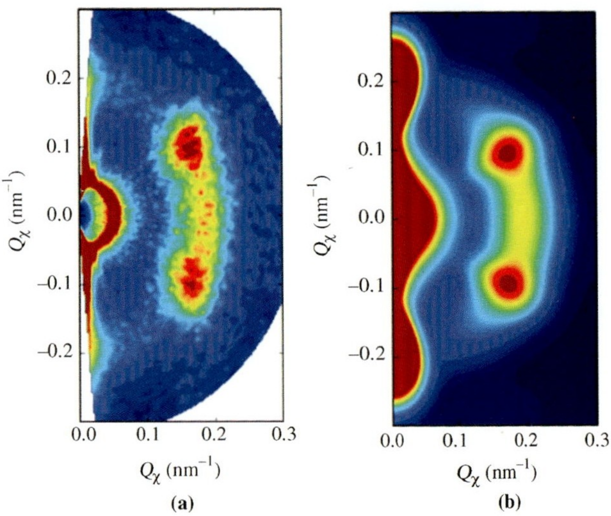


**Figure IV.6.11** Rotational SANS pattern (a) from a 141 nm thick dPS-*b*-PMMA film annealed for 15 h at 147°C.